Mathematics for Engineers

Mathematics for Engineers offers a comprehensive treatment of the core mathematical topics required for a modern engineering degree. The book begins with an introduction to the basics of mathematical reasoning and builds up the level of complexity as it progresses. The approach of the book is to build understanding through engagement, with numerous exercises and illuminating examples throughout the text designed to foster a practical understanding of the topics under discussion.

Features

- Replete with examples, exercises, and applications
- Suitable for engineers but also for other students of the quantitative sciences
- Written in an engaging and accessible style while preserving absolute rigor.

Mathematics for Engineers

Francesc Pozo Montero
Núria Parés Mariné
Yolanda Vidal Seguí

CRC Press
Taylor & Francis Group
Boca Raton London New York

CRC Press is an imprint of the
Taylor & Francis Group, an **informa** business
A CHAPMAN & HALL BOOK

Designed cover image: Francesc Pozo Montero

First edition published 2025
by CRC Press
2385 NW Executive Center Drive, Suite 320, Boca Raton FL 33431

and by CRC Press
4 Park Square, Milton Park, Abingdon, Oxon, OX14 4RN

CRC Press is an imprint of Taylor & Francis Group, LLC

ISBN: 978-1-032-50544-2 (hbk)
ISBN: 978-1-032-50805-4 (pbk)
ISBN: 978-1-003-39973-5 (ebk)

DOI: 10.1201/9781003399735

Typeset in Latin Modern font
by KnowledgeWorks Global Ltd.

Publisher's note: This book has been prepared from camera-ready copy provided by the authors.

Access the Support Material: www.Routledge.com/9781032505442

With gratitude to our students and colleagues, whose curiosity and insights enriched this work, and to our families and friends, whose support and encouragement made this book possible.

Contents

Preface

The book you hold in your hands (whether in physical or digital format) is the result of over twenty-five years of teaching mathematics in engineering degrees. The book follows the basic scheme of a core calculus course in one variable of the new degrees adapted to the European Higher Education Area, but includes three additional chapters: one dedicated to mathematical reasoning and proof methods, another dedicated to numerical sets, with special emphasis on complex numbers, and a final chapter intended as an introduction to linear algebra, in which matrices and systems of linear equations are presented.

Throughout these years of teaching and research, we have imparted substantial knowledge but without any doubt, it is these students who have taught us how to approach our teaching to be better teachers! In this sense, this book aims to reflect the way of studying of the students that we can base on the necessary compromise between the quantity of knowledge we want to convey and the quality of that knowledge or how that knowledge is acquired. Formalisms sometimes put obstacles to understanding. But without formalisms, new knowledge cannot be built.

This book reflects a rigorous and precise mathematical approach combined with a clear didactic intention. In each chapter of the book, we wanted to justify everything that is exposed, proving most of the theorems and propositions, but without that generating problems or hindering reading. The basic scheme followed by each of the chapters is the following:

- An introduction, which briefly presents the contents of the chapter.
- The development of theoretical concepts and procedures, with a significant number of examples and demonstrations of theorems.
- A series of Proposed Exercises, interspersed in the text, to check the level of achievement of the material. The solution to the Proposed Exercises is found in an Annex (www.Routledge.com/9781032505442).
- Solved Problems at the end of each chapter, organized according to theme and degree of difficulty.

In particular, and as a demonstration of the effort we have put into transmitting mathematics in the most didactic and clear way possible, the book includes more than 300 figures and more than 200 solved problems.

We sincerely believe that this book can be an excellent tool to guide and complete the study of a calculus course in one variable in a face-to-face study, but it can also be used as basic support material in calculus courses taught at a distance.

We cannot conclude this prologue without thanking those who have helped us, in one way or another, to make this book a reality. Thus, we thank Daniel Amposta Navarro, Ignacio Arruga Cantalapiedra, Sergio Arruga Cantalapiedra, Antonio de la Casa Gómez, and Jesús Martínez Fernández.

Barcelona, August 27, 2024.

Francesc Pozo Montero
Núria Parés Mariné
Yolanda Vidal Seguí

Author Biographies

Francesc Pozo Montero received his degree in Mathematics from the University of Barcelona in 2000 and completed his PhD in Applied Mathematics at the Universitat Politècnica de Catalunya (UPC) in 2005. Since 2000, he has been a member of the Department of Mathematics at UPC, where he currently holds the position of Full Professor. He also serves as the coordinator of the Control, Data, and Artificial Intelligence research group and is recognized as a Senior Member of IEEE.

Francesc Pozo Montero's research interests include condition monitoring, control systems, data-driven modeling, system identification, and structural health monitoring, with a particular emphasis on applications related to wind turbines.

He is an Editorial Board Member for several prestigious international journals, including *Structural Control and Health Monitoring* (Wiley), *International Journal of Distributed Sensor Networks* (Wiley), *Mathematical Problems in Engineering* (Wiley), *Mathematics* (MDPI), *Sensors* (MDPI), *Algorithms* (MDPI), *Journal of Vibration and Control* (Sage), *Frontiers in Built Environment* (Frontiers), *Frontiers in Energy Research* (Frontiers), and *Energies* (MDPI).

Francesc Pozo Montero has made substantial contributions to his field, authoring over 70 high-impact journal articles, participating in 23 competitive R&D&I projects, and writing 34 book chapters and 12 books. He has successfully supervised six PhD candidates, filed one invention patent, and established a collaboration contract with an industry partner. His work has also been presented in more than 130 conference papers.

Núria Parés Mariné holds a degree in Mathematics (1999) and a PhD in Applied Mathematics (2005) from the Universitat Politècnica de Catalunya (UPC), Barcelona, Spain. As a Full Professor at UPC, she focuses on developing and advancing numerical methods applied to engineering. Over the years, her research has concentrated on three primary areas: result certification, model reduction techniques, and machine learning. Through her innovative approach, she has successfully collaborated with internationally renowned researchers, contributing significantly to the academic and scientific community.

Núria Parés Mariné has made substantial contributions to her field, authoring over 25 high-impact journal articles and participating in 19 competitive R&D&I projects. Her prolific output also includes writing three book chapters and ten books. She has guided and supervised three PhD candidates to successful completion, showcasing her dedication to mentoring the next generation of researchers. In addition, she has been involved in technology transfer, establishing a collaboration contract with an industry partner and bridging the gap between academia and real-world applications.

Her work has been widely recognized and presented in more than 65 conference papers, illustrating her commitment to disseminating knowledge and sharing advancements with the broader scientific community. Prof. Parés continues to be a leading figure in her research areas, driving innovation and excellence in applied mathematics. Her ongoing efforts in integrating machine learning with traditional numerical methods are paving the way for new, groundbreaking approaches in engineering and related disciplines.

Yolanda Vidal Seguí holds a degree in Mathematics (1999) and a PhD in Applied Mathematics (2005) from the Universitat Politècnica de Catalunya (UPC), Barcelona, Spain. As an Associate Professor at UPC and an IEEE Senior Member, she is deeply involved in multidisciplinary research, with a particular focus on the application of her expertise to wind turbines. Her research areas encompass condition monitoring, structural health monitoring, fault diagnosis and prognosis, predictive maintenance, machine learning, deep learning, artificial intelligence, and mathematical modeling.

Yolanda Vidal Seguí serves on the Editorial Board of several prestigious international journals, including Engineering Applications of AI (Elsevier), Wind Energy (Wiley), Wind Energy Science (Copernicus), Journal of Vibration and Control (SAGE), IET Renewable Power Generation (IET), Mathematics (MDPI), Sensors (MDPI), Energies (MDPI), Frontiers in Built Environment, and Frontiers in Energy Research.

Her prolific contributions to the field are demonstrated by over 70 high-impact journal articles, 23 competitive R+D+I projects, 17 book chapters, and 10 books. She has supervised 7 PhD theses, with 4 currently ongoing. Additionally, she holds 1 invention patent and has secured a collaboration contract with an industry partner. Dr. Vidal has also presented 120 conference papers, further showcasing her dedication and impact in her areas of expertise.

Part I

Mathematical Reasoning

1

Introduction to Mathematical Reasoning

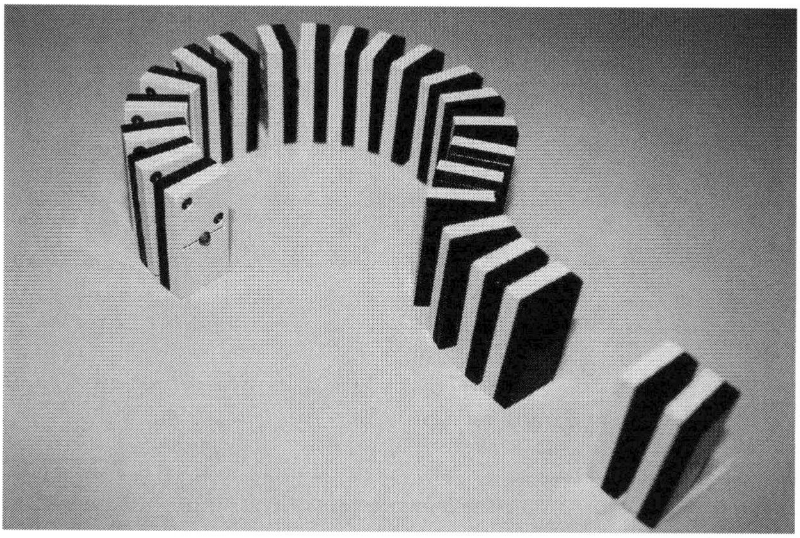

«If people do not believe that mathematics is simple, it is only because they do not realize how complicated life is.» (John von Neumann)

In this first introductory chapter, the different mechanisms that mathematics uses to prove various types of properties are presented. Among the different strategies presented (direct proof, contrapositive, or proof by contradiction), mathematical induction stands out, which allows proving properties about natural numbers.

1.1 Mathematical Reasoning

According to the 22nd edition of the Spanish Language Dictionary, mathematics can be defined as follows:

Definition 1.1 (Mathematics). Deductive science that studies the properties of abstract entities, such as numbers, geometric figures, symbols, and their relationships.

DOI: 10.1201/9781003399735-1

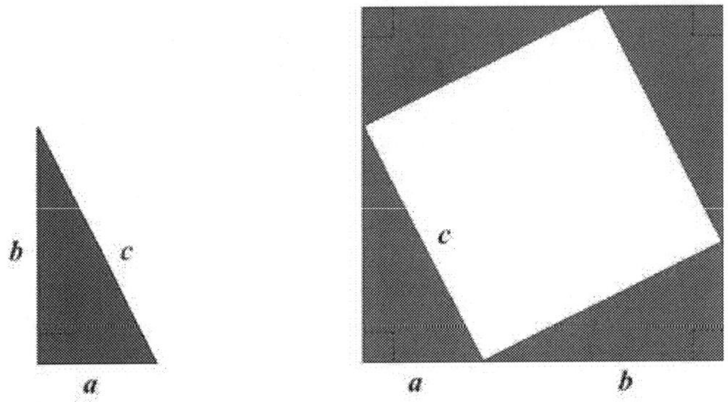

FIGURE 1.1
In a right triangle with hypotenuse c and legs a and b, we have $a^2 + b^2 = c^2$.

For example, the Pythagorean theorem (Figure 1.1) tells us that if we have a right triangle with sides a and b and hypotenuse c, then the sum of the squares of the legs is equal to the square of the hypotenuse, that is:

$$a^2 + b^2 = c^2.$$

In this chapter, we will see some of the tools that mathematics uses to prove these types of properties.

Mathematical properties or statements are called different names depending on their importance. The most important properties are called **theorems** (such as the Pythagorean theorem), while lemmas, corollaries, or propositions collect less important properties. Let's see some more formal definitions.

Definition 1.2 (Theorem). A proposition that can be logically proven from axioms or other theorems already proven, using accepted rules of inference.

Definition 1.3 (Lemma). A proposition that must be proven before establishing a theorem.

Definition 1.4 (Corollary). A proposition that does not require a particular proof but is easily deduced from what has been proven before.

Definition 1.5 (Proposition). A statement of a truth that has been proven or that is being tried to be proven.

The importance of theorems, lemmas, corollaries, or propositions is that they collect properties that **have been proven to be true** (following the logical-deductive method characteristic of mathematics).

But what is a mathematical proof?

Definition 1.6 (Proof). A mathematical proof is a **coherent sequence of steps** that allows one to ensure the truth of a statement.

The starting point of a proof is a set of statements that are considered true and are called **premises**. These premises can be hypotheses, axioms, or other propositions or theorems previously proven. Considering the premises as true, one must arrive at the original statement by **applying certain logical rules or applying previously proven properties**.

1.2 Is It Easy to Make a Mathematical Proof?

In general, proving a mathematical property can be very complex. The key questions we must ask ourselves are:

- Where do I start?
- What steps should I follow?

That is, the difficulty lies in knowing how to choose the premises well and then knowing which logical reasoning to follow to arrive at the original statement.

In this book, we will present and review basic concepts of calculus in one variable and algebra, delving into reasoning and proofs.

1.3 Methods of Proof

Although there is generally no single procedure for proving theorems, there are different **methods of proof** that are commonly used in mathematics. The five basic types of reasoning are:

- Direct proof.
- Proof by contrapositive.
- Proof by contradiction or reductio ad absurdum.
- Proof by counterexample.
- Proof by induction.

Next, we will see several examples of simple proofs obtained from these four methods.

Suppose we want to prove that a statement is true. We will call this statement Q. For example, let's imagine that we want to prove that if a and b are two real numbers, then the square of their sum is equal to the square of the first term plus the square of the second term plus twice the first term times the second term, that is:

$$(a+b)^2 = a^2 + b^2 + 2ab.$$

To prove Q, we generally need to start with some premises and logical rules. We will call these properties P. In the case of wanting to prove that $(a+b)^2 = a^2+b^2+2ab$, the premises will be the ability to apply the properties of the sum and the product of real numbers. The fact of proving Q from P is denoted by:

$$P \Rightarrow Q$$

and is read as "P implies Q" or "Q is deduced from P."

Remember that Q is what we want to prove and that P is what we assume to be true, the rules of the game.

Similarly to chess, where each piece has a permitted movement, not all steps are allowed in a mathematical proof. We can only use previously proven properties (the properties included in P).

1.4 Direct Proof

In direct proof, we assume P is true, and through a coherent sequence of steps, we show that Q is true.

Example 1.1 (Formula for the square of a sum). *We want to prove that if a and b are two real numbers:*

$$(a+b)^2 = a^2 + 2ab + b^2.$$

The premises, in this case, are the properties of the sum, product, and power of real numbers that we will be applying step by step. When presenting a proof, it is essential to explain in each step which property has been used.

Proof.

$$\begin{aligned}
(a+b)^2 &= (a+b)\cdot(a+b) && \text{[definition of squaring]}\\
&= (a+b)\cdot a + (a+b)\cdot b && \text{[distributive property of multiplication over addition]}\\
&= a\cdot(a+b) + b\cdot(a+b) && \text{[commutative property of multiplication]}\\
&= a\cdot a + a\cdot b + b\cdot a + b\cdot b && \text{[distributive property of multiplication over addition]}\\
&= a^2 + a\cdot b + b\cdot a + b^2 && \text{[definition of squaring]}\\
&= a^2 + a\cdot b + a\cdot b + b^2 && \text{[commutative property of multiplication]}\\
&= a^2 + 1\cdot a\cdot b + 1\cdot a\cdot b + b^2 && \text{[multiplicative identity]}\\
&= a^2 + 2\cdot a\cdot b + b^2 && \text{[distributive property of multiplication over addition]}
\end{aligned}$$

In the previous steps, we have also used the commutative property of addition. □

Example 1.2 (Solutions of the quadratic equation). *Given the quadratic equation $ax^2 + bx + c = 0$, if $a \neq 0$ and $b^2 - 4ac \geq 0$, then its two solutions can be calculated as:*

$$x = \frac{-b \pm \sqrt{b^2 - 4ac}}{2a}.$$

The hypotheses are that we have a quadratic equation ($a \neq 0$) and that:

$$b^2 - 4ac \geq 0.$$

In addition, we will use the properties of addition, multiplication, and division of real numbers, as well as the basic properties of working with equalities.

Proof. Since $a \neq 0$, we can divide both sides of the equation by a:

$$x^2 + \frac{b}{a}x + \frac{c}{a} = 0. \tag{1.1}$$

The second step in the proof is called completing the square. The idea is to convert the expression with two terms:

$$x^2 + \frac{b}{a}x$$

into a term of the form:

$$(x + C)^2.$$

Observing that:

$$\left(x + \frac{b}{2a}\right)^2 = x^2 + \frac{b}{a}x + \left(\frac{b}{2a}\right)^2$$

we have:

$$x^2 + \frac{b}{a}x = \left(x + \frac{b}{2a}\right)^2 - \left(\frac{b}{2a}\right)^2.$$

Substituting this equality into equation (1.1), we obtain:

$$\left(x + \frac{b}{2a}\right)^2 - \left(\frac{b}{2a}\right)^2 + \frac{c}{a} = 0.$$

Isolating the terms that involve x, we have:

$$\left(x + \frac{b}{2a}\right)^2 = \frac{b^2}{4a^2} - \frac{c}{a}.$$

Taking the square root of both sides (which is possible since we have assumed that $b^2 - 4ac \geq 0$):

$$x + \frac{b}{2a} = \pm\sqrt{\frac{b^2 - 4ac}{4a^2}}$$

and finally, solving for x, we obtain:

$$x = -\frac{b}{2a} \pm \frac{\sqrt{b^2 - 4ac}}{2\sqrt{a^2}}.$$

Now we have one technical detail left to consider, which depends on the sign of a, since $\sqrt{a^2} = |a|$. Let us assume that $a > 0$, then:

$$x = -\frac{b}{2a} \pm \frac{\sqrt{b^2 - 4ac}}{2\sqrt{a^2}} = -\frac{b}{2a} \pm \frac{\sqrt{b^2 - 4ac}}{2a} = \frac{-b \pm \sqrt{b^2 - 4ac}}{2a}.$$

On the other hand, if $a < 0$, then:

$$x = -\frac{b}{2a} \pm \frac{\sqrt{b^2 - 4ac}}{2\sqrt{a^2}} = -\frac{b}{2a} \pm \frac{\sqrt{b^2 - 4ac}}{-2a} = \frac{-b \mp \sqrt{b^2 - 4ac}}{2a}.$$

□

Proposed Exercise 1.1. Find the two solutions of the second-degree equation $2x^2+5x+6=0$ without using the formula, that is, using the technique of completing the square.

Example 1.3 (Pythagorean Theorem). *Given a right triangle with hypotenuse c and legs a and b, then:*

$$a^2 + b^2 = c^2.$$

Proof. Note that in Figure 1.1 we have a large square (which we call CG) of side $a+b$ that is formed by 4 right triangles with sides a and b (which we call T) and a small square of side c (which we call CP). Therefore, we have that:

$$A_{\mathrm{CG}} = 4A_{\mathrm{T}} + A_{\mathrm{CP}},$$

where A_{CG}, A_{T}, and A_{CP} represent the area of the large square, each of the triangles, and the small square, respectively. That is:

$$(a+b)^2 = 4\frac{ab}{2} + c^2 \iff (a+b)^2 = 2ab + c^2$$

Expanding the square, we have:

$$a^2 + b^2 + 2ab = 2ab + c^2$$

and, simplifying, we arrive at the Pythagorean Theorem:

$$a^2 + b^2 = c^2.$$

□

Example 1.4. *Let a be an integer, that is, $a \in \mathbb{Z}$. If a is an even number, then a^2 is also an even number.*

Proof. If a is an even number, then a can be expressed as twice an integer, that is:

$$a = 2k, \ k \in \mathbb{Z}$$

Then, its square:

$$a^2 = (2k)^2 = 2 \cdot \underbrace{(2k^2)}_{\tilde{k}}$$

can also be expressed as twice an integer:

$$\tilde{k} = 2k^2 \in \mathbb{Z}.$$

Thus, we can conclude that a^2 is also an even number. □

Proposed Exercise 1.2 (Direct Proof). If a and b are two positive real numbers, then their arithmetic mean is greater than or equal to their geometric mean. That is:

$$\sqrt{ab} \leq \frac{a+b}{2}.$$

1.5 Proof by Contrapositive

Recall that in a proof, the goal is to demonstrate that $P \Rightarrow Q$. In a proof by contrapositive, the following property is used:

$$\boxed{P \Rightarrow Q} \quad \text{is equivalent to} \quad \boxed{\neg Q \Rightarrow \neg P}$$

where $\neg Q$ and $\neg P$ represent the negation of Q and P, respectively. This means that it is equivalent to prove either of the following statements:

- $P \Rightarrow Q$ (if we assume P is true, then Q is true.)
- $\neg Q \Rightarrow \neg P$ (the negation of Q implies the negation of P, that is, if we assume that Q is not true, then P will not be true either.)

Example 1.5. *Consider the following sentence:*

«If it rains, the patio gets wet»

This sentence can be written logically as $P \Rightarrow Q$. In this case, the premise P is «it rains» and the consequence Q is «the patio gets wet», that is, raining implies that the patio gets wet. This sentence is equivalent to:

«If the patio does not get wet, then it has not rained»

That is, the patio not getting wet implies that it has not rained.

We will see in the following example that, at times, it is easier to prove some propositions using the proof by contrapositive than the direct proof.

Example 1.6. *Let a be an integer, that is, $a \in \mathbb{Z}$. If a^2 is, an even number, then a is also an even number.*

Proof. Instead of proving:

$$a^2 \text{ is even} \Longrightarrow a \text{ is even}$$

we will prove:

$$a \text{ is not even (odd)} \Longrightarrow a^2 \text{ is not even (odd)}$$

Indeed, that a is odd means that there exists an integer k such that a can be expressed as $a = 2k + 1$. That is:

$$\begin{aligned} a \text{ odd} \quad \Longrightarrow& \quad a = 2k+1, \ k \in \mathbb{Z} \\ \Longrightarrow& \quad a^2 = (2k+1)^2 = 4k^2 + 4k + 1 \\ \Longrightarrow& \quad a^2 = 2(2k^2 + 2k) + 1 \\ \Longrightarrow& \quad a^2 = 2\tilde{k} + 1, \ \tilde{k} \in \mathbb{Z} \\ \Longrightarrow& \quad a^2 \text{ is odd.} \end{aligned}$$

□

If we add the result of Example 1.4 to this last example, what we get is the following proposition:

Proposition 1.1. Let a be an integer, that is $a \in \mathbb{Z}$. a is an even number if and only if, a^2 is an even number. That is:

$$a^2 \text{ is even} \iff a \text{ is even}$$

Proof. To prove a biconditional statement, as in this case, it is necessary to prove the implication from left to right:

$$a^2 \text{ is even} \Longrightarrow a \text{ is even},$$

and the implication from right to left:

$$a \text{ is even} \Longrightarrow a^2 \text{ is even}.$$

These two implications have already been proven in Examples 1.4 and 1.6. □

Proposed Exercise 1.3 (Proof by Contrapositive). Let $a \geq 0$ be a real number. If $\forall \varepsilon > 0$, $0 \leq a < \varepsilon$, then $a = 0$.

1.6 Proof by Contradiction or Reduction to Absurdity

The implication $P \Rightarrow Q$ means that if P is true, then Q is also true. Therefore, if this statement is true, P and $\neg Q$ cannot occur simultaneously.

Example 1.7. *If raining implies that the patio gets wet, it cannot be the case that it has rained and the patio has not gotten wet.*

Proofs by contradiction are based on assuming P and $\neg Q$ to be true at the same time and through a sequence of steps arriving at an absurd or contradictory result. This way, the initial hypothesis must be false, and therefore, if P is true, then Q must also be true.

Example 1.8. *The equation $x^2 = 2$ has no solution in the set $\mathbb{Q}$ of rational numbers.*

Proof. Suppose there is a rational number $x \in \mathbb{Q}$ that is a solution to the equation $x^2 = 2$. That $x \in \mathbb{Q}$ means that there are two integers p and q, such that $x = \frac{p}{q}$. Additionally, we can assume that $\frac{p}{q}$ is in simplest form; that is, p and q are coprime. Then:

$$x^2 = 2 \Longrightarrow \frac{p^2}{q^2} = 2 \Longrightarrow p^2 = 2q^2.$$

Therefore, p^2 is an even number, and as we have seen in Proposition 1.1, this implies that p is also an even number. That is, $p = 2\tilde{p}$ where $\tilde{p}$ is an integer.

Thus, the equation becomes:

$$(2\tilde{p})^2 = 2q^2 \Longrightarrow 4\tilde{p}^2 = 2q^2 \Longrightarrow 2\tilde{p}^2 = q^2.$$

Therefore, if $q^2 = 2\tilde{p}^2$, q^2 must be an even number. Now, if q^2 is an even number, then q must also be even. Hence, we conclude that both p and q are even numbers, which contradicts the fact that p and q are coprime.

Since we reached a contradiction, we can deduce that the equation $x^2 = 2$ has no solution in $\mathbb{Q}$, as we wanted to prove. □

Here is another example of using proof by contradiction.

Theorem 1.1 (Euclid's Theorem on the Infinitude of Prime Numbers). **The set of prime numbers is infinite.**

Proof. We present below an adaptation to the current notation of Euclid's own proof of his theorem in the collection of books *Elements* published around 300 B.C.

Suppose the set of prime numbers is finite; that is, there are certain numbers:

$$p_1, p_2, \ldots, p_n$$

which are the *only* prime numbers. Now consider the following number:

$$p = p_1 \cdot p_2 \cdots p_n + 1$$

which, by its very construction, is clearly greater than 1 and different from all the primes p_i in the set. The number p can be prime or composite. If it is prime, we have a prime number that is not in the set, which leads us to a contradiction. On the other hand, if it is a composite number, there will be some factor q that divides p. Suppose q is one of the p_i. It can then be deduced that if q divides p and by definition also divides $p_1 \cdot p_2 \cdots p_n$, q must also divide the difference, that is:

$$q \text{ divides } p - p_1 \cdot p_2 \cdots p_n = 1,$$

but no prime number can divide 1, which also leads to a contradiction. The consequence of these two contradictions is that the set of prime numbers cannot be finite. □

1.7 Proof by Counterexample

Strictly speaking, a proof by counterexample serves to demonstrate that a property is not true. For example, consider the so-called Mersenne numbers:

$$M_n = 2^n - 1, \ n \in \mathbb{N},$$

named after the French philosopher Marin Mersenne. If we calculate the value of the Mersenne numbers for the first prime numbers:

$$M_2 = 3$$
$$M_3 = 7$$
$$M_5 = 31$$

we can observe that we also obtain prime numbers. In this way, we could (incorrectly) establish the following theorem:

Theorem 1.2. If p is a prime number, then $M_p = 2^p - 1$ is also a prime number.

At this point, we must either be able to prove the theorem or find a counterexample—a single instance where the theorem is not true—if we believe the theorem is not valid.

In this case, simply consider the prime number $p = 11$ to realize that:

$$M_{11} = 2^{11} - 1 = 2047 = 23 \cdot 89,$$

that is, it is not a prime number and the theorem is not valid.

Proposed Exercise 1.4 (Proof by Counterexample). Demonstrate that if A and B are two 3×3 matrices, it is not true that $AB = BA$.

1.8 Proof by Induction

We will introduce proof by induction through an example and using the analogy of domino pieces.

Imagine we want to prove that:

$$1 + 2 + \cdots + n = \frac{n(n+1)}{2}$$

is true for every natural number n ($\forall n \in \mathbb{N}$).

The first thing we can do is select a few natural numbers:

$$n = 1, 2, 7, \ldots$$

and see that the equality holds for these values. However, this is not sufficient to ensure that the property is true for all $n \in \mathbb{N}$.

The problem we face is proving a property that refers to *all* natural numbers.

How can we prove that a property is true for every natural number without having to check the property for infinite numbers?

Imagine, we have an infinite row of domino pieces, aligned and ready to be knocked down.

We know from experience that if the pieces are well placed, by knocking down the first piece, all of them will fall.

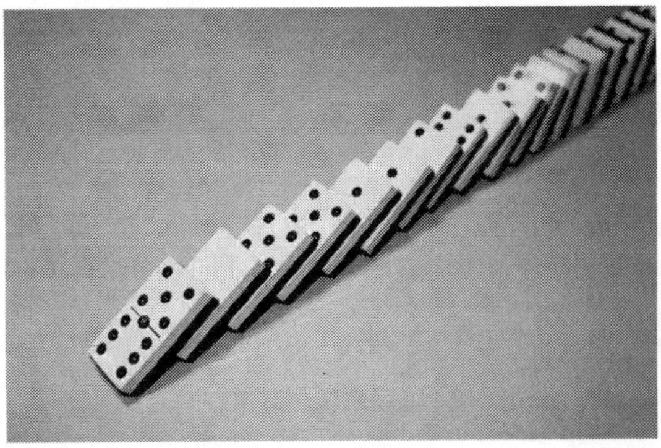

How can we prove that all the pieces will fall?

1. We know from experience that if we hit one domino piece, it will fall.

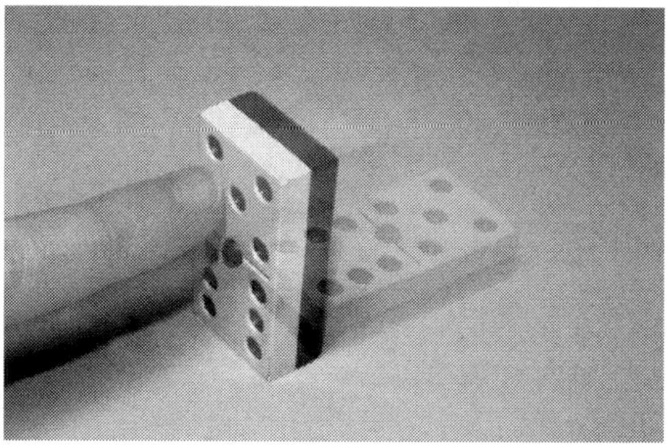

2. We also know that if one piece falls, and the next one is well placed, the next one will also fall.

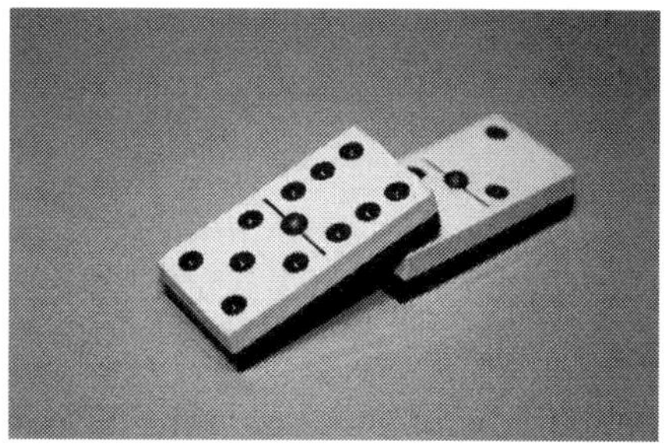

Intuition tells us that the pieces will keep falling one after another infinitely. Therefore, if we can prove:

1. That the first domino piece falls:

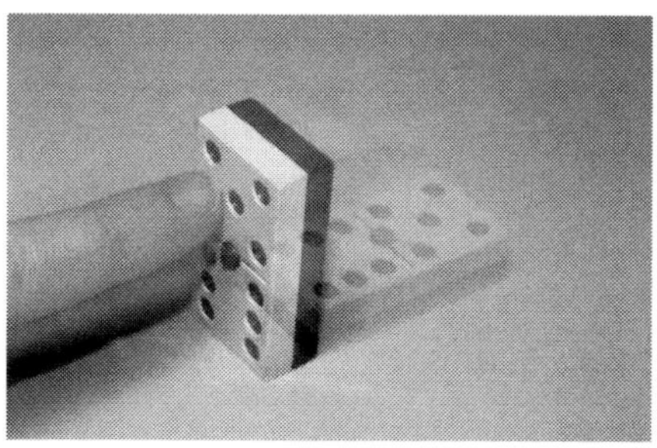

2. And that if one piece falls, the next one will also fall:

 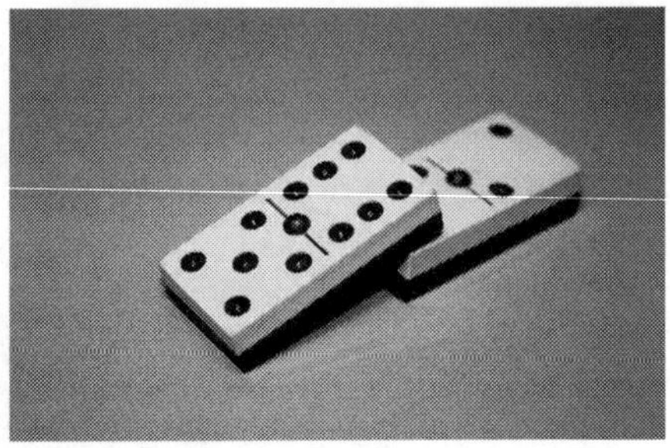

Then we can be sure that all the pieces will fall:

The problem we face is to prove a property, let us call it $\mathscr{P}$, that refers to all natural numbers. That is, we we want to prove that:

$$\mathscr{P}(n) \text{ is true } \forall n \in \mathbb{N}.$$

To see that $\mathscr{P}(n)$ is true $\forall n \in \mathbb{N}$ it is sufficient to see that:

(a) *Base case*: $\mathscr{P}(1)$ is true.

(b) *Inductive case*: assuming that $\mathscr{P}(n)$ is true (inductive hypothesis), see that $\mathscr{P}(n+1)$ is true. That is, see that:

$$\mathscr{P}(n) \text{ true } \Longrightarrow \mathscr{P}(n+1) \text{ true.}$$

Example 1.9. *Let us prove that for every natural number n the following holds:*

$$1 + 2 + \cdots + n = \frac{n(n+1)}{2}.$$

Proof. To see that the property is true for every natural number, we need to prove two things:

(a) **Base case**. That the formula is true when $n = 1$.

(b) **Inductive case**. That if the formula is true for a generic natural number k, then the formula must also be true for $k + 1$.

Indeed:

(a) **Base case.** The property is true for $n = 1$, since:

$$1 = \frac{1 \cdot 2}{2}.$$

(b) **Inductive case.** Suppose the property is true for $n = k$, that is, suppose it is true that:

$$1 + 2 + \cdots + k = \frac{k(k+1)}{2} \quad \text{(inductive hypothesis)}$$

We need to prove that it is also true for $n = k + 1$. Therefore, we want to prove that:

$$1 + 2 + \cdots + k + (k+1) = \frac{(k+1)(k+2)}{2}.$$

Using the inductive hypothesis, we know how to calculate the value of the first k terms of the sum we need to perform:

$$\begin{aligned}
\underbrace{1 + 2 + \cdots + k}_{\frac{k(k+1)}{2}} + (k+1) &= \frac{k(k+1)}{2} + (k+1) \\
&= \frac{k(k+1) + 2(k+1)}{2} \\
&= \frac{k^2 + k + 2k + 2}{2} \\
&= \frac{k^2 + 3k + 2}{2} \\
&= \frac{(k+1)(k+2)}{2}
\end{aligned}$$

Therefore, the property is proven for $\forall n \in \mathbb{N}$.

□

Proposed Exercise 1.5. Use the principle of induction to see that the following equality is true for every natural number:

$$1 + 3 + 5 + \cdots + (2n - 1) = n^2$$

Proposed Exercise 1.6. Use the principle of induction to see that the following equality is true for every natural number:

$$1^2 + 2^2 + 3^2 + \cdots + n^2 = \frac{1}{6} n(n+1)(2n+1)$$

1.9 Solved Problems (Direct Proof)

Problem 1.1. If $a \in \mathbb{N}$ is a divisor of $b \in \mathbb{N}$ and, in turn, b is a divisor of $c \in \mathbb{N}$, prove that a is a divisor of c.

Solution to the problem 1.1. *If a is a divisor of b, then there is a natural number k such that $b = ak$. Similarly, if b is a divisor of c, then there is another natural number k' such that $c = bk'$. Therefore,*

$$c = bk' = akk' = ak'', \; k'' = kk' \in \mathbb{N},$$

which means that a is a divisor of c.

Problem 1.2. If $a \in \mathbb{N}$ is a divisor of $b \in \mathbb{N}$ and, in turn, a is a divisor of $c \in \mathbb{N}$, prove that a is a divisor of $b + c$.

Solution to the problem 1.2. *If a is a divisor of b, then there is a natural number k such that $b = ak$. Similarly, if a is a divisor of c, then there is another natural number k' such that $c = ak'$. Therefore,*

$$b + c = ak + ak' = a(k + k') = ak'', \; k'' = k + k' \in \mathbb{N},$$

which means that a is a divisor of $b + c$.

Problem 1.3. If a is an integer divisible by 4, prove that a can be expressed as the difference of two perfect squares.

Solution to the problem 1.3. *In this case, the proof is* constructive, *in the sense that we will find an exact expression for a as the difference of two perfect squares. Since a is divisible by four, we propose the following form:*

$$\begin{aligned} a &= \left(\frac{a}{4} + k\right)^2 - \left(\frac{a}{4} + s\right)^2 \\ &= \left(\frac{1}{2}k - \frac{1}{2}s\right) a + (k^2 - s^2). \end{aligned}$$

In this way, it is necessary that

$$\begin{aligned} \frac{1}{2}k - \frac{1}{2}s &= 1 \\ k^2 - s^2 &= 0. \end{aligned}$$

Solving the system, we obtain

$$\begin{aligned} k &= 1, \\ s &= -1, \end{aligned}$$

that is,

$$a = \left(\frac{a}{4} + 1\right)^2 - \left(\frac{a}{4} - 1\right)^2,$$

which is the difference of two perfect squares.

Problem 1.4. If a and b are real numbers, prove that $a^2 + b^2 \geq 2ab$.

Solution to the problem 1.4. *The inequality*

$$a^2 + b^2 \geq 2ab$$

is equivalent to

$$a^2 + b^2 - 2ab \geq 0$$

The expression $a^2 + b^2 - 2ab$ *is equivalent to* $(a-b)^2$. *Since it is the square of a difference, it is clearly non-negative. That is,*

$$(a-b)^2 \geq 0,$$

as we wanted to demonstrate.

Problem 1.5. Prove that the sum of two rational numbers is a rational number.

Solution to the problem 1.5. *A rational number is one that can be expressed as a quotient of integers, where the denominator is not zero. If* $a, b \in \mathbb{Q}$ *are two rational numbers, then there exist integers* $p, q, p', q' \in \mathbb{Z},\ q, q' \neq 0$ *such that*

$$a = \frac{p}{q},\ b = \frac{p'}{q'}.$$

The sum of these two numbers is

$$a + b = \frac{p}{q} + \frac{p'}{q'} = \frac{pq' + p'q}{qq'}.$$

Since the numbers $pq' + p'q, qq' \in \mathbb{Z}$ *are integers and, additionally,* $qq' \neq 0$, *it is proven that* $a + b \in \mathbb{Q}$.

Problem 1.6. If r_1 and r_2 are the distinct roots of the polynomial $p(x) = x^2 + bx + c$, prove that $r_1 + r_2 = -b$ and $r_1 r_2 = c$.

Solution to the problem 1.6. *The proof in this case is constructive. If* r_1 *and* r_2 *are the two roots of the polynomial* $p(x) = x^2 + bx + c$, *then* $p(x)$ *can be expressed as* $p(x) = (x - r_1)(x - r_2)$. *Expanding this expression:*

$$p(x) = x^2 - (r_1 + r_2)x + r_1 r_2.$$

Therefore, the coefficient of the linear term b *is equal to* $-r_1 - r_2$ *and the constant term* c *is equal to* $r_1 r_2$.

1.10 Solved Problems (Proof by Contrapositive)

Problem 1.7. If x and y are two integers such that $x+y$ is even, prove that x and y have the same parity.

Solution to the problem 1.7. *We will prove this using the contrapositive method. We will show that if two integers x and y have different parity, then the sum $x+y$ is odd. Indeed, suppose x is even and y is odd. This means that there exist two integers $k, k' \in \mathbb{Z}$ such that*

$$\begin{aligned} x &= 2k \\ y &= 2k' + 1 \end{aligned}$$

Then,

$$x + y = 2k + 2k' + 1 = 2(k + k') + 1 = 2k'' + 1, \; k'' = k + k' \in \mathbb{Z},$$

which implies that the sum $x+y$ is odd.

Problem 1.8. If n is a positive integer such that $n \mod 4$ is 2 or 3, prove that n is not a perfect square. If m and n are two integers, $n \mod m$ is the remainder obtained when dividing n by m.

Solution to the problem 1.8. *We will prove this using the contrapositive method. We will show that the integers that are perfect squares, when divided by 4, never have a remainder of 2 or 3. Suppose n is a perfect square of an even number, that is,*

$$n = (2k)^2, \; k \in \mathbb{Z}.$$

Then, it is evident that $n = 4k^2$, that is, n is divisible by 4. Therefore, the remainder of the division of n by 4 is zero.

Suppose now that n is a perfect square of an odd number, that is,

$$n = (2k+1)^2, \; k \in \mathbb{Z}.$$

In this case, we have that

$$n = 4k^2 + 4k + 1 = 4(k^2 + k) + 1 = 4k'' + 1, \; k'' = k^2 + k \in \mathbb{Z}.$$

If we now divide n by 4, the remainder obtained is 1. Therefore, the remainders we have obtained are 0 or 1. That is, we never have a remainder of 2 or 3, as we wanted to prove.

Problem 1.9. If x and y are two integers whose product is even, prove that at least one of them must be even.

Solution to the problem 1.9. *We will prove this using the contrapositive method. We will show that if two integers are odd, then their product is an odd number. Indeed, if x and y are two odd numbers, there exist two integers $k, k' \in \mathbb{Z}$ such that*

$$x = 2k + 1, \; y = 2k' + 1.$$

Then,

$$\begin{aligned} xy &= (2k+1)(2k'+1) = 4kk' + 2k + 2k' + 1 = 2(2kk' + k + k') + 1 \\ &= 2k'' + 1, \; k'' = 2kk' + k + k' \in \mathbb{Z}. \end{aligned}$$

That is, xy is an odd number, as we wanted to prove.

Problem 1.10. If x and y are two integers whose product is odd, prove that x and y must be odd.

Solution to the problem 1.10. *We will prove this using the contrapositive method. We will show that given two integers x and y, if not both are odd, then their product is even. Thus, we have two cases to consider: (i) both numbers are even and (ii) one of the numbers is odd and the other is even. In the first case, we can assume that there are two integers $k, k' \in \mathbb{Z}$ such that*

$$x = 2k, \ \ y = 2k'.$$

Then,

$$xy = 2k \cdot 2k' = 4kk' = 2(2kk') = 2k'', \ \ k'' = 2kk' \in \mathbb{Z},$$

which implies that the product is even.

If we consider the second case, we can assume that there are two integers $k, k' \in \mathbb{Z}$ such that

$$x = 2k, \ \ y = 2k' + 1.$$

Then,

$$xy = (2k)(2k' + 1) = 4kk' + 2k = 2(2kk' + k) = 2k'', \ \ k'' = 2kk' + k \in \mathbb{Z},$$

which implies that the product is also even, as we wanted to prove.

Problem 1.11. If n is a positive integer such that $n \mod 3 = 2$, prove that n is not a perfect square.

Solution to the problem 1.11. *We will prove this using the contrapositive method. We will show that if a natural number is a perfect square, then the remainder of the division of this number by 3 is never 2. In this case, we consider that natural numbers are classified according to their remainder when divided by 3. This remainder can be 0 (multiples of 3), 1, or 2. Then, if n is a perfect square, there will be a natural number k such that*

$$\begin{aligned} n &= (3k)^2 \ \ or \\ n &= (3k+1)^2 \ \ or \\ n &= (3k+2)^2. \end{aligned}$$

In the first case,

$$n = (3k)^2 = 9k^2 = 3(3k^2) = 3k'', \ \ k'' = 3k^2 \in \mathbb{Z},$$

which implies that the remainder when dividing n by 3 is zero. In the second case,

$$\begin{aligned} n &= (3k+1)^2 = 9k^2 + 6k + 1 \\ &= 3(3k^2 + 2k) + 1 = 3k'' + 1, \ \ k'' = 3k^2 + 2k \in \mathbb{Z}, \end{aligned}$$

which implies that the remainder when dividing n by 3 is 1. Finally,

$$\begin{aligned} n &= (3k+2)^2 = 9k^2 + 12k + 4 \\ &= 3(3k^2 + 4k + 1) + 1 = 3k'' + 1, \ \ k'' = 3k^2 + 4k + 1 \in \mathbb{Z}, \end{aligned}$$

which implies that the remainder when dividing n by 3 is also 1, as we wanted to prove.

Problem 1.12. If a and b are real numbers such that the product ab is an irrational number, prove that a and/or b must be irrational numbers.

Solution to the problem 1.12. *We will prove this using the contrapositive method. We will show that if two real numbers are rational, then the product is also rational. Let $a, b \in \mathbb{Q}$ be two rational numbers, then there exist integers $p, q, p', q' \in \mathbb{Z}$, $q, q' \neq 0$ such that*

$$a = \frac{p}{q}, \ b = \frac{p'}{q'}.$$

The product of these two numbers is

$$ab = \frac{p}{q} \cdot \frac{p'}{q'} = \frac{pp'}{qq'}.$$

Since the numbers $pp', qq' \in \mathbb{Z}$ are integers and, additionally, $qq' \neq 0$, it is proven that $ab \in \mathbb{Q}$, as we wanted to prove.

1.11 Solved Problems (Proof by Contradiction or Reduction to Absurdity)

Problem 1.13. Prove that there are infinitely many prime numbers.

Solution to the problem 1.13. *We present below a different proof than the one already presented in the proof of Theorem 1.1. Nevertheless, as in that case, we will use the proof by reduction to absurdity. Suppose the set of prime numbers is finite and we will reach a contradiction. Let*

$$p_1 < p_2 < \cdots < p_n$$

be the set of prime numbers ordered from smallest to largest. Now consider the number

$$p = p_1 p_2 \cdots p_n + 1,$$

formed by the product of all prime numbers plus 1. The number p cannot be prime because it is not any of the primes in the set $p \neq p_i$, $i = 1, \ldots, n$ and because it would be a new *prime number. Thus, p must be a composite number, which means that there exists at least one prime number from the list that is a divisor of p. Suppose this prime number is p_k. Thus, we can write that*

$$p_k | p,$$

that is, p_k is a divisor of p. However, it is evident that p_k is also a divisor of $p_1 p_2 \cdots p_n$, that is,

$$p_k | p_1 p_2 \cdots p_n.$$

Therefore, p_k must also be a divisor of the difference, that is,

$$p_k | 1,$$

which contradicts the fact that p_k is a prime number.

Problem 1.14. Prove that the square root of 2 is an irrational number.

Solution to the problem 1.14. *The solution to this problem can be found in the resolution of Example 1.8.*

Problem 1.15. Prove that the Diophantine equation $x^2 - y^2 = 1$ has no positive integer solutions.

Solution to the problem 1.15. *A Diophantine equation is an algebraic equation with integer coefficients, involving two or more unknowns, and we are only interested in its integer solutions. One of the most famous families of Diophantine equations is the following:*

$$x^n + y^n = z^n,$$

where n is an integer and x, y, z are also positive integers. When $n = 1$, it is evident that we can find infinitely many solutions; when $n = 2$, a possible solution is $x = 3, y = 4$, and $z = 5$, which represents, for example, the sides of a right triangle. Finally, when $n \geq 3$, the French mathematician Pierre de Fermat conjectured in 1637 that the Diophantine equation has no solution (x, y, and z positive integers). The so-called Fermat's Last Theorem was not proven until 1995 (358 years later) by the British mathematician Andrew Wiles.

Returning to our problem, we will use proof by reduction to absurdity. Suppose there are two positive integers that are solutions to the equation, and we will reach a contradiction. The equation

$$x^2 - y^2 = 1$$

can be rewritten as

$$(x + y)(x - y) = 1.$$

Since x and y are two positive integers, the quantities $x+y$ and $x-y$ are integers. Therefore, for the product of two integers to be 1, they must be 1 or -1. But $x+y$ cannot be -1 because it must necessarily be a positive quantity. Therefore, it must be that

$$x + y = 1,$$
$$x - y = 1,$$

but this implies that $y = 0$, which contradicts the positivity of the number y.

Problem 1.16. Prove that the Diophantine equation $x^2 - y^2 = 10$ has no positive integer solutions.

Solution to the problem 1.16. *We will use proof by reduction to absurdity. Suppose there are two positive integers that are solutions to the equation, and we will reach a contradiction. The equation*

$$x^2 - y^2 = 10$$

can be rewritten as

$$(x + y)(x - y) = 10.$$

Since x and y are two positive integers, the quantities $x+y$ and $x-y$ are integers. Since x and y are positive integers, the expression $x+y$ is greater than $x-y$, and $x+y$ is clearly positive. Additionally, for the product of two integers to be 10, they must be 10 *and* 1*, or* 5 *and* 2*. Thus, we have two systems of two linear equations with two unknowns. The first of these systems is:*

$$x + y = 10,$$
$$x - y = 1,$$

which has the solution $x = 11/2$ and $y = 9/2$, which contradicts the fact that both numbers are integers. On the other hand, the second of these systems is:

$$x + y = 5,$$
$$x - y = 2,$$

which has the solution $x = 7/2$ and $y = 3/2$, which also contradicts the fact that both numbers are integers.

Problem 1.17. Prove that the equation $x^3 + x + 1 = 0$ has no rational solutions.

Solution to the problem 1.17. *We will use proof by reduction to absurdity. Suppose there is a rational number $\frac{p}{q} \in \mathbb{Q}$, $p, q \in \mathbb{Z}$, $q \neq 0$ (considering that p and q are also coprime) that is a solution to the equation $x^3 + x + 1 = 0$, and we will reach a contradiction. Indeed, if $\frac{p}{q}$ is a solution to the equation, it satisfies that*

$$\frac{p^3}{q^3} + \frac{p}{q} + 1 = 0,$$

which is equivalent to

$$p^3 + pq^2 + q^3 = 0.$$

The integers p and q cannot both be even because they are coprime. Thus, we can consider three cases: (i) both are odd; (ii) p is even and q is odd; and (iii) p is odd and q is even. In the first case, we have that p^3 is odd, pq^2 is also odd, and q^3 is also odd. Thus, the sum of three odd numbers is odd and, therefore, cannot be 0*. In the second case, we have that p^3 is even, pq^2 is also even, but q^3 is odd and, therefore, the sum is also an odd number and cannot be* 0*. Finally, in the third case, we have that p^3 is odd, pq^2 is even, and q^3 is also even and, therefore, the sum is an odd number and cannot be* 0*. Therefore, in all cases, we reach a contradiction.*

Problem 1.18. Prove that the equation $x^5 + x^4 + x^3 + x^2 + 1 = 0$ has no rational solutions.

Solution to the problem 1.18. *We will use proof by reduction to absurdity. Suppose there is a rational number $\frac{p}{q} \in \mathbb{Q}$, $p, q \in \mathbb{Z}$, $q \neq 0$ (considering that p and q are also coprime) that is a solution to the equation $x^5 + x^4 + x^3 + x^2 + 1 = 0$, and we will reach a contradiction. Indeed, if $\frac{p}{q}$ is a solution to the equation, it satisfies that*

$$\frac{p^5}{q^5} + \frac{p^4}{q^4} + \frac{p^3}{q^3} + \frac{p^2}{q^2} + 1 = 0,$$

which is equivalent to –multiplying by $q^5 \neq 0$–

$$p^5 + p^4q + p^3q^2 + p^2q^3 + q^5 = 0.$$

The integers p and q cannot both be even because they are coprime. Thus, we can consider three cases: (i) both are odd; (ii) p is even and q is odd; and (iii) p is odd and q is even. In the first case, we have that all the terms are odd. Thus, the sum of five odd numbers is odd and, therefore, cannot be 0. In the second case, we have that all terms except q^5 are even and, therefore, the sum is also an odd number and cannot be 0. Finally, in the third case, we have that all terms except p^5 are even and, therefore, the sum is an odd number and cannot be 0. Therefore, in all cases, we reach a contradiction.

Problem 1.19. If a is a rational number and b is an irrational number, prove that $a + b$ is an irrational number.

Solution to the problem 1.19. *We will use proof by reduction to absurdity. Suppose $a + b$ is a rational number and we will reach a contradiction. In this case, we will use the property from Problem 1.5. Indeed, if $a + b$ is rational and $-a$ is also a rational number, their sum must also be rational. But the sum is equal to b, which we have said is irrational, and therefore, we reach a contradiction.*

1.12 Solved Problems (Proof by Induction)

Problem 1.20. Prove that for all $n \in \mathbb{N}$, the following holds:

$$(1 + 2 + \cdots + n)^2 = 1^3 + 2^3 + \cdots + n^3.$$

Solution to the problem 1.20. *To see that the property is true for all natural numbers, we will use the principle of induction. Therefore, we need to see:*

(a) That it is true for $n = 1$.

(b) That if it is true for $n = k$, it is also true for $n = k + 1$.

First, let's check the base case, that is, that the property is true for $n = 1$:

$$1^2 = 1^3$$

Now, suppose the property is true for $n = k$, that is, suppose that:

$$(1 + 2 + \cdots + k)^2 = 1^3 + 2^3 + \cdots + k^3. \tag{1.2}$$

Let's see that the property is true for $n = k + 1$, that is, we want to prove if it is true that:

$$(1 + 2 + \cdots + k + (k + 1))^2 = 1^3 + 2^3 + \cdots + k^3 + (k + 1)^3.$$

Using the property $(a + b)^2 = a^2 + 2ab + b^2$, we have that:

$$(\underbrace{1 + 2 + \cdots + k}_{a} + \underbrace{(k + 1)}_{b})^2 = (1 + \cdots + k)^2 + 2(1 + \cdots + k)(k + 1) + (k + 1)^2.$$

If we consider that:

$$1 + 2 + \cdots + k = \frac{k(k+1)}{2},$$

and the inductive hypothesis of equation (1.2), we have that:

$$\begin{aligned}(1 + \cdots + k + (k+1))^2 &= (1 + \cdots + k)^2 + (k+1)^2 + 2(1 + \cdots + k)(k+1)\\ &= 1^3 + \cdots + k^3 + (k+1)^2 + 2\frac{k(k+1)}{2}(k+1)\\ &= 1^3 + \cdots + k^3 + (k+1)^2 + k(k+1)^2\\ &= 1^3 + \cdots + k^3 + (k+1)^2(1+k)\\ &= 1^3 + \cdots + k^3 + (k+1)^3,\end{aligned}$$

as we wanted to demonstrate.

Problem 1.21. Prove that for all $n \in \mathbb{N}$, the following holds:

$$\frac{1}{1\cdot 2} + \frac{1}{2\cdot 3} + \cdots + \frac{1}{n\cdot(n+1)} = \frac{n}{n+1}$$

Solution to the problem 1.21. *To see that the property is true for all natural numbers, we will use the principle of induction.*

(a) ***Base case****. The base case is true because:*

$$\frac{1}{1\cdot 2} = \frac{1}{2}.$$

(b) ***Inductive case****. Now, suppose the property is true for $n = k$, that is:*

$$\frac{1}{1\cdot 2} + \frac{1}{2\cdot 3} + \cdots + \frac{1}{k\cdot(k+1)} = \frac{k}{k+1} \tag{1.3}$$

Let's see that it is also true for $n = k+1$, that is, we want to prove that:

$$\frac{1}{1\cdot 2} + \frac{1}{2\cdot 3} + \cdots + \frac{1}{k\cdot(k+1)} + \frac{1}{(k+1)\cdot(k+2)} = \frac{k+1}{k+2}.$$

Using the inductive hypothesis of equation (1.3), we have that:

$$\begin{aligned}\frac{1}{1\cdot 2} + \frac{1}{2\cdot 3} + \cdots + \frac{1}{k(k+1)} + \frac{1}{(k+1)(k+2)} &= \frac{k}{k+1} + \frac{1}{(k+1)(k+2)}\\ &= \frac{1}{k+1}\left(k + \frac{1}{k+2}\right)\\ &= \frac{1}{k+1}\frac{k(k+2)+1}{k+2}\\ &= \frac{1}{k+1}\frac{k^2+2k+1}{k+2}\\ &= \frac{1}{k+1}\frac{(k+1)^2}{k+2}\\ &= \frac{k+1}{k+2},\end{aligned}$$

as we wanted to demonstrate.

Problem 1.22. Prove that for every $n \in \mathbb{N}$, the following holds:

$$1 + 2 + 2^2 + 2^3 + \cdots + 2^n = 2^{n+1} - 1$$

Solution to the problem 1.22. *To prove the property by induction, we need to prove*

(a) That the property is true for $n = 1$.

(b) That if the property is true for $n = k$, it is also true for $n = k + 1$.

Base case. *For $n = 1$, we have that $1 + 2 = 3 = 2^2 - 1$.*
Inductive case. *Suppose now that the property is true for $n = k$, that is:*

$$1 + 2 + 2^2 + 2^3 + \cdots + 2^k = 2^{k+1} - 1.$$

Let's see that it is true for $n = k + 1$, that is, we want to see that:

$$1 + 2 + 2^2 + 2^3 + \cdots + 2^k + 2^{k+1} = 2^{k+2} - 1.$$

Using the inductive hypothesis:

$$\begin{aligned} 1 + 2 + 2^2 + 2^3 + \cdots + 2^k + 2^{k+1} &= 2^{k+1} - 1 + 2^{k+1} \\ &= 2 \cdot 2^{k+1} - 1 \\ &= 2^{k+2} - 1. \end{aligned}$$

Problem 1.23. From the following equalities:

$$1 + \frac{1}{2} = 2 - \frac{1}{2}$$
$$1 + \frac{1}{2} + \frac{1}{4} = 2 - \frac{1}{4}$$
$$1 + \frac{1}{2} + \frac{1}{4} + \frac{1}{8} = 2 - \frac{1}{8}$$

deduce if there exists a *general law* for any $n \in \mathbb{N}$. If so, state it and prove it by induction.

Solution to the problem 1.23. *Let's determine the value of the expression:*

$$f(n) = 1 + \frac{1}{2} + \frac{1}{4} + \frac{1}{8} + \cdots + \frac{1}{2^n}.$$

To do so, we will assign different values to n:

$$\begin{aligned} f(1) &= 1 + \frac{1}{2} = 2 - \frac{1}{2} \\ f(2) &= 1 + \frac{1}{2} + \frac{1}{4} = 2 - \frac{1}{4} \\ f(3) &= 1 + \frac{1}{2} + \frac{1}{4} + \frac{1}{8} = 2 - \frac{1}{8} \\ f(4) &= 1 + \frac{1}{2} + \frac{1}{4} + \frac{1}{8} + \frac{1}{16} = \frac{31}{16} = 2 - \frac{1}{16} \\ f(5) &= 1 + \frac{1}{2} + \frac{1}{4} + \frac{1}{8} + \frac{1}{16} + \frac{1}{32} = \frac{63}{32} = 2 - \frac{1}{32} \end{aligned}$$

Therefore, we can deduce that:

$$f(n) = 2 - \frac{1}{2^n}.$$

To prove the property by induction, we need to prove

(a) *That the property is true for* $n = 1$.

(b) *That if the property is true for* $n = k$, *it is also true for* $n = k + 1$.

Base case. *For* $n = 1$, *we have that:*

$$1 + \frac{1}{2} = \frac{3}{2} = 2 - \frac{1}{2}.$$

Inductive case. *Now, suppose the formula is true for* $n = k$, *that is:*

$$1 + \frac{1}{2} + \cdots + \frac{1}{2^k} = 2 - \frac{1}{2^k}. \tag{1.4}$$

Let's see that the formula is true for $n = k + 1$, *that is, we want to prove that:*

$$1 + \frac{1}{2} + \cdots + \frac{1}{2^k} + \frac{1}{2^{k+1}} = 2 - \frac{1}{2^{k+1}}.$$

Using the inductive hypothesis:

$$\begin{aligned} 1 + \frac{1}{2} + \cdots + \frac{1}{2^k} + \frac{1}{2^{k+1}} &= 2 - \frac{1}{2^k} + \frac{1}{2^{k+1}} \\ &= 2 - \frac{1}{2^k} + \frac{1}{2 \cdot 2^k} \\ &= 2 - \frac{1}{2^k}\left(1 - \frac{1}{2}\right) \\ &= 2 - \frac{1}{2^k}\frac{1}{2} \\ &= 2 - \frac{1}{2^{k+1}}, \end{aligned}$$

as we wanted to demonstrate.

Problem 1.24. Prove that:

$$n^3 + (n+1)^3 + (n+2)^3$$

is a multiple of 9 for all $n \in \mathbb{N}$.

Solution to the problem 1.24. *To prove the property by induction, we need to prove*

(a) *That the property is true for* $n = 1$.

(b) *That if the property is true for* $n = k$, *it is also true for* $n = k + 1$.

Base case. *For* $n = 1$, *we have:*

$$1^3 + 2^3 + 3^3 = 1 + 8 + 27 = 36 = 4 \cdot 9,$$

which is a multiple of 9.

Inductive case. *Suppose that $n^3 + (n+1)^3 + (n+2)^3$ is a multiple of 9 for $n = k$, that is, there exists a natural number p such that:*

$$k^3 + (k+1)^3 + (k+2)^3 = 9p.$$

Let's see that it is also true for $n = k+1$, that is, there exists a natural number $\tilde{p}$ such that:

$$(k+1)^3 + (k+2)^3 + (k+3)^3 = 9\tilde{p}. \tag{1.5}$$

If we expand the cube $(k+3)^3$, we have that:

$$(k+3)^3 = k^3 + 9k^2 + 27k + 27.$$

Therefore:

$$(k+1)^3 + (k+2)^3 + (k+3)^3 = (k+1)^3 + (k+2)^3 + k^3 + 9k^2 + 27k + 27.$$

Using the inductive hypothesis of equation (1.2), we have that:

$$\begin{aligned}(k+1)^3 + (k+2)^3 + (k+3)^3 &= k^3 + (k+1)^3 + (k+2)^3 + 9k^2 + 27k + 27\\ &= 9p + 9k^2 + 27k + 27\\ &= 9(p + k^2 + 3k + 3) = 9\tilde{p},\end{aligned}$$

where:

$$\tilde{p} = p + k^2 + 3k + 3.$$

Therefore, the property is true.

Problem 1.25. Prove that:

$$\frac{1}{2} + \frac{1}{4} + \cdots + \frac{1}{2^n} < 1$$

for all $n \in \mathbb{N}$.

Solution to the problem 1.25. *To prove the inequality by induction, we need to prove*

(a) That the inequality is true for $n = 1$.

(b) That if the inequality is true for $n = k$, it is also true for $n = k+1$.

Base case. *For $n = 1$, we have that, indeed, $\frac{1}{2} < 1$.*

Inductive case. *Now, suppose the inequality is true for $n = k$, that is:*

$$\frac{1}{2} + \frac{1}{4} + \cdots + \frac{1}{2^{k-1}} + \frac{1}{2^k} < 1. \tag{1.6}$$

Let's see that the inequality is true for $n = k+1$, that is, we want to prove that:

$$\frac{1}{2} + \frac{1}{4} + \cdots + \frac{1}{2^k} + \frac{1}{2^{k+1}} < 1.$$

Multiplying equation (1.6) by $\frac{1}{2}$, we have that:

$$\frac{1}{4} + \cdots + \frac{1}{2^k} + \frac{1}{2^{k+1}} < \frac{1}{2},$$

and, therefore:

$$\frac{1}{2} + \frac{1}{4} + \cdots + \frac{1}{2^k} + \frac{1}{2^{k+1}} < \frac{1}{2} + \frac{1}{2} = 1,$$

as we wanted to demonstrate.

Problem 1.26. Prove that $2^n \leq n!$ for all $n \in \mathbb{N},\ n \geq 4$.

Solution to the problem 1.26. *To prove the inequality by induction, we need to prove*

(a) *That the inequality is true for* $n = 4$.

(b) *That if the inequality is true for* $n = k$, *it is also true for* $n = k + 1$.

Base case. *For* $n = 4$, *we have that, indeed:*

$$2^4 = 16 < 24 = 4!.$$

Inductive case. *Suppose that* $2^k \leq k!$. *Let's see if it is true for* $n = k + 1$, *that is, we want to prove if:*

$$2^{k+1} \leq (k + 1)!.$$

Using the inductive hypothesis:

$$2^{k+1} = 2^k \cdot 2 \leq k! \cdot 2.$$

Then, since $2 < k + 1$, *as* $k \geq 4$,

$$2^{k+1} \leq k! \cdot 2 < k!(k + 1) = (k + 1)!,$$

as we wanted to demonstrate.

Problem 1.27. Prove that $2n + 1 \leq 2^n$ for all $n \in \mathbb{N},\ n \geq 3$.

Solution to the problem 1.27. *To prove the inequality by induction, we need to prove*

(a) *That the inequality is true for* $n = 3$.

(b) *That if the inequality is true for* $n = k$, *it is also true for* $n = k + 1$.

Base case. *For* $n = 3$, *we have that, indeed:*

$$2 \cdot 3 + 1 = 7 \leq 8 = 2^3.$$

Inductive case. *Suppose that* $2k + 1 \leq 2^k$. *Let's see if it is true for* $n = k + 1$, *that is, we want to prove if:*

$$2(k + 1) + 1 \leq 2^{k+1}.$$

Indeed, using the inductive hypothesis:

$$2(k + 1) + 1 = 2k + 2 + 1 = 2k + 1 + 2 \leq 2^k + 2 \leq 2^k + 2^k = 2 \cdot 2^k = 2^{k+1}.$$

In the last inequality, we considered $2 \leq 2^k$ *(since* $k \geq 3$*).*

Problem 1.28. Prove that $n^2 \leq 2^n$ for all $n \in \mathbb{N},\ n \geq 4$.

Solution to the problem 1.28. *To prove the inequality by induction, we need to prove*

(a) *That the inequality is true for* $n = 4$.

(b) *That if the inequality is true for $n = k$, it is also true for $n = k + 1$.*

Base case. *For $n = 4$, we have that, indeed:*

$$4^2 = 16 \leq 16 = 2^4.$$

Inductive case. *Suppose that $k^2 \leq 2^k$. Let's see if it is true for $n = k + 1$, that is, we want to prove if:*

$$(k + 1)^2 \leq 2^{k+1}.$$

Using the inductive hypothesis:

$$(k + 1)^2 = k^2 + 2k + 1 \leq 2^k + 2k + 1 \leq 2^k + 2^k = 2 \cdot 2^k = 2^{k+1},$$

where in the last inequality we considered that $2k + 1 \leq 2^k$, since $k \geq 4$.

Problem 1.29. Consider the sequence of numbers defined by the recursive formula:

$$a_n = 2a_{n-1} - a_{n-1}^2,$$

where a_0 and a_1 are any real numbers. Prove that:

$$a_n = 1 - (1 - a_0)^{2^n}.$$

Solution to the problem 1.29. *To prove the property by induction, we need to prove*

(a) *That the property is true for $n = 1$.*

(b) *That if the property is true for $n = k$, it is also true for $n = k + 1$.*

Base case. *For $n = 1$, we have that:*

$$a_1 = 2a_0 - a_0^2 = 1 - 1 + 2a_0 - a_0^2 = 1 - (1 - a_0)^2,$$

and therefore, the base case is true.

Inductive case. *Suppose now that it is true for $n = k$, that is:*

$$a_k = 1 - (1 - a_0)^{2^k}. \tag{1.7}$$

Let's see that it is true for $n = k + 1$, that is, we want to prove that:

$$a_{k+1} = 1 - (1 - a_0)^{2^{k+1}}.$$

If we use the definition of the term a_{k+1} and the inductive hypothesis of equation (1.2), we have that:

$$\begin{aligned}
a_{k+1} &= 2a_k - a_k^2 = 2(1 - (1 - a_0)^{2^k}) - (1 - (1 - a_0)^{2^k})^2 \\
&= 2 - 2(1 - a_0)^{2^k} - 1 - ((1 - a_0)^{2^k})^2 + 2(1 - a_0)^{2^k} \\
&= 2 - 1 - ((1 - a_0)^{2^k})^2 \\
&= 1 - (1 - a_0)^{2^k \cdot 2} \\
&= 1 - (1 - a_0)^{2^{k+1}},
\end{aligned}$$

Problem 1.30. The Fibonacci numbers are defined recursively as follows:

$$f_1 = f_2 = 1, \quad f_n = f_{n-1} + f_{n-2} \ \forall n \geq 2,$$

so that the first Fibonacci numbers are:

$$1,\ 1,\ 2,\ 3,\ 5,\ 8,\ 13,\ 21,\ 34,\ 55,\ 89,\ 144,\ 233,\ 377,\ 610,\ \ldots$$

Prove the following properties of these numbers:

(i) f_{3n} is even, $\forall n \in \mathbb{N}$.

(ii) f_{4n} is divisible by three, $\forall n \in \mathbb{N}$.

Solution to the problem 1.30. *To prove the properties of parts (i) and (ii) by induction, we need to prove*

(a) That the property is true for $n = 1$.

(b) That if the property is true for $n = k$*, it is also true for* $n = k + 1$.

(i) Let's see that f_{3n} *is even for all* $n \in \mathbb{N}$.

Base case. *The base case is* $n = 1$*. In this case, we have that* $f_3 = 2$*, which is even.*

Inductive case. *Suppose that* f_{3n} *is even for* $n = k$*, that is,* $f_{3k} = 2p$ *where* $p \in \mathbb{N}$*. Let's see that* $f_{3(k+1)}$ *is also even, that is, we want to prove that there exists* $\tilde{p} \in \mathbb{N}$ *such that:*

$$f_{3(k+1)} = 2\tilde{p}.$$

By the definition of the Fibonacci numbers:

$$f_{3(k+1)} = f_{3(k+1)-1} + f_{3(k+1)-2} = f_{3k+2} + f_{3k+1}.$$

Similarly, using the definition of the Fibonacci numbers:

$$f_{3k+2} = f_{3k+1} + f_{3k},$$

and therefore:

$$f_{3(k+1)} = f_{3k+2} + f_{3k+1} = f_{3k+1} + f_{3k} + f_{3k+1} = 2f_{3k+1} + f_{3k}.$$

Then, using the inductive hypothesis:

$$f_{3(k+1)} = 2f_{3k+1} + f_{3k} = 2(f_{3k+1} + p),$$

and therefore, it is an even number (taking $\tilde{p} = p + f_{3k+1}$*).*

(ii) We need to prove that f_{4n} *is divisible by three, or in other words, that it is a multiple of three.*

Base case. *The base case is* $f_4 = 3$*, which is a multiple of three.*

Inductive case. *Suppose now that* $f_{4k} = 3p, p \in \mathbb{N}$*. Let's see that* $f_{4(k+1)}$ *is also a* *multiple of 3. Indeed:*

$$\begin{aligned} f_{4(k+1)} &= f_{4k+3} + f_{4k+2} \\ &= f_{4k+2} + f_{4k+1} + f_{4k+2} \\ &= 2f_{4k+2} + f_{4k+1} \\ &= 2(f_{4k+1} + f_{4k}) + f_{4k+1} \\ &= 2f_{4k+1} + 2f_{4k} + f_{4k+1} \\ &= 3f_{4k+1} + 2f_{4k} \\ &= 3(f_{4k+1} + 2p), \end{aligned}$$

which is a multiple of 3.

Problem 1.31. Calculate the value of the expression:

$$\frac{2+4+\cdots+2n}{1+3+\cdots+(2n-1)}$$

for all $n \in \mathbb{N}$, and prove the resulting formula by induction.

Solution to the problem 1.31. *Let's determine the value of the expression:*

$$f(n) = \frac{2+4+\cdots+2n}{1+3+\cdots+(2n-1)}.$$

To do so, we will assign different values to n:

$$\begin{aligned} f(1) &= \frac{2}{1} \\ f(2) &= \frac{2+4}{1+3} = \frac{6}{4} = \frac{2\cdot 3}{2^2} = \frac{3}{2} \\ f(3) &= \frac{2+4+6}{1+3+5} = \frac{12}{9} = \frac{3\cdot 4}{3^2} = \frac{4}{3} \\ f(4) &= \frac{2+4+6+8}{1+3+5+7} = \frac{20}{16} = \frac{4\cdot 5}{4^2} = \frac{5}{4} \\ f(5) &= \frac{2+4+6+8+10}{1+3+5+7+9} = \frac{30}{25} = \frac{5\cdot 6}{5^2} = \frac{6}{5} \end{aligned}$$

Therefore, we can deduce that:

$$f(n) = \frac{n(n+1)}{n^2} = \frac{n+1}{n}$$

To prove this, we have two options. We can separately prove that:

$$2+4+\cdots+2n = n(n+1)$$

and that:

$$1+3+5+\cdots+(2n-1) = n^2,$$

or we can prove that:

$$f(n) = \frac{n+1}{n}.$$

We will do it separately because it will be simpler. Therefore, we need to prove that:

(i) $2+4+\cdots+2n = n(n+1)$

(ii) $1+3+5+\cdots+(2n-1) = n^2$

(i) ***Base case***. *When* $n=1$, *indeed* $2 = 1\cdot 2$.

Inductive case. *Suppose the property is true for* $n=k$, *that is:*

$$2+4+\cdots+2k = k(k+1).$$

Let's see that the property is true for $n=k+1$. *That is, we want to prove that:*

$$2+4+\cdots+2k+2(k+1) = (k+1)(k+2).$$

Therefore:

$$2+4+\cdots+2k+2(k+1) = k(k+1)+2(k+1) = (k+1)(k+2),$$

as we wanted to demonstrate.

(ii) ***Base case***. *When* $n=1$, *indeed* $1 = 1^2$.

Inductive case. *Suppose the property is true for* $n=k$, *that is:*

$$1+3+\cdots+(2k-1) = k^2.$$

Let's see that the property is true for $n=k+1$. *That is, we we want to prove that:*

$$1+3+\cdots+(2k-1)+(2k+1) = (k+1)^2.$$

Therefore:

$$1+3+\cdots+(2k-1)+(2k+1) = k^2+2k+1 = (k+1)^2,$$

as we wanted to demonstrate.

Therefore, using (i) and (ii), we have that:

$$f(n) = \frac{2+4+\cdots+2n}{1+3+\cdots+(2n-1)} = \frac{n(n+1)}{n^2} = \frac{n+1}{n},$$

as we wanted to demonstrate.

Part II

The Set of Complex Numbers

2

Numerical Sets. The Set of Complex Numbers

«Number is the ruler of forms and ideas, and the cause of gods and daemons.» Pythagoras, cited in the book *Life of Pythagoras* by Iamblichus of Chalcis, according to the English translation by Thomas Taylor (1818).

In this second chapter, the most important sets of numbers are presented, such as natural numbers, integers, rational numbers, real numbers, and complex numbers. For the first four sets, in addition to presenting them from a constructivist perspective, a brief review of their mathematical structure is provided. For the set of complex numbers, for which no prior knowledge is assumed, the mathematical operations that can be performed with them and their different representations (standard form, polar form, exponential form) are defined.

2.1 Numerical Sets

2.1.1 The Set of Natural Numbers

Numbers, such as 2, -3, the fraction $\frac{1}{2}$, or π, are human creations that have responded to different needs. For example, since ancient times, humans have had the need to count:

DOI: 10.1201/9781003399735-2

- the sheep in a flock,
- the number of trees in a small plantation,
- the number of sons and daughters.

In this way, the concept of natural numbers arises, which we represent with the symbol $\mathbb{N}$, and which is a set formed by the numbers $1, 2, 3, \ldots$ This is indicated as follows:

$$\mathbb{N} = \{1, 2, 3, \ldots\}$$

To denote that the number 3 is a natural number, we write it as $3 \in \mathbb{N}$, which reads as «3 belongs to the set of natural numbers».

It could be considered that the number 0 is also part of the set of natural numbers. In this case, we would have:

$$\mathbb{N} = \{0, 1, 2, 3, \ldots\}$$

and the set that does not contain zero would be denoted as $\mathbb{N}^* = \{1, 2, 3, \ldots\}$.

As a numerical set, the set of natural numbers has two internal binary operations, addition and multiplication:

$$+ : \mathbb{N} \times \mathbb{N} \to \mathbb{N}(a, b) \to a + b$$

$$\cdot : \mathbb{N} \times \mathbb{N} \to \mathbb{N}(a, b) \to a \cdot b$$

In this context, the term internal refers to the fact that the result of adding or multiplying two natural numbers is another natural number. These two operations have the following properties:

(A1) **Addition is associative**. That is, given $a, b, c \in \mathbb{N}$, we have:

$$(a + b) + c = a + (b + c)$$

(A2) **Addition is commutative**. That is, given $a, b \in \mathbb{N}$, we have:

$$a + b = b + a$$

(A3) **Existence of an additive identity**. That is, given $a \in \mathbb{N}$, there exists the element $0 \in \mathbb{N}$ such that:

$$a + 0 = 0 + a = a$$

(A4) **Multiplication is associative**. That is, given $a, b, c \in \mathbb{N}$, we have:

$$(a \cdot b) \cdot c = a \cdot (b \cdot c)$$

(A5) **Existence of a multiplicative identity**. That is, given $a \in \mathbb{N}$, there exists the element $1 \in \mathbb{N}$ such that:

$$a \cdot 1 = 1 \cdot a = a$$

(A6) **Distributive property of multiplication over addition**. Given $a, b, c \in \mathbb{N}$, we have:

$$a \cdot (b + c) = a \cdot b + a \cdot c$$
$$(a + b) \cdot c = a \cdot c + b \cdot c$$

(A7) **Zero is the absorbing element in multiplication**. That is, given $a \in \mathbb{N}$, we have:

$$a \cdot 0 = 0 \cdot a = 0$$

(A8) **Multiplication is commutative**. That is, given $a, b \in \mathbb{N}$, we have:

$$a \cdot b = b \cdot a.$$

The set of natural numbers endowed with the two internal operations of addition and multiplication and with properties A1–A7 that we have just described is called a **semiring**. If property A8 is also satisfied, we say it is a **commutative semiring**.

In addition to these properties, the set of natural numbers has other characteristics:

- Every natural number has a *successor*. We will see that this property is also shared with integers. For example, the successor of the number 1 is the number 2, the successor of the number 2 is the number 3, and so on.
- Every natural number, except 0 (or except 1 if we consider $\mathbb{N}^*$), has a *predecessor*. This element is called the *first element*.
- There are no numbers between two consecutive natural numbers. This property is also satisfied by the set of integers.
- There are infinitely many natural numbers.

Although not explicitly stated, these last properties of natural numbers are used (although implicitly) in proofs by mathematical induction presented in the chapter "Introduction to Mathematical Reasoning."

2.1.2 The Set of Integers

Clearly, the natural number 2 can represent the fact «I have two sheep in my flock». However, if as a result of a commercial agreement, I go from *having* two sheep to *owing* two sheep, how do we represent this fact? In such a case, we can turn to *integers*, represented by the symbol $\mathbb{Z}$, which comprise the natural numbers, the natural numbers with *negative signs*, and zero. This set can be represented as follows:

$$\mathbb{Z} = \{\ldots, -3, -2, -1, 0, 1, 2, 3, \ldots\}.$$

As a numerical set, the set of integers inherits the two binary operations of natural numbers, addition and multiplication:

$$\begin{aligned} + : \mathbb{Z} \times \mathbb{Z} &\to \mathbb{Z} \\ (a, b) &\to a + b \end{aligned}$$

$$\begin{aligned} \cdot : \mathbb{Z} \times \mathbb{Z} &\to \mathbb{Z} \\ (a, b) &\to a \cdot b \end{aligned}$$

and to properties A1–A8, a new property is added:

(A9) **Existence of additive inverses**. That is, given $a \in \mathbb{Z}$, there exists an element $-a \in \mathbb{Z}$ such that:

$$a + (-a) = (-a) + a = 0.$$

The set of integers endowed with the two internal operations of addition and multiplication and with properties A1–A9 that we have just described is called a **commutative ring**. Since integers include natural numbers, we can write:

$$\mathbb{N} \subset \mathbb{Z}.$$

The human need for numerical sets can also be posed in terms of equations. For example, the equation:

$$x + 2 = 0$$

can only be solved if we have the set of integers $\mathbb{Z}$, as the solution is the integer -2, i.e., $x = -2 \in \mathbb{Z}$. In general, the equation:

$$x + a = 0, \text{ where } a \text{ is a natural number}$$

has as its solution the integer $-a$, i.e., $x = -a \in \mathbb{Z}$.

2.1.3 The Set of Rational Numbers

The next numerical set corresponds to rational numbers, which is formed by fractions and is represented by the symbol $\mathbb{Q}$. Its creation is based on the need to represent parts of a whole, such as dividing agricultural land into three equal parts as a result of an inheritance. In this sense, this set can be represented as follows:

$$\mathbb{Q} = \left\{ \frac{a}{b} \text{ such that } a, b \in \mathbb{Z},\ b \neq 0 \right\},$$

that is, it is the set formed by the quotients or fractions of integers, in which it is essential that the denominator b is different from zero. Rational numbers are $\frac{1}{2}$ or $\frac{2}{-3}$, i.e., numbers whose decimal expression has a repeating pattern. For example:

$$\frac{1}{2} = 0.5,$$
$$\frac{2}{-3} = -0.6666\ldots = -0.\bar{6},$$

or:

$$\frac{211}{990} = 0.21313\ldots = 0.2\overline{13}.$$

But we can also write:

$$1 = \frac{1}{1} = \frac{2}{2},$$
$$-3 = \frac{-3}{1} = \frac{3}{-1},$$

i.e., natural numbers and integers are also rational numbers.

In this sense, we can write:

$$\mathbb{N} \subset \mathbb{Z} \subset \mathbb{Q},$$

which reads «the set of natural numbers is a subset of the set of integers, which in turn is a subset of the set of rational numbers». That is, every natural number is also an integer and every integer is, in turn, a rational number.

Just as to solve equations of the form $x + a = 0$ where a is a natural number, we need to create the set of integers; the equation:

$$2x = 3$$

can only be solved with the help of the set of rational numbers $\mathbb{Q}$, as the solution is the rational number $\frac{3}{2}$, i.e., $x = \frac{3}{2} \in \mathbb{Q}$. In general, the equation:

$$bx = a, \text{ where } a, b \text{ are integers and } b \neq 0$$

has as its solution the rational number $\frac{a}{b}$, i.e., $x = \frac{a}{b} \in \mathbb{Q}$.

The set of rational numbers, seen as a mathematical structure, also inherits the operations of addition and multiplication. That is:

$$\begin{aligned} + : \mathbb{Q} \times \mathbb{Q} &\to \mathbb{Q} \\ \left(\frac{a}{b}, \frac{c}{d}\right) &\mapsto \frac{a}{b} + \frac{c}{d} = \frac{ad + bc}{bd} \end{aligned}$$

$$\begin{aligned} \cdot : \mathbb{Q} \times \mathbb{Q} &\to \mathbb{Q} \\ \left(\frac{a}{b}, \frac{c}{d}\right) &\mapsto \frac{ac}{bd} \end{aligned}$$

and to properties A1–A9, a new property is added:

(A10) **Existence of multiplicative inverses.** That is, given $a \in \mathbb{Q}$, there exists an element $a^{-1} \in \mathbb{Q}$ such that:

$$a \cdot a^{-1} = a^{-1} \cdot a = 1.$$

The set of rational numbers endowed with the two internal operations of addition and multiplication and with properties A1–A10 that we have described is called a **commutative field**.

2.1.4 The Set of Real Numbers

At this point, we can ask ourselves, are there equations that cannot be solved in the set of rational numbers?

Imagine we want to calculate the length of the hypotenuse of a right triangle with legs of length 1, as shown in Figure 2.1.

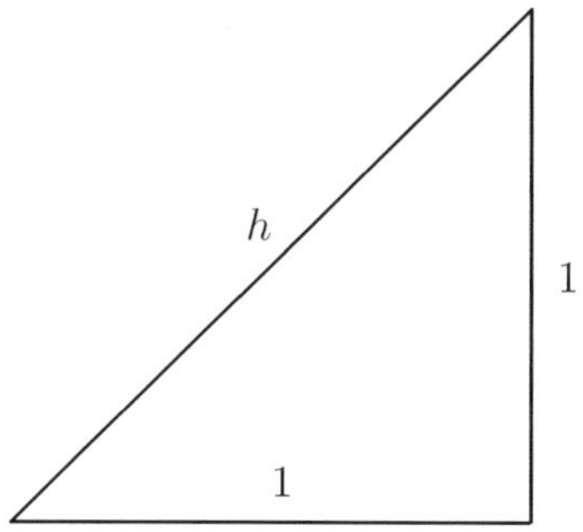

FIGURE 2.1
Right triangle with equal legs of length 1.

According to the Pythagorean theorem, we have:

$$h^2 = 1^2 + 1^2 = 2.$$

The Greeks already demonstrated that the equation:

$$h^2 = 2,$$

as simple as it may seem, has no solution that can be expressed as a fraction. However, we all know that one of the solutions to this equation is the number $\sqrt{2}$. The other solution is $-\sqrt{2}$.

It can be proven that the number $\sqrt{2}$ is *not* a rational number (in fact, you will find the proof in the chapter «Introduction to Mathematical Reasoning»). Numbers that are not rational are called *irrational numbers* and are usually represented by the symbol $\mathbb{R}\backslash\mathbb{Q}$ or $\mathbb{I}$.

Other *irrational* numbers include the number π (which is approximately 3.1416), the number e (which is approximately 2.7183), and the square roots of prime numbers $\sqrt{2}, \sqrt{3}, \ldots$

Recall that rational numbers, when expressed in decimal form, are either numbers with a finite number of decimals or numbers with an infinite number of decimals but with a repeating pattern. For example:

$$\frac{3}{4} = 0.75, \quad \frac{5}{33} = 0.1515\ldots = 0.\overline{15}.$$

In contrast, irrational numbers, when expressed in decimal form, have an infinite number of decimals with no periodicity. For example, the number π has infinite decimals with no kind of periodicity:

$$\pi = 3.14159265358979323846264338327950288419716939937 51\ldots$$

The set of *real* numbers, which we represent with the letter $\mathbb{R}$, is defined as the *union* of the sets of rational numbers and irrational numbers, that is:

$$\mathbb{R} = \mathbb{Q} \cup [\mathbb{R}\backslash\mathbb{Q}] = \mathbb{Q} \cup \mathbb{I}.$$

Remember that we use the symbol $\cup$ to denote the union of sets. The previous expression reads «the set of real numbers is the union of the set of rational numbers and the set of irrational numbers».

Observation 2.1. Finally, we have:

$$\mathbb{N} \subset \mathbb{Z} \subset \mathbb{Q} \subset \mathbb{R}.$$

This inclusion relationship can be seen in Figure 2.2.

At this point, it is interesting to note that real numbers, such as $\sqrt{2}$ or π, are not just a mathematical invention but also appear in nature. For example:

- What is the length of the diagonal of a square with a side equal to one meter?
- What is the area of a circle with a radius equal to one meter?

From the perspective of mathematical structure, the set of real numbers also satisfies properties A1–A10, making it a **field**.

Property 2.1. *Real numbers can be represented as points on a line called the* real line *(see Figure 2.3). The point corresponding to zero is called the origin. Conversely, for each point on the line, there is one and only one associated real number.*

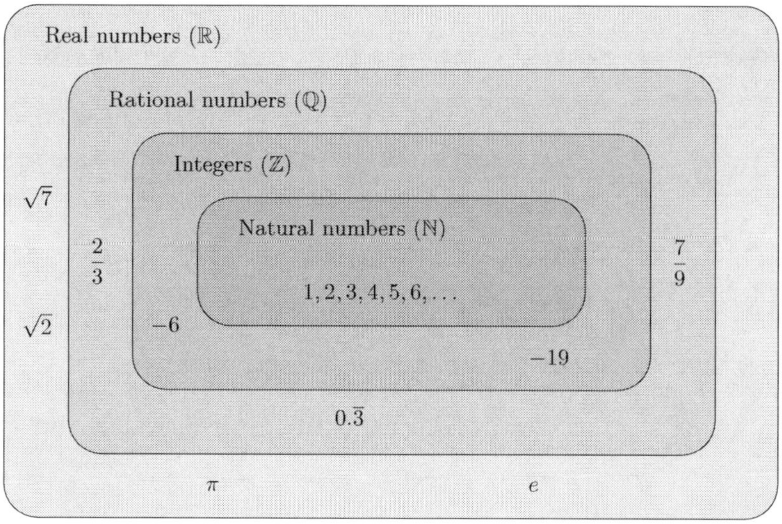

FIGURE 2.2
Inclusion relationship among the different numerical sets.

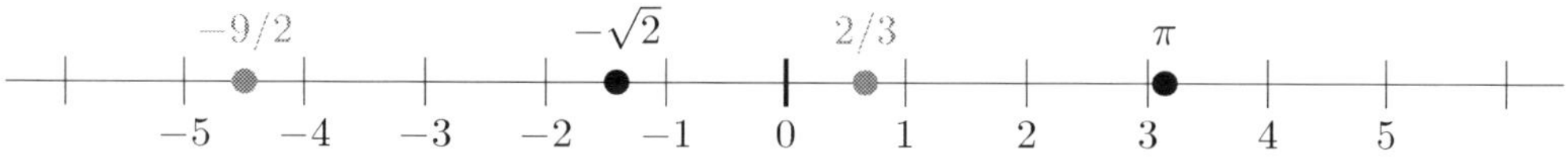

FIGURE 2.3
Representation of the real line.

Proposed Exercise 2.1. Express the numbers $1.3\overline{26}$ and $0.\overline{9}$ in fraction form.

2.2 The Set of Complex Numbers

Do all quadratic equations have a solution?

We have already seen in the previous sections that in order to solve linear equations of the form $ax + b = 0$, where a and b are integers, we need more than just integers; we need rational numbers. Similarly, we have seen that to solve some quadratic equations, such as $x^2 = 2$, we need real numbers (which include irrational numbers).

The problem now arises when we want to solve the equation:

$$x^2 = -1.$$

We know that there is *no* real number whose square is a negative number. Therefore, it could be said that $x^2 = -1$ has no solutions.

Recall that to solve the equation $x^2 = 2$ we imagined the number $\sqrt{2}$; thus, analogously, to solve the equation $x^2 = -1$ we create a new number:

$$\sqrt{-1}.$$

This number is assigned the symbol i. Therefore, to obtain solutions for the equation $x^2 = -1$ and for *all* quadratic equations, $ax^2 + bx + c = 0$, we need to create a new set: the set of complex numbers.

This allows us to make a first definition of what the set of complex numbers will be:

Definition 2.1. The set of complex numbers is defined as:

$$\mathbb{C} = \{z \text{ is a solution of the equation } ax^2 + bx + c = 0 \text{ where } a, b, c \in \mathbb{R}\},$$

that is, $\mathbb{C}$ is the set of numbers where every quadratic equation has a solution.

What we propose is a significant creative leap, as it involves creating, out of nothing, a solution for a type of equation, not being content with the existence of equations without a solution. The advantages of this creative leap will outweigh the initial discomfort.

2.2.1 Introductory Example

Throughout history, various problems have arisen that had no solution within the set of real numbers. Below, some historical problems are presented, in which the need for a new numerical set, the set of complex numbers, begins to be intuited.

Suppose we want to solve the following problem, posed by the Greek mathematician Diophantus of Alexandria in 275 BC:

Consider a right triangle like the one in Figure 2.4.

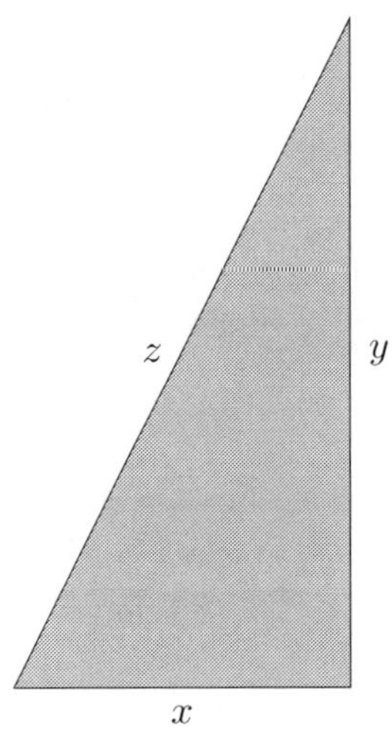

FIGURE 2.4
Right triangle with sides x, y, and z.

Find the sides of the triangle (x, y, and z) such that its perimeter $x + y + z$ is equal to 12 and the area is equal to 7.

In solving this problem, the following equation appears:

$$6x^2 - 43x + 84 = 0,$$

where x is the length of one of the sides of the triangle.

The formula to solve a quadratic equation of the form $ax^2 + bx + c = 0$ is:

$$x = \frac{-b \pm \sqrt{b^2 - 4ac}}{2a} = \frac{-b \pm \sqrt{\Delta}}{2a},$$

where the expression $\Delta = b^2 - 4ac$ is called the *discriminant.* In the case of the example we are considering, we can see that the discriminant is negative:

$$\Delta = b^2 - 4ac = (-43)^2 - 4 \cdot 6 \cdot 84 = 1849 - 2016 = -167 < 0,$$

and therefore, there are no real solutions. That is, we cannot construct any triangle with an area of 7 and a perimeter of 12.

Now consider another problem, posed by the Italian physician and mathematician Girolamo Cardano in 1545.

> Divide the number 10 into two parts, so that the product of these parts is 40. That is, we want to determine two numbers x and y such that their sum is 10 and their product is 40.

In this case, the two equations presented by the problem are:

$$\begin{aligned} x + y &= 10 \\ xy &= 40 \end{aligned}$$

where, if we solve for y from the first equation and substitute it into the second:

$$x(10 - x) = 40 \quad \text{or equivalently,} \quad x^2 - 10x + 40 = 0.$$

This equation also has no real solutions. However, in this case, Cardano himself said that the two solutions were:

$$5 + \sqrt{-15} \quad \text{and} \quad 5 - \sqrt{-15}.$$

Gauss was the first to call such expressions *complex numbers.*

As we have seen before, the set of real numbers is insufficient to find solutions to quadratic equations. For this reason, we make the following definition:

Definition 2.2. We define the number i such that $i^2 = -1$ or, in other words, $i = \sqrt{-1}$. The number i is called the *imaginary unit.*

René Descartes was the first to use the term *imaginary unit* in 1637.

Note now that, since the imaginary number i satisfies $i = \sqrt{-1}$, the solutions to the equation:

$$x^2 - 10x + 40 = 0$$

can be denoted as:

$$5 \pm \sqrt{-15} = 5 \pm \sqrt{15 \cdot (-1)} = 5 \pm \sqrt{15}\sqrt{-1} = 5 \pm \sqrt{15}\,i.$$

Thus, even if the discriminant of a quadratic equation is negative, we can always find the solutions using the imaginary unit i. Recall that the number i allows us to work with roots of negative numbers.

Consequently, if we allow solutions to be complex numbers, that is, numbers of the form:

$$a + bi$$

all quadratic equations will have a solution.

TABLE 2.1
Real part and imaginary part.

$z = a + bi$	Re(z)	Im(z)
$z_1 = 2 + 3i$	2	3
$z_2 = 3 - 4i$	3	-4
$z_3 = -2/3 + 8i$	$-2/3$	8
$z_4 = \sqrt{2} + 10i$	$\sqrt{2}$	10

Definition 2.3. We define the set of *complex numbers*, represented by the symbol $\mathbb{C}$, as the set consisting of expressions of the form $a + bi$, where a and b are real numbers and i is such that its square is equal to -1. Formally:

$$\mathbb{C} = \{a + bi \mid a, b \in \mathbb{R},\ i^2 = -1\}.$$

2.3 Representation of Complex Numbers: Standard Form

We have already seen that a complex number is a number of the form:

$$a + bi,$$

where a and b are real numbers.

Examples of complex numbers are:

$$z_1 = 2 + 3i\ , \qquad z_2 = 3 - 4i\ ,$$
$$z_3 = -\frac{2}{3} + 8i\ , \qquad z_4 = \sqrt{2} + 10i\ .$$

From the previous expressions, we can observe that a complex number is formed by the sum of two clearly differentiated parts:

$$\underbrace{a}_{\text{real part}} + \underbrace{b}_{\text{imaginary part}} i,$$

which we define as the *real part* (a) and the *imaginary part* (b).

In Table 2.1, we can find the real part and the imaginary part of the four complex numbers considered earlier.

Property 2.2. *Equality of two complex numbers Two complex numbers are equal if, and only if, they have the same real part and the same imaginary part. That is,* $z_1 = z_2$ *if, and only if:*

$$Re(z_1) = Re(z_2) Im(z_1) = Im(z_2)$$

- If a complex number has no real part (its real part is zero), it is called a *pure imaginary* complex number. For example, $i, 2i, \sqrt{3}i$ are pure imaginary complex numbers.
- If, on the other hand, a complex number has no imaginary part (its imaginary part is zero), it is called a *real* number. For example, $2, -1, e$ are real numbers.

Property 2.3 (Inclusion of $\mathbb{R}$ in $\mathbb{C}$). *Every real number $a \in \mathbb{R}$ is associated with the complex number $a + 0 \cdot i$, that is, it is a complex number with a zero imaginary part.*

Real numbers can be represented as points on a line (the real line). However, there is no place for $\sqrt{-1}$ on the real line, that is, there is no place for i. How can we then represent complex numbers graphically?

To solve this problem, Gauss asked what would happen if a new direction (a vertical axis, like the y-axis) were introduced and if to position i, a point located above the number line at a unit distance was used (*see* Figure 2.5). All the new numbers needed to solve any equation would now be combinations of i and the usual numbers.

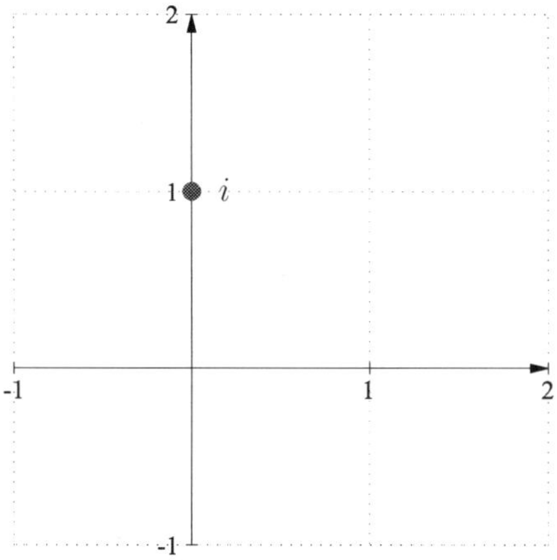

FIGURE 2.5
Representation of the complex plane. In particular, representation of the number i in the complex plane.

We have already seen that complex numbers consist of a real part and an imaginary part. In this way, as Gauss pointed out, each complex number can be represented by a unique point in the plane, which is obtained as follows: the real part represents the *abscissa* (colloquially, the x-coordinate) and the imaginary part is the *ordinate* (colloquially, the y-coordinate). *See* Figure 2.6.

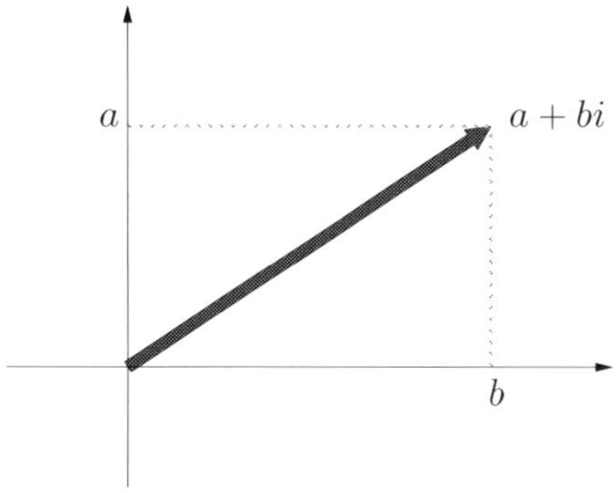

FIGURE 2.6
Representation in the plane of the complex number $a + bi$.

We can observe this correspondence in Table 2.2.

TABLE 2.2
Correspondence between complex numbers and points in the plane.

$z = a + bi$	$\mathrm{Re}(z)$	$\mathrm{Im}(z)$	Point
$z_1 = 2 + 3i$	2	3	$(2, 3)$
$z_2 = 3 - 4i$	3	-4	$(3, -4)$
$z_3 = -2/3 + 8i$	$-2/3$	8	$(-2/3, 8)$
$z_4 = \sqrt{2} + 10i$	$\sqrt{2}$	10	$(\sqrt{2}, 10)$

In this way, the complex number $a + bi$ will be identified with the point (a, b) in the plane, allowing us a new definition of the set of complex numbers.

Definition 2.4. We define the set of *complex numbers*, represented by the symbol $\mathbb{C}$, as the set of points in the plane, that is, as the set formed by the Cartesian product of $\mathbb{R}$ with $\mathbb{R}$. That is:

$$\mathbb{C} = \mathbb{R} \times \mathbb{R} = \{(a, b) \mid a, b \in \mathbb{R}\}$$

2.4 Operations with Complex Numbers

As with any set, we need to know what operations we can perform with its elements. In this case, for the set of complex numbers, we will see how we can add, subtract, multiply, divide, etc.

2.4.1 Addition and Subtraction of Complex Numbers in Standard Form

To add complex numbers, we add the real parts and the imaginary parts separately.

Given $z_1 = 1 + 2i$ and $z_2 = 4 + 3i$, then:

$$\begin{aligned} z_1 + z_2 &= (1 + 2i) + (4 + 3i) \\ &= (1 + 4) + (2 + 3)i \\ &= 5 + 5i. \end{aligned}$$

Notice that we treat the number i as if it were the variable of a polynomial, grouping together terms without i and terms with i.

In the complex plane, numbers are added in the same way as vectors in the plane. That is, the parallelogram law is used, as shown in Figure 2.7.

The subtraction of two complex numbers $z_1 - z_2$, where $z_1 = a + bi$ and $z_2 = c + di$, can essentially be seen as an addition. Thus, the difference:

$$z_1 - z_2 = z_1 + (-z_2)$$

is simply the *addition* of the complex numbers z_1 and $-z_2$, where:

$$-z_2 = -(c + di) = -c - di.$$

Thus, we can write:

$$\begin{aligned} z_1 - z_2 &= z_1 + (-z_2) \\ &= (a + bi) + (-c - di) \\ &= (a - c) + (b - d)i. \end{aligned}$$

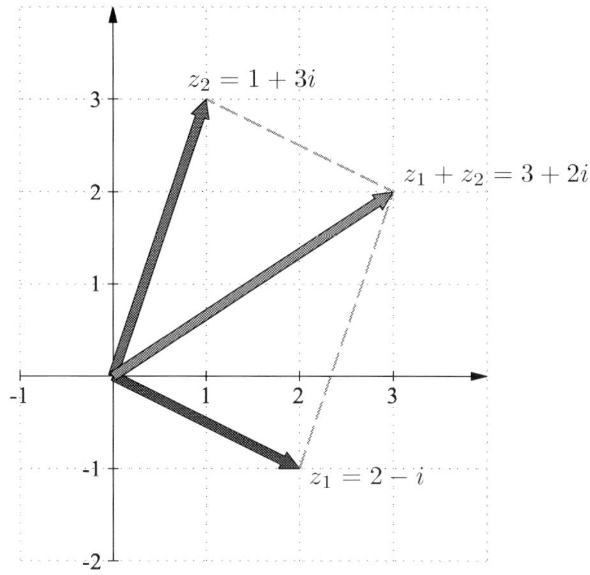

FIGURE 2.7
Addition of two complex numbers using the parallelogram law.

For example, given $z_1 = 2 - i$ and $z_2 = 1 + 3i$, then:

$$\begin{aligned} z_1 - z_2 &= z_1 + (-1) \cdot z_2 \\ &= (2 - i) + (-1) \cdot (1 + 3i) \\ &= (2 - i) + (-1 - 3i) \\ &= (2 - 1) + (-1 - 3)i \\ &= 1 - 4i. \end{aligned}$$

A graphical example of the difference of two complex numbers can be seen in Figure 2.8.

Definition 2.5 (Opposite)**.** The complex number $-z$ is called the *opposite* of the complex number z. From a geometric point of view, the complex number $-z$ is the symmetric with respect to the origin of the complex number z, as shown in Figure 2.9.

2.4.2 Product of Complex Numbers in Standard Form

To multiply a complex number by a real number, we multiply the real and imaginary parts of the complex number by the real number.

Given $z = 1 + 2i$, to multiply it by a, a real number, we do:

$$a \cdot z = a \cdot (1 + 2i) = a \cdot 1 + a \cdot 2i = a + 2ai.$$

Remember that real numbers are included within complex numbers. Specifically, any real number a can be written as $a + 0i$. Therefore, we have just seen how to multiply a complex number by another complex number with a zero imaginary part. Next, we will see how to multiply two complex numbers where both may have non-zero imaginary parts. But first, let's recall how to multiply polynomials $p(x) = a + bx$ and $q(x) = c + dx$.

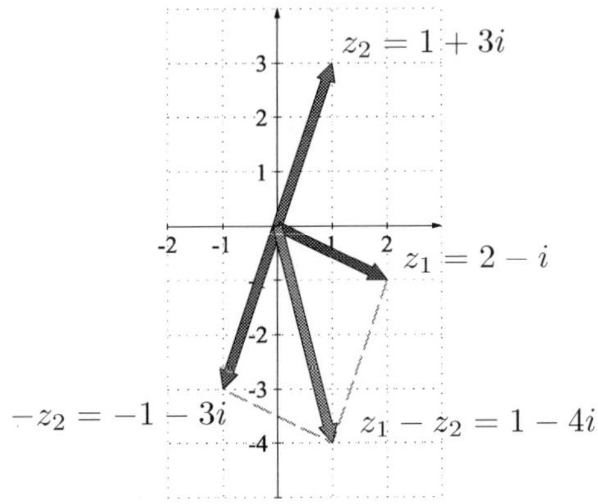

FIGURE 2.8
Subtraction of two complex numbers. To calculate the difference between the complex numbers $z_1 - z_2$, we consider the sum of the numbers $z_1 + (-z_2)$, where $-z_2$ is the *opposite* of the complex number z_2.

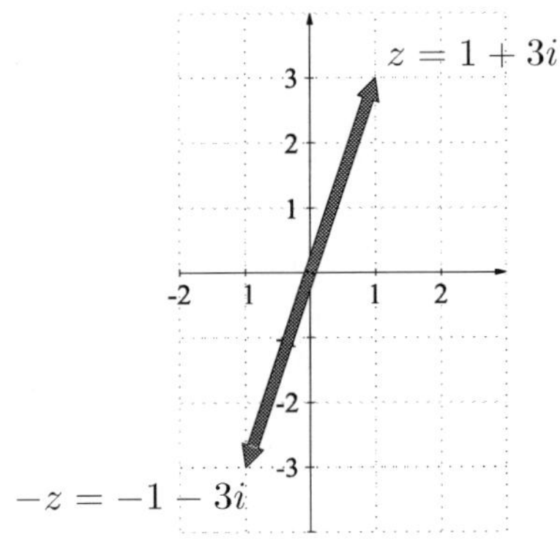

FIGURE 2.9
The complex number z and its opposite $-z$.

Two polynomials are multiplied by applying the distributive property, that is:

$$
\begin{aligned}
p(x) \cdot q(x) &= (a + bx) \cdot (c + dx) \\
&= ac + adx + bcx + bdx^2 \\
&= ac + (ad + bc)x + bdx^2.
\end{aligned}
$$

To multiply complex numbers, we must proceed in the same way, noting that $i^2 = -1$. Given $z_1 = 1 + 2i$ and $z_2 = 4 + 3i$:

$$
\begin{aligned}
z_1 \cdot z_2 &= (1 + 2i) \cdot (4 + 3i) \\
&= 1 \cdot (4 + 3i) + 2i \cdot (4 + 3i) \\
&= 1 \cdot 4 + 1 \cdot 3i + 2 \cdot 4i + 2 \cdot 3i^2 \\
&= 1 \cdot 4 + 1 \cdot 3i + 2 \cdot 4i - 2 \cdot 3 \qquad (\text{using } i^2 = -1) \\
&= (1 \cdot 4 - 2 \cdot 3) + (1 \cdot 3 + 2 \cdot 4)i \\
&= -2 + 11i.
\end{aligned}
$$

2.4.3 The Conjugate of a Complex Number

Definition 2.6 (Conjugate). The conjugate of a complex number, $z = a + bi$, is another complex number with the same real part, but with the imaginary part having the opposite sign. The conjugate of a complex number z is denoted as $\overline{z} = a - bi$.

Given $z = 1 + 2i$, its conjugate is:

$$\overline{z} = 1 - 2i.$$

We will use the conjugate of a complex number primarily for dividing complex numbers, as we will see later.

In the complex plane, the conjugate of a complex number is obtained by reflecting it across the x-axis (*see* Figure 2.10).

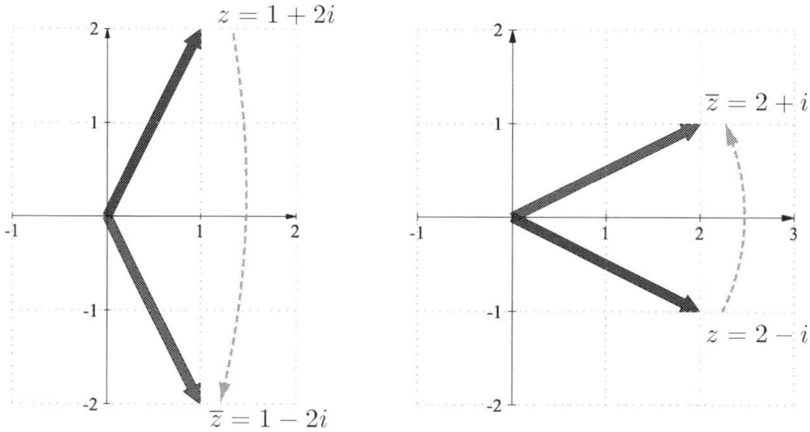

FIGURE 2.10
The conjugate of a complex number is obtained by reflecting it across the x-axis.

Note that for any complex number, if we calculate the conjugate of the conjugate, we recover the original complex number, that is:

$$z = 1 + 2i \quad \longrightarrow \quad \overline{z} = 1 - 2i \quad \longrightarrow \quad \overline{\overline{z}} = 1 + 2i = z.$$

Property 2.4. *Regarding the conjugate of a complex number, the following properties hold, considering $z = a + bi$:*

(1) $z + \overline{z} = 2a = 2\Re(z)$.

(2) $z - \overline{z} = 2bi = 2i\Im(z)$.

(3) $z \cdot \overline{z} = a^2 + b^2$.

PROOF. Let $z = a + bi$ be a complex number and $\overline{z} = a - bi$ its conjugate. Then:

(1)
$$z + \overline{z} = (a + bi) + (a - bi) = 2a = 2\Re(z).$$

(2)
$$z - \overline{z} = (a + bi) - (a - bi) = 2bi = 2i\Im(z).$$

(3)
$$z \cdot \overline{z} = (a + bi)(a - bi) = a^2 + b^2.$$

As we wanted to show. □

2.4.3.1 The Product of a Complex Number by Its Conjugate

Another important property is that if we multiply a complex number z by its conjugate $\overline{z}$, the result is a real and positive number (or zero, in the case that $z = 0$). Indeed, if we consider $z = a + bi$ and its conjugate $\overline{z} = a - bi$, we have:

$$z \cdot \overline{z} = (a + bi)(a - bi) = a^2 - b^2 i^2 = a^2 + b^2 \in \mathbb{R}.$$

Later, we will see that this will be of great interest for the division of complex numbers.

Example 2.1. *If we have* $z = 1 + 2i$ *and* $\overline{z} = 1 - 2i$*:*

$$\begin{aligned} z \cdot \overline{z} &= (1 + 2i) \cdot (1 - 2i) \\ &= 1^2 - 1 \cdot 2i + 1 \cdot 2i - 2^2 i^2 \\ &= 1^2 - 1 \cdot 2i + 1 \cdot 2i + 2^2 \qquad (\textit{using } i^2 = -1) \\ &= 1^2 + 2^2 \geq 0. \end{aligned}$$

2.4.4 Division of Complex Numbers in Standard Form

In the division of complex numbers, we will consider two cases:

(1) The complex number dividing is a *real* number. For example:

$$\frac{z_1}{z_2} = \frac{1 + 2i}{5}.$$

(2) The complex number dividing is a complex number with a non-zero *imaginary* part. For example:

$$\frac{z_1}{z_2} = \frac{1 + 2i}{4 + 3i}.$$

We observe that in the **first** case, the denominator is a real number; we have:

$$\frac{z_1}{z_2} = \frac{1 + 2i}{5} = \frac{1}{5} + \frac{2}{5} i.$$

In the more general case where the dividing complex number has a non-zero imaginary part, we will follow the following strategy: multiply and divide by the conjugate of the complex number in the denominator.

Thus, if we multiply both the numerator and the denominator by $\overline{z}_2$, the new denominator will be a real number and, therefore, the division will be easy to perform:

$$\begin{aligned} \frac{z_1}{z_2} &= \frac{1 + 2i}{3 - 4i} = \frac{1 + 2i}{3 - 4i} \cdot \frac{3 + 4i}{3 + 4i} = \frac{(1 + 2i)(3 + 4i)}{(3 - 4i)(3 + 4i)} \\ &= \frac{3 - 8 + 6i + 4i}{3^2 + 4^2} = \frac{-5 + 10i}{25} = -\frac{5}{25} + \frac{10}{25} i = -\frac{1}{5} + \frac{2}{5} i. \end{aligned}$$

2.5 Polar Form: An Alternative Representation of Complex Numbers

A complex number, which we have so far represented in *standard* form $z = a + bi$, and which we have drawn as the point (a, b) in the plane, can also be represented in another way.

Cartesian coordinates or orthogonal coordinates, such as those in standard form, represent points using a pair of numbers: the x-coordinate, or abscissa (horizontal displacement), and the y-coordinate, or ordinate (vertical displacement). This type of coordinates can be very useful, for example, to indicate a location in the Eixample of Barcelona, as shown in Figure 2.11.

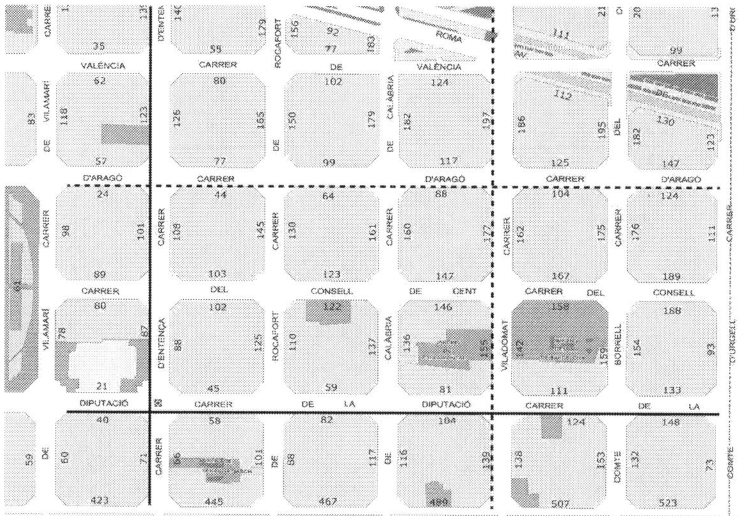

FIGURE 2.11
Cartesian coordinates in the Eixample of Barcelona. Considering the intersection of Diputació and Entença streets as the origin of coordinates, the intersection of Aragó and Viladomat streets can be found three streets to the right and two up, that is, at point $(3, 2)$. [Source: City Council of Barcelona, 2013.]

However, this is not the only way to represent points. As we have said, Cartesian coordinates can be useful for representing the Earth's surface on a plane, but nevertheless, ships use a two-dimensional radar system that places points on circles centered at the origin of coordinates, as shown in Figure 2.12.

Thus, the representation of a point (a, b) in these new coordinates is based on (Figure 2.13):

- The distance of the point (a, b) to the origin of coordinates, which we call **magnitude** or **modulus** and is usually represented as r.
- The angle formed by the positive x-axis with the ray that joins the points $(0, 0)$ and (a, b), which can also be called **argument** and is represented as θ.

Definition 2.7 (Modulus). To each complex number $z = a + bi$, we associate a positive real number called the **modulus** of z, denoted by $|z|$, which we define as:

$$|z| = r = \sqrt{a^2 + b^2}.$$

Definition 2.8 (Principal Argument). If $z = a + bi$ is a complex number, the **principal argument** of z, written as $\arg(z)$, is defined as the unique real number $\theta \in (-\pi, \pi]$ such

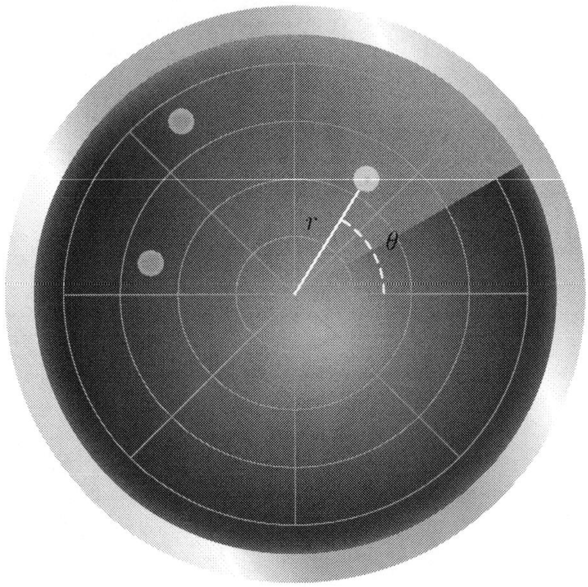

FIGURE 2.12
Polar coordinates on a radar.

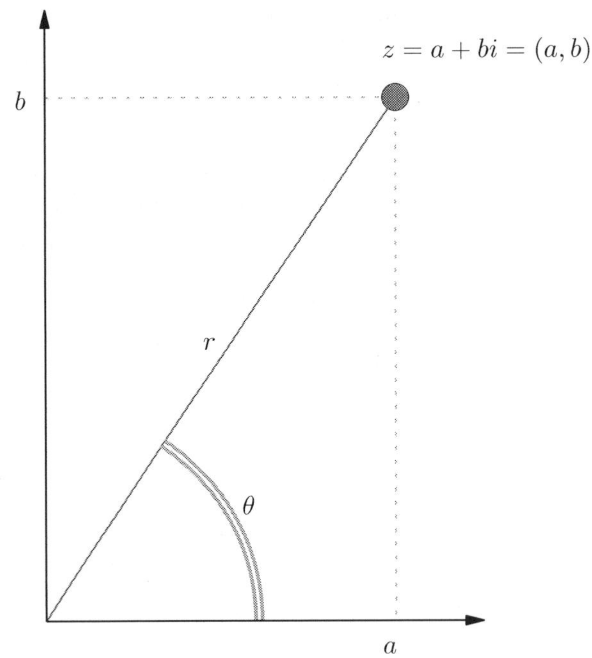

FIGURE 2.13
From Cartesian coordinates to polar coordinates. The complex number $a + bi$ can be expressed in polar form using the distance of the point to the origin, r, and the angle it forms with the positive x-axis, θ. That is, $a + bi = r_\theta$.

that:

$$a = |z|\cos(\theta)$$
$$b = |z|\sin(\theta).$$

Definition 2.9 (Argument). The **argument** of a complex number $z \in \mathbb{C}$ is defined as any element of the set:

$$\{\arg(z) + 2\pi k \mid k \in \mathbb{Z}\}.$$

Definition 2.10 (Polar Form). The following way of representing a complex number is called **polar form** or modulus-argument form:

$$r_\theta$$

This way of representing a complex number, r_θ, is called the *polar form*. In some books, you may also find the notation $r\angle\theta$.

Property 2.1. Let $z, u \in \mathbb{C}$. Then:

(1) The conjugate of the sum is the sum of conjugates:

$$\overline{z+u} = \overline{z} + \overline{u}.$$

(2) The conjugate of the product is the product of conjugates:

$$\overline{zu} = \overline{z} \cdot \overline{u}.$$

(3) The modulus of a number and its conjugate are equal:

$$|\overline{z}| = |z|.$$

(4) The product of a number by its conjugate is the square of the modulus of the number:

$$z \cdot \overline{z} = |z|^2.$$

(5) The modulus of the product is the product of moduli, and the argument of the product is the sum of arguments:

$$|zu| = |z||u|,$$
$$\arg(zu) = \arg(z) + \arg(u).$$

(6) The modulus of the quotient is the quotient of moduli, and the argument of the quotient is the difference of arguments:

$$\left|\frac{z}{u}\right| = \frac{|z|}{|u|},$$
$$\arg\left(\frac{z}{u}\right) = \arg(z) - \arg(u).$$

(7) The modulus of the sum is less than or equal to the sum of moduli:

$$|z+u| \leq |z| + |u|.$$

Proof. Consider $z = z_1 + z_2 i$ and $u = u_1 + u_2 i$. Then:

(1)

$$\begin{aligned}
\overline{z+u} &= \overline{z_1 + z_2 i + u_1 + u_2 i} \\
&= \overline{(z_1 + u_1) + (z_2 + u_2)i} \\
&= (z_1 + u_1) - (z_2 + u_2)i \\
&= (z_1 - z_2 i) + (u_1 - u_2 i) \\
&= \overline{z_1 + z_2 i} + \overline{u_1 + u_2 i} \\
&= \overline{z} + \overline{u}.
\end{aligned}$$

(2)

$$\begin{aligned}
\overline{zu} &= \overline{(z_1 + z_2 i)(u_1 + u_2 i)} \\
&= \overline{(z_1 u_1 - z_2 u_2) + (z_1 u_2 + z_2 u_1)i} \\
&= (z_1 u_1 - z_2 u_2) - (z_1 u_2 + z_2 u_1)i \\
&= (z_1 - z_2 i)(u_1 - u_2 i) \\
&= \overline{z_1 + z_2 i} \cdot \overline{u_1 + u_2 i} \\
&= \overline{z} \cdot \overline{u}.
\end{aligned}$$

(3)

$$\begin{aligned}
|\overline{z}| &= |\overline{z_1 + z_2 i}| \\
&= |z_1 - z_2 i| \\
&= \sqrt{z_1^2 + (-z_2)^2} \\
&= \sqrt{z_1^2 + z_2^2} \\
&= |z_1 + z_2 i| \\
&= |z|.
\end{aligned}$$

(4)

$$\begin{aligned}
z \cdot \overline{z} &= (z_1 + z_2 i)(\overline{z_1 + z_1 i}) \\
&= (z_1 + z_2 i)(z_1 - z_2 i) \\
&= z_1^2 + z_2^2 \\
&= \left(\sqrt{z_1^2 + z_2^2}\right)^2 \\
&= |z|^2.
\end{aligned}$$

(5) For this proof, we present two solutions. The first uses the definition of modulus as the square root of the product of a complex number by its conjugate:

$$|zu| = \sqrt{zu\overline{zu}} = \sqrt{zu\overline{z} \cdot \overline{u}} = \sqrt{z\overline{z}u\overline{u}} = \sqrt{z\overline{z}}\sqrt{u\overline{u}} = |z||u|.$$

The second is based on expressing the complex number in standard form:

$$\begin{aligned}
|zu| &= |(z_1 + z_2 i)(u_1 + u_2 i)| \\
&= |(z_1 u_1 - z_2 u_2) + (z_1 u_2 + z_2 u_1) i| \\
&= \sqrt{(z_1 u_1 - z_2 u_2)^2 + (z_1 u_2 + z_2 u_1)^2} \\
&= \sqrt{z_1^2 u_1^2 z_2 u_2^2 + z_1^2 u_2^2 + z_2^2 u_1^2} \\
&= \sqrt{(z_1^2 + z_2^2)(u_1^2 + u_2^2)} \\
&= \sqrt{z_1^2 + z_2^2}\sqrt{u_1^2 + u_2^2} \\
&= |z||u|.
\end{aligned}$$

The proof that the argument of the product is the sum of the arguments can be found in part (1) of Proposition 2.1.

(6)

$$\begin{aligned}
\left|\frac{z}{u}\right| &= \left|\frac{z}{u}\frac{\overline{u}}{\overline{u}}\right| \\
&= \left|\frac{z\overline{u}}{|u|^2}\right| \\
&= \frac{1}{|u|^2}|z\overline{u}| \\
&= \frac{1}{|u|^2}|z||\overline{u}| \\
&= \frac{1}{|u|^2}|z||u| \\
&= \frac{|z|}{|u|}.
\end{aligned}$$

The proof that the argument of the quotient is the difference of the arguments will be provided later (part (1) of Proposition 2.1).

(7) Instead of proving the expression:

$$|z + u| \leq |z| + |u|,$$

we will prove its square. Similarly, we will use the following properties of complex numbers in the proof:

$$\begin{aligned}
z + \overline{z} &= 2\Re(z) \\
|z| &\geq |\Re(z)|
\end{aligned}$$

In effect:

$$\begin{aligned}
|z + u|^2 &= (z + u)(\overline{z + u}) = (z + u)(\overline{z} + \overline{u}) = z\overline{z} + z\overline{u} + \overline{z}u + u\overline{u} \\
&= |z|^2 + |u|^2 + z\overline{u} + \underbrace{\overline{z}u}_{\overline{z\overline{u}}} = |z|^2 + |u|^2 + 2\Re(z\overline{u}) \leq |z|^2 + |u|^2 + 2|z\overline{u}| \\
&= |z|^2 + |u|^2 + 2|z||\overline{u}| = |z|^2 + |u|^2 + 2|z||u| = (|z| + |u|)^2
\end{aligned}$$

which is what we wanted to prove.

□

Example 2.2. *Consider the complex numbers* $3, 2i, -4,$ *and* $-5i$. *What is their representation in polar form? See Figure 2.14.*

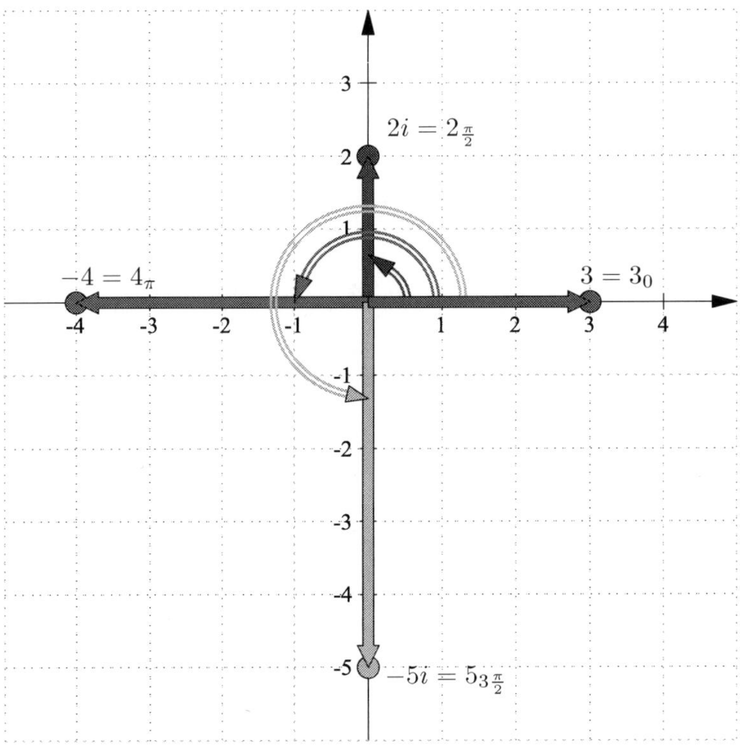

FIGURE 2.14
Polar form of the complex numbers 3, $2i$, -4, and $-5i$.

SOLUTION.

- The complex number 3 has an imaginary part equal to zero. This implies that it is located on the positive part of the x-axis (in fact, it is a positive real number). Consequently, the angle formed by the positive x-axis with the ray joining the origin and this point is precisely 0. On the other hand, the distance from this point to the origin is also 3. Therefore, we can write:

$$3 = 3 + 0i = 3_0.$$

- The complex number $2i$ has a real part equal to zero. Its representation as a point would be $(0, 2)$, which implies that it is located on the positive part of the y-axis. On the other hand, the angle formed by the positive x-axis with the ray joining the origin and this point is a right angle, that is, it has a value of $\frac{\pi}{2}$ rad or 90°. Finally, the distance from this point to the origin is 2. Therefore, we can write:

$$2i = 0 + 2i = 2_{\frac{\pi}{2}} = 2_{90^\circ}.$$

- The complex number -4 has an imaginary part equal to zero and a negative real part. This implies that it is located on the negative part of the x-axis (in fact, it is a negative real number). Consequently, the angle formed by the positive x-axis with the ray joining the origin and this point is a straight angle, that is, it has a value of π rad or 180°. On the other hand, the distance from this point to the origin is 4. Therefore, we can write:

$$-4 = -4 + 0i = 4_{\pi} = 4_{180^\circ}.$$

- The complex number $-5i$ has a real part equal to zero and a negative imaginary part. Its representation as a point would be $(0, -5)$, which implies that it is located on the negative part of the y-axis. On the other hand, the angle formed by the positive x-axis with the ray joining the origin and this point is the sum of a right angle and a straight angle, that is, it has a value of $\frac{3\pi}{2}$ rad or 270° (if we measure the angles counterclockwise). The distance from this point to the origin is 5. Therefore, we can write:

$$-5i = 0 - 5i = 5_{\frac{3\pi}{2}} = 5_{270^\circ}.$$

In this example, we have expressed in polar form the complex numbers located *on* the axes. In the next section, we will see how to generalize all this. □

2.5.1 From Standard to Polar Form

Suppose we have a complex number in standard form:

$$a + bi,$$

which can be graphically represented as the point (a, b) in the plane. By applying the Pythagorean theorem, we know that the distance from the point to the origin is:

$$r = \sqrt{a^2 + b^2}.$$

Similarly, we know that the relationship between the point and the angle formed by the positive x-axis with the ray that connects the points $(0, 0)$ and (a, b) is:

$$\tan(\theta) = \frac{b}{a}.$$

Therefore, θ can be obtained with the expression:

$\theta = \arctan\left(\frac{b}{a}\right)$, if a (real part) is positive, or

$\theta = \arctan\left(\frac{b}{a}\right) + \pi$, if a is negative and b (imaginary part) is positive, or

$\theta = \arctan\left(\frac{b}{a}\right) - \pi$, if a is negative and b is also negative.

In all cases, we will obtain an angle θ in the interval $(-\pi, \pi]$.

How do we calculate the arctangent?

To obtain the angle θ, we can use a calculator, any symbolic calculation software like *Wiris*, *Maple*, or even WolframAlpha.

If the arctangent is calculated with a calculator, it is important to know whether it is in `rad` (radians) mode or `deg` (degrees) mode.

Example 2.3 (of calculating the arctangent). *Consider the complex numbers:*

$$z_1 = 1 + i$$
$$z_2 = -1 - i$$

The complex number z_1 is in the first quadrant *since it has positive real and imaginary parts:*

$$Re(z_1) = 1 > 0, \quad Im(z_1) = 1 > 0.$$

On the other hand, the complex number z_2 is in the third quadrant *since it has negative real and imaginary parts:*

$$Re(z_2) = -1 < 0, \quad Im(z_2) = -1 < 0.$$

The modulus of both complex numbers is:

$$r_1 = \sqrt{1^2 + 1^2} = \sqrt{2}, \quad r_2 = \sqrt{(-1)^2 + (-1)^2} = \sqrt{2}.$$

But what is their angle or argument?

- *In the case of $z_1 = 1 + i$, since the real part is positive, the formula is:*

$$\theta = \arctan\left(\frac{1}{1}\right) = \underbrace{\arctan(1)}_{calculator} = \frac{\pi}{4} \; rad$$

- *In the case of $z_2 = -1-i$, since the real and imaginary parts are negative, the formula is:*

$$\theta = \arctan\left(\frac{-1}{-1}\right) - \pi = \underbrace{\arctan(1)}_{calculator} - \pi = \frac{\pi}{4} - \pi = -\frac{3\pi}{4} \; rad$$

Thus, the complex numbers $z_1 = 1+i$ and $z_2 = -1-i$ can be represented in polar coordinates as:

$$z_1 = 1 + i = \sqrt{2}_{\frac{\pi}{4}}$$
$$z_2 = -1 - i = \sqrt{2}_{-\frac{3\pi}{4}}$$

Proposed Exercise 2.2. Express the complex numbers $3 + 4i$ and $3 - 4i$ in polar form.

The Polar Representation is Not Unique

The polar representation of a complex number is not unique. For example, if we consider the complex number:

$$z = 1_{\frac{\pi}{4}},$$

and add any multiple of 2π radians to the argument (i.e., we make full rotations), we obtain the same number. That is:

$$1_{\frac{\pi}{4}} = 1_{\frac{\pi}{4}+2\pi} = 1_{\frac{\pi}{4}+4\pi} = 1_{\frac{\pi}{4}+6\pi}.$$

In general, we can write (Figure 2.15):

$$r_\theta = r_{\theta+2\pi k}, \quad k \in \mathbb{Z}.$$

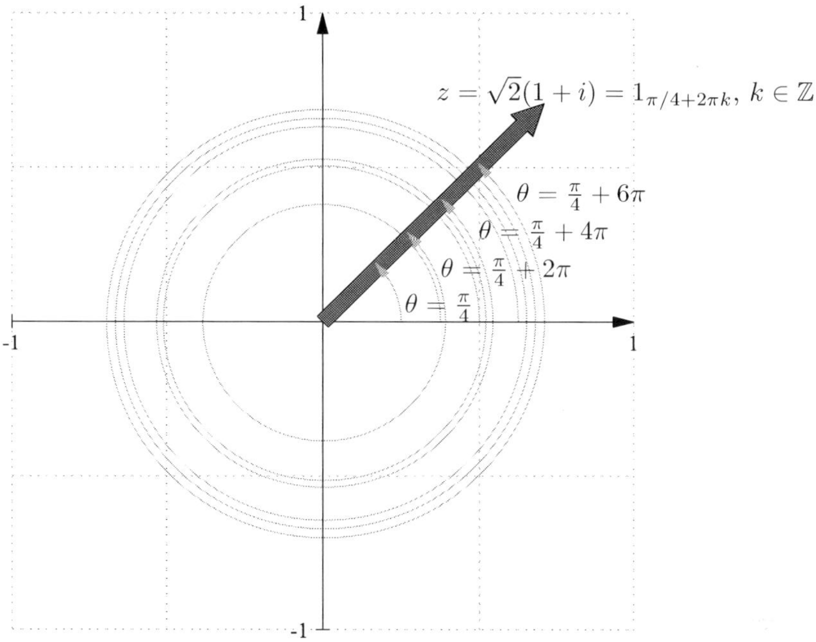

FIGURE 2.15
The polar representation of a complex number is not unique because if we have a number r_θ and add any multiple of 2π to the angle, we get the same number. In this figure, $z = 1_{\pi/4} = 1_{\pi/4+2\pi} = 1_{\pi/4+4\pi} = 1_{\pi/4+6\pi}$.

To unify the representation of complex numbers in polar form, it is common to consider the angle θ that satisfies:

$$-\pi < \theta \leq \pi,$$

that is, of all the angles that can represent a complex number, we must choose the one between $-\pi$ and $+\pi$ radians. To do this, it may be necessary to subtract or add multiples of 2π radians. This angle, as we have defined earlier, is called the **principal argument**.

When converting a complex number from standard form to polar form, it is very important to represent the complex number in the plane. This way, we will know which quadrant it is in and can avoid errors when calculating the argument.

2.5.2 From Polar to Standard Form

Suppose we have a complex number $a + bi$ expressed in terms of its magnitude and the angle it forms with the positive x-axis, that is, $z = r_\theta$, as in Figure 2.13.

Then, using the definitions of the trigonometric functions, sine and cosine, we have:

$$\sin(\theta) = \frac{b}{r} \quad \text{and} \quad \cos(\theta) = \frac{a}{r}.$$

Therefore:

$$a = r\cos(\theta), \quad b = r\sin(\theta).$$

That is, the number r_θ in standard form is:

$$z = r_\theta = \underbrace{r\cos(\theta)}_{\text{real part}} + \underbrace{r\sin(\theta)}_{\text{imaginary part}}\ i.$$

TABLE 2.3
Sine, cosine, and tangent of the most common angles in the first quadrant

θ	0	$\pi/6$	$\pi/4$	$\pi/3$	$\pi/2$
sin	0	$\frac{1}{2}$	$\frac{\sqrt{2}}{2}$	$\frac{\sqrt{3}}{2}$	1
cos	1	$\frac{\sqrt{3}}{2}$	$\frac{\sqrt{2}}{2}$	$\frac{1}{2}$	0
tan	0	$\frac{1}{\sqrt{3}}$	1	$\sqrt{3}$	∞

The equivalence between the polar and standard form of a complex number is:

$$\begin{aligned} r_\theta &= r\cos(\theta) + r\sin(\theta)i \\ &= r(\cos(\theta) + i\sin(\theta)) \end{aligned}$$

Definition 2.11 (Trigonometric Form)**.** The representation $r(\cos(\theta) + i\sin(\theta))$ of a complex number is called the *trigonometric form.*

As a reminder, Table 2.3 shows the sine, cosine, and tangent of the most common angles.

Example 2.4 (of representing a complex number in standard form)**.** *Consider the complex number in polar form:*

$$\sqrt{3}_{\frac{\pi}{6}}.$$

What is its representation in standard form?

SOLUTION. Applying the formula directly, we have:

$$\begin{aligned} \sqrt{3}_{\frac{\pi}{6}} &= \sqrt{3}\cos\left(\frac{\pi}{6}\right) + \sqrt{3}\sin\left(\frac{\pi}{6}\right) i \\ &= \sqrt{3}\frac{\sqrt{3}}{2} + \sqrt{3}\frac{1}{2}i \\ &= \frac{3}{2} + \frac{\sqrt{3}}{2}i \end{aligned}$$

For this calculation, the values of cosine and sine were looked up in Table 2.3. For other angles, you can use a calculator or the instructions applied below.

For angles in the second, third, and fourth quadrants, use the following rules:

- If $\theta \in (\pi/2, \pi)$ (second quadrant), then:

$$\begin{aligned} \sin(\theta) &= \sin(\pi - \theta) \\ \cos(\theta) &= -\cos(\pi - \theta) \end{aligned}$$

- If $\theta \in (-\pi, -\pi/2)$ (third quadrant), then:

$$\begin{aligned} \sin(\theta) &= -\sin(\pi + \theta) \\ \cos(\theta) &= -\cos(\pi + \theta) \end{aligned}$$

- If $\theta \in (-\pi/2, 0)$ (fourth quadrant), then:

$$\begin{aligned} \sin(\theta) &= -\sin(-\theta) \\ \cos(\theta) &= \cos(-\theta) \end{aligned}$$

Figure 2.16 will help understand these rules. □

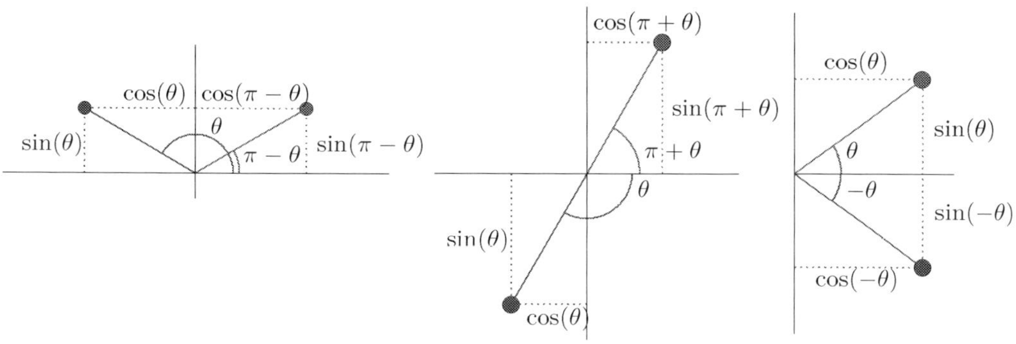

FIGURE 2.16
Trigonometric relationships of angles in the second, third, and fourth quadrants. Trigonometric relationships to calculate the sine and cosine of angles in the second, third, and fourth quadrants based on the sine and cosine of angles in the first quadrant.

Proposed Exercise 2.3. Express the following complex numbers in standard form: $1_0, 2_\pi, 1_{\pi/3}$.

2.5.3 Arithmetic Operations with Complex Numbers in Polar Form

Addition of complex numbers in polar form

To add or subtract two complex numbers, we first need to express them in standard form, that is, write them in the form $z = a + bi$ and then add them as indicated in the previous section.

Therefore, addition and subtraction operations will *always* be done in standard form. If desired, the final result can then be converted to polar form.

Example 2.5. *We have* $z_1 = 2_\pi$ *and* $z_2 = 5_{-\frac{\pi}{2}}$*, two complex numbers in polar form, and we want to calculate:*

$$z_1 + z_2, \quad \textit{and} \quad z_1 - z_2$$

The first thing we do is convert the two numbers to standard form:

$$z_1 = 2\underbrace{\cos(\pi)}_{-1} + 2\underbrace{\sin(\pi)}_{0}\, i = -2,$$
$$z_2 = 5\underbrace{\cos\left(-\frac{\pi}{2}\right)}_{0} + 5\underbrace{\sin\left(-\frac{\pi}{2}\right)}_{-1} i = -5i$$

Then:

$$z_1 + z_2 = -2 + (-5i) = -2 - 5i$$
$$z_1 - z_2 = -2 - (-5i) = -2 + 5i$$

Multiplication and Division of Complex Numbers in Polar Form

To multiply two numbers in polar form, multiply the magnitudes and add the arguments. When we talk about the exponential form, we will see why this is the way to multiply complex numbers.

Example 2.6. *We have* $z_1 = 2_\pi$ *and* $z_2 = 5_{-\frac{\pi}{2}}$, *two complex numbers in polar form, and we want to calculate:*

$$z_1 \cdot z_2.$$

Then:

$$\begin{aligned} z_1 \cdot z_2 &= 2_\pi \cdot 5_{-\frac{\pi}{2}} \\ &= (2 \cdot 5)_{\pi+(-\frac{\pi}{2})} \\ &= 10_{\frac{\pi}{2}} \end{aligned}$$

In a general case, therefore, we can write:

$$r_{\theta_1} \cdot s_{\theta_2} = (r \cdot s)_{\theta_1+\theta_2}$$

Similarly, to divide two complex numbers in polar form, divide the magnitudes and subtract the arguments.

Example 2.7. *We have* $z_1 = 2_\pi$ *and* $z_2 = 5_{-\frac{\pi}{2}}$, *two complex numbers in polar form, and we want to calculate:*

$$\frac{z_1}{z_2}.$$

Then:

$$\begin{aligned} \frac{z_1}{z_2} &= \frac{2_\pi}{5_{-\frac{\pi}{2}}} \\ &= \left(\frac{2}{5}\right)_{\pi-(-\frac{\pi}{2})} \\ &= 0.4_{\frac{3\pi}{2}} \end{aligned}$$

In a general case, therefore, we can write:

$$\frac{r_{\theta_1}}{s_{\theta_2}} = \left(\frac{r}{s}\right)_{\theta_1-\theta_2}$$

2.6 The Exponential of a Complex Number

The exponential form of a complex number allows us to represent complex numbers in a very compact way. Additionally, we can express the trigonometric functions sine and cosine in terms of complex numbers.

But let's start by recalling the two ways of representing complex numbers we have seen so far:

- the standard form, $z = a + bi$, or
- the polar form, $z = r_\theta = r(\cos(\theta) + i\sin(\theta))$.

In this subsection, we will see a third way to represent complex numbers: the exponential form.

Consider the real exponential function, $\exp(x) = e^x$. This function can be rewritten as a sum of infinite terms:

$$e^x = 1 + x + \frac{x^2}{2!} + \frac{x^3}{3!} + \frac{x^4}{4!} + \cdots$$

where:

$$n! = n \cdot (n-1) \cdot (n-2) \cdot 2 \cdot 1$$

is the factorial of the natural number n. This sum of infinite terms is called the *power series expansion*. Although it is an infinite sum, in practice, only a finite number of terms are usually used. For example, if we consider only six terms and $x = 1$, we have:

$$\begin{aligned} e^1 &\approx 1 + 1 + \frac{1}{2!} + \frac{1}{3!} + \frac{1}{4!} + \frac{1}{5!} \\ &= 2 + 0.5 + 0.16667 + 0.04167 + 0.00833 \\ &= 2.71666 \end{aligned}$$

The result is already good enough compared to the actual value, which is $e = 2.71828$ (to five decimal places).

We can also express the trigonometric functions sine and cosine as an infinite sum of terms:

$$\begin{aligned} \sin(x) &= x - \frac{x^3}{3!} + \frac{x^5}{5!} - \frac{x^7}{7!} + \cdots \\ \cos(x) &= 1 - \frac{x^2}{2!} + \frac{x^4}{4!} - \frac{x^6}{6!} + \cdots \end{aligned}$$

We return to the expression of the exponential e^x but replacing x with i, and finally reorganizing the terms:

$$\begin{aligned} e^{\theta i} &= 1 + (\theta i) + \frac{(\theta i)^2}{2!} + \frac{(\theta i)^3}{3!} + \frac{(\theta i)^4}{4!} + \frac{(\theta i)^5}{5!} + \cdots \\ &= 1 + \theta i - \frac{\theta^2}{2!} - \frac{\theta^3}{3!} i + \frac{\theta^4}{4!} + \frac{\theta^5}{5!} i + \cdots \\ &= \left[1 - \frac{\theta^2}{2!} + \frac{\theta^4}{4!} - \cdots\right] + \left[\theta - \frac{\theta^3}{3!} + \frac{\theta^5}{5!} - \cdots\right] i \\ &= \cos(\theta) + \sin(\theta) i \end{aligned}$$

Definition 2.12 (Euler's Formula). The exponential of a pure imaginary complex number is, therefore:

$$e^{\theta i} = \cos(\theta) + \sin(\theta) i.$$

Since every complex number can be expressed as $z = r(\cos(\theta) + \sin(\theta)i)$ and, as we have just seen, $e^{\theta i} = \cos(\theta) + \sin(\theta)i$, this allows us to represent complex numbers in another way: $z = re^{\theta i}$, which we call the exponential form.

Definition 2.13 (Exponential of a Complex Number). The exponential of a complex number $z = a + bi \in \mathbb{C}$ that is not purely imaginary is:

$$\exp(z) = e^{a+bi} = e^a e^{bi} = e^a(\cos(b) + i\sin(b)).$$

Definition 2.14 (Exponential Form). The exponential form of a complex number is:

$$z = re^{\theta i}$$

where $r = |z|$ is the modulus of z and θ is the argument of z.

The polar representation of a complex number is, therefore, equivalent to the exponential representation of the same complex number. Indeed:

$$re^{\theta i} = r(\cos(\theta) + i\sin(\theta)) = r_\theta.$$

See Figure 2.17.

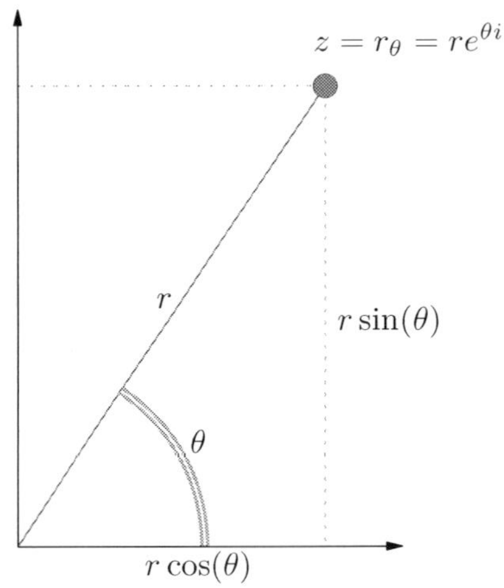

FIGURE 2.17
Representation of a complex number in polar and exponential form.

Example 2.8. *Table 2.4 shows the equivalences between the different representations (standard, polar, and exponential) of the following complex numbers:*

$$z_1 = 1 + i, \quad z_2 = i, \quad z_3 = 1 + \sqrt{3}i, \quad z_4 = 1 - \sqrt{3}i$$

Proposition 2.1. Complex numbers in exponential form have the following properties:

(1) $e^{(\theta_1+\theta_2)i} = e^{\theta_1 i}e^{\theta_2 i}$

(2) $e^0 = 1$

(3) $e^{-\theta i} = \frac{1}{e^{\theta i}}$

TABLE 2.4
Relationship between standard, polar, and exponential forms.

Standard	Polar	Exponential
$z_1 = 1 + i$	$z_1 = \sqrt{2}_{\frac{\pi}{4}}$	$z_1 = \sqrt{2}e^{\frac{\pi}{4}i}$
$z_2 = i$	$z_2 = 1_{\frac{\pi}{2}}$	$z_2 = e^{\frac{\pi}{2}i}$
$z_3 = 1 + \sqrt{3}i$	$z_3 = 2_{\frac{\pi}{3}}$	$z_3 = 2e^{\frac{\pi}{3}i}$
$z_4 = 1 - \sqrt{3}i$	$z_4 = 2_{-\frac{\pi}{3}}$	$z_4 = 2e^{-\frac{\pi}{3}i}$

(4) $e^{\theta i + 2\pi k i} = e^{\theta i}$

Proof. (1) First, we need to recall the formulas for the sine and cosine of the sum of angles. Indeed:

$$\sin(\theta_1 + \theta_2) = \sin(\theta_1)\cos(\theta_2) + \cos(\theta_1)\sin(\theta_2)$$
$$\cos(\theta_1 + \theta_2) = \cos(\theta_1)\cos(\theta_2) - \sin(\theta_1)\sin(\theta_2)$$

Then, applying the definition of complex exponential:

$$\begin{aligned} e^{\theta_1 i}e^{\theta_2 i} &= (\cos(\theta_1) + i\sin(\theta_1)) \cdot (\cos(\theta_2) + i\sin(\theta_2)) \\ &= [\cos(\theta_1)\cos(\theta_2) - \sin(\theta_1)\sin(\theta_2)] + i\,[\sin(\theta_1)\cos(\theta_2) + \cos(\theta_1)\sin(\theta_2)] \\ &= \cos(\theta_1 + \theta_2) + i\sin(\theta_1 + \theta_2) \\ &= e^{(\theta_1 + \theta_2)i} \end{aligned}$$

(2) Applying the definition directly:

$$e^0 = e^{0\cdot i} = \underbrace{\cos(0)}_{1} + i\underbrace{\sin(0)}_{0} = 1.$$

(3) In this case, also applying the definition, we have:

$$\begin{aligned} \frac{1}{e^{\theta i}} &= \frac{1}{\cos(\theta) + i\sin(\theta)} \\ &= \frac{1}{\cos(\theta) + i\sin(\theta)} \cdot \frac{\cos(\theta) - i\sin(\theta)}{\cos(\theta) - i\sin(\theta)} \\ &= \frac{\cos(\theta) - i\sin(\theta)}{\cos^2(\theta) + \sin^2(\theta)} \\ &= \frac{\cos(\theta) - i\sin(\theta)}{1} \\ &= \cos(\theta) - i\sin(\theta) \\ &= \cos(-\theta) + i\sin(-\theta) \\ &= e^{-\theta i}. \end{aligned}$$

(4) The proof of this part is a direct consequence of the periodicity of the sine and cosine functions. Indeed:

$$\begin{aligned} e^{\theta i+2\pi ki} &= \cos(\theta+2\pi k)+i\sin(\theta+2\pi k) \\ &= \cos(\theta)+i\sin(\theta) \\ &= e^{\theta i}, \end{aligned}$$

where we have taken into account that:

$$\cos(\theta+2\pi k)=\cos(\theta), \quad \sin(\theta+2\pi k)=\sin(\theta)$$

for $k \in \mathbb{Z}$. □

□

2.6.1 Operations of Complex Numbers in Exponential Form

The operations of complex numbers in exponential form are equivalent to the operations we have already seen in polar form. These are:

- **Multiplication**. Multiply the magnitudes and add the arguments:

$$\begin{aligned} 2e^{\frac{\pi}{3}i}\cdot 3e^{\frac{\pi}{2}i} &= (2\cdot 3)e^{\left(\frac{\pi}{3}+\frac{\pi}{2}\right)i} \\ &= 6e^{\frac{5\pi}{6}i}. \end{aligned}$$

In general:

$$r_1e^{\theta_1 i}\cdot r_2e^{\theta_2 i}=r_1r_2e^{(\theta_1+\theta_2)i}.$$

- **Division**. Divide the magnitudes and subtract the arguments:

$$\begin{aligned} \frac{2e^{\frac{\pi}{3}i}}{3e^{\frac{\pi}{2}i}} &= \frac{2}{3}e^{\left(\frac{\pi}{3}-\frac{\pi}{2}\right)i} \\ &= \frac{2}{3}e^{-\frac{\pi}{6}i}. \end{aligned}$$

In general:

$$\frac{r_1e^{\theta_1 i}}{r_2e^{\theta_2 i}}=\frac{r_1}{r_2}e^{(\theta_1-\theta_2)i}.$$

- **Conjugation**. The conjugate of a complex number expressed in exponential form is the complex number with the same magnitude and opposite argument:

$$\overline{2e^{\frac{\pi}{3}i}}=2e^{-\frac{\pi}{3}i}.$$

In general:

$$\overline{re^{\theta i}}=re^{-\theta i}.$$

- **Addition and Subtraction.** To add or subtract complex numbers in exponential form, we need to express them in standard form:

$$\begin{aligned} r_1e^{\theta_1 i} + r_2e^{\theta_2 i} &= r_1(\cos(\theta_1) + i\sin(\theta_1)) + r_2(\cos(\theta_2) + i\sin(\theta_2)) \\ &= (r_1\cos(\theta_1) + r_2\cos(\theta_2)) + i\,(r_1\sin(\theta_1) + r_2\sin(\theta_2)) \end{aligned}$$

The introduction of the exponential form also allows us to calculate powers of complex numbers very easily:

- **Exponentiation.** To raise a complex number in exponential form to a power, we use the definition of the product of complex numbers in polar form:

$$\begin{aligned} \left(re^{\theta i}\right)^2 &= re^{\theta i}\cdot re^{\theta i} = r^2e^{(\theta+\theta)i} \\ &= r^2e^{2\theta i} \end{aligned}$$

That is, we have multiplied the magnitudes ($r\cdot r = r^2$) and added the arguments ($\theta+\theta = 2\theta$). Thus, we have:

$$\left(re^{\theta i}\right)^2 = r^2e^{2\theta i}$$

Similarly, to raise a complex number in exponential form to the power n, we raise the magnitude to n and multiply the argument by n. Indeed, we can see that:

$$\begin{aligned} \left(re^{\theta i}\right)^n &= \underbrace{re^{\theta i}\cdot re^{\theta i}\cdots re^{\theta i}}_{n \text{ times}} = \underbrace{(r\cdots r)}_{n\text{ times}}\, e^{\overbrace{(\theta+\cdots+\theta)}^{n\text{ times}}\,i} \\ &= r^ne^{n\theta i} \end{aligned}$$

That is, we have reached the expression:

$$\left(re^{\theta i}\right)^n = r^ne^{n\theta i}$$

which we can also write as:

$$[r(\cos(\theta) + i\sin(\theta)]^n = r^n(\cos(n\theta) + i\sin(n\theta)).$$

This last equality is known as the *De Moivre's Theorem*, named after the French mathematician Abraham De Moivre (1667-1754), and can be used to calculate the formulas for the sine and cosine of angles, such as $2\theta, 3\theta, \ldots$ in terms of $\sin(\theta)$ and $\cos(\theta)$.

Example 2.9. *Calculate an expression for* $\sin(2\theta)$ *in terms of* $\sin(\theta)$ *and* $\cos(\theta)$.
SOLUTION. *By De Moivre's theorem with* $r = 1$ *and* $n = 2$, *we have:*

$$[\cos(\theta) + i\sin(\theta)]^2 = \underbrace{\cos(2\theta) + i\sin(2\theta)}_{z}$$

We need to calculate $\sin(2\theta)$, *that is, the imaginary part of the complex number* z:

$$Im(\cos(2\theta) + i\sin(2\theta)) = \sin(2\theta).$$

Therefore, we proceed to develop the expression:

$$[\cos(\theta) + i\sin(\theta)]^2$$

from which we will
then consider its imaginary part. Indeed:

$$\begin{aligned}
[\cos(\theta) + i\sin(\theta)]^2 &= \cos^2(\theta) + i^2\sin^2(\theta) + 2i\cos(\theta)\sin(\theta) \\
&= \underbrace{\cos^2(\theta) - \sin^2(\theta)}_{Re(z)} + i\underbrace{2\cos(\theta)\sin(\theta)}_{Im(z)}.
\end{aligned}$$

Thus, not only do we obtain:

$$\sin(2\theta) = 2\cos(\theta)\sin(\theta),$$

but also, by equating the real parts, we can establish that:

$$\cos(2\theta) = \cos^2(\theta) - \sin^2(\theta).$$

Proposed Exercise 2.4. Prove by induction that De Moivre's theorem is true for any natural number $n \in \mathbb{N}$.

Proposed Exercise 2.5. Following the example 2.9, calculate an expression for $\sin(3\theta)$ in terms of $\sin(\theta)$ and $\cos(\theta)$.

2.7 Trigonometric Functions in Terms of the Complex Exponential

If we use the complex exponential, we can obtain expressions for sine and cosine in terms of complex numbers. Let us then proceed in a *constructive* manner to establish these relationships.

- First, consider the exponential of the complex numbers θi and $-\theta i$:

$$\begin{aligned}
e^{\theta i} &= \cos(\theta) + i\sin(\theta), \\
e^{-\theta i} &= \cos(-\theta) + i\sin(-\theta).
\end{aligned}$$

- Second, recall that:

$$\begin{aligned}
\cos(-\theta) &= \cos(\theta), \\
\sin(-\theta) &= -\sin(\theta),
\end{aligned}$$

  as seen in Figure 2.18.

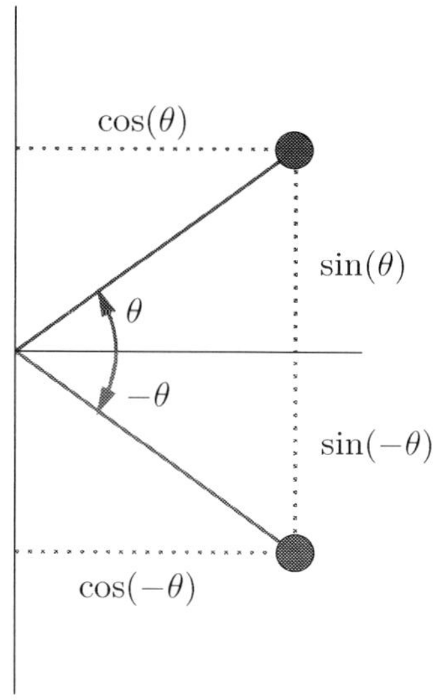

FIGURE 2.18
Trigonometric relationships: $\cos(-\theta) = \cos(\theta), \sin(-\theta) = -\sin(\theta)$.

- This allows us to write again:

$$\begin{aligned} e^{\theta i} &= \cos(\theta) + i\sin(\theta) \\ e^{-\theta i} &= \cos(-\theta) + i\sin(-\theta) \\ &= \cos(\theta) - i\sin(\theta). \end{aligned}$$

- If we add these two expressions:

$$\begin{aligned} e^{\theta i} + e^{-\theta i} &= \cos(\theta) + i\sin(\theta) + \cos(\theta) - i\sin(\theta) \\ &= 2\cos(\theta) \end{aligned}$$

  and therefore, dividing the whole equality by 2, we obtain the value of $\cos(\theta)$:

$$\cos(\theta) = \frac{e^{\theta i} + e^{-\theta i}}{2}.$$

- If instead of adding, we subtract the two expressions:

$$\begin{aligned} e^{\theta i} - e^{-\theta i} &= \cos(\theta) + i\sin(\theta) - (\cos(\theta) - i\sin(\theta)) \\ &= \cos(\theta) + i\sin(\theta) - \cos(\theta) + i\sin(\theta) \\ &= 2i\sin(\theta). \end{aligned}$$

  Dividing again, but now by $2i$, we obtain:

$$\sin(\theta) = \frac{e^{\theta i} - e^{-\theta i}}{2i}.$$

The expressions for the trigonometric functions sine, cosine, and tangent in terms of the complex exponential are:

$$\sin(\theta) = \frac{e^{\theta i} - e^{-\theta i}}{2i}$$
$$\cos(\theta) = \frac{e^{\theta i} + e^{-\theta i}}{2}$$
$$\tan(\theta) = \frac{i\left(e^{-\theta i} - e^{\theta i}\right)}{e^{-\theta i} + e^{\theta i}}$$

Hyperbolic functions are very interesting functions that appear in many branches of science. Their origin traces back to the German mathematician Johann Heinrich Lambert (1728-1777) when he was comparing the area of a semicircular region with the area of a region bounded by a hyperbola. Recall the definitions of the hyperbolic functions:

$$\sinh(x) = \frac{e^x - e^{-x}}{2},$$
$$\cosh(x) = \frac{e^x + e^{-x}}{2},$$
$$\tanh(x) = \frac{e^x - e^{-x}}{e^x + e^{-x}},$$

which are read as *hyperbolic sine*, *hyperbolic cosine*, and *hyperbolic tangent.*

We can observe the similarities between these functions and the trigonometric functions. In particular, we can establish the following equivalences:

Proposition 2.2. The trigonometric functions and hyperbolic functions are related as follows:

(1) $\sin(\theta) = -i\sinh(\theta i)$

(2) $\cos(\theta) = \cosh(\theta i)$

(3) $\sin(\theta i) = i\sinh(\theta)$

(4) $\cos(\theta i) = \cosh(\theta)$

Proof. Suppose $\theta \in \mathbb{R}$, then:

(1)

$$\begin{aligned} -i\sinh(\theta i) &= -i\frac{e^{\theta i} - e^{-\theta i}}{2} \\ &= -i\frac{e^{\theta i} - e^{-\theta i}}{2} \cdot \frac{i}{i} \\ &= \frac{e^{\theta i} - e^{-\theta i}}{2i} \\ &= \sin(\theta) \end{aligned}$$

(2)

$$\begin{aligned} \cosh(\theta i) &= \frac{e^{\theta i} + e^{-\theta i}}{2} \\ &= \cos(\theta) \end{aligned}$$

(3)

$$\begin{aligned}\sin(\theta i) &= \frac{e^{(\theta i)i} - e^{-(\theta i)i}}{2i} \\ &= \frac{e^{-\theta} - e^{\theta}}{2i} \\ &= \frac{e^{-\theta} - e^{\theta}}{2i} \cdot \frac{i}{i} \\ &= \frac{e^{-\theta} - e^{\theta}}{-2} \cdot i \\ &= i\frac{e^{\theta} - e^{-\theta}}{2} \\ &= i\sinh(\theta)\end{aligned}$$

(4)

$$\begin{aligned}\cos(\theta i) &= \frac{e^{(\theta i)i} + e^{-(\theta i)i}}{2} \\ &= \frac{e^{-\theta} + e^{\theta}}{2} \\ &= \frac{e^{\theta} + e^{-\theta}}{2} \\ &= \cosh(\theta)\end{aligned}$$

as we wanted to demonstrate.

□

Example 2.10. *Calculate the standard form of* $\sin(i)$.

SOLUTION. Considering the equality $\sin(\theta i) = i\sinh(\theta)$ with $\theta = 1$, we have:

$$\sin(i) = i\sinh(1) = \left(\frac{e - \frac{1}{e}}{2}\right) i,$$

which is a purely imaginary number. □

2.8 Roots of Complex Numbers

In the set of real numbers, the square root of a positive number has two solutions. For example, the *two* square roots of the number 1 are -1 and $+1$, since:

$$\begin{aligned}1^2 &= 1, \\ (-1)^2 &= 1.\end{aligned}$$

If we try to find a cube root, there is only *one* solution. For example, the cube root of -1 is -1:

$$(-1)^3 = -1.$$

This can be seen in Figure 2.19:

- When we calculate square roots, for example, of 2, we seek the numbers that, when squared, give 2. These numbers are, as shown, $-\sqrt{2}$ and $\sqrt{2}$. Square roots of negative numbers cannot be calculated.
- When we calculate cube roots, for example, of 2, we seek the numbers that, when cubed, give 2. In this case, we have a single value, $\sqrt[3]{2}$ (approximately 1.26), as seen in Figure 2.19. All numbers, both positive and negative, have a cube root.

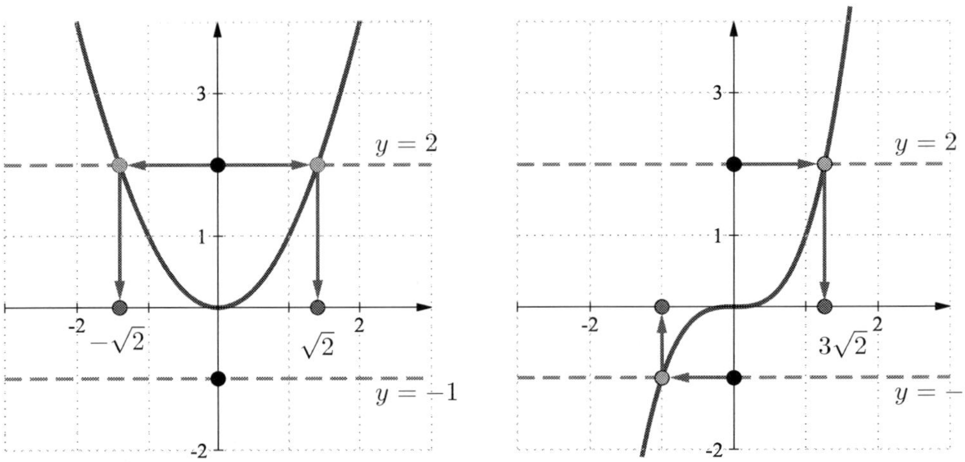

FIGURE 2.19
Square roots of a real number (left): to determine the square roots of a number s, we look for the intersections of the parabola $y = x^2$ with the line $y = s$. Cube roots of a real number (right): to determine the cube roots of a number s, we look for the intersections of the function $y = x^3$ with the line $y = s$.

In the set of real numbers:

- Even-indexed roots (square root, fourth root, etc.) do not have a solution if the number is negative.
- Even-indexed roots (square root, fourth root, etc.) have two solutions if the number is positive.
- Odd-indexed roots (cube root, fifth root, etc.) always have one solution.

In the set of complex numbers, there are differences compared to the set of real numbers:

- A square root always has two solutions.
- A cube root always has three solutions.
- And, in general, an n-th root always has n solutions.

Consider, for example, the problem of finding the complex numbers z that satisfy the equation:

$$z^3 = 1.$$

In this case, the solutions to this equation, that is, the numbers z that satisfy the above equation, are called *cube roots of unity*. To solve the above equation, the first thing we need to consider is that the number 1 is a complex number that in polar coordinates has a modulus of 1 and an argument of 0:

$$1 = 1_0 = 1e^{0i}.$$

Additionally, we can use other equivalent arguments, such as $2\pi, 4\pi, 6\pi$, or any other multiple of 2π, since adding a full turn does not change the position of the vector representing this complex number and, therefore, does not change its value. Thus, we can also write:

$$\begin{aligned} 1 &= 1_0 = 1_{2\pi} = 1_{4\pi} = 1_{6\pi} = \cdots \\ &= e^{0i} = e^{2\pi i} = e^{4\pi i} = e^{6\pi i} = \cdots \\ &= e^{2k\pi i}, \quad k = 0, 1, 2, 3, \ldots \end{aligned}$$

Thus, the equation $z^3 = 1$ is equivalent to:

$$z^3 = e^{2\pi k i}, \quad k = 0, 1, 2, 3, \ldots$$

To solve this equation, cube roots are taken on both sides of the equation, or equivalently, both sides of the equation are raised to the power of $\frac{1}{3}$. If we do this, we obtain the following equation:

$$\left(z^3\right)^{\frac{1}{3}} = z = e^{\frac{2k\pi i}{3}}, \quad k = 0, 1, 2, 3, \ldots$$

Finally, to determine the solutions, we only need to assign values to the variable k. In this way, we will find the values of the different cube roots. In particular, since we know we must find exactly 3 cube roots, we substitute the value of k with 0, 1, and 2:

$$\begin{aligned} k = 0: &\qquad z = e^{0i} = 1 \\ k = 1: &\qquad z = e^{\frac{2\pi}{3}i} = \cos\left(\frac{2\pi}{3}\right) + i\sin\left(\frac{2\pi}{3}\right) = -\frac{1}{2} + \frac{\sqrt{3}}{2}i \\ k = 2: &\qquad z = e^{\frac{4\pi}{3}i} = \cos\left(\frac{4\pi}{3}\right) + i\sin\left(\frac{4\pi}{3}\right) = -\frac{1}{2} - \frac{\sqrt{3}}{2}i \end{aligned}$$

If we give more values to k, such as $k = 3$, $k = 4$, or $k = 5$, we do not obtain new roots, as these would repeat.

In a similar way to how the cube roots of unity are calculated, the roots of any complex number can be calculated.

Proposition 2.3. The equation $z^n = re^{\theta i}$ has n solutions which are:

$$z = \sqrt[n]{r}\, e^{\left(\frac{\theta}{n} + \frac{2\pi}{n}k\right)i}, \quad k = 0, 1, 2, 3, \ldots, n-1$$

Note that, since r is a positive real number, $\sqrt[n]{r}$ refers to the positive n-th root of r.

Example 2.11 (Cube roots of -8). *Now consider the problem of finding the complex numbers z that satisfy the equation:*

$$z^3 = -8.$$

In this case, the solutions to this equation, that is, the numbers z that satisfy the above equation, are called cube roots *of -8.*

To solve the above equation, the first thing to consider is that the number -8 is a complex number that in polar coordinates has a modulus of 8 and an argument of π:

$$-8 = 8_\pi = 8e^{\pi i}.$$

As before, we can use other equivalent arguments, such as $\pi + 2\pi$, $\pi + 4\pi$, or $\pi + 6\pi$, since we are adding full turns. Thus, we can also write:

$$\begin{aligned} -8 &= 8_{\pi} = 8_{\pi+2\pi} = 8_{\pi+4\pi} = 8_{\pi+6\pi} = \cdots \\ &= 8e^{\pi i} = 8e^{(\pi+2\pi)i} = 8e^{(\pi+4\pi)i} = 8e^{(\pi+6\pi)i} = \cdots \\ &= 8e^{(\pi+2k\pi)i}, \quad k = 0, 1, 2, 3, \ldots \end{aligned}$$

Thus, the equation $z^3 = -8$ is equivalent to:

$$z^3 = 8e^{(\pi+2k\pi)i}, \quad k = 0, 1, 2, 3, \ldots$$

To solve this equation, cube roots are taken on both sides of the equation, or equivalently, both sides of the equation are raised to the power of $\frac{1}{3}$. If we do this, we obtain this equation:

$$\left(z^3\right)^{\frac{1}{3}} = z = \sqrt[3]{8}e^{\left(\frac{\pi}{3}+\frac{2k\pi}{3}\right)i}, \quad k = 0, 1, 2, 3, \ldots$$

Finally, to determine the solutions, we only need to assign values to the variable k. In this way, we will find the values of the different cube roots. In particular, since we know we must find exactly 3 cube roots, we substitute the value of k with 0, 1, and 2:

$$\begin{aligned} k = 0: \quad & z = \sqrt[3]{8}e^{\frac{\pi}{3}i} = 2\left(\cos\left(\frac{\pi}{3}\right) + i\sin\left(\frac{\pi}{3}\right)\right) \\ & = 2\left(\frac{1}{2} + \frac{\sqrt{3}}{2}i\right) = 1 + \sqrt{3}i \\ k = 1: \quad & z = \sqrt[3]{8}e^{\pi i} = 2\left(\cos(\pi) + i\sin(\pi)\right) \\ & = 2(-1) = -2 \\ k = 2: \quad & z = \sqrt[3]{8}e^{\frac{5\pi}{3}i} = 2\left(\cos\left(\frac{5\pi}{3}\right) + i\sin\left(\frac{5\pi}{3}\right)\right) \\ & = 2\left(\frac{1}{2} - \frac{\sqrt{3}}{2}i\right) = 1 - \sqrt{3}i \end{aligned}$$

Example 2.12 (Square, cube, and fourth roots of -1). *First, we express the complex number -1 in exponential form:*

$$-1 = e^{\pi i} = e^{(\pi+2\pi k)i}, \quad k = 0, 1, 2, 3, \ldots$$

Therefore, we want to solve the equation:

$$z^n = e^{(\pi+2\pi k)i}, \quad k = 0, 1, 2, 3, \ldots$$

for $n = 2$, 3, and 4. We apply the n-th root on both sides of the equality:

$$\left(z^n\right)^{\frac{1}{n}} = z = e^{\frac{(\pi+2\pi k)i}{n}} = e^{\left(\frac{\pi}{n}+\frac{2\pi k}{n}\right)i}, \quad k = 0, 1, 2, 3, \ldots$$

To determine the square roots, we substitute $n = 2$ and take 0 and 1 as values of k:

$$\begin{aligned} k = 0: \quad & z = e^{\frac{\pi}{2}i} = \cos\left(\frac{\pi}{2}\right) + i\sin\left(\frac{\pi}{2}\right) = i \\ k = 1: \quad & z = e^{\left(\frac{\pi}{2}+\pi\right)i} = e^{\frac{3\pi}{2}i} = \cos\left(\frac{3\pi}{2}\right) + i\sin\left(\frac{3\pi}{2}\right) = -i \end{aligned}$$

Similarly, to determine the cube roots, we substitute $n = 3$ and take 0, 1, and 2 as values of k:

$$k=0: \quad z = e^{\frac{\pi}{3}i} = \cos\left(\frac{\pi}{3}\right) + i\sin\left(\frac{\pi}{3}\right) = \frac{1}{2} + \frac{\sqrt{3}}{2}i$$

$$k=1: \quad z = e^{\left(\frac{\pi}{3}+\frac{2\pi}{3}\right)i} = e^{\pi i} = -1$$

$$k=2: \quad z = e^{\left(\frac{\pi}{3}+\frac{4\pi}{3}\right)i} = e^{\frac{5\pi}{3}i} = \cos\left(\frac{5\pi}{3}\right) + i\sin\left(\frac{5\pi}{3}\right) = \frac{1}{2} - \frac{\sqrt{3}}{2}i$$

Finally, to determine the fourth roots, we substitute $n = 4$ and take 0, 1, 2, and 3 as values of k:

$$k=0: \quad z = e^{\frac{\pi}{4}i} = \cos\left(\frac{\pi}{3}\right) + i\sin\left(\frac{\pi}{3}\right) = \frac{1}{2} + \frac{\sqrt{3}}{2}i = \frac{\sqrt{2}}{2} + \frac{\sqrt{2}}{2}i$$

$$k=1: \quad z = e^{\left(\frac{\pi}{4}+\frac{2\pi}{4}\right)i} = e^{\frac{3\pi}{4}i} = \cos\left(\frac{3\pi}{4}\right) + i\sin\left(\frac{3\pi}{4}\right) = -\frac{\sqrt{2}}{2} + \frac{\sqrt{2}}{2}i$$

$$k=2: \quad z = e^{\left(\frac{\pi}{4}+\frac{4\pi}{4}\right)i} = e^{\frac{5\pi}{4}i} = \cos\left(\frac{5\pi}{4}\right) + i\sin\left(\frac{5\pi}{4}\right) = -\frac{\sqrt{2}}{2} - \frac{\sqrt{2}}{2}i$$

$$k=3: \quad z = e^{\left(\frac{\pi}{4}+\frac{6\pi}{4}\right)i} = e^{\frac{7\pi}{4}i} = \cos\left(\frac{7\pi}{4}\right) + i\sin\left(\frac{7\pi}{4}\right) = \frac{\sqrt{2}}{2} - \frac{\sqrt{2}}{2}i$$

In Figure 2.20, we have represented the square, cube, and fourth roots of the number $z = -1$. Notice that all the roots have the same modulus (they are located on a circumference) and are located at the vertices of a regular polygon with n vertices. Thus, the square roots are always aligned, the cube roots are the vertices of an equilateral triangle, and the fourth roots are the vertices of a square. In Figure 2.21, we have represented the fifth and sixth roots of $z = -1$, which are also the vertices of a regular polygon.

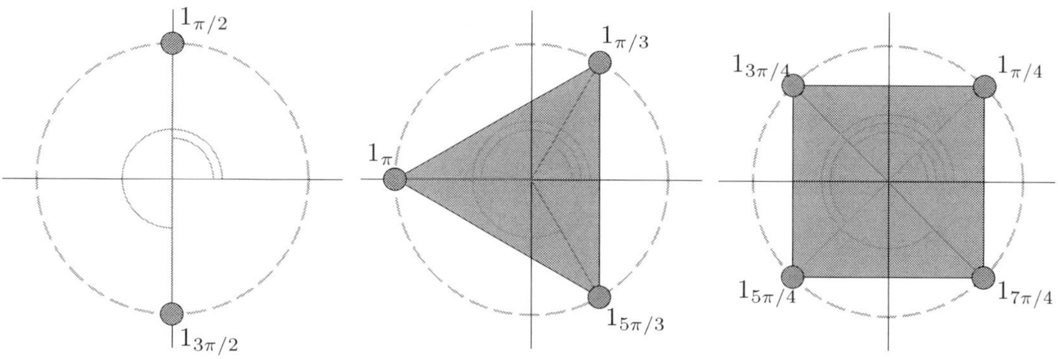

FIGURE 2.20
Square, cube, and fourth roots of the number $z = -1$.

2.9 The Logarithm of a Complex Number

In the set of real numbers, the exponential function is bijective and therefore invertible, with its inverse being the logarithm (natural or neperian). That is:

$$\exp(x) = e^x = y$$

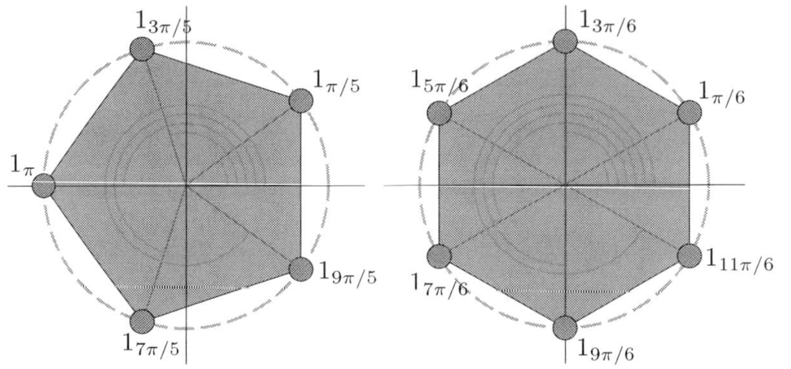

FIGURE 2.21
Fifth and sixth roots of the number $z = -1$.

is equivalent to:

$$x = \ln(y).$$

Recall that the domain of the exponential function is $\mathrm{Dom}(\exp) = \mathbb{R}$ and its range is the interval $(0, +\infty)$, while the domain of the logarithm function is $\mathrm{Dom}(\ln) = (0, +\infty)$ and its range is $\mathbb{R}$.

However, when defining a function equivalent to the logarithm in the set of complex numbers, we must consider that the complex exponential function is no longer bijective, since:

$$re^{\theta i} = re^{(\theta + 2\pi k)i}, \ k \in \mathbb{Z}.$$

Therefore, if $x + iy$ is the logarithm of $z = re^{\theta i}$, it must be true that:

$$e^{x+iy} = e^x e^{iy} = re^{\theta i},$$

which implies that:

$$e^x = r \ \Rightarrow \ x = \ln(r)$$

and also that:

$$y = \theta + 2\pi k, \ k \in \mathbb{Z}.$$

Thus, we can establish the following definition:

Definition 2.15. (Complex logarithm). Let $z = re^{\theta i} \in \mathbb{C}$. The complex logarithm of z, $\ln(z)$, is defined as:

$$\ln(z) = \ln(r) + (\theta + 2\pi k)i, \ k \in \mathbb{Z}.$$

From the above definition, it follows that the logarithm of a complex number is not a single number but a family of infinite complex numbers, all with the same real part equal to $\ln(r)$ and with imaginary parts differing by multiples of 2π.

Definition 2.16. (Principal value of the logarithm). Let $z = re^{\theta i} \in \mathbb{C}$. The principal value of the complex logarithm of z, $\ln(z)$, is defined as:

$$\ln(z) = \ln(r) + \theta i.$$

That is, the logarithm of z has been considered with $k = 0$.

Remark 2.1. The complex logarithm is defined for all complex numbers except for the complex number $z = 0$.

Example 2.13. *Express* $\ln(i)$ *in standard form.*

SOLUTION. *First, to calculate the logarithm, we must express the complex number in polar or exponential form. In this case, we have:*

$$i = e^{\frac{\pi}{2}i}.$$

Therefore:

$$\ln(i) = \ln\left(e^{\frac{\pi}{2}i}\right) = \ln(1) + \left(\frac{\pi}{2} + 2\pi k\right)i = \left(\frac{\pi}{2} + 2\pi k\right)i,\ k \in \mathbb{Z}.$$

In this case, the principal value of the logarithm is:

$$\ln(i) = \frac{\pi}{2}i$$

Note that the principal value of $\ln(i)$ *is a pure imaginary complex number with a modulus of* $|\ln(i)| = \frac{\pi}{2}$ *and a principal argument of* $\frac{\pi}{2}$ *radians.*

Example 2.14. *Express* $i^{\ln(i)}$ *in standard form.*

SOLUTION. *First, let's express this complex number differently:*

$$i^{\ln(i)} = e^{\ln(i)\cdot\ln(i)} = e^{(\ln(i))^2}.$$

Now, using the result from Example 2.13, we know that:

$$\ln(i) = \left(\frac{\pi}{2} + 2\pi k\right)i,\ k \in \mathbb{Z}.$$

Therefore:

$$e^{(\ln(i))^2} = e^{\left(\left(\frac{\pi}{2}+2\pi k\right)i\right)^2},\ k \in \mathbb{Z} = e^{-\left(\frac{\pi}{2}+2\pi k\right)^2},\ k \in \mathbb{Z}$$

Note that the complex number $i^{\ln(i)} = e^{-\left(\frac{\pi}{2}+2\pi k\right)^2}$, $k \in \mathbb{Z}$, *is a complex number with a zero imaginary part and a modulus of:*

$$\left|e^{-\left(\frac{\pi}{2}+2\pi k\right)^2}\right| = e^{-\left(\frac{\pi}{2}+2\pi k\right)^2}$$

and with a principal argument of:

$$arg\left(e^{-\left(\frac{\pi}{2}+2\pi k\right)^2}\right) = 0.$$

Property 2.2. On the complex logarithm Let $z, u \in \mathbb{C}$. Then:

$\ln(z \cdot u) = \ln(z) + \ln(u) + 2\pi ki,\ k \in \mathbb{Z}.$

(2) $\ln(z/u) = \ln(z) - \ln(u) + 2\pi ki,\ k \in \mathbb{Z}.$

(3) $\ln(z^n) = n\ln(z) + 2\pi ki,\ k \in \mathbb{Z}.$

PROOF. Consider the exponential form of z and u, i.e., $z = re^{\alpha i}$ and $u = se^{\beta i}$. Then:

(1)

$$\begin{aligned}
\ln(z) + \ln(u) + 2\pi ki &= \ln\left(re^{\alpha i}\right) + \ln\left(se^{\beta i}\right) + 2\pi ki \\
&= \ln(r) + \alpha i + 2\pi k_1 i + \ln(s) + \beta i + 2\pi k_2 i + 2\pi ki \\
&= \ln(rs) + (\alpha + \beta)i + 2\pi(k_1 + k_2 + k)i \\
&= \ln(rse^{(\alpha+\beta)i}) \\
&= \ln(re^{\alpha i} \cdot se^{\beta i}) \\
&= \ln(z \cdot u)
\end{aligned}$$

(2)

$$\begin{aligned}
\ln(z/u) &= \ln\left(\frac{re^{\alpha i}}{se^{\beta i}}\right) \\
&= \ln\left(re^{\alpha i} \cdot \frac{1}{s}e^{-\beta i}\right) \\
&= \ln\left(re^{\alpha i}\right) + \ln\left(s^{-1}e^{-\beta i}\right) + 2\pi ki \\
&= \ln(z) + \ln\left(s^{-1}\right) - \beta i + 2\pi ki \\
&= \ln(z) - \ln(s) - \beta i + 2\pi ki \\
&= \ln(z) - (\ln(s) + \beta i) + 2\pi ki \\
&= \ln(z) - \ln(u) + 2\pi ki
\end{aligned}$$

(3)

$$\begin{aligned}
\ln\left(z^n\right) &= \ln\left(\left(re^{\alpha i}\right)^n\right) \\
&= \ln\left(r^n e^{n\alpha i}\right) \\
&= \ln\left(r^n\right) + n\alpha i + 2\pi ki \\
&= n\ln(r) + n\alpha i + 2\pi ki \\
&= n\left(\ln(r) + \alpha i\right) + 2\pi ki \\
&= n\ln\left(re^{\alpha i}\right) + 2\pi ki \\
&= n\ln\left(z\right) + 2\pi ki
\end{aligned}$$

Trigonometric functions in terms of the complex exponential and the complex logarithm also allow us to solve equations in $\mathbb{C}$ such as:

$$\sin(z) = 3$$

or even:

$$\tan(x) = \frac{i}{2}$$

Let's solve the equation $\sin(z) = 3$ in the following example.

Example 2.15 (Trigonometric equation). *Solve the equation* $\sin(z) = 3$.

SOLUTION. Recalling the expression of the sine in terms of the complex exponential, we can write:

$$\begin{aligned}
\sin(z) = 3 &\Leftrightarrow \frac{e^{zi} - e^{-zi}}{2i} = 3 \\
&\Leftrightarrow e^{zi} - e^{-zi} = 6i \\
&\Leftrightarrow e^{zi} - \frac{1}{e^{zi}} = 6i \\
&\Leftrightarrow e^{2zi} - 1 = 6e^{zi}i \\
&\Leftrightarrow e^{2zi} - 6e^{zi}i - 1 = 0 \\
&\Leftrightarrow -e^{2zi}i^2 - 6e^{zi}i - 1 = 0 \\
&\Leftrightarrow -(e^{zi}i)^2 - 6e^{zi}i - 1 = 0
\end{aligned}$$

If we now define $x = e^{zi}i$, the previous equation becomes:

$$-x^2 - 6x - 1 = 0,$$

or equivalently:

$$x^2 + 6x + 1 = 0.$$

The solutions to this equation are:

$$x = \frac{-6 \pm \sqrt{36 - 4}}{2} = -3 \pm 2\sqrt{2}.$$

Therefore:

$$x = e^{zi}i \Leftrightarrow e^{zi}i = -3 \pm 2\sqrt{2} \Leftrightarrow e^{zi} = \left(3 \mp 2\sqrt{2}\right) i$$

Applying logarithms:

$$zi = \ln\left(\left(3 \mp 2\sqrt{2}\right) i\right) = \ln\left(3 \mp 2\sqrt{2}\right) + \ln(i) = \ln\left(3 \mp 2\sqrt{2}\right) + \left(\frac{\pi}{2} + 2\pi k\right) i, \quad k \in \mathbb{Z}$$

Finally:

$$z = \left(\frac{\pi}{2} + 2\pi k\right) - i \cdot \ln\left(3 \mp 2\sqrt{2}\right), \quad k \in \mathbb{Z}.$$

Proposed Exercise 2.6. Solve the equation $\tan(x) = \frac{i}{2}$.

2.10 Applications of Complex Numbers to AC Circuits

An alternating current (AC) electrical circuit, such as the one in Figure 2.22, consisting of resistors, capacitors, and inductors, can be analyzed using the equation:

$$V = ZI,$$

where V is the voltage, Z is the impedance, and I is the electric current. Both V, Z, and I are complex numbers.

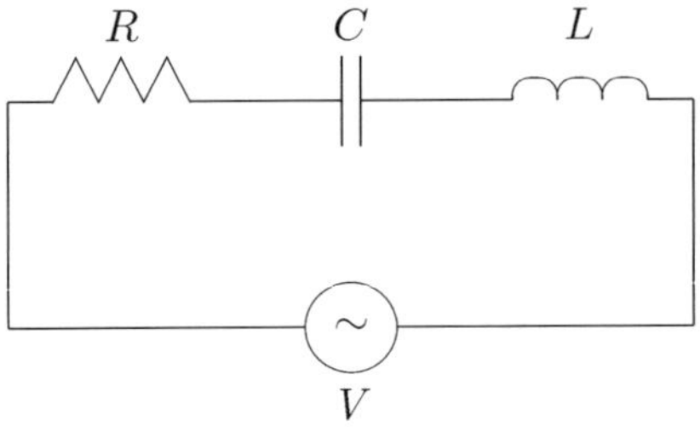

FIGURE 2.22
RLC electrical circuit.

Each component of an electrical circuit has an associated complex *impedance*. For two components in series, the resulting impedance, Z_R, is calculated with the formula:

$$Z_R = Z_1 + Z_2.$$

For two components in parallel, the resulting impedance is determined by:

$$\frac{1}{Z_R} = \frac{1}{Z_1} + \frac{1}{Z_2}.$$

Therefore, in the case of circuits with parallel components, it may be easier to calculate the inverse of the impedance, the *admittance*, $Y = 1/Z$. Thus, the resulting admittance of a circuit with parallel elements is:

$$Y_R = Y_1 + Y_2 \text{ and } V = I/Y.$$

The real part of Z is called *resistance* or *resistive part* and the imaginary part is called *reactance* or *reactive part*, so it can be written as:

$$Z = R + Si,$$

where R and S are measured in ohms.

The impedances of the components of an electrical circuit are found in Table 2.5, where ω is the angular frequency of the input signal (or source) and $\omega = 2\pi f$, where f is the frequency (in hertz), R is the resistance (in ohms), C is the capacitance of the capacitor (in farads), and L is the inductance (in henrys).

Example 2.16. *Calculate the impedance of the circuit in Figure 2.22, with a frequency of* 20 *kHz, where* $L = 2$ *mH,* $C = 100$ μ*F, and* $R = 2000$ Ω*. Assume that the voltage has an amplitude of* 300 *V and also calculate the electric current* I *and the relative phase shift.*

SOLUTION. The impedances of the resistor, capacitor, and inductor are:

$$R, \ \frac{-i}{\omega C}, \ i\omega L,$$

TABLE 2.5
Impedances of electrical circuit components..

Resistor	$Z = R$	No reactive part
Capacitor	$Z = 1/(i\omega C) = -i/(\omega C)$	Purely reactive
Inductor	$Z = i\omega L$	Purely reactive

respectively. Since we are given the frequency, f, in hertz, the value of $\omega = 2\pi f$ is:

$$\omega = 2\pi f = 2\pi \cdot 20 \cdot 10^3 = 4 \cdot 10^4 \cdot \pi.$$

Considering that the elements of the circuit are in series, the resulting impedance is the sum of the impedance of each component. Indeed:

$$\begin{aligned} Z &= R - \frac{i}{\omega C} + i\omega L \\ &= 2000 - \frac{i}{4 \cdot 10^4 \cdot \pi \cdot 100 \cdot 10^{-6}} + i \cdot 4 \cdot 10^4 \cdot 2 \cdot 10^{-3} \\ &= 2000 - \frac{i}{4\pi} + i \cdot 80 \\ &= 2000 + i\left(80 - \frac{1}{4\pi}\right) \\ &\approx 2000 + 79.92i \end{aligned}$$

In polar coordinates, $Z = 2000 + 79.92i$ can be expressed as $Z = 2001.60_{2.29^\circ}$. So, if we consider the relationship $V = ZI$ and assume that $V = 300_{0^\circ}$:

$$I = \frac{V}{Z} = \frac{300_{0^\circ}}{2001.60_{2.29^\circ}} = 0.15_{-2.29^\circ},$$

and therefore, we obtain an electric current with a magnitude of 0.15 A with a phase delay of 2.29°.

Proposed Exercise 2.7. The circuit in Figure 2.23 can be used as a low-pass filter (which allows low frequencies to pass and attenuates high frequencies). If we consider that $R = 300\ \Omega$, $L = 2$ mH, and $C = 10\ \mu$F, calculate the admittance of the circuit assuming a frequency of 2 kHz. If we also assume that the electric current has an amplitude of 12 A, calculate the magnitude and phase shift of the voltage.

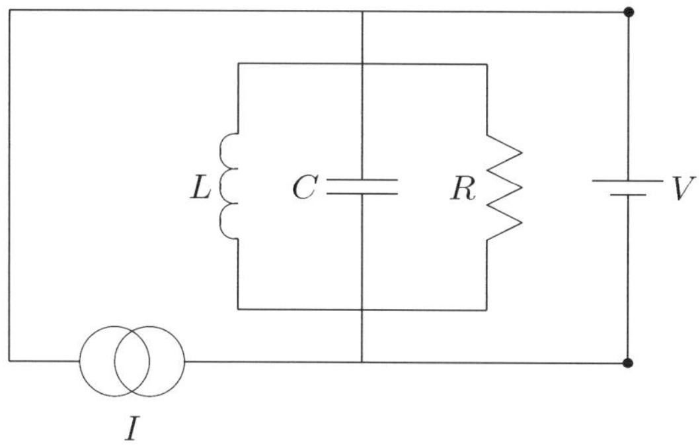

FIGURE 2.23
Electrical circuit.

Solved Problems (Complex Numbers)

Problem 2.1. Apply the formula to find the solutions of a quadratic equation to the equation:

$$x^2 - 10x + 40 = 0$$

and verify that the solutions are $5 + \sqrt{-15}$ and $5 - \sqrt{-15}$ (even though there is a negative number under the square root).

Solution to the problem 2.1. *In this exercise, we are asked to consider the equation:*

$$x^2 - 10x + 40 = 0$$

and to find the solutions:

$$5 + \sqrt{-15} \quad and \quad 5 - \sqrt{-15},$$

using the formula for the solutions of a quadratic equation.

If we apply the formula (noting that in this case, $a = 1$, $b = -10$, *and* $c = 40$*) we have:*

$$\begin{aligned} x &= \frac{-b \pm \sqrt{b^2 - 4ac}}{2a} = \frac{10 \pm \sqrt{100 - 160}}{2} \\ &= \frac{10 \pm \sqrt{-60}}{2} = 5 \pm \frac{\sqrt{-15}\sqrt{4}}{2} = 5 \pm \frac{\sqrt{-15} \cdot 2}{2} = 5 \pm \sqrt{-15}. \end{aligned}$$

Thus, we have recovered the two solutions we wanted.

Problem 2.2. Calculate the solutions of the following quadratic equation:

$$x^2 - 2x + 5 = 0.$$

Do you notice any common characteristic in its solutions?

Solution to the problem 2.2. *In this exercise, we need to calculate the solutions of the equation*

$$x^2 - 2x + 5 = 0;$$

knowing that if we encounter a negative number under the square root, we can use the fact that $\sqrt{-1} = i$.

To calculate the solutions, we use the formula for the solutions of a quadratic equation of the form $ax^2 + bx + c = 0$. *In our case,* $a = 1$, $b = -2$, *and* $c = 5$*; therefore:*

$$\begin{aligned} x &= \frac{-b \pm \sqrt{b^2 - 4ac}}{2a} = \frac{2 \pm \sqrt{4 - 20}}{2} \\ &= \frac{2 \pm \sqrt{-16}}{2} = 1 \pm \frac{\sqrt{16}\sqrt{-1}}{2} = 1 \pm \frac{4i}{2} = 1 \pm 2i. \end{aligned}$$

Thus, the two solutions are $1 + 2i$ *and* $1 - 2i$.

We can observe that the two solutions are identical except for the constant accompanying i*, which changes sign.*

Problem 2.3. Calculate the solutions of the following quadratic equation:

$$x^2 - 6x + 25 = 0.$$

Do you notice any common characteristic in the solutions of the equations that have been solved?

Solution to the problem 2.3. *This exercise is similar to Problem 2.2, but with different coefficients for the equation to be solved.*

As we did in Problem 2.2, we will use the formula to calculate the solutions of a quadratic equation of the form $ax^2 + bx + c = 0$. *Now,* $a = 1$, $b = -6$, *and* $c = 25$; *therefore:*

$$\begin{aligned} x &= \frac{-b \pm \sqrt{b^2 - 4ac}}{2a} = \frac{6 \pm \sqrt{36 - 100}}{2} \\ &= \frac{6 \pm \sqrt{-64}}{2} = 3 \pm \frac{\sqrt{64}\sqrt{-1}}{2} = 3 \pm \frac{8i}{2} = 3 \pm 4i. \end{aligned}$$

Thus, the two solutions are $3 + 4i$ *and* $3 - 4i$.

As in Problem 2.2, we can observe that the two solutions are identical except for the constant accompanying i, *which changes sign. That is, the solutions are of the form:*

$$x = c_1 + c_2 i \qquad \text{and} \qquad x = c_1 - c_2 i,$$

where the constants c_1 *and* c_2 *depend on the equation we are considering.*

Problem 2.4. Although the *irrational* number $\sqrt{2}$ cannot be expressed as a fraction, it can be located on a number line (in fact, it is found between 1.4 and 1.5). How can it be exactly located?

Solution to the problem 2.4. *Consider a square with a side length of* 1. *If we apply the Pythagorean theorem, we know that its diagonal has a length equal to:*

$$d = \sqrt{1^2 + 1^2} = \sqrt{2}.$$

Therefore, to locate the number $\sqrt{2}$ *on the real line, we only need a ruler, a set square, and a compass.*

Using the ruler and the set square, we draw two coordinate axes and draw the square $[0, 1] \times [0, 1]$. *Once the square is drawn, we draw its diagonal. Finally, with the compass point at* $(0, 0)$, *we transfer the diagonal of the square onto the* x*-axis. See Figure 2.24.*

Problem 2.5. Consider a right triangle with a perimeter of 12 and an area of 7. Verify that the length x of one of the sides of the triangle satisfies the equation:

$$6x^2 - 43x + 84 = 0.$$

Graph the parabola:

$$y = 6x^2 - 43x + 84$$

and verify that it does not intersect the x-axis.

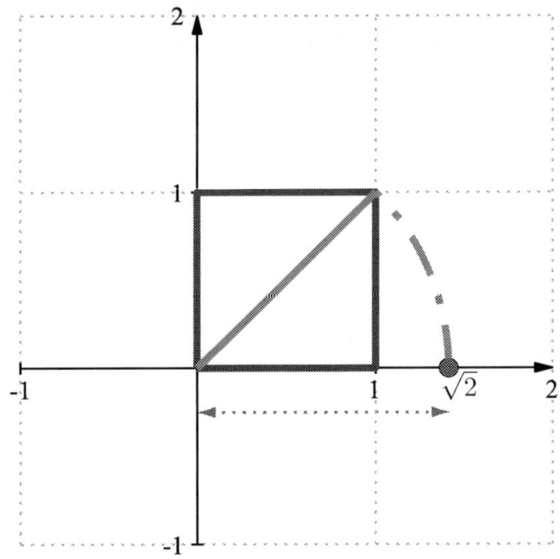

FIGURE 2.24
Square with side length 1 (Problem 2.0).

Solution to the problem 2.5. *Consider the right triangle in Figure 2.25.*
We know that the perimeter of the triangle must be 12, meaning:

$$x + y + z = 12.$$

Additionally, we know that the area of the triangle must be 7, meaning:

$$(1/2)xy = 7$$

(recall that the area of a triangle is half the base (x) times the height (y)).

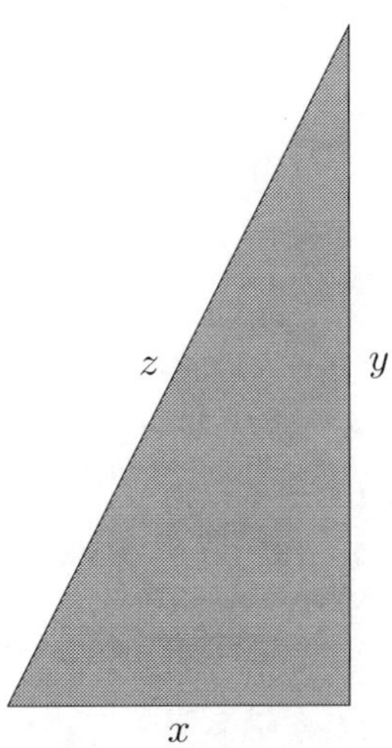

FIGURE 2.25
Right triangle with sides x, y, z (Problem 2.0).

Also, since the triangle is a right triangle, we can apply the Pythagorean theorem, which states:

$$z^2 = x^2 + y^2.$$

Therefore, the three equations that the side lengths must satisfy are:

$$\begin{aligned}&(1)\ \ x + y + z = 12\\&(2)\ \ xy = 14\\&(3)\ \ z^2 = x^2 + y^2.\end{aligned}$$

If we isolate z from equation (1) and square the equality, we get:

$$z = 12 - x - y \quad \longrightarrow \quad z^2 = (12 - x - y)^2.$$

Now we can equate this result with equation (3), yielding:

$$(12 - x - y)^2 = x^2 + y^2.$$

Next, we need to expand the left side of the previous equation and use equation (2). Indeed, let's expand the left-hand side:

$$\begin{aligned}(12 - x - y)^2 &= (12 - (x + y))^2 = 12^2 + (x + y)^2 - 2 \cdot 12(x + y)\\&= 12^2 + x^2 + y^2 + 2xy - 24x - 24y.\end{aligned}$$

If we equate this with the right-hand side, we obtain:

$$12^2 + x^2 + y^2 + 2xy - 24x - 24y = x^2 + y^2,$$

where we can simplify the $x^2 + y^2$ terms to obtain:

$$12^2 + 2xy - 24x - 24y = 0.$$

Now, to use equation (2), we multiply the entire equation by x and substitute all xy products with 14.

$$12^2x + 2xxy - 24x^2 - 24xy = 0 \quad \longrightarrow \quad 12^2x + 28x - 24x^2 - 24 \cdot 14 = 0.$$

Finally, we group terms:

$$144x + 28x - 24x^2 - 336 = 0 \quad \longrightarrow \quad 24x^2 - 172x + 336 = 0,$$

and dividing by 4, we obtain the desired result:

$$6x^2 - 43x + 84 = 0.$$

Next, we will calculate the values of x that satisfy the previous equation. To do this, we apply the formula for the solutions of a quadratic equation of the form $ax^2 + bx + c = 0$.

$$\begin{aligned}x &= \frac{-b \pm \sqrt{b^2 - 4ac}}{2a} = \frac{43 \pm \sqrt{1849 - 2016}}{12}\\&= \frac{43 \pm \sqrt{-167}}{12} = \frac{43}{12} \pm \frac{\sqrt{-167}}{12} = 5 \pm \frac{\sqrt{-15 \cdot 2}}{2} = 5 \pm \sqrt{-15}.\end{aligned}$$

Since there are no real solutions to the equation, the parabola:

$$y = 6x^2 - 43x + 84$$

does not intersect the x-axis, as shown in Figure 2.26. This implies that a triangle with the characteristics described in the problem statement cannot be constructed.

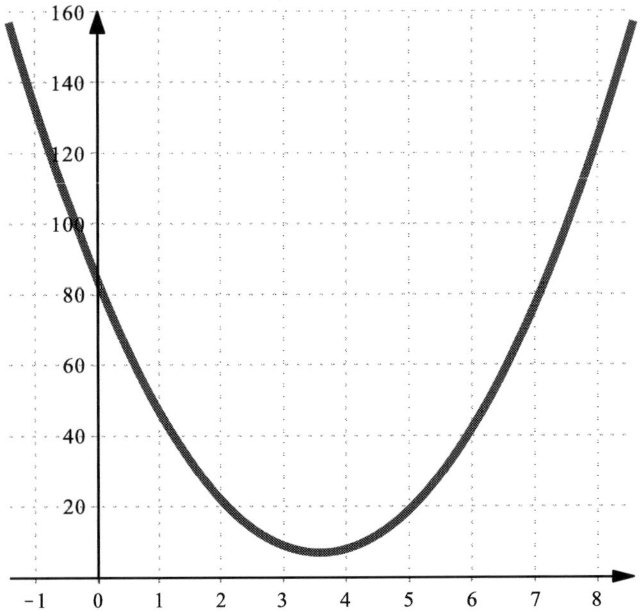

FIGURE 2.26
Parabola $y = 6x^2 - 43x + 84$ (Problem 2.0).

Problem 2.6. Plot the following numbers in the complex plane:

$$z_1 = 3, \quad z_2 = 3i, \quad z_3 = -2i, \quad z_4 = 1 + 2i, \quad z_5 = -2 + i.$$

Solution to the problem 2.6. *Recall that a complex number $a + bi$ can be associated with a point in the complex plane, the point (a, b). In this way, we can associate the given points with:*

$$\begin{aligned} z_1 &= 3 && \longrightarrow && (3, 0) \\ z_2 &= 3i && \longrightarrow && (0, 3) \\ z_3 &= -2i && \longrightarrow && (0, -2) \\ z_4 &= 1 + 2i && \longrightarrow && (1, 2) \\ z_5 &= -2 + i && \longrightarrow && (-2, 1). \end{aligned}$$

In Figure 2.27, we have plotted these points in the complex plane.

Problem 2.7. Graphically add the vectors $(2, -1)$ and $(1, 3)$.

Solution to the problem 2.7. *To graph the sum of the given vectors $(2, -1)$ and $(1, 3)$, the first thing we need to do is represent them graphically (see the left graph in Figure 2.28). These points are associated with the complex points $2 - i$ and $1 + 3i$.*

The second step is to draw two lines parallel to the given vectors so that the two vectors and the two parallels form a parallelogram (as seen in the center graph of Figure 2.28).

Finally, the sum of the two vectors originates at the point $(0, 0)$ and ends at the intersection of the two drawn lines (see the right graph of Figure 2.28).

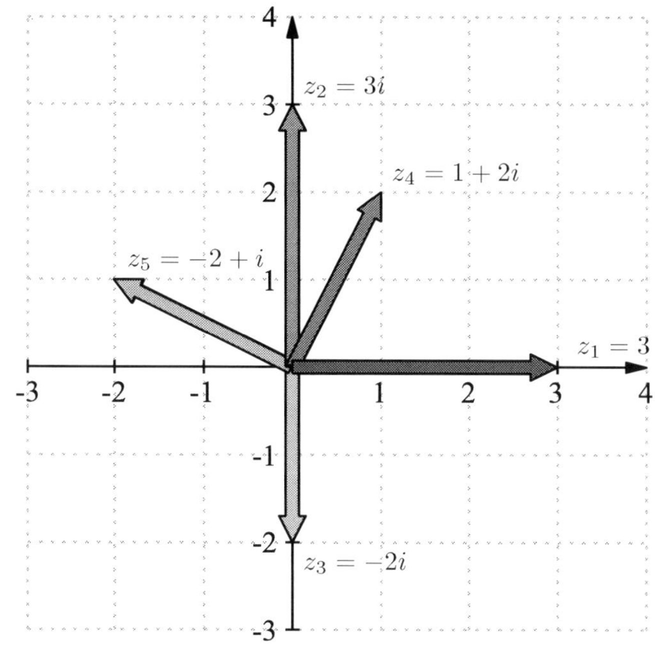

FIGURE 2.27
Points $(3,0)$, $(0,3)$, $(0,-2)$, $(1,2)$, and $(-2,1)$ in the complex plane (Problem 2.0).

Problem 2.8. Given $z_1 = 3 + 4i$ and $z_2 = 1 - 2i$, calculate $z_1 + z_2$. Plot these three complex numbers in the complex plane and verify that the parallelogram law holds.

Solution to the problem 2.8. *Consider the numbers* $z_1 = 3 + 4i$ *and* $z_2 = 1 - 2i$. *We need to calculate their sum:*

$$z_1 + z_2.$$

Recall that to add two complex numbers, we add the real parts and the imaginary parts separately (as if we were treating i as a variable and operating with the numbers).

$$\begin{aligned} z_1 + z_2 &= (3 + 4i) + (1 - 2i) = (3 + 1) + (4 - 2)i \\ &= 4 + 2i. \end{aligned}$$

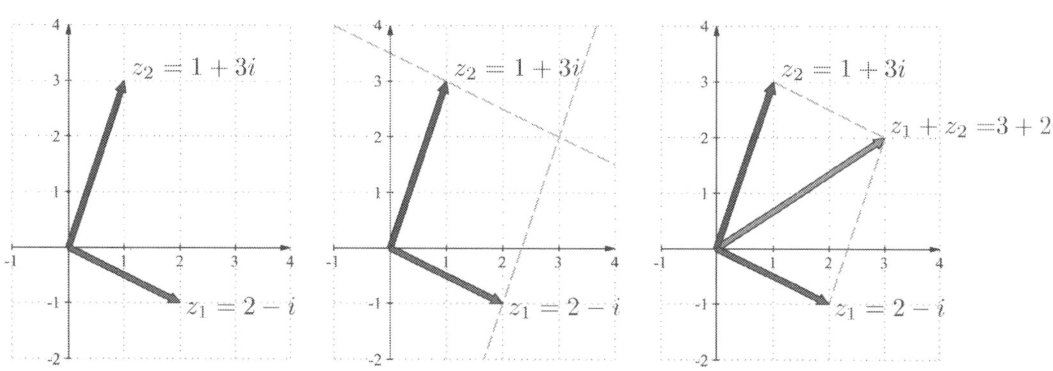

FIGURE 2.28
Graphical sum of the vectors $(2,-1)$ and $(1,3)$ (Problem 2.0).

To plot these complex numbers in the complex plane, remember that every complex number of the form $a+bi$ is associated with the point (a,b) in the complex plane. Therefore, we need to plot the points:

Complex Number	$\longrightarrow$	*Point in the Plane*
$z_1 = 3+4i$	$\longrightarrow$	$(3,4)$
$z_2 = 1-2i$	$\longrightarrow$	$(1,-2)$
$z_1+z_2 = 4+2i$	$\longrightarrow$	$(4,2)$.

Note in Figure 2.29 that the parallelogram law holds.

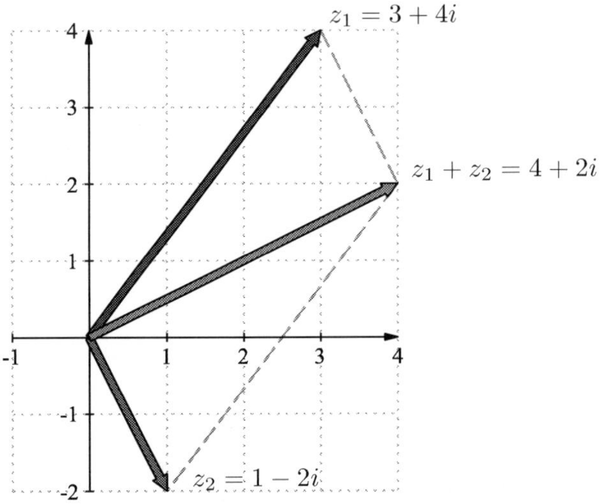

FIGURE 2.29
Sum of the vectors $z_1 = 3+4i$ and $z_2 = 1-2i$ (Problem 2.0).

Problem 2.9. Given $z_1 = 3+4i$ and $z_2 = 1-2i$, calculate $z_1 - z_2$. Plot these three complex numbers in the complex plane.

Solution to the problem 2.9. *Consider the numbers $z_1 = 3+4i$ and $z_2 = 1-2i$. We need to calculate their difference:*

$$z_1 - z_2.$$

Recall that to subtract two complex numbers, we subtract the real parts and the imaginary parts separately (as if we were treating i as a variable and operating with the numbers).

$$\begin{aligned} z_1 - z_2 &= (3+4i)-(1-2i) = (3-1)+(4-(-2))i \\ &= 2+6i. \end{aligned}$$

To plot these complex numbers in the complex plane, remember that every complex number of the form $a+bi$ is associated with the point (a,b) in the complex plane. Therefore, we need to plot the points:

Complex Number	$\longrightarrow$	*Point in the Plane*
$z_1 = 3+4i$	$\longrightarrow$	$(3,4)$
$z_2 = 1-2i$	$\longrightarrow$	$(1,-2)$
$z_1-z_2 = 2+6i$	$\longrightarrow$	$(2,6)$.

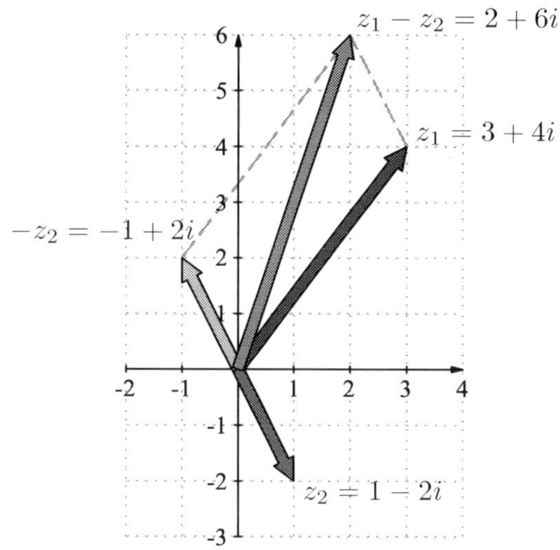

FIGURE 2.30
Plot of the numbers $z_1 = 3 + 4i$, $z_2 = 1 - 2i$, and $z_1 - z_2$ in the complex plane (Problem 2.0).

Note that to subtract two complex numbers z_1 and z_2, we are essentially adding the numbers z_1 and $-z_2$, as shown in Figure 2.30.

Problem 2.10. Given $z_1 = 3 + 4i$ and $z_2 = 1 - 2i$, calculate $z_1 \cdot z_2$.

Solution to the problem 2.10. *Recall that the product of two complex numbers is done in the same way as the product of two polynomials of the form:*

$$(a + bx) \cdot (c + dx) = a(c + dx) + bx(c + dx) = ac + adx + bcx + bdx^2.$$

The only thing to keep in mind is that when i^2 appears, we will substitute it by -1.

$$\begin{aligned} z_1 \cdot z_2 &= (3 + 4i) \cdot (1 - 2i) \\ &= 3 \cdot (1 - 2i) + 4i \cdot (1 - 2i) \\ &= 3 \cdot 1 + 3 \cdot (-2)i + 4i \cdot 1 + 4i \cdot (-2i) \\ &= 3 - 6i + 4i - 8i^2 \\ &= (3 + 8) + (-6 + 4)i \qquad \textit{(using } i^2 = -1 \textit{)} \\ &= 11 - 2i. \end{aligned}$$

Figure 2.31 shows z_1, z_2, and their product.

Problem 2.11. Given $z = 4 + 3i$, calculate $\bar{z}$. Then calculate $\bar{\bar{z}}$. What do you observe?

Solution to the problem 2.11. *Consider $z = 4 + 3i$. The conjugate of this number is obtained by changing the sign of the imaginary part, that is:*

$$z = 4 + 3i \quad \longrightarrow \quad \bar{z} = 4 - 3i.$$

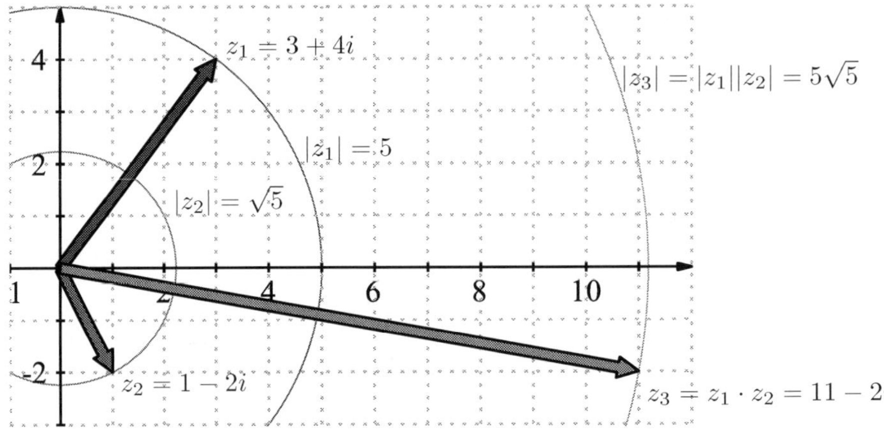

FIGURE 2.31
Plot of the numbers $z_1 = 3+4i$, $z_2 = 1-2i$, and $z_1 \cdot z_2$ in the complex plane (Problem 2.0).

Notice that if we calculate the conjugate again, that is, if we calculate $\bar{\bar{z}}$, *we obtain:*

$$z = 4+3i \quad \longrightarrow \quad \bar{z} = 4-3i \quad \longrightarrow \quad \bar{\bar{z}} = 4-(-3)i = 4+3i.$$

That is, $\bar{\bar{z}} = z$.

Problem 2.12. Given $z = 1+5i$, calculate $\bar{z}$. Then calculate $z \cdot \bar{z}$. What do you observe?

Solution to the problem 2.12. *We have* $z = 1+5i$ *and, therefore,* $\bar{z} = 1-5i$. *Now let's calculate* $z \cdot \bar{z}$.

$$\begin{aligned} z \cdot \bar{z} &= (1+5i) \cdot (1-5i) \\ &= 1 \cdot 1 + 1 \cdot (-5)i + 5i \cdot 1 + 5i \cdot (-5i) \\ &= 1 - 5i + 5i - 25i^2 \\ &= 1 - 25(-1) \\ &= 1 + 25 = 26. \end{aligned}$$

We observe that $z \cdot \bar{z}$ *is a positive real number and, furthermore, that:*

$$z \cdot \bar{z} = Re(z)^2 + Im(z)^2 = (1)^2 + (5^2) = 1 + 25 = 26.$$

Problem 2.13. Find the conjugates of the following complex numbers and verify that $z \cdot \bar{z}$ is real and positive (or zero) in all cases: $2-5i$, $-4+2i$, -5, $6i$, $x+yi$. Plot the numbers and their conjugates in the complex plane.

Solution to the problem 2.13. *Let's first consider* $z_1 = 2-5i$. *The conjugate of a complex number is obtained by changing the sign of the imaginary part, that is:*

$$z_1 = 2-5i \quad \longrightarrow \quad \overline{z_1} = 2+5i.$$

Therefore:

$$\begin{aligned} z_1 \cdot \overline{z_1} &= (2-5i)\cdot(2+5i) = 2\cdot 2 + 2\cdot 5i - 5i\cdot 2 - 5i\cdot 5i \\ &= 4 + 10i - 10i - 25i^2 \\ &= 4 - 25(-1) = 4 + 25 = 29. \end{aligned}$$

Note that multiplying z_1 by its conjugate gives:

$$z_1 \cdot \overline{z_1} = 4 + 25 = Re(z_1)^2 + Im(z_1)^2 = 2^2 + (-5)^2 = 4 + 25 = 29.$$

The plot of z_1 and $\overline{z_1}$ can be seen in Figure 2.32.

Now let's do the same with $z_2 = -4+2i$. The conjugate of a complex number is obtained by changing the sign of the imaginary part, that is:

$$z_2 = -4 + 2i \quad\longrightarrow\quad \overline{z_2} = -4 - 2i.$$

Therefore:

$$\begin{aligned} z_2 \cdot \overline{z_2} &= (-4+2i)\cdot(-4-2i) = (-4)\cdot(-4) + (-4)\cdot(-2i) + 2i\cdot(-4) + 2i\cdot(-2i) \\ &= 16 + 8i - 8i - 4i^2 \\ &= 16 - 4(-1) = 16 + 4 = 20. \end{aligned}$$

Note that multiplying z_2 by its conjugate gives, just as in the previous example:

$$z_2 \cdot \overline{z_2} = 16 + 4 = Re(z_2)^2 + Im(z_2)^2 = (-4)^2 + 2^2 = 16 + 4 = 20.$$

The plot of z_2 and $\overline{z_2}$ can be seen in Figure 2.32.

Now consider $z_3 = -5$, which is a real number since the imaginary part of z_3 is zero. In this case:

$$z_3 = -5 \quad\longrightarrow\quad \overline{z_3} = z_3 = -5.$$

Therefore:

$$z_3 \cdot \overline{z_3} = (-5)\cdot(-5) = 25.$$

Note that we could have also used the formula:

$$z_3 \cdot \overline{z_3} = Re(z_3)^2 + Im(z_3)^2 = (-5)^2 + 0^2 = 25.$$

The plot of z_3 and $\overline{z_3}$ can be seen in Figure 2.32.

Now consider $z_4 = 6i$, which is a pure imaginary number since the real part of z_4 is zero. In this case:

$$z_4 = 6i \quad\longrightarrow\quad \overline{z_4} = -z_4 = -6i.$$

Therefore:

$$z_4 \cdot \overline{z_4} = (6i)\cdot(-6i) = -36i^2 = 36.$$

Note that, even in this case, we could have used the formula:

$$z_4 \cdot \overline{z_4} = Re(z_4)^2 + Im(z_4)^2 = 0^2 + 6^2 = 36.$$

The plot of z_4 and $\overline{z_4}$ can be seen in Figure 2.32.

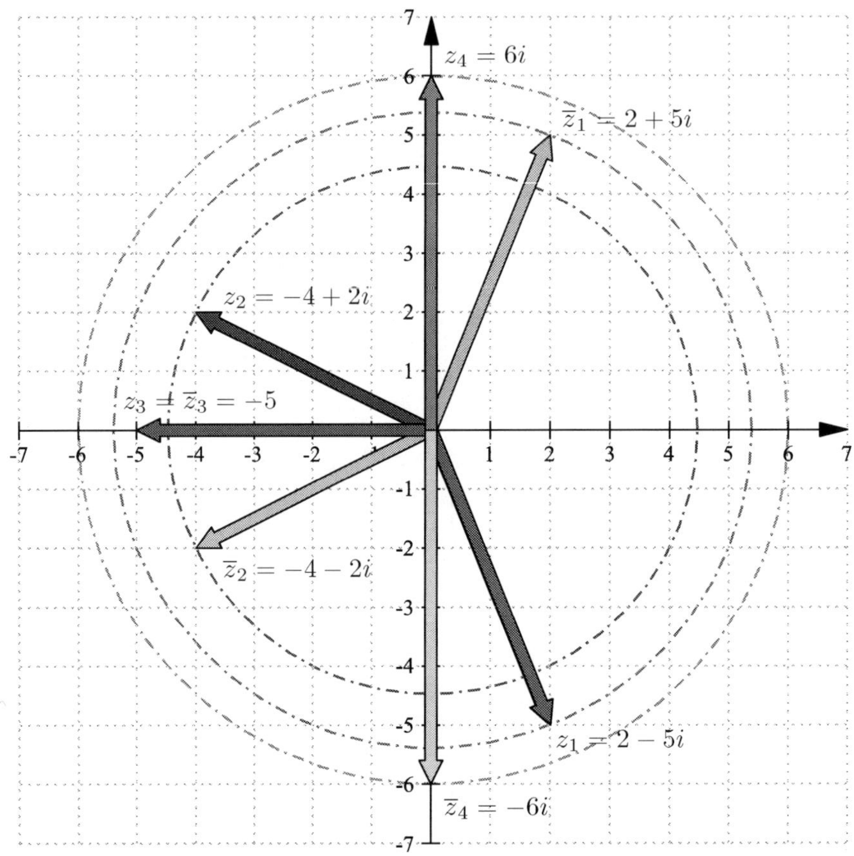

FIGURE 2.32
Plot of the numbers $2-5i$, $-4+2i$, -5, and $6i$ in the complex plane (Problem 2.0).

Finally, let's perform the exercise for a generic complex number $z_5 = x + yi$. *In this case:*

$$z_5 = x + yi \quad \longrightarrow \quad \overline{z_5} = x - yi.$$

Therefore:

$$\begin{aligned} z_5 \cdot \overline{z_5} &= (x+yi)\cdot(x-yi) = x\cdot x + x\cdot(-yi) + yi\cdot x + yi\cdot(-yi) \\ &= x^2 - xyi + xyi - y^2i^2 \\ &= x^2 - y^2i^2 = x^2 + y^2. \end{aligned}$$

Note, as seen before, that indeed:

$$z_5 \cdot \overline{z_5} = x^2 + y^2 = Re(z_5)^2 + Im(z_5)^2.$$

Problem 2.14. Verify with an example that when a complex number is added and multiplied by its conjugate, the result is a real number in each case.

Solution to the problem 2.14. *Instead of verifying this with an example, let's see it in general. Let* $z = a + bi$, *then its conjugate is* $\overline{z} = a - bi$.

Therefore, given that:

$$z + \overline{z} = (a + bi) + (a - bi) = 2a \qquad y$$
$$z \cdot \overline{z} = (a + bi) \cdot (a - bi) = a^2 - abi + abi - (bi)^2 = a^2 + b^2,$$

we see that both $z + \overline{z}$ and $z \cdot \overline{z}$ are real numbers.

Problem 2.15. Given $z_1 = 3 + 4i$ and $z_2 = 1 - 2i$, find $\frac{z_1}{z_2}$.

Solution to the problem 2.15. *Recall that when dealing with a quotient of complex numbers, the first step is to eliminate the imaginary terms in the denominator. To do this, we multiply the denominator by its conjugate, which also requires multiplying the numerator by the same number.*

$$\frac{z_1}{z_2} = \frac{3 + 4i}{1 - 2i} = \frac{3 + 4i}{1 - 2i} \cdot \frac{1 + 2i}{1 + 2i} = \frac{(3 + 4i)(1 + 2i)}{(1 - 2i)(1 + 2i)}$$

Calculating the numerator and denominator separately:

$$(3 + 4i)(1 + 2i) = 3 + 6i + 4i + 8i^2 = 3 + 10i - 8 = -5 + 10i$$

$$(1 - 2i)(1 + 2i) = 1 + 2i - 2i - 4i^2 = 1 + 4 = 5$$

Thus:

$$\frac{z_1}{z_2} = \frac{-5 + 10i}{5} = -1 + 2i$$

Problem 2.16. Calculate the following complex numbers:

$$\frac{1}{i}, \quad \frac{1}{a + bi}, \quad \frac{1}{2 - 3i} \cdot \frac{1}{1 + i}, \quad \frac{1 + 2i}{3 - 4i} + \frac{2 - i}{5i}$$

Solution to the problem 2.16. *In this exercise, we need to perform divisions of complex numbers. As in Problem 2.15, recall that the first step is to eliminate the imaginary terms in the denominator. We start with the first number, $\frac{1}{i}$:*

$$\frac{1}{i} = \frac{1}{i} \cdot \frac{-i}{-i} = \frac{-i}{-i^2} = \frac{-i}{1} = -i$$

Similarly, for the next quotient, $\frac{1}{a+bi}$:

$$\frac{1}{a + bi} = \frac{1}{a + bi} \cdot \frac{a - bi}{a - bi} = \frac{a - bi}{(a + bi)(a - bi)} = \frac{a - bi}{a^2 - b^2(-1)} = \frac{a - bi}{a^2 + b^2}$$

For the next calculation, we have two options because we have a product of complex numbers in the denominator. The first option is to multiply the denominator and then multiply by the conjugate of the result. The second option is to multiply by the conjugates of both denominators and then operate. The first option yields:

$$\begin{aligned} \frac{1}{2 - 3i} \cdot \frac{1}{1 + i} &= \frac{1}{(2 - 3i)(1 + i)} = \frac{1}{2 + 2i - 3i - 3i^2} = \frac{1}{2 - i + 3} = \frac{1}{5 - i} \\ &= \frac{1}{5 - i} \cdot \frac{5 + i}{5 + i} = \frac{5 + i}{(5 - i)(5 + i)} = \frac{5 + i}{25 + 5i - 5i - i^2} = \frac{5 + i}{26}. \end{aligned}$$

Let's see how it would be done using the second option:

$$\begin{aligned}\frac{1}{2-3i}\cdot\frac{1}{1+i} &= \frac{1}{2-3i}\cdot\frac{2+3i}{2+3i}\cdot\frac{1}{1+i}\cdot\frac{1-i}{1-i} = \frac{2+3i}{(2-3i)(2+3i)}\cdot\frac{1-i}{(1+i)(1-i)}\\ &= \frac{2+3i}{4+9}\cdot\frac{1-i}{1+1} = \frac{(2+3i)(1-i)}{26}\\ &= \frac{2-2i+3i-3i^2}{26} = \frac{2+i+3}{26} = \frac{5+i}{26}.\end{aligned}$$

Note that in this exercise we used the fact that a number multiplied by its conjugate is the square of the real part plus the square of the imaginary part. That is, we used:

$$(2-3i)(2+3i) = 4+6i-6i-9i^2 = 4+9 = 2^2+(-3)^2$$

and:

$$(1+i)(1-i) = 1-i+i-i^2 = 1+1 = 1^2+1^2.$$

Finally, let's calculate the result of the following operation:

$$\frac{1+2i}{3-4i}+\frac{2-i}{5i}$$

In this case, we first need to calculate each term separately and then add the results.

$$\begin{aligned}\frac{1+2i}{3-4i}+\frac{2-i}{5i} &= \frac{1+2i}{3-4i}\cdot\frac{3+4i}{3+4i}+\frac{2-i}{5i}\cdot\frac{-5i}{-5i} = \frac{(1+2i)(3+4i)}{(3-4i)(3+4i)}+\frac{(2-i)(-5i)}{-25i^2}\\ &= \frac{(1+2i)(3+4i)}{3^2+4^2}+\frac{(2-i)(-5i)}{25} = \frac{3+4i+6i+8i^2}{25}+\frac{-10i+5i^2}{25}\\ &= \frac{3+10i-8}{25}+\frac{-10i-5}{25} = \frac{-5+10i}{25}+\frac{-10i-5}{25}\\ &= \frac{-1+2i}{5}+\frac{-2i-1}{5} = \frac{-1+2i-2i-1}{5} = \frac{-2}{5} = -\frac{2}{5}.\end{aligned}$$

Problem 2.17. Given the complex number $-1-\sqrt{3}i$, calculate its distance from the origin and the angle it forms with the positive x-axis. What do you observe?

Solution to the problem 2.17. *First, consider the number $z=-1-\sqrt{3}i$. To calculate its distance from the origin and the angle it forms with the positive x-axis, we start with a diagram. The number $z=-1-\sqrt{3}i$ is associated with the point $(-1,-\sqrt{3})$, as seen in Figure 2.33; thus, it is located in the third quadrant.*

We calculate the distance to the origin using the Pythagorean theorem. Note that the point $(-1,-\sqrt{3})$ forms a triangle with legs of lengths 1 and $\sqrt{3}$, and we want to find the length of the hypotenuse. Thus:

$$r=\sqrt{1^2+(\sqrt{3})^2}=\sqrt{1+3}=\sqrt{4}=2.$$

To find the angle it forms with the x-axis, we first calculate the angle we marked in the triangle in Figure 2.33. Because it is a right triangle, we have:

$$\sin(\theta)=\frac{1}{2}\Longrightarrow\theta=\frac{\pi}{6}.$$

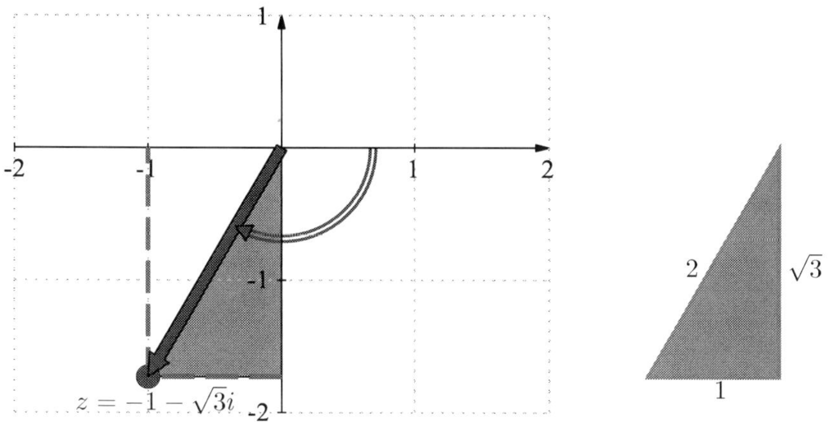

FIGURE 2.33
Representation of the number $-1 - \sqrt{3}i$ in the complex plane (left) and the triangle formed by the components of the number (right) (Problem 2.0).

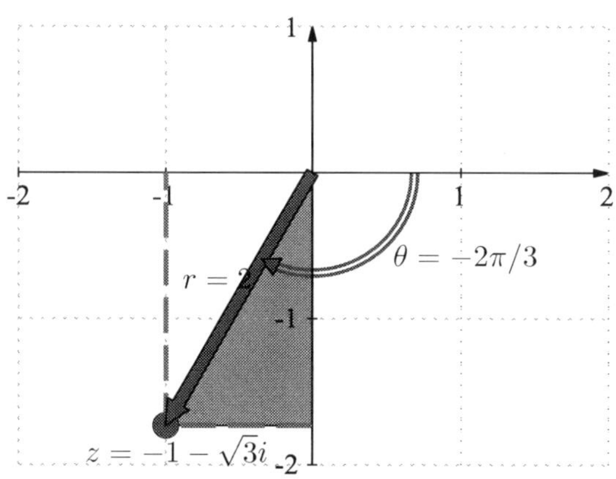

FIGURE 2.34
Representation of the number $-1 - \sqrt{3}i$, its modulus, and its angle in the complex plane (Problem 2.0).

The angle we seek is the marked angle plus $90° = \pi/2$ *and, since it goes clockwise, we must change the sign. Thus, the angle we are looking for is:*

$$-\left(\frac{\pi}{6} + \frac{\pi}{2}\right) = -\frac{\pi + 3\pi}{6} = -\frac{4\pi}{6} = -\frac{2\pi}{3}.$$

Finally, we make another diagram of the results we obtained (see Figure 2.34).

Problem 2.18. Express the following complex numbers in polar coordinates: $3 + 2i$, $-2 - 5i$, $-4 + 2i$, $4 - 2i$.

Solution to the problem 2.18. *First, consider the number $z = 3 + 2i$. When converting a number from rectangular form to polar form, it is important to make a diagram to avoid mistakes. Therefore, the first step is to plot $z = 3 + 2i$ in the complex plane. This number is associated with the point $(3, 2)$, as shown in Figure 2.35; thus, it is located in the first quadrant.*

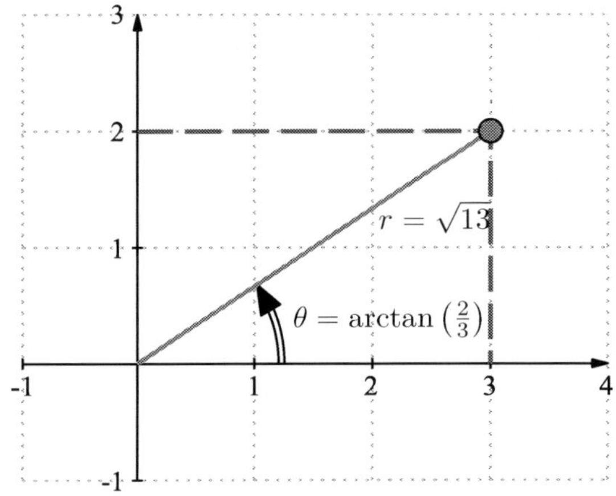

FIGURE 2.35
Representation of the number $z = 3 + 2i$, its modulus, and its angle in the complex plane (Problem 2.0).

From Figure 2.35, we can see that the argument (or angle) of this complex number must be between $0°$ and $90° = \pi/2$ rad.

Now we calculate the modulus and the argument. The modulus (the distance from the point $(3, 2)$ to the origin) is calculated as:

$$r = \sqrt{3^2 + 2^2} = \sqrt{9 + 4} = \sqrt{13}.$$

To calculate the argument θ, note that from Figure 2.35, we have:

$$\tan(\theta) = \frac{\sin(\theta)}{\cos(\theta)} = \frac{2/r}{3/r} = \frac{2}{3} = \frac{Im(z)}{Re(z)}.$$

Thus, the angle θ is obtained as:

$$\theta = \arctan\left(\frac{2}{3}\right).$$

Calculating this value with a calculator (in radians) gives:

$$\arctan\left(\frac{2}{3}\right) = 0.588 \ rad,$$

which is the value we are looking for, as:

$$0 \leq 0.588 \leq \frac{\pi}{2} = 1.570796327.$$

Therefore:

$$z = 3 + 2i = \sqrt{13}_{0.588 \ rad} = \sqrt{13}_{0.588 \ rad \cdot \frac{180°}{\pi \ rad}} = \sqrt{13}_{33.690°}.$$

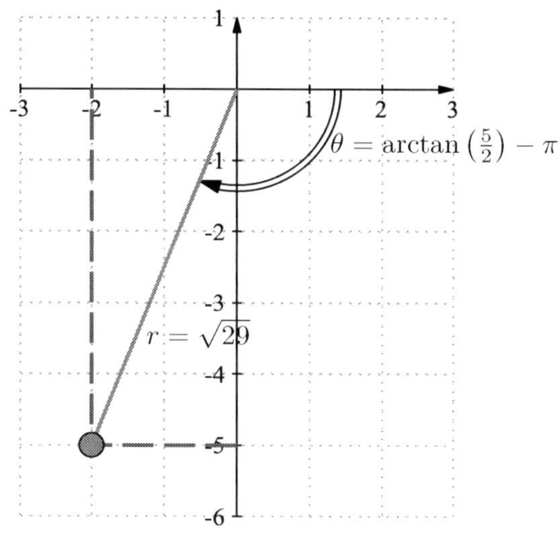

FIGURE 2.36
Representation of the number $z = -2 - 5i$ in the complex plane (Problem 2.0).

For the remaining points, we proceed in the same way. First, we make a diagram to determine which quadrant we are in, then we calculate the modulus and the argument.

Consider $z = -2 - 5i$. *This complex number is associated with the point* $(-2, -5)$, *as shown in Figure 2.36, so it is located in the third quadrant.*

From Figure 2.36, we can see that the argument (or angle) of this complex number must be between $180° = \pi$ *rad and* $270° = 3\pi/2$ *rad. Note that we can also say that the angle is between* $-90° = -\pi/2$ *rad and* $-180° = -\pi$ *rad.*

The modulus is:

$$r = \sqrt{(-2)^2 + (-5)^2} = \sqrt{4 + 25} = \sqrt{29}.$$

To calculate the argument θ, *note that from Figure 2.36, we have:*

$$\tan(\theta) = \frac{\sin(\theta)}{\cos(\theta)} = \frac{-5/r}{-2/r} = \frac{5}{2}.$$

Thus, the angle θ *is obtained as:*

$$\theta = \arctan\left(\frac{5}{2}\right).$$

Calculating this value with a calculator (in radians) gives:

$$\arctan\left(\frac{5}{2}\right) = 1.190 \text{ rad},$$

which is not the value we are looking for, as:

$$1.190 \leq \pi = 3.141592654.$$

See Figure 2.37, where we have represented the angle we want to calculate and the angle given by the calculator.

The angle obtained is the equivalent of the first quadrant, as shown in Figure 2.37. To recover the angle, we want from the angle given by the calculator, we must add or subtract

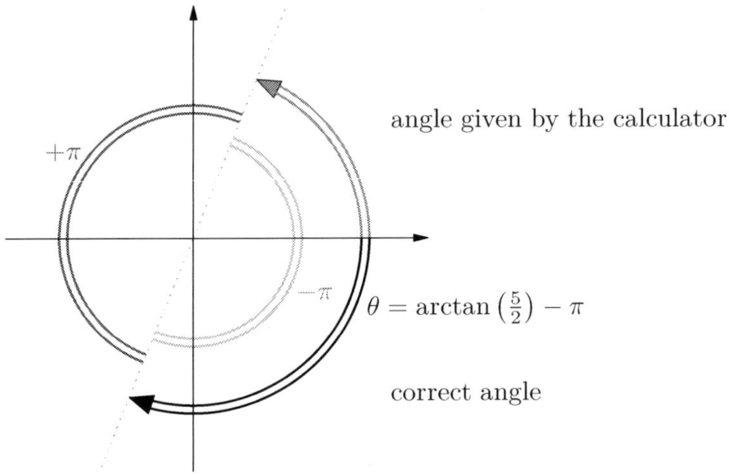

FIGURE 2.37
Correct angle of the number $z = -2 - 5i$ and angle calculated with the calculator (Problem 2.0).

π. Although the result will be different, the value obtained will be the same angle. In this case, we decided to subtract π:

$$\theta = 1.190 - \pi = -1.951,$$

and the complex number is:

$$z = -2 - 5i = \sqrt{19}_{-1.951\ rad} = \sqrt{19}_{-1.951\ rad \cdot \frac{180^\circ}{\pi\ rad}} = \sqrt{19}_{-111.801^\circ}.$$

Now consider $z = -4 + 2i$. This complex number is associated with the point $(-4, 2)$, as shown in Figure 2.38, so it is located in the second quadrant.

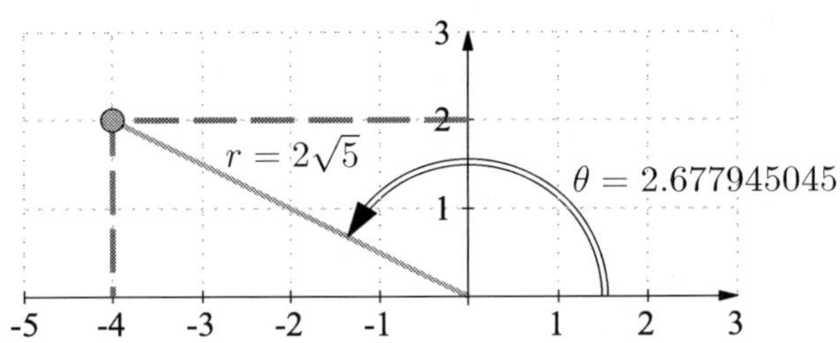

FIGURE 2.38
Representation of the number $z = -4 + 2i$ in the complex plane, its modulus, and its angle (Problem 2.0).

From Figure 2.38, we can see that the argument (or angle) of this complex number must be between $90^\circ = \pi/2$ rad and $180^\circ = \pi$ rad.

The modulus is:

$$r = \sqrt{(-4)^2 + 2^2} = \sqrt{16 + 4} = \sqrt{20} = \sqrt{4}\sqrt{5} = 2\sqrt{5}.$$

To calculate the argument θ, note that from Figure 2.38, we have:

$$\tan(\theta) = \frac{\sin(\theta)}{\cos(\theta)} = \frac{2/r}{-4/r} = \frac{2}{-4} = -\frac{1}{2}.$$

Thus, the angle θ *is obtained as:*

$$\theta = \arctan\left(-\frac{1}{2}\right).$$

Calculating this value with a calculator (in radians) gives:

$$\arctan\left(-\frac{1}{2}\right) = -0.464\ rad,$$

which is not the value we are looking for, as:

$$-0.464 < 0.$$

See Figure 2.39, where we have represented the angle we want to calculate and the angle given by the calculator.

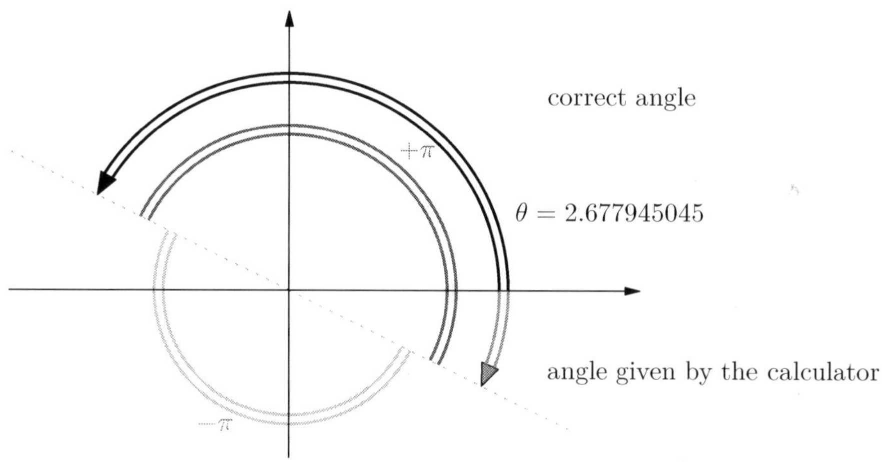

FIGURE 2.39
Correct angle of the number $z = -4 + 2i$ and angle calculated with the calculator (Problem 2.0).

The angle obtained is the equivalent of the fourth quadrant, as shown in Figure 2.39. To recover the correct angle, from the angle given by the calculator, we must add or subtract π. *In this case, since the calculator gives us a negative value, we add* π:

$$\theta = -0.464 + \pi = 2.678,$$

and the complex number is:

$$z = -4 + 2i = 2\sqrt{5}_{2.678\ rad} = 2\sqrt{5}_{2.678\ rad \cdot \frac{180^\circ}{\pi\ rad}} = 2\sqrt{5}_{153.435^\circ}.$$

Finally, consider the number $z = 4 - 2i$. *This complex number is associated with the point* $(4, -2)$, *as shown in Figure 2.40; thus, it is located in the fourth quadrant.*

From Figure 2.40, we can see that the argument (or angle) of this complex number must be between $270^\circ = 3\pi/2$ *rad and* $360^\circ = 2\pi$ *rad, or between* $0^\circ = 0$ *rad and* $-90^\circ = -\pi/2$ *rad.*

The modulus is:

$$r = \sqrt{4^2 + (-2)^2} = \sqrt{16 + 4} = \sqrt{20} = \sqrt{4}\sqrt{5} = 2\sqrt{5}.$$

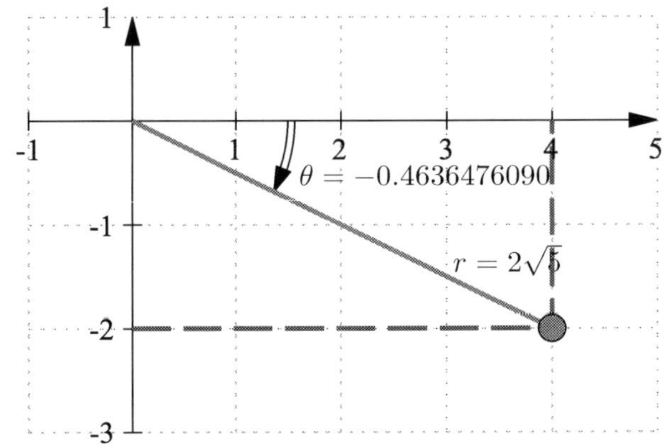

FIGURE 2.40
Representation of the number $(4, -2)$ in the complex plane (Problem 2.0).

To calculate the argument θ, note that from Figure 2.40, we have:

$$\tan\theta = \frac{\sin\theta}{\cos\theta} = \frac{-2/r}{4/r} = \frac{-2}{4} = -\frac{1}{2}.$$

Thus, the angle θ is obtained as:

$$\theta = \arctan\left(-\frac{1}{2}\right).$$

Calculating this value with a calculator (in radians) gives:

$$\arctan\left(-\frac{1}{2}\right) = -0.464\ rad,$$

which is the angle we are looking for, as:

$$-\frac{\pi}{2} = -1.571 < -0.464 < 0.$$

Therefore:

$$z = 4 - 2i = 2\sqrt{5}\angle -0.464\ rad = 2\sqrt{5}\angle\left(-0.464\ rad \cdot \frac{180^\circ}{\pi\ rad}\right) = 2\sqrt{5}\angle -26.565^\circ.$$

Problem 2.19. Express the following complex numbers in polar coordinates:

$$-i, \quad -1-i, \quad 1-\sqrt{3}i, \quad -\sqrt{3}+i, \quad \sqrt{3}+i.$$

Solution to the problem 2.19. *As in Problem 2.18, remember that when converting a number from rectangular form to polar form, it is important to make a diagram to avoid mistakes. Additionally, in simple cases, we can directly give the result.*

First, consider the number $z = -i$, which is associated with the point $(0, -1)$, as shown in Figure 2.41.

In this case, note that the distance to the origin is 1 and that the angle is $-90^\circ = -\pi/2$ rad. Therefore, in polar form:

$$z = -i = 1_{-\pi/2}.$$

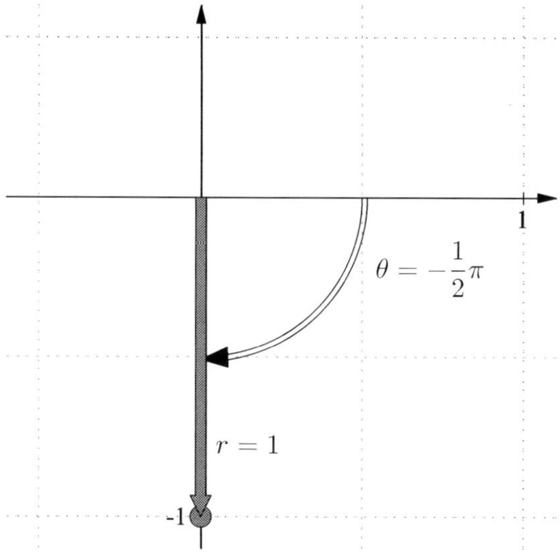

FIGURE 2.41
Representation of the number $z = -i$ in the complex plane (Problem 2.0).

Now consider the number $z = -1 - i$, *which is associated with the point* $(-1, -1)$, *as shown in Figure 2.42.*

In this case, the distance to the origin is $\sqrt{2}$ *because it is the diagonal of a square with side* 1. *Moreover, the angle is* $(-90 - 45)^\circ = -135^\circ = -3\pi/4$ *rad, as can be easily verified. Therefore, in polar form:*

$$z = -1 - i = \sqrt{2}_{-3\pi/4}.$$

Now consider the number $z = 1 - \sqrt{3}i$, *which is associated with the point* $(1, -\sqrt{3})$. *As shown in Figure 2.43, it is located in the fourth quadrant.*

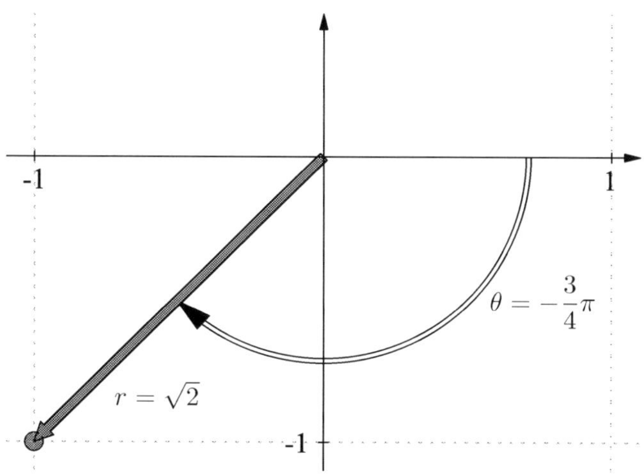

FIGURE 2.42
Representation of the number $z = -1 - i$ in the complex plane (Problem 2.0).

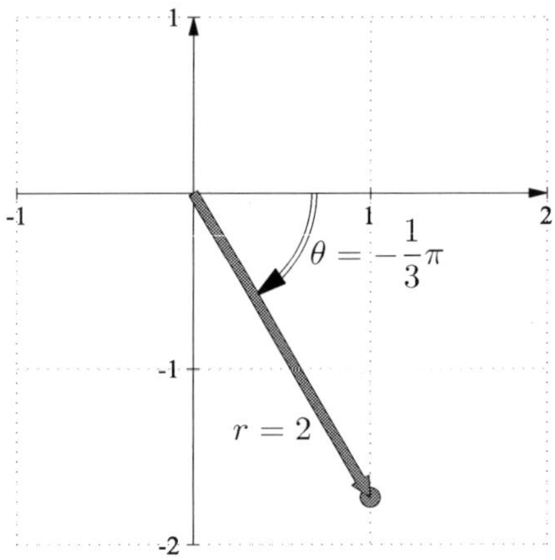

FIGURE 2.43
Representation of the number $z = 1 - \sqrt{3}i$ in the complex plane (Problem 2.0).

From Figure 2.43, we can see that the argument (or angle) of this complex number must be between $270° = 3\pi/2$ *rad and* $360° = 2\pi$ *rad, or between* $0° = 0$ *rad and* $-90° = -\pi/2$ *rad.*

The modulus is:

$$r = \sqrt{1^2 + (-\sqrt{3})^2} = \sqrt{1+3} = \sqrt{4} = 2.$$

To calculate the argument θ*, note that from Figure 2.43, we have:*

$$\tan(\theta) = \frac{\sin(\theta)}{\cos(\theta)} = \frac{-\sqrt{3}/r}{1/r} = -\sqrt{3}.$$

Thus, the angle θ *is obtained as:*

$$\theta = \arctan\left(-\sqrt{3}\right).$$

Calculating this value with a calculator (in radians) gives:

$$\arctan\left(-\frac{1}{2}\right) = -1.047 \; rad = -\frac{\pi}{3} \; rad.$$

This angle is what we are looking for, as:

$$-\frac{\pi}{2} < -\frac{\pi}{3} < 0.$$

Therefore:

$$z = 1 - \sqrt{3}i = 2_{-\pi/3 \; rad} = 2_{-\pi/3 \; rad \cdot \frac{180°}{\pi \; rad}} = 2_{-60°}.$$

To finish the exercise, we need to convert the numbers $z_2 = -\sqrt{3} + i$ *and* $z_3 = \sqrt{3} + i$ *to polar form. We will do this easily using the information we just calculated. Let* $z_1 = 1 - \sqrt{3}i$*. We just saw that:*

$$z_1 = 1 - \sqrt{3}i = 2_{-\pi/3 \; rad} = 2_{-60°}.$$

Now, we draw the points z_1*,* z_2*, and* z_3 *in the complex plane.*

We can see that the three points lie on the same circle. Thus, the three points have the same modulus, which is equal to 2.

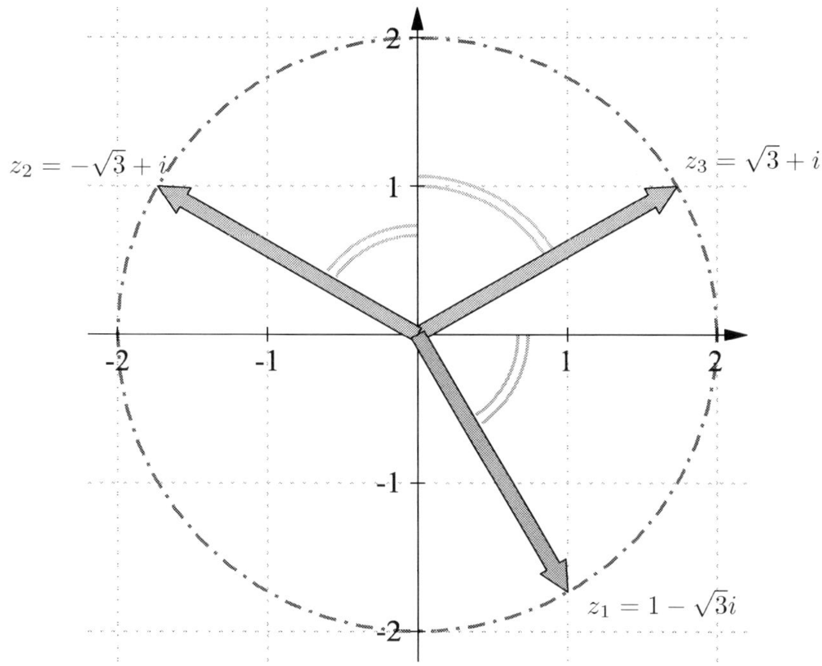

FIGURE 2.44
Representation of the numbers $z_1 = 1 - \sqrt{3}i$, $z_2 = -\sqrt{3} + i$, and $z_3 = \sqrt{3} + i$ in the complex plane (Problem 2.0).

Now, we just need to calculate the angles. In Figure 2.44, we have marked the three angles, which are equal (with an opening of $60°$*). Therefore, the angle associated with* z_2 *is* $90° + 60° = 150°$*. Similarly, the angle associated with* z_3 *is* $90° - 60° = 30°$*.*

Thus:

$$z_1 = 1 - \sqrt{3}i = 2_{-60°} = 2_{-\pi/3\ rad}$$
$$z_2 = -\sqrt{3} + i = 2_{150°} = 2_{-\pi/3\ rad}$$
$$z_3 = \sqrt{3} + i = 2_{30°} = 2_{5\pi/6\ rad}.$$

A good exercise to practice is to redo this exercise without the visual aid and check that you get the same result.

Problem 2.20. Express the complex numbers $-2 - 5i$ and $-2 + 5i$ in polar form.

Solution to the problem 2.20. *Consider the complex numbers* $z_1 = -2 - 5i$ *and* $z_2 = -2 + 5i$*. Note that* z_2 *is the conjugate of* z_1*, meaning:*

$$z_2 = \overline{z_1}.$$

With this information, we already know that the modulus of both numbers will be the same, which is:

$$r = \sqrt{2^2 + 5^2} = \sqrt{4 + 25} = \sqrt{29}.$$

To correctly calculate the angles, we start with a diagram in the complex plane.

As seen in Figure 2.45, the angles of z_1 *and* z_2 *are equal but with opposite signs. Therefore, if we denote the angle of* z_1 *as* θ_1 *and the angle of* z_2 *as* θ_2*, we have:*

$$\theta_2 = -\theta_1.$$

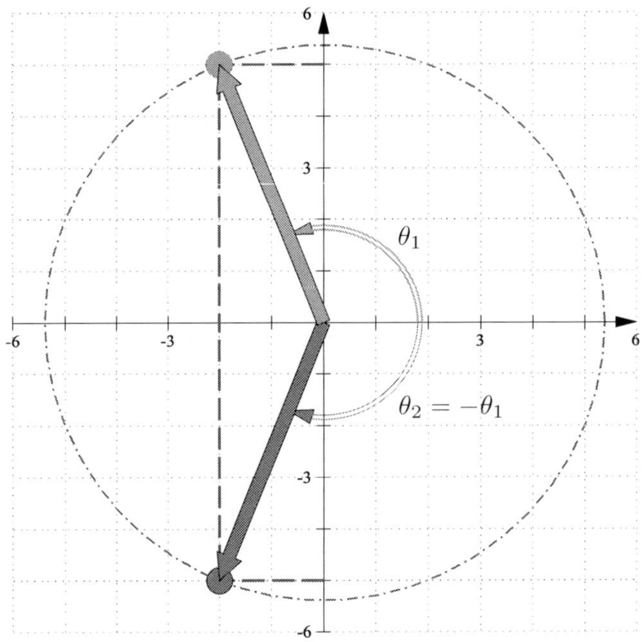

FIGURE 2.45
Representation of the numbers $z_1 = -2-5i$ and $z_2 = -2+5i$ in the complex plane (Problem 2.0).

Note that this is equivalent to saying:

$$\theta_2 = 2\pi - \theta_1,$$

since we can always add (or subtract) 2π *to angles without changing the associated complex number.*

Therefore, let's calculate θ_1*. Note that since* z_1 *is in the third quadrant, its argument (or angle) must be between* $180° = \pi$ *rad and* $270° = 3\pi/2$ *rad, or between* $-90° = -\pi/2$ *rad and* $-180° = -\pi$ *rad. From Figure 2.45, we see that:*

$$\tan(\theta_1) = \frac{\sin(\theta_1)}{\cos(\theta_1)} = \frac{-5/r}{-2/r} = \frac{5}{2}.$$

Therefore, the angle θ_1 *is obtained as:*

$$\theta_1 = \arctan\left(\frac{5}{2}\right).$$

Calculating this value with a calculator (in radians) gives:

$$\arctan\left(\frac{5}{2}\right) = 1.190 \; rad = 1.190 \; rad \cdot \frac{180°}{\pi \; rad} = 68.199°.$$

Therefore, this is not the angle we are looking for, as we must subtract (or add) π*, meaning:*

$$\theta_1 = (1.190 - \pi) \; rad = -1.951 \; rad = -111.801°.$$

Finally:

$$\theta_2 = -\theta_1 = 1.951 \; rad = 111.801°.$$

Therefore:

$$z_1 = -2 - 5i = \sqrt{29}_{-1.951\ rad}$$
$$z_2 = -2 + 5i = \sqrt{29}_{1.951\ rad}.$$

Problem 2.21. Write the following complex numbers:

(a) In polar form: $1 - i$, $i - \sqrt{3}$, -2.45, $-\frac{i}{3}$, $\frac{1}{2} + \frac{\sqrt{3}}{2}i$.

(b) In rectangular form: 3_π, $8e^{\frac{\pi}{3}i}$, $\sqrt{2}e^{\frac{\pi}{4}i}$, $5e^{\frac{3\pi}{5}i}$.

Solution to the problem 2.21. *(a) To express a complex number in polar form, we need to find its polar coordinates, which are the modulus ($r = |z|$) and the argument ($\theta = \arg z$).*

- $\boxed{z = 1 - i}$
Before starting the problem, it is very important to plot the complex number in the complex plane. In this case, as seen in Figure 2.46, $|z| = \sqrt{2}$ and $\theta = -\frac{\pi}{4}$. However, let's calculate them using the formulas:

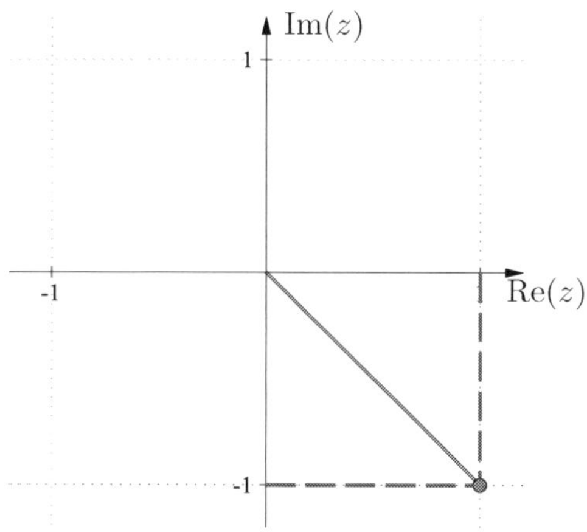

FIGURE 2.46
Representation of the number $1 - i$ in the complex plane (Problem 2.0).

Given a complex number in rectangular form ($z = a+bi$), its modulus is defined as $|z| = \sqrt{a^2 + b^2}$. Thus, the modulus of z in this case is:

$$r = |z| = \sqrt{1^2 + (-1)^2} = \sqrt{2}.$$

The argument θ can be found using the expressions $a = |z| \cdot \cos\theta$ and $b = |z| \cdot \sin\theta$. We find and express it in radians:

$$1 = r\cos\theta \Longrightarrow \cos\theta = \frac{1}{r} = \frac{1}{\sqrt{2}} \Longrightarrow \theta = \begin{cases} 45^\circ \\ or \\ -45^\circ \end{cases}$$

$$-1 = r\sin\theta \Longrightarrow \sin\theta = \frac{-1}{r} = -\frac{1}{\sqrt{2}} \Longrightarrow \theta = \begin{cases} -45^\circ \\ or \\ -135^\circ \end{cases}$$

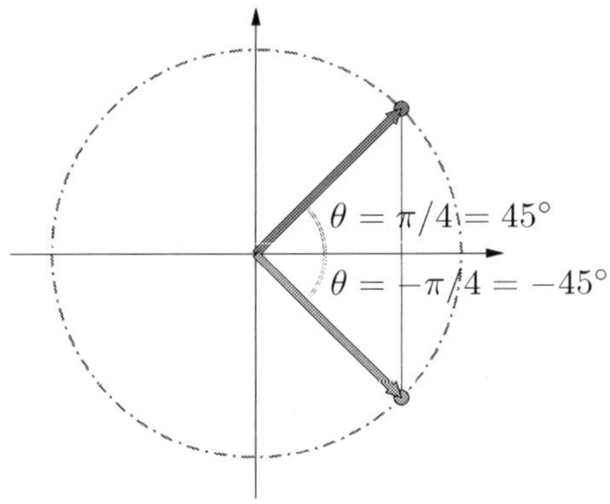

FIGURE 2.47
Angles satisfying $\cos\theta = \frac{1}{r} = \frac{1}{\sqrt{2}}$ (Problem 2.0).

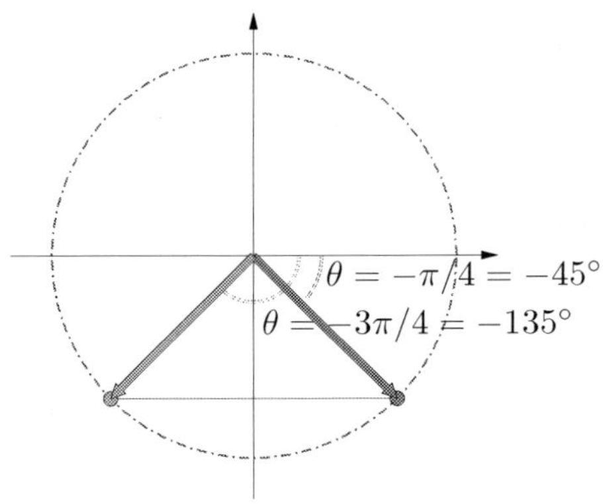

FIGURE 2.48
Angles satisfying $\sin\theta = -\frac{1}{\sqrt{2}}$ (Problem 2.0).

Therefore, $\theta = -45°$, which in radians is:

$$-45° \cdot \frac{\pi\, rad}{180°} = -\frac{\pi}{4} rad.$$

Note that with the diagram, once we see that $\theta = 45°$ or $\theta = -45°$, we could directly say that $\theta = -45°$ without needing to solve the sine equation.

The polar form of a complex number is r_θ. Therefore, $z = 1 - i$ in polar form is:

$$z = \sqrt{2}_{-\frac{\pi}{4}}.$$

Verification:

$$z = \sqrt{2}_{-\frac{\pi}{4}} = \sqrt{2}\left(\cos\left(-\frac{\pi}{4}\right) + i\sin\left(-\frac{\pi}{4}\right)\right) = 1 - 1i = 1 - i.$$

- $\boxed{z = i - \sqrt{3}}$

Let's represent this number in the complex plane, as shown in Figure 2.49.

$$z = i - \sqrt{3} = 0 - \tfrac{1}{3}i$$
$$r = |z| = \sqrt{0^2 + \left(-\tfrac{1}{3}\right)^2} = 2$$

Now, let's find the argument:

$$1 = r\sin\theta \Longrightarrow \theta = \begin{cases} 60^\circ \\ or \\ 150^\circ \end{cases}$$

Since we know z is in the second quadrant, $\theta = 150^\circ$, which in radians is:

$$150^\circ \cdot \frac{\pi\,rad}{180^\circ} = \frac{5\pi}{6}\,rad.$$

Therefore, $z = i - \sqrt{3}$ in polar form is:

$$z = 2_{\frac{5\pi}{6}}.$$

Verification:

$$z = 2_{\frac{5\pi}{6}} = 2\left(\cos\frac{5\pi}{6} + i\sin\frac{5\pi}{6}\right) = -1.732 + 1i = -\sqrt{3} + i.$$

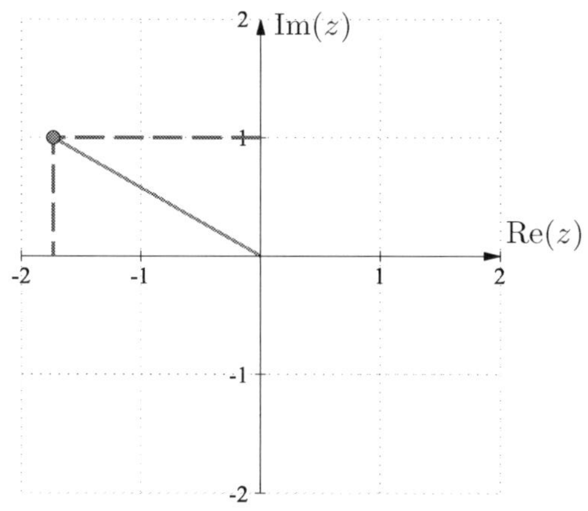

FIGURE 2.49
Representation of the number $z = i - \sqrt{3}$ in the complex plane (Problem 2.0).

We proceed similarly for the remaining cases:

- $\boxed{z = -2.45}$

Let's represent this number in the complex plane, as shown in Figure 2.50.

$$\begin{aligned} z &= -2.45 = -2.45 + 0i \\ r &= |z| = \sqrt{(-2.45)^2 + 0^2} \\ &= 2.45 = \frac{245}{100} = \frac{49}{20} \\ -2.45 &= 2.45\cos\theta \Longrightarrow \theta = \pi\,rad \end{aligned}$$

Therefore:

$$z = \left(\frac{49}{20}\right)_\pi = 2.45_\pi.$$

Verification:

$$z = 2.45_\pi = 2.45\,(\cos\pi + i\sin\pi) = 2.45(-1 + 0i) = -2.45.$$

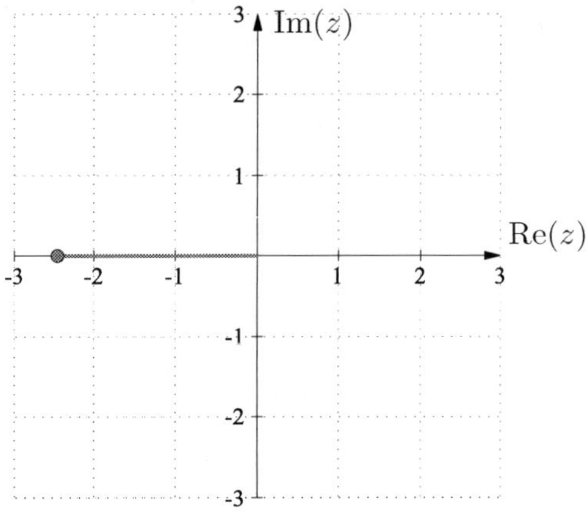

FIGURE 2.50
Representation of the number $z = -2.45$ in the complex plane (Problem 2.0).

- $\boxed{z = -\dfrac{i}{3}}$

  Let's represent this number in the complex plane, as shown in Figure 2.51.

$$z = -\tfrac{i}{3} = 0 - \tfrac{1}{3}i$$
$$r = |z| = \sqrt{0^2 + \left(-\tfrac{1}{3}\right)^2} = \tfrac{1}{3}$$
$$-\tfrac{1}{3} = \tfrac{1}{3}\sin\theta \Longrightarrow \theta = -\tfrac{\pi}{2}\,rad$$

Therefore:

$$z = \left(\frac{1}{3}\right)_{-\frac{\pi}{2}}$$

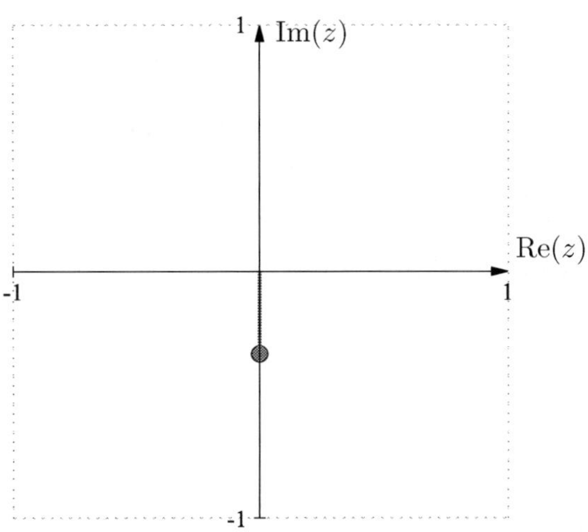

FIGURE 2.51
Representation of the number $z = -\frac{i}{3}$ in the complex plane (Problem 2.0).

Verification:

$$z = \left(\frac{1}{3}\right)_{-\frac{\pi}{2}} = \frac{1}{3}\left(\cos\left(-\frac{\pi}{2}\right) + i\sin\left(-\frac{\pi}{2}\right)\right) = \frac{1}{3}(0 - i) = -\frac{i}{3}.$$

- $\boxed{z = \frac{1}{2} + \frac{\sqrt{3}}{2}i}$

Let's represent this number in the complex plane, as shown in Figure 2.52.

$$z = \tfrac{1}{2} + \tfrac{\sqrt{3}}{2}i$$
$$r = |z| = \sqrt{\left(\tfrac{1}{2}\right)^2 + \left(\tfrac{\sqrt{3}}{2}\right)^2} = 1$$
$$\tfrac{1}{2} = 1\cos\theta \Longrightarrow \theta = \tfrac{\pi}{3}\,rad$$

Therefore:

$$z = 1_{\frac{\pi}{3}}$$

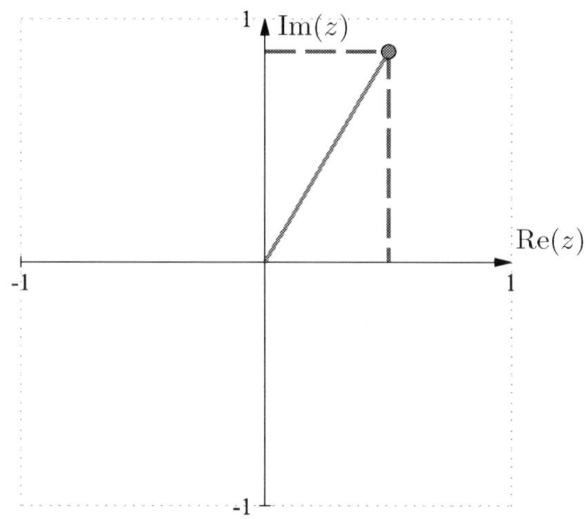

FIGURE 2.52
Representation of the number $z = \frac{1}{2} + \frac{\sqrt{3}}{2}i$ in the complex plane (Problem 2.0).

Verification:

$$z = 1_{\frac{\pi}{3}} = 1\left(\cos\frac{\pi}{3} + i\sin\frac{\pi}{3}\right) = \frac{1}{2} + \frac{\sqrt{3}}{2}i.$$

(b) *To express a complex number in rectangular form, we need to find its real and imaginary parts, knowing that $z = r(\cos\theta + i \cdot \sin\theta)$.*

- $\boxed{z = 3_\pi} \quad \Longrightarrow \quad r = 3, \quad \theta = \pi$

Therefore:

$$z = 3(\cos\pi + i \cdot \sin\pi) = -3 + 0i = -3.$$

- $\boxed{z = 8e^{\frac{\pi}{3}i}} \quad \Longrightarrow \quad r = 8, \quad \theta = \frac{\pi}{3}$

Therefore:

$$z = 8\left(\cos\frac{\pi}{3} + i \cdot \sin\frac{\pi}{3}\right) = 4 + 8\frac{\sqrt{3}}{2}i = 4 + 4\sqrt{3}i.$$

- $\boxed{z = \sqrt{2}e^{\frac{\pi}{4}i}} \implies r = \sqrt{2}, \quad \theta = \frac{\pi}{4}$

  Therefore:

$$z = \sqrt{2}\left(\cos\frac{\pi}{4} + i \cdot \sin\frac{\pi}{4}\right) = \sqrt{2}\left(\frac{\sqrt{2}}{2} + \frac{\sqrt{2}}{2}i\right) = 1 + 1i.$$

- $\boxed{z = 5e^{\frac{3\pi}{5}i}} \implies r = 5, \quad \theta = \frac{3\pi}{5}$

  Therefore:

$$z = 5\left(\cos\frac{3\pi}{5} + i \cdot \sin\frac{3\pi}{5}\right) = -5\cos\left(\frac{3}{5}\pi\right) + 5\sin\left(\frac{3}{5}\pi\right)i$$
$$= -1.5451 + 4.7553i$$

Problem 2.22. Given the complex number $2_{\pi/3}$, calculate a and b from Figure 2.53.

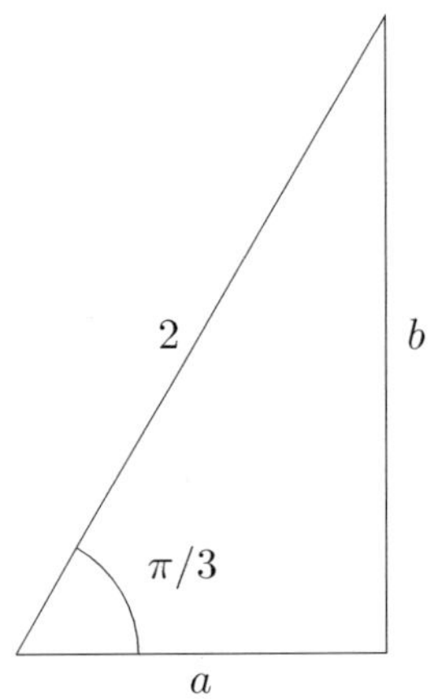

FIGURE 2.53
Right triangle with the real and imaginary components of the number $2_{\pi/3}$ (Problem 2.22).

Solution to the problem 2.22. *To calculate a and b, we need to use the trigonometric relationships of sine and cosine. Considering that we have a right triangle, we know that:*

$$\cos\left(\frac{\pi}{3}\right) = \frac{a}{2} \implies a = 2\cos\left(\frac{\pi}{3}\right) = 2 \cdot \frac{1}{2} = 1.$$

Similarly:

$$\sin\left(\frac{\pi}{3}\right) = \frac{b}{2} \implies b = 2\sin\left(\frac{\pi}{3}\right) = 2 \cdot \frac{\sqrt{3}}{2} = \sqrt{3}.$$

Let's now verify the Pythagorean theorem:

$$2^2 = a^2 + b^2 = 1^2 + (\sqrt{3})^2 = 1 + 3 = 4.$$

Problem 2.23. Express the following complex numbers in binomial form:

$$1_0, \quad 2_\pi, \quad 1_{\pi/3}, \quad 4_{-\pi/6}, \quad 2_{3\pi/4}.$$

Solution to the problem 2.23. *To convert numbers from polar form to binomial form, drawing is not strictly necessary, although it is always recommended for checking results. To perform this exercise, we must consider the following formula that transforms a complex number from binomial form to polar form:*

$$r_\theta = r\cos(\theta) + r\sin(\theta)i.$$

Accordingly, we have:

$$\begin{aligned}
z_1 &= 1_0 = 1\cos(0) + 1\sin(0)i = 1 + 0i = 1, \\
z_2 &= 2_\pi = 2\cos(\pi) + 2\sin(\pi)i = 2\cdot(-1) + 0i = -2, \\
z_3 &= 1_{\pi/3} = 1\cos(\pi/3) + 1\sin(\pi/3)i = 1\cdot\frac{1}{2} + 1\cdot\frac{\sqrt{3}}{2}i = \frac{1}{2} + \frac{\sqrt{3}}{2}i, \\
z_4 &= 4_{-\pi/6} = 4\cos(-\pi/6) + 4\sin(-\pi/6)i = 4\cdot\frac{\sqrt{3}}{2} + 4\cdot\frac{-1}{2}i = 2\sqrt{3} - 2i, \\
z_5 &= 2_{3\pi/4} = 2\cos(3\pi/4) + 2\sin(3\pi/4)i = 2\cdot\frac{-\sqrt{2}}{2} + 2\cdot\frac{\sqrt{2}}{2}i = -\sqrt{2} + \sqrt{2}i.
\end{aligned}$$

To verify the results, we have plotted the numbers in the complex plane (shown in Figure 2.54).

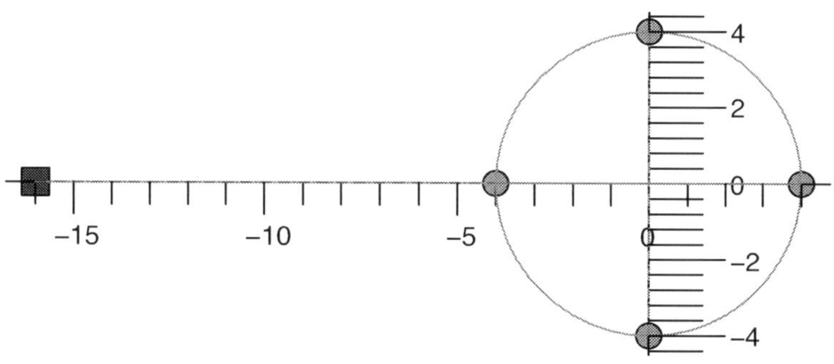

FIGURE 2.54
Representation of the numbers 1_0, 2_π, $1_{\pi/3}$, $4_{-\pi/6}$, and $2_{3\pi/4}$ in the complex plane (Problem 2.23).

Problem 2.24. Express the result of the division $6_{120^\circ} : 3_{30^\circ}$ in binomial form.

Solution to the problem 2.24. *To divide two complex numbers in polar form, divide the magnitudes and subtract the arguments:*

$$z = 6_{120^\circ} : 3_{30^\circ} = \frac{6}{3}_{120^\circ - 30^\circ} = 2_{90^\circ}.$$

Convert the result to binomial form:

$$z = 2_{90^\circ} = 2\cos 90 + 2\sin 90i = 2i.$$

Problem 2.25. Given $z_1 = 3_{\pi/6}$ and $z_2 = 2_{\pi/4}$, calculate: $z_1 + z_2, z_1 - z_2, z_1 \cdot z_2, z_1/z_2$.

Solution to the problem 2.25. *Remember that the sum and subtraction of two complex numbers must always be done in binomial form. However, the product and division of complex numbers can be done in binomial, polar, or exponential form.*

Let's start by calculating the sum and difference of z_1 and z_2. We need to express the numbers in binomial form, as it is essential for adding them:

$$z_1 = 3_{\pi/6} = 3\cos(\pi/6) + 3\sin(\pi/6)i = \frac{3\sqrt{3}}{2} + 3\frac{1}{2}i = \frac{3\sqrt{3}}{2} + \frac{3}{2}i,$$

$$z_2 = 2_{\pi/4} = 2\cos(\pi/4) + 2\sin(\pi/4)i = \frac{2\sqrt{2}}{2} + \frac{2\sqrt{2}}{2}i = \sqrt{2} + \sqrt{2}i.$$

Therefore:

$$z_1 + z_2 = \frac{3\sqrt{3}}{2} + \frac{3}{2}i + \sqrt{2} + \sqrt{2}i = \frac{3\sqrt{3}}{2} + \sqrt{2} + \left(\frac{3}{2} + \sqrt{2}\right) i,$$

$$z_1 - z_2 = \frac{3\sqrt{3}}{2} + \frac{3}{2}i - \left(\sqrt{2} + \sqrt{2}i\right) = \frac{3\sqrt{3}}{2} - \sqrt{2} + \left(\frac{3}{2} - \sqrt{2}\right) i.$$

The results can be seen in Figure 2.55.

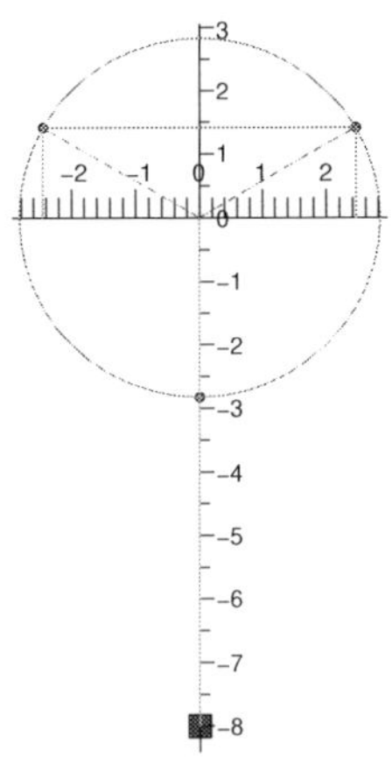

FIGURE 2.55
Representation of the numbers $z_1 = 3_{\pi/6}$, $z_2 = 2_{\pi/4}$, $z_1 + z_2$, $-z_2$, and $z_1 - z_2$ (Problem 2.25).

Now, let's calculate the product and quotient of z_1 and z_2:

$$z_1 \cdot z_2 = 3_{\pi/6} \cdot 2_{\pi/4} = (3 \cdot 2)_{\pi/6+\pi/4} = 6_{5\pi/12},$$

$$\frac{z_1}{z_2} = \frac{3_{\pi/6}}{2_{\pi/4}} = \left(\frac{3}{2}\right)_{\pi/6-\pi/4} = 1.5_{-\pi/12}.$$

Note that to calculate the product of two complex numbers, multiply the magnitudes *and* add the arguments. *Similarly, to calculate the division of two complex numbers,* divide the magnitudes *and* subtract the arguments.

Problem 2.26. Express the following complex numbers in exponential form:

(a) $z = 1 + i\sqrt{3}$

(b) $z = \sqrt{3} + i$

Solution to the problem 2.26. *The exponential form of a complex number is:* $re^{\theta i}$. *Thus, we need to find the magnitude,* r, *and the argument,* θ.

(a) $z = 1 + i\sqrt{3}$

We represent this number in the complex plane, as shown in Figure 2.56.

$$r = |z| = \sqrt{1^2 + (\sqrt{3})^2} = 2,$$
$$\theta = \tan^{-1}\left(\frac{\sqrt{3}}{1}\right) = \frac{\pi}{3}.$$

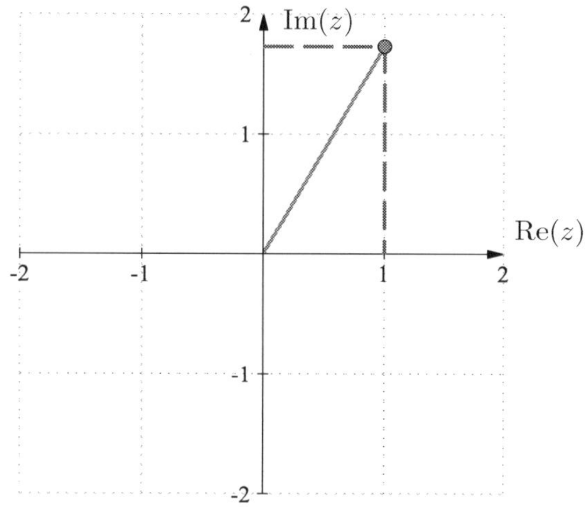

FIGURE 2.56
Representation of the number $z = 1 + i\sqrt{3}$ in the complex plane (Problem 2.26).

Since z *is in the first quadrant, we have:*

$$z = 2e^{\frac{\pi}{3}i}.$$

(b) $z = \sqrt{3} + i$

We represent this number in the complex plane, as shown in Figure 2.57.

$$r = |z| = \sqrt{(\sqrt{3})^2 + 1^2} = 2,$$
$$\theta = \tan^{-1}\left(\frac{1}{\sqrt{3}}\right) = \frac{\pi}{6}.$$

Since z *is in the first quadrant, we have:*

$$z = 2e^{\frac{\pi}{6}i}.$$

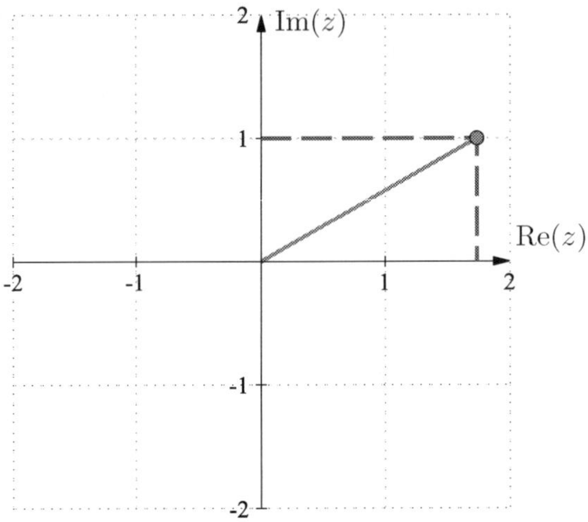

FIGURE 2.57
Representation of the number $z = \sqrt{3} + i$ in the complex plane (Problem 2.26).

Problem 2.27. Verify the equivalences in the following table.

$z = a + bi$	Polar	Exponential
$z_1 = 1 + i$	$z_1 = \sqrt{2}_{\frac{\pi}{4}}$	$z_1 = \sqrt{2}e^{\frac{\pi}{4}i}$
$z_2 = i$	$z_2 = 1_{\frac{\pi}{2}}$	$z_2 = e^{\frac{\pi}{2}i}$
$z_3 = 1 + \sqrt{3}i$	$z_3 = 2_{\frac{\pi}{3}}$	$z_3 = 2e^{\frac{\pi}{3}i}$
$z_4 = 1 - \sqrt{3}i$	$z_4 = 2_{-\frac{\pi}{3}}$	$z_4 = 2e^{-\frac{\pi}{3}i}$

Solution to the problem 2.27. *The correspondence between polar and exponential form is straightforward, since:*

$$r_\theta = re^{\theta i},$$

is simply a matter of definition.

To verify the correspondence between binomial and polar forms, we have two options: either consider the numbers in binomial form and convert them to polar, or the reverse. As it is simpler to go from polar to binomial rather than the other way around, we will proceed in this manner.

$$\begin{aligned}
z_1 &= \sqrt{2}_{\frac{\pi}{4}} = \sqrt{2}\cos(\pi/4) + \sqrt{2}\sin(\pi/4)i = \sqrt{2}\sqrt{2}/2 + \sqrt{2}\sqrt{2}/2i = 1 + 1i, \\
z_2 &= 1_{\frac{\pi}{2}} = 1\cos(\pi/2) + 1\sin(\pi/2)i = 0 + 1i = i, \\
z_3 &= 2_{\frac{\pi}{3}} = 2\cos(\pi/3) + 2\sin(\pi/3)i = 2 \cdot 1/2 + 2 \cdot \sqrt{3}/2i = 1 + \sqrt{3}i, \\
z_4 &= 2_{-\frac{\pi}{3}} = 2\cos(-\pi/3) + 2\sin(-\pi/3)i = 2 \cdot 1/2 - 2 \cdot \sqrt{3}/2i = 1 - \sqrt{3}i.
\end{aligned}$$

The four complex numbers are represented in Figure 2.58:

Problem 2.28. Given $z = \pi i$, calculate e^z and convert this number to polar form. What magnitude and angle result?

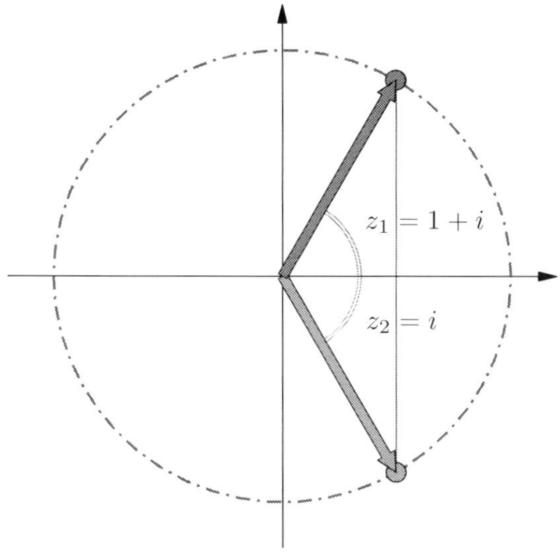

FIGURE 2.58
Representation of the numbers $z_1 = \sqrt{2}_{\frac{\pi}{4}}$, $z_2 = 1_{\frac{\pi}{2}}$, $z_3 = 2_{\frac{\pi}{3}}$, and $z_4 = 2_{-\frac{\pi}{3}}$ in the complex plane (Problem 2.27).

Solution to the problem 2.28. *Using the definition of a pure imaginary number, we have:*

$$e^{\pi i} = \cos(\pi) + \sin(\pi)i = -1 + 0i = -1.$$

Since the result is a negative real number, converting it to polar form is straightforward. The magnitude of $e^{\pi i}$ *is* 1 *and the angle is* π*. Thus:*

$$e^{\pi i} = -1 = 1_{\pi}.$$

Problem 2.29. Given $z = 2 + \pi i$, calculate e^z (considering that $e^{2+\pi i} = e^2 e^{\pi i}$) and convert this number to polar form. What magnitude and angle result?

Solution to the problem 2.29. *Given* $z = 2 + \pi i$*, let's calculate* e^z*:*

$$e^z = e^{2+\pi i} = e^2 e^{\pi i}.$$

Using the result from Problem 2.28, we have:

$$e^z = e^2 e^{\pi i} = -e^2.$$

Therefore, we obtain a negative real number again. Thus:

$$e^z = -e^2 = e^2_{\pi}.$$

The magnitude of e^z *is* e^2 *and its angle is* π *(or* $-\pi$*).*

Problem 2.30. Using the fact that:

$$\tan(\theta) = \frac{\sin(\theta)}{\cos(\theta)},$$

derive the expression of the tangent in terms of the complex exponential.

Solution to the problem 2.30. *We know that we can express the sine and cosine of an angle* θ *in terms of the complex exponential as:*

$$\sin(\theta) = \frac{e^{\theta i} - e^{-\theta i}}{2i} \qquad \textit{and} \qquad \cos(\theta) = \frac{e^{\theta i} + e^{-\theta i}}{2}.$$

Therefore, we have:

$$\begin{aligned}\tan(\theta) &= \frac{\sin(\theta)}{\cos(\theta)} = \frac{e^{\theta i} - e^{-\theta i}}{2i} : \frac{e^{\theta i} + e^{-\theta i}}{2} = \frac{(e^{\theta i} - e^{-\theta i})2}{(e^{\theta i} + e^{-\theta i})2i} = \frac{e^{\theta i} - e^{-\theta i}}{e^{\theta i} + e^{-\theta i}} \cdot \frac{1}{i} \\ &= \frac{e^{\theta i} - e^{-\theta i}}{e^{\theta i} + e^{-\theta i}} \cdot \frac{1}{i} \cdot \frac{i}{i} = \frac{e^{\theta i} - e^{-\theta i}}{e^{\theta i} + e^{-\theta i}} \cdot \frac{i}{i^2} = \frac{e^{\theta i} - e^{-\theta i}}{e^{\theta i} + e^{-\theta i}} \cdot \frac{i}{-1} = -i\frac{e^{\theta i} - e^{-\theta i}}{e^{\theta i} + e^{-\theta i}}.\end{aligned}$$

Problem 2.31. Find an expression for $\sin(3\theta)$ in terms of $\sin(\theta)$ and $\cos(\theta)$. Hint: *Recall the binomial theorem.*

Solution to the problem 2.31. *We need to find an expression for* $\sin(3\theta)$ *in terms of* $\sin(\theta)$ *and* $\cos(\theta)$.

By De Moivre's theorem with $r = 1$ *and* $n = 3$*, we have:*

$$[\cos(\theta) + i\sin(\theta)]^3 = \cos(3\theta) + i\sin(3\theta).$$

Thus, if we denote:

$$z = \cos(3\theta) + i\sin(3\theta),$$

we want to calculate the imaginary part of the complex number z*:*

$$Im(\cos(3\theta) + i\sin(3\theta)) = \sin(3\theta).$$

We know that:

$$z = \cos(3\theta) + i\sin(3\theta) = [\cos(\theta) + i\sin(\theta)]^3.$$

We expand the right side of the previous equation using the binomial theorem:

$$(a+b)^3 = a^3 + 3a^2b + 3ab^2 + b^3,$$

yielding:

$$\begin{aligned} z &= [\cos(\theta) + i\sin(\theta)]^3 \\ &= \cos^3(\theta) + 3\cos^2(\theta)i\sin(\theta) + 3\cos(\theta)(i\sin(\theta))^2 + (i\sin(\theta))^3 \\ &= \cos^3(\theta) + 3i\cos^2(\theta)\sin(\theta) + 3i^2\cos(\theta)\sin^2(\theta) + i^3\sin^3(\theta) \\ &= \cos^3(\theta) + 3i\cos^2(\theta)\sin(\theta) - 3\cos(\theta)\sin^2(\theta) + ii^2\sin^3(\theta) \\ &= \cos^3(\theta) + 3i\cos^2(\theta)\sin(\theta) - 3\cos(\theta)\sin^2(\theta) - i\sin^3(\theta) \\ &= \cos^3(\theta) - 3\cos(\theta)\sin^2(\theta) + \left(3\cos^2(\theta)\sin(\theta) - \sin^3(\theta)\right)i. \end{aligned}$$

Thus, the imaginary part of z *is:*

$$Im(z) = 3\cos^2(\theta)\sin(\theta) - \sin^3(\theta),$$

so:

$$\sin(3\theta) = Im(\cos(3\theta) + i\sin(3\theta)) = 3\cos^2(\theta)\sin(\theta) - \sin^3(\theta).$$

This expression can be simplified further using:

$$\cos^2(\theta) + \sin^2(\theta) = 1.$$

Using this equality, we can derive the following two expressions:

$$\begin{aligned}\sin(3\theta) &= 3\cos^2(\theta)\sin(\theta) - \sin^3(\theta) = 3(1 - \sin^2(\theta))\sin(\theta) - \sin^3(\theta)\\ &= 3\sin(\theta) - 3\sin^3(\theta) - \sin^3(\theta) = 3\sin(\theta) - 4\sin^3(\theta),\end{aligned}$$

and:

$$\begin{aligned}\sin(3\theta) &= 3\cos^2(\theta)\sin(\theta) - \sin^3(\theta) = 3\cos^2(\theta)\sin(\theta) - \sin(\theta)\sin^2(\theta)\\ &= 3\cos^2(\theta)\sin(\theta) - \sin(\theta)(1 - \cos^2(\theta))\\ &= 3\cos^2(\theta)\sin(\theta) - \sin(\theta) + \cos^2(\theta)\sin(\theta) = 4\cos^2(\theta)\sin(\theta) - \sin(\theta).\end{aligned}$$

Problem 2.32. In physics and engineering, a *phasor* is a representation (in vector form) of a sinusoidal function with an amplitude A, phase θ, and frequency ω, which do not vary over time. More specifically, the following identification is made:

$$A\cos(\omega t + \theta) = A_\theta.$$

Thus, A_θ represents a vector in the plane, anchored at the origin of coordinates, with magnitude A and forming an angle of θ radians with the positive x-axis. Often, in addition to the notation A_θ, the equivalent notation $A\angle\theta$ is also used. Note finally that the phasor A_θ is just the complex number A_θ expressed in polar form. Identify the phasors (magnitude and phase) from Figure 2.59.

Solution to the problem 2.32. *In Figure 2.59, we have three vectors in the plane originating at $(0,0)$. Each of these vectors is characterized by:*

1. *Phase (θ): the angle formed by the positive x-axis with the vector;*

2. *Magnitude or amplitude (A): the length of the vector (or distance from the tip of the vector to the origin $(0,0)$).*

Therefore, we can identify each of these points in the plane with the phasor A_θ. We only need to calculate, for each vector, its phase and amplitude.

Let's start with vector $\vec{a}$, which is associated with the point $(1,1)$. The magnitude of this vector can be calculated using the Pythagorean theorem:

$$A = \sqrt{1^2 + 1^2} = \sqrt{2}.$$

Additionally, as shown in Figure 2.59, the angle formed by this vector with the positive x-axis is $45° = \pi/4$ rad. Thus, this vector is associated with the phasor $\sqrt{2}_{\pi/4}$.

Next, consider vector $\vec{b}$, associated with the point $(-1,-1)$. Its magnitude is obtained by calculating the distance from the point $(-1,-1)$ to $(0,0)$:

$$A = \sqrt{(-1)^2 + (-1)^2} = \sqrt{2}.$$

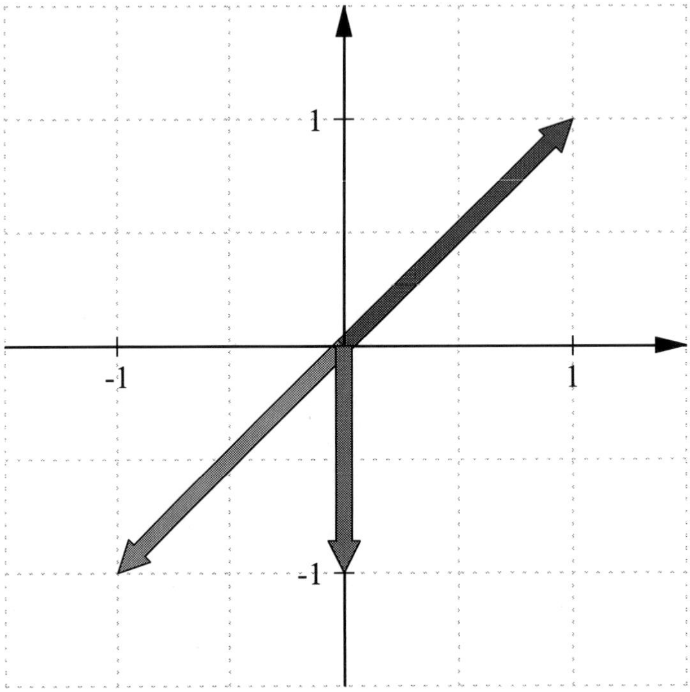

FIGURE 2.59
Phasors whose magnitudes and phases need to be identified (Problem 2.32).

Furthermore, as shown in Figure 2.59, the angle formed by this vector with the x-axis is:

$$180° + 45° = 225° = 5\pi/4 \text{ rad.}$$

Therefore, this vector is associated with the phasor $\sqrt{2}_{5\pi/4}$.

Finally, consider vector $\vec{c}$*, associated with the point* $(0,-1)$*. Its magnitude is obtained by calculating the distance from* $(0,-1)$ *to* $(0,0)$*:*

$$A = \sqrt{0^2 + (-1)^2} = \sqrt{1} = 1.$$

Additionally, as shown in Figure 2.59, the angle formed by this vector with the x-axis is $-90° = -\pi/2$ *rad. Thus, this vector is associated with the phasor* $1_{-\pi/2}$.

In summary, the phasors associated with the given vectors are:

Associated endpoint of the vector	$\longrightarrow$	*Phasor*
$(1,1)$	$\longrightarrow$	$\sqrt{2}_{\pi/4}$
$(-1,-1)$	$\longrightarrow$	$\sqrt{2}_{5\pi/4}$
$(0,-1)$	$\longrightarrow$	$1_{-\pi/2}$

Problem 2.33. Consider the phasor that represents the sinusoidal function:

$$\sqrt{2}\cos(\omega t + \frac{\pi}{4}).$$

What are the phasors obtained by multiplying this phasor by i, i^2, i^3, and i^4, respectively? Represent the input phasor and the four output phasors.

Solution to the problem 2.33. *The phasor representing the sinusoidal function $\sqrt{2}\cos(\omega t+\frac{\pi}{4})$ is given by the vector:*

$$\sqrt{2}\cos\left(\omega t+\frac{\pi}{4}\right)=\sqrt{2}_{\frac{\pi}{4}}.$$

Before proceeding with the multiplication by i, i^2, i^3, and i^4, we will calculate these successive powers of the number i:

$$\begin{aligned} i &= i, \\ i^2 &= -1, \\ i^3 &= -i, \\ i^4 &= 1, \end{aligned}$$

which in polar form can be written as:

$$\begin{aligned} i &= i = 1_{\frac{\pi}{2}}, \\ i^2 &= -1 = 1_{\pi}, \\ i^3 &= -i = 1_{-\frac{\pi}{2}}, \\ i^4 &= 1 = 1_0. \end{aligned}$$

Now, performing the multiplication of complex numbers in polar form:

$$1_{\frac{\pi}{2}}\cdot\sqrt{2}_{\frac{\pi}{4}}=\left(1\cdot\sqrt{2}\right)_{\frac{\pi}{2}+\frac{\pi}{4}}=\sqrt{2}_{\frac{3\pi}{4}},$$

$$1_{\pi}\cdot\sqrt{2}_{\frac{\pi}{4}}=\left(1\cdot\sqrt{2}\right)_{\pi+\frac{\pi}{4}}=\sqrt{2}_{\frac{5\pi}{4}},$$

$$1_{-\frac{\pi}{2}}\cdot\sqrt{2}_{\frac{\pi}{4}}=\left(1\cdot\sqrt{2}\right)_{-\frac{\pi}{2}+\frac{\pi}{4}}=\sqrt{2}_{-\frac{\pi}{4}},$$

$$1_0\cdot\sqrt{2}_{\frac{\pi}{4}}=\left(1\cdot\sqrt{2}\right)_{0+\frac{\pi}{4}}=\sqrt{2}_{\frac{\pi}{4}}.$$

These four phasors represent the following sinusoidal functions:

$$\sqrt{2}_{\frac{3\pi}{4}}=\sqrt{2}\cos\left(\omega t+\frac{3\pi}{4}\right),$$

$$\sqrt{2}_{\frac{5\pi}{4}}=\sqrt{2}\cos\left(\omega t+\frac{5\pi}{4}\right),$$

$$\sqrt{2}_{-\frac{\pi}{4}}=\sqrt{2}\cos\left(\omega t-\frac{\pi}{4}\right),$$

$$\sqrt{2}_{\frac{\pi}{4}}=\sqrt{2}\cos\left(\omega t+\frac{\pi}{4}\right),$$

which can be seen represented in Figure 2.60 (considering, for example, a frequency $\omega=1$).

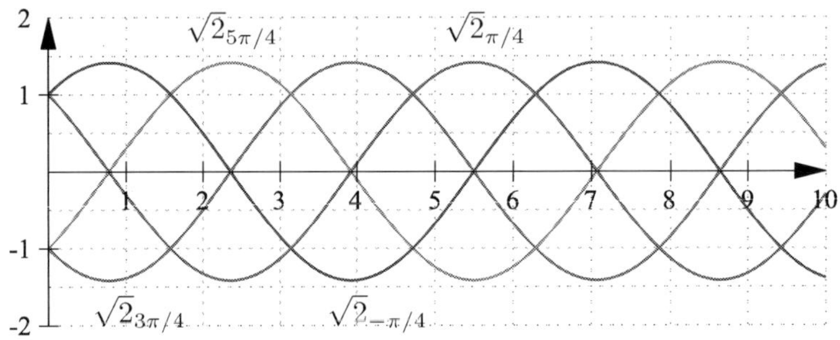

FIGURE 2.60
Functions represented by the phasors $\sqrt{2}_{\frac{\pi}{4}}$, $\sqrt{2}_{\frac{3\pi}{4}}$, $\sqrt{2}_{\frac{5\pi}{4}}$, and $\sqrt{2}_{-\frac{\pi}{4}}$ (Problem 2.0).

Problem 2.34. Observe closely these two functions, represented in Figure 2.61:

$$F(t) = \cos(t)$$
$$x(t) = \frac{1}{2}\cos\left(t - \frac{\pi}{2}\right)$$

What similarities and differences can be observed?

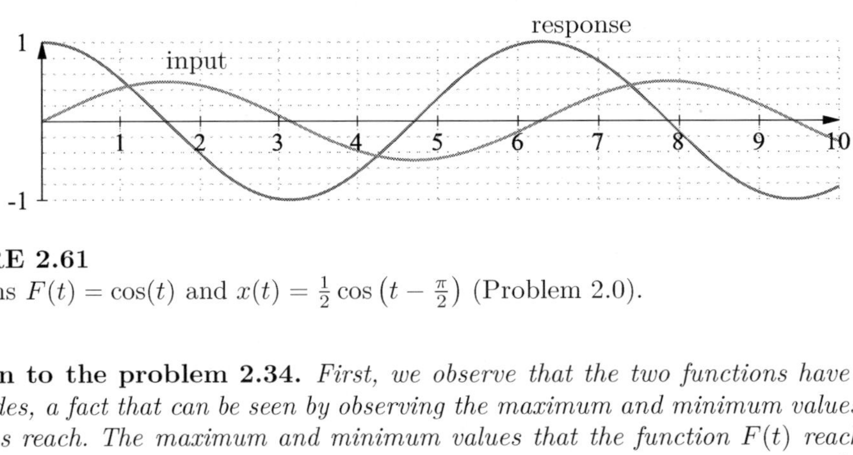

FIGURE 2.61
Functions $F(t) = \cos(t)$ and $x(t) = \frac{1}{2}\cos\left(t - \frac{\pi}{2}\right)$ (Problem 2.0).

Solution to the problem 2.34. *First, we observe that the two functions have different amplitudes, a fact that can be seen by observing the maximum and minimum values that the functions reach. The maximum and minimum values that the function $F(t)$ reaches are 1 and -1, respectively, while the maximum and minimum values that the function $x(t)$ reaches are $1/2$ and $-1/2$. Therefore, the amplitude of the function $x(t)$ is half the amplitude of $F(t)$.*

Now let's see if they have the same frequency. Consider the function $F(t)$; we see that at time $t = 0$ the value of the function is equal to 1. To determine the frequency, we will look at when the function again reaches the same value. We see that this occurs at $t = 2\pi$, that is:

$$F(0) = F(2\pi) = 1.$$

If we now consider the function $x(t)$, we see that at $t = 0$ the function is equal to 0 and that this value is found again at $t = 2\pi$. Therefore, both functions have the same frequency.

Finally, we need to compare the phases of the two functions. We observe that the function $F(t)$ reaches its first maximum at $t = 0$, while the function $x(t)$ reaches it at $t = \pi/2$. Therefore, if we shift the function $x(t)$ $-\pi/2$ to the left, the two functions will have the same phase.

In conclusion, both functions have the same frequency, the amplitude of $x(t)$ is half the amplitude of $F(t)$, and finally, the function $x(t)$ is phase-shifted $-\pi/2$ with respect to $F(t)$.

Problem 2.35. Represent the three cube roots of unity in the complex plane. Connect the three roots with straight lines. What geometric figure is obtained?

Solution to the problem 2.35. *The three cube roots of unity are:*

$$z_1 = e^{0i} = 1$$

$$z_2 = e^{2\pi/3i} = -\ frac12 + \frac{\sqrt{3}}{2}i$$

$$z_3 = e^{4\pi/3i} = -\frac{1}{2} - \frac{\sqrt{3}}{2}i.$$

The representation of these three roots in the complex plane can be seen in Figure 2.62.

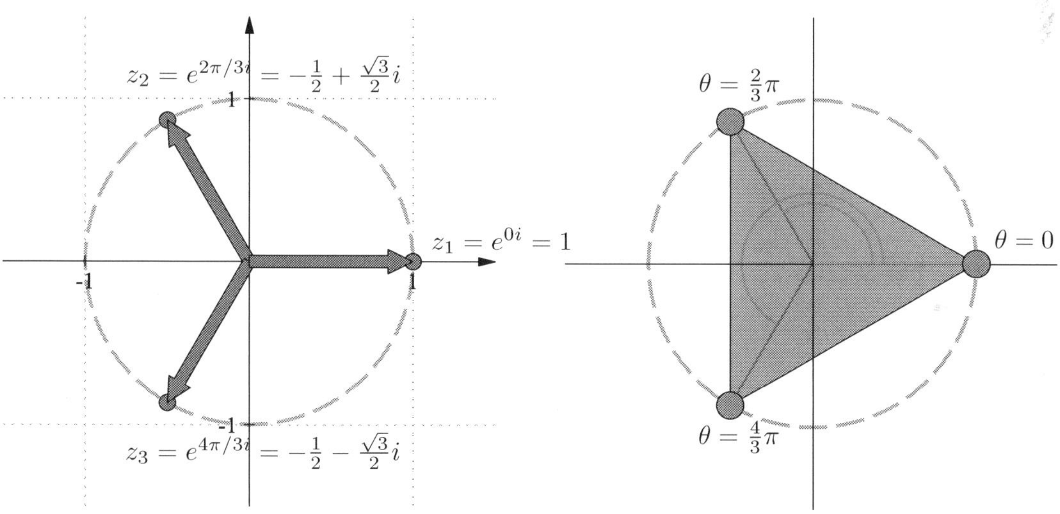

FIGURE 2.62
Representation of the three cube roots of unity in the complex plane (Problem 2.0).

As can be seen, the three roots form an equilateral triangle. Furthermore, the vertices of this triangle lie on the circumference centered at $(0, 0)$ with radius 1.

Problem 2.36. Verify that the solutions of the previous problem satisfy $z^3 = 1$, that is, raise the roots obtained to the cube and verify that they yield 1.

Solution to the problem 2.36. *In this exercise we are asked to verify that the following complex numbers:*

$$z = e^{0i} = 1$$
$$z = e^{2\pi/3i} = -\frac{1}{2} + \frac{\sqrt{3}}{2}i$$
$$z = e^{4\pi/3i} = -\frac{1}{2} - \frac{\sqrt{3}}{2}i$$

satisfy:

$$z^3 = 1.$$

Therefore, we need to raise the given numbers to the cube and verify that the result is 1. *This can be done both in binomial form and in exponential form. We will do it both ways, although we will see that it is simpler to do it in exponential form.*

Let's start by performing the calculations in exponential form:

$$z = e^{0i} \quad \Longrightarrow \quad z^3 = (e^{0i})^3 = e^{3\cdot 0i} = e^{0i} = 1$$
$$z = e^{2\pi/3i} \quad \Longrightarrow \quad z^3 = (e^{2\pi/3i})^3 = e^{3\cdot 2\pi/3i} = e^{2\pi i} = e^{0i} = 1$$
$$z = e^{4\pi/3i} \quad \Longrightarrow \quad z^3 = (e^{4\pi/3i})^3 = e^{3\cdot 4\pi/3i} = e^{4\pi i} = e^{0i} = 1$$

Note that in the calculations we performed, we first used the property:

$$(a^b)^c = a^{c\cdot b}.$$

Furthermore, we also used the fact that the angles 2π *and* 4π *are equal to the angle* 0, *and therefore we substituted them in the expression of the exponential. That is, we used:*

$$r_\theta = r_{\theta+2\pi} = r_{\theta+4\pi},$$

which, in exponential form, is:

$$re^{\theta i} = re^{\theta i + 2\pi i} = re^{\theta i + 4\pi i}.$$

In particular, we applied this for $r = 1$ *and* $\theta = 0$, *that is, we used the fact that:*

$$e^{0i} = e^{2\pi i} = e^{4\pi i}.$$

Now we will perform the calculations in binomial form. To do this, since we need to calculate the cube of a sum, we must use the formula:

$$(a+b)^3 = a^3 + 3a^2b + 3ab^2 + b^3.$$

- $\boxed{z = 1}$

$$z^3 = 1^3 = 1$$

- $\boxed{z = -\frac{1}{2} + \frac{\sqrt{3}}{2}i}$

$$
\begin{aligned}
z^3 &= \left(-\frac{1}{2} + \frac{\sqrt{3}}{2}i\right)^3 \\
&= \left(-\frac{1}{2}\right)^3 + 3\left(-\frac{1}{2}\right)^2\left(\frac{\sqrt{3}}{2}i\right) + 3\left(-\frac{1}{2}\right)\left(\frac{\sqrt{3}}{2}i\right)^2 + \left(\frac{\sqrt{3}}{2}i\right)^3 \\
&= -\frac{1}{8} + 3\frac{1}{4}\frac{\sqrt{3}}{2}i - 3\frac{1}{2}\frac{3}{4}i^2 + \frac{3\sqrt{3}}{8}i^3 = -\frac{1}{8} + \frac{3\sqrt{3}}{8}i - \frac{9}{8}(-1) + \frac{3\sqrt{3}}{8}ii^2 \\
&= -\frac{1}{8} + \frac{3\sqrt{3}}{8}i + \frac{9}{8} - \frac{3\sqrt{3}}{8}i = 1.
\end{aligned}
$$

- $\boxed{z = -\frac{1}{2} - \frac{\sqrt{3}}{2}i}$

$$
\begin{aligned}
z^3 &= \left(-\frac{1}{2} - \frac{\sqrt{3}}{2}i\right)^3 \\
&= \left(-\frac{1}{2}\right)^3 + 3\left(-\frac{1}{2}\right)^2\left(-\frac{\sqrt{3}}{2}i\right) + 3\left(-\frac{1}{2}\right)\left(-\frac{\sqrt{3}}{2}i\right)^2 + \left(-\frac{\sqrt{3}}{2}i\right)^3 \\
&= -\frac{1}{8} - 3\frac{1}{4}\frac{\sqrt{3}}{2}i - 3\frac{1}{2}\frac{3}{4}i^2 - \frac{3\sqrt{3}}{8}i^3 = -\frac{1}{8} - \frac{3\sqrt{3}}{8}i - \frac{9}{8}(-1) - \frac{3\sqrt{3}}{8}ii^2 \\
&= -\frac{1}{8} - \frac{3\sqrt{3}}{8}i + \frac{9}{8} + \frac{3\sqrt{3}}{8}i = 1.
\end{aligned}
$$

Problem 2.37. Calculate the square roots, fourth roots, fifth roots, and sixth roots of unity. Represent them in the complex plane. What geometric figures are obtained?

Solution to the problem 2.37. *First, express the complex number* 1 *in exponential form:*

$$1 = e^{0i} = e^{2\pi k i}, \quad k = 0, 1, 2, 3, \ldots$$

Thus, what we want to solve is:

$$z^n = e^{2\pi k i}, \quad k = 0, 1, 2, 3, \ldots$$

for $n = 2, 4, 5,$ *and* 6. *Apply the nth root to both sides of the equation:*

$$(z^n)^{\frac{1}{n}} = z = e^{\frac{2\pi k i}{n}} = e^{2\pi k/n\, i}, \quad k = 0, 1, 2, 3, \ldots$$

To determine the square roots, we substitute $n = 2$:

$$z = e^{2\pi k/2\, i} = e^{\pi k\, i}, \quad k = 0, 1, 2, 3, \ldots$$

and take the values $k = 0, 1$:

$$
\begin{aligned}
k = 0: &\qquad z_1 = e^{0i} = \cos\Big(0\Big) + i\sin\Big(0\Big) = 1 \\
k = 1: &\qquad z_2 = e^{\pi i} = \cos\Big(\pi\Big) + i\sin\Big(\pi\Big) = -1.
\end{aligned}
$$

Similarly, to determine the fourth roots, we substitute $n = 4$:

$$z = e^{2\pi k/4\,i} = e^{\pi/2k\,i}, \quad k = 0, 1, 2, 3, \ldots$$

and take the values $k = 0, 1, 2, 3$:

$$\begin{aligned}
k = 0: &\quad z_1 = e^{0i} = 1 \\
k = 1: &\quad z_2 = e^{\pi/2i} = \cos\left(\frac{\pi}{2}\right) + i\sin\left(\frac{\pi}{2}\right) = 0 + i = i \\
k = 2: &\quad z_3 = e^{\pi i} = \cos\left(\pi\right) + i\sin\left(\pi\right) = -1 \\
k = 3: &\quad z_4 = e^{3\pi/2i} = \cos\left(\frac{3\pi}{2}\right) + i\sin\left(\frac{3\pi}{2}\right) = 0 - i = -i.
\end{aligned}$$

Now, let's obtain the fifth roots. First, we substitute $n = 5$:

$$z = e^{2\pi k/5\,i} = e^{2\pi/5k\,i}, \quad k = 0, 1, 2, 3, \ldots$$

and take the values $k = 0, 1, 2, 3, 4$:

$$\begin{aligned}
k = 0: &\quad z_1 = e^{0i} = 1 \\
k = 1: &\quad z_2 = e^{2\pi/5i} = \cos\left(\frac{2\pi}{5}\right) + i\sin\left(\frac{2\pi}{5}\right) = 0.309 + 0.951i \\
k = 2: &\quad z_3 = e^{4\pi/5i} = \cos\left(\frac{4\pi}{5}\right) + i\sin\left(\frac{4\pi}{5}\right) = -0.809 + 0.588i \\
k = 3: &\quad z_4 = e^{6\pi/5i} = \cos\left(\frac{6\pi}{5}\right) + i\sin\left(\frac{6\pi}{5}\right) = -0.809 - 0.588i = \overline{z_3} \\
k = 4: &\quad z_5 = e^{8\pi/5i} = \cos\left(\frac{8\pi}{5}\right) + i\sin\left(\frac{8\pi}{5}\right) = 0.309 - 0.951i = \overline{z_2}.
\end{aligned}$$

Finally, let's obtain the sixth roots. First, we substitute $n = 6$:

$$z = e^{2\pi k/6\,i} = e^{2\pi/5k\,i}, \quad k = 0, 1, 2, 3, \ldots$$

and take the values $k = 0, 1, 2, 3, 4, 5$:

$$\begin{aligned}
k = 0: &\quad z_1 = e^{0i} = 1 \\
k = 1: &\quad z_2 = e^{2\pi/6i} = \cos\left(\frac{2\pi}{6}\right) + i\sin\left(\frac{2\pi}{6}\right) = \frac{1}{2} + \frac{\sqrt{3}}{2}i \\
k = 2: &\quad z_3 = e^{4\pi/6i} = \cos\left(\frac{4\pi}{6}\right) + i\sin\left(\frac{4\pi}{6}\right) = -\frac{1}{2} + \frac{\sqrt{3}}{2}i \\
k = 3: &\quad z_4 = e^{6\pi/6i} = e^{\pi i} = -1 \\
k = 4: &\quad z_5 = e^{8\pi/6i} = \cos\left(\frac{8\pi}{6}\right) + i\sin\left(\frac{8\pi}{6}\right) = -\frac{1}{2} - \frac{\sqrt{3}}{2}i = \overline{z_3} \\
k = 5: &\quad z_6 = e^{10\pi/6i} = \cos\left(\frac{10\pi}{6}\right) + i\sin\left(\frac{10\pi}{6}\right) = \frac{1}{2} - \frac{\sqrt{3}}{2}i = \overline{z_2}.
\end{aligned}$$

In Figure 2.63, we have represented the square roots, fourth roots, fifth roots, and sixth roots of the number $z = 1$. Note that all the roots have the same modulus (they are located on a circle) and are situated at the vertices of a regular polygon with n sides. Thus, the square roots are always aligned, the fourth roots are the vertices of a square, etc.

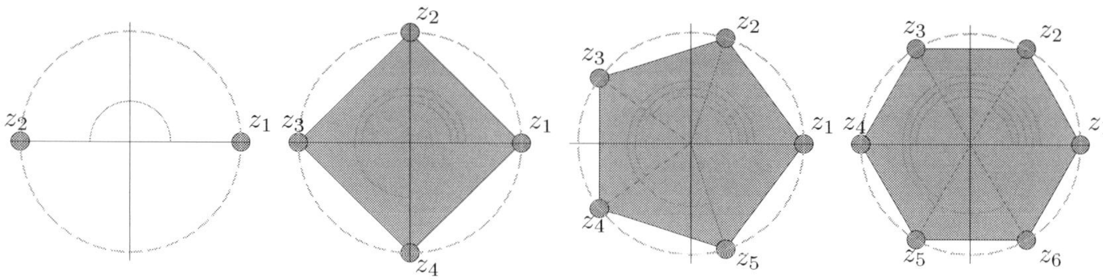

FIGURE 2.63
Square roots, fourth roots, fifth roots, and sixth roots of unity (Problem 2.0).

Problem 2.38. Verify that the sum of the fourth roots of unity is 0.

Solution to the problem 2.38. *The fourth roots of unity, as we calculated in Problem 2.37, are:*

$$z_1 = 1$$
$$z_2 = i$$
$$z_3 = -1$$
$$z_4 = -i.$$

We verify that $z_1 + z_2 + z_3 + z_4 = 0$*:*

$$1 + i - 1 - i = 0.$$

Problem 2.39. Express the fourth roots of $z = 4$ in binomial form.

Solution to the problem 2.39. *To calculate the fourth roots of a complex number, we need to convert the complex number to polar or exponential form; in this case, we have:*

$$z = 4 = 4_0 = 4e^{0i}.$$

There are four fourth roots of a complex number. The first root has a modulus equal to the fourth root of the modulus of z and an argument equal to the argument of z divided by 4, that is, if we denote by z_1 the first root:

$$z_1 = \sqrt[4]{4}e^{\frac{0}{4}i} = \sqrt{2}e^{0i}.$$

This number in binomial form is directly: $z_1 = \sqrt{2}$.

To find the other three roots, we need to remember that the fourth roots are the vertices of a square centered at the origin, that is, all have the same modulus and the angles differ by $\frac{2\pi}{4} = \frac{\pi}{2}$ *radians.*

Therefore, the other three roots are:

$$z_2 = \sqrt{2}e^{0i+\frac{\pi}{2}i} = \sqrt{2}e^{\frac{\pi}{2}i}$$
$$z_3 = \sqrt{2}e^{\frac{\pi}{2}i+\frac{\pi}{2}i} = \sqrt{2}e^{\pi i}$$
$$z_4 = \sqrt{2}e^{\pi i+\frac{\pi}{2}i} = \sqrt{2}e^{\frac{3\pi}{2}i} = \sqrt{2}e^{(\frac{3\pi}{2}-2\pi)i} = \sqrt{2}e^{-\frac{\pi}{2}i},$$

which in binomial form are:

$$z_1 = \sqrt{2}, \quad z_2 = \sqrt{2}i, \quad z_3 = -\sqrt{2}, \quad z_4 = -\sqrt{2}i.$$

The graphical representation in the complex plane is shown in Figure 2.64.

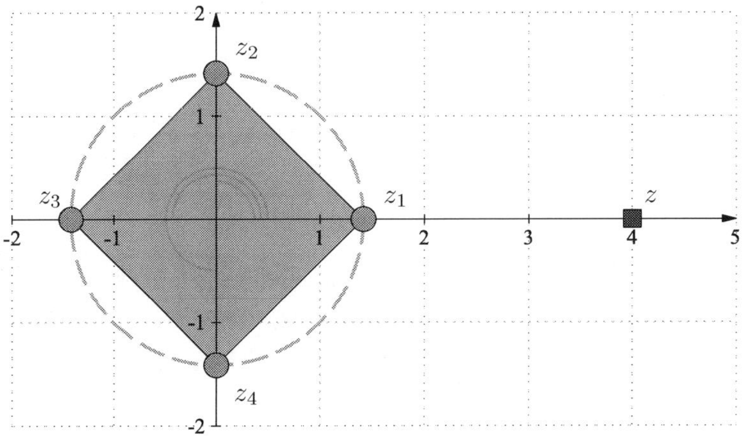

FIGURE 2.64
Fourth roots of $z = 4$ (Problem 2.0).

Problem 2.40. Calculate the solutions to the following equations:

$$z^4 + 16 = 0, \quad z^6 = i, \quad z^3 + 8i = 0.$$

Solution to the problem 2.40. *Let's first solve the equation $z^4 + 16 = 0$. Note that solving this equation is equivalent to finding the values of z that make:*

$$z^4 = -16.$$

Therefore, we need to calculate the fourth roots of the complex number -16.

This exercise can be done very simply using the following reasoning:

$$z^4 = -16 = 16(-1) \Longrightarrow z^4 = \sqrt[4]{16} \cdot \sqrt[4]{-1} = 2\sqrt[4]{-1}.$$

Therefore, we can obtain the fourth roots of -16 from the fourth roots of -1, which we have already calculated previously, by multiplying by two. Thus, the four fourth roots are:

$$z_1 = 2\left(\frac{\sqrt{2}}{2} + \frac{\sqrt{2}}{2}i\right) = \sqrt{2} + \sqrt{2}i$$

$$z_2 = 2\left(-\frac{\sqrt{2}}{2} + \frac{\sqrt{2}}{2}i\right) = -\sqrt{2} + \sqrt{2}i$$

$$z_3 = 2\left(-\frac{\sqrt{2}}{2} - \frac{\sqrt{2}}{2}i\right) = -\sqrt{2} - \sqrt{2}i$$

$$z_4 = 2\left(\frac{\sqrt{2}}{2} - \frac{\sqrt{2}}{2}i\right) = \sqrt{2} - \sqrt{2}i.$$

However, we can also do the same exercise step by step. To do this, the first thing we will do is convert the number -16 *to exponential form:*

$$-16 = 16e^{\pi i} = 16e^{(\pi+2\pi k)i}, \quad k = 0, 1, 2, 3, \ldots$$

Therefore, what we want to solve is:

$$z^4 = 16e^{(\pi+2\pi k)i}, \quad k = 0, 1, 2, 3, \ldots$$

Now, let's apply the fourth root to both sides of the equation:

$$\left(z^4\right)^{\frac{1}{4}} = z = \left(16e^{(\pi+2\pi k)i}\right)^{\frac{1}{4}} = \sqrt[4]{16}e^{(\pi+2\pi k)/4i} = 2e^{(\pi/4+\pi/2k)i}, \quad k = 0, 1, 2, 3, \ldots$$

To determine the fourth roots, we take the values $k = 0, 1, 2, 3$*:*

$$\begin{aligned}
k = 0: &\quad z_1 = 2e^{\pi/4i} = 2\cos\left(\frac{\pi}{4}\right) + 2\sin\left(\frac{\pi}{4}\right)i = \sqrt{2} + \sqrt{2}i \\
k = 1: &\quad z_2 = 2e^{(\pi/4+\pi/2)i} = 2e^{3\pi/4i} = 2\cos\left(\frac{3\pi}{4}\right) + 2\sin\left(3\frac{\pi}{4}\right)i = -\sqrt{2} + \sqrt{2}i \\
k = 2: &\quad z_3 = 2e^{(\pi/4+\pi)i} = 2e^{5\pi/4i} = 2\cos\left(\frac{5\pi}{4}\right) + 2\sin\left(\frac{5\pi}{4}\right)i = -\sqrt{2} - \sqrt{2}i \\
k = 3: &\quad z_4 = 2e^{(\pi/4+3\pi/2)i} = 2e^{7\pi/4i} = 2\cos\left(\frac{7\pi}{4}\right) + 2\sin\left(7\frac{\pi}{4}\right)i = \sqrt{2} - \sqrt{2}i.
\end{aligned}$$

We see that we obtain the same results as with the simpler method of doing the exercise.

Another equally simple way to calculate the fourth roots is to remember that the four roots are the vertices of a square. Consider the number -16*, which has a modulus of* 16 *and an angle of* π*. We know that the first root can be obtained by taking the fourth root of the modulus and dividing the argument by* 4*, so we have:*

$$z_1 = \sqrt[4]{16}e^{\pi/4i} = \sqrt{2} + \sqrt{2}i.$$

If we plot this point (see Figure 2.65), we see that it lies on the bisector of the first quadrant (line that divides the first quadrant into two equal halves, the line $y = x$*). We now draw the circle with radius* 2 *and place the four vertices of the square on it.*

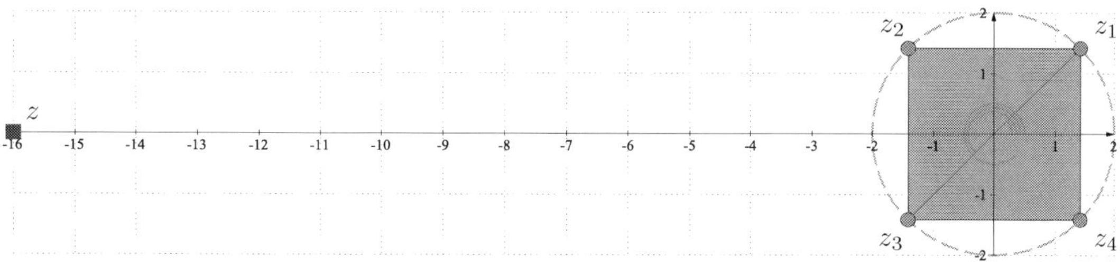

FIGURE 2.65
Fourth roots of $z = 16e^{\pi/4i}$ (Problem 2.0).

Graphically, we see that if $z_1 = \sqrt{2} + \sqrt{2}i$*, to obtain* z_2 *it is only necessary to change the sign of the* x *coordinate, that is, the real part. Similarly, we see that* $z_3 = -z_1$*. Finally, to obtain* z_4*, we need to change the sign of the* y *coordinate, that is, the imaginary part.*

Thus:

$$z_1 = \sqrt{2} + \sqrt{2}i$$
$$z_2 = -\sqrt{2} + \sqrt{2}i$$
$$z_3 = -z_1 = -\sqrt{2} + \sqrt{2}i$$
$$z_4 = -\sqrt{2} + \sqrt{2}i.$$

Let's now solve the equation $z^6 = i$*. In this case, we are asked to find the sixth roots of the number* i*. The first thing to do is convert this number to exponential form:*

$$i = e^{\pi/2i}.$$

Therefore, since i *has a modulus of* 1*, the sixth roots will have a modulus of* $\sqrt[6]{1} = 1$*. That is, the complex numbers we are looking for are located on the circle with radius* 1*.*

Remember that the first root is obtained by dividing the initial angle by 6*, that is:*

$$z_1 = e^{\pi/2/6i} = e^{\pi/12i}.$$

The other roots are found by placing them at the vertices of a regular hexagon as shown in Figure 2.66. That is, the angle difference between each root is $2\pi/6 = \pi/3$*.*

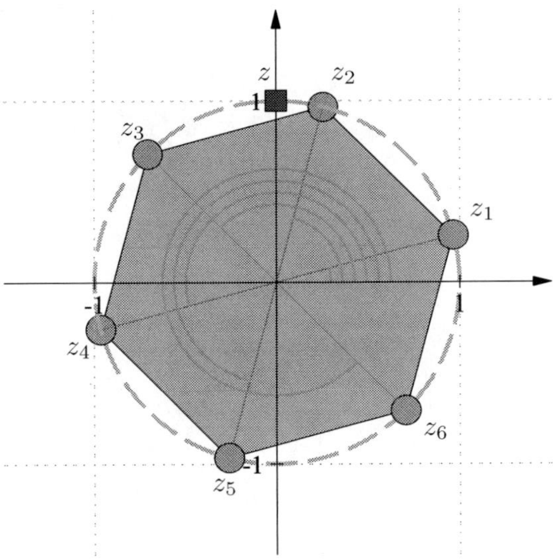

FIGURE 2.66
Sixth roots of $z = i$ (Problem 2.0).

In this way:

$$z_1 = e^{\pi/12i}$$
$$z_2 = e^{(\pi/12+\pi/3)i} = e^{5\pi/12i}$$
$$z_3 = e^{(\pi/12+2\pi/3)i} = e^{3\pi/4i}$$
$$z_4 = e^{(\pi/12+3\pi/3)i} = e^{13\pi/12i}$$
$$z_5 = e^{(\pi/12+4\pi/3)i} = e^{17\pi/12i}$$
$$z_6 = e^{(\pi/12+5\pi/3)i} = e^{7\pi/4i}.$$

Finally, let's find the solutions to the equation $z^3 + 8i = 0$. *In this case, we need to solve:*

$$z^3 = -8i,$$

that is, we need to find the cube roots of:

$$-8i = 8e^{-\pi/2i}.$$

We know that the modulus of the three roots will be $\sqrt[3]{8} = 2$; *therefore, they are located on the circle centered at* $(0,0)$ *and radius* 2. *The first root is obtained by dividing the angle by* 3, *that is:*

$$z_1 = 2e^{-\pi/2/3i} = 2e^{-\pi/6i}.$$

In this case, since the roots are located on a triangle, the angle difference between the points is $2\pi/3$. *This results in:*

$$\begin{aligned} z_1 &= 2e^{-\pi/6i} \\ z_2 &= 2e^{(-\pi/6+2\pi/3)i} = e^{\pi/2i} \\ z_3 &= 2e^{(-\pi/6+4\pi/3)i} = e^{7\pi/6i}. \end{aligned}$$

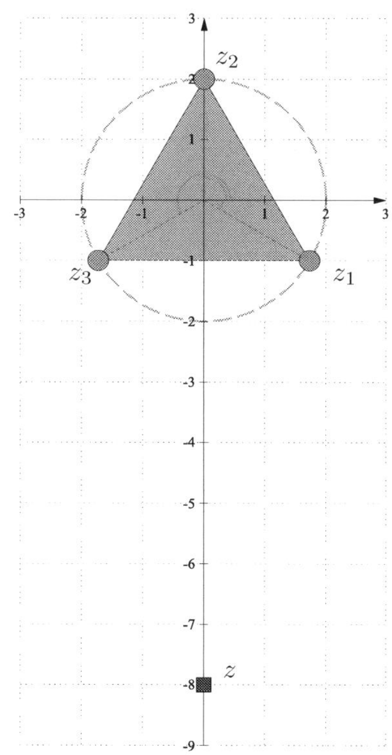

FIGURE 2.67
Cube roots of $z = -8i$ (Problem 2.0).

As seen in Figure 2.67, if we add $2\pi/3$ *to the angle of* z_1, *we obtain an angle of* $\pi/2$, *that is, the second root is on the positive* y*-axis. This means it is a pure positive imaginary number; therefore,* $z_2 = 2i$. *Also, in Figure 2.67, we see that to obtain* z_3 *we only need to* *change the sign of the real part of* z_1. *Thus:*

$$z_1 = \sqrt{3} - i$$
$$z_2 = 2i$$
$$z_3 = -\sqrt{3} - i.$$

Problem 2.41. One of the cube roots of a complex number is $1_{\frac{\pi}{3}}$. Calculate this complex number and the other two roots.

Solution to the problem 2.41. *There are three cube roots of a complex number. We know one:*

$$z_1 = 1_{\frac{\pi}{3}}.$$

To find the other two roots, keep in mind that the cube roots are the vertices of an equilateral triangle centered at the origin. Thus, all the moduli are equal, while the angles differ by $\frac{2\pi}{3}$ *radians (resulting from dividing the* 2π *radians of a circle by 3, since a triangle has 3 vertices).*

Therefore, the other two roots are:

$$z_2 = 1_{\frac{\pi}{3}+\frac{2}{3}\pi} = 1_{\pi},$$
$$z_3 = 1_{\pi+\frac{2}{3}\pi} = 1_{\frac{5}{3}\pi} = 1_{\frac{5}{3}\pi-2\pi} = 1_{-\frac{\pi}{3}}.$$

It is important to remember that the principal argument of a complex number is such that:

$$-\pi < \theta \leq \pi.$$

To find the complex number, simply raise any of the three roots to the cube:

$$z = z_1^3 = z_2^3 = z_3^3.$$

We use the exponential form to make it clearer:

$$z = z_1^3 = \left(1e^{\frac{\pi}{3}i}\right)^3 = 1^3 e^{3\cdot\frac{\pi}{3}i} = 1e^{\pi i} = 1_{\pi}.$$

We verify that the result is the same with the other two roots:

$$z = z_2^3 = \left(1e^{\pi i}\right)^3 = 1^3 e^{3\pi i} = 1e^{3\pi i} = 1e^{3\pi i - 2\pi i} = 1e^{\pi i} = 1_{\pi}$$
$$z = z_3^3 = \left(1e^{-\frac{\pi}{3}i}\right)^3 = 1^3 e^{-3\frac{\pi}{3}i} = 1e^{-\pi i} = 1e^{\pi i} = 1_{\pi}.$$

Therefore:

$$z = 1_{\pi} \quad ; \quad z_1 = 1_{\frac{\pi}{3}} \quad ; \quad z_2 = 1_{\pi} \quad ; \quad z_3 = 1_{-\frac{\pi}{3}}.$$

The graphical representation in the complex plane is shown in Figure 2.68.

Problem 2.42. If z_1, z_2, and z_3 are the cube roots of $z = 1 + i$, calculate:

$$z_1^6 + z_2^{12} + z_3^{24}.$$

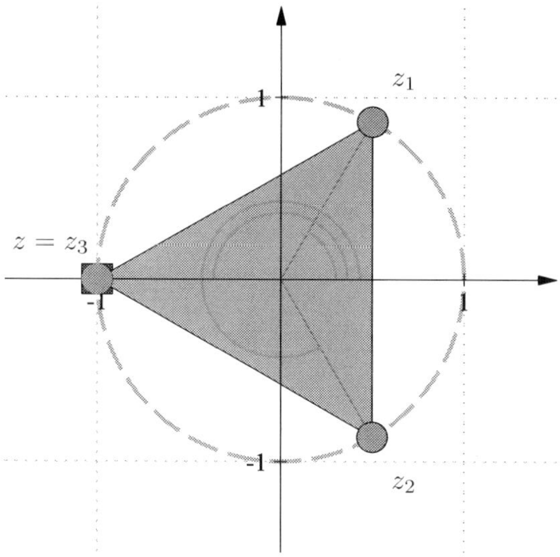

FIGURE 2.68
Cube roots of the number $z = 1_\pi$ (Problem 2.0).

Solution to the problem 2.42. *If z_1, z_2, and z_3 are the cube roots of $z = 1 + i$, then $z_1^3 = z_2^3 = z_3^3 = z$. Therefore:*

$$z_1^6 = \left(z_1^3\right)^2 = z^2,$$
$$z_2^{12} = \left(z_2^3\right)^4 = z^4,$$
$$z_3^{24} = \left(z_3^3\right)^8 = z^8.$$

Convert $z = 1 + i$ to exponential form $z = 1 + i = \sqrt{2}e^{\frac{\pi}{4}i}$ and then:

$$z_1^6 + z_2^{12} + z_3^{24} = z^2 + z^4 + z^8 = 2e^{\frac{\pi}{2}i} + 4e^{\pi i} + 16e^{0i}.$$

Convert to binomial form to sum and obtain the result as follows:

$$z_1^6 + z_2^{12} + z_3^{24} = (0 + 2i) + (-4 + 0i) + (16 + 0i) = 12 + 2i.$$

If we do not realize that it is not necessary to calculate the three roots to solve the problem, the procedure would be as follows. First, express z in exponential form to be able to calculate the roots:

$$z = 1 + i = \sqrt{2}e^{\frac{\pi}{4}i}.$$

Then the three cube roots are:

$$z_1 = \sqrt[6]{2}e^{\frac{\frac{\pi}{4}}{3}i} = \sqrt[6]{2}e^{\frac{\pi}{12}i},$$
$$z_2 = \sqrt[6]{2}e^{\left(\frac{\pi}{12}+\frac{2}{3}\pi\right)i} = \sqrt[6]{2}e^{\frac{9}{12}\pi i} = \sqrt[6]{2}e^{\frac{3}{4}\pi i},$$
$$z_3 = \sqrt[6]{2}e^{\left(\frac{3}{4}\pi+\frac{2}{3}\pi\right)i} = \sqrt[6]{2}e^{\frac{17}{12}\pi i-2\pi i} = \sqrt[6]{2}e^{-\frac{7}{12}\pi i}.$$

Therefore:

$$z_1^6 + z_2^{12} + z_3^{24} = \left(\sqrt[6]{2}e^{\frac{\pi}{12}i}\right)^6 + \left(\sqrt[6]{2}e^{\frac{3}{4}\pi i}\right)^{12} + \left(\sqrt[6]{2}e^{-\frac{7}{12}\pi i}\right)^{24} =$$
$$= 2e^{\frac{6\pi}{12}i} + 2^2e^{12\cdot\frac{3}{4}\pi i} + 2^4e^{-24\cdot\frac{7}{12}\pi i} =$$
$$= 2e^{\frac{\pi}{2}i} + 4e^{9\pi i} + 16e^{-14\pi i} = 2e^{\frac{\pi}{2}i} + 4e^{\pi i} + 16e^{0i},$$

and we see that we arrive at the same result as before.

Problem 2.43. Calculate the following expressions in exponential form:

$$\frac{(3-2i)(2+3i)}{3-4i}, \quad i^{29}, \quad (1+i)^4.$$

Solution to the problem 2.43. *To simplify the expressions given in exponential form, the first thing we need to do is convert the numbers from binomial to exponential form.*

Let's start by calculating the expression:

$$\frac{(3-2i)(2+3i)}{3-4i}.$$

To do this, we need to convert the three numbers that appear into exponential form.

$$\begin{aligned}
z_1 = 3-2i \quad &\longrightarrow \quad r_1 = \sqrt{3^2+(-2)^2} = \sqrt{9+4} = \sqrt{13}\\
&\longrightarrow \quad \tan(\theta_1) = \tfrac{-2}{3} \Longrightarrow \theta_1 = -0.5880 \ rad\\
z_2 = 2+3i \quad &\longrightarrow \quad r_2 = \sqrt{2^2+3^2} = \sqrt{4+9} = \sqrt{13}\\
&\longrightarrow \quad \tan(\theta_2) = \tfrac{3}{2} \Longrightarrow \theta_2 = 0.9828 \ rad\\
z_3 = 3-4i \quad &\longrightarrow \quad r_3 = \sqrt{3^2+(-4)^2} = \sqrt{9+16} = \sqrt{25} = 5\\
&\longrightarrow \quad \tan(\theta_3) = \tfrac{-4}{3} \Longrightarrow \theta_3 = -0.9273 \ rad
\end{aligned}$$

Thus, we have:

$$\begin{aligned}
\frac{(3-2i)(2+3i)}{3-4i} &= \frac{z_1 \cdot z_2}{z_3} = \frac{\sqrt{13}e^{-0.5880i} \cdot \sqrt{13}e^{0.9828i}}{5e^{-0.9273i}}\\
&= \frac{\sqrt{13}^2}{5}e^{(-0.5880+0.9828+0.9273)i} = \frac{13}{5}e^{1.3221i}.
\end{aligned}$$

Finally, we can convert the result to binomial form:

$$\begin{aligned}
\frac{13}{5}e^{1.3221i} &= \frac{13}{5}\cos(1.3221) + \frac{13}{5}\sin(1.3221)i\\
&= 0.64 + 2.52i.
\end{aligned}$$

Note that by approximating the angles to four decimal places, we have obtained an approximate result. In fact, it is a good exercise to perform the same calculations in binomial form and verify that:

$$\frac{(3-2i)(2+3i)}{3-4i} = \frac{16}{25} - \frac{63}{25}i = 0.64 + 2.52i.$$

Now let's calculate i^{29}. The number i is associated with the point $(0,1)$ in the complex plane and, therefore, has a modulus of 1 and an angle of $\pi/2 = 90°$. Using De Moivre's formula, we obtain:

$$i^{29} = (1e^{\pi/2i})^{29} = (e^{\pi/2i})^{29} = e^{29\cdot\pi/2i}.$$

Note that the angle obtained is:

$$\frac{29}{2}\pi = \frac{29}{4}\cdot 2\pi = 7.25 \cdot 2\pi,$$

which represents more than 7 full rotations. To determine the angle between 0 *and* 2π*, we subtract 7 full rotations:*

$$\frac{29}{2}\pi - 7 \cdot 2\pi = \left(\frac{29}{2} - 14\right)\pi = \frac{29-28}{2}\pi = \frac{\pi}{2}.$$

Consequently, the result is:

$$e^{\pi/2i} = \cos(\pi/2) + \sin(\pi/2)i = i.$$

Finally, let's calculate $(1+i)^4$*. First, we will convert the number* $1+i$ *to exponential form. Note that the complex number* $1+i$ *is associated with the point* $(1,1)$*, which has a modulus of* $\sqrt{2}$ *and an angle of* $45° = \pi/4$ *rad. Therefore:*

$$(1+i)^4 = \left(\sqrt{2}e^{\pi/4i}\right)^4 = \left(\sqrt{2}\right)^4 e^{4\cdot\pi/4i} = 2^2 e^{\pi i} = 4e^{\pi i}.$$

Note that the angle π *indicates that it is a negative real number:*

$$(1+i)^4 = 4e^{\pi i} = -4.$$

Remember that to avoid mistakes, it is very useful to always represent the complex numbers with which you are working, especially in the calculation of angles.

Problem 2.44. Calculate:

$$\frac{(4+7i)+(1-i)(2-i)}{(5+i)^2}, \quad \frac{10-((1-4i)-(2-3i))}{3i+(2+\sqrt{3i})(2-\sqrt{3i})}, \quad \left(\frac{1-i}{1+i}\right)^6.$$

Solution to the problem 2.44. *When performing calculations with complex numbers, the binomial form is used for addition and subtraction, and generally the polar or exponential form is used for multiplication and division, although these operations can also be done in binomial form.*

In the first case we have:

- $z_1 = \dfrac{(4+7i)+(1-i)(2-i)}{(5+i)^2}$

  Since we have simple sums and products, to avoid converting from one form to another, we will simplify the numerator and the denominator in binomial form:

$$z_1 = \frac{(4+7i)+(1-i)(2-i)}{(5+i)^2} = \frac{(4+7i)+(2-1)+(-1-2)i}{25+10i+i^2} =$$
$$= \frac{(4+7i)+(1-3i)}{25+10i+(-1)} = \frac{5+4i}{24+10i}.$$

  To divide complex numbers in binomial form, we multiply and divide by the conjugate of the denominator:

$$z_1 = \frac{5+4i}{24+10i} \cdot \frac{24-10i}{24-10i} = \frac{(5\cdot 24 - 4(-10)) + (5(-10)+4\cdot 24)i}{24^2-(10i)^2} =$$
$$= \frac{(120+40)+(-50+96)i}{24^2+10^2} = \frac{160+46i}{676} = \frac{40}{169} + \frac{23}{338}i.$$

  Thus:

$$z_1 = \frac{40}{169} + \frac{23}{338}i.$$

- $z_2 = \dfrac{10 - ((1-4i) - (2-3i))}{3i + (2+\sqrt{3}i)(2-\sqrt{3}i)}$

  We start by solving the numerator in binomial form. Regarding the denominator, we note that there is a product of sums and differences, which we know is the difference of squares. This operation eliminates the imaginary part of the denominator, making the exercise easier to solve:

$$z_2 = \frac{10 - ((1-4i) - (2-3i))}{3i + (2+\sqrt{3}i)(2-\sqrt{3}i)} = \frac{10 - (-1-i)}{3i + 2^2 - (\sqrt{3}i)^2} = \frac{11+i}{3i+4-3i} =$$

$$= \frac{11+i}{4} = \frac{11}{4} + \frac{1}{4}i.$$

Therefore:

$$z_2 = \frac{11}{4} + \frac{1}{4}i.$$

- $z_3 = \left(\dfrac{1-i}{1+i}\right)^6$

  In this case, since we need to raise to the power of 6, it is convenient to perform the calculations in polar or exponential form. Therefore, the first thing we do is convert the numerator and the denominator to exponential form to calculate the quotient and the power:

$$1 - i = \sqrt{2}e^{-\frac{\pi}{4}i}$$

$$1 + i = \sqrt{2}e^{\frac{\pi}{4}i}$$

So:

$$z_3 = \left(\frac{1-i}{1+i}\right)^6 = \left(\frac{\sqrt{2}e^{-\frac{\pi}{4}i}}{\sqrt{2}e^{\frac{\pi}{4}i}}\right)^6 = \left(\frac{\sqrt{2}}{\sqrt{2}}e^{\left(-\frac{\pi}{4}-\frac{\pi}{4}\right)i}\right)^6$$

$$= \left(1e^{-\frac{\pi}{2}i}\right)^6 = 1e^{-\frac{6\pi}{2}i} = 1e^{-3\pi i} = 1e^{-3\pi i + 2\pi i + 2\pi i}$$

$$= 1e^{\pi i} = 1(\cos\pi + i\cdot\sin\pi) = -1.$$

Therefore:

$$z_3 = -1.$$

Problem 2.45. Perform the indicated operations:

$$\frac{2-3\mathrm{i}}{(2+3\mathrm{i})(1+\mathrm{i})} - \frac{3+i}{4(2-i)}, \quad (1-2i)^{52}, \quad \left(\frac{1+\mathrm{i}\sqrt{3}}{1-\mathrm{i}}\right)^{12}.$$

Solution to the problem 2.45.

- $z = \dfrac{2-3\mathrm{i}}{(2+3\mathrm{i})(1+\mathrm{i})} - \dfrac{3+i}{4(2-i)}$

Since these are simple sums and products, we simplify the denominator of the first fraction in binomial form:

$$z = \frac{2-3i}{(2+3i)(1+i)} - \frac{3+i}{4(2-i)} = \frac{2-3i}{2+5i+3i^2} - \frac{3+i}{4(2-i)} =$$

$$= \frac{2-3i}{2+5i+3(-1)} - \frac{3+i}{4(2-i)} = \frac{2-3i}{-1+5i} - \frac{3+i}{4(2-i)}.$$

To divide the complex numbers that compose each fraction, we multiply and divide by the conjugate of the denominator:

$$z = \frac{(2-3i)(-1-5i)}{(-1+5i)(-1-5i)} - \frac{(3+i)(2+i)}{4(2-i)(2+i)}$$

$$= \frac{(2(-1)-(-3)(-5)) + (2(-5)+(-3)(-1))i}{(-1)^2-(5i)^2} - \frac{(3\cdot 2 - 1\cdot 1) + (3\cdot 1 + 1\cdot 2)i}{4(2^2-(i)^2)}$$

$$= \frac{(-2-15)+(-10+3)i}{(-1)^2+5^2} - \frac{(6-1)+(3+2)i}{4(2^2+1)} = \frac{-17-7i}{26} - \frac{5+5i}{20}$$

$$= \frac{-17-7i}{26} - \frac{1+i}{4}.$$

Finally, we reduce the fractions to a common denominator and add them:

$$z = \frac{-17-7i}{2\cdot 13} - \frac{1+i}{2\cdot 2} = \frac{(-17-7i)\cdot 2}{(2\cdot 13)\cdot 2} - \frac{(1+i)\cdot 13}{(2\cdot 2)\cdot 13}$$

$$= \frac{(-34-14i)-(13+13i)}{52} = -\frac{47}{52} - \frac{27}{52}i.$$

Thus:

$$z = -\frac{47}{52} - \frac{27}{52}i.$$

- $z = (1-2i)^{52}$

In this case, since we need to raise to the power of 52, it is more convenient to operate in polar or exponential form. We convert $(1-2i)$ to exponential form:

$$1-2i = \sqrt{5}e^{-1.10715i}.$$

We raise the result to the power of 52:

$$z = (\sqrt{5}e^{-1.10715i})^{52} = \sqrt{5}^{52}e^{52\cdot(-1.10715i)} = 5^{26}e^{-0.32567\pi i}$$

$$= 1.49012\cdot 10^{18}e^{-1.02312i}.$$

Therefore:

$$z = 1.49012\cdot 10^{18}e^{-1.02312i} = 7.7598\cdot 10^{17} - 1.2721\cdot 10^{18}i.$$

- $z = \left(\dfrac{1+\mathrm{i}\sqrt{3}}{1-\mathrm{i}}\right)^{12}$

In this case, we need to solve the division of complex numbers and then raise the result to the power of 12. Since for the second operation it is more convenient to operate in polar or exponential form, we directly convert the numerator and denominator to exponential form before calculating the division:

$$1+i\sqrt{3} = 2e^{\frac{\pi}{3}i}$$
$$1-i = \sqrt{2}e^{-\frac{\pi}{4}i}.$$

Thus:

$$z = \left(\frac{2e^{\frac{\pi}{3}i}}{\sqrt{2}e^{-\frac{\pi}{4}i}}\right)^{12}$$

We calculate the division:

$$z = \left(\frac{2}{\sqrt{2}}e^{\left(\frac{\pi}{3}-\left(-\frac{\pi}{4}\right)\right)i}\right)^{12} = \left(\sqrt{2}e^{\frac{7\pi}{12}i}\right)^{12}.$$

We raise to the power of 12:

$$z_3 = \left(\sqrt{2}e^{\frac{7\pi}{12}i}\right)^{12} = \sqrt{2}^{12}e^{12\frac{7\pi}{12}i} = 2^6e^{7\pi i} = 2^6e^{\pi i} = -64.$$

Thus:

$$z_3 = -64.$$

Problem 2.46. Calculate:

$$\sqrt[5]{1-\sqrt{3}i}, \quad (1-i)^{\frac{5}{4}}, \quad \frac{\sqrt[3]{-1+i}}{\sqrt{1+\sqrt{3}i}}.$$

Solution to the problem 2.46.

- $z = \sqrt[5]{1-i\sqrt{3}}$

We must express the number $1-i\sqrt{3}$ in polar or exponential form to calculate its fifth roots. Expressed in polar form:

$$z = \sqrt[5]{1-i\sqrt{3}} = \sqrt[5]{2_{-\frac{\pi}{3}}}.$$

The first root will have a modulus of the fifth root of two and an argument of $-\frac{\pi}{3}$ divided by 5:

$$z_1 = \sqrt[5]{2}_{-\frac{\left(\frac{\pi}{3}\right)}{5}} = \sqrt[5]{2}_{-\frac{\pi}{15}}.$$

The other four roots will have the same modulus, and their angles will differ by $\frac{2\pi}{5}$ radians. Then:

$$z_2 = \sqrt[5]{2}_{\frac{5\pi}{15}} = \sqrt[5]{2}_{\frac{\pi}{3}},$$
$$z_3 = \sqrt[5]{2}_{\frac{11}{15}\pi},$$
$$z_4 = \sqrt[5]{2}_{\frac{17}{15}\pi},$$
$$z_5 = \sqrt[5]{2}_{\frac{23}{15}\pi}.$$

Therefore, the five solutions are:

$$z_1 = \sqrt[5]{2}_{-\frac{\pi}{15}},$$
$$z_2 = \sqrt[5]{2}_{\frac{\pi}{3}},$$
$$z_3 = \sqrt[5]{2}_{\frac{11}{15}\pi},$$
$$z_4 = \sqrt[5]{2}_{\frac{17}{15}\pi},$$
$$z_5 = \sqrt[5]{2}_{\frac{23}{15}\pi}.$$

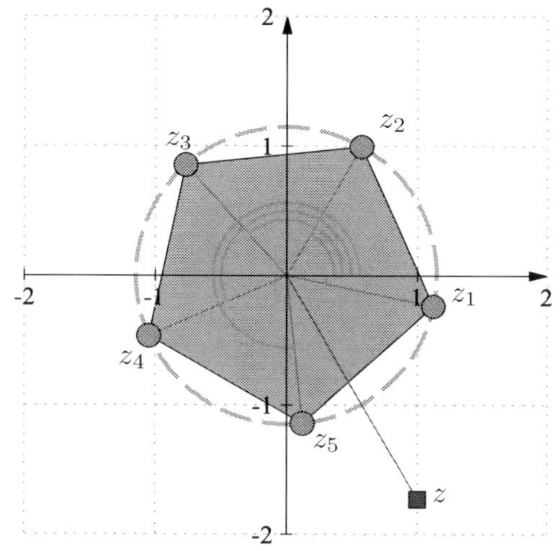

FIGURE 2.69
Fifth roots of the number $1 - i\sqrt{3}$ (Problem 2.46).

- $z = (1-i)^{\frac{5}{4}}$

  Observe that a number raised to the power of $\frac{5}{4}$ is the fourth roots of the number raised to the power of 5:

  $$z = (1-i)^{\frac{5}{4}} = (1-i)^{5\cdot\frac{1}{4}} = \left((1-i)^5\right)^{\frac{1}{4}} = \sqrt[4]{(1-i)^5}.$$

  Therefore, we need to calculate $(1-i)$ raised to the power of 5 and then calculate the fourth roots of the result. To raise to the power of 5, we operate in polar or exponential form. We convert $(1-i)$ to exponential form:

  $$1 - i = \sqrt{2}e^{-\frac{\pi}{4}i}.$$

  We raise to the power of 5:

  $$\left(\sqrt{2}e^{-\frac{\pi}{4}i}\right)^5 = 2^{\frac{5}{2}}e^{-\frac{5\pi}{4}i}.$$

  Next, we calculate the first fourth root:

  $$z_1 = \sqrt[4]{2^{\frac{5}{2}}}e^{-\frac{5\pi}{4}\cdot\frac{1}{4}i} = 2^{\frac{5}{8}}e^{-\frac{5\pi}{16}i}.$$

The other three roots will have the same modulus, and their arguments will differ by $\frac{2\pi}{4} = \frac{\pi}{2}$ *radians. Then:*

$$z_2 = 2^{\frac{5}{8}} e^{-\frac{5\pi}{16} + \frac{2\pi}{4} i} = 2^{\frac{5}{8}} e^{-\frac{5\pi}{16} + \frac{8\pi}{16} i} = 2^{\frac{5}{8}} e^{\frac{3\pi}{16} i}$$

$$z_3 = 2^{\frac{5}{8}} e^{\frac{3\pi}{16} + \frac{2\pi}{4} i} = 2^{\frac{5}{8}} e^{\frac{3\pi}{16} + \frac{8\pi}{16} i} = 2^{\frac{5}{8}} e^{\frac{11\pi}{16} i}$$

$$z_4 = 2^{\frac{5}{8}} e^{\frac{11\pi}{16} + \frac{2\pi}{4} i} = 2^{\frac{5}{8}} e^{\frac{11\pi}{16} + \frac{8\pi}{16} i} = 2^{\frac{5}{8}} e^{\frac{19\pi}{16} i}$$

Finally:

$$z_1 = \left(2^{\frac{5}{8}}\right)_{-\frac{5}{16}\pi}$$

$$z_2 = \left(2^{\frac{5}{8}}\right)_{\frac{3}{16}\pi}$$

$$z_3 = \left(2^{\frac{5}{8}}\right)_{\frac{11}{16}\pi}$$

$$z_4 = \left(2^{\frac{5}{8}}\right)_{\frac{19}{16}\pi}.$$

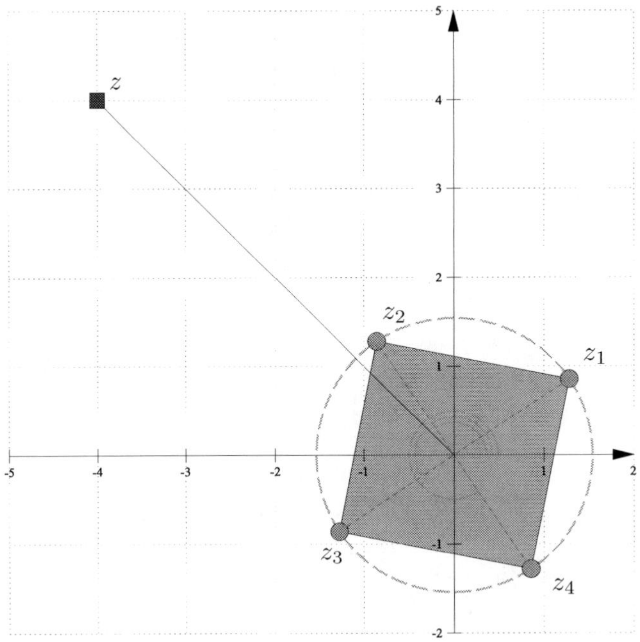

FIGURE 2.70
Fourth roots of $2^{\frac{5}{2}} e^{-\frac{5\pi}{4} i}$ (Problem 2.46).

- $z = \dfrac{\sqrt[3]{-1+i}}{\sqrt{1+\sqrt{3}i}}$

  In this case, we need to calculate the roots of two numbers, so we must operate in polar or exponential form. We convert the numbers to exponential form:

$$-1 + i = \sqrt{2} e^{\frac{3\pi}{4} i}$$

$$1 + \sqrt{3}i = 2e^{\frac{\pi}{3} i}.$$

We calculate the roots:

$$\sqrt[3]{\sqrt{2}e^{\frac{3\pi}{4}i}} = \sqrt[3]{\sqrt{2}}e^{\left(\frac{\pi}{4}+\frac{2\pi k}{3}\right)i}, \quad k=0,1,2$$

$$\sqrt{2e^{\frac{\pi}{3}i}} = \sqrt{2}e^{\left(\frac{\pi}{6}+\pi l\right)i}, \quad l=0,1.$$

Next, we calculate the division:

$$z = \frac{2^{\frac{1}{6}}e^{\left(\frac{\pi}{4}+\frac{2\pi k}{3}\right)i}}{2^{\frac{1}{2}}e^{\left(\frac{\pi}{6}+\pi l\right)i}} = 2^{\frac{1}{6}-\frac{1}{2}}e^{\left(\frac{\pi}{4}+\frac{2\pi k}{3}-\left(\frac{\pi}{6}+\pi l\right)\right)i}$$

$$= 2^{-\frac{4}{12}}e^{\left(\frac{\pi}{4}+\frac{2\pi k}{3}-\frac{\pi}{6}-\pi l\right)i}$$

$$= 2^{-\frac{1}{3}}e^{\left(\frac{\pi}{12}+\frac{2\pi k}{3}-\pi l\right)i}, \quad k=0,1,2, \quad l=0,1.$$

Note that we have an expression with three possible values of k and two possible values of l. Therefore, there will be $3 \cdot 2 = 6$ solutions. We calculate the solutions:

for $k=0$ and $l=0$: $z_1 = 2^{-\frac{1}{3}}e^{\frac{\pi}{12}i}$,

for $k=0$ and $l=1$: $z_2 = 2^{-\frac{1}{3}}e^{\left(\frac{\pi}{12}-\pi\right)i} = 2^{-\frac{1}{3}}e^{-\frac{11\pi}{12}i}$,

for $k=1$ and $l=0$: $z_3 = 2^{-\frac{1}{3}}e^{\left(\frac{\pi}{12}+\frac{2\pi}{3}\right)i}, = 2^{-\frac{1}{3}}e^{\frac{9\pi}{12}i}$

for $k=1$ and $l=1$: $z_4 = 2^{-\frac{1}{3}}e^{\left(\frac{\pi}{12}+\frac{2\pi}{3}-\pi\right)i} = 2^{-\frac{1}{3}}e^{-\frac{3\pi}{12}i} = 2^{-\frac{1}{3}}e^{-\frac{\pi}{4}i}$,

for $k=2$ and $l=0$: $z_5 = 2^{-\frac{1}{3}}e^{\left(\frac{\pi}{12}+\frac{2\pi\cdot 2}{3}\right)i} = 2^{-\frac{1}{3}}e^{\frac{17\pi}{12}i}$,

for $k=2$ and $l=1$: $z_6 = 2^{-\frac{1}{3}}e^{\left(\frac{\pi}{12}+\frac{2\pi\cdot 2}{3}-\pi\right)i} = 2^{-\frac{1}{3}}e^{\frac{5\pi}{12}i}$.

Problem 2.47. Solve in $\mathbb{C}$ the following equations:

(a) $x^2+ix+1=0$;

(b) $x^4+x^2+1=0$;

(c) $z^3-iz+(-1+i)=0$;

(d) $z^3+(1+i)z^2+(-2+i)z-2i=0$.

Solution to the problem 2.47. *(a)* $x^2+ix+1=0$

In this section, we need to solve a quadratic equation. To do this, we will use the formula that allows us to calculate the solutions of an equation of the form $ax^2+bx+c=0$. In this case, $a=1$, $b=i$, and $c=1$; therefore:

$$\begin{aligned} x &= \frac{-b\pm\sqrt{b^2-4ac}}{2a} = \frac{-i\pm\sqrt{-1-4}}{2} \\ &= \frac{-i\pm\sqrt{-5}}{2} = \frac{-i\pm\sqrt{5}\sqrt{-1}}{2} = \frac{-i\pm\sqrt{5}i}{2} = \frac{-1\pm\sqrt{5}}{2}i. \end{aligned}$$

Thus, the two solutions are $\frac{-1+\sqrt{5}}{2}i$ and $\frac{-1-\sqrt{5}}{2}i$.

(b) $x^4+x^2+1=0$

In this section, we need to solve a quartic equation, but we can convert it into a quadratic equation by substituting $y = x^2$. *This way, the equation becomes:*

$$x^4 + x^2 + 1 = (x^2)^2 + (x^2) + 1 = y^2 + y + 1 = 0.$$

To solve it, we will use the formula that allows us to calculate the solutions of an equation of the form $ax^2 + bx + c = 0$. *In this case,* $a = 1$, $b = 1$, *and* $c = 1$; *therefore:*

$$\begin{aligned} x &= \frac{-b \pm \sqrt{b^2 - 4ac}}{2a} = \frac{-1 \pm \sqrt{1-4}}{2} \\ &= \frac{-1 \pm \sqrt{-3}}{2} = \frac{-1 \pm \sqrt{3}\sqrt{-1}}{2} = \frac{-1 \pm \sqrt{3}i}{2} = -\frac{1}{2} \pm \frac{\sqrt{3}}{2}i. \end{aligned}$$

Thus, the two solutions are $y_1 = -\frac{1}{2} + \frac{\sqrt{3}}{2}i$ *and* $y_2 = -\frac{1}{2} - \frac{\sqrt{3}}{2}i$.

Next, we undo the substitution:

$$x_1 = \sqrt{y_1} = \sqrt{-\frac{1}{2} + \frac{\sqrt{3}}{2}i},$$

$$x_2 = \sqrt{y_2} = \sqrt{-\frac{1}{2} - \frac{\sqrt{3}}{2}i}.$$

To calculate these roots, we will operate in exponential form:

$$x_1 = \sqrt{-\frac{1}{2} + \frac{\sqrt{3}}{2}i} = \sqrt{1e^{\frac{2\pi}{3}i}},$$

$$x_2 = \sqrt{-\frac{1}{2} - \frac{\sqrt{3}}{2}i} = \sqrt{1e^{\frac{4\pi}{3}i}}.$$

We calculate the roots of x_1:

$$x_{11} = \sqrt{1e^{\frac{2\pi}{3}i}} = \sqrt{1}e^{\frac{2\pi}{3}\frac{1}{2}i} = 1e^{\frac{\pi}{3}i},$$

$$x_{12} = 1e^{\left(\frac{\pi}{3} + \frac{2\pi}{2}\right)i} = 1e^{\frac{4\pi}{3}i}.$$

We calculate the roots of x_2:

$$x_{21} = \sqrt{1e^{\frac{4\pi}{3}i}} = \sqrt{1}e^{\frac{4\pi}{3}\frac{1}{2}i} = 1e^{\frac{2\pi}{3}i},$$

$$x_{22} = 1e^{\left(\frac{2\pi}{3} + \frac{2\pi}{2}\right)i} = 1e^{\frac{5\pi}{3}i}.$$

Thus, the solutions of the equation $x^4 + x^2 + 1 = 0$ *are:*

$$x_{11} = 1e^{\frac{\pi}{3}i} = \frac{1}{2} + \frac{\sqrt{3}}{2}i,$$

$$x_{12} = 1e^{\frac{4\pi}{3}i} = -\frac{1}{2} - \frac{\sqrt{3}}{2}i,$$

$$x_{21} = 1e^{\frac{2\pi}{3}i} = -\frac{1}{2} + \frac{\sqrt{3}}{2}i,$$

$$x_{22} = 1e^{\frac{5\pi}{3}i} = \frac{1}{2} - \frac{\sqrt{3}}{2}i.$$

(c) $z^3 - iz + (-1+i) = 0$

In this case, we need to solve a cubic equation, so we cannot use the formula used in the previous sections, as it only works for equations of the form $ax^2 + bx + c = 0$. We can try to solve the equation using Ruffini's rule:

$$\begin{array}{c|cccc}
 & 1 & 0 & -i & (-1+i) \\
1 & & 1 & 1 & (1-i) \\
\hline
 & 1 & 1 & (1-i) & 0 \\
i & & i & (i-1) & \\
\hline
 & 1 & (1+i) & 0 &
\end{array}$$

We can factor the polynomial as:

$$z^3 - iz + (-1+i) = (z + (1+i))(z-i)(z-1) = 0$$

Therefore, the solutions of the equation are:

$$\begin{aligned} z_1 &= -1 - i, \\ z_2 &= i, \\ z_3 &= 1. \end{aligned}$$

(d) $z^3 + (1+i)z^2 + (-2+i)z - 2i = 0$
To solve this equation, we will use Ruffini's rule, as in the previous section:

$$\begin{array}{c|cccc}
 & 1 & (1+i) & (-2+i) & -2i \\
1 & & 1 & (2+i) & 2i \\
\hline
 & 1 & (2+i) & 2i & 0 \\
-2 & & -2 & -2i & \\
\hline
 & 1 & i & 0 &
\end{array}$$

We can factor the polynomial as:

$$z^3 + (1+i)z^2 + (-2+i)z - 2i = (z+i)(z+2)(z-1) = 0$$

Therefore, the solutions of the equation are:

$$\begin{aligned} z_1 &= -i, \\ z_2 &= -2, \\ z_3 &= 1. \end{aligned}$$

Problem 2.48. Calculate:

(a) $z = \dfrac{3}{i} + \dfrac{2i}{2+i} + \dfrac{2+3i}{1+2i}$;

(b) The real numbers $a, b \in \mathbb{R}$, knowing that:

$$(-5 + ai)(b + 19i) = 3 + 2i;$$

(c) The real number $a \in \mathbb{R}$ such that $z = \dfrac{3-2ai}{4-3i}$ is located on the bisector of the first quadrant.

Solution to the problem 2.48. *(a)* $z = \frac{3}{i} + \frac{2i}{2+i} + \frac{2+3i}{1+2i}$

To sum the fractions, we simplify them by multiplying and dividing by the conjugate of the denominator:

$$
\begin{aligned}
z &= \frac{3}{i}\frac{-i}{-i} + \frac{2i}{2+i}\frac{2-i}{2-i} + \frac{2+3i}{1+2i}\frac{1-2i}{1-2i} \\
&= \frac{-3i}{-(i^2)} + \frac{4i-2(i)^2}{2^2-(i)^2} + \frac{2+3i-4i-6(i)^{2\,2}}{(1)^2-2^2(i} \\
&= \frac{-3i}{-(-1)} + \frac{4i-2(-1)}{4-(-1)} + \frac{2-i-6(-1)}{1-4(-1)} \\
&= -3i + \frac{2-4i}{5} + \frac{8-i}{5}.
\end{aligned}
$$

Next, we reduce the fractions to a common denominator and add them:

$$
\begin{aligned}
z &= -3i + \frac{2-4i}{5} + \frac{8-i}{5} \\
&= \frac{-15i}{5} + \frac{2-4i}{5} + \frac{8-i}{5} \\
&= \frac{10-12i}{5} \\
&= 2 - \frac{12}{5}i.
\end{aligned}
$$

(b) The real numbers $a, b \in \mathbb{R}$*, knowing that:*

$$(-5+ai)(b+19i) = 3+2i$$

First, we convert the product $(-5+ai)(b+19i)$ *to binomial form:*

$$
\begin{aligned}
(-5+ai)(b+19i) &= -5b + abi - 95i + 19a(i)^2 \\
&= -5b + abi - 95i - 19a \\
&= (-19a - 5b) + (ab - 95)i.
\end{aligned}
$$

The real part and the imaginary part of the number $(-19a-5b)+(ab-95)i$ *must match the real part and the imaginary part of the number* $3+2i$*. Therefore:*

$$-19a - 5b = 3 \ \Leftrightarrow\ ab - 95 = 2.$$

We need to solve the above system of equations

. By operating on the first equation:

$$
\begin{aligned}
-19a - 5b = 3 &\ \Leftrightarrow\ -19a = 3 + 5b \\
&\ \Leftrightarrow\ a = -\frac{3}{19} - \frac{5}{19}b.
\end{aligned}
$$

Substituting a *into the second equation:*

$$
\begin{aligned}
ab - 95 = 2 &\ \Leftrightarrow\ \left(-\frac{3}{19} - \frac{5}{19}b\right) \cdot b - 95 = 2 \\
&\ \Leftrightarrow\ -\frac{3}{19}b - \frac{5}{19}b^2 - 95 = 2 \\
&\ \Leftrightarrow\ -\frac{5}{19}b^2 - \frac{3}{19}b - 97 = 0.
\end{aligned}
$$

To calculate b, we need to solve the above quadratic equation. To do this, we will use the formula that allows us to calculate the solutions of an equation of the form $ax^2 + bx + c = 0$*. In this case,* $a = -\frac{5}{19}$, $b = -\frac{3}{19}$*, and* $c = -97$*; therefore:*

$$\begin{aligned} x &= \frac{-b \pm \sqrt{b^2 - 4ac}}{2a} = \frac{-(-\frac{3}{19}) \pm \sqrt{(-\frac{3}{19})^2 - 4(-\frac{5}{19})(-97)}}{2(-\frac{5}{19})} \\ &= \frac{\frac{3}{19} \pm \sqrt{\frac{9}{361} - \frac{1940}{19}}}{-\frac{10}{19}} = \frac{\frac{3}{19} \pm \sqrt{\frac{9}{361} - \frac{36860}{361}}}{-\frac{10}{19}} = \frac{\frac{3}{19} \pm \sqrt{\frac{-36851}{361}}}{-\frac{10}{19}} \\ &= \frac{\frac{3}{19} \pm \sqrt{\frac{36851}{361}}i}{-\frac{10}{19}} = \frac{\frac{3}{19}}{-\frac{10}{19}} \pm \frac{\frac{\sqrt{36851}}{19}}{-\frac{10}{19}}i = -\frac{3}{10} \pm \left(-\frac{\sqrt{36851}}{10}i\right). \end{aligned}$$

Thus, the two solutions are:

$$-\frac{3}{10} + \frac{\sqrt{36851}}{10}i, \qquad -\frac{3}{10} - \frac{\sqrt{36851}}{10}i.$$

The statement asks us to find the real numbers $a, b \in \mathbb{R}$ *that satisfy:*

$$(-5 + ai)(b + 19i) = 3 + 2i,$$

but the solutions we have found for b are complex numbers. Therefore, we can conclude that there are no real a and b that satisfy the above property.

(c) *The real number* $a \in \mathbb{R}$ *such that* $z = \dfrac{3 - 2ai}{4 - 3i}$ *is located on the bisector of the first quadrant.*

We start by simplifying the fraction by multiplying and dividing by the conjugate of the denominator:

$$\begin{aligned} z &= \frac{3 - 2ai}{4 - 3i} = \frac{3 - 2ai}{4 - 3i}\frac{4 + 3i}{4 + 3i} \\ &= \frac{12 - 8ai + 9i - 6a(i)^2}{4^2 - 3^2(i)^2} \\ &= \frac{12 + (9 - 8a)i - 6a(-1)}{16 - 9(-1)} \\ &= \frac{12 + 6a + (9 - 8a)i}{25} \\ &= \frac{12 + 6a}{25} + \frac{9 - 8a}{25}i. \end{aligned}$$

The bisector of the first quadrant is the line that passes through the origin and forms an angle of $\frac{\pi}{4}$ *radians with the x-axis. Therefore, the complex numbers located on the bisector of the first quadrant are of the form:*

$$re^{\frac{\pi}{4}i}$$

Expressed in binomial form:

$$r \cdot \cos(\frac{\pi}{4}) + r \cdot \sin(\frac{\pi}{4})i = r\frac{\sqrt{2}}{2} + r\frac{\sqrt{2}}{2}i$$

Therefore, they are all numbers whose real and imaginary parts are equal and positive. So, we need to impose that the real and imaginary parts of the number:

$$\frac{12+6a}{25}+\frac{9-8a}{25}i$$

are equal and positive for it to be located on the bisector of the first quadrant:

$$\begin{aligned} Re(z) &= Im(z); \\ \frac{12+6a}{25} &= \frac{9-8a}{25}; \\ 12+6a &= 9-8a; \end{aligned}$$

$$14a=-3 \quad \Rightarrow \quad a=-\frac{3}{14}.$$

Therefore, the value of a for $z=\dfrac{3-2ai}{4-3i}$ *to be located on the bisector of the first quadrant is:*

$$a=-\frac{3}{14}.$$

Note that with this value of a, we get $Re(z)=Im(z)=\frac{3}{7}>0$.

Problem 2.49. Prove that if z is a complex number, the quotient:

$$\frac{z-\overline{z}}{z+\overline{z}}$$

is a pure imaginary number.

Solution to the problem 2.49. *If z is a complex number, we can express it in binomial form as follows:*

$$z=a+bi,$$

where a is the real part and b is the imaginary part, with a and $b \in \mathbb{R}$.

Its conjugate is:

$$\overline{z}=a-bi.$$

Then:

$$\frac{z-\overline{z}}{z+\overline{z}}=\frac{(a+bi)-(a-bi)}{(a+bi)+(a-bi)}=\frac{2bi}{2a}=\frac{b}{a}i=0+\frac{b}{a}i.$$

We can see that the real part is 0 and, therefore, it is proven that the quotient:

$$\frac{z-\overline{z}}{z+\overline{z}}$$

is a pure imaginary number.

Problem 2.50. Determine a necessary and sufficient condition for two complex numbers $z_1, z_2 \in \mathbb{C}$ to satisfy the equality:

$$|z_1+z_2|=|z_1|+|z_2|.$$

Solution to the problem 2.50. *Since we have to add the two numbers, it is convenient to express them in binomial form, that is, we consider* $z_1 = a_1 + b_1$ *and* $z_2 = a_2 + b_2 i$.

Then, the equality $|z_1 + z_2| = |z_1| + |z_2|$ *leads to:*

$$\sqrt{(a_1 + a_2)^2 + (b_1 + b_2)^2} = \sqrt{a_1^2 + b_1^2} + \sqrt{a_2^2 + b_2^2},$$

and squaring both sides, we have:

$$(a_1 + a_2)^2 + (b_1 + b_2)^2 = a_1^2 + b_1^2 + a_2^2 + b_2^2 + 2\sqrt{a_1^2 + b_1^2}\sqrt{a_2^2 + b_2^2}.$$

Therefore:

$$\begin{aligned} -2\sqrt{(a_1^2 + b_1^2)(a_2^2 + b_2^2)} &= a_1^2 + b_1^2 + a_2^2 + b_2^2 - (a_1 + a_2)^2 - (b_1 + b_2)^2 \\ &= -2a_1 a_2 - 2b_1 b_2. \end{aligned}$$

Then, dividing by -2 *and squaring again:*

$$(a_1^2 + b_1^2)(a_2^2 + b_2^2) = (a_1 a_2 + b_1 b_2)^2,$$

from which, simplifying, we obtain:

$$\begin{aligned} 0 &= (a_1^2 + b_1^2)(a_2^2 + b_2^2) - (a_1 a_2 + b_1 b_2)^2 \\ &= a_1^2 a_2^2 + a_1^2 b_2^2 + b_1^2 a_2^2 + b_1^2 b_2^2 - \left(a_1^2 a_2^2 + 2a_1 a_2 b_1 b_2 + b_1^2 b_2^2\right) \\ &= a_1^2 b_2^2 + b_1^2 a_2^2 - 2a_1 a_2 b_1 b_2 = \left(a_1 b_2 - a_2 b_1\right)^2. \end{aligned}$$

Therefore, we arrive at the condition that a_1, a_2, b_1 *and* b_2 *must satisfy:*

$$a_1 b_2 - a_2 b_1 = 0,$$

which is equivalent to imposing that:

$$\begin{vmatrix} a_1 & b_1 \\ a_2 & b_2 \end{vmatrix} = 0.$$

This condition tells us that the two vectors (a_1, b_1) *and* (b_1, b_2) *must be parallel, that is, the two numbers must be aligned. Thus, for the condition:*

$$|z_1 + z_2| = |z_1| + |z_2|,$$

the vectors determining z_1 *and* z_2 *must be parallel.*

Problem 2.51. (a) Determine two complex numbers z_1 and z_2, knowing that their quotient is 3, that the sum of their arguments is $\pi/3$ and that the sum of their magnitudes is 8.

(b) One of the cube roots of a certain complex number z is $3i$; calculate this complex number z and the rest of its cube roots.

Solution to the problem 2.51. *(a) We need to determine two complex numbers* z_1 *and* z_2*, knowing that their quotient is 3, that the sum of their arguments is* $\pi/3$ *and that the sum of their magnitudes is 8. From the first condition, we know that:*

$$\frac{z_1}{z_2} = \frac{r_1 e^{\theta 1 i}}{r_2 e^{\theta 2 i}} = \frac{r_1}{r_2} e^{(\theta 1 - \theta 2)i}.$$

The quotient of the two numbers is equal to 3, which in exponential form is $3e^{2\pi k i}$*,* $k \in \mathbb{Z}$*; therefore:*

$$\begin{aligned}
&\frac{r_1}{r_2} e^{(\theta 1 - \theta 2)i} = 3e^{2\pi k i}, \quad k \in \mathbb{Z};\\
&\frac{r_1}{r_2} = 3;\\
&(\theta 1 - \theta 2) = 2\pi k, \quad k \in \mathbb{Z};
\end{aligned}$$

Thus, the magnitudes and arguments of the numbers must satisfy the following conditions:

$$\begin{aligned}
&\frac{r_1}{r_2} = 3,\\
&\theta 1 - \theta 2 = 2\pi k, \quad k \in \mathbb{Z},\\
&\theta 1 + \theta 2 = \frac{\pi}{3},\\
&r_1 + r_2 = 8.
\end{aligned}$$

By operating on the first equation:

$$\frac{r_1}{r_2} = 3; r_1 = 3r_2.$$

Substituting in the fourth equation:

$$\begin{aligned}
&r_1 + r_2 = 8;\\
&3r_2 + r_2 = 8;\\
&4r_2 = 8;\\
&r_2 = 2;\\
&r_1 = 3r_2; r_1 = 6.
\end{aligned}$$

By operating on the second equation:

$$\begin{aligned}
&\theta 1 - \theta 2 = 2\pi k, \quad k \in \mathbb{Z};\\
&\theta 1 = \theta 2 + 2\pi k, \quad k \in \mathbb{Z}.
\end{aligned}$$

Substituting in the third equation:

$$\begin{aligned}
&\theta 1 + \theta 2 = \frac{\pi}{3};\\
&\theta 2 + 2\pi k + \theta 2 = \frac{\pi}{3}, \quad k \in \mathbb{Z};\\
&2 \cdot \theta 2 = \frac{\pi}{3} - 2\pi k, \quad k \in \mathbb{Z};\\
&\theta 2 = \frac{\pi}{6} - \pi k, \quad k \in \mathbb{Z};\\
&\theta 1 = \theta 2 + 2\pi k = \frac{\pi}{6} - \pi k + 2\pi k = \frac{\pi}{6} + \pi k, \quad k \in \mathbb{Z}.
\end{aligned}$$

Therefore, the two complex numbers that satisfy the conditions mentioned in the problem statement are:

$$z_1 = 6e^{\frac{\pi}{6}+\pi k} \quad \textit{and} \quad z_2 = 2e^{\frac{\pi}{6}-\pi k}, \quad k \in \mathbb{Z}.$$

(b) *Knowing that one of the cube roots of a certain complex number z is $3i$, we need to calculate this complex number z and the rest of its cube roots. We know that:*

$$\sqrt[3]{z} = 3i.$$

We calculate the complex number z:

$$z = (3i)^3 = 3^3(i)^3 = -27i = 27e^{-\frac{\pi}{2}}.$$

Next, we can calculate the cube roots of z, but it is not necessary since we already know one of them and we know that the other two will have the same magnitude, while the angles will differ by $\frac{2\pi}{3}$ radians. Since the first root is:

$$z_1 = 3i = 3e^{\frac{\pi}{2}i},$$

the rest of the roots will be:

$$z_2 = 3e^{\left(\frac{\pi}{2}+\frac{2\pi}{3}\right)i} = 3e^{\frac{7\pi}{6}i},$$
$$z_3 = 3e^{\left(\frac{7\pi}{6}+\frac{2\pi}{3}\right)i} = 3e^{\frac{11\pi}{6}i}.$$

Therefore, the complex number is $z = 27e^{-\frac{\pi}{2}i} = -27i$ and its cube roots are:

$$z_1 = 3e^{\frac{\pi}{2}i} = 3i, \quad z_2 = 3e^{\frac{7}{6}\pi i} = -\frac{3}{2}\sqrt{3} - \frac{3}{2}i, \quad \textit{and} \quad z_3 = 3e^{\frac{11}{6}\pi i} = \frac{3}{2}\sqrt{3} - \frac{3}{2}i.$$

Problem 2.52. Calculate $z \in \mathbb{C}$ such that z, z^{-1}, and $1 - z$ have the same magnitude.

Solution to the problem 2.52. *Consider z expressed in polar form as $z = r_\theta$.*

We have two unknowns: the magnitude and the argument. We need, therefore, two equations to solve the system. We obtain these from the conditions given in the problem statement: that z, z^{-1}, and $1 - z$ have the same magnitude.

First, we impose that $|z| = |z^{-1}|$. We have $r = \frac{1}{r}$, from which $r^2 = 1$. Since $r \geq 0$, we have $r = 1$, and therefore:

$$z = 1_\theta = \cos\theta + i\sin\theta.$$

Next, we impose that $|z| = |1 - z|$. Since $|z| = 1$ and $z = \cos\theta + i\sin\theta$, we have:

$$\begin{aligned} 1 &= |-1 + (\cos\theta + i\sin\theta)| = |(-1+\cos\theta) + i\sin\theta| \\ &= \sqrt{(-1+\cos\theta)^2 + (\sin\theta)^2}. \end{aligned}$$

Squaring the above expression:

$$\begin{aligned} 1 &= (-1+\cos\theta)^2 + (\sin\theta)^2 = 1 - 2\cos\theta + (\cos\theta)^2 + (\sin\theta)^2 \\ &= 2 - 2\cos\theta, \end{aligned}$$

Thus, $\cos\theta = \frac{1}{2}$.

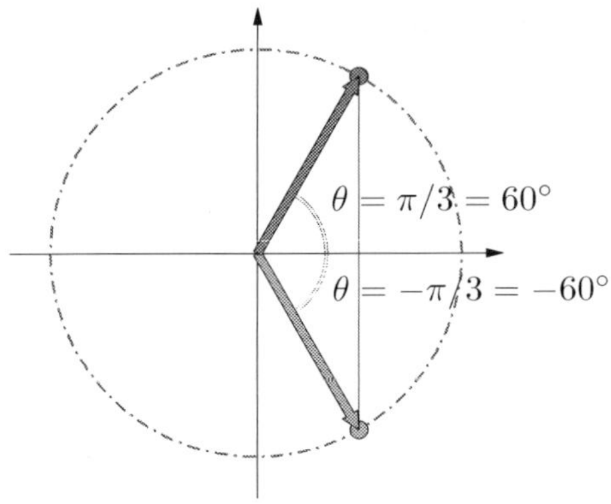

FIGURE 2.71
Angles that satisfy $\cos\theta = \frac{1}{2}$ (Problem 2.52).

We see that there are two principal arguments that satisfy $\cos\theta = \frac{1}{2}$ *(see Figure 2.71):*

$$\theta_1 = \frac{\pi}{3} \text{ and } \theta_2 = -\frac{\pi}{3}.$$

Therefore, there are two solutions:

$$z_1 = 1_{\frac{\pi}{3}} \quad \text{and} \quad z_2 = 1_{-\frac{\pi}{3}}.$$

If we evaluate z^{-1} *and* $1-z$ *for each of the values found for* z *and calculate their magnitudes, we will verify that all are* $r = 1$*, like the magnitude of* z *itself.*

Problem 2.53. Determine the complex numbers z such that:

(a) $\bar{z} = z^{-1}$;

(b) $|1 - z|^2 \leq 1 - |z|^2$;

(c) $|z - \alpha| = |z - \beta|$, $(\alpha, \beta \in \mathbb{R})$.

Solution to the problem 2.53. *(a) Let* $z = a + bi = r_\theta$*, then imposing that* $\bar{z} = z^{-1}$ *in binomial form is equivalent to imposing that:*

$$a - bi = \frac{1}{a + bi}$$

Multiplying the equality by $a + bi$*, we have:*

$$(a - bi)(a + bi) = 1 \iff a^2 - (bi)^2 = a^2 + b^2 = 1 \iff r^2 = 1.$$

Since $r \geq 0$*, the equation* $r^2 = 1$ *tells us that* $r = 1$*.*

Therefore, the numbers that satisfy $\bar{z} = z^{-1}$ *are the numbers with magnitude* 1*, that is, they are the numbers of the form* $z = 1_\theta$*.*

(b) Let $z = a + bi$*, and we impose that* $|1 - z|^2 \leq 1 - |z|^2$*. Since:*

$$|1 - z|^2 = |1 - a - bi|^2 = (1 - a)^2 + b^2,$$
$$1 - |z|^2 = 1 - (a^2 + b^2) = 1 - a^2 - b^2,$$

we want to satisfy the inequality:

$$(1-a)^2 + b^2 \le 1 - a^2 - b^2$$
$$\iff 1 + a^2 - 2a + b^2 \le 1 - a^2 - b^2$$
$$\iff 2a^2 + 2b^2 \le 2a \iff b^2 \le a - a^2.$$

If $a - a^2 = a(1-a) < 0$, the equation $b^2 \le a - a^2$ has no real solution. This occurs when $a < 0$ and when $a > 1$. That is, the equation only has a solution if the real part is within the interval $[0, 1]$.

In this case, it is also necessary that $b^2 \le a - a^2$. Since this inequality can be expressed as:

$$b^2 + (a - 1/2)^2 < 1/4$$

(by completing the square), the representation of the complex numbers that satisfy this inequality is a circle centered at the point $(1/2, 0)$ and with radius $1/2$, as shown in Figure 2.72.

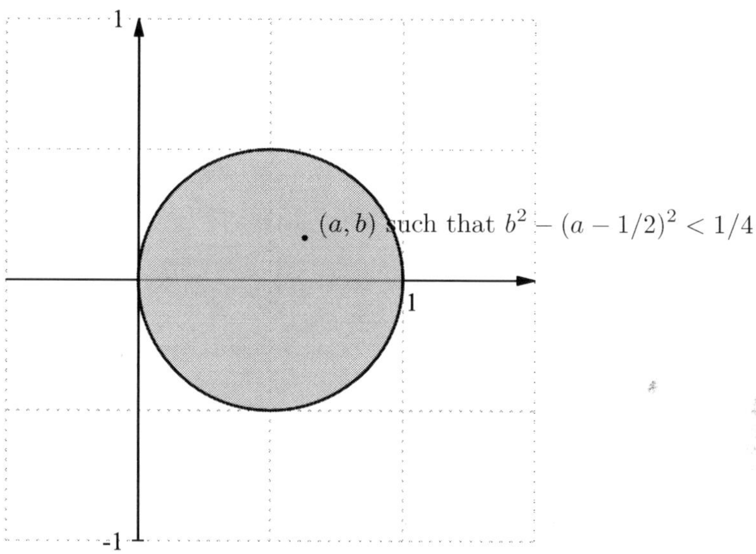

FIGURE 2.72
Region of the complex plane that contains the numbers that satisfy $|1 - z|^2 \le 1 - |z|^2$ (Problem 2.53).

(c) *Since the magnitudes of complex numbers are always positive, imposing that $|z - \alpha| = |z - \beta|$ is equivalent to imposing that $|z - \alpha|^2 = |z - \beta|^2$. Let $z = a + bi$, then we have:*

$$|z - \alpha|^2 = |a + bi - \alpha|^2 = (a - \alpha)^2 + b^2,$$
$$|z - \beta|^2 = |a + bi - \beta|^2 = (a - \beta)^2 + b^2,$$

and, therefore, if we impose that $|z - \alpha|^2 = |z - \beta|^2$, we have:

$$(a - \alpha)^2 + b^2 = (a - \beta)^2 + b^2 \iff (a - \alpha)^2 - (a - \beta)^2 = 0.$$

Simplifying the above expression:

$$0 = (a-\alpha)^2 - (a-\beta)^2 = a^2 + \alpha^2 - 2a\alpha - a^2 - \beta^2 + 2a\beta$$
$$= 2a(\beta - \alpha) + \alpha^2 - \beta^2.$$

Therefore, we get:

$$2a(\beta - \alpha) = \beta^2 - \alpha^2.$$

If $\beta = \alpha$, *then the above expression becomes* $0 = 0$, *that is, in the case where* $\alpha = \beta$, *the equality given is always true.*

If $\beta \neq \alpha$, *we can divide by this term and we obtain that:*

$$a = \frac{1}{2}\frac{\beta^2 - \alpha^2}{\beta - \alpha} = \frac{1}{2}\frac{(\beta-\alpha)(\beta+\alpha)}{\beta - \alpha} = \frac{1}{2}(\beta + \alpha).$$

Therefore, the answer to the problem is:

if $\alpha = \beta$ *the equality is true for all* $z \in \mathbb{C}$

if $\alpha \neq \beta$ $z = \frac{1}{2}(\beta + \alpha) + bi, b \in \mathbb{R}$.

Part III

Differential Calculus

3

Functions, Limits, and Continuity

«My special pleasure in mathematics rested particularly on its purely speculative part.» (Bernard Bolzano)

This third chapter is fundamental, as it contains a key concept in mathematics: the concept of *limit*. The limit of a function at a point characterizes the behavior of a function in the vicinity of that point. Although limits already appeared in the development of calculus in the 17th and 18th centuries, the current definition and notation, which we present in this book, correspond to Bernard Bolzano, a Czech mathematician (1781–1848), who already used the so-called *epsilon-delta* technique. Before that, some aspects of real functions of a real variable, such as the domain and the range, will be recalled. The concept of a continuous function at a point will also be discussed. The idea of the limit will also be used in Chapters 4 and 5 when defining the derivative and the Riemann integral.

DOI: 10.1201/9781003399735-3

3.1 Real Functions of a Real Variable

3.1.1 Concept of Function. Injectivity. Surjectivity. Bijectivity

Definition 3.1 (Function). A function of one variable is a correspondence between two sets:

$$\begin{aligned} f : A &\longrightarrow B \\ x &\longrightarrow f(x) = y \end{aligned}$$

In Figure 3.1, you can see an example of a function (left) and also a curve that is not the graph of any function (right).

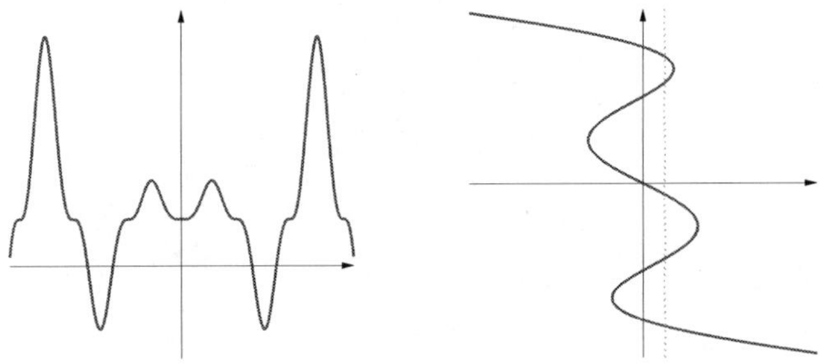

FIGURE 3.1
Example of a function (left) and example of a curve in the plane that does not represent the graph of any function (right).

In this book, we will only consider real functions of a real variable, that is:

$$f : A \subset \mathbb{R} \longrightarrow \mathbb{R}.$$

Definition 3.2 (Injective Function). A function $f : A \subset \mathbb{R} \longrightarrow B \subset \mathbb{R}$ is injective if two different numbers $x_1, x_2 \in A$, $x_1 \neq x_2$, have different images $f(x_1) \neq f(x_2)$. Or, in other words, if the equality of two images $f(x_1) = f(x_2)$ implies the equality of the points $x_1 = x_2$.

Example 3.1. *The function $f : \mathbb{R} \longrightarrow \mathbb{R}$ defined as $f(x) = x^2$ is not injective, since the points $x_1 = 1$ and $x_2 = -1$ have the same image $f(1) = f(-1) = 1$.*

Example 3.2. *The function $f : \mathbb{R} \longrightarrow \mathbb{R}$ defined as $f(x) = 2x + 7$ is injective, since if $f(x_1) = f(x_2)$, this implies that $2x_1 + 7 = 2x_2 + 7$, which in turn implies that $x_1 = x_2$.*

Remark 3.1. The injectivity of a function depends on the set A over which it is defined. For example, the function $f : [0, +\infty) \longrightarrow \mathbb{R}$ defined as $f(x) = x^2$ is indeed injective.

Proposed Exercise 3.1. Prove that the function $f : [0, +\infty) \longrightarrow \mathbb{R}$ defined as $f(x) = x^2$ is injective.

Definition 3.3 (Surjective Function). A function $f : A \subset \mathbb{R} \longrightarrow B \subset \mathbb{R}$ is surjective if for every element $y \in B$ there exists an element $x \in A$ such that:

$$f(x) = y.$$

Remark 3.2. The surjectivity of a function depends on the set B over which it is defined. *See* Example 3.3.

Example 3.3. *The function $f : [0, +\infty) \longrightarrow \mathbb{R}$ defined as $f(x) = x^2$ is not surjective because there are values in $\mathbb{R}$, such as $y = -1$ or $y = -2$ for which there is no number $x \in [0, +\infty)$ such that $f(x) = x^2 = -1$ or $f(x) = x^2 = -2$.*

Example 3.4. *If we consider instead the function $f : \mathbb{R} \longrightarrow [0, +\infty) \subset \mathbb{R}$ defined as $f(x) = x^2$, now we can say that it is surjective, since for every $y \in [0, +\infty)$ there is at least one number $x \in \mathbb{R}$ such that:*

$$f(x) = x^2 = y > 0.$$

In fact, in this case, there are two numbers, which are:

$$x = \pm\sqrt{y}.$$

3.1.2 Important Sets

Definition 3.4 (Domain). The domain of a function $y = f(x)$ is the set of values x for which $f(x)$ is defined. It is generally denoted by $\mathrm{Dom}(f)$. In the case of real functions of a real variable, the domain of a function is all the real numbers that have a real number as their image.

In Table 3.1 the domains of some functions are summarized.

Definition 3.5 (Image). The image or range of the function $f : A \subset \mathbb{R} \to B \subset \mathbb{R}$ is defined as the set of values $f(x) \in B$ for all $x \in A$. That is, it is the set:

$$\mathrm{Im}(f) = f(A) = \{y \in B \mid y = f(x),\ x \in A\}.$$

You can see in Figure 3.2 an example of what we consider domain and image.

Definition 3.6 (Equality of Functions). Two functions f and g are equal if and only if:

(i) $\mathrm{Dom}(f) = \mathrm{Dom}(g)$ and

(ii) $f(x) = g(x),\ \forall x \in \mathrm{Dom}(f) = \mathrm{Dom}(g)$.

Example 3.5. *As an example, in Table 3.2 you can find the image of some simple functions.*

TABLE 3.1
Domain of some important functions. Note that $\mathbb{R}[x]$ refers to the set of polynomials with real coefficients.

Function	**Domain**
$f(x) = p(x) \in \mathbb{R}[x]$	$\mathrm{Dom}(f) = \mathbb{R}$
$f(x) = \frac{p(x)}{q(x)},\ p(x), q(x) \in \mathbb{R}[x]$	$\mathrm{Dom}(f) = \mathbb{R} - \{x \mid q(x) = 0\}$
$f(x) = \sqrt[2n]{g(x)}$	$\mathrm{Dom}(f) = \{x \mid g(x) \geq 0\}$
$f(x) = \sqrt[2n+1]{g(x)}$	$\mathrm{Dom}(f) = \mathrm{Dom}(g)$
$f(x) = \log(g(x))$	$\mathrm{Dom}(f) = \{x \mid g(x) > 0\}$
$f(x) = \sin(g(x))$	$\mathrm{Dom}(f) = \mathrm{Dom}(g)$
$f(x) = \cos(g(x))$	$\mathrm{Dom}(f) = \mathrm{Dom}(g)$
$f(x) = \tan(g(x))$	$\mathrm{Dom}(f) = \mathrm{Dom}(g) - \{x \mid \cos(g(x)) = 0\}$
$f(x) = e^{g(x)}$	$\mathrm{Dom}(f) = \mathrm{Dom}(g)$

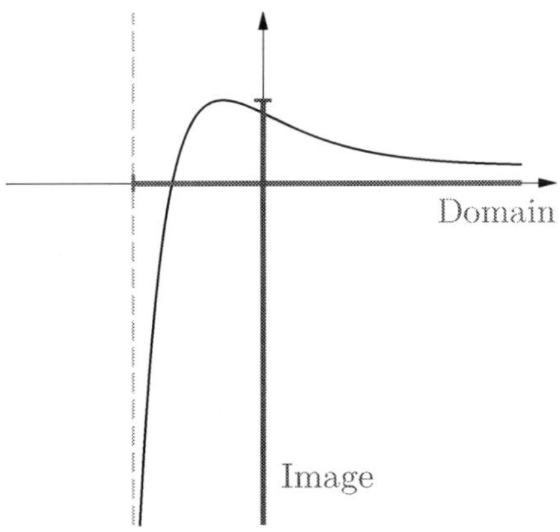

FIGURE 3.2
In this figure, the domain of the function is represented (on the x-axis) and its image (on the y-axis).

Definition 3.7 (Graph of the Function). Given a function $f : A \subset \mathbb{R} \to B \subset \mathbb{R}$, we call the *graph* of the function the set of points in the plane defined as:

$$G_f = \{(x, f(x)) \mid x \in A\} \subset A \times B \subset \mathbb{R}^2.$$

We must not confuse the graph of the function with the function itself, although sometimes we identify them.

Definition 3.8 (Preimage). Given a function $f : A \subset \mathbb{R} \to B \subset \mathbb{R}$, and given an element $y \in B$, the set of preimages of y, denoted by $f^{-1}(y)$, is defined as the set of elements $x \in A$ such that $f(x) = y$, that is:

$$f^{-1}(y) = \{x \in A \mid f(x) = y\}.$$

The set of preimages can be an empty set, a set with a single element, with several elements, or with infinite elements. An example of this can be seen in Figure 3.3.

TABLE 3.2
Image or range of some functions.

Function	Image
$f(x) = 2x + 1$	$\text{Im}(f) = \mathbb{R}$
$f(x) = \sin(x)$	$\text{Im}(f) = [-1, 1]$
$f(x) = \frac{1}{x+1}$	$\text{Im}(f) = \mathbb{R} - \{0\}$

Example 3.6. *Consider the function $f : \mathbb{R} \to \mathbb{R}$ such that $f(x) = x^2$. In this case, we have that:*

$$f^{-1}(-1) = \emptyset$$

since there is no real number x such that $f(x) = x^2 = -1$;

$$f^{-1}(0) = 0$$

since $f(0) = 0$ and, finally:

$$f^{-1}(1) = \{-1, 1\}$$

since $f(-1) = f(1) = 1$.

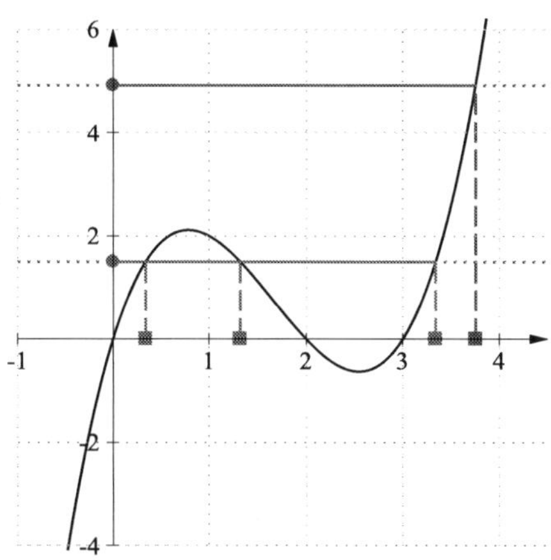

FIGURE 3.3
Graph of the function $f(x) = x^3 - 5x^2 + 6x$. It can be seen that there are points that have three preimages and other points that have a single preimage.

Example 3.7. *If we want to calculate the set of preimages of a point y of the function $f(x) = 2x - 1$, we take $y = f(x)$ and solve for x in terms of y. Then:*

$$y = 2x - 1 \quad \Longleftrightarrow \quad x = \frac{y+1}{2} \quad \Longleftrightarrow \quad f^{-1}(y) = \frac{y+1}{2}$$

3.1.3 Invertible Functions

If for some $y \in f(A)$, the set $f^{-1}(y)$ consists of more than one element, then $f^{-1} : f(A) \longrightarrow A$ is not a function.

Otherwise, if $f^{-1}(y)$ always consists of a single element, f^{-1} is called the **inverse function** of f.

Example 3.8 (of invertible functions). *Here are some invertible functions and their inverses.*

- *The inverse of the function* $f(x) = 2x + 1$ *is the function* $f^{-1}(x) = \frac{x-1}{2}$, *since:*

$$y = 2x + 1 \quad \Longleftrightarrow \quad x = \frac{y-1}{2} \quad \Longleftrightarrow \quad f^{-1}(y) = \frac{y-1}{2}$$

- *The inverse of the function* $f(x) = e^x + 1$ *is the function* $f^{-1}(x) = \ln(x-1)$, *since:*

$$y = e^x + 1 \quad \Longleftrightarrow \quad x = \ln(y-1) \quad \Longleftrightarrow \quad f^{-1}(y) = \ln(y-1)$$

Note that although in the development of the inverse function, it depends on y, *finally the inverse is expressed as a function of the variable* x.

Example 3.9 (of a non-invertible function). *The function* $f(x) = x^2$ *is a non-invertible function because, by isolating the* y *from the expression* $y = x^2$ *we obtain:*

$$y = x^2 \quad \Longleftrightarrow \quad x = \pm\sqrt{y}$$

Therefore:

- *If* $y < 0$, $f^{-1}(y) = \emptyset$;
- *if* $y = 0$, $f^{-1}(y) = \{0\}$; *and*
- *if* $y > 0$, $f^{-1}(y)$ *is a set formed by two elements* $\{-\sqrt{y}, \sqrt{y}\}$.

3.1.4 Restriction of a Function

Sometimes it may be interesting to reduce the domain of a function to calculate an inverse function or make the function bijective. This will be achieved by restricting the function.

Definition 3.9. If $X \subset A$, the **restriction** of f to X is the application:

$$f_{|X} : X \longrightarrow B$$

Example 3.10. *Consider the function* $f : \mathbb{R} \longrightarrow [0, \infty)$ *such that* $f(x) = x^2$. *Calculate the preimage of all points* $y \in [0, \infty)$ *and the inverse function, if it exists.*

Solution.

Given $y \geq 0$, the set of preimages will be the points $x \in \mathbb{R}$ such that $f(x) = y$, that is, such that $x^2 = y$.

Therefore:

$$\begin{aligned} y = 0 \quad &\Longrightarrow \quad f^{-1}(y) = 0 \\ y > 0 \quad &\Longrightarrow \quad f^{-1}(y) = \pm\sqrt{y} \end{aligned}$$

Since the preimage of points $y > 0$ is not unique, the function is not invertible.

Then:

$$f^{-1} : [0, \infty) \longrightarrow \mathbb{R}, \quad f^{-1}(y) = \pm\sqrt{y}$$

is not a function, as can be seen in Figure 3.4. □

In Example 3.11, the domain is restricted to allow for the existence of the inverse function.

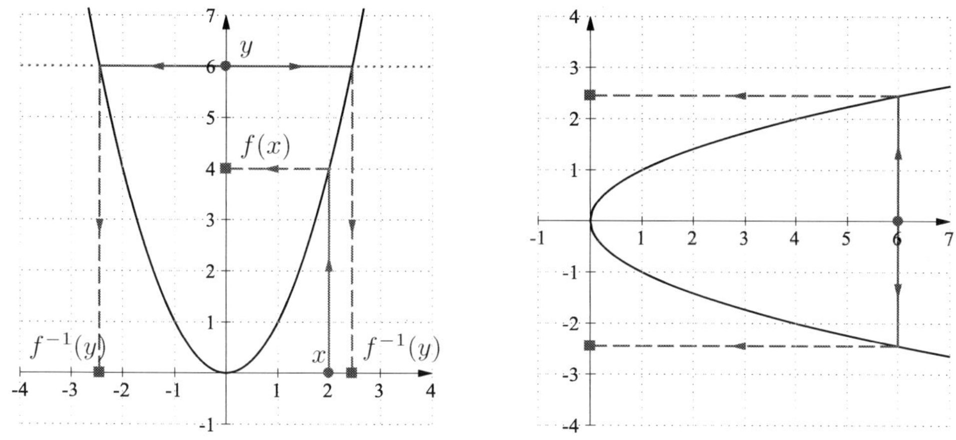

FIGURE 3.4
Graph of the function $f(x) = x^2$ (left) and graph of its inverse (right). As can be seen, the inverse is not a function since, given a value, it corresponds to two images.

Example 3.11. *Consider the function $f : \mathbb{R} \longrightarrow [0, \infty)$, $f(x) = x^2$, restricted to the interval $X = [0, \infty)$. Calculate the preimage of all points $y \in [0, \infty)$, and calculate the inverse function, if it exists.*

SOLUTION.
Now we are considering the function:

$$f_{|X} : [0, \infty) \longrightarrow [0, \infty)$$

Given $y \geq 0$, its preimage will be the points $x \in [0, \infty)$ such that $f_{|X}(x) = y$, that is:

$$x^2 = y.$$

In this case, the equation $x^2 = y$ has two real solutions:

$$x = \pm\sqrt{y},$$

but only $x = +\sqrt{y}$ is in the domain $[0, \infty)$.

Therefore, given $y \geq 0$, its preimage $f_{|X}^{-1}(y) = +\sqrt{y}$ is unique and the inverse function exists and is:

$$f_{|X}^{-1} : [0, \infty) \longrightarrow \mathbb{R}, \quad f_{|X}^{-1}(y) = \sqrt{y}$$

The restricted function $f_{|X}(x)$ and its inverse are represented in Figure 3.5. □

3.2 Basic Functions

Here are some of the functions that will be used most throughout the book. For each function, its name, explicit definition, domain, image, and inverse function, if it exists, are given. Table 3.3 contains a summary of this information.

TABLE 3.3
Definition, domain, image, and inverse of the most important functions.

Function	$f(x)$	Domain	Image	Inverse
Identity	$f(x) = x$	$\mathbb{R}$	$\mathbb{R}$	$f^{-1}(x) = x$
Constant	$f(x) = c,\ c \in \mathbb{R}$	$\mathbb{R}$	$\{c\}$	$\nexists$
Sign	$f(x) = \begin{cases} -1 & x < 0 \\ 0 & x = 0 \\ 1 & x > 0 \end{cases}$	$\mathbb{R}$	$\{-1, 0, 1\}$	$\nexists$
Unit step at a, $a \in \mathbb{R}$	$f(x) = \begin{cases} 0 & x < a \\ 1 & x \geq a \end{cases}$	$\mathbb{R}$	$\{0, 1\}$	$\nexists$
Floor	$f(x) = \lfloor x \rfloor$	$\mathbb{R}$	$\mathbb{Z}$	$\nexists$
Ceiling	$f(x) = \lceil x \rceil$	$\mathbb{R}$	$\mathbb{Z}$	$\nexists$
Exponential	$f(x) = e^x$	$\mathbb{R}$	$(0, +\infty)$	$f^{-1}(x) = \ln(x)$
Logarithm	$f(x) = \ln(x)$	$(0, +\infty)$	$\mathbb{R}$	$f^{-1}(x) = e^x$
Square root	$f(x) = \sqrt{x}$	$[0, +\infty)$	$[0, +\infty)$	$f^{-1}(x) = x^2,\ x > 0.$
Sine	$f(x) = \sin(x)$	$\mathbb{R}$	$[-1, 1]$	$f^{-1}_{\mid[-\pi/2,\pi/2]}(x) = \arcsin(x)$
Cosine	$f(x) = \cos(x)$	$\mathbb{R}$	$[-1, 1]$	$f^{-1}_{\mid[0,\pi]}(x) = \arccos(x)$
Tangent	$f(x) = \tan(x)$	$\mathbb{R} - \{\pi/2 + k\pi,\ k \in \mathbb{Z}\}$	$\mathbb{R}$	$f^{-1}_{\mid(-\pi/2,\pi/2)}(x) = \arctan(x)$

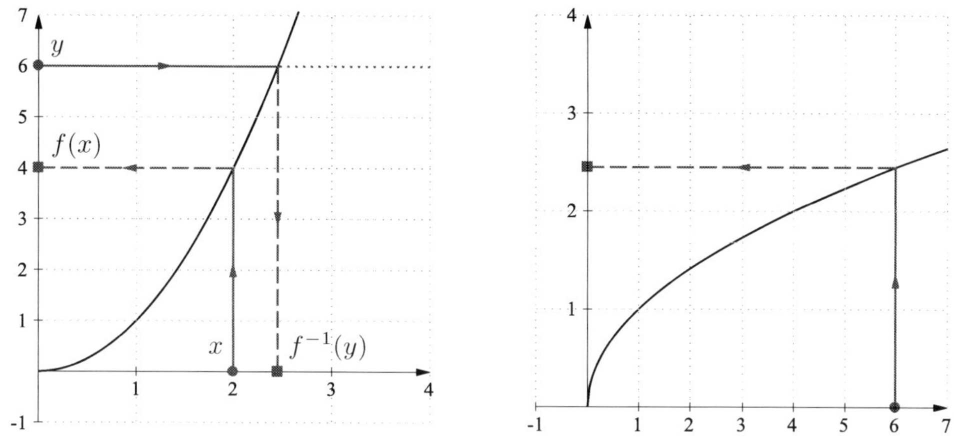

FIGURE 3.5
Graph of the function $f(x) = x^2$ restricted over the set $[0, +\infty)$ (left) and graph of its inverse function (right).

- **Identity**. $f(x) = x$. The domain is $\mathbb{R}$, the image is $\mathbb{R}$, and the inverse is itself. *See* Figure 3.6 (left).
- **Constant**. $f(x) = c$, $c \in \mathbb{R}$. The domain is $\mathbb{R}$, the image is $\{c\}$, and there is no inverse. *See* Figure 3.6 (right).

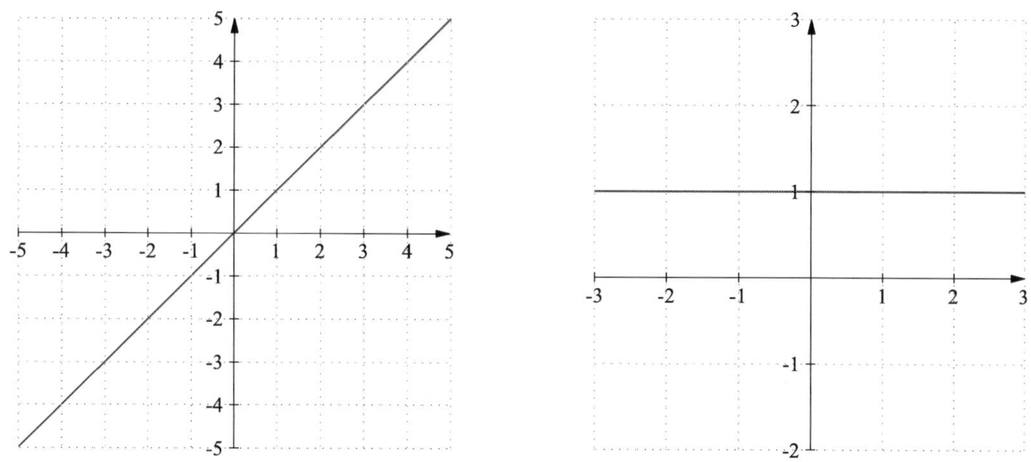

FIGURE 3.6
Graphs of the identity function (left) and the constant function (right).

- **Sign Function**. $f(x) = -1$ if $x < 0$; $f(0) = 0$; $f(x) = 1$ if $x > 0$. The domain is $\mathbb{R}$, and the image is $\{-1, 0, 1\}$. There is no inverse. *See* Figure 3.7 (left).
- **Unit Step Function at a Point**. $f(x) = 0$ if $x < a$, $a \in \mathbb{R}$; $f(x) = 1$ if $x \geq a$. The domain is $\mathbb{R}$, the range is $0, 1$, and there is no inverse. *See* Figure 3.7 (right).
- **Floor Function**. $f(x) = m$ if $x \in [m, m+1)$, $m \in \mathbb{Z}$. The domain is $\mathbb{R}$, the image is $\mathbb{Z}$. There is no inverse. *See* Figure 3.8 (left).

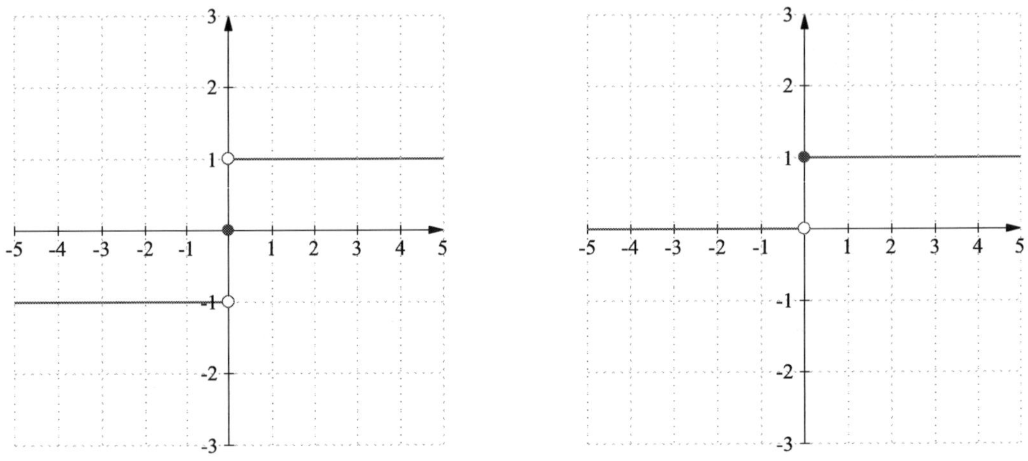

FIGURE 3.7
Graphs of the sign function (left) and the unit step function (right).

- **Ceiling Function.** $f(x) = m$ if $x \in (m-1, m]$, $m \in \mathbb{Z}$. The domain is $\mathbb{R}$, the image is $\mathbb{Z}$. There is no inverse. *See* Figure 3.8 (right).

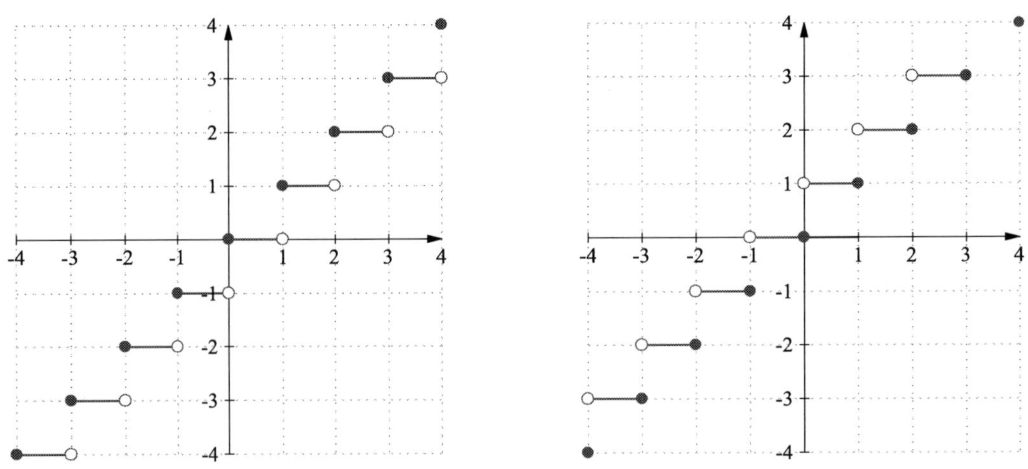

FIGURE 3.8
Graphs of the floor function (left) and the ceiling function (right).

- **Exponential Function.** $f(x) = e^x$. The domain is $\mathbb{R}$, the image is $(0, +\infty)$, and the inverse is the natural logarithm function. *See* Figure 3.9 (left).
- **Logarithm Function.** $f(x) = \ln x$. The domain is $(0, +\infty)$, the image is $\mathbb{R}$. The inverse function is the exponential function. *See* Figure 3.9 (right).
- **Square root function.** $f(x) = \sqrt{x}$. The domain is $[0, +\infty)$, the range is $[0, +\infty)$. The inverse function is x^2 (restricting the domain of x^2 to $[0, +\infty)$). *See* Figure 3.10 (left).

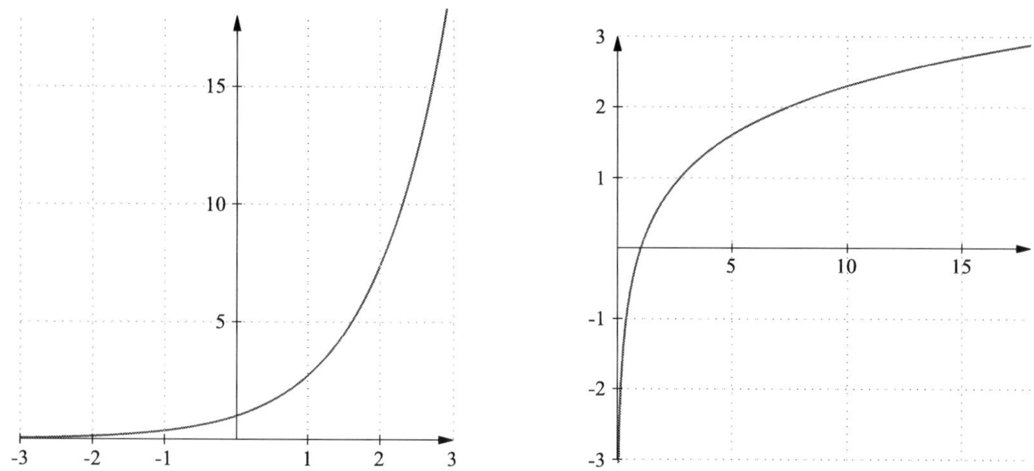

FIGURE 3.9
Graphs of the exponential function (left) and the natural logarithm function (right).

- **Sine function**. $f(x) = \sin(x)$. The domain is $\mathbb{R}$, the range is $[-1, 1]$. The inverse function is:

$$f^{-1}_{|[-\pi/2,\pi/2]}(x) = \arcsin(x)$$

(restricting the domain of the function $\sin(x)$ to $[-\pi/2, \pi/2]$). *See* Figure 3.10 (right).

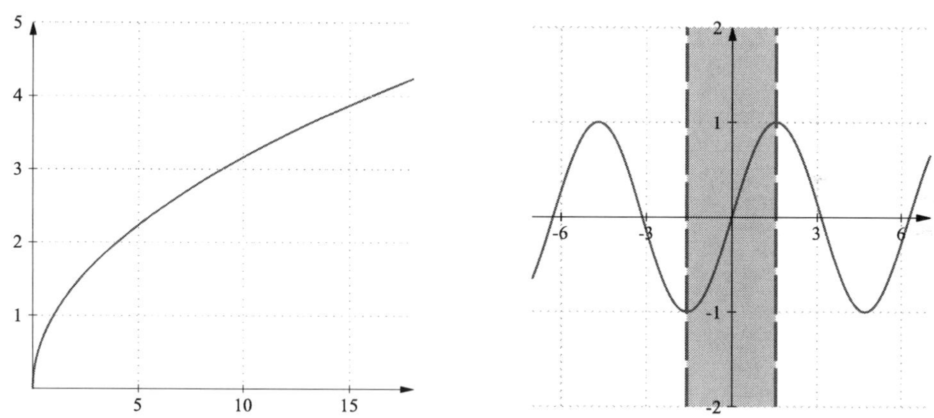

FIGURE 3.10
Graphs of the square root function (left) and the sine function (right). The shaded region corresponds to the region where the sine function is invertible.

- **Cosine function**. $f(x) = \cos(x)$. The domain is $\mathbb{R}$, the range is $[-1, 1]$. The inverse function is:

$$f^{-1}_{|[0,\pi]}(x) = \arccos(x)$$

(restricting the domain of the function $\cos(x)$ to $[0, \pi]$). *See* Figure 3.11 (left).

- **Tangent function**. $f(x) = \tan(x)$. The domain is all real numbers, except the points where the cosine is zero; the range is $\mathbb{R}$. The inverse function is:

$$f^{-1}_{|(-\pi/2,\pi/2)}(x) = \arctan(x)$$

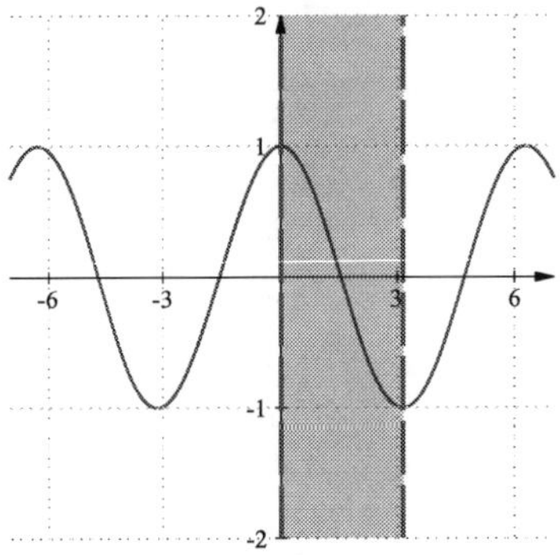

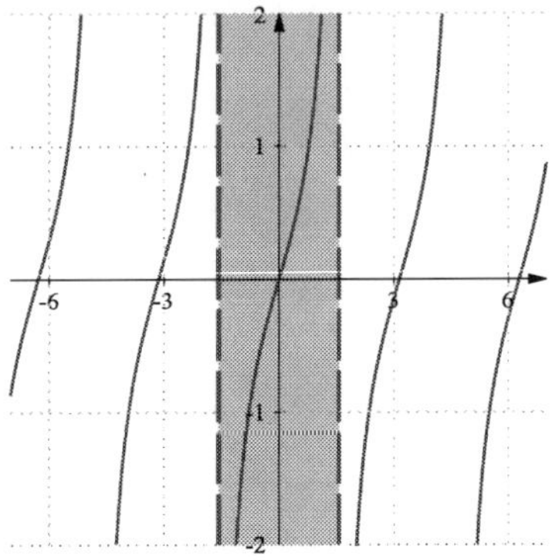

FIGURE 3.11
Graphs of the cosine function (left) and the tangent function (right). The shaded regions correspond to the regions where the cosine and tangent functions are invertible, respectively.

(restricting the domain of the function $\tan(x)$ to $(-\pi/2, \pi/2)$). *See* Figure 3.11 (right).

3.3 Algebraic Operations with Functions

Consider $f : A \to \mathbb{R}$ and $g : B \to \mathbb{R}$, where A and B are the domains of definition of f and g, respectively. Let $D = A \cap B \subset \mathbb{R}$ be the intersection of the two domains. We define the following operations:

- Addition/Subtraction:

$$\begin{aligned} f \pm g : A \cap B &\to \mathbb{R} \\ x &\mapsto (f \pm g)(x) = f(x) \pm g(x) \end{aligned}$$

- Scalar multiplication:

$$\begin{aligned} \lambda f : A &\to \mathbb{R} \\ x &\mapsto (\lambda f)(x) = \lambda \cdot f(x), \end{aligned}$$

  where $\lambda \in \mathbb{R}$.

- Multiplication:

$$\begin{aligned} f \cdot g : A \cap B &\to \mathbb{R} \\ x &\mapsto (f \cdot g)(x) = f(x) \cdot g(x) \end{aligned}$$

- Quotient:

$$\begin{aligned} f/g : A \cap B - \{x \mid g(x) = 0\} &\to \mathbb{R} \\ x &\mapsto (f/g)(x) = \frac{f(x)}{g(x)} \end{aligned}$$

3.4 Composite Function

Definition 3.10 (Composite function). Consider the functions $f : A \to B$ and $g : C \to D$. We define the *composite function* of f and g, denoted as $(g \circ f)(x)$, as:

$$(g \circ f)(x) = g(f(x))$$

Note that for this composition to make sense, $B \subset C$. Also note that the resulting function is such that:

$$(g \circ f) : A \to D$$

whose domain is defined as:

$$\text{Dom}(g \circ f) = \{x \in \text{Dom}(f) \subset A \mid f(x) \in \text{Dom}(g) \subset C\}$$

Example 3.12. *Consider* $h(x) = \sqrt{2x+2}$ *and* $k(x) = 3x + 2$. *Calculate* $h \circ k$ *and* $k \circ h$ *and their domains.*

SOLUTION.

- The function $h \circ k$ is defined as:

$$(h \circ k)(x) = h(k(x)) = h(3x+2) = \sqrt{2(3x+2)+2} = \sqrt{6x+6}.$$

On the other hand:

$$\begin{aligned}\text{Dom}(h \circ k) &= \{x \in \mathbb{R} \mid 6x + 6 \geq 0\} \\ &= \{x \in \mathbb{R} \mid 6(x+1) \geq 0\} \\ &= \{x \in \mathbb{R} \mid x + 1 \geq 0\} \\ &= \{x \in \mathbb{R} \mid x \geq -1\} \\ &= [-1, \infty)\end{aligned}$$

- The function $k \circ h$ is defined as:

$$(k \circ h)(x) = k(h(x)) = k(\sqrt{2x+2}) = 3\sqrt{2x+2} + 2.$$

On the other hand:

$$\begin{aligned}\text{Dom}(k \circ h) &= \{x \in \mathbb{R} \mid 2x + 2 \geq 0\} \\ &= \{x \in \mathbb{R} \mid 2(x+1) \geq 0\} \\ &= \{x \in \mathbb{R} \mid x + 1 \geq 0\} \\ &= \{x \in \mathbb{R} \mid x \geq -1\} \\ &= [-1, \infty)\end{aligned}$$

□

Remark 3.3. Note that, in general, the composition of functions is not commutative. That is:

$$(h \circ k) \neq (k \circ h).$$

3.5 Inverse Function

Definition 3.11 (Inverse function). Let $f : A \longrightarrow B$ be an invertible (or, equivalently, bijective) function, that is, its inverse function $f^{-1} : B \longrightarrow A$ exists. Then:

$$f \circ f^{-1} = I, \quad \text{and} \quad f^{-1} \circ f = I$$

where $I(x) = x$ is the identity function.

Remark 3.4. In the previous definition, the two identity functions are not the same function. In effect, the function:

$$f \circ f^{-1} = I$$

has domain B, while the function:

$$f^{-1} \circ f = I$$

has domain A.

Remark 3.5. The graphs of the functions f and f^{-1} are symmetric with respect to the bisector of the first and third quadrants. *See* Figure 3.12.

Example 3.13. *Let $g : \mathbb{R} \longrightarrow \mathbb{R}$ such that $g(x) = x + 1$. Calculate the inverse function and verify that $g \circ g^{-1} = I$ and that $g^{-1} \circ g = I$.*

SOLUTION.

Let $y \in \mathbb{R}$. Isolating x from the expression $y = x + 1$ we obtain $x = y - 1$. Therefore, $g^{-1}(y) = y - 1$ or, equivalently, $g^{-1}(x) = x - 1$.

Then:

$$(g^{-1} \circ g)(x) = g^{-1}(g(x)) = g^{-1}(x + 1) = (x + 1) - 1 = x = I(x)$$
$$(g \circ g^{-1})(x) = g(g^{-1}(x)) = g(x - 1) = (x - 1) + 1 = I(x)$$

□

Proposed Exercise 3.2. Calculate the inverses of the functions:

(a) $f(x) = 4x - 7$

(b) $f(x) = \dfrac{2x + 5}{x - 1}$

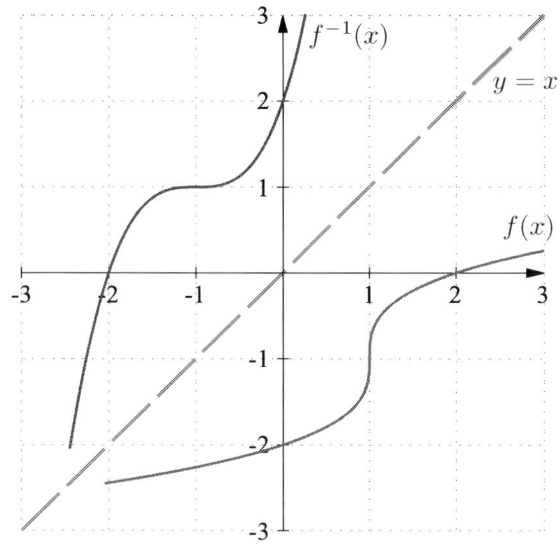

FIGURE 3.12
The graphs of the functions f and f^{-1} are symmetric with respect to the bisector of the first and third quadrants.

(c) $f(x) = \ln(x+1)$

Verify that $f \circ f^{-1} = I$ and that $f^{-1} \circ f = I$.

3.6 Limit of a Function at a Point

3.6.1 Intuitive Idea

The goal of the concept of a limit is to understand the behavior of the function **near** a certain point, without it being important what the function does at that point.

Let's consider, for example, and to illustrate the concept of a limit, the function:

$$f(x) = \frac{x^2 - 4}{x - 2}$$

We are interested in understanding the behavior of the function around the point $x = 2$. First of all, observe that the function is not defined for $x = 2$.

To calculate the limit intuitively, let's make a table of values taking points close to 2 from the left (smaller than 2) and from the right (larger than 2) and look at the behavior of the images (*see* Table 3.4):

We see that when x tends to 2 ($x \to 2$) (*see* Figure 3.13), the images of these points tend to 4 ($f(x) \to 4$). This can be denoted by:

$$\lim_{x \to 2} f(x) = 4$$

TABLE 3.4
If we consider the function $f(x) = \frac{x^2-4}{x-2}$, when the points approach 2, their respective images approach 4.

x	$f(x)$	x	$f(x)$
1.99	3.99	2.01	4.01
1.999	3.999	2.001	4.001
1.9999	3.9999	2.0001	4.0001
1.99999	3.99999	2.00001	4.00001
1.999999	3.999999	2.000001	4.000001

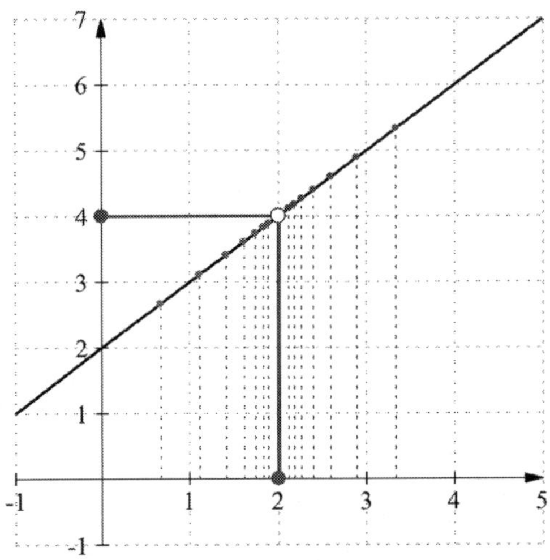

FIGURE 3.13
Graph of the function $f(x) = \frac{x^2-4}{x-2}$. Although the function is not defined at $x = 2$, it can be seen how the images of points close to 2 approach 4.

Now let's consider the point $x = 1$ and the function:

$$f(x) = \begin{cases} -1, & x < 1 \\ 1, & x \geq 1 \end{cases}$$

If we take points close to 1 from the left and from the right and look at the images, we can generate a set of values as in Table 3.5.

When considering values approaching 1, the images do not tend to a single value (*see* Figure 3.14), and therefore, we will say that the limit does not exist.

3.6.2 Accumulation Point

To be able to calculate the limit of a function when x tends to a point a, it is necessary that this point a be what we call an *accumulation point*.

TABLE 3.5
If we consider the function $f(x) = -1$ if $x < 1$ and $f(x) = 1$ if $x \geq 1$, when the points approach 1, their respective images do not tend to a single value.

x	$f(x)$	x	$f(x)$
0.99	-1	1.01	1
0.999	-1	1.001	1
0.9999	-1	1.0001	1
0.99999	-1	1.00001	1
0.999999	-1	1.000001	1

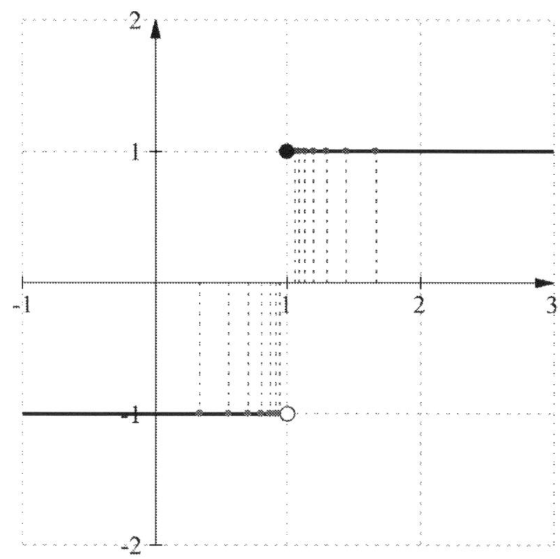

FIGURE 3.14
Graph of the function $f(x) = -1$ if $x < 1$ and $f(x) = 1$ if $x \geq 1$. It can be seen how the images of points close to 1 are different if these points are smaller or larger than 1.

Definition 3.12 (Accumulation Point). Given a point $a \in \mathbb{R}$ and a set $X \subset \mathbb{R}$, we say that a is an **accumulation point** of the set X if for any open set S_a that includes a, we have

$$(S_a - \{a\}) \cap X \neq \emptyset.$$

Definition 3.13 (Accumulation Point (Alternative Definition)). Given a point $a \in \mathbb{R}$ and a set $X \subset \mathbb{R}$, we say that a is an **accumulation point** of the set X if for every $\delta > 0$, we have

$$((a - \delta, a + \delta) - \{a\}) \cap X \neq \emptyset.$$

Definition 3.14 (Isolated Point). Given a point $a \in \mathbb{R}$ and a set $X \subset \mathbb{R}$, we say that a is an **isolated point** of the set X if it is **not** an accumulation point.

Example 3.14 (Accumulation Point). *Consider, for example, the function* $f(x) = \ln(x)$ *whose domain is the interval* $(0, +\infty)$. *The point* $a = 0$ *is an accumulation point because any open interval containing the point* $a = 0$ *will have a non-empty intersection with the interval* $(0, +\infty)$.

On the other hand, the point $a = -1$ *is an isolated point because the interval* $(-3/2, -1/2)$ *that contains it has an empty intersection with the interval* $(0, +\infty)$. *That is,*

$$((-3/2, -1/2) - \{-1\}) \cap (0, +\infty) = \emptyset.$$

Therefore, it will make sense to calculate

$$\lim_{x \to 0^+} \ln(x),$$

while it will not make sense to calculate

$$\lim_{x \to -1} \ln(x).$$

Remark 3.6. For the rest of the chapter, for simplicity, we will not explicitly refer to the need for the points we tend towards to be accumulation points. That is, we will assume that in the calculation of the limit

$$\lim_{x \to a} f(x)$$

the point a will be an accumulation point of the domain of f.

3.6.3 Formal Definition

Let's begin this section by defining, informally, the concept of the limit of a function at a point.

Definition 3.15 (Informal Definition of Limit of a Function at a Point). Let $f : A \subset \mathbb{R} \to \mathbb{R}$ be a real function of a real variable and $a \in \mathbb{R}$ a point that does not necessarily belong to the domain of f. It is said that the function f has a **limit** $L \in \mathbb{R}$ **at the point** a if $|f(x) - L|$ is arbitrarily small when x is sufficiently close to a, without x ever being equal to a.

The previous definition is *informal* because it includes the expressions "arbitrarily small" and "sufficiently close." Let's see below a new definition, somewhat more complex, but which avoids all ambiguity.

Definition 3.16 (Limit of a Function at a Point). Let $f : A \subset \mathbb{R} \to \mathbb{R}$ be a real function of a real variable and $a \in \mathbb{R}$ a point that does not necessarily belong to the domain of f. It is said that the function f has a **limit** $L \in \mathbb{R}$ **at the point** a if for every positive number ε there exists a positive number δ (which generally depends on ε), such that for all points x that satisfy $0 < |x - a| < \delta$, it holds that $|f(x) - L| < \varepsilon$.

In other words, the function f has a **limit** $L \in \mathbb{R}$ **at the point** a if:

$$\forall \varepsilon > 0 \; \exists \delta > 0 \text{ such that } \forall x \text{ that satisfies } 0 < |x - a| < \delta, \text{ then } |f(x) - L| < \varepsilon.$$

Remark 3.7. In the previous definition, it is important to highlight that we need $0 < |x - a|$ to prevent x from being equal to a.

The limit of a function f at a point a is denoted by:

$$L = \lim_{x \to a} f(x)$$

or even:

$$L = \lim_{a} f(x).$$

3.6.4 Geometric Interpretation of the Formal Definition of Limit

The definition of the limit of a function at a point can seem very confusing, but after a little analysis, it can be perfectly understood. In the following lines, we will try to approach this formal definition through some images.

The formal definition basically has three parts:

- "For all $\varepsilon > 0$." We can also say "given a positive ε value" or "considering any positive number ε." In this case, with this positive number ε, we generate an interval containing the limit value L, of the form $[L-\varepsilon, L+\varepsilon]$. This is the interval where we want the images of x to *fall*. It is important to note at this point that the expression $|f(x)-L| < \varepsilon$ is equivalent to $f(x) \in (L-\varepsilon, L+\varepsilon)$. Let's see it:

$$\begin{aligned} |f(x)-L| < \varepsilon \quad &\iff \quad -\varepsilon < f(x)-L < \varepsilon \\ &\iff \quad L-\varepsilon < f(x) < L+\varepsilon \\ &\iff \quad f(x) \in (L-\varepsilon, L+\varepsilon) \end{aligned}$$

  A graphical representation of this interval can be seen in Figure 3.15.

- "There exists $\delta > 0$." We can also say "we are able to find a positive number δ." This value is related to ε in the sense that it depends on it. In this case, with this positive number δ, we generate an interval centered at the point a, with amplitude 2δ and excluding the point a itself, that is:

$$(a-\delta, a) \cup (a, a+\delta)$$

  or:

$$(a-\delta, a+\delta) - \{a\}.$$

  This is the interval whose images we need to be within the interval $(L-\varepsilon, L+\varepsilon)$. It is important to note at this point that the expression $0 < |x-a| < \delta$ is equivalent to $x \in (a-\delta, a+\delta),\ x \neq a$. Let's see it in two parts. On one hand:

$$\begin{aligned} |x-a| < \delta \quad &\iff \quad -\delta < x-a < \delta \\ &\iff \quad a-\delta < x < a+\delta \\ &\iff \quad x \in (a-\delta, a+\delta) \end{aligned}$$

  On the other hand, $0 < |x-a|$ means that $x \neq a$.

  A graphical representation of this interval can be seen in Figure 3.15.

- "Such that for all x that satisfies $0 < |x-a| < \delta$ **then** $|f(x)-L| < \varepsilon$." This last part is an implication, that is, it is the key part of the formal definition of limit. A graphical representation of this implication can be seen in Figure 3.15.

3.6.5 Calculating Limits Using the Definition

Example 3.15 (Limit of the Constant Function)**.** *Consider the function $f(x) = k,\ k \in \mathbb{R}$ and $a \in \mathbb{R}$. Prove that:*

$$\lim_{x \to a} f(x) = k$$

Solution. In this case, for every positive number ε it holds that:

$$|f(x)-L| = |k-k| = 0 < \varepsilon,$$

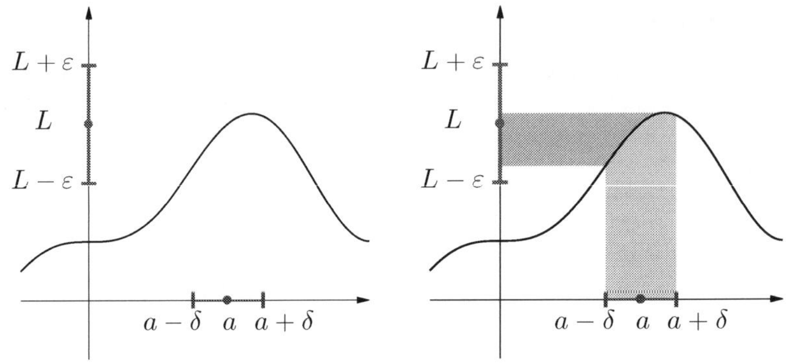

FIGURE 3.15
Graphical representation of the interval $(L-\varepsilon, L+\varepsilon)$ on the vertical axis and the interval $(a-\delta, a+\delta)$ on the horizontal axis (left). For the limit of $f(x)$ when x tends to a to be L, it is necessary that the images of the points in the interval $(a-\delta, a+\delta) - \{a\}$ fall within the interval $(L-\varepsilon, L+\varepsilon)$ (right).

regardless of the value of $\delta > 0$. Therefore, it is always true that if:

$$0 < |x-a| < \delta,$$

then $|f(x) - L| = |k-k| = 0 < \varepsilon$. □

Example 3.16 (Limit of the Identity Function). *Consider the function* $f(x) = x$ *and* $a \in \mathbb{R}$. *Prove that:*

$$\lim_{x \to a} f(x) = a$$

SOLUTION. We will prove this limit following the two steps below:

- **Step 1**. Relate the difference $|f(x) - L|$ with $|x-a|$.
- **Step 2**. Select δ as a function of ε.

For this case, we have $f(x) = x$ and $L = a$. Therefore:

- **Step 1**.
$$|f(x) - L| = |x-a| = 1 \cdot |x-a| < \varepsilon$$
- **Step 2**. If $|x-a| < \varepsilon$, then $|f(x) - L| < \varepsilon$. Therefore, we select $\delta = \varepsilon$.

These two preliminary steps serve to select δ as a function of ε. The proof is completed in the following lines.

We need to prove that for every $\varepsilon > 0$ there exists $\delta > 0$, such that if we take $x \in \mathbb{R}$ that satisfies $0 < |x-a| < \delta$, then $|f(x) - a| < \varepsilon$.

Consider the given $\varepsilon > 0$ and take $\delta = \varepsilon$.

Then, for all $x \in \mathbb{R}$ such that $0 < |x-a| < \delta$, given that $f(x) = x$, we have that:

$$|f(x) - a| = |x-a| < \delta = \varepsilon$$

as we wanted to demonstrate. *See* Figure 3.16. □

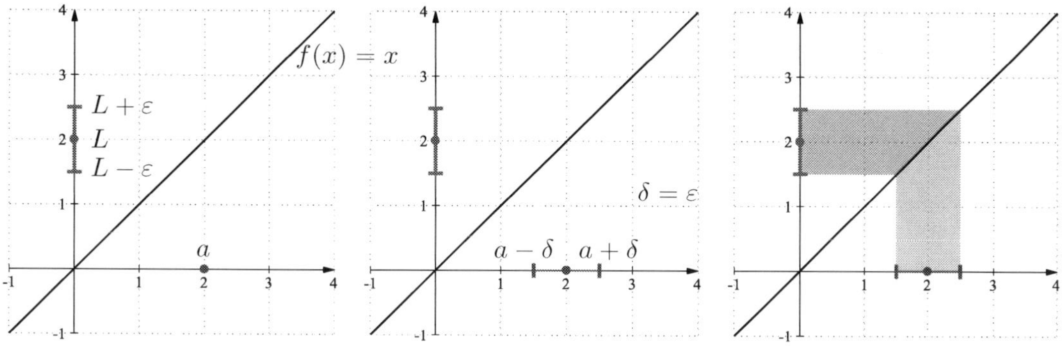

FIGURE 3.16
Graphical representation of the proof of the limit of $f(x)$ when x tends to a. Note that in this case, $\delta = \varepsilon$.

Example 3.17. *Consider the function* $f(x) = \frac{x}{2}$ *and* $a \in \mathbb{R}$. *Prove that:*

$$\lim_{x \to a} f(x) = \frac{a}{2}$$

SOLUTION. We will prove this limit also following the two steps below:

- **Step 1**. Relate the difference $|f(x) - L|$ with $|x - a|$.
- **Step 2**. Select δ as a function of ε.

For this case, we have $f(x) = \frac{x}{2}$ and $L = \frac{a}{2}$. Therefore:

- **Step 1**.

$$|f(x) - L| = \left| \frac{x}{2} - \frac{a}{2} \right| = \frac{1}{2} \cdot |x - a| < \varepsilon$$

- **Step 2**. If $|x - a| < \varepsilon$, then $|f(x) - L| < \varepsilon$. Therefore, we select $\delta = 2\varepsilon$.

These two preliminary steps serve to select δ as a function of ε. The proof is completed in the following lines.

We need to prove that for every $\varepsilon > 0$ there exists $\delta > 0$, such that if we take $x \in \mathbb{R}$ that satisfies $0 < |x - a| < \delta$, then:

$$\left| f(x) - \frac{a}{2} \right| < \varepsilon.$$

Consider the given $\varepsilon > 0$ and take $\delta = 2\varepsilon$.

Then, for all $x \in \mathbb{R}$ such that $0 < |x - a| < \delta$, given that $f(x) = \frac{x}{2}$, we have:

$$|x - a| < \delta \Longrightarrow \left| \frac{x}{2} - \frac{a}{2} \right| = \frac{|x - a|}{2} < \frac{\delta}{2} \Longrightarrow \left| f(x) - \frac{a}{2} \right| < \frac{\delta}{2} = \varepsilon$$

as we wanted to demonstrate. *See* Figure 3.17. □

Example 3.18. *Consider the function* $f(x) = 2x$ *and* $a \in \mathbb{R}$. *Prove that:*

$$\lim_{x \to a} f(x) = 2a$$

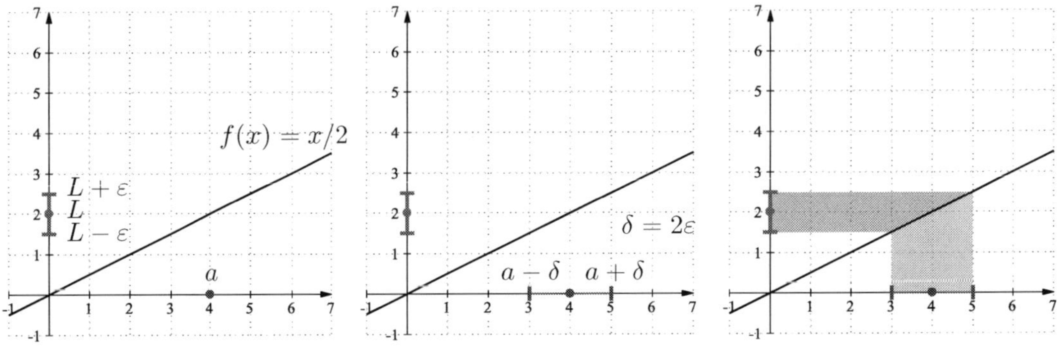

FIGURE 3.17
Graphical representation of the proof of the limit of $f(x)$ when x tends to a. Note that in this case, $\delta = 2\varepsilon$.

SOLUTION. We will prove this limit following the two steps below:

- **Step 1**. Relate the difference $|f(x) - L|$ with $|x - a|$.
- **Step 2**. Select δ as a function of ε.

For this case, we have $f(x) = 2x$ and $L = 2a$. Therefore:

- **Step 1**.
$$|f(x) - L| = |2x - 2a| = 2 \cdot |x - a| < \varepsilon$$
- **Step 2**. If $|x - a| < \varepsilon$, then $|f(x) - L| < \varepsilon$. Therefore, we select $\delta = \dfrac{\varepsilon}{2}$.

These two preliminary steps serve to select δ as a function of ε. The proof is completed in the following lines.

We need to see that for every $\varepsilon > 0$ there exists $\delta > 0$, such that if we take:

$$x \in (a - \delta, a + \delta), \quad x \neq a,$$

then:

$$f(x) \in (2a - \varepsilon, 2a + \varepsilon).$$

Consider the given $\varepsilon > 0$ and take $\delta = \varepsilon/2$.
Then, if $x \in (a - \delta, a + \delta)$, $f(x) = 2x$ and:

$$\begin{aligned} a - \delta < x < a + \delta &\Longrightarrow 2a - 2\delta < 2x < 2a + 2\delta \\ &\Longrightarrow 2a - 2\varepsilon/2 < 2x < 2a + 2\varepsilon/2 \\ &\Longrightarrow 2a - \varepsilon < f(x) < 2a + \varepsilon \\ &\Longrightarrow f(x) \in (2a - \varepsilon, 2a + \varepsilon), \end{aligned}$$

as we wanted to demonstrate. *See* Figure 3.18. □

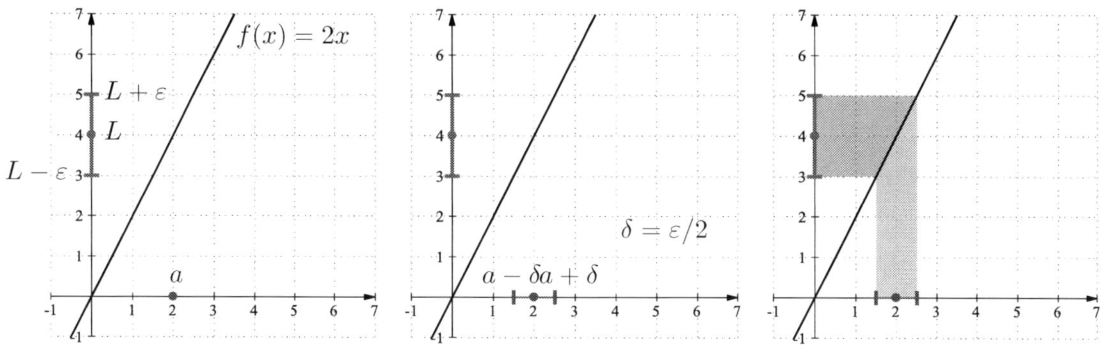

FIGURE 3.18
Graphical representation of the proof of the limit of $f(x) = 2x$ when x tends to a. Note that in this case, $\delta = \varepsilon/2$.

Example 3.19 (Function That Does Not Have a Limit). *Consider the function:*

$$f(x) = \begin{cases} x, & \textit{if } x < 2 \\ 2x, & \textit{if } x \geq 2 \end{cases}$$

and $a = 2$. *Prove that:*

$$\lim_{x \to 2} f(x) \textit{ does not exist}$$

SOLUTION. If we observe Figure 3.19, we can see that the trend of the function as x approaches $a = 2$ depends on the side from which we come. From the left, the trend is to 2 and from the right, it is to 4. Can we say that the limit is 4? Can we say that the limit is 2? We will see that neither is true, based on the formal definition of a limit. That is, we will consider a specific value of ε for which no value of δ makes the images of the points $(a - \delta, a + \delta) - \{a\}$ fall within $(L - \varepsilon, L + \varepsilon)$.

Let's take $\varepsilon = 1$.

- The limit is not $L = 4$ because we can find points as close as we want to 2 such that $|f(x) - 4| > 1$. For example, $f(1.9) = 1.9$ which does not belong to the interval:

$$(L - \varepsilon, L + \varepsilon) = (4 - 1, 4 + 1) = (3, 5).$$

- Similarly, the limit is not $L = 2$ because we can find points as close as we want to 2 such that $|f(x) - 2| > 1$. For example, $f(2.1) = 4.2$ which does not belong to the interval:

$$(L - \varepsilon, L + \varepsilon) = (2 - 1, 2 + 1) = (1, 3).$$

In conclusion, for any candidate limit L, we can always find points as close as we want to 2 such that $|f(x) - L| > 1$, therefore, the limit does not exist. □

Example 3.20. *Consider the function* $f(x) = x^2$ *and the point* $a = 2$. *Prove that:*

$$\lim_{x \to 2} f(x) = 4$$

SOLUTION.

To prove this limit, we need to prove that $\forall \varepsilon > 0$, $\exists \delta > 0$, such that:

$$\forall x \in \mathbb{R}\,,\; 0 < |x - 2| < \delta \Longrightarrow |x^2 - 4| < \varepsilon.$$

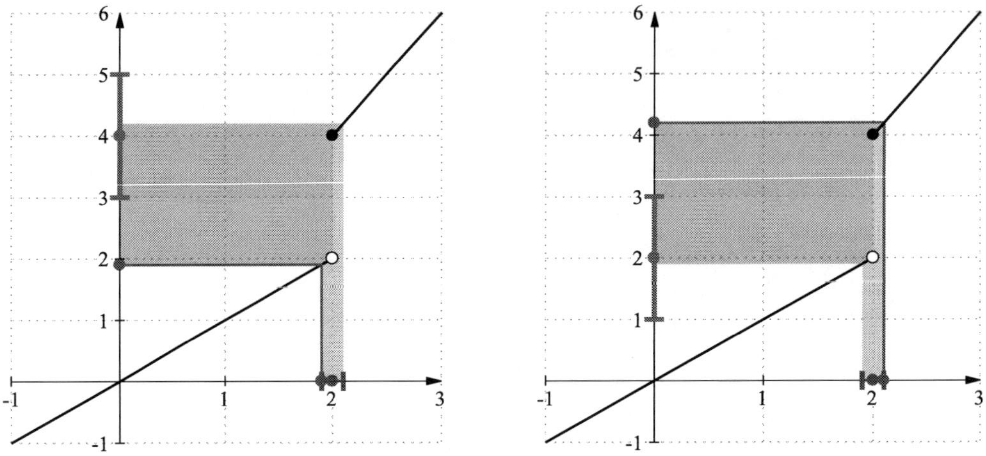

FIGURE 3.19
Graphical representation of a function that does not have a limit at a point. No matter how close the points are to 2, there are always images that escape the interval $(L - \varepsilon, L + \varepsilon) = (1, 3)$.

In a first approximation to the proof of this limit, we will only see that for $\varepsilon = 3$ we can consider $\delta = 1/2$. Indeed:

$$\begin{aligned}
|x-2| < \frac{1}{2} \quad &\Longrightarrow \quad -\frac{1}{2} < x - 2 < \frac{1}{2} \\
&\Longrightarrow \quad 2 - \frac{1}{2} < x < \frac{1}{2} + 2 \\
&\Longrightarrow \quad \frac{3}{2} < x < \frac{5}{2} \\
&\Longrightarrow \quad \frac{9}{4} < x^2 < \frac{25}{4} \\
&\Longrightarrow \quad x^2 \in (2.25, 6.25) \\
&\Longrightarrow \quad x^2 \in (1, 7) \\
&\Longrightarrow \quad 1 < x^2 < 7 \\
&\Longrightarrow \quad 1 - 4 < x^2 - 4 < 7 - 4 \\
&\Longrightarrow \quad -3 < x^2 - 4 < 3 \\
&\Longrightarrow \quad |x^2 - 4| < 3.
\end{aligned}$$

For a more general proof, we can proceed as follows when choosing the appropriate value of δ. Suppose a given $\varepsilon > 0$. Then, we want to find a $\delta > 0$ such that the images of the points $(2 - \delta, 2 + \delta) - \{2\}$ are within the interval $(4 - \varepsilon, 4 + \varepsilon)$. If we consider the preimages of the endpoints of this latter interval, and considering that the function x^2 is strictly increasing, we obtain:

$$\text{preimage of } 4 - \varepsilon = \sqrt{4 - \varepsilon} < 2$$
$$\text{preimage of } 4 + \varepsilon = \sqrt{4 + \varepsilon} > 2$$

It will suffice to choose δ as the smallest of the distances between the number 2 and these two preimages. That is:

$$\delta = \min\{2 - \sqrt{4-\varepsilon}, \sqrt{4+\varepsilon} - 2\}$$

□

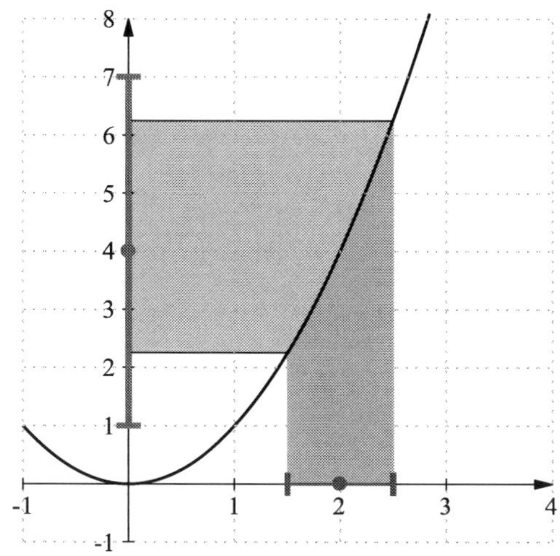

FIGURE 3.20
Graphical representation of the limit of the function x^2 when x tends to 2. In this case, when $\varepsilon = 3$, it is enough to consider $\delta = \frac{1}{2}$.

Remark 3.8. The limit of a function at a point aims to characterize the behavior of the images of the points close to the point a.

Remark 3.9. Calculating limits using the definition is very difficult; therefore, alternative techniques must be sought to calculate them.

3.6.6 Properties of the Limit of a Function at a Point

Proposition 3.1. The limit of a function at a point, if it exists, is unique.

PROOF. The proof will be done by contradiction.

Suppose a function $f(x)$ has two limits $L_1 \neq L_2$ when x tends to a. Suppose, without loss of generality, that $L_2 > L_1$. Define:

$$\varepsilon = \frac{L_2 - L_1}{2} > 0.$$

Then, by the definition of limit, we know that:

- Given ε, there exists $\delta_1 > 0$ such that if $x \in (a - \delta_1, a + \delta_1) - \{a\}$, then:

$$f(x) \in (L_1 - \varepsilon, L_1 + \varepsilon)$$

- Given ε, there exists $\delta_2 > 0$ such that if $x \in (a - \delta_2, a + \delta_2) - \{a\}$, then:

$$f(x) \in (L_2 - \varepsilon, L_2 + \varepsilon)$$

Given ε, if we now consider $\delta = \min\{\delta_1, \delta_2\}$, we have that, simultaneously:

$$f(x) \in (L_1 - \varepsilon, L_1 + \varepsilon)$$

and:

$$f(x) \in (L_2 - \varepsilon, L_2 + \varepsilon).$$

But $f(x)$ cannot belong simultaneously to both intervals, since the intervals:

$$(L_1 - \varepsilon, L_1 + \varepsilon) \cap (L_2 - \varepsilon, L_2 + \varepsilon) = \emptyset$$

are disjoint. With this, we arrive at a contradiction, which implies that the limit of a function, if it exists, is unique. □

Proposition 3.2. If a function has a limit at a point, then there exists a neighborhood of that point in which the function is bounded.

PROOF. The proof is a direct consequence of the very definition of limit. Consider the function $f(x)$ of which we know that:

$$\lim_{x \to a} f(x) = L$$

This means that for every $\varepsilon > 0$ there exists $\delta > 0$, such that if $0 < |x - a| < \delta$ then:

$$|f(x) - L| < \varepsilon.$$

We have already seen that this last inequality is equivalent to:

$$f(x) \in (L - \varepsilon, L + \varepsilon).$$

This last implies that:

$$|f(x)| < \max\{|L - \varepsilon|, |L + \varepsilon|\},$$

which demonstrates that the function is bounded. This idea can be illustrated in Figure 3.21.

Proposition 3.3. If $f : A \to \mathbb{R}$ is a function such that:

$$\lim_{x \to a} f(x) = L \neq 0,$$

then there exists a reduced neighborhood of the point a in which the function preserves the sign.

PROOF. The proof is also a direct consequence of the very definition of limit. Consider the function $f(x)$ of which we know that:

$$\lim_{x \to a} f(x) = L \neq 0$$

Suppose $L > 0$. Now consider $\varepsilon = \frac{L}{2} > 0$.

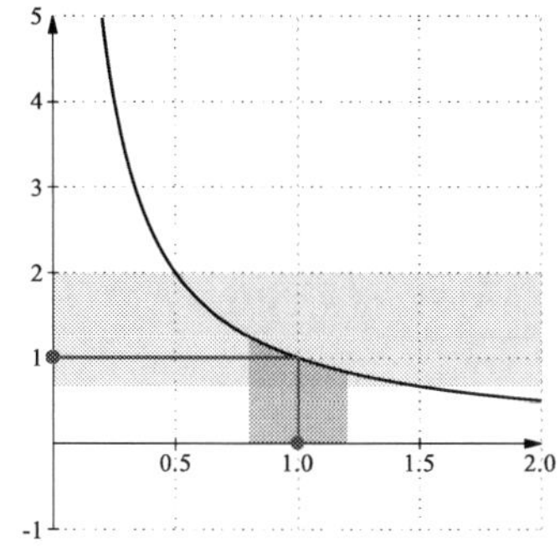

FIGURE 3.21
If a function has a limit at a point, then there exists a neighborhood of that point in which the function is bounded.

We know that there exists $\delta > 0$ such that if $0 < |x - a| < \delta$ then $|f(x) - L| < \varepsilon$. This last inequality is equivalent to:

$$f(x) \in (L - \varepsilon, L + \varepsilon) = \left(L - \frac{L}{2}, L + \frac{L}{2}\right) = \left(\frac{L}{2}, \frac{3L}{2}\right).$$

This last implies that there is a neighborhood of the point a, the interval $(a - \delta, a + \delta)$, in which all images are positive. In particular, greater than $\frac{L}{2}$.

The case $L < 0$ is proved similarly.

This idea can be illustrated in Figure 3.22.

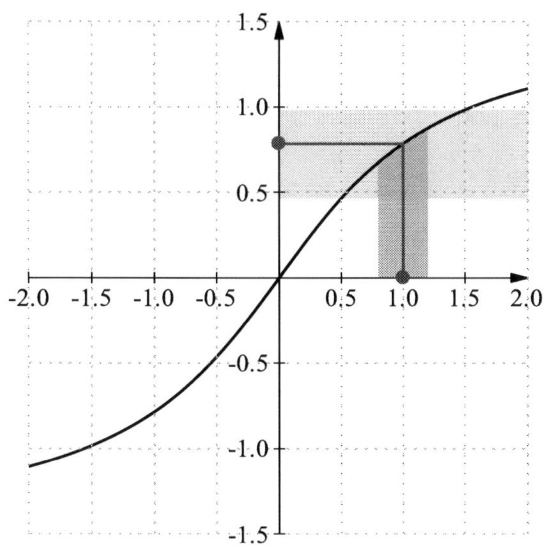

FIGURE 3.22
If the limit of a function is different from zero, then there exists a neighborhood of that point in which the function preserves the sign.

3.7 Calculation of Limits in Algebraic Form

Proposition 3.4 (Algebra of Function Limits). If $f : A \to \mathbb{R}$ and $g : A \to \mathbb{R}$ are two real functions of a real variable, a is a point in A, and these functions satisfy:

$$\lim_{x \to a} f = L_1$$
$$\lim_{x \to a} g = L_2$$

then the following hold:

(a) Limit of the sum:

$$\lim_{x \to a} (f+g)(x) = L_1 + L_2$$

(b) Limit of the scalar product:

$$\lim_{x \to a} (\lambda f)(x) = \lambda L_1,$$

for any $\lambda \in \mathbb{R}$;

Proposition 3.5 (Algebra of Function Limits (cont.)). **(c) Limit of the product:**

$$\lim_{x \to a} (fg)(x) = L_1 L_2$$

(d) Limit of the quotient:

$$\lim_{x \to a} \left(\frac{f}{g}\right)(x) = \frac{L_1}{L_2},$$

if $L_2 \neq 0$;

(e) Limit of the absolute value:

$$\lim_{x \to a} |f| = |L_1|.$$

Proof. The proofs of parts (a), (b), (c), (d), and (e) are presented below.

(a) Consider $\varepsilon > 0$. Then, given that:

$$\lim_{x \to a} f(x) = L_1,$$

we know that for the positive number $\frac{\varepsilon}{2}$, there exists a positive number δ_1 such that:

$$\text{if } 0 < |x-a| < \delta_1 \text{ then } |f(x) - L_1| < \frac{\varepsilon}{2}$$

Similarly, given that:

$$\lim_{x \to a} g(x) = L_2,$$

we know that for the positive number $\frac{\varepsilon}{2}$, there exists a positive number δ_2 such that:

$$\text{if } 0 < |x-a| < \delta_2 \text{ then } |g(x) - L_2| < \frac{\varepsilon}{2}$$

Now consider $\delta = \min\{\delta_1, \delta_2\}$. We will show that if $0 < |x-a| < \delta$, then $|f(x) + g(x) - (L_1 + L_2)| < \varepsilon$, that is:

$$\lim_{x \to a} (f+g)(x) = L_1 + L_2.$$

Indeed, if $0 < |x-a| < \delta$, then:

$$\begin{aligned} |f(x)+g(x)-(L_1+L_2)| &= |(f(x)-L_1)+(g(x)-L_2)| \\ &\leq \underbrace{|f(x)-L_1|}_{<\frac{\varepsilon}{2} \text{ if } 0<|x-a|<\delta\leq\delta_1} + \underbrace{|g(x)-L_2|}_{<\frac{\varepsilon}{2} \text{ if } 0<|x-a|<\delta\leq\delta_2} \\ &< \frac{\varepsilon}{2} + \frac{\varepsilon}{2} \\ &= \varepsilon, \end{aligned}$$

as we wanted to show.

(b) If we consider $\lambda = 0$, then $\lambda \cdot f$ is equal to the constant function:

$$f(x) = 0.$$

The limit, in this case, is equal to 0, as already demonstrated in Example 3.15.

Now suppose that $\lambda \neq 0$. Consider $\varepsilon > 0$. Then, given that:

$$\lim_{x \to a} f(x) = L_1,$$

we know that for the positive number $\frac{\varepsilon}{|\lambda|}$, there exists a positive number δ such that:

$$\text{if } 0 < |x-a| < \delta \text{ then } |f(x) - L_1| < \frac{\varepsilon}{|\lambda|}$$

Now, if $0 < |x-a| < \delta$ we have:

$$\begin{aligned} |(\lambda \cdot f)(x) - \lambda \cdot L_1| &= |\lambda \cdot f(x) - \lambda \cdot L_1| \\ &= |\lambda(f(x) - L_1)| \\ &= |\lambda||f(x) - L_1| \\ &< |\lambda|\frac{\varepsilon}{|\lambda|} \\ &= \varepsilon \end{aligned}$$

(c) Suppose $g(x)$ is the constant function equal to zero in a neighborhood of the point a. Then, the product function $f(x)g(x)$ is the constant function equal to zero in a neighborhood of a and its limit is 0, as we have seen in part (b).

Now suppose that $g(x)$ is not the constant function equal to zero. Then, by Proposition 3.2, there exists a neighborhood of the point a in which the function is bounded. Let $\bar{g}$ be an upper bound of the absolute value of the function, that is:

$$|g(x)| \leq \bar{g} \neq 0.$$

Suppose also that $L_1 \neq 0$.

Consider $\varepsilon > 0$. Then, given that:

$$\lim_{x \to a} f(x) = L_1$$

we know that for the positive number $\frac{\varepsilon}{2\bar{g}}$, there exists a positive number δ_1 such that:

$$\text{if } 0 < |x - a| < \delta_1 \text{ then } |f(x) - L_1| < \frac{\varepsilon}{2\bar{g}}$$

Similarly, given that:

$$\lim_{x \to a} g(x) = L_2,$$

we know that for the positive number $\frac{\varepsilon}{2|L_1|}$, there exists a positive number δ_2 such that:

$$\text{if } 0 < |x - a| < \delta_2 \text{ then } |g(x) - L_2| < \frac{\varepsilon}{2|L_1|}$$

Now consider $\delta = \min\{\delta_1, \delta_2\}$. We will show that if $0 < |x - a| < \delta$, then $|f(x)g(x) - L_1L_2| < \varepsilon$, that is:

$$\begin{aligned}
|f(x)g(x) - L_1L_2| &= |f(x)g(x) - L_1L_2 + L_1g(x) - L_1g(x)| \\
&= |(f(x) - L_1)g(x) + L_1(g(x) - L_2)| \\
&\leq |(f(x) - L_1)g(x)| + |L_1(g(x) - L_2)| \\
&= |f(x) - L_1||g(x)| + |L_1||g(x) - L_2| \\
&\leq |f(x) - L_1|\bar{g} + |L_1||g(x) - L_2| \\
&< \frac{\varepsilon}{2\bar{g}}\bar{g} + |L_1|\frac{\varepsilon}{2|L_1|} \\
&< \frac{\varepsilon}{2} + \frac{\varepsilon}{2} \\
&= \varepsilon
\end{aligned}$$

Now suppose that $L_1 = 0$.

Consider $\varepsilon > 0$. Then, given that:

$$\lim_{x \to a} f(x) = 0$$

we know that for the positive number $\frac{\varepsilon}{\bar{g}}$, there exists a positive number δ such that:

$$\text{if } 0 < |x - a| < \delta \text{ then } |f(x)| < \frac{\varepsilon}{\bar{g}}$$

We will show that if $0 < |x - a| < \delta$ then $|f(x)g(x)| < \varepsilon$, that is:

$$\begin{aligned}
|f(x)g(x)| &= |f(x)||g(x)| \\
&< \frac{\varepsilon}{\bar{g}}\bar{g} \\
&= \varepsilon
\end{aligned}$$

(d) We will prove the limit:

$$\lim_{x \to a} \frac{1}{g(x)} = \frac{1}{L_2},$$

as long as $L_2 \neq 0$. Using this proof along with the proof of part (c), we will fully demonstrate the limit of the quotient.

Suppose that $g(x)$ is not the constant function equal to zero. Then, by Proposition 3.2, there exists a neighborhood of the point a in which the function is bounded. Let $\underline{g}$ be a lower bound of the absolute value of the function, that is:

$$0 \neq \underline{g} \leq |g(x)|$$

Given that:

$$\lim_{x \to a} g(x) = L_2$$

we know that for the positive number $|\underline{g}L_2|\varepsilon$, there exists a positive number δ such that:

$$\text{if } 0 < |x-a| < \delta \text{ then } |g(x) - L_2| < |\underline{g}L_2|\varepsilon$$

We will show that if $0 < |x-a| < \delta$ then:

$$\left| \frac{1}{g(x)} - \frac{1}{L_2} \right| < \varepsilon,$$

that is:

$$\begin{aligned} \left| \frac{1}{g(x)} - \frac{1}{L_2} \right| &= \left| \frac{L_2 - g(x)}{g(x)L_2} \right| \\ &= |L_2 - g(x)| \left| \frac{1}{g(x)} \cdot \frac{1}{L_2} \right| \\ &= |g(x) - L_2| \left| \frac{1}{g(x)} \right| \left| \frac{1}{L_2} \right| \\ &\leq |L_2 - g(x)| \left| \frac{1}{\underline{g}} \right| \left| \frac{1}{L_2} \right| \\ &< |\underline{g}L_2|\varepsilon \left| \frac{1}{\underline{g}} \right| \left| \frac{1}{L_2} \right| \\ &= \varepsilon \end{aligned}$$

(e) Suppose that $L_1 > 0$. Then, by Proposition 3.6, there exists a neighborhood of the point a in which the function is positive. Therefore:

$$\lim_{x \to a} |f(x)| = \lim_{x \to a} f(x) = L_1 = |L_1|.$$

Now suppose that $L_1 < 0$. Then, by Proposition 3.6, there exists a neighborhood of the point a in which the function is negative. Therefore:

$$\lim_{x \to a} |f(x)| = \lim_{x \to a} -f(x) = -\lim_{x \to a} f(x) = -L_1 = |L_1|.$$

Now suppose that $L_1 = 0$. Consider $\varepsilon > 0$. Then, given that:

$$\lim_{x \to a} f(x) = 0,$$

there exists a positive number δ such that:

$$\text{if } 0 < |x - a| < \delta \text{ then } |f(x)| < \varepsilon$$

We will show that if $0 < |x - a| < \delta$ then $||f|(x) - |0|| < \varepsilon$, that is:

$$\begin{aligned} ||f|(x) - |0|| &= ||f(x)| - |0|| \\ &= ||f(x)|| \\ &= |f(x)| \\ &< \varepsilon \end{aligned}$$

3.7.1 Calculation of Limits of Polynomials and Rational Functions

Using the properties of the algebra of limits and the two basic limits:

$$\begin{aligned} \lim_{x \to a} x &= a \\ \lim_{x \to a} 1 &= 1 \end{aligned}$$

some limits can be calculated easily.

Example 3.21. *Calculate the following limits:*

(a) $\lim_{x \to a} 2x$

(b) $\lim_{x \to a} x + 1$

(c) $\lim_{x \to a} x^2$

(d) $\lim_{x \to a} |x|$

(e) $\lim_{x \to a} \frac{x^2}{x+1}$

SOLUTION.

(a) $\lim_{x \to a} 2x = 2\left(\lim_{x \to a} x\right) = 2a$

(b) $\lim_{x \to a} x + 1 = \left(\lim_{x \to a} x\right) + \left(\lim_{x \to a} 1\right) = a + 1$

(c) $\lim_{x \to a} x^2 = \left(\lim_{x \to a} x\right)\left(\lim_{x \to a} x\right) = a \cdot a = a^2$

(d) $\lim_{x \to a} |x| = \left|\lim_{x \to a} x\right| = |a|$

(e) $\lim_{x \to a} \frac{x^2}{x+1} = \frac{\lim_{x \to a} x^2}{\lim_{x \to a} x + 1} = \frac{a^2}{a+1}, \quad a \neq -1.$

□

Example 3.22. *Verify that* $\lim_{x \to a} k_2 x^2 + k_1 x = k_2 a^2 + k_1 a$.

SOLUTION. We know that $\lim_{x \to a} x = a$. Then, by Proposition 3.4 (b):

$$\lim_{x \to a} k_1 \cdot x = k_1 \cdot a$$

Similarly, by Proposition 3.4 (c):

$$\lim_{x \to a} x^2 = \lim_{x \to a} x \cdot x = a \cdot a = a^2$$

and again by Proposition 3.4 (b):

$$\lim_{x \to a} k_2 x^2 = k_2 a^2$$

Finally, by Proposition 3.4 (a) we have:

$$\lim_{x \to a} k_2 x^2 + k_1 x = k_2 a^2 + k_1 a$$

In summary:

$$\begin{aligned}
\lim_{x \to a} (k_2 x^2 + k_1 x) &= \left(\lim_{x \to a} k_2 x^2 \right) + \left(\lim_{x \to a} k_1 x \right) \\
&= k_2 \left(\lim_{x \to a} x^2 \right) + k_1 \left(\lim_{x \to a} x \right) = \\
&= k_2 \left(\lim_{x \to a} x \right) \left(\lim_{x \to a} x \right) + k_1 \left(\lim_{x \to a} x \right) \\
&= k_2 a^2 + k_1 a.
\end{aligned}$$

□

Example 3.23. *Using Example 3.22, demonstrate that if* $a \neq 0$*:*

$$\lim_{x \to a} \frac{x^2 - 2x}{x} = a - 2.$$

SOLUTION. From Example 3.22 we know that:

$$\lim_{x \to a} x^2 - 2x = a^2 - 2a$$

Then, for $a \neq 0$, the final limit is calculated using Proposition 3.4 (d):

$$\lim_{x \to a} \frac{x^2 - 2x}{x} = \frac{\lim_{x \to a} x^2 - 2x}{\lim_{x \to a} x} = \frac{a^2 - 2a}{a} = \frac{a(a-2)}{a} = a - 2.$$

□

Proposed Exercise 3.3. Using the properties of limits and considering that:

$$\lim_{x\to a} x = a, \qquad \lim_{x\to a} 1 = 1$$

calculate:

(a) $\lim_{x\to 1}(x^3 - 2x + 4)$

(b) $\lim_{x\to 3} \frac{x^2+3}{x-5}$

Proposed Exercise 3.4 (Polynomial Limits)**.** Considering that:

$$\lim_{x\to a} x = a, \qquad \lim_{x\to a} 1 = 1$$

demonstrate by induction, using the properties of the algebra of limits, that if:

$$p(x) = b_n x^n + b_{n-1}x^{n-1} + b_{n-2}x^{n-2} + \cdots + b_1 x + b_0$$

then:

$$\lim_{x\to a} p(x) = b_n a^n + b_{n-1}a^{n-1} + b_{n-2}a^{n-2} + \ldots + b_1 a + b_0.$$

That is, if $p(x)$ is a polynomial of degree n:

$$\lim_{x\to a} p(x) = p(a).$$

Proposed Exercise 3.5 (Limit of the Quotient of Polynomials)**.** Using Proposed 3.4, demonstrate that if $p(x)$ is a polynomial and $q(x)$ is a polynomial such that $q(a) \neq 0$, then:

$$\lim_{x\to a} \frac{p(x)}{q(x)} = \frac{p(a)}{q(a)}.$$

We have just seen that using the properties of the algebra of function limits from Proposition 3.4 and the two basic limits:

$$\lim_{x\to a} x = a, \qquad \lim_{x\to a} 1 = 1,$$

all limits of polynomial functions and rational functions (quotients of polynomials) can be calculated, as long as the denominator does not vanish at the point where the limit is calculated.

Furthermore, the value of the limit is obtained simply by evaluating the polynomials at the given point. That is:

$$\lim_{x\to a} p(x) = p(a), \qquad \lim_{x\to a} \frac{p(x)}{q(x)} = \frac{p(a)}{q(a)}, \qquad q(a) \neq 0,$$

where $p(x), q(x) \in \mathbb{R}[x]$ are polynomials with real coefficients.

3.7.2 Calculation of Limits of Roots of Polynomials

Example 3.24 (Roots of Polynomials). *Consider $f(x) = \sqrt{x}$ and $a \geq 0$. Demonstrate that if the limit exists, it must be equal to:*

$$\lim_{x \to a} \sqrt{x} = \sqrt{a}$$

SOLUTION. Let L denote the limit we want to calculate and suppose it exists:

$$L = \lim_{x \to a} \sqrt{x}.$$

Then, using Proposition 3.4 (c) for the limit of a product of functions, we have:

$$L^2 = \left(\lim_{x \to a} \sqrt{x}\right)\left(\lim_{x \to a} \sqrt{x}\right) = \lim_{x \to a} (\sqrt{x})^2 = \lim_{x \to a} x = a$$

Therefore, if $L^2 = a \Longrightarrow L = \sqrt{a}$, as we wanted to show. □

Proposed Exercise 3.6. Demonstrate that if the following limits exist, their value must be:

(a) $\lim_{x \to a} \sqrt[3]{x} = \sqrt[3]{a}$

(b) $\lim_{x \to a} \sqrt[4]{x+2} = \sqrt[4]{a+2},\ a \geq -2$

Proposition 3.6. Let $p(x)$ be a polynomial. The limit of the nth root of $p(x)$ as x approaches a is the nth root of $p(a)$, provided that $p(a)$ is positive and n is even, or for any value of $p(a)$ if n is odd.

3.7.3 Calculation of Limits of Rational Functions of the Type «0/0»

We will illustrate in the following example how to resolve the indeterminacy that arises when calculating the limit of a rational function, where both the numerator and the denominator simultaneously vanish.

Example 3.25 (Rational Function of the Type «0/0»). *Calculate:*

$$\lim_{x \to 1} \frac{x^2 - 1}{x - 1}$$

SOLUTION. First, we evaluate the polynomials to calculate the limit:

$$\lim_{x \to 1} \frac{x^2 - 1}{x - 1} = «\frac{0}{0}» \text{ (indeterminacy)}$$

Since the denominator vanishes at $x = 1$, we cannot calculate the limit directly using the algebra of limits. However, when we calculate a limit, we take values of x close to 1, but never $x = 1$. Therefore, for $x \neq 1$:

$$\frac{(x+1)(x-1)}{x-1} = x + 1.$$

Therefore:

$$\lim_{x\to 1}\frac{x^2-1}{x-1}=\lim_{x\to 1}\frac{(x+1)(x-1)}{x-1}=\lim_{x\to 1}(x+1)=2$$

□

In general, we can state the following result.

Proposition 3.7 (Limit of Rational Function of the Type «0/0»)**.** Suppose that:

$$f(x)=\frac{p(x)}{q(x)}$$

is a rational function, where $p(x), q(x) \in \mathbb{R}[x]$ are polynomials and, moreover, $p(a) = q(a) = 0$. Then the limit:

$$\lim_{x\to a}\frac{p(x)}{q(x)}$$

is obtained by dividing the numerator and the denominator by the binomial $(x-a)$ and calculating the limit of the resulting rational function. That is:

$$\lim_{x\to a}\frac{p(x)}{q(x)}=\lim_{x\to a}\frac{\dfrac{p(x)}{x-a}}{\dfrac{q(x)}{x-a}}=\lim_{x\to a}\frac{\tilde{p}(x)}{\tilde{q}(x)},$$

where:

$$\tilde{p}(x)=\frac{p(x)}{x-a}\in\mathbb{R}[x],\qquad \tilde{q}(x)=\frac{q(x)}{x-a}\in\mathbb{R}[x].$$

Remark 3.10. It is worth noting that in Proposition 3.7, $\tilde{p}(x)$ and $\tilde{q}(x)$ are polynomials since, according to the remainder theorem, if $p(a) = q(a) = 0$, then the polynomials $p(x)$ and $q(x)$ are exactly divisible by the binomial $x-a$, that is, $p(x)$ and $q(x)$ can be expressed as:

$$p(x)=\tilde{p}(x)(x-a)$$
$$q(x)=\tilde{q}(x)(x-a)$$

3.7.4 Calculating Limits of Functions through Rationalization

We will illustrate in the following example a problem similar to that of rational functions. In this case, however, it will not involve a quotient of polynomials but a quotient of polynomials and expressions with radicals.

Example 3.26 (Rationalization)**.** *Calculate:*

$$\lim_{x\to 4}\frac{\sqrt{x}-2}{x-4}$$

SOLUTION. By the algebra of limits, we know that the limits of polynomials and roots of polynomials are calculated by evaluating the function at the point (provided the denominator does not vanish). In this case:

$$\lim_{x\to 4} \frac{\sqrt{x}-2}{x-4} = \frac{\sqrt{4}-2}{4-4} = \frac{0}{0} \text{ (indeterminate form)}$$

However, we can simplify the problem by multiplying by the conjugate of the numerator, since this is where we find the expression with radicals. Indeed:

$$\lim_{x\to 4} \frac{\sqrt{x}-2}{x-4} = \lim_{x\to 4} \frac{\sqrt{x}-2}{x-4} \cdot \frac{\sqrt{x}+2}{\sqrt{x}+2} = \lim_{x\to 4} \frac{x-4}{(x-4)(\sqrt{x}+2)} = \lim_{x\to 4} \frac{1}{\sqrt{x}+2} = \frac{1}{4}$$

□

In general, we can establish the following result.

Proposition 3.8 (Rationalization). Suppose that $f(x) = \frac{r(x)}{s(x)}$ is a function, where $r(x), s(x)$ are expressions with radicals and that, furthermore, $r(a) = s(a) = 0$. Then, the limit:

$$\lim_{x\to a} \frac{r(x)}{s(x)}$$

is calculated by multiplying by the conjugate of the expression with radicals to eliminate the indeterminate form.

Rationalization

Example 3.27. *Calculate the following limits:*

(a) $\lim_{x\to 4}(x^2 - 3x + 1)$

(b) $\lim_{x\to 4} \frac{x^2 - 3x + 1}{2x - 1}$

(c) $\lim_{x\to 4} \frac{x^3 - 5x^2 - 8x + 48}{x^4 - 4x^3 - 13x^2 + 40x + 48}$

(d) $\lim_{x\to -3} \frac{x+3}{x^3 + 6x^2 + 9x}$

(e) $\lim_{x\to 1} \frac{x^2 - 2x + 1}{x^4 + 5x^3 - x^2 - 17x + 12}$

(f) $\lim_{x\to -1} \frac{\sqrt{x+5} - x - 3}{x+1}$

(g) $\lim_{x\to 1} \frac{\sqrt{x+8} - 3}{x-1}$

(h) $\lim_{x\to -3} \frac{x^2 + 5x + 6}{\sqrt{x+6} - \sqrt{3}}$

SOLUTION.

(a) $\lim_{x\to 4}(x^2 - 3x + 1) = 4^2 - 3\cdot 4 + 1 = 5$

(b) $\lim_{x\to 4}\dfrac{x^2 - 3x + 1}{2x - 1} = \dfrac{4^2 - 3\cdot 4 + 1}{2\cdot 4 - 1} = \dfrac{5}{7}$

(c)

$$\begin{aligned}\lim_{x\to 4}\frac{x^3 - 5x^2 - 8x + 48}{x^4 - 4x^3 - 13x^2 + 40x + 48} &= \frac{4^3 - 5\cdot 4^2 - 8\cdot 4 + 48}{4^4 - 3\cdot 4^3 - 14\cdot 4^2 + 40\cdot 4 + 48}\\ &= \frac{0}{0}\ \text{(indeterminate form)}\\ &= \lim_{x\to 4}\frac{(x-4)^2(x+3)}{(x-4)^2(x^2+4x+3)}\\ &= \lim_{x\to 4}\frac{x+3}{x^2+4x+3}\\ &= \frac{4+3}{4^2+4\cdot 4+3} = \frac{7}{35} = \frac{1}{5}\end{aligned}$$

(d)

$$\begin{aligned}\lim_{x\to -3}\frac{x+3}{x^3+6x^2+9x} &= \frac{-3+3}{(-3)^3+6\cdot(-3)^2+9\cdot(-3)} = \frac{0}{0}\ \text{(indeterminate form)}\\ &= \lim_{x\to -3}\frac{(x+3)}{(x+3)(x^2+3x)}\\ &= \lim_{x\to -3}\frac{1}{x^2+3x} = \frac{1}{(-3)^2+3\cdot(-3)} = \frac{1}{0}\ \text{(does not exist)}\end{aligned}$$

(e)

$$\begin{aligned}\lim_{x\to 1}\frac{x^2-2x+1}{x^4+5x^3-x^2-17x+12} &= \frac{0}{0}\ \text{(indeterminate form)}\\ &= \lim_{x\to 1}\frac{(x-1)(x-1)}{(x-1)(x^3+6x^2+5x-12)}\\ &= \frac{0}{0}\ \text{(indeterminate form)}\\ &= \lim_{x\to 1}\frac{x-1}{(x-1)(x^2+7x+12)}\\ &= \lim_{x\to 1}\frac{1}{x^2+7x+12} = \frac{1}{20}\end{aligned}$$

(f)

$$\begin{aligned}
\lim_{x\to-1}\frac{\sqrt{x+5}-x-3}{x+1} &= \frac{\sqrt{-1+5}-(-1)-3}{-1+1} = \frac{0}{0} \text{ (indeterminate form)}\\
&= \lim_{x\to-1}\frac{\sqrt{x+5}-(x+3)}{x+1}\cdot\frac{\sqrt{x+5}+x+3}{\sqrt{x+5}+x+3}\\
&= \lim_{x\to-1}\frac{(\sqrt{x+5})^2-(x+3)^2}{(x+1)(\sqrt{x+5}+x+3)}\\
&= \lim_{x\to-1}\frac{x+5-x^2-6x-9}{(x+1)(\sqrt{x+5}+x+3)}\\
&= \lim_{x\to-1}\frac{-x^2-5x-4}{(x+1)(\sqrt{x+5}+x+3)} = \frac{0}{0} \text{ (indeterminate form)}\\
&= \lim_{x\to-1}\frac{(x+1)(-x-4)}{(x+1)(\sqrt{x+5}+x+3)}\\
&= \lim_{x\to-1}\frac{-x-4}{\sqrt{x+5}+x+3} = \frac{1-4}{\sqrt{4}+2} = -\frac{3}{4}
\end{aligned}$$

(g)

$$\begin{aligned}
\lim_{x\to1}\frac{\sqrt{x+8}-3}{x-1} &= \frac{\sqrt{1+8}-3}{1-1} = \frac{0}{0} \text{ (indeterminate form)}\\
&= \lim_{x\to1}\frac{\sqrt{x+8}-3}{x-1}\cdot\frac{\sqrt{x+8}+3}{\sqrt{x+8}+3}\\
&= \lim_{x\to1}\frac{x+8-9}{(x-1)(\sqrt{x+8}+3)}\\
&= \lim_{x\to1}\frac{x-1}{(x-1)(\sqrt{x+8}+3)}\\
&= \lim_{x\to1}\frac{1}{\sqrt{x+8}+3} = \frac{1}{\sqrt{1+8}+3} = \frac{1}{6}
\end{aligned}$$

(h)

$$\begin{aligned}
\lim_{x\to-3}\frac{x^2+5x+6}{\sqrt{x+6}-\sqrt{3}} &= \frac{(-3)^2+5(-3)+6}{\sqrt{-3+6}-\sqrt{3}} = \frac{0}{0} \text{ (indeterminate form)}\\
&= \lim_{x\to-3}\frac{x^2+5x+6}{\sqrt{x+6}-\sqrt{3}}\cdot\frac{\sqrt{x+6}+\sqrt{3}}{\sqrt{x+6}+\sqrt{3}}\\
&= \lim_{x\to-3}\frac{(x^2+5x+6)(\sqrt{x+6}+\sqrt{3})}{x+6-3}\\
&= \lim_{x\to-3}\frac{(x^2+5x+6)(\sqrt{x+6}+\sqrt{3})}{x+3}\\
&= \frac{((-3)^2+5(-3)+6)(\sqrt{(-3)+6}+\sqrt{3})}{-3+3} = \frac{0}{0} \text{ (indeterminate form)}\\
&= \lim_{x\to-3}\frac{(x+3)(x+2)(\sqrt{x+6}+\sqrt{3})}{x+3}\\
&= \lim_{x\to-3}(x+2)(\sqrt{x+6}+\sqrt{3})\\
&= (-3+2)(\sqrt{-3+6}+\sqrt{3})\\
&= -2\sqrt{3}
\end{aligned}$$

□

3.7.5 Calculating Limits through Criteria

With the properties of the algebra of limits of functions and knowing how to calculate polynomial limits at a point, we can only calculate the limits of functions that can be obtained by operating with polynomials.

To calculate limits of general functions, we need other tools. Therefore, we will introduce two criteria that will help us calculate limits. These criteria are the Squeeze Theorem and the Zero Product Theorem.

Proposition 3.9 (Squeeze Theorem). If $f, g, h : A \subset \mathbb{R} \to \mathbb{R}$ are three real functions of a real variable such that:

(1) $f(x) \leq h(x) \leq g(x), \quad \forall x \in [a-r, a+r] - \{a\}$ for some $r > 0$.

(2) $\lim_{x \to a} f(x) = \lim_{x \to a} g(x) = L$.

Then we can conclude that:

$$\lim_{x \to a} h(x) = L$$

That is, when a function is "squeezed" around a point a between two functions that have the same limit at this point, the limit as x approaches a of the squeezed function must coincide. See an illustration of this criterion in Figure 3.23.

Proof Consider $\varepsilon > 0$. Then, since $\lim_{x \to a} f(x) = L$, we know that, given the positive number ε, there exists a positive number δ_1 such that:

$$\text{if } 0 < |x-a| < \delta_1 \text{ then } |f(x) - L| < \varepsilon \text{ or, equivalently, } L - \varepsilon < f(x) < L + \varepsilon$$

Similarly, since $\lim_{x \to a} g(x) = L$, we know that given the positive number ε, there exists a positive number δ_2 such that:

$$\text{if } 0 < |x-a| < \delta_2 \text{ then } |g(x) - L| < \varepsilon \text{ or, equivalently, } L - \varepsilon < g(x) < L + \varepsilon$$

Now consider: $\delta = \min\{\delta_1, \delta_2\}$. We will demonstrate that if $0 < |x-a| < \delta$ then $|h(x) - L| < \varepsilon$, that is:

$$\lim_{x \to a} h(x) = L.$$

Indeed, if $0 < |x-a| < \delta$, then:

$$L - \varepsilon < f(x) \leq h(x) \leq g(x) < L + \varepsilon$$

which implies that:

$$L - \varepsilon < h(x) < L + \varepsilon,$$

that is, $|h(x) - L| < \varepsilon$, as we wanted to demonstrate. □

Example 3.28. *Demonstrate, using the Squeeze Theorem, that:*

$$\lim_{x \to \frac{\pi}{2}} \sin(x) = 1$$

SOLUTION. Consider the functions:

$$h(x) = \sin(x)$$
$$f(x) = 1 - \left(x - \frac{\pi}{2}\right)^2$$
$$g(x) = \left(x - \frac{\pi}{2}\right)^2 + 1$$

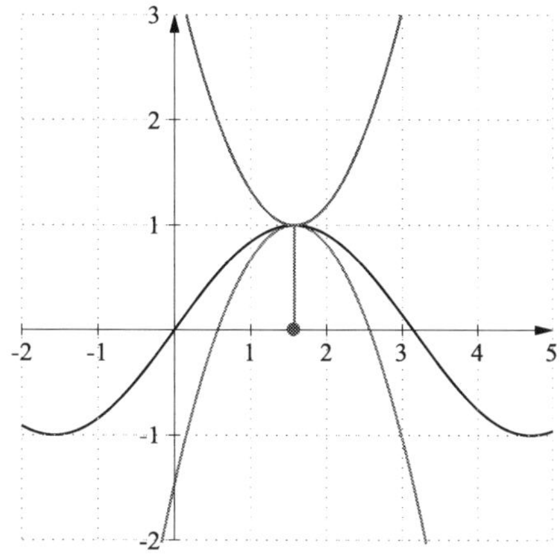

FIGURE 3.23
Illustration of the Squeeze Theorem for the functions in Example 3.28.

It can be shown that:

$$f(x) \leq h(x) \leq g(x)$$

for all real numbers and, furthermore:

$$\lim_{x \to \frac{\pi}{2}} f(x) = \lim_{x \to \frac{\pi}{2}} g(x) = 1$$

Therefore:

$$\lim_{x \to \frac{\pi}{2}} \sin(x) = 1$$

□

Remark 3.11. The difficulty of applying the Squeeze Theorem lies in finding the two bounding functions and being able to prove that $f(x) \leq h(x) \leq g(x)$.

Remark 3.12. It is important to note that it is sufficient that:

$$f(x) \leq h(x) \leq g(x), \ \forall x \in [a - r, a + r]$$

that is, in the limit of Example 3.28 we could also have considered:

$$f(x) = 1 - \left(x - \frac{\pi}{2}\right)^2 (4 - x)$$
$$g(x) = \left(x - \frac{\pi}{2}\right)^2 + 1$$

since, for example, taking $r = 1$ we have that:

$$f(x) \le h(x) \le g(x), \ \forall x \in \left[\frac{\pi}{2} - 1, \frac{\pi}{2} + 1\right]$$

as can be seen in Figure 3.24.

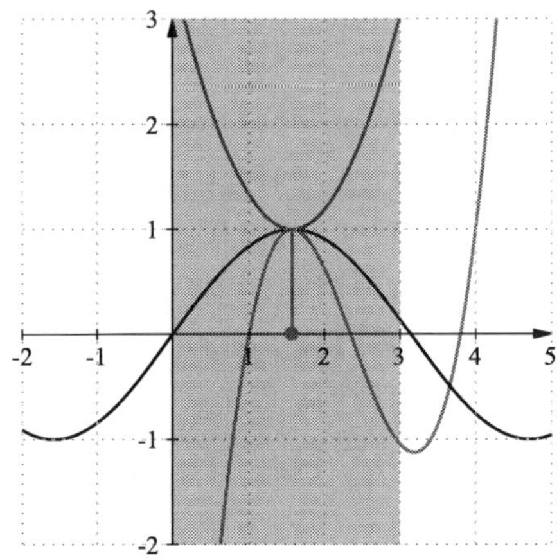

FIGURE 3.24
In the application of the Squeeze Theorem, it is sufficient that the inequalities $f(x) \le h(x) \le g(x)$ are satisfied in a neighborhood of the point a.

Proposition 3.10 (Zero Product Theorem). If $f, g : A \subset \mathbb{R} \to \mathbb{R}$ are two real functions of a real variable such that:

(1) $\lim_{x \to a} f(x) = 0$;

(2) g is bounded in a neighborhood of the point a, that is, there exists a number $r > 0$ and a positive number $k > 0$ such that $|g(x)| < k, \ \forall x \in (a - r, a + r) - \{a\}$.

Then it is verified that:

$$\lim_{x \to a} f(x) \cdot g(x) = 0.$$

Example 3.29. *Calculate the limit of:*

$$f(x) = (x - 2) \sin\left(\frac{1}{x - 2}\right)$$

as x approaches 2.

SOLUTION. Given that:

$$\lim_{x \to 2} (x - 2) = 0$$

and the function $\sin\left(\frac{1}{x-2}\right)$ is bounded since $-1 \le \sin\left(\frac{1}{x-2}\right) \le 1$ (where it is defined), then:

$$\lim_{x \to 2} (x - 2) \sin\left(\frac{1}{x - 2}\right) = 0.$$

□

Example 3.30. *Calculate the limit of $f(x) = x\sin(x)$ as x approaches 0.*

SOLUTION. Given that:

$$\lim_{x\to 0} x = 0$$

and the function $\sin(x)$ is bounded since $-1 \leq \sin(x) \leq 1$, then:

$$\lim_{x\to 0} x\sin(x) = 0.$$

□

As we have seen, calculating limits through the definition is a complex task. Therefore, we have used the algebra of limits to calculate other limits such as:

- polynomials,
- rational functions,
- functions with radicals, etc.

The Squeeze Theorem also allows us to demonstrate other limits, but as we have seen, proving limits through the Squeeze Theorem is also complex.

Next, we present the limits of other basic functions whose proofs are beyond the scope of this book. For example:

$$\begin{aligned} \lim_{x\to a} e^x &= e^a \\ \lim_{x\to a} \ln(x) &= \ln(a) \quad \text{if } a > 0 \\ \lim_{x\to a} \sin(x) &= \sin(a) \\ \lim_{x\to a} \cos(x) &= \cos(a) \end{aligned}$$

3.8 One-Sided Limits

3.8.1 Intuitive Idea

Calculating one-sided limits will allow us to characterize the behavior of a function around a point when the trends to the left and right are different. Let's see this with an example.

Consider the function:

$$f(x) = \begin{cases} 7 - x, & x < 3 \\ x + 3, & x \geq 3 \end{cases}$$

In this case, we are interested in understanding the behavior of the function around the point $x = 3$.

To intuitively study the behavior of the function, let's create a table of values (*see* Table 3.6) taking points close to 3 from the left (less than 3) and from the right (greater than 3).

We can clearly see that $f(x)$ does not have a limit at $x = 3$ because, as we take points close to 3, the images do not converge to a single value.

However, if we only take points close to 3 from the left, the images converge to 4, and if we only take points close to 3 from the right, the images converge to 6.

Therefore, we could describe the behavior around $x = 3$ by saying that approaching from the left we tend to 4, and approaching from the right we tend to 6.

TABLE 3.6
If we consider the function $f(x) = 7 - x$ if $x < 3$ and $f(x) = x + 3$ if $x \geq 3$, when the points approach 3, their respective images do not tend to a single value.

x	$f(x)$	x	$f(x)$
2.9	4.1	3.1	6.1
2.99	4.01	3.01	6.01
2.999	4.001	3.001	6.001
2.9999	4.0001	3.0001	6.0001
2.99999	4.00001	3.00001	6.00001

3.8.2 Formal Definition

Definition 3.17 (Right-Hand Limits of a Function at a Point)**.** Let $f : A \subset \mathbb{R} \to \mathbb{R}$ be a real function of a real variable and $a \in \mathbb{R}$ a point that does not necessarily belong to the domain of f. It is said that the function f has a **right-hand limit $L^+ \in \mathbb{R}$ at the point** a, if:

$\forall \varepsilon > 0 \quad \exists \delta > 0$ such that $\forall x \in A$ that satisfies $a < x < a+\delta$, we have that $|f(x) - L^+| < \varepsilon$.

See Figure 3.25.

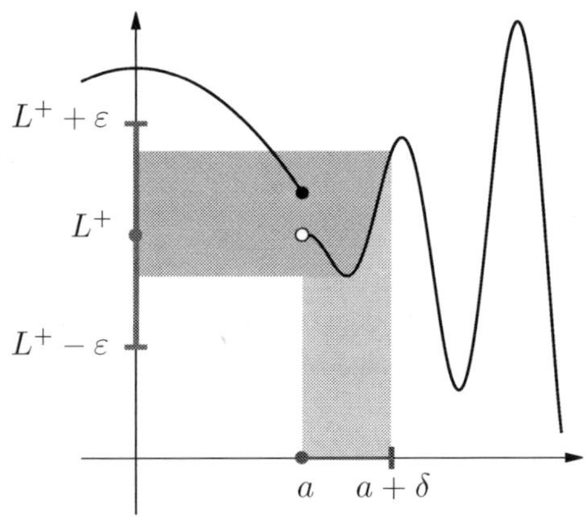

FIGURE 3.25
For the limit of $f(x)$ as x approaches a from the right to be L^+, the images of the points in the interval $(a, a + \delta)$ must fall within the interval $(L - \varepsilon, L + \varepsilon)$.

The left-hand limit is defined analogously.

Definition 3.18 (Left-Hand Limits of a Function at a Point)**.** Let $f : A \subset \mathbb{R} \to \mathbb{R}$ be a real function of a real variable and $a \in \mathbb{R}$ a point that does not necessarily belong to the domain of f. It is said that the function f has a **left-hand limit $L^- \in \mathbb{R}$ at the point** a, if:

$\forall \varepsilon > 0 \quad \exists \delta > 0$ such that $\forall x \in A$ that satisfies $a - \delta < x < a$, we have that $|f(x) - L^-| < \varepsilon$.

See Figure 3.26.

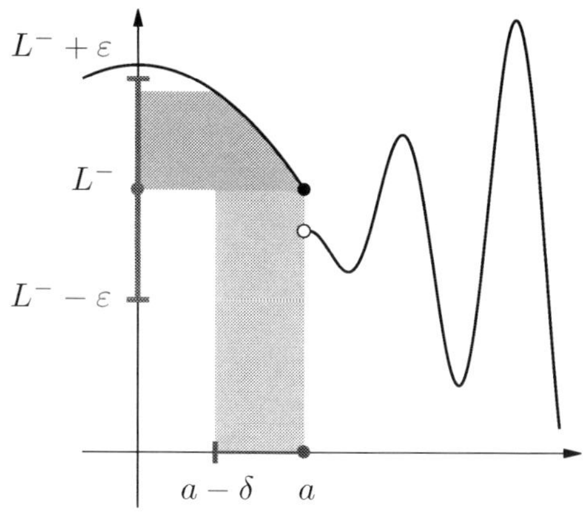

FIGURE 3.26
For the limit of $f(x)$ as x approaches a from the left to be L^-, the images of the points in the interval $(a - \delta, a)$ must fall within the interval $(L - \varepsilon, L + \varepsilon)$.

The right-hand limit and the left-hand limit are known as **one-sided limits** of a function and are denoted by:

$$L^+ = \lim_{a^+} f = \lim_{x \to a^+} f(x)$$

$$L^- = \lim_{a^-} f = \lim_{x \to a^-} f(x)$$

Example 3.31. *Calculate the one-sided limits at $x = 1$ of the function:*

$$f(x) = \begin{cases} 2 - x, & x < 1 \\ x + 1, & x \geq 1 \end{cases}$$

Solution. The left-hand limit is calculated considering values of x smaller than 1, that is, $x < 1$. Therefore, the first segment of the function will be considered. For the right-hand limit, the second segment will be considered, which is when $x > 1$. Therefore:

$$\lim_{x \to 1^-} f(x) = \lim_{x \to 1^-} (2 - x) = 1$$

$$\lim_{x \to 1^+} f(x) = \lim_{x \to 1^+} (x + 1) = 2$$

The equality of the one-sided limits will imply the existence of the limit, and vice versa, as established in Proposition 3.11. □

Proposition 3.11. If $f : A \subset \mathbb{R} \to \mathbb{R}$ is a real function of a real variable and $a \in A$, the following statements are equivalent:

(1) $\lim_{x \to a^+} f = \lim_{x \to a^-} f = L$;

(2) $\lim_{x \to a} f = L$.

Example 3.32. *Calculate the limit at point $x = 2$ of the function:*

$$f(x) = \begin{cases} (x - 2) \sin\left(\dfrac{1}{x - 2}\right), & x < 2 \\[2ex] \dfrac{x^2 - 3x + 2}{x - 1}, & x \geq 2 \end{cases}$$

SOLUTION.

$$\lim_{x\to 2^-} f(x) = \lim_{x\to 2^-} \underbrace{(x-2)}_{\to 0} \underbrace{\sin\left(\frac{1}{x-2}\right)}_{\text{acotada}} = 0$$

$$\lim_{x\to 2^+} f(x) = \lim_{x\to 2^+} \frac{x^2-3x+2}{x-1} = \frac{0}{1} = 0$$

Since:

$$\lim_{x\to 2^-} f(x) = \lim_{x\to 2^+} f(x) = 0 \Longrightarrow \lim_{x\to 2} f(x) = 0.$$

□

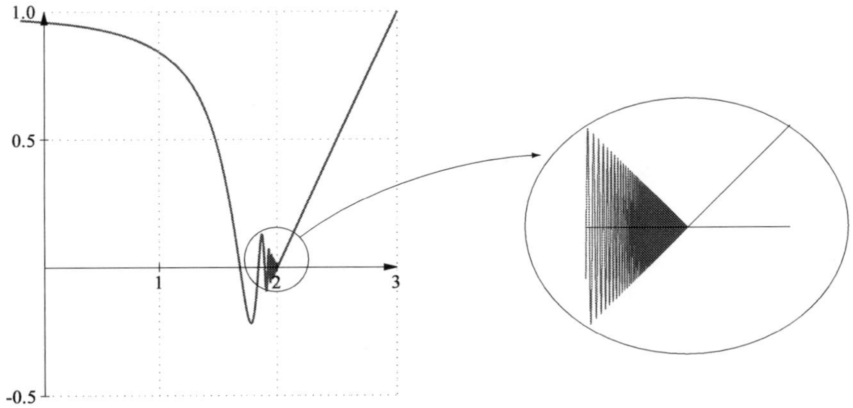

FIGURE 3.27
Graphical representation of the function in Example 3.32.

3.9 Extension of the Concept of Limit (Infinite Limit, Limit at Infinity)

The cases we have considered so far correspond to the limit of a function at a point where the result was a real number. In this section, the concept of a limit is extended to more cases, summarized in Table 3.7.

TABLE 3.7
Summary of different types of limits that can be encountered. In particular, horizontal asymptotes (HA) occur when calculating a limit at infinity with a finite result. On the other hand, vertical asymptotes (VA) correspond to an infinite limit at a point.

	Limit of a function at a point	Limit at infinity
Finite limit	$\lim_{x\to a} f(x) = L \in \mathbb{R},\ a \in \mathbb{R}$	$\lim_{x\to\pm\infty} f(x) = L \in \mathbb{R}$ (HA)
Infinite limit	$\lim_{x\to a} f(x) = \pm\infty,\ a \in \mathbb{R}$ (VA)	$\lim_{x\to\pm\infty} f(x) = \pm\infty$

3.9.1 Limit of a Function at Infinity

The objective we now set is to extend the concept of a limit to be able to define and calculate a limit of the form:

$$\lim_{x \to +\infty} f(x) = L$$

If we consider the function:

$$f(x) = \frac{1}{x} + 1$$

and take increasingly larger points ($x \to +\infty$), then we can observe in Table 3.8 how the images tend to 1.

TABLE 3.8
If we consider the function $f(x) = \frac{1}{x} + 1$ and take increasingly larger points ($x \to +\infty$), we can observe how the images tend to 1.

x	$f(x)$
1	2
10	1.1
100	1.01
1000	1.001
10000	1.0001
100000	1.00001

In Figure 3.28, the graphical representation of this function can be seen.

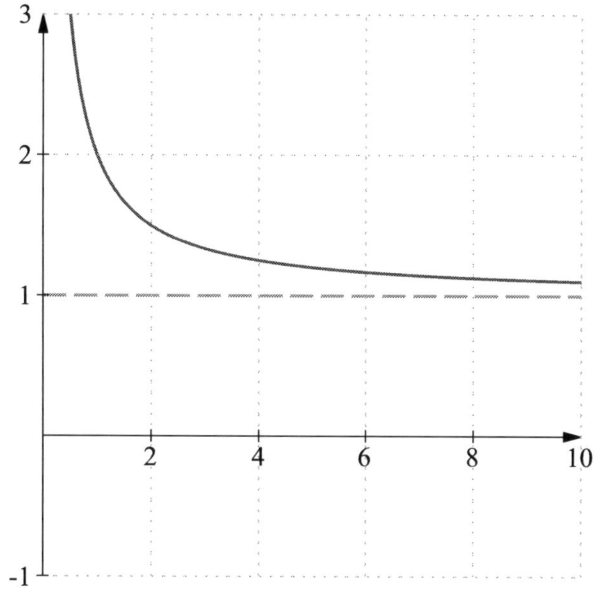

FIGURE 3.28
Graphical representation of the function $f(x) = \frac{1}{x} + 1$.

Let's formalize what we mean by the limit of a function at infinity.

Definition 3.19 (Limit of a Function at $+\infty$). A function f is said to have a limit $L \in \mathbb{R}$ as x tends to $+\infty$ if, for every positive number ε, there exists a number λ such that for all points x satisfying $x > \lambda$, we have $|f(x) - L| < \varepsilon$.

In other words, a function f has a **limit** $L \in \mathbb{R}$ as x tends to $+\infty$ if:

$$\forall \varepsilon > 0 \; \exists \lambda \text{ such that if } x > \lambda, \text{ then } |f(x) - L| < \varepsilon.$$

Remark 3.13. In other words, the value of ε serves to create a neighborhood or interval: $(L - \varepsilon, L + \varepsilon)$ around the supposed limit value. Then, once this neighborhood is set, we must ensure that there exists a point λ such that all points greater than λ have their images within this neighborhood. This idea is represented in Figure 3.29.

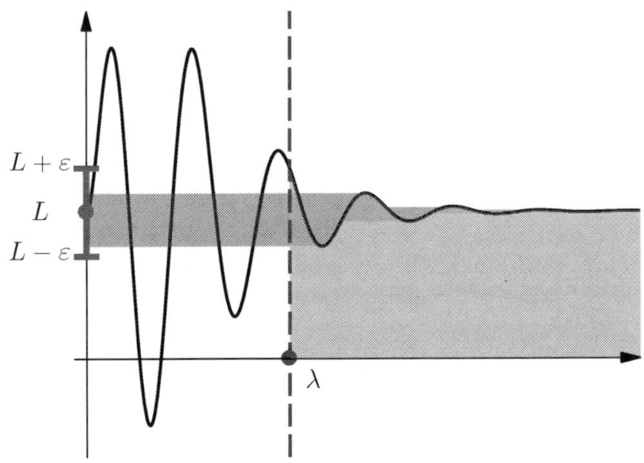

FIGURE 3.29
Graphical representation of the limit at infinity. Given a value of ε, there exists a value λ such that for all x greater than λ, its image is within the interval $[L - \varepsilon, L + \varepsilon]$.

Similarly, the limit of f when x tends to $-\infty$, $\lim_{x \to -\infty} f(x)$, is defined analogously.

Definition 3.20 (Limit of a Function at $-\infty$). A function f is said to have a limit $L \in \mathbb{R}$ as x tends to $-\infty$ if, for every positive number ε, there exists a number λ such that for all points x satisfying $x < \lambda$, we have $|f(x) - L| < \varepsilon$.

In other words, a function f has a **limit** $L \in \mathbb{R}$ as x tends to $-\infty$ if:

$$\forall \varepsilon > 0 \; \exists \lambda \text{ such that if } x < \lambda, \text{ then } |f(x) - L| < \varepsilon.$$

Observation 3.1. Horizontal Asymptote If a function has a limit L as x tends to $+\infty$, then the function has a horizontal asymptote at $y = L$.

Similarly, if a function has a limit L as x tends to $-\infty$, then the function has a horizontal asymptote at $y = L$.

Example 3.33. *Prove, using the definition of limit at infinity, that:*

$$\lim_{x \to +\infty} \frac{1}{x} = 0$$

SOLUTION. The proof of this limit will be done in the following two steps:

- **Step 1**. Relate the difference $|f(x) - L|$ to x.
- **Step 2**. Select λ as a function of ε.

For this case, we have $f(x) = \frac{1}{x}$ and $L = 0$. Therefore:

- **Step 1**.
$$|f(x) - L| = \left|\frac{1}{x} - 0\right| = \left|\frac{1}{x}\right| = \frac{1}{x} < \varepsilon \Leftrightarrow x > \frac{1}{\varepsilon}$$
- **Step 2**. If $x > \frac{1}{\varepsilon}$, then $|f(x) - L| < \varepsilon$. Therefore, we select $\lambda = \frac{1}{\varepsilon}$.

Note that in the proof we considered:

$$\left|\frac{1}{x}\right| = \frac{1}{x},$$

since when x tends to infinity, we can consider x as positive. □

Example 3.34. *Prove, using the definition of limit at infinity, that:*

$$\lim_{x \to +\infty} \frac{1}{x^p} = 0, \ p > 0$$

SOLUTION. The proof of this limit will be done in the following two steps:

- **Step 1**. Relate the difference $|f(x) - L|$ to x.
- **Step 2**. Select λ as a function of ε.

For this case, we have $f(x) = \frac{1}{x^p}$, $p > 0$ and $L = 0$. Therefore:

- **Step 1**.
$$|f(x) - L| = \left|\frac{1}{x^p} - 0\right| = \left|\frac{1}{x^p}\right| = \frac{1}{x^p} < \varepsilon \Leftrightarrow x^p > \frac{1}{\varepsilon} \Leftrightarrow x > \frac{1}{\sqrt[p]{\varepsilon}}$$
- **Step 2**. If $x > \frac{1}{\sqrt[p]{\varepsilon}}$, then $|f(x) - L| < \varepsilon$. Therefore, we select $\lambda = x > \frac{1}{\sqrt[p]{\varepsilon}}$.

Note that in the proof we considered:

$$\left|\frac{1}{x^p}\right| = \frac{1}{x^p},$$

since when x tends to infinity, we can consider x as positive. □

Remark 3.14. For the calculation of the limit:

$$\lim_{x \to -\infty} f(x),$$

the variable change $y = -x$ can be considered, and the limit:

$$\lim_{y \to +\infty} f(-y),$$

where the variable now tends to $+\infty$.

Example 3.35. *Calculate the limit:*

$$\lim_{x \to +\infty} \frac{x^3 + x^2 + 1}{2x^3 + x + 1}$$

SOLUTION. Knowing that:

$$\lim_{x \to +\infty} \frac{1}{x^p} = 0, \quad p > 0$$

and dividing by the highest degree of x, we have:

$$\lim_{x \to +\infty} \frac{x^3 + x^2 + 1}{2x^3 + x + 1} = \lim_{x \to +\infty} \frac{\frac{x^3}{x^3} + \frac{x^2}{x^3} + \frac{1}{x^3}}{\frac{2x^3}{x^3} + \frac{x}{x^3} + \frac{1}{x^3}} = \lim_{x \to +\infty} \frac{1 + \frac{1}{x} + \frac{1}{x^3}}{2 + \frac{1}{x^2} + \frac{1}{x^3}} = \frac{1 + 0 + 0}{2 + 0 + 0} = \frac{1}{2}$$

□

Example 3.36. *Calculate the limit:*

$$\lim_{x \to +\infty} \frac{x^2 + 3}{2x^4 + 5}$$

SOLUTION. Knowing that:

$$\lim_{x \to +\infty} \frac{1}{x^p} = 0, \quad p > 0$$

and dividing by the highest degree of x, we have:

$$\lim_{x \to +\infty} \frac{x^2 + 3}{2x^4 + 5} = \lim_{x \to +\infty} \frac{\frac{x^2}{x^4} + \frac{3}{x^4}}{\frac{2x^4}{x^4} + \frac{5}{x^4}} = \lim_{x \to +\infty} \frac{\frac{1}{x^2} + \frac{3}{x^4}}{2 + \frac{5}{x^4}} = \frac{0 + 0}{2 + 0} = \frac{0}{2} = 0$$

□

As we will see next, the properties of the algebra of limits of functions are also valid when the *point* where the limit is calculated is $\pm\infty$.

Proposition 3.12 (Algebra of Limits of Functions at Infinity)**.** If $f : A \to \mathbb{R}$ and $g : A \to \mathbb{R}$ are two real functions of a real variable, and if these functions are such that:

$$\lim_{x\to\infty} f = L_1$$

$$\lim_{x\to\infty} g = L_2$$

then the following holds:

(a) Limit of the sum:

$$\lim_{x\to\infty}(f+g)(x) = L_1 + L_2$$

(b) Limit of the scalar product:

$$\lim_{x\to\infty}(\lambda f)(x) = \lambda L_1,$$

for any $\lambda \in \mathbb{R}$;

(c) Limit of the product:

$$\lim_{x\to\infty}(fg)(x) = L_1 L_2;$$

(d) Limit of the quotient:

$$\lim_{x\to\infty}\left(\frac{f}{g}\right)(x) = \frac{L_1}{L_2},$$

if $L_2 \neq 0$;

(e) Limit of the absolute value:

$$\lim_{x\to\infty}|f| = |L_1|.$$

Proposed Exercise 3.7. Knowing that $\lim_{x\to+\infty}\frac{1}{x} = 0$ and using the algebra of limits, calculate the following limit:

$$\lim_{x\to+\infty}\frac{1}{x} + \frac{1}{x^2} + 1.$$

Similarly, the *squeeze theorem* and the *boundedness criterion* are also valid when the *point* where the limit is calculated is $\pm\infty$.

Example 3.37. *Calculate the limit of:*

$$f(x) = \frac{\sin(x)}{x}$$

at infinity.

SOLUTION.

$$\lim_{x \to +\infty} \frac{\sin(x)}{x} = \lim_{x \to +\infty} \frac{1}{x} \cdot \sin(x) = 0 \cdot \text{bounded} = 0$$

□

Proposed Exercise 3.8. Knowing that:

$$\lim_{x \to +\infty} \frac{1}{x} = \lim_{x \to +\infty} \frac{1}{x^3} = 0,$$

reason using the squeeze theorem that:

$$\lim_{x \to +\infty} \frac{1}{x^2} = 0$$

3.9.2 Infinite Limits

Up until this point, we have only considered the possibility that the result of calculating a limit is a real number. Otherwise, we have said that the limit did not exist. However, there is a case where the function does not have a limit, but the behavior of the function around a point will be that of growing or decreasing indefinitely.

We want to study, for example, the behavior of the function

$$f(x) = \frac{1}{x^2}$$

around the point $x = 0$. We have represented the graph of this function in Figure 3.30.

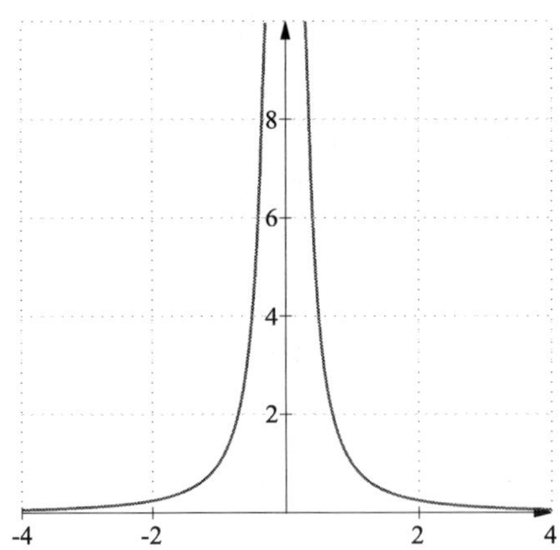

FIGURE 3.30
Graphical representation of the function $f(x) = \frac{1}{x^2}$.

Let's observe that, clearly, the limit of the function $f(x)$ does not exist when x tends to 0, that is:

$$\nexists \lim_{x \to 0} f(x).$$

But as we approach both from the right and from the left to $x = 0$, the values tend to become indefinitely larger. We will say that the values tend to positive infinity, $+\infty$. In other words, if we take points close to 0 from the left and from the right, the values do not tend to a finite number L and, therefore:

$$\lim_{x \to 0} \frac{1}{x^2} \neq L$$

but we do see (Table 3.9) that the values become increasingly larger (they approach $+\infty$).

TABLE 3.9
If we consider the function $f(x) = \frac{1}{x^2}$ and take points close to 0 from the left and the right, the values become increasingly larger.

x	$f(x)$
1	1
0.1	100
0.01	10000
0.001	1000000
0.0001	100000000

x	$f(x)$
−1	1
−0.1	100
−0.01	10000
−0.001	1000000
−0.0001	100000000

Let's formalize this idea.

Definition 3.21 (Infinite limit at a point). Let $f : A \subset \mathbb{R} \to \mathbb{R}$ be a real function of a real variable and $a \in \mathbb{R}$ a point that does not necessarily belong to the domain of f. It is said that the function f has a **limit** $+\infty$ **at the point** a, if for every positive number k, there is a positive number δ (which generally depends on k) such that for all points x that satisfy $0 < |x - a| < \delta$, we have $f(x) > k$.

In other words, the function f has a **limit** $+\infty$ **at the point** a if:

$$\forall k > 0 \ \exists \delta > 0 \text{ such that if } 0 < |x - a| < \delta, \text{ then } f(x) > k.$$

In this case, it is written:

$$\lim_{x \to a} f(x) = +\infty.$$

Remark 3.15. In other words, the value of k serves to indicate a lower bound for the values of all the points in the interval $(a - \delta, a + \delta) - \{a\}$. This idea is represented in Figure 3.31.

Analogously, the limit negative infinity at a point is defined.

Definition 3.22 (Negative infinite limit at a point). Let $f : A \subset \mathbb{R} \to \mathbb{R}$ be a real function of a real variable and $a \in \mathbb{R}$ a point that does not necessarily belong to the domain of f. It is said that the function f has a **limit** $-\infty$ **at the point** a, if for every positive number k, there is a positive number δ (which generally depends on k) such that for all points x that satisfy $0 < |x - a| < \delta$, we have $f(x) < -k$.

In other words, the function f has a **limit** $-\infty$ **at the point** a if:

$$\forall k > 0 \ \exists \delta > 0 \text{ such that if } 0 < |x - a| < \delta, \text{ then } f(x) < -k.$$

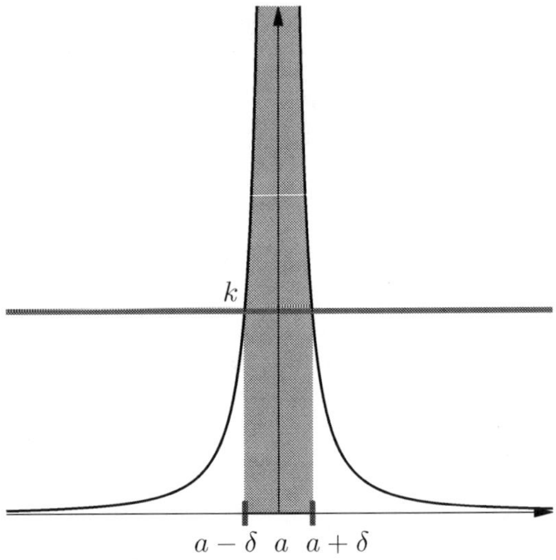

FIGURE 3.31
Graphical representation of an infinite limit at a point. Given a value k, there exists a value δ such that the values of all points in the interval $(a-\delta, a+\delta) - \{a\}$ are greater than k.

In this case, it is written:

$$\lim_{x \to a} f(x) = -\infty.$$

Observation 3.2 (Vertical Asymptote). If a function has a limit $+\infty$ or $-\infty$ when x tends to a, then it has a vertical asymptote at $x = a$.

Example 3.38. *Prove using the definition of an infinite limit that:*

$$\lim_{x \to 0} \frac{1}{x^2} = +\infty$$

SOLUTION. We will prove this limit by relating the positive real numbers k and δ. That is, given the value k, we will look for the value of δ that satisfies the definition of the limit. In this case, consider that $f(x) = \frac{1}{x^2}$ and $a = 0$.

Indeed:

$$\begin{aligned} f(x) > k \quad &\Leftrightarrow \quad \frac{1}{x^2} > k, \quad x \neq 0 \\ &\Leftrightarrow \quad \frac{1}{k} > x^2, \quad x \neq 0 \\ &\Leftrightarrow \quad |x| < \frac{1}{\sqrt{k}}, \quad x \neq 0 \\ &\Leftrightarrow \quad 0 < |x - 0| < \frac{1}{\sqrt{k}} \end{aligned}$$

If $0 < |x - 0| < \frac{1}{\sqrt{k}}$, then $f(x) > k$. Therefore, we select $\delta = \frac{1}{\sqrt{k}}$. □

Definition 3.23 (One-sided infinite limit at a point)**.** Let $f : A \subset \mathbb{R} \to \mathbb{R}$ be a real function of a real variable and $a \in \mathbb{R}$ a point that does not necessarily belong to the domain of f. It is said that the function f has a **limit** $+\infty$ **at the point** a **from the right**, if for every positive number k, there is a positive number δ (which generally depends on k) such that for all points x that satisfy $a < x < a + \delta$, we have $f(x) > k$.

In other words, the function f has a **limit** $+\infty$ **at the point** a **from the right** if:

$$\forall k > 0\ \exists \delta > 0 \text{ such that if } a < x < a + \delta, \text{ then } f(x) > k.$$

In this case, it is written:

$$\lim_{x \to a^+} f(x) = +\infty.$$

Proposed Exercise 3.9. Define, following the example of Definition 3.23, the following concepts:

- Limit $+\infty$ when x tends to a from the left.
- Limit $-\infty$ when x tends to a from the right.
- Limit $-\infty$ when x tends to a from the left.

Now consider the function $f(x) = \frac{1}{x}$ and study its behavior around the point $x = 0$, which can be observed in Figure 3.32.

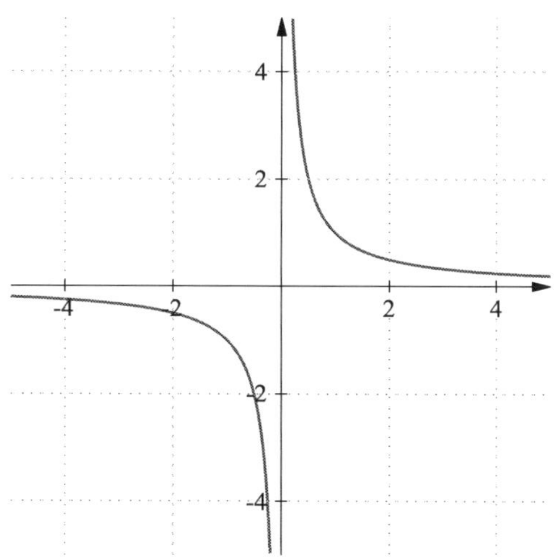

FIGURE 3.32
Graphical representation of the function $f(x) = \frac{1}{x}$.

In this case, it is easy to see that:

$$\lim_{x \to 0^-} f(x) = -\infty$$
$$\lim_{x \to 0^+} f(x) = +\infty$$

and, therefore:

$$\nexists \lim_{x \to 0} f(x)$$

Proposition 3.13. Let $f, g : A \subset \mathbb{R} \to \mathbb{R}$ be two functions such that:

$$\lim_{x\to a} f(x) = L, \quad L > 0$$
$$\lim_{x\to a} g(x) = 0$$

with $g(x)$ also being positive in a neighborhood of the point a. Then:

$$\lim_{x\to a} \frac{f(x)}{g(x)} = +\infty.$$

Example 3.39. *Calculate the limit:*

$$\lim_{x\to 2} \frac{x-1}{(x-2)^2}$$

SOLUTION. The function whose limit we need to calculate is the division of two functions. The function of the numerator is $f(x) = x - 1$ and the function of the denominator is $g(x) = (x-2)^2$. We have:

$$\lim_{x\to 2} f(x) = \lim_{x\to 2}(x-1) = 1$$

$$\lim_{x\to 2} g(x) = \lim_{x\to 2}(x-2)^2 = 0$$

Moreover, it is evident that $g(x) = (x-2)^2 > 0$ for all $x \in \mathbb{R} - \{2\}$. Therefore, applying Proposition 3.13, we have:

$$\lim_{x\to 2} \frac{x-1}{(x-2)^2} = +\infty.$$

□

Proposed Exercise 3.10. Calculate the limit:

$$\lim_{x\to 0} \frac{-1}{x^2 \sin\left(\frac{1}{x}\right)^2}$$

Proposed Exercise 3.11. Calculate the limit:

$$\lim_{x\to -\infty} \frac{1}{e^x}$$

3.9.3 Infinite Limit at Infinity

Next, we will define what we understand by infinite limit at infinity, that is:

$$\lim_{x\to +\infty} f(x) = +\infty.$$

The definition is, in a sense, a mix between Definitions 3.19 and 3.21.

Definition 3.24 (Infinite limit at infinity). Let $f : A \subset \mathbb{R} \to \mathbb{R}$ be a real function of a real variable. It is said that the function f has a **limit** $+\infty$ **at infinity**, if for every positive number k, there is a number λ (which generally depends on k) such that for all points x that satisfy $x > \lambda$, we have $f(x) > k$.

In other words, the function f has a **limit** $+\infty$ **at infinity** if:

$$\forall k > 0 \ \exists \lambda \text{ such that if } x > \lambda, \text{ then } f(x) > k.$$

In this case, it is written:

$$\lim_{x \to +\infty} f(x) = +\infty.$$

Proposed Exercise 3.12. Define, following the example of Definition 3.24, the following concepts:

- Limit $+\infty$ when x tends to negative infinity.
- Limit $-\infty$ when x tends to positive infinity.
- Limit $-\infty$ when x tends to negative infinity.

Proposed Exercise 3.13. Calculate the following limits at $+\infty$:

(a) $\lim_{x\to+\infty} x^2$

(b) $\lim_{x\to+\infty} e^x$

(c) $\lim_{x\to+\infty} \ln(x)$

(d) $\lim_{x\to+\infty} \sqrt{x}$

3.9.4 Algebra of Infinite Limits

In the following section, we will describe how to calculate limits of combinations of functions with known limits, including infinite limits. We will do this through the algebra of infinite limits.

Proposition 3.14 (Algebra of infinite limits). Consider $f, g, h : A \to \mathbb{R}$ three real functions of a real variable such that:

$$\lim_{x \to \overline{a}} f(x) = L \in \mathbb{R}$$
$$\lim_{x \to \overline{a}} g(x) = +\infty$$
$$\lim_{x \to \overline{a}} h(x) = +\infty$$

where $\overline{a} \in \overline{\mathbb{R}} = [-\infty, +\infty]$, that is, $\overline{a}$ can be a real number, but it can also be $+\infty$ or $-\infty$. Then:

- Limit of the sum/difference.

$$\lim_{x \to \overline{a}} (f(x) \pm g(x)) = L \pm (+\infty) = \pm\infty$$
$$\lim_{x \to \overline{a}} (g(x) + h(x)) = +\infty + (+\infty) = +\infty$$

- Limit of the product.

$$\lim_{x \to \overline{a}} f(x) \cdot (\pm g(x)) = L \cdot (\pm\infty) = \text{sgn}(L) \cdot (\pm\infty)$$

$$\lim_{x \to \overline{a}} g(x) \cdot h(x) = (+\infty) \cdot (+\infty) = +\infty$$

- Limit of the quotient.

$$\lim_{x \to \overline{a}} \frac{f(x)}{\pm g(x)} = \frac{L}{\pm\infty} = 0$$

As a summary, the limits that can be calculated using the algebra of infinite limits are found in Table 3.10.

TABLE 3.10
Algebra of infinite limits.

	L and $+\infty$	$+\infty$ and $+\infty$
Sum	$L \pm (+\infty) = \pm\infty$	$+\infty + (+\infty) = +\infty$
Product	$L \cdot (\pm\infty) = \text{sgn}(L) \cdot (\pm\infty)$	$(+\infty) \cdot (+\infty) = +\infty$
Quotient	$\frac{L}{\pm\infty} = 0$	

Example 3.40. *Calculate, using the algebra of infinite limits, the following limits:*

(a) $\lim_{x \to \frac{\pi}{2}} (5\cos(4x) + |\tan(x)|)$

(b) $\lim_{x \to 0} 2\frac{1}{x^2}$

(c) $\lim_{x \to 0} -\frac{1}{x^2}$

(d) $\lim_{x \to 1} 2(x+1)\frac{1}{(x-1)^2}$

(e) $\lim_{x \to 1} -2(x+1)\frac{1}{(x-1)^2}$

(f) $\lim_{x \to \frac{\pi}{2}} \frac{20\sin(x)}{\frac{1}{\cos^2(x)}}$

(g) $\lim_{x \to 0} \frac{1}{x^2} + \frac{1}{\sin^2(x)}$

(h) $\lim_{x \to 0} \frac{1}{x^2} \cdot \frac{1}{x^2}$

SOLUTION.

(a) $\lim_{x \to \frac{\pi}{2}} \left(\underbrace{5\cos(4x)}_{5} + \underbrace{|\tan(x)|}_{+\infty} \right) = 5 + (+\infty) = +\infty$

(b) $\lim_{x \to 0} 2\frac{1}{x^2} = 2 \cdot (+\infty) = +\infty$

(c) $\lim_{x \to 0} -\frac{1}{x^2} = -1 \cdot (+\infty) = -\infty$

(d) $\lim_{x \to 1} \underbrace{2(x+1)}_{4} \underbrace{\frac{1}{(x-1)^2}}_{+\infty} = 1 \cdot (+\infty) = +\infty$

(e) $\lim_{x\to 1} \underbrace{-2(x+1)}_{-4} \underbrace{\frac{1}{(x-1)^2}}_{+\infty} = -1 \cdot (+\infty) = -\infty$

(f) $\lim_{x\to \frac{\pi}{2}} \frac{20\sin(x)}{\frac{1}{\cos^2(x)}} = \frac{20}{+\infty} = 0$

(g) $\lim_{x\to 0} \frac{1}{x^2} + \frac{1}{\sin^2(x)} = (+\infty) + (+\infty) = +\infty$

(h) $\lim_{x\to 0} \frac{1}{x^2} \cdot \frac{1}{x^2} = (+\infty) \cdot (+\infty) = +\infty$

□

3.10 Indeterminate Forms

The algebra of infinite limits allows us to calculate the limit of some combinations of functions. However, there are other combinations that the algebra of limits does not resolve and that are called *indeterminate forms*.

Definition 3.25. *(Indeterminate Limit or Indeterminate Form).* We say that a limit presents an indeterminate form if the result depends on the functions in the specific example.

For example, if when calculating a limit, we encounter a difference of functions whose limits are both $+\infty$:

$$\boxed{(+\infty) - (+\infty)}$$

the result cannot be determined directly. In fact, we will see that the result will depend on the specific functions and that, in this case, it could be a constant, positive infinity, or negative infinity.

Indeed:

$$\begin{aligned}\lim_{x\to+\infty} (x - 2x) &= \text{«}(+\infty) - (+\infty)\text{»} \text{ (indeterminate form)}\\ &= \lim_{x\to+\infty} -x\\ &= -\infty\end{aligned}$$

$$\begin{aligned}\lim_{x\to+\infty} (2x - x) &= \text{«}(+\infty) - (+\infty)\text{»} \text{ (indeterminate form)}\\ &= \lim_{x\to+\infty} x\\ &= +\infty\end{aligned}$$

$$\begin{aligned}\lim_{x\to+\infty} (x - x) &= \text{«}(+\infty) - (+\infty)\text{»} \text{ (indeterminate form)}\\ &= \lim_{x\to+\infty} 0\\ &= 0\end{aligned}$$

Proposition 3.15 (Infinite limits not resolved by the algebra of limits). Consider $f_1, f_2, g, h, u : A \subset \mathbb{R} \to \mathbb{R}$ five real functions of a real variable such that:

$$\lim_{x \to \overline{a}} f_1(x) = 0$$
$$\lim_{x \to \overline{a}} f_2(x) = 0$$
$$\lim_{x \to \overline{a}} g(x) = +\infty$$
$$\lim_{x \to \overline{a}} h(x) = +\infty$$
$$\lim_{x \to \overline{a}} u(x) = 1$$

where $\overline{a} \in \overline{\mathbb{R}} = [-\infty, +\infty]$, that is, $\overline{a}$ can be a real number, but it can also be $+\infty$ or $-\infty$. Then, the following limits are indeterminate:

- $\lim_{x\to\overline{a}}(g(x) - h(x)) = «(+\infty) - (+\infty)»$
- $\lim_{x\to\overline{a}} f(x) \cdot (\pm g(x)) = «0 \cdot (\pm\infty)»$
- $\lim_{x\to\overline{a}} \frac{\pm g(x)}{\pm h(x)} = «\frac{\pm\infty}{\pm\infty}»$
- $\lim_{x\to\overline{a}} \frac{f_1(x)}{f_2(x)} = «\frac{0}{0}»$
- $\lim_{x\to\overline{a}} u(x)^{\pm g(x)} = «1^{\pm\infty}»$
- $\lim_{x\to\overline{a}} f_1(x)^{f_2(x)} = «0^0»$
- $\lim_{x\to\overline{a}} (\pm g(x))^{f_1(x)} = «(\pm\infty)^0»$

Before presenting how to solve these types of indeterminate forms, we will introduce in the following sections the concept of local equivalence of functions, as well as the orders of infinitude.

3.11 Local Equivalence of Functions

In this section, we develop a concept of great practical interest: that of locally equivalent functions, which will allow certain manipulations in the expression of the limit and to calculate the limit of a function from other functions with known limits.

Definition 3.26. *(Locally Equivalent Functions).* If $f, g : A \subset \mathbb{R} \to \mathbb{R}$ are two real functions of a real variable, f and g are said to be locally equivalent at point a if:

$$\lim_{x \to a} \frac{f(x)}{g(x)} = 1$$

When f and g are locally equivalent at point a, we write:

$$f \sim_a g$$

Proposition 3.16. If $f(x)$ and $g(x)$ are two functions such that:

$$\lim_{x\to a} f(x) = \lim_{x\to a} g(x) = L \neq 0,$$

then they are equivalent.

Proof. Let $f(x)$ and $g(x)$ be two functions such that:

$$\lim_{x\to a} f(x) = \lim_{x\to a} g(x) = L \neq 0.$$

Then, using the algebra of limits, the limit of the quotient is the quotient of limits:

$$\lim_{x\to a} \frac{f(x)}{g(x)} = \frac{L}{L} = 1$$

and, therefore, $f(x)$ and $g(x)$ are equivalent at $x = a$. □

Observation 3.3. It is important to note that if two functions f and g have the same limit $L = 0$ as x tends to a, Proposition 3.16 cannot guarantee that the two functions are equivalent at a. For example:

$$x \sim_0 x$$

but:

$$x^2 \not\sim_0 x$$

Proposition 3.17. If two functions $f, g : A \subset \mathbb{R} \to \mathbb{R}$ are locally equivalent at point a, then either both have the same limit or neither of them has a limit.

Proof. Let $f(x)$ and $g(x)$ be two functions equivalent at $x = a$ and assume that f has a limit L. Since f and g are equivalent:

$$\lim_{x\to a} \frac{f(x)}{g(x)} = 1$$

and we have:

$$\lim_{x\to a} g(x) = \lim_{x\to a} \left(g(x) \cdot \frac{f(x)}{g(x)} \right) = \lim_{x\to a} f(x) = L,$$

which proves the existence of the limit of g. □

Proposition 3.18. Local equivalence of functions is an equivalence relation. That is, given three functions $f, g, h : A \subset \mathbb{R} \to \mathbb{R}$ and a point a, we have:

(i) **Reflexivity**.

$$f \sim_a f,$$

that is, every function is locally equivalent to itself at any point a.

(ii) **Symmetry**.

$$f \sim_a g \Rightarrow g \sim_a f,$$

that is, if f is locally equivalent to g at point a, then g is locally equivalent to f at the same point a.

(iii) **Transitivity.**
$$f \sim_a g,\ g \sim_a h \Rightarrow f \sim_a h,$$
that is, if f is locally equivalent to g at point a, and if g is locally equivalent to h at the same point a, then f is locally equivalent to h at that point a.

Proof. (i) Given that:
$$\lim_{x\to a} \frac{f(x)}{f(x)} = \lim_{x\to a} 1 = 1,$$
then:
$$f \sim_a f.$$

(ii) We know that:
$$\lim_{x\to a} \frac{f(x)}{g(x)} = 1.$$
Then:
$$\lim_{x\to a} \frac{g(x)}{f(x)} = \lim_{x\to a} \frac{1}{\frac{f(x)}{g(x)}} = \frac{1}{\lim_{x\to a} \frac{f(x)}{g(x)}} = \frac{1}{1} = 1,$$
and, therefore:
$$g \sim_a f.$$

(iii) In this case, we know that:
$$\lim_{x\to a} \frac{f(x)}{g(x)} = 1$$
$$\lim_{x\to a} \frac{g(x)}{h(x)} = 1$$
Then:
$$\lim_{x\to a} \frac{f(x)}{h(x)} = \lim_{x\to a} \frac{f(x)}{h(x)} \cdot \frac{g(x)}{g(x)} = \lim_{x\to a} \frac{f(x)}{g(x)} \cdot \frac{g(x)}{h(x)} = 1 \cdot 1 = 1,$$
which implies $f \sim_a h$, as we wanted to prove.

□

3.11.1 Equivalent Infinitesimals and Infinites

Definition 3.27 (Infinitesimal)**.** The function $f(x)$ is called an *infinitesimal* when x tends to a if:
$$\lim_{x\to a} f(x) = 0.$$

Definition 3.28 (Equivalent Infinitesimals)**.** We say that $f(x)$ and $g(x)$ are two *equivalent infinitesimals* at point a if they are both infinitesimals at point a and, furthermore, they are locally equivalent at the same point a. That is, if:

$$\lim_{x\to a} f(x) = 0$$
$$\lim_{x\to a} g(x) = 0$$
$$\lim_{x\to a} \frac{f(x)}{g(x)} = 1$$

The following is a list of equivalent infinitesimals at the origin.

$$\begin{aligned} \sin(x) &\sim_0 x & 1-\cos(x) &\sim_0 \frac{x^2}{2} \\ \arcsin(x) &\sim_0 x & e^x - 1 &\sim_0 x \\ \tan(x) &\sim_0 x & c^x - 1 &\sim_0 x\ln(c),\ c\in\mathbb{R} \\ \arctan(x) &\sim_0 x & (1+x)^k - 1 &\sim_0 kx,\ k\in\mathbb{Z} \\ \ln(1+x) &\sim_0 x & \log_c(1+x) &\sim_0 \frac{x}{\ln(c)},\ c>0 \end{aligned}$$

Observation 3.4. That two functions are locally equivalent at a, that is, that the limit when x tends to a of the quotient of these two functions is 1, means that their behavior around point a is *similar*. This idea can be seen in Figures 3.33 to 3.36.

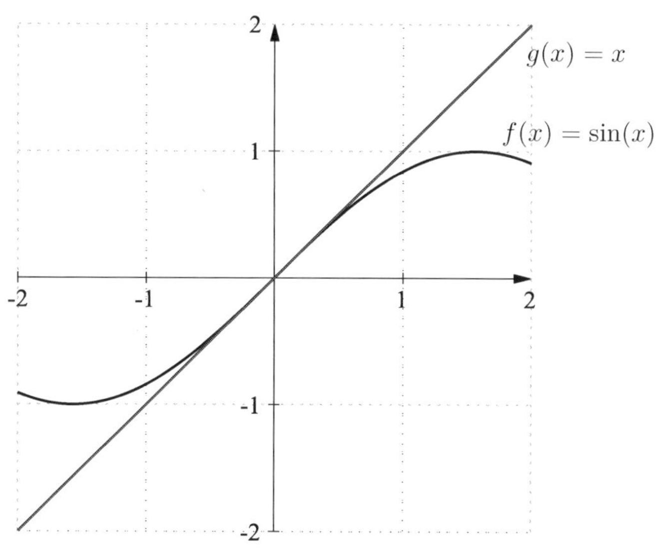

FIGURE 3.33
The function $\sin(x)$ is an equivalent infinitesimal to the function x at the origin.

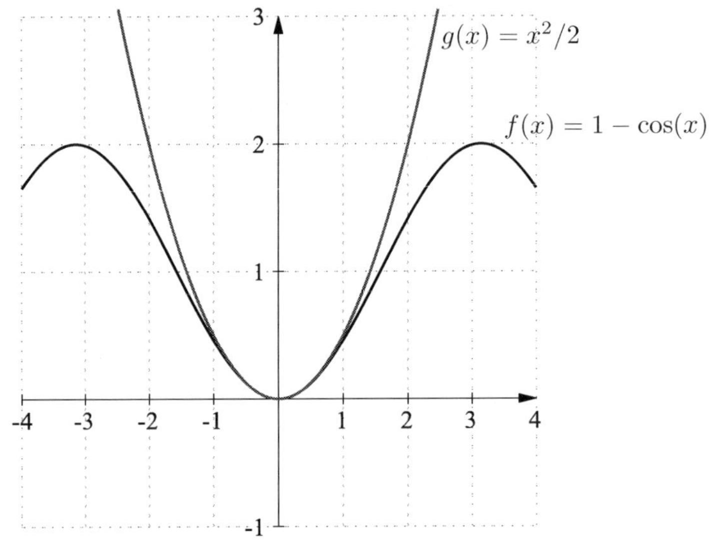

FIGURE 3.34
The function $1 - \cos(x)$ is an equivalent infinitesimal to the function $\frac{x^2}{2}$ at the origin.

Definition 3.29 (Infinite). The function $f(x)$ is called *infinite* when x tends to a if:

$$\lim_{x \to a} f(x) = \pm\infty.$$

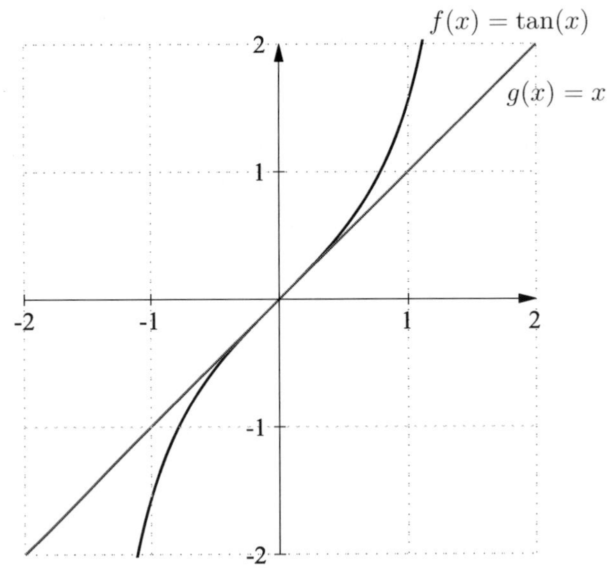

FIGURE 3.35
The function $\tan(x)$ is an equivalent infinitesimal to the function x at the origin.

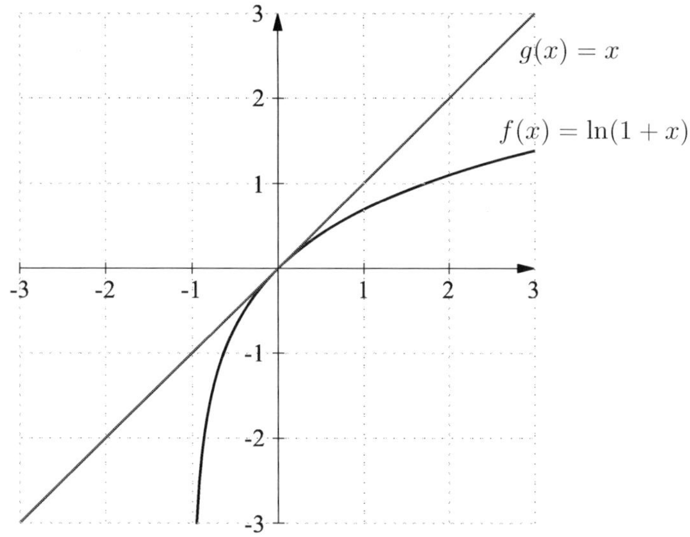

FIGURE 3.36
The function $\ln(1 + x)$ is an equivalent infinitesimal to the function x at the origin.

Remark 3.16. In Definitions 3.27-3.29, point a can be any point on the real line, $+\infty$ or $-\infty$.

In relation to the list of equivalent infinitesimals at the origin, we can also encounter equivalent infinitesimals that are not expressed exactly as in that list.

Proposition 3.19. If $\lim_{x \to a} f(x) = 0$, then:

$$\sin(f(x)) \sim_a f(x)$$

PROOF. The proof is based on the change of variable $y = f(x)$. When $x \to a$, then $y = f(x) \to 0$. Therefore:

$$\lim_{x \to a} \frac{\sin(f(x))}{f(x)} = \lim_{y \to 0} \frac{\sin(y)}{y} = 1,$$

which means that:

$$\sin(f(x)) \sim_a f(x),$$

as we wanted to demonstrate. □

Definition 3.30 (Equivalent Infinites)**.** We say that $f(x)$ and $g(x)$ are two *infinites* at point a if they are two infinites at point a and, furthermore, they are locally equivalent at the same point a. That is, if:

$$\lim_{x \to a} f(x) = 0$$
$$\lim_{x \to a} g(x) = 0$$
$$\lim_{x \to a} \frac{f(x)}{g(x)} = 1$$

Example 3.41 (Equivalent Infinites). *Two polynomials of the same degree and with the same leading coefficient are equivalent infinites at infinity. For example:*

$$x^3 + 2x + 3 \sim_{\infty} x^3$$

3.11.2 Orders of Infinity

Definition 3.31. Let $f(x)$ and $g(x)$ be two infinites. We say that $g(x)$ is an *infinite of higher order* than $f(x)$ at point a if:

$$\lim_{x \to a} \frac{f(x)}{g(x)} = 0,$$

and we denote it as:

$$f(x) \ll g(x)$$

Remark 3.17. In relation to Definition 3.31, it is equivalent to say that $g(x)$ is an *infinite of higher order* than $f(x)$ at point a or $f(x)$ is an *infinite of lower order* than $g(x)$ at point a.

Below is an ordered list of infinites at infinity:

$$\boxed{\ln(x) \ll \sqrt[n]{x} \ll \sqrt[n-1]{x} \ll x \ll x^n \ll x^{n+1} \ll 2^x \ll e^x \ll 3^x}$$

3.12 Resolving Indeterminate Forms

In this section, we will study how to resolve different types of indeterminate forms. Among other methods, we will use local equivalence of functions (equivalent infinitesimals and infinites) and orders of infinity. The following list, while not exhaustive, aims to cover as many cases as possible.

$\boxed{(+\infty) - (+\infty)}$

This type of indeterminate form is usually resolved by manipulating the expression, for example, by summing rational functions, multiplying by the conjugate of functions with radicals, or factoring out the common term.

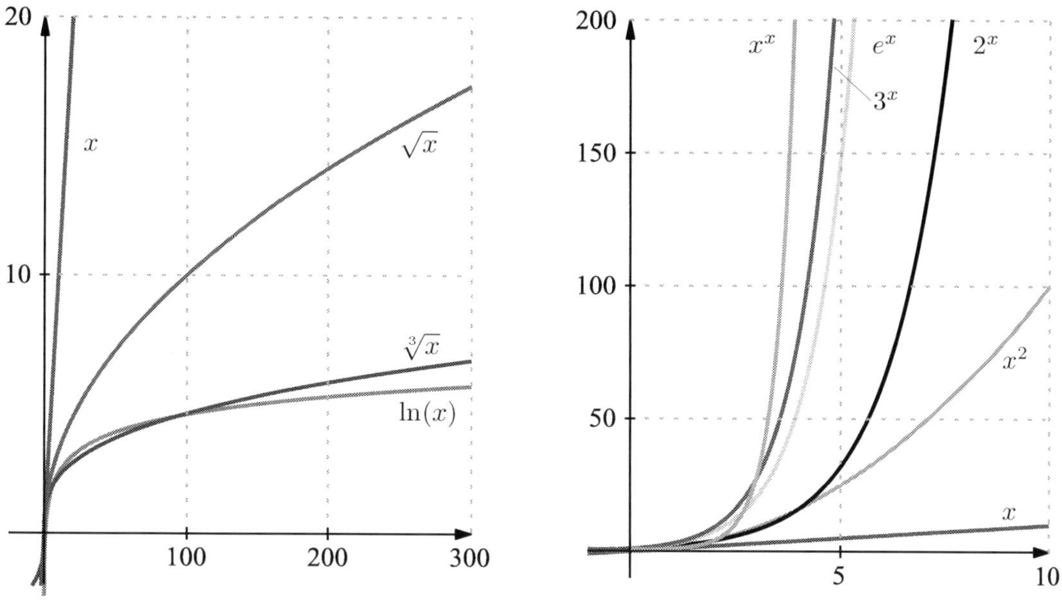

FIGURE 3.37
Graphical representation of orders of infinity. On the left are functions that tend to infinity *more slowly* than the identity function. On the right are functions that tend to infinity *faster* than the identity function.

- Simplifying the expression of the function is sometimes sufficient to resolve the indeterminate form:

$$\begin{aligned}\lim_{x\to+\infty}(x-2x) &= << (+\infty)-(+\infty) >> \text{ (indeterminate form)}\\ &= \lim_{x\to+\infty}(-x)\\ &= -\infty\end{aligned}$$

- In this case, it is necessary to factor out a common term to resolve the indeterminate form:

$$\begin{aligned}\lim_{x\to+\infty}(x-\sqrt{x}) &= << (+\infty)-(+\infty) >> \text{ (indeterminate form)}\\ &= \lim_{x\to+\infty}\sqrt{x}(\sqrt{x}-1)\\ &= (+\infty)\cdot(+\infty)\\ &= +\infty\end{aligned}$$

- In this example, it is necessary to multiply by the conjugate of an expression with radicals to resolve the indeterminate form:

$$
\begin{aligned}
\lim_{x\to+\infty}\left(\sqrt{x^2-2}-\sqrt{x^2+x}\right) &= «(+\infty)-(+\infty)» \text{ (indeterminate form)}\\
&= \lim_{x\to+\infty}\left(\sqrt{x^2-2}-\sqrt{x^2+x}\right)\frac{\sqrt{x^2-2}+\sqrt{x^2+x}}{\sqrt{x^2-2}+\sqrt{x^2+x}}\\
&= \lim_{x\to+\infty}\frac{x^2-2-x^2-x}{\sqrt{x^2-2}+\sqrt{x^2+x}}\\
&= \lim_{x\to+\infty}\frac{-x-2}{\sqrt{x^2-2}+\sqrt{x^2+x}}\\
&= \lim_{x\to+\infty}\frac{-\frac{x}{x}-\frac{2}{x}}{\sqrt{\frac{x^2}{x^2}-\frac{2}{x^2}}+\sqrt{\frac{x^2}{x^2}+\frac{x}{x^2}}}\\
&= \lim_{x\to+\infty}\frac{-1-\frac{2}{x}}{\sqrt{1-\frac{2}{x^2}}+\sqrt{1+\frac{1}{x}}}=\frac{-1}{1+1}=-\frac{1}{2}
\end{aligned}
$$

$\boxed{\dfrac{\pm\infty}{\pm\infty}}$

This type of indeterminate form is usually resolved by comparing the infinities. In the case of polynomial functions, dividing by the highest degree term; in the case of exponential functions, dividing by the highest base exponential.

- In this case, being a limit of the quotient of two polynomials, the indeterminate form is resolved by dividing by the highest degree term:

$$
\begin{aligned}
\lim_{x\to+\infty}\frac{3x^4+2x+1}{5x^4+2x+1} &= << \frac{\pm\infty}{\pm\infty} >> \text{ (indeterminate form)}\\
&= \lim_{x\to+\infty}\frac{3\left(x^4+\frac{2}{3}x+\frac{1}{3}\right)}{5\left(x^4+\frac{2}{5}x+\frac{1}{5}\right)}\\
&= \frac{3}{5}\underbrace{\lim_{x\to+\infty}\frac{x^4+\frac{2}{3}x+\frac{1}{3}}{x^4+\frac{2}{5}x+\frac{1}{5}}}_{\to 1}=\frac{3}{5}
\end{aligned}
$$

- In this limit, we use the orders of infinity to resolve the indeterminate form:

$$
\begin{aligned}
\lim_{x\to+\infty}\frac{3^x}{2^x} &= << \frac{\pm\infty}{\pm\infty} >> \text{ (indeterminate form)}\\
&= +\infty \text{ since } 2^x \ll 3^x
\end{aligned}
$$

- In this example, it is sufficient to divide by the highest base exponential to resolve the indeterminate form:

$$\begin{aligned}
\lim_{x\to+\infty} \frac{3^{x+2}+2^x}{3^{x-2}} &= << \frac{\pm\infty}{\pm\infty} >> \text{ (indeterminate form)} \\
&= \lim_{x\to+\infty} \frac{3^2\cdot 3^x + 2^x}{3^{-2}\cdot 3^x} \\
&= \lim_{x\to+\infty} \frac{\frac{3^2\cdot 3^x}{3^x} + \frac{2^x}{3^x}}{\frac{3^{-2}\cdot 3^x}{3^x}} \\
&= \lim_{x\to+\infty} \frac{3^2 + \frac{2^x}{3^x}}{3^{-2}} = \frac{3^2}{3^{-2}} = 3^4 = 81
\end{aligned}$$

$\boxed{\frac{0}{0}}$

This type of indeterminate form is usually resolved by factorization if they are rational functions without radicals, by multiplying by the conjugate if they are rational functions with radicals, or by using equivalent infinitesimals.

- In this case, the indeterminate form is resolved by using equivalent infinitesimals:

$$\begin{aligned}
\lim_{x\to 0} \frac{1-\cos(x)}{x^2} &= << \frac{0}{0} >> \text{ (indeterminate form)} \\
&= \lim_{x\to 0} \frac{1-\cos(x)}{x^2} \cdot \underbrace{\frac{\frac{x^2}{2}}{1-\cos(x)}}_{\to 1} \\
&= \lim_{x\to 0} \frac{\frac{x^2}{2}}{x^2} = \lim_{x\to 0} \frac{1}{2} = \frac{1}{2}
\end{aligned}$$

- In this limit, equivalent infinitesimals are also used to resolve the indeterminate form:

$$\begin{aligned}
\lim_{x\to 0} \frac{x^2 \sin\left(\frac{1}{x}\right)}{\sin(x)} &= << \frac{0}{0} >> \text{ (indeterminate form)} \\
&= \lim_{x\to 0} \frac{x^2 \sin\left(\frac{1}{x}\right)}{\sin(x)} \cdot \underbrace{\frac{\sin(x)}{x}}_{\to 1} \\
&= \lim_{x\to 0} x \cdot \underbrace{\sin\left(\frac{1}{x}\right)}_{\text{bounded}} = 0
\end{aligned}$$

- In this case, equivalent infinitesimals are also used to resolve the indeterminate form:

$$\begin{aligned}
\lim_{x\to 0} \frac{\sin(x)}{x^3} &= << \frac{0}{0} >> \text{ (indeterminate form)} \\
&= \lim_{x\to 0} \frac{\sin(x)}{x} \frac{1}{x^2} \\
&= \lim_{x\to 0} \frac{1}{x^2} = +\infty
\end{aligned}$$

- In this example, an equivalent trigonometric expression for $\sin(2x)$ and equivalent infinitesimals are used to resolve the indeterminate form:

$$\begin{aligned}\lim_{x\to 0}\frac{\sin(2x)}{x} &= << \frac{0}{0} >> \text{ (indeterminate form)}\\ &= \lim_{x\to 0}\frac{2\sin(x)\cos(x)}{x}\\ &= \lim_{x\to 0}\frac{\sin(x)}{x}2\cos(x)\\ &= \lim_{x\to 0}2\cos(x) = 2\cos(0) = 2\end{aligned}$$

Proposed Exercise 3.14. Calculate the limit:

$$\lim_{x\to 0}\frac{3-4\cos(x)+\cos(2x)}{x}$$

knowing that $\cos(2x) = 2\cos^2(x) - 1$.

$0\cdot(\pm\infty)$

This type of indeterminate form is usually resolved by transforming this indeterminate form into an indeterminate form of the type $\frac{0}{0}$ or $\frac{\pm\infty}{\pm\infty}$.

- To resolve the indeterminate form in this example, we manipulate it to obtain a quotient of polynomials, i.e., an indeterminate form of the type $\frac{\pm\infty}{\pm\infty}$:

$$\begin{aligned}\lim_{x\to +\infty}(x+7)\sqrt{\frac{1}{4x^2+3}} &= << 0\cdot(\pm\infty) >> \text{ (indeterminate form)}\\ &= \lim_{x\to +\infty}\sqrt{\frac{(x+7)^2}{4x^2+3}}\\ &= \lim_{x\to +\infty}\sqrt{\frac{x^2+14x+49}{4x^2+3}}\\ &= \lim_{x\to +\infty}\sqrt{\frac{1+\frac{14}{x}+\frac{49}{x^2}}{4+\frac{3}{x^2}}} = \sqrt{\frac{1}{4}} = \frac{1}{2}\end{aligned}$$

0^0

This type of indeterminate form is usually resolved by applying logarithms, which transforms the limit into another indeterminate form.

- To calculate this limit, we apply logarithms and calculate the new limit:

$$L = \lim_{x \to +\infty} \left(\frac{1}{e^{2x}} \right)^{-\frac{1}{x}} = << 0^0 >> \text{ (indeterminate form)}$$

Then:

$$\begin{aligned}
\ln(L) &= \ln \left(\lim_{x \to +\infty} \left(\frac{1}{e^{2x}} \right)^{-\frac{1}{x}} \right) \\
&= \lim_{x \to +\infty} \ln \left[\left(\frac{1}{e^{2x}} \right)^{-\frac{1}{x}} \right] \\
&= \lim_{x \to +\infty} -\frac{1}{x} \ln \left(e^{-2x} \right) \\
&= \lim_{x \to +\infty} -\frac{1}{x}(-2x) = \lim_{x \to +\infty} 2 = 2
\end{aligned}$$

Therefore:

$$\ln(L) = 2 \Leftrightarrow L = e^2$$

$(\pm\infty)^0$

This type of indeterminate form is usually resolved by applying logarithms, which transforms the limit into another indeterminate form.

- To calculate this limit, we apply logarithms and calculate the new limit:

$$L = \lim_{x \to +\infty} x^{\frac{\sin(x)}{x}} = << (\pm\infty)^0 >> \text{ (indeterminate form)}$$

Then:

$$\begin{aligned}
\ln(L) &= \ln \left(\lim_{x \to +\infty} x^{\frac{\sin(x)}{x}} \right) \\
&= \lim_{x \to +\infty} \ln \left[x^{\frac{\sin(x)}{x}} \right] \\
&= \lim_{x \to +\infty} \frac{\sin(x)}{x} \ln(x) \\
&= \lim_{x \to +\infty} \underbrace{\sin(x)}_{\text{bounded}} \underbrace{\frac{\ln(x)}{x}}_{\to 0} = 0
\end{aligned}$$

Therefore:

$$\ln(L) = 0 \Leftrightarrow L = e^0 = 1$$

$\boxed{1^{\pm\infty}}$

To resolve this type of indeterminate form, we use:

$$\lim_{x\to\pm\infty}\left(1+\frac{1}{x}\right)^{x}=e.$$

In fact, if $\lim_{x\to a} f(x)=\pm\infty$, we have:

$$\lim_{x\to a}\left(1+\frac{1}{f(x)}\right)^{f(x)}=e$$

where point a can be a real number, $+\infty$, or $-\infty$.

Therefore, when calculating a limit of this type, we must try to express the limit in this form. Here are some examples of calculating this type of limit.

Example 3.42. *Calculate the limit:*

$$\lim_{x\to+\infty}\left(4-\frac{3x}{x+5}\right)^{x^2}$$

SOLUTION.

$$\begin{aligned}
\lim_{x\to+\infty}\left(4-\frac{3x}{x+5}\right)^{x^2} &= << 1^{+\infty} >> \text{ (indeterminate form)}\\
&= \lim_{x\to+\infty}\left(1+3-\frac{3x}{x+5}\right)^{x^2}\\
&= \lim_{x\to+\infty}\left(1+\frac{3x+15-3x}{x+5}\right)^{x^2}\\
&= \lim_{x\to+\infty}\left(1+\frac{15}{x+5}\right)^{x^2}\\
&= \lim_{x\to+\infty}\left(1+\frac{1}{\frac{x+5}{15}}\right)^{x^2}\\
&= \lim_{x\to+\infty}\left(1+\frac{1}{\frac{x+5}{15}}\right)^{\frac{x+5}{15}\frac{15}{x+5}x^2}\\
&= e^{\left(\underbrace{\lim_{x\to+\infty}\frac{15}{x+5}x^2}_{\to+\infty}\right)} = +\infty
\end{aligned}$$

□

Example 3.43. *Calculate the limit:*

$$\lim_{x\to+\infty}\left(\frac{x+2}{x+1}\right)^{x}$$

SOLUTION.

$$\begin{aligned}
\lim_{x\to+\infty}\left(\frac{x+2}{x+1}\right)^x &= << 1^{+\infty} >> \text{ (indeterminate form)}\\
&= \lim_{x\to+\infty}\left(1+\frac{x+2}{x+1}-1\right)^x\\
&= \lim_{x\to+\infty}\left(1+\frac{x+2-(x+1)}{x+1}\right)^x\\
&= \lim_{x\to+\infty}\left(1+\frac{1}{x+1}\right)^x\\
&= \lim_{x\to+\infty}\left(1+\frac{1}{x+1}\right)^{(x+1)\frac{1}{x+1}x}\\
&= e^{\left(\underbrace{\lim_{x\to+\infty}\frac{x}{x+1}}_{\to 1}\right)} = e
\end{aligned}$$

□

Example 3.44 (Equivalent Infinitesimals). *Calculate the following limit using equivalent infinitesimals:*

$$\lim_{x\to 0}\frac{1-\cos(x)}{x^k},\quad k=1,2,3,4$$

SOLUTION. Since $1-\cos(x)\sim_0 \frac{x^2}{2}$, we have:

$$\begin{aligned}
\lim_{x\to 0}\frac{1-\cos(x)}{x^k} &= << \frac{0}{0} >> \text{ (indeterminate form)}\\
&= \lim_{x\to 0}\underbrace{\frac{1-\cos(x)}{x^2/2}}_{\to 1}\frac{x^2/2}{x^k}\\
&= \lim_{x\to 0}\frac{1}{2}\frac{x^2}{x^k}\\
&= \begin{cases} 0 & , \quad k=1\\ \frac{1}{2} & , \quad k=2\\ \nexists & , \quad k=3\\ +\infty & , \quad k=4\end{cases}
\end{aligned}$$

□

Example 3.45 (Equivalent Infinitesimals). *Calculate the following limit:*

$$\lim_{x\to 0}\frac{2\ln(x+1)}{\tan(x)}.$$

SOLUTION. Since $x\sim_0 \tan(x)$ and $x\sim_0 \ln(1+x)$, we have:

$$\begin{aligned}
\lim_{x\to 0}\frac{2\ln(x+1)}{\tan(x)} &= \lim_{x\to 0} 2\underbrace{\frac{\ln(x+1)}{x}}_{\to 1}\underbrace{\frac{x}{\tan(x)}}_{\to 1}\\
&= 2
\end{aligned}$$

□

Example 3.46 (Equivalent Infinitesimals). *Calculate the following limit:*

$$\lim_{x\to 0}\frac{3-4\cos(x)+\cos(2x)}{x},$$

knowing that $\cos(2x)=2\cos^2(x)-1$.

SOLUTION.

$$\begin{aligned}\lim_{x\to 0}\frac{3-4\cos(x)+\cos(2x)}{x} &= \lim_{x\to 0}\frac{3-4\cos(x)+2\cos^2(x)-1}{x}\\ &= \lim_{x\to 0}\frac{2\cos^2(x)-4\cos(x)+2}{x}\\ &= \lim_{x\to 0}\frac{2\left(\cos(x)-1\right)^2}{x}\\ &= \lim_{x\to 0}\frac{2\left(\cos(x)-1\right)^2}{x}\cdot\frac{\left(\frac{x^2}{2}\right)^2}{(1-\cos(x))^2}\\ &= \lim_{x\to 0}\frac{2x^4}{4x}=\lim_{x\to 0}\frac{x^3}{2}=0\end{aligned}$$

□

Example 3.47 (Equivalent Infinitesimals). *Calculate the following limit:*

$$\lim_{x\to 0}\frac{1-\cos(1-\cos(x))}{\sin^2(x^2)}$$

SOLUTION. In this limit, we will use:

$$\begin{aligned}1-\cos(f(x)) &\sim_0 \frac{f(x)^2}{2}\\ \sin(f(x)) &\sim_0 f(x)\end{aligned}$$

whenever $f(x)\to 0$ as $x\to 0$. Indeed:

$$\begin{aligned}\lim_{x\to 0}\frac{1-\cos(1-\cos(x))}{\sin^2(x^2)} &= \lim_{x\to 0}\frac{1-\cos(1-\cos(x))}{\sin^2(x^2)}\cdot\frac{\frac{(1-\cos(x))^2}{2}}{1-\cos(1-\cos(x))}\cdot\frac{\sin(x^2)}{x^2}\cdot\frac{\sin(x^2)}{x^2}\\ &= \lim_{x\to 0}\frac{(1-\cos(x))^2}{2x^4}\\ &= \lim_{x\to 0}\frac{(1-\cos(x))^2}{2x^4}\cdot\frac{\frac{x^2}{2}}{1-\cos(x)}\cdot\frac{\frac{x^2}{2}}{1-\cos(x)}\\ &= \lim_{x\to 0}\frac{1}{2}\cdot\frac{1}{x^4}\cdot\frac{x^4}{4}=\lim_{x\to 0}\frac{1}{8}=\frac{1}{8}\end{aligned}$$

□

Example 3.48 (Orders of Infinity). *Calculate the limit:*

$$\lim_{x\to +\infty}\frac{\ln(x)}{x}.$$

SOLUTION. By the orders of infinity, we have:

$$\ln(x) \ll x.$$

Therefore:

$$\lim_{x \to +\infty} \frac{\ln(x)}{x} = 0.$$

□

Example 3.49 (Orders of Infinity). *Calculate the limit:*

$$\lim_{x \to +\infty} \frac{x^5}{e^x}.$$

SOLUTION. By the orders of infinity, we have:

$$x^5 \ll e^x.$$

Therefore:

$$\lim_{x \to +\infty} \frac{x^5}{e^x} = 0.$$

□

Example 3.50 (Orders of Infinity). *Calculate the limit:*

$$\lim_{x \to +\infty} \frac{\ln(x)}{\sqrt{x}}.$$

SOLUTION. By the orders of infinity, we have:

$$\ln(x) \ll \sqrt{x}.$$

Therefore:

$$\lim_{x \to +\infty} \frac{\ln(x)}{\sqrt{x}} = 0.$$

□

Example 3.51 (Orders of Infinity). *Calculate the limit:*

$$\lim_{x \to +\infty} \frac{x}{\ln(x)}.$$

SOLUTION. By the orders of infinity, we have:

$$\ln(x) \ll \sqrt{x}.$$

Therefore:

$$\begin{aligned}
\lim_{x \to +\infty} \frac{x}{\ln(x)} &= \lim_{x \to +\infty} \frac{1}{\frac{\ln(x)}{x}} \\
&= \frac{1}{\lim_{x \to +\infty} \frac{\ln(x)}{x}} \\
&= \frac{1}{0^+} = +\infty
\end{aligned}$$

since $\frac{\ln(x)}{x} > 0$ for large values of x. □

Example 3.52 (Number e). *Calculate the limit:*

$$L = \lim_{x \to +\infty} \left(1 + \frac{1}{x(\sin(10x)+2)}\right)^{x(\sin(10x)+2)}.$$

SOLUTION. If we define $f(x) = x(\sin(10x)+2)$, we have:

$$\lim_{x \to +\infty} f(x) = \lim_{x \to +\infty} x \underbrace{(\sin(10x)+2)}_{>0 \text{ and bounded}} = +\infty,$$

and, therefore:

$$L = \lim_{x \to +\infty} \left(1 + \frac{1}{f(x)}\right)^{f(x)} = e.$$

In Figure 3.38, this function is graphically represented and its limit as x tends to $+\infty$ can be seen. □

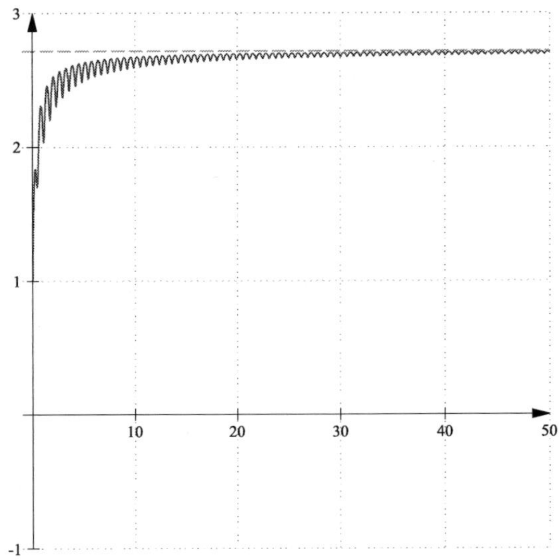

FIGURE 3.38
Graph of the function $f(x) = \left(1 + \frac{1}{x(\sin(10x)+2)}\right)^{x(\sin(10x)+2)}$, where it can be seen that as x tends to $+\infty$, the function presents a horizontal asymptote at $y = e$.

Example 3.53 (Number e). *Calculate the limit:*

$$\lim_{x \to \infty} \left(\frac{x+2}{x+1}\right)^{\dfrac{ax^2-1}{x+3}}$$

SOLUTION. These limits are calculated by introducing the number e:

$$\lim_{x\to+\infty}\left(\frac{x+2}{x+1}\right)^{\frac{ax^2-1}{x+3}} = \lim_{x\to+\infty}\left(1+\frac{1}{x+1}\right)^{\frac{(x+1)(ax^2-1)}{(x+1)(x+3)}}$$
$$= \lim_{x\to+\infty}\left[\left(1+\frac{1}{x+1}\right)^{x+1}\right]^{\frac{ax^2-1}{(x+1)(x+3)}}$$
$$= e^{\lim\limits_{x\to+\infty}\frac{ax^2-1}{(x+1)(x+3)}} = e^a$$

□

Proposed Exercise 3.15. Calculate the limit:

$$\lim_{x\to e}\frac{\ln(\ln(x))}{x-e}$$

Proposed Exercise 3.16. Calculate the limit:

$$\lim_{x\to\pi}\frac{1+\cos(x)}{(x-\pi)^2}$$

by making the change of variable $y = x - \pi$ and remembering that $\cos(y+\pi) = -\cos(y)$.

Proposed Exercise 3.17. Calculate the limit:

$$\lim_{x\to+\infty}\frac{x^5-\sqrt{x}+\ln(x)}{e^x+\cos(x)+\sqrt{x}}$$

by dividing by the highest order infinite.

3.13 Continuity of a Function at a Point

Intuitively, we can say that a function is continuous if we can graphically represent it in one stroke. Obviously, this notion is insufficient and is too closely tied to the graphical representation of the function. To formalize this idea, we first define the concept of continuity of a function at a point, and later, we will define the continuity of a function on intervals.

Definition 3.32. If $f : A \subset \mathbb{R} \to \mathbb{R}$ is a real function of a real variable, the function f is said to be continuous at $x = a$ if it satisfies:

$$\lim_{x\to a} f(x) = f(a)$$

Polynomials, the exponential function, logarithm, sine and cosine functions, and functions with radicals are continuous at all points in their domain, as they all satisfy:

$$\lim_{x\to a} f(x) = f(a)$$

Example 3.54. *The function* $f(x) = \frac{1}{x}$ *is not continuous at* $x = 0$ *because* $\lim_{x\to 0} f(x)$ *does not exist, as can be seen in Figure 3.39.*

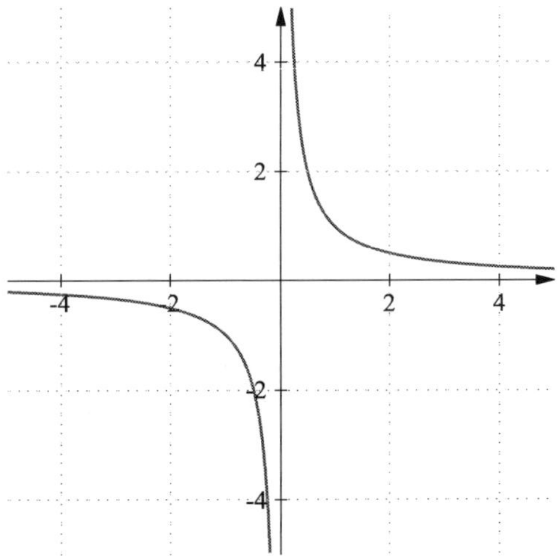

FIGURE 3.39
The function $f(x) = \frac{1}{x}$ is not continuous at $x = 0$.

Example 3.55. *The function:*

$$f(x) = x \sin\left(\frac{1}{x}\right)$$

is not continuous at $x = 0$ *because although:*

$$\lim_{x\to 0} f(x) = 0,$$

the function is not defined at $x = 0$*. The graph of the function and a detail of it can be seen in Figure 3.40.*

Proposition 3.20 (Continuity Criteria)**.** If two functions $f, g : A \subset \mathbb{R} \to \mathbb{R}$ are continuous at the point $x = a \in A$, then the following properties hold:

(a) **Continuity of the Sum Function.** The function $f + g$ is continuous at a.

(b) **Continuity of the Product Function.** The function $f \cdot g$ is continuous at a.

(c) **Continuity of the Scalar Multiple Function.** The function $\lambda \cdot f$ is continuous at a for all $\lambda \in \mathbb{R}$.

(d) **Continuity of the Quotient Function.** The function $\frac{f}{g}$ is continuous at a if $g(a) \neq 0$.

Proof. The proof of these properties is based on the algebra of limits. Indeed:

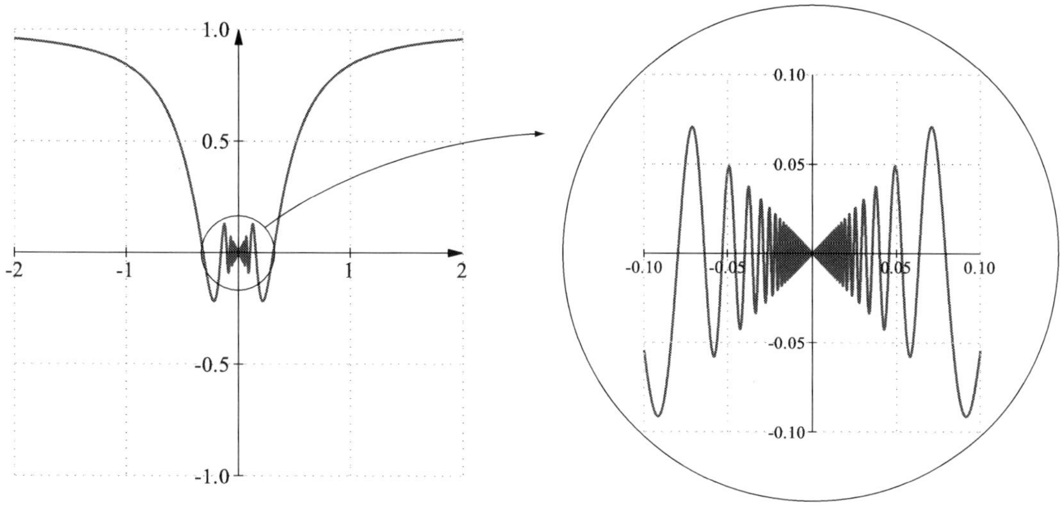

FIGURE 3.40
Graphical representation of the function $f(x) = x \sin\left(\frac{1}{x}\right)$ (left) and detail of the graph (right).

(a) **Continuity of the Sum Function:**

$$\begin{aligned}\lim_{x\to a}(f+g)(x) &= \lim_{x\to a}(f(x)+g(x)) \\ &= \underbrace{\lim_{x\to a} f(x)}_{f(a)} + \underbrace{\lim_{x\to a} g(x)}_{g(a)} \\ &= f(a)+g(a) \\ &= (f+g)(a)\end{aligned}$$

(b) **Continuity of the Product Function:**

$$\begin{aligned}\lim_{x\to a}(f\cdot g)(x) &= \lim_{x\to a}(f(x)\cdot g(x)) \\ &= \underbrace{\lim_{x\to a} f(x)}_{f(a)} \cdot \underbrace{\lim_{x\to a} g(x)}_{g(a)} \\ &= f(a)\cdot g(a) \\ &= (f\cdot g)(a)\end{aligned}$$

(c) **Continuity of the Scalar Multiple Function:**

$$\begin{aligned}\lim_{x\to a}(\lambda\cdot f)(x) &= \lim_{x\to a}\lambda\cdot f(x) \\ &= \lambda\cdot\lim_{x\to a} f(x) \\ &= \lambda\cdot f(a) \\ &= (\lambda\cdot f)(a)\end{aligned}$$

(d) **Continuity of the Quotient Function:**

$$\begin{aligned}\lim_{x\to a}(f/g)(x) &= \lim_{x\to a}(f(x)/g(x))\\ &= \underbrace{\lim_{x\to a} f(x)}_{f(a)} / \underbrace{\lim_{x\to a} g(x)}_{g(a)}\\ &= f(a)/g(a),\ g(a)\neq 0\\ &= (f/g)(a)\end{aligned}$$

□

Example 3.56. *The function:*

$$f(x) = x + 2e^x + \sqrt{x^2+1} + \frac{x+3}{x^4+1}$$

is continuous over all $\mathbb{R}$ *because it is formed by sums, products, and quotients of continuous functions where the denominators do not vanish.*

3.13.1 Continuity Field

Definition 3.33. If $f : A \subset \mathbb{R} \to \mathbb{R}$ is a real function of a real variable, the function f is said to be continuous on $X \subset A$ if f is continuous at $x = a$ for every $a \in X$.

Example 3.57. *The function:*

$$f(x) = \frac{x+1}{(x^2-1)(x-3)}$$

is continuous on $\mathbb{R}\backslash\{-1, 1, 3\}$*, since except for these three points, the function is a quotient of polynomials where the denominator is not zero. The graph of this function can be found in Figure 3.41.*

Example 3.58. *The functions* $f(x) = \lfloor x \rfloor$ *and* $g(x) = \lceil x \rceil$ *(the floor and ceiling functions, respectively) are continuous on* $X = \mathbb{R}\backslash\mathbb{Z}$*. In Figure 3.42, the graphs of these two functions are shown.*

Example 3.59. *The functions* $f(x) = x - \lfloor x \rfloor$ *and* $g(x) = \lceil x \rceil - x$ *are continuous on* $X = \mathbb{R}\backslash\mathbb{Z}$*. In Figure 3.43, we can see the graphs of these two functions.*

3.13.2 Lateral Continuity

Definition 3.34. If $f : A \subset \mathbb{R} \to \mathbb{R}$ is a real function of a real variable, the function f is said to be continuous from the **right** at $x = a$ if the following equality holds:

$$\lim_{x\to a^+} f(x) = f(a)$$

Similarly, **left** continuity is defined as follows.

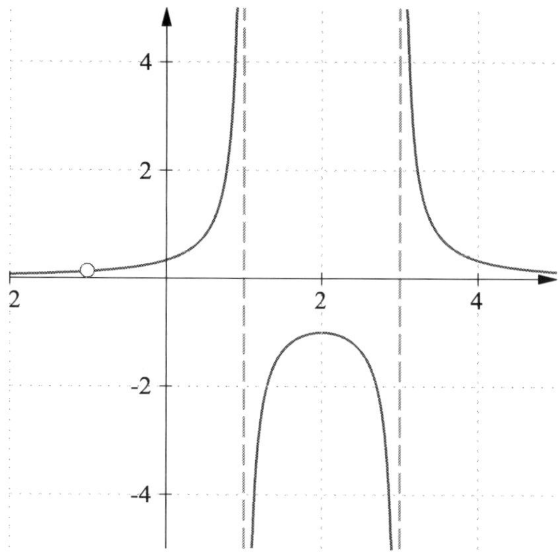

FIGURE 3.41
Graphical representation of the function $\frac{x+1}{(x^2-1)(x-3)}$. The function is not continuous at $x = -1$, $x = 1$ and $x = 3$.

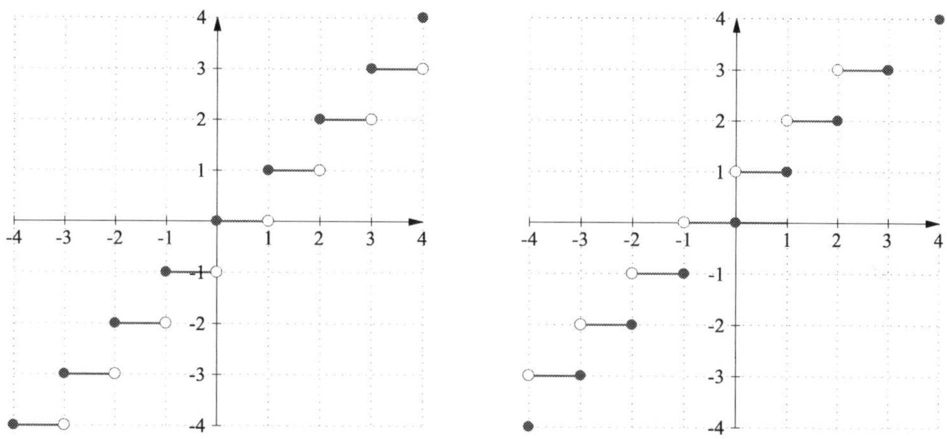

FIGURE 3.42
Graph of the floor function (left) and the ceiling function (right). The points where these two functions are not continuous are all integers.

Definition 3.35. If $f : A \subset \mathbb{R} \to \mathbb{R}$ is a real function of a real variable, the function f is said to be continuous from the **left** at $x = a$ if the following equality holds:

$$\lim_{x \to a^-} f(x) = f(a)$$

Proposition 3.21. If $f : A \subset \mathbb{R} \to \mathbb{R}$ is a real function of a real variable, the function f is said to be continuous at $x = a$ if and only if it is simultaneously continuous from the right and from the left.

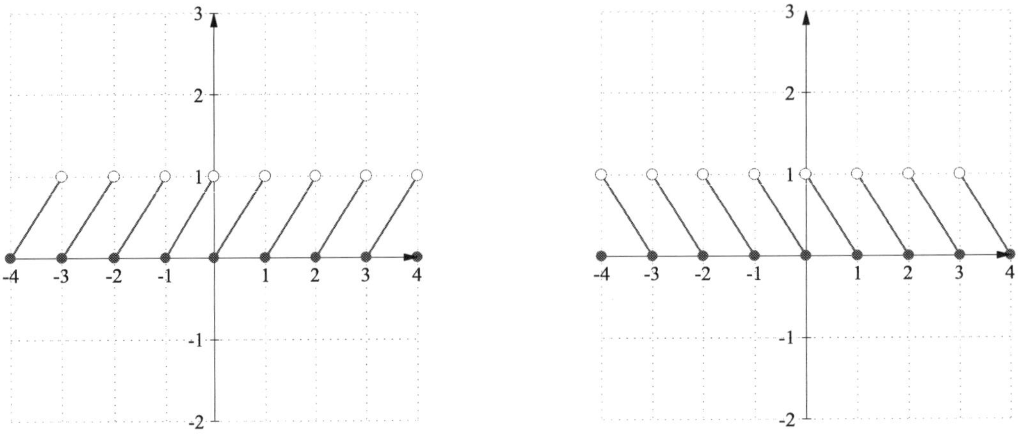

FIGURE 3.43
Graph of the function $f(x) = x - \lfloor x \rfloor$ (left) and $g(x) = \lceil x \rceil - x$ (right). The points where these two functions are not continuous are all integers.

3.13.3 Continuity in Open and Closed Intervals

Definition 3.36 (Continuity in an open interval)**.** If $f : [a, b] \to \mathbb{R}$, the function f is said to be continuous in the open interval (a, b) if f is continuous at the point x for every $x \in (a, b)$. In this case, we can write:

$$f \in \mathcal{C}^0(a, b).$$

Remark 3.18. Continuity in an open interval includes continuity over the *entire* set of real numbers. For example, polynomials, quotients of polynomials with a non-zero denominator, and trigonometric functions are continuous functions over the entire set of real numbers $\mathbb{R}$. In these cases, we will write:

$$\begin{aligned} &p(x) \in \mathcal{C}^0(\mathbb{R}),\ p(x) \in \mathbb{R}[x] \\ &\frac{p(x)}{q(x)} \in \mathcal{C}^0(\mathbb{R}),\ p(x), q(x) \in \mathbb{R}[x],\ q(x) \neq 0 \\ &\sin(x) \in \mathcal{C}^0(\mathbb{R}) \\ &\cos(x) \in \mathcal{C}^0(\mathbb{R}) \end{aligned}$$

Definition 3.37 (Continuity in a closed interval)**.** If $f : [a, b] \to \mathbb{R}$, the function f is said to be continuous in the closed interval $[a, b]$ if:

(i) f is continuous at the point x, for every $x \in (a, b)$;

(ii) f is continuous from the right at the point $x = a$;

(iii) f is continuous from the left at the point $x = b$.

In this case, we can write:

$$f \in C^0[a, b].$$

Example 3.60. *Consider the function:*

$$f(x) = \begin{cases} \frac{x}{4} & x < 0 \\ 2 - x & 0 \le x \le 4 \\ \cos(x) & x > 4 \end{cases}$$

The function is continuous on $\mathbb{R} - \{0, 4\}$.

Additionally, the function is continuous from the right at $x = 0$ *and continuous from the left at* $x = 4$, *as shown in Figure 3.44.*

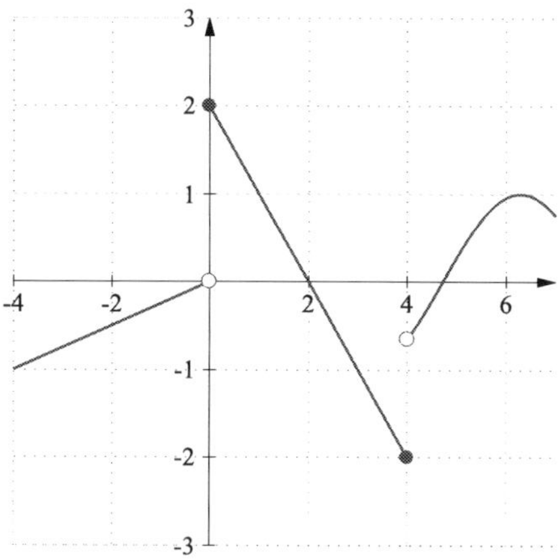

FIGURE 3.44
Graph of the function from Example 3.60.

Example 3.61. *The functions* $f(x) = \lfloor x \rfloor$ *and* $\tilde{f}(x) = x - \lfloor x \rfloor$ *are continuous from the right throughout* $\mathbb{R}$. *On the other hand, the functions* $g(x) = \lceil x \rceil$ *and* $\tilde{g}(x) = \lceil x \rceil - x$ *are continuous from the left throughout* $\mathbb{R}$.

Example 3.62. *The function* $g(x) = sgn(\sin(\pi x))$ *is neither continuous from the right nor from the left in the set of integers* $\mathbb{Z}$, *if we define the function sgn as follows:*

$$sgn(x) = \begin{cases} 1 & x > 0 \\ 0 & x = 0 \\ -1 & x < 0 \end{cases}$$

See Figure 3.45.

3.13.4 Classification of Discontinuities of a Function

The points where a function is not continuous can be classified in three ways: *removable discontinuity*, *jump discontinuity*, or *essential discontinuity*.

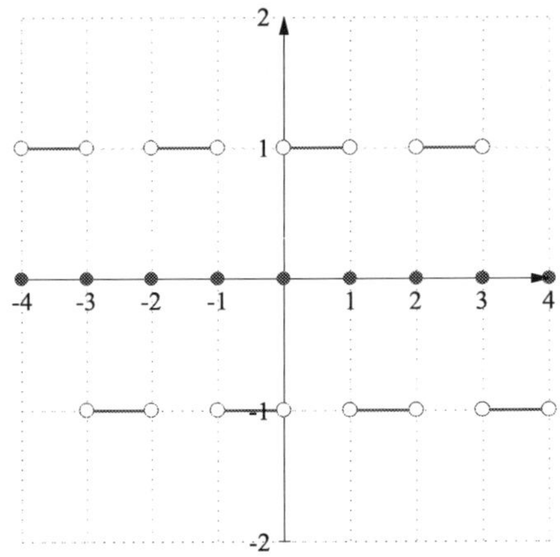

FIGURE 3.45
Graph of the function $g(x) = \text{sgn}(\sin(\pi x))$.

Definition 3.38 (Removable discontinuity). We say that the function $f : A \subset \mathbb{R} \to \mathbb{R}$ has a *removable discontinuity* at $x = a$ when either:

$$\lim_{x \to a} f(x) = L \neq f(a),$$

or:

$$\lim_{x \to a} f(x) = L, \quad \text{but } f(a) \text{ does not exist}$$

that is, when the limit of the function exists but it does not coincide with $f(a)$, either because it is different or because a is not in the domain of the function. *See* Figure 3.46.

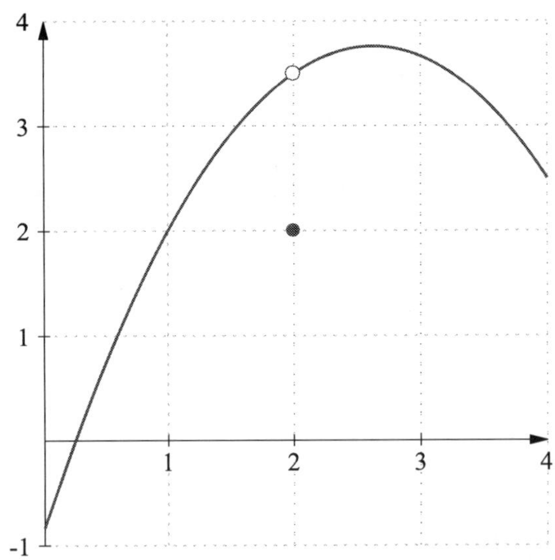

FIGURE 3.46
The function in the graph has a removable discontinuity at the point $x = 2$.

Definition 3.39 (Jump discontinuity). We say that the function $f : A \subset \mathbb{R} \to \mathbb{R}$ has a *jump discontinuity* (or finite jump or first kind discontinuity) at $x = a$ when:

$$\lim_{x\to a^-} f(x) = L_1 \neq L_2 = \lim_{x\to a^+} f(x),$$

that is, when the one-sided limits exist but are different. *See* Figure 3.47.

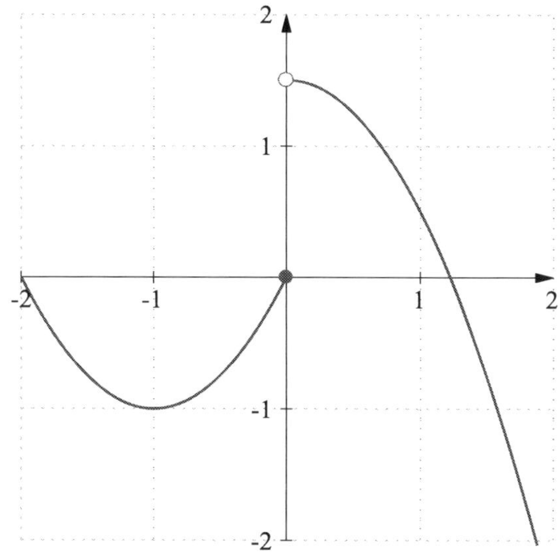

FIGURE 3.47
The function in the graph has a jump discontinuity at the point $x = 0$.

Definition 3.40 (Essential discontinuity). We say that the function $f : A \subset \mathbb{R} \to \mathbb{R}$ has an *essential discontinuity* (or asymptotic or infinite jump or second kind discontinuity) at $x = a$ when either $\lim_{x\to a^-} f(x)$ does not exist, or $\lim_{x\to a^+} f(x)$ does not exist, or neither one-sided limit exists. *See* Figure 3.48.

Remark 3.19 (Asymptotic or essential discontinuity). We can find books that distinguish between an *asymptotic* discontinuity and an *essential* discontinuity as follows:

(i) We say that the function $f : A \subset \mathbb{R} \to \mathbb{R}$ has an *asymptotic discontinuity* at $x = a$ when $\lim_{x\to a^-} f(x) = \pm\infty$, or $\lim_{x\to a^+} f(x) = \pm\infty$, or both limits are infinite.

(ii) We say that the function $f : A \subset \mathbb{R} \to \mathbb{R}$ has an *essential discontinuity* at $x = a$ when $\nexists \lim_{x\to a^-} f(x)$ or $\nexists \lim_{x\to a^+} f(x)$.

Example 3.63 (essential discontinuity). *Consider the function:*

$$f(x) = \begin{cases} x, & x \in \mathbb{Q} \\ -x, & x \in \mathbb{R} - \mathbb{Q} \end{cases}$$

The only point where this function is continuous is $x = 0$. At all other points, the function has an essential (non-asymptotic) discontinuity.

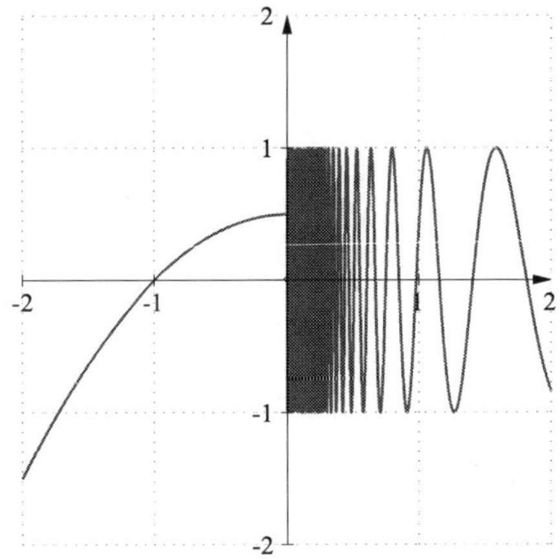

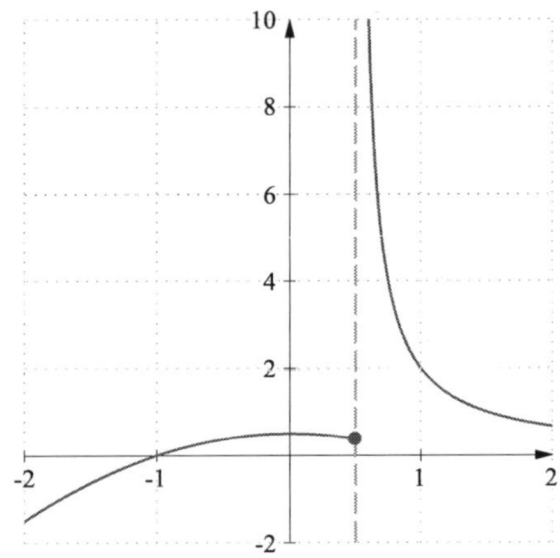

FIGURE 3.48
The function on the left has an essential discontinuity at the point $x = 0$ since the right-hand limit does not exist. The function on the right also has an essential (asymptotic) discontinuity at a point since the right-hand limit at that point is plus infinity.

3.14 Theorems on Continuous Functions

Below are some key theorems on continuous functions.

Proposition 3.22. If $f : [a, b] \to \mathbb{R}$ is **continuous** on the interval $[a, b]$, then f is bounded on that interval. That is, there exists $k \in \mathbb{R}$ such that $|f(x)| < k$ for all $x \in [a, b]$.

Proof. The proof of this proposition is based on Cantor's nesting theorem, which we state below:

Theorem 3.1 (Cantor's nesting theorem). Let $\{C_n\}$ be a sequence of non-empty closed sets in $\mathbb{R}$ such that:

$$C_n \subset C_{n-1},\ n = 2, 3, \ldots$$

with C_1 bounded. Then, the intersection:

$$C = \bigcap_{n=1}^{\infty} C_n,$$

is a non-empty closed set.

We will prove this by contradiction. Suppose the function is continuous on the interval $[a, b]$ but not bounded.

If $f(x)$ is not bounded, there will be values in the interval $[a, b]$ such that $f(x) > k$ for any arbitrary number $k \in \mathbb{R}$. If the interval $[a, b]$ is divided into two equal parts, in at least one of these two parts there will be values for which $f(x) > k$. Dividing this part again into two equal parts, there will be values for which $f(x) > k$. Repeating this scheme, we obtain a sequence of nested intervals, whose width tends to zero, in which there are values for which $f(x) > k$.

According to Cantor's nesting theorem 3.1, there exists a value α that is common to all these intervals. On one hand, since the function is continuous at $\alpha \in [a, b]$, it is possible to find a neighborhood where $|f(x) - f(\alpha)| < \varepsilon$. But in this neighborhood, there are values for which $f(x) > k$, which contradicts the fact that $|f(x) - f(\alpha)| < \varepsilon$.

Therefore, the function must be bounded, contrary to our initial assumption. □

Theorem 3.2 (Weierstrass's theorem (the maximum and minimum theorem)). If $f : [a, b] \to \mathbb{R}$ is **continuous** on the interval $[a, b]$, then f attains an **absolute minimum** and an **absolute maximum** on that interval.

Proof. We will prove the existence of the maximum. The proof for the existence of the minimum would be done similarly.

According to Proposition 3.22, the function $f(x)$ is bounded on the interval $[a, b]$ since it is continuous on that interval. Let k be the least upper bound, that is, if $f(x) < K$ on the interval $[a, b]$, then $k \leq K$.

If there exists a point $M \in [a, b]$ such that $f(M) = k$, the theorem is proved. If this is not the case, it must hold that $k - f(x) > 0$ for all $x \in [a, b]$. Then, if we define the function:

$$g(x) = \frac{1}{k - f(x)},$$

this function is continuous on the interval $[a, b]$ and by applying Proposition 3.22 again, the function $g(x)$ will be bounded. That is, there exists a real number $\tilde{k} \in \mathbb{R}$ such that:

$$g(x) = \frac{1}{k - f(x)} < \tilde{k}.$$

Therefore:

$$\frac{1}{k - f(x)} < \tilde{k} \Rightarrow k - f(x) > \frac{1}{\tilde{k}} \Rightarrow f(x) < k - \frac{1}{\tilde{k}},$$

which contradicts the fact that k was the least upper bound. Thus, the possibility that the maximum is not attained by the function is excluded. □

Theorem 3.3 (Bolzano's theorem). If $f : [a, b] \to \mathbb{R}$ is **continuous** on the interval $[a, b]$ and such that:

$$f(a) \cdot f(b) < 0,$$

then there exists at least one point $c \in (a, b)$ such that $f(c) = 0$.

Proof. We will assume, without loss of generality, that $f(a) > 0$ and that $f(b) < 0$. We consider the midpoint of this interval, that is:

$$x_1 = \frac{a + b}{2}.$$

If $f(x_1) = 0$, the theorem is proved. Otherwise, if for example $f(x_1) < 0$, we now consider the interval $\left[a, \frac{a+b}{2}\right] = [a, x_1]$. At the endpoints of this interval, the function still changes sign. We repeat the process, considering the midpoint of this new interval:

$$x_2 = \frac{a + x_1}{2} = \frac{a + \frac{a+b}{2}}{2}.$$

If $f(x_2) = 0$, the theorem is proved. Otherwise, we repeat the process again. According to Cantor's nesting theorem 3.1, there exists a point c that belongs to all the intervals. If $f(c) > 0$, $f(x)$ must be positive in a neighborhood of c, which contradicts the fact that $f(x)$ takes values of opposite signs at the endpoints of this interval. If $f(c) < 0$, $f(x)$ must be negative in a neighborhood of c, which again contradicts the fact that $f(x)$ takes values of opposite signs at the endpoints of this interval. Therefore, it can only be that $f(c) = 0$. □

Observation 3.5. Bolzano's theorem guarantees that continuous functions that change sign at their endpoints have at least one point where they intersect the x-axis. This can be seen in Figure 3.49.

Bolzano's theorem allows us to demonstrate the existence of solutions to equations.

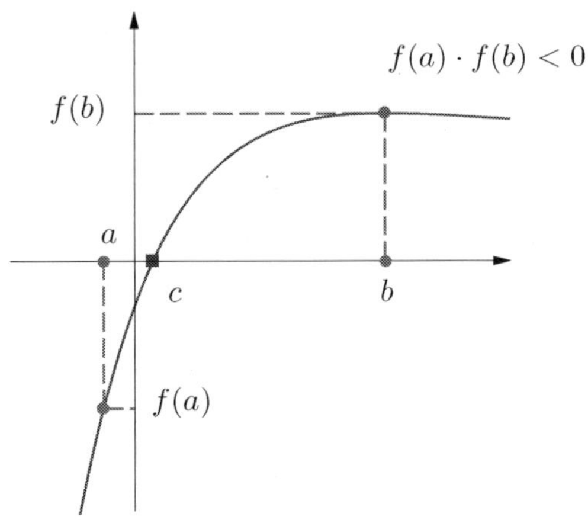

FIGURE 3.49
If the function changes sign at its endpoints and is continuous, there exists a point c where the function intersects the x-axis.

Example 3.64. *Demonstrate that the equation:*

$$\cos(x) - 2x + 1 = 0$$

has at least one solution in the interval $\left[0, \frac{\pi}{2}\right]$.

SOLUTION. Consider the function $f(x) = \cos(x) - 2x + 1$, which is continuous in the interval $\left[0, \frac{\pi}{2}\right]$. Additionally:

$$f(0) = \cos(0) - 2 \cdot 0 + 1 = 1 + 1 = 2 > 0$$
$$f\left(\frac{\pi}{2}\right) = \cos\left(\frac{\pi}{2}\right) - 2 \cdot \frac{\pi}{2} + 1 = 0 - \pi + 1 = 1 - \pi < 0$$

Therefore, according to Bolzano's theorem 3.3, there exists a point $c \in \left[0, \frac{\pi}{2}\right]$ such that $f(c) = 0$. That is, the equation has at least one solution.

The following theorem is, in a way, a generalization of Bolzano's theorem.

Theorem 3.4 (Intermediate value theorem). If $f : [a, b] \to \mathbb{R}$ is **continuous** on the interval $[a, b]$ and such that:

$$f(a) < f(b),$$

then for any $y \in \mathbb{R}$ such that:

$$f(a) < y < f(b)$$

there exists a $t \in (a, b)$ with $f(t) = y$.

Proof. Consider a new function $g(x)$ defined as:

$$g(x) = f(x) - y.$$

Since $f(x)$ is continuous on $[a, b]$, $g(x)$ is also continuous as it is the difference between a continuous function and a constant function. Additionally:

$$g(a) = f(a) - y < 0$$
$$g(b) = f(b) - y > 0$$

Therefore, by Bolzano's theorem, there exists $t \in [a, b]$ such that $g(t) = 0$. But this is equivalent to saying that there exists $t \in [a, b]$ such that:

$$f(t) = g(t) + y = y,$$

just as we wanted to prove. □

This idea is illustrated in Figure 3.50.

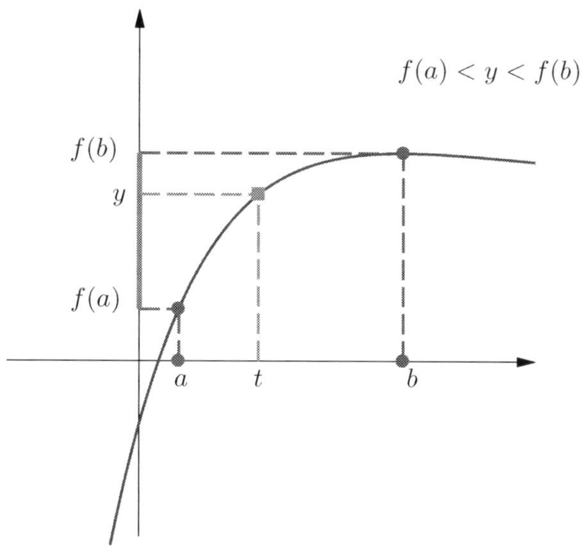

FIGURE 3.50
If the function is continuous and $f(a) < f(b)$, the function attains all intermediate values.

3.15 Solved Problems (Functions)

Problem 3.1. Determine the domain of the functions with the following analytical definitions:

(a) $f(x) = x^2 + 1$

(b) $f(x) = \dfrac{x}{x+1}$

(c) $f(x) = \sqrt{x^2 - 1}$

(d) $f(x) = \sqrt{-x}$

(e) $f(x) = \ln(x)$

(f) $f(x) = \sqrt{\ln(x)}$

(g) $f(x) = e^{x/2} - 6\log(x+2) + 3$

Solution to the problem 3.1. *(a) The function $f(x) = x^2 + 1$ is a polynomial. Therefore, $Dom(f) = \mathbb{R}$.*

(b) The function:

$$f(x) = \frac{x}{x+1}$$

is a ratio of polynomials, and the denominator $q(x) = x + 1$ is zero at $x = -1$. Therefore:

$$\begin{aligned} Dom(f) &= \mathbb{R} - \{x \mid q(x) = 0\} = \mathbb{R} - \{x \mid x + 1 = 0\} = \\ &= \mathbb{R} - \{x \mid x = -1\} = \mathbb{R} - \{-1\}. \end{aligned}$$

(c) The function $f(x) = \sqrt{x^2 - 1}$ is of the form:

$$f(x) = \sqrt[2n]{g(x)}, \quad n \in \mathbb{Z},$$

where $n = 1$ and $g(x) = x^2 - 1$. Therefore, its domain is:

$$\begin{aligned} Dom(f) &= \{x \mid x^2 - 1 \geq 0\} = \{x \mid x^2 \geq 1\} = \{x \mid |x| \geq 1\} \\ &= \{x \mid x \geq 1 \text{ or } x \leq -1\} = \mathbb{R} - (-1, 1). \end{aligned}$$

(d) The function $f(x) = \sqrt{-x}$ is of the form:

$$f(x) = \sqrt[2n]{g(x)}, \quad n \in \mathbb{Z},$$

where $n = 1$ and $g(x) = -x$. Therefore, its domain is:

$$Dom(f) = \{x \mid -x \geq 0\} = \{x \mid x \leq 0\} = (-\infty, 0].$$

(e) The function $f(x) = \ln(x)$ can be rewritten as:

$$f(x) = \frac{\log(x)}{\log(e)} = \frac{1}{\log(e)} \log(x),$$

thus, it is the product of a scalar by the function $f_1(x) = \log(x)$. *Hence:*

$$Dom(f) = Dom(f_1),$$

since $f_1 = \lambda f$.
The function $f_1(x) = \log(x)$ *is of the form* $f(x) = \log(g(x))$, *where* $g(x) = x$. *Therefore, its domain is:*

$$Dom(f_1) = \{x \mid x > 0\} = (0, +\infty).$$

Thus:

$$Dom(f) = Dom(f_1) = (0, +\infty).$$

(f) *The function* $f(x) = \sqrt{\ln(x)}$ *is of the form* $f(x) = \sqrt[2n]{g(x)}$, $n \in \mathbb{Z}$, *where* $n = 1$ *and* $g(x) = \ln(x)$. *Therefore, its domain is:*

$$Dom(f) = \{x \mid \ln(x) \geq 0\} = \{x \mid x \geq e^0 = 1\}$$

$$= \{x \mid x \geq 1\} = [1, +\infty).$$

(g) *The function* $f(x) = e^{x/2} - 6\log(x+2) + 3$ *is the sum of the functions* $f_1 = e^{x/2}$, $f_2 = -6\log(x+2)$ *and* $f_3 = 3$, *thus its domain is:*

$$Dom(f) = Dom(f_1) \cap Dom(f_2) \cap Dom(f_3).$$

The function $f_1(x) = e^{x/2}$ *is of the form* $f(x) = e^{g(x)}$, *where* $g(x) = \frac{x}{2}$. *Therefore, its domain is:*

$$Dom(f_1) = Dom\left(\frac{x}{2}\right) = \mathbb{R}.$$

The function $f_2(x) = -6\log(x+2)$ *is the product of a scalar by the function:*

$$f_{21}(x) = \log(x+2),$$

thus, $Dom(f_2) = Dom(f_{21})$, *since* $f_{21} = \lambda f_2$.
The function $f_{21}(x) = \log(x+2)$ *is of the form* $f(x) = \log(g(x))$, *where* $g(x) = x+2$. *Therefore, its domain is:*

$$Dom(f_2) = \{x \mid x + 2 > 0\} = \{x \mid x > -2\} = (-2, +\infty).$$

The function $f_3(x) = 3$ *is a zero-degree polynomial, thus:*

$$Dom(f_3) = \mathbb{R}.$$

Finally:

$$Dom(f) = \underbrace{Dom(f_1)}_{\mathbb{R}} \cap \underbrace{Dom(f_2)}_{>-2} \cap \underbrace{Dom(f_3)}_{\mathbb{R}} = \mathbb{R} \cap (-2, +\infty) \cap \mathbb{R} = (-2, +\infty).$$

Problem 3.2. Determine the domain of the functions with the following analytical definitions:

(a) $f(x) = \sqrt{\ln\left(\frac{x-3}{x-2}\right)}$

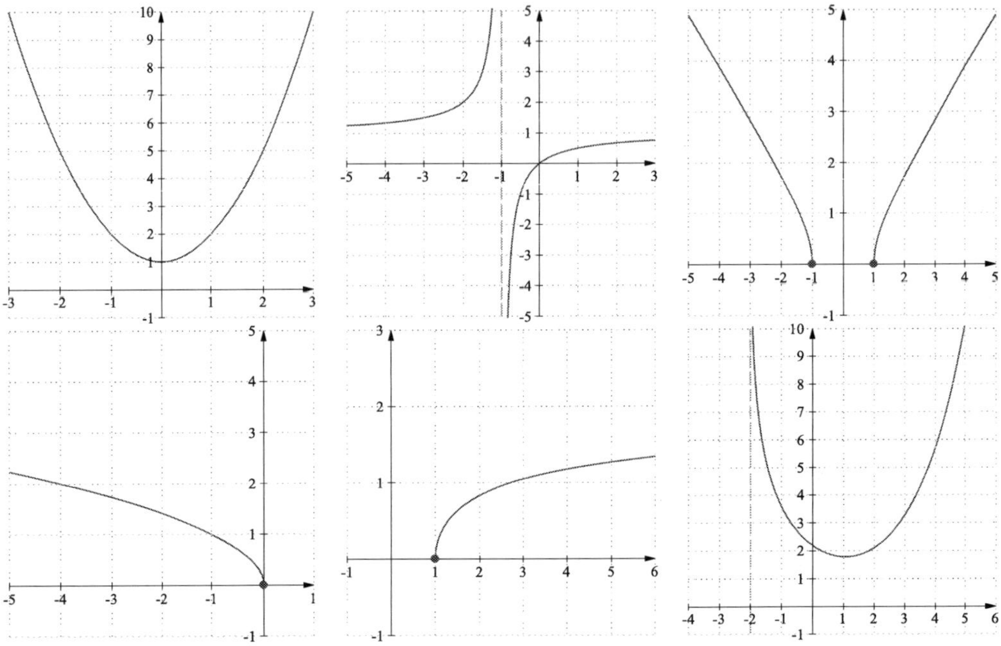

FIGURE 3.51
From left to right and top to bottom: graph of the functions in Problem 3.1: $f(x) = x^2 + 1$, $f(x) = \frac{x}{x+1}$, $f(x) = \sqrt{x^2-1}$, $f(x) = \sqrt{-x}$, $f(x) = \sqrt{\ln(x)}$ and $f(x) = e^{x/2} - 6\log(x+2) + 3$.

(b) $f(x) = \dfrac{1}{\sqrt[3]{x-2}}$

Solution to the problem 3.2. *(a) The function:*

$$f(x) = \sqrt{\ln\left(\frac{x-3}{x-2}\right)}$$

is of the form $f(x) = \sqrt[2n]{g(x)}$, $n \in \mathbb{Z}$, *where* $n = 1$ *and:*

$$g(x) = \ln\left(\frac{x-3}{x-2}\right).$$

In turn, $g(x) = \ln(h(x))$ *where* $h(x)$ *is a ratio of polynomials with the denominator* $q(x) = x - 2$. *Therefore, the domain of* $f(x)$ *is:*

$$\begin{aligned} Dom(f) &= \left\{x \mid \ln\left(\frac{x-3}{x-2}\right) \geq 0\right\} \cap \{x \mid x-2 \neq 0\} = \left\{x \mid \frac{x-3}{x-2} \geq e^0 = 1\right\} \\ &= \{x \text{ such that if } x-2>0, x-3 \geq x-2 \text{ or if } x-2<0, x-3 \leq x-2\} \\ &= \{x \text{ such that if } x-2>0, -3 \geq -2 \text{ or if } x-2<0, -3 \leq -2\} \\ &= \{x \text{ such that } x-2<0\} = \{x \text{ such that } x<2\} = (-\infty, 2). \end{aligned}$$

(b) The function:

$$f(x) = \frac{1}{\sqrt[3]{x-2}}$$

is a ratio of the functions $f_1(x) = 1$ *and* $f_2(x) = \sqrt[3]{x-2}$, *so:*

$$f(x) = \frac{f_1(x)}{f_2(x)}.$$

The function $f_1(x) = 1$ *is a zero-degree polynomial, so* $Dom(f_1) = \mathbb{R}$.
The function $f_2(x) = \sqrt[3]{x-2}$ *is of the form:*

$$f(x) = \sqrt[2n+1]{g(x)}, \quad n \in \mathbb{Z},$$

where $g(x) = x - 2$ *and* $n = 1$. *Therefore, its domain is:*

$$Dom(f_2) = Dom(x-2) = \mathbb{R}.$$

Thus, the domain of f *is:*

$$Dom(f) = Dom\left(\frac{f_1}{f_2}\right) = Dom(f_1) \cap Dom(f_2) - \{x \mid f_2(x) = 0\}$$

$$= \underbrace{Dom(1)}_{\mathbb{R}} \cap \underbrace{Dom(\sqrt[3]{x-2})}_{\mathbb{R}} - \{x \mid \sqrt[3]{x-2} = 0\}$$

$$= \mathbb{R} \cap \mathbb{R} - \{x \mid x - 2 = 0\} = \mathbb{R} - \{x \mid x = 2\}$$

$$= \mathbb{R} - \{2\}.$$

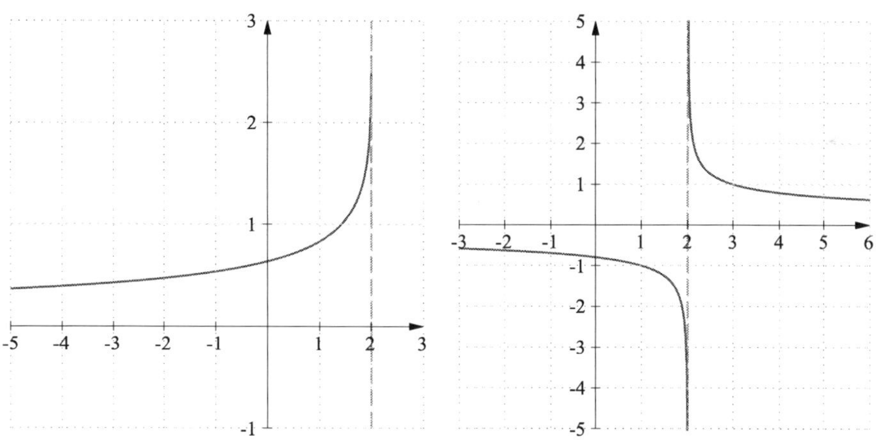

FIGURE 3.52
Graph of the functions $f(x) = \sqrt{\ln\left(\frac{x-3}{x-2}\right)}$ (left) and $f(x) = \frac{1}{\sqrt[3]{x-2}}$ (right) (Problem 3.2).

Problem 3.3. Are the functions f_1 and f_2 equal?

(a) $f_1(x) = \dfrac{2x^2 + x}{x}$; $\quad f_2(x) = 2x + 1$.

(b) $f_1(x) = \dfrac{2x^2 + x}{x}$; $f_2(x) = 2x + 1$, $x \neq 0$.

Solution to the problem 3.3. *(a) The functions f_1 and f_2 are not equal because they do not have the same domain. In particular, $x = 0 \notin Dom(f_1)$, whereas $f_2(0) = 1$.*

(b) The two functions are equal since $Dom(f_1) = Dom(f_2)$ and the range is the same for all points in the domain.

Problem 3.4. Express each of the following functions as a composition of two functions (u, g) such that $f(x) = g(u(x))$:

(a) $f(x) = (x^2 + 5x + 1)^5$

(b) $f(x) = \cos^3(x)$

(c) $f(x) = \sin(x^3)$

(d) $f(x) = \sqrt{5x^2 - x}$

Solution to the problem 3.4. *(a) For $g(x) = x^5$ and $u(x) = x^2 + 5x + 1$, $f(x) = g(u(x)) = (x^2 + 5x + 1)^5$.*

(b) For $g(x) = x^3$ and $u(x) = \cos(x)$, $f(x) = g(u(x)) = (\cos(x))^3 = \cos^3(x)$.

(c) For $g(x) = \sin(x)$ and $u(x) = x^3$, $f(x) = g(u(x)) = \sin(x^3)$.

(d) For $g(x) = \sqrt{x}$ and $u(x) = 5x^2 - x$, $f(x) = g(u(x)) = \sqrt{5x^2 - x}$.

Problem 3.5. Let $f(x) = \sin(x)$ and $g(x)$ be such that $\mathrm{Dom}(g) = (0, +\infty)$. What is the domain of the composite function $g(f(x))$?

Solution to the problem 3.5. *The composite function $(g \circ f)$ has the domain:*

$$\begin{aligned} Dom(g \circ f) &= \{x \in \underbrace{Dom(f)}_{\mathbb{R}} \mid f(x) \in \underbrace{Dom(g)}_{>0}\} \\ &= \{x \in \mathbb{R} \mid f(x) > 0\} \\ &= \{x \in \mathbb{R} \mid \sin(x) > 0\} \\ &= \cdots \cup (-4\pi, -3\pi) \cup (-2\pi, -\pi) \cup (0, \pi) \cup (2\pi, 3\pi) \cup \cdots \end{aligned}$$

Problem 3.6. Let $f(x) = \dfrac{x^2 + 2}{\sin(x^2) + 2} - 3$.

(a) Determine $\mathrm{Dom}(f)$.

(b) Determine if f is invertible.

Solution to the problem 3.6. *(a) The function* $f(x) = \dfrac{x^2+2}{\sin(x^2)+2} - 3$ *is a ratio and sum of the functions:*

$$f_1(x) = x^2 + 2, \quad f_2(x) = \frac{1}{\sin(x^2)+2}, \quad f_3(x) = -3,$$

such that $f(x) = f_1(x)f_2(x) + f_3(x)$*. Thus:*

$$Dom(f) = Dom(f_1) \cap Dom(f_2) \cap Dom(f_3).$$

The function $f_1(x) = x^2 + 2$ *is polynomial. Therefore,* $Dom(f_1) = \mathbb{R}$*.*
The function $f_2(x)$ *is the inverse of the function* $\sin(x^2) + 2$*, thus:*

$$\begin{aligned} Dom(f_2) &= \mathbb{R} - \{x \mid \sin(x^2) + 2 = 0\} \\ &= \mathbb{R} - \{x \mid \sin(x^2) = -2\} \end{aligned}$$

However, since $-1 \leq \sin(x^2) \leq 1$*, then:*

$$Dom(f_2) = \mathbb{R}.$$

Thus:

$$\begin{aligned} Dom(f) &= Dom(f_1) \cap Dom(f_2) \cap Dom(f_3) \\ &= \mathbb{R} \cap \mathbb{R} \cap \mathbb{R} = \mathbb{R}. \end{aligned}$$

(b) In Figure 3.53, it can be seen that $f(-2) = f(2)$ *and therefore,* $f(x)$ *is not invertible.*

Problem 3.7. For the function f defined in each case, find its inverse if it exists and, if so, specify its domain:

(a) $f(x) = 2x - 1$

(b) $f(x) = \dfrac{1}{x-5}$

(c) $f(x) = x^2$

(d) $f(x) = x^3 + 2$

(e) $f(x) = \ln\left(\dfrac{x}{2}\right)$

(f) $f(x) = \sin(x)$

(g) $f(x) = x^2,\ x \leq 0$

(h) $f(x) = \sqrt{1-x^2}\, x \in (-1, 1)$

Solution to the problem 3.7. *(a)*

$$y = 2x - 1 \iff x = \frac{y+1}{2} \iff f^{-1}(y) = \frac{y+1}{2}.$$

Therefore:

$$f^{-1}(x) = \frac{x+1}{2} \quad \text{and} \quad Dom(f^{-1}) = \mathbb{R}.$$

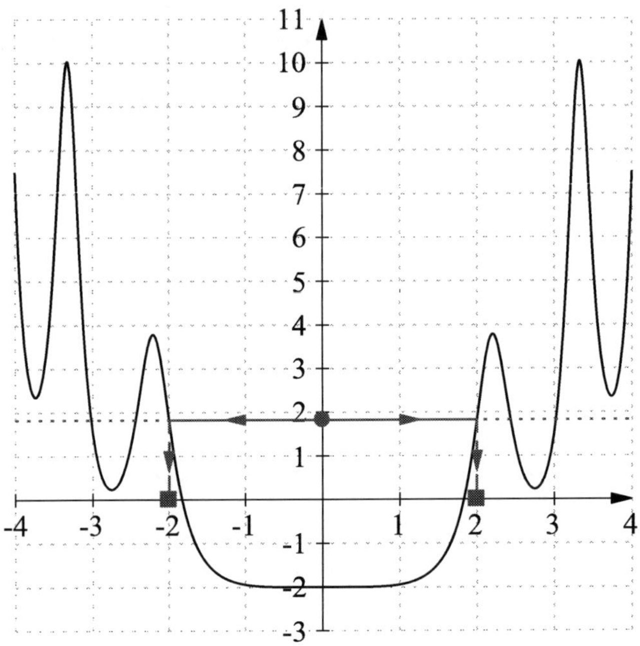

FIGURE 3.53
Graph of the function $f(x) = \frac{x^2+2}{\sin(x^2)+2} - 3$ (Problem 3.6). This function is not invertible since $y = 2$ has more than one preimage.

(b)

$$y = \frac{1}{x-5} \iff x = 5 + \frac{1}{y} \iff f^{-1}(y) = 5 + \frac{1}{y}.$$

Therefore:

$$f^{-1}(x) = 5 + \frac{1}{x} \quad \textit{and} \quad \textit{Dom}(f^{-1}) = \mathbb{R} - \{0\}.$$

(c)

$$y = x^2 \iff x = \pm\sqrt{y} \iff f^{-1}(y) = \pm\sqrt{y}.$$

Thus:

- *if* $y < 0$, $f^{-1}(y) = \emptyset$;
- *if* $y = 0$, $f^{-1}(y) = \{0\}$;
- *if* $y > 0$, $f^{-1}(y)$ *is a set of two elements* $\{-\sqrt{y}, \sqrt{y}\}$.

Therefore, $f(x)$ *does not have an inverse. For example,* $f^{-1}(1) = \{-1, 1\}$.

(d)

$$y = x^3 + 2 \iff x = \sqrt[3]{y-2} \iff f^{-1}(y) = \sqrt[3]{y-2}.$$

Therefore:

$$f^{-1}(x) = \sqrt[3]{x-2} \quad \textit{and} \quad \textit{Dom}(f^{-1}) = \mathbb{R}.$$

(e)

$$y = \ln\left(\frac{x}{2}\right) \iff x = 2e^y \iff f^{-1}(y) = 2e^y.$$

Therefore:

$$f^{-1}(x) = 2e^x \quad \textit{and} \quad Dom(f^{-1}) = \mathbb{R}.$$

(f) *For* $k \in \mathbb{Z}$*:*

$$y = \sin(x) \iff x = \arcsin(y) + 2\pi k \iff f^{-1}(y) = \arcsin(y) + 2\pi k.$$

Therefore, $f(x)$ *does not have an inverse. For example,* $f^{-1}(1) = \{2\pi k,\ k \in \mathbb{Z}\}$.

(g)

$$y = x^2 \textit{ if } x \leq 0 \iff x = -\sqrt{y} \iff f^{-1}(y) = -\sqrt{y}.$$

Therefore:

$$f^{-1}(x) = -\sqrt{x} \quad \textit{and} \quad Dom(f^{-1}) = [0, +\infty).$$

(h)

$$y = \sqrt{1 - x^2} \textit{ if } x \in (-1, 1) \iff x = \pm\sqrt{1 - y^2} \iff f^{-1}(y) = \pm\sqrt{1 - y^2}.$$

Thus:

- *if* $y < -1$, $f^{-1}(y) = \emptyset$*;*
- *if* $-1 \leq y \leq 1$, $f^{-1}(y)$ *is a set of two elements* $\{-\sqrt{1 - y^2}, \sqrt{1 - y^2}\}$*;*
- *if* $y > 1$, $f^{-1}(y) = \emptyset$.

Therefore, $f(x)$ *does not have an inverse. For example,* $f^{-1}(0) = \{-1, 1\}$.

Problem 3.8. Consider the function $f(x) = |x^3 - 8|$. Calculate:

(a) $f^{-1}(19)$

(b) $f^{-1}(0)$

(c) $f^{-1}(92)$

Solution to the problem 3.8. *(a) First, let's calculate the preimage of* 19*:*

$$\begin{aligned}
f^{-1}(19) &= \{x \in \mathbb{R} \textit{ such that } f(x) = 19\} = \{x \in \mathbb{R} \textit{ such that } |x^3 - 8| = 19\} \\
&= \{x \in \mathbb{R} \textit{ such that } x^3 - 8 = 19 \textit{ or } x^3 - 8 = -19\} \\
&= \{x \in \mathbb{R} \textit{ such that } x^3 = 27 \textit{ or } x^3 = -11\} \\
&= \{x \in \mathbb{R} \textit{ such that } x = \sqrt[3]{27} = 3 \textit{ or } x = \sqrt[3]{-11} = -\sqrt[3]{11}\} \\
&= \{3, -\sqrt[3]{11}\}.
\end{aligned}$$

Therefore, $f^{-1}(19) = \{-\sqrt[3]{11}, 3\}$.

$$\begin{aligned}
f^{-1}(0) &= \{x \in \mathbb{R} \textit{ such that } f(x) = 0\} = \{x \in \mathbb{R} \textit{ such that } |x^3 - 8| = 0\} \\
&= \{x \in \mathbb{R} \textit{ such that } x^3 = 8\} = \{2\}.
\end{aligned}$$

Therefore, $f^{-1}(0) = \{2\}$.

(c) Finally, we can also calculate the preimage of 92*:*

$$\begin{aligned} f^{-1}(92) &= \{x \in \mathbb{R} \text{ such that } f(x) = 92\} = \{x \in \mathbb{R} \text{ such that } |x^3 - 8| = 92\} \\ &= \{x \in \mathbb{R} \text{ such that } x^3 - 8 = 92 \text{ or } x^3 - 8 = -92\} \\ &= \{x \in \mathbb{R} \text{ such that } x^3 = 100 \text{ or } x^3 = -84\} \\ &= \{x \in \mathbb{R} \text{ such that } x = \sqrt[3]{100} \text{ or } x = \sqrt[3]{-84}\}. \end{aligned}$$

Therefore, $f^{-1}(92) = \{\sqrt[3]{100}, \sqrt[3]{-84}\}$*. The graph of the function and these three preimages can be seen in Figure 3.54.*

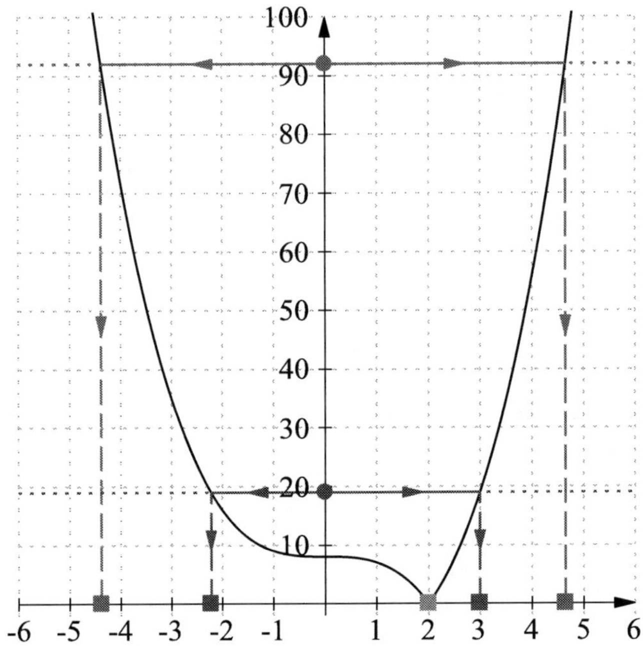

FIGURE 3.54
Graph of the function $f(x) = |x^3 - 8|$ (Problem 3.8), showing the preimages of the values $y = 19$, $y = 92$, and $y = 0$.

Problem 3.9. Consider the function $f(x) = |x^3|$. Determine if $f(x)$ is invertible if we restrict it to $x \geq 0$.

Solution to the problem 3.9. *The function* $f(x)$ *can be rewritten as:*

$$f(x) = \begin{cases} x^3, & x \geq 0 \\ -x^3, & x < 0 \end{cases}$$

therefore:

$$\begin{cases} \text{If } y < 0,\ f^{-1}(y) = \emptyset \\ \text{If } y \geq 0,\ f^{-1}(y) \text{ is a set of two elements } \{-\sqrt[3]{y}, \sqrt[3]{y}\}. \end{cases}$$

We see that $f(x)$ is not invertible over its entire domain, as for example $f(-1) = f(1)$, as seen in Figure 3.55.

To ensure the preimage of each point is at most one element, we must restrict the values of the function $f(x) = |x^3|$ to either positive or negative numbers.
Thus, $f(x)$ is invertible if restricted to $x \geq 0$.

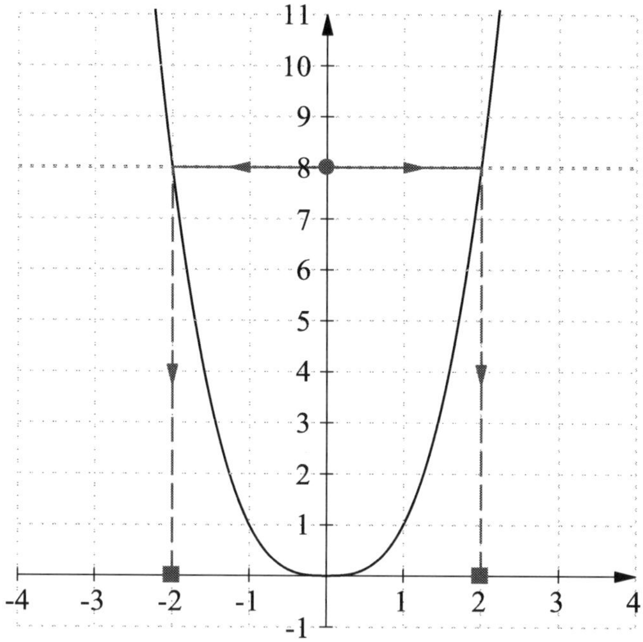

FIGURE 3.55
Graph of the function $f(x) = |x^3|$ (Problem 3.9), showing the preimage of $y = 8$. To make this function invertible, it must be restricted to positive real numbers.

Problem 3.10. Consider the function $f(x) = |x^2 - 2|$. Determine if $f(x)$ is invertible if we restrict it to $x \geq 0$.

Solution to the problem 3.10. *The function $f(x)$ can be rewritten as:*

$$f(x) = \begin{cases} -x^2 + 2, & x^2 - 2 < 0 \\ x^2 - 2, & x^2 - 2 \geq 0 \end{cases}$$

therefore:

$$f(x) = \begin{cases} -x^2 + 2, & |x| < \sqrt{2} \\ x^2 - 2, & |x| \geq \sqrt{2} \end{cases}$$

thus:

$$f(x) = \begin{cases} x^2 - 2, & x < -\sqrt{2} \\ -x^2 + 2, & -\sqrt{2} \le x \le \sqrt{2} \\ x^2 - 2, & x > \sqrt{2}. \end{cases}$$

Then:

$$\begin{cases} \textit{If } y < 0, & f^{-1}(y) = \emptyset \\ \textit{If } y = 0, & f^{-1}(y) \textit{ is a set of two elements } \{-\sqrt{2}, \sqrt{2}\} \\ \textit{If } 0 < y < 2, & f^{-1}(y) \textit{ is a set of four elements:} \\ & \{-\sqrt{-y+2}, -\sqrt{y+2}, \sqrt{-y+2}, \sqrt{y+2}\} \\ \textit{If } y = 2, & f^{-1}(y) \textit{ is a set of three elements } \{-2, 0, 2\} \\ \textit{If } y > 2, & f^{-1}(y) \textit{ is a set of two elements } \{-\sqrt{y+2}, \sqrt{y+2}\}. \end{cases}$$

We see that $f(x)$ is not invertible over its entire domain because for all $y \ge 0$, $f^{-1}(y)$ has two, three, or four preimages. It is also not invertible if restricted to $x \ge 0$ because $f(0) = f(2) = 2$.

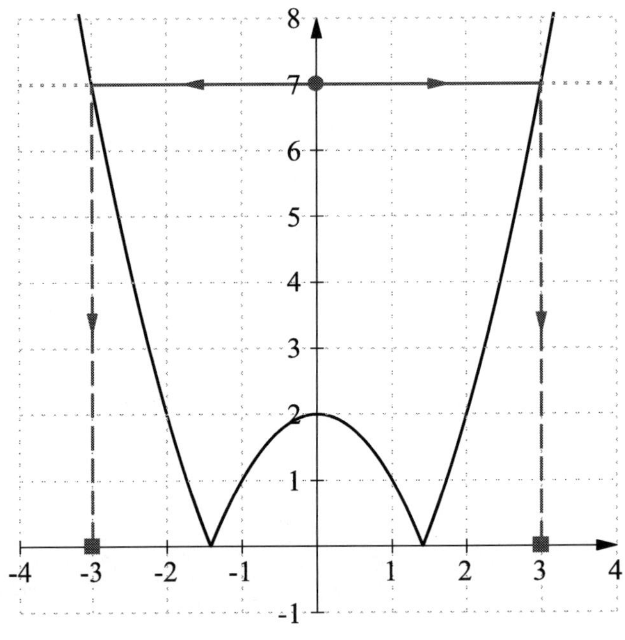

FIGURE 3.56
Graph of the function $f(x) = |x^2 - 2|$ (Problem 3.10), showing the preimage of $y = 7$. The function is not invertible if restricted to positive real numbers, as for example, $f(0) = f(2) = 2$.

3.16 Solved Problems (Limits)

Problem 3.11. Using the definition of the limit of a function at a point, prove that:

$$\lim_{x \to a} 3x = 3a.$$

Solution to the problem 3.11. *We will prove this limit by following these two steps:*

- ***Step 1.*** *Relate the difference* $|f(x) - L|$ *with* $|x - a|$.
- ***Step 2.*** *Select* δ *as a function of* ε.

In this case, we have $f(x) = 3x$ *and* $L = 3a$. *Therefore:*

- ***Step 1.***
$$|f(x) - L| = |3x - 3a| = 3 \cdot |x - a| < \varepsilon$$
- ***Step 2.*** *If* $|x - a| < \varepsilon$, *then* $|f(x) - L| < \varepsilon$. *Therefore, we select* $\delta = \dfrac{\varepsilon}{3}$.

These two preliminary steps serve to select δ *as a function of* ε. *The proof is completed in the following lines.*

We need to prove that for every $\varepsilon > 0$, *there exists* $\delta > 0$ *such that if we take:*

$$x \in (a - \delta, a + \delta), \quad x \neq a,$$

then:

$$f(x) \in (3a - \varepsilon, 3a + \varepsilon).$$

Consider $\varepsilon > 0$ *given and take* $\delta = \varepsilon/3$.

Then, if $x \in (a - \delta, a + \delta)$, $f(x) = 3x$ *and:*

$$\begin{aligned} a - \delta < x < a + \delta &\Longrightarrow 3a - 3\delta < 3x < 3a + 3\delta \\ &\Longrightarrow 3a - 3\varepsilon/3 < 3x < 3a + 3\varepsilon/3 \\ &\Longrightarrow 3a - \varepsilon < f(x) < 3a + \varepsilon \\ &\Longrightarrow f(x) \in (3a - \varepsilon, 3a + \varepsilon), \end{aligned}$$

as we wanted to show.

Problem 3.12. Using the definition of the limit of a function at a point, prove that:

$$\lim_{x \to a} 3x + 2 = 3a + 2.$$

Solution to the problem 3.12. *We will prove this limit by following these two steps:*

- ***Step 1.*** *Relate the difference* $|f(x) - L|$ *with* $|x - a|$.
- ***Step 2.*** *Select* δ *as a function of* ε.

In this case, we have $f(x) = 3x + 2$ *and* $L = 3a + 2$. *Therefore:*

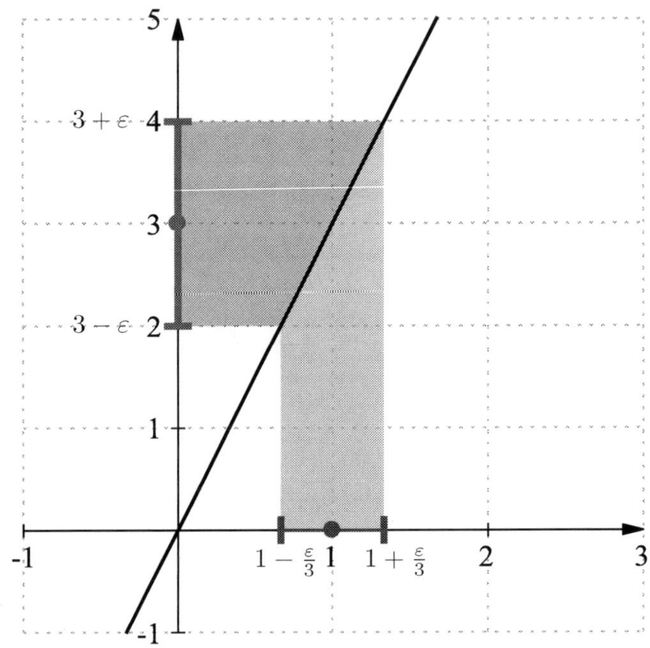

FIGURE 3.57
Graphical representation of the relationship between ε and δ for the calculation of the limit of the function $f(x) = 3x$ (Problem 3.11).

- ***Step 1.***
$$|f(x) - L| = |3x + 2 - (3a + 2)| = 3 \cdot |x - a| < \varepsilon$$
- ***Step 2.*** *If* $|x - a| < \varepsilon$, *then* $|f(x) - L| < \varepsilon$. *Therefore, we select* $\delta = \dfrac{\varepsilon}{3}$.

These two preliminary steps serve to select δ *as a function of* ε. *The proof is completed in the following lines.*

We need to prove that for every $\varepsilon > 0$, *there exists* $\delta > 0$ *such that if we take:*

$$x \in (a - \delta, a + \delta), \quad x \neq a,$$

then:

$$f(x) \in (3a + 2 - \varepsilon, 3a + 2 + \varepsilon).$$

Consider $\varepsilon > 0$ *given and take* $\delta = \varepsilon/3$.

Then, if $x \in (a - \delta, a + \delta)$, $f(x) = 3x + 2$ *and:*

$$\begin{aligned} a - \delta < x < a + \delta &\Longrightarrow 3a - 3\delta < 3x < 3a + 3\delta \\ &\Longrightarrow 3a - 3\varepsilon/3 < 3x < 3a + 3\varepsilon/3 \\ &\Longrightarrow 3a + 2 - \varepsilon < \underbrace{3x + 2}_{f(x)} < 3a + 2 + \varepsilon \\ &\Longrightarrow f(x) \in (3a + 2 - \varepsilon, 3a + 2 + \varepsilon), \end{aligned}$$

as we wanted to show.

Problem 3.13. State the formal definition of $\lim_{x \to +\infty} f(x) = +\infty$. Using this definition, prove that:

$$\lim_{x \to +\infty} x^2 = +\infty.$$

Solution to the problem 3.13. $\lim_{x \to +\infty} f(x) = +\infty \iff \forall k > 0,\ \exists \lambda$ *such that if* $x > \lambda$ *then,* $f(x) > k$.

Let's now prove that:

$$\lim_{x \to +\infty} x^2 = +\infty.$$

For $k > 0$ *we take* $\lambda = \sqrt{k}$. *We need to prove that* $x > \lambda \Longrightarrow f(x) > k$:

$$x > \lambda \quad \Longrightarrow \quad x > \sqrt{k} \quad \Longrightarrow \quad f(x) = x^2 > (\sqrt{k})^2 \quad \Longrightarrow \quad f(x) > k$$

as we wanted to show. This idea is represented in Figure 3.58.

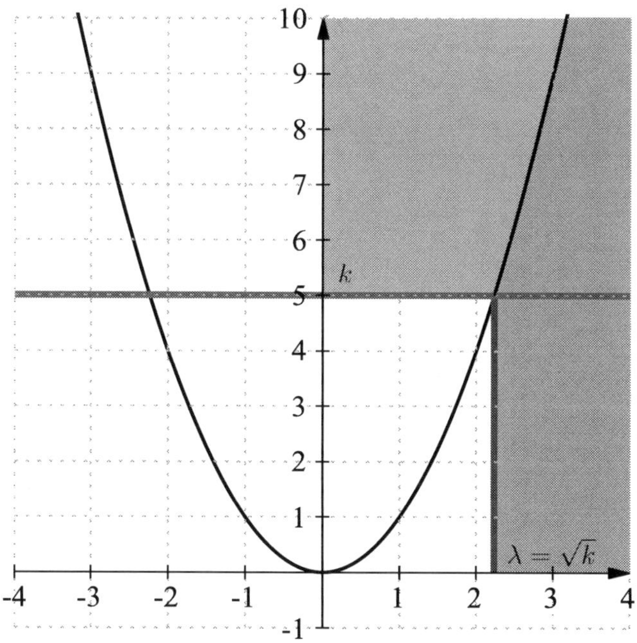

FIGURE 3.58
Graphical representation of the limit of the function $f(x) = x^2$ as x tends to $+\infty$. It can be seen that for values $x > \sqrt{k}$, the respective images $f(x)$ are greater than k (Problem 3.13).

Problem 3.14. Define the concept of *infinitesimals*. Give an example of three *different* functions that are equivalent infinitesimals as $x \to 2$.

Solution to the problem 3.14. *We say that the function* $f : A \subset \mathbb{R} \to \mathbb{R}$ *is an* infinitesimal *at the point* $x = a$ *if the limit of the function as* x *tends to* a *is* 0, *that is:*

$$\lim_{x \to a} f(x) = 0.$$

To find three examples of equivalent infinitesimals, we recall the infinitesimals at zero:

$$x \sim_0 \sin(x) \sim_0 \ln(1+x) \sim_0 \tan(x).$$

Then, making the change of variable $x = y - 2$, we will have:

$$y - 2 \sim_2 \sin(y-2) \sim_2 \ln(y-1) \sim_2 \tan(y-2).$$

Problem 3.15. Calculate the following limits:

(a) $\lim\limits_{x\to 2} \dfrac{x-2}{x^2-4}$

(b) $\lim\limits_{x\to 0} \dfrac{x}{\sqrt{1+2x}-\sqrt{1-2x}}$

(c) $\lim\limits_{x\to 0} \dfrac{\sqrt{1+2x}-\sqrt{1-2x}}{-x^2+3x+\pi}$

Solution to the problem 3.15. *(a) This is a quotient of polynomials that are zero at $x = 2$. So, if we factor by the binomial $x - 2$, we have:*

$$\lim_{x\to 2} \frac{x-2}{x^2-4} = «\frac{0}{0}» \text{ (indeterminate)}$$

$$= \lim_{x\to 2} \frac{x-2}{(x-2)(x+2)}$$

$$= \lim_{x\to 2} \frac{1}{x+2} = \frac{1}{4}$$

(b) In this case, both the numerator and the denominator are zero at $x = 0$, but the denominator is not polynomial. Therefore, we multiply and divide by the conjugate of the denominator:

$$\lim_{x\to 0} \frac{x}{\sqrt{1+2x}-\sqrt{1-2x}} = «\frac{0}{0}» \text{ (indeterminate)}$$

$$= \lim_{x\to 0} \frac{x}{\sqrt{1+2x}-\sqrt{1-2x}} \cdot \frac{\sqrt{1+2x}+\sqrt{1-2x}}{\sqrt{1+2x}+\sqrt{1-2x}}$$

$$= \lim_{x\to 0} \frac{x(\sqrt{1+2x}+\sqrt{1-2x})}{1+2x-(1-2x)}$$

$$= \lim_{x\to 0} \frac{\cancel{x}(\sqrt{1+2x}+\sqrt{1-2x})}{4\cancel{x}}$$

$$= \lim_{x\to 0} \frac{\sqrt{1+2x}+\sqrt{1-2x}}{4} = \frac{1}{2}.$$

(c) In this case, there is no indeterminate form and the limit can be calculated directly.

$$\lim_{x\to 0} \frac{\sqrt{1+2x}-\sqrt{1-2x}}{-x^2+3x+\pi} = \frac{\sqrt{1}-\sqrt{1}}{\pi} = 0.$$

Problem 3.16. Calculate the following limits:

(a) $\lim\limits_{x\to+\infty}\left(\dfrac{3x+1}{x}\right)^{x}$

(b) $\lim\limits_{x\to1}\dfrac{x-1}{\sqrt{x+3}\cdot(x^2-1)}$

(c) $\lim\limits_{x\to0^-}\sin(x)\cdot\dfrac{1}{x^2}$

(d) $\lim\limits_{x\to1}\left(\dfrac{2x}{x+1}\right)^{\dfrac{1}{x-1}}$

(e) $\lim\limits_{x\to1}\dfrac{\ln(x)}{2x^2-8x+6}$

(f) $\lim\limits_{x\to+\infty}\dfrac{\sin(x)+x+\sqrt{x}}{e^x}$

(g) $\lim\limits_{x\to-\infty}\dfrac{e^x}{x}$

Solution to the problem 3.16. *(a) In this case, there is no indeterminate form and the limit can be calculated directly:*

$$\lim_{x\to+\infty}\left(\frac{3x+1}{x}\right)^{x}=\lim_{x\to+\infty}\left(3+\underbrace{\frac{1}{x}}_{\to0}\right)^{x}=3^{+\infty}=+\infty.$$

(b) To resolve the indeterminate form, we will factor the polynomial that appears in the denominator and simplify the term $x-1$*:*

$$\begin{aligned}\lim_{x\to1}\frac{x-1}{\sqrt{x+3}\cdot(x^2-1)}&=\frac{0}{2\cdot0}=«\frac{0}{0}» \text{ (indeterminate form)}\\&=\lim_{x\to1}\frac{x-1}{\sqrt{x+3}\cdot(x+1)\cdot(x-1)}\\&=\lim_{x\to1}\frac{1}{\sqrt{x+3}\cdot(x+1)}\\&=\frac{1}{2\cdot2}=\frac{1}{4}\end{aligned}$$

(c) To resolve the indeterminate form, we will use $\sin(x)\sim_0 x$*, that is,*

$$\lim_{x\to0}\frac{\sin(x)}{x}=1.$$

$$\lim_{x\to 0^-} \sin(x)\cdot\frac{1}{x^2} = 0\cdot\frac{1}{0^+} = «0\cdot(+\infty)» \text{ (indeterminate form)}$$

$$= \lim_{x\to 0^-} \underbrace{\frac{\sin(x)}{x}}_{\to 1}\cdot\frac{1}{x}$$

$$= \lim_{x\to 0^-} \frac{1}{x} = \frac{1}{0^-} = -\infty.$$

(d) *To resolve the indeterminate form, we will rewrite the limit so that the number e appears:*

$$\lim_{x\to 1}\left(\frac{2x}{x+1}\right)^{\frac{1}{x-1}} = \left(\frac{2}{2}\right)^{\frac{1}{0}} = «1^{\infty}» \text{ (indeterminate form)}$$

$$= \lim_{x\to 1}\left(\frac{x+1+x-1}{x+1}\right)^{\frac{1}{x-1}} = \lim_{x\to 1}\left(1+\frac{x-1}{x+1}\right)^{\frac{1}{x-1}}$$

$$= \lim_{x\to 1}\left(1+\frac{1}{\frac{x+1}{x-1}}\right)^{\frac{x+1}{x-1}\cdot\frac{x-1}{x+1}\cdot\frac{1}{x-1}}$$

$$= \lim_{x\to 1}\left[\left(1+\frac{1}{\frac{x+1}{x-1}}\right)^{\frac{x+1}{x-1}}\right]^{\frac{x-1}{x+1}\cdot\frac{1}{x-1}}$$

$$= \lim_{x\to 1}\left[\left(1+\frac{1}{\frac{x+1}{x-1}}\right)^{\frac{x+1}{x-1}}\right]^{\frac{1}{x+1}} = e^{\lim_{x\to 1}\frac{1}{x+1}} = e^{1/2}$$

where we have used:

$$\lim_{x\to 1}\left(1+\frac{1}{\frac{x+1}{x-1}}\right)^{\frac{x+1}{x-1}} = e,$$

since if $x\to 1$:

$$\frac{x+1}{x-1}\to\infty.$$

(e) *To resolve the indeterminate form, we will make the change of variable $x = 1 + t$ in order to use the infinitesimal* $\ln(t+1) \sim_0 t$.

$$\lim_{x\to 1} \frac{\ln(x)}{2x^2 - 8x + 6} = \frac{\ln 1}{2 - 8 + 6} = «\frac{0}{0}» \text{ (indeterminate form)}$$

$$= \lim_{x\to 1} \frac{\ln(x)}{2(x^2 - 4x + 3)} = \lim_{x\to 1} \frac{\ln(x)}{2(x-1)(x-3)} = \lim_{t\to 0} \frac{\ln(1+t)}{2t(t-2)}$$

$$= \lim_{t\to 0} \underbrace{\frac{\ln(1+t)}{t}}_{\to 1} \cdot \frac{1}{2(t-2)} = \lim_{t\to 0} \frac{1}{2(t-2)} = \frac{1}{2\cdot(-2)} = -\frac{1}{4}.$$

(f) *This indeterminate form is resolved using orders of infinity:*

$$\lim_{x\to+\infty} \frac{\sin(x) + x + \sqrt{x}}{e^x} = «\frac{+\infty}{+\infty}» \text{ (indeterminate form).}$$

We have:

$$\sin(x) << \sqrt{x} << x << e^x.$$

Therefore:

$$\lim_{x\to+\infty} \frac{\sin(x)}{e^x} = \lim_{x\to+\infty} \frac{\sqrt{x}}{e^x} = \lim_{x\to+\infty} \frac{x}{e^x} = 0.$$

Then:

$$\lim_{x\to+\infty} \frac{\sin(x) + x + \sqrt{x}}{e^x} = \lim_{x\to+\infty} \frac{\sin(x)/e^x + x/e^x + \sqrt{x}/e^x}{e^x/e^x}$$

$$= \lim_{x\to+\infty} \frac{\overbrace{\sin(x)/e^x}^{\to 0} + \overbrace{x/e^x}^{\to 0} + \overbrace{\sqrt{x}/e^x}^{\to 0}}{1} = \frac{0}{1} = 0.$$

(g) *In this case, there is no indeterminate form and the limit can be calculated directly.*

$$\lim_{x\to-\infty} \frac{e^x}{x} = \lim_{x\to-\infty} e^x \cdot \frac{1}{x} = e^{-\infty} \cdot \frac{1}{-\infty} = 0 \cdot 0 = 0.$$

Problem 3.17. Calculate the following limits:

(a) $\lim_{x\to 3} \frac{x^2 - 2x - 3}{x^2 - 4x + 3}$

(b) $\lim_{x\to 0} \frac{\sqrt{x+8} - 2\sqrt{2}}{x}$

(c) $\lim_{x\to 0} \frac{e^x - e^{-x}}{x}$

(d) $\lim_{x\to+\infty} \frac{e^{-x}}{\sqrt{x}}$

(e) $\lim\limits_{x\to+\infty}\left(\dfrac{7x-5}{7x-7}\right)^{10+x}$

(f) $\lim\limits_{x\to 0}\dfrac{\sin(x)}{\tan(x)}$

Solution to the problem 3.17. *(a) This is a quotient of polynomials that are zero at $x = 3$. So, if we factor by the binomial $x - 3$, we have:*

$$\begin{aligned}\lim_{x\to 3}\frac{x^2-2x-3}{x^2-4x+3} &= «\frac{0}{0}» \text{ (indeterminate form)}\\ &= \lim_{x\to 3}\frac{(x+1)(x-3)}{(x-1)(x-3)}\\ &= \lim_{x\to 3}\frac{x+1}{x-1} = \frac{4}{2} = 2\end{aligned}$$

where, to simplify the rational expression, we have used that $x \neq 3$, since we are calculating a limit.

(b) In this case, both the numerator and the denominator are zero at $x = 0$, but the numerator is not polynomial. Therefore, we multiply and divide by the conjugate of the numerator:

$$\lim_{x\to 0}\frac{\sqrt{x+8}-2\sqrt{2}}{x} = «\frac{0}{0}» \text{ (indeterminate form)}$$

$$\begin{aligned}\lim_{x\to 0}\frac{\sqrt{x+8}-2\sqrt{2}}{x} &= \lim_{x\to 0}\frac{\sqrt{x+8}-2\sqrt{2}}{x}\,\frac{\sqrt{x+8}+2\sqrt{2}}{\sqrt{x+8}+2\sqrt{2}}\\ &= \lim_{x\to 0}\frac{\left(\sqrt{x+8}\right)^2-\left(2\sqrt{2}\right)^2}{x\left(\sqrt{x+8}+2\sqrt{2}\right)}\\ &= \lim_{x\to 0}\frac{x+\not{8}-\not{8}}{x\left(\sqrt{x+8}+2\sqrt{2}\right)}\\ &= \lim_{x\to 0}\frac{1}{\sqrt{x+8}+2\sqrt{2}}\\ &= \frac{1}{\sqrt{8}+2\sqrt{2}} = \frac{1}{2\sqrt{2}+2\sqrt{2}} = \frac{1}{4\sqrt{2}} = \frac{\sqrt{2}}{8}.\end{aligned}$$

(c) To resolve the indeterminate form, we will use $e^x - 1 \sim_0 x$, that is,

$$\lim_{x\to 0}\frac{e^x-1}{x} = 1.$$

$$\lim_{x\to 0}\frac{e^x-e^{-x}}{x} = «\frac{0}{0}» \text{ (indeterminate form)}$$

$$= \lim_{x\to 0}\frac{e^x-1+1-e^{-x}}{x}$$

$$= \lim_{x\to 0}\frac{e^x-1}{x}+\frac{1-e^{-x}}{x}$$

$$= \lim_{x\to 0}\underbrace{\frac{e^x-1}{x}}_{\to 1}+\underbrace{\frac{e^{-x}-1}{-x}}_{\to 1} = 2$$

To calculate the last summand, we have made the change of variable $t=-x$ in order to use the infinitesimal $e^t-1\sim_0 t$. In fact:

$$\lim_{x\to 0}\frac{e^{-x}-1}{-x} = \lim_{t\to 0}\frac{e^t-1}{t} = 1$$

(d) *In this case, there is no indeterminate form and the limit can be calculated directly.*

$$\lim_{x\to+\infty}\frac{e^{-x}}{\sqrt{x}} = \frac{0}{\infty} = 0.$$

(e) *To resolve the indeterminate form, we will rewrite the limit so that the number e appears:*

$$\lim_{x\to+\infty}\left(\frac{7x-5}{7x-7}\right)^{10+x} = «1^{+\infty}» \text{ (indeterminate form)}$$

$$= \lim_{x\to+\infty}\left(\frac{7x-7}{7x-7}+\frac{2}{7x-7}\right)^{10+x}$$

$$= \lim_{x\to+\infty}\left(1+\frac{1}{\frac{7x-7}{2}}\right)^{\frac{7x-7}{2}\frac{2}{7x-7}(10+x)}$$

$$= \lim_{x\to+\infty}\underbrace{\left[\left(1+\frac{1}{\frac{7x-7}{2}}\right)^{\frac{7x-7}{2}}\right]}_{\to e}^{\frac{2(10+x)}{7x-7}} = e^{\lim_{x\to+\infty}\frac{\frac{2x}{x}+\frac{20}{x}}{\frac{7x}{x}-\frac{7}{x}}} = e^{\frac{2}{7}}$$

where we have used:

$$\lim_{x\to+\infty}\left(1+\frac{1}{\frac{7x-7}{2}}\right)^{\frac{7x-7}{2}} = e,$$

since if $x\to+\infty$, $\frac{7x-7}{2}\to\infty$.

(f) *This limit will be resolved using equivalent infinitesimals. Indeed:*

$$\lim_{x\to 0}\frac{\sin x}{\tan x} = «\frac{0}{0}» \text{ (indeterminate form)} = \lim_{x\to 0}\frac{\overbrace{\frac{\sin x}{x}}^{\to 1}}{\underbrace{\frac{\tan x}{x}}_{\to 1}} = \frac{1}{1} = 1.$$

Problem 3.18. Calculate the following limits:

(a) $\lim_{x\to 3}\frac{x^2-9}{\sqrt{x^4-17}-(3x-1)}$

(b) $\lim_{x\to 0^+}\left(\frac{\ln(x)+1}{\ln(x)}\right)^{\ln(x)}$

(c) $\lim_{x\to 3}\frac{x^2-4x+3}{x^2-5x+6}$

(d) $\lim_{x\to 0}\frac{\sqrt{x+27}-3\sqrt{3}}{x}$

(e) $\lim_{x\to +\infty}\frac{e^{-x}}{\ln x}$

Solution to the problem 3.18. (a) *In this case, both the numerator and the denominator are zero at $x = 3$, but the denominator is not polynomial. Therefore, we multiply and divide by the conjugate of the denominator:*

$$\begin{aligned}
\lim_{x\to 3}\frac{x^2-9}{\sqrt{x^4-17}-(3x-1)} &= «\frac{0}{0}» \text{ (indeterminate form)}\\
&= \lim_{x\to 3}\frac{x^2-9}{\sqrt{x^4-17}-(3x-1)}\,\frac{\sqrt{x^4-17}+(3x-1)}{\sqrt{x^4-17}+(3x-1)}\\
&= \lim_{x\to 3}\frac{(x^2-9)(\sqrt{x^4-17}+(3x-1))}{x^4-17-(3x-1)^2}\\
&= \lim_{x\to 3}\frac{(x^2-9)(\sqrt{x^4-17}+3x-1)}{x^4-17-(9x^2-6x+1)}\\
&= \lim_{x\to 3}\frac{(x^2-9)(\sqrt{x^4-17}+3x-1)}{x^4-9x^2+6x-18}.
\end{aligned}$$

Next, we will factor the polynomial that appears in the denominator and simplify the term $x-3$:

$$\lim_{x\to 3}\frac{(x^2-9)(\sqrt{x^4-17}+3x-1)}{x^4-9x^2+6x-18} = \lim_{x\to 3}\frac{(x+3)(x-3)(\sqrt{x^4-17}+3x-1)}{(x-3)(x^3+3x^2+6)}$$

$$\lim_{x\to 3}\frac{(x^2-9)(\sqrt{x^4-17}+3x-1)}{x^4-9x^2+6x-18}=\lim_{x\to 3}\frac{(x+3)(\sqrt{x^4-17}+3x-1)}{x^3+3x^2+6}$$
$$=\frac{6(8+9-1)}{27+27+6}=\frac{96}{60}=\frac{8}{5}$$

(b) *To resolve the indeterminate form, we rewrite the limit so that the number e appears:*

$$\lim_{x\to 0^+}\left(\frac{\ln(x)+1}{\ln(x)}\right)^{\ln(x)}=«1^{-\infty}» \text{ (indeterminate form)}$$
$$=\underbrace{\lim_{x\to 0^+}\left(1+\frac{1}{\ln(x)}\right)^{\ln(x)}}_{\to e}=e$$

where we have used:

$$\lim_{x\to 0^+}\left(1+\frac{1}{\ln(x)}\right)^{\ln(x)}=e,$$

since if $x\to 0^+$, $\ln(x)\to-\infty$.

(c) *This is a quotient of polynomials that are zero at* $x=3$. *So, if we factor by the binomial* $x-3$, *we have:*

$$\lim_{x\to 3}\frac{x^2-4x+3}{x^2-5x+6}=«\frac{0}{0}» \text{ (indeterminate form)}$$
$$=\lim_{x\to 3}\frac{(x-1)(x-3)}{(x-2)(x-3)}=\lim_{x\to 3}\frac{x-1}{x-2}=\frac{2}{1}=2$$

(d) *In this case, both the numerator and the denominator are zero at* $x=0$, *but the numerator is not polynomial. Therefore, we multiply and divide by the conjugate of the numerator:*

$$\lim_{x\to 0}\frac{\sqrt{x+27}-3\sqrt{3}}{x}=«\frac{0}{0}» \text{ (indeterminate form)}$$
$$=\lim_{x\to 0}\frac{\sqrt{x+27}-3\sqrt{3}}{x}\frac{\sqrt{x+27}+3\sqrt{3}}{\sqrt{x+27}+3\sqrt{3}}$$
$$=\lim_{x\to 0}\frac{\left(\sqrt{x+27}\right)^2-\left(3\sqrt{3}\right)^2}{x\left(\sqrt{x+27}+3\sqrt{3}\right)}$$
$$=\lim_{x\to 0}\frac{x+\cancel{27}-\cancel{27}}{x\left(\sqrt{x+27}+3\sqrt{3}\right)}$$
$$=\lim_{x\to 0}\frac{1}{\sqrt{x+27}+3\sqrt{3}}$$
$$=\frac{1}{\sqrt{27}+3\sqrt{3}}=\frac{1}{3\sqrt{3}+3\sqrt{3}}=\frac{1}{6\sqrt{3}}=\frac{\sqrt{3}}{18}.$$

(e) In this case, there is no indeterminate form and the limit can be calculated directly.

$$\lim_{x\to+\infty} \frac{e^{-x}}{\ln x} = 0.$$

Problem 3.19. Verify the following limits:

(a) $\lim_{x\to 0} \frac{\sqrt{2+x}-\sqrt{2}}{x} = \frac{1}{2\sqrt{2}}$

(b) $\lim_{x\to 0} x \sin\left(\frac{1}{x^2}\right) = 0$

(c) $\lim_{x\to 2^+} \left(\frac{2x+1}{x+3}\right)^{\frac{1}{x-2}} = e^{\frac{1}{5}}$

(d) $\nexists \lim_{x\to 0} \frac{1}{x}$ or $\begin{cases} \lim_{x\to 0-} \frac{1}{x} = -\infty \\ \lim_{x\to 0+} \frac{1}{x} = +\infty \end{cases}$

(e) $\lim_{x\to 0} \frac{1}{x^2} = +\infty$

(f) $\lim_{x\to+\infty} \frac{x+\cos(3x)}{x-\cos(3x)} = 1$

Solution to the problem 3.19. *(a) In this case, both the numerator and the denominator are zero at $x = 0$, but the numerator is not polynomial. Therefore, we multiply and divide by the conjugate of the numerator:*

$$\begin{aligned}
\lim_{x\to 0} \frac{\sqrt{2+x}-\sqrt{2}}{x} &= «\frac{0}{0}» \text{ (indeterminate form)} \\
&= \lim_{x\to 0} \frac{\sqrt{2+x}-\sqrt{2}}{x} \cdot \frac{\sqrt{2+x}+\sqrt{2}}{\sqrt{2+x}+\sqrt{2}} = \lim_{x\to 0} \frac{(2+x)-2}{x\left(\sqrt{2+x}+\sqrt{2}\right)} \\
&= \lim_{x\to 0} \frac{x}{x\left(\sqrt{2+x}+\sqrt{2}\right)} = \lim_{x\to 0} \frac{1}{\sqrt{2+x}+\sqrt{2}} = \frac{1}{\sqrt{2}+\sqrt{2}} \\
&= \frac{1}{2\sqrt{2}} = \frac{\sqrt{2}}{4}.
\end{aligned}$$

(b) This limit will be solved using the "zero times bounded" criterion. Indeed:

$$\lim_{x\to 0} \underbrace{x}_{\to 0} \underbrace{\sin\left(\frac{1}{x^2}\right)}_{bounded} = 0$$

(c) *To resolve the indeterminate form, we rewrite the limit to make the number e appear:*

$$\lim_{x\to 2^+}\left(\frac{2x+1}{x+3}\right)^{\frac{1}{x-2}} = «1^{+\infty}» \text{ (indeterminate form)}$$

$$= \lim_{x\to 2^+}\left(\frac{x+3+x-2}{x+3}\right)^{\frac{1}{x-2}} = \lim_{x\to 2^+}\left(1+\frac{x-2}{x+3}\right)^{\frac{1}{x-2}}$$

$$= \lim_{x\to 2^+}\left(1+\frac{1}{\frac{x+3}{x-2}}\right)^{\frac{x+3}{x-2}\cdot\frac{1}{x+3}} = \lim_{x\to 2^+}\underbrace{\left[\left(1+\frac{1}{\frac{x+3}{x-2}}\right)^{\frac{x+3}{x-2}}\right]}_{\to e}{}^{\frac{1}{x+3}}$$

$$= e^{\lim\limits_{x\to 2^+}\frac{1}{x+3}} = e^{\frac{1}{5}}$$

where we have used that:

$$\lim_{x\to 2^+}\left(1+\frac{1}{\frac{x+3}{x-2}}\right)^{\frac{x+3}{x-2}} = e,$$

since if $x\to 2^+$, $\frac{x+3}{x-2}\to\infty$.

(d) *In this case, we verify that the limit does not exist because the limit from the left is not equal to the limit from the right:*

$$\nexists \lim_{x\to 0}\frac{1}{x}$$

because:

$$\lim_{x\to 0^-}\frac{1}{x} = -\infty \quad \textit{and} \quad \lim_{x\to 0^+}\frac{1}{x} = +\infty.$$

(e) *In this case, there is no indeterminate form and the limit can be calculated directly:*

$$\lim_{x\to 0}\frac{1}{x^2} = +\infty.$$

(f) *This limit will be solved using the "zero times bounded" criterion:*

$$\lim_{x\to+\infty}\frac{x+\cos(3x)}{x-\cos(3x)} = \lim_{x\to+\infty}\frac{\frac{x}{x}+\overbrace{\frac{\cos(3x)}{x}}^{\to 0}}{\frac{x}{x}-\underbrace{\frac{\cos(3x)}{x}}_{\to 0}} = \frac{1+0}{1-0} = 1.$$

Problem 3.20. Verify the following limits at infinity:

(a) $\lim\limits_{x\to+\infty}\frac{2-x-5x^2}{3x^2+1} = -\frac{5}{3}$

(b) $\lim_{x\to+\infty}\left(3+\dfrac{2-2x}{x-4}\right)^{\frac{x+1}{2}}=e^{-3}$

(c) $\lim_{x\to+\infty}\dfrac{7x-3}{4x^2+2}=0$

(d) $\lim_{x\to+\infty}\sqrt{x^2+1}-x-2=-2$

(e) $\lim_{x\to+\infty}\left(\dfrac{x+5}{x+2}\right)^{x+3}=e^3$

(f) $\lim_{x\to+\infty}\dfrac{x^2+x+2}{2x+3}+\dfrac{1-x^3}{x^2+5}=-\infty$

(g) $\lim_{x\to+\infty}\sqrt{3x^2+5x-9}-\sqrt{3x^2-x+1}=\sqrt{3}$

Solution to the problem 3.20. *(a) This is a quotient of polynomials that tend to infinity. So, if we divide both polynomials by* x^2 *we have:*

$$\lim_{x\to+\infty}\frac{2-x-5x^2}{3x^2+1}=\lim_{x\to+\infty}\frac{\frac{2}{x^2}-\frac{x}{x^2}-\frac{5x^2}{x^2}}{\frac{3x^2}{x^2}+\frac{1}{x^2}}=\frac{0-0-5}{3+0}=-\frac{5}{3}.$$

(b) To resolve the indetermination, we will rewrite the limit to make the number e appear:

$$\lim_{x\to+\infty}\left(3+\frac{2-2x}{x-4}\right)^{\frac{x+1}{2}}=\lim_{x\to+\infty}\left(1+\frac{2x-8+2-2x}{x-4}\right)^{\frac{x+1}{2}}$$

$$=\lim_{x\to+\infty}\left(1+\frac{-6}{x-4}\right)^{\frac{x+1}{2}}=\lim_{x\to+\infty}\left(1+\frac{1}{\frac{4-x}{6}}\right)^{\frac{4-x}{6}\cdot\frac{6}{4-x}\cdot\frac{x+1}{2}}$$

$$=\lim_{x\to+\infty}\left[\left(1+\frac{1}{\frac{4-x}{6}}\right)^{\frac{4-x}{6}}\right]^{\frac{6}{4-x}\cdot\frac{x+1}{2}}$$

$$=e^{\lim_{x\to+\infty}\frac{3(x+1)}{4-x}}=e^{-3}$$

where we have used that:

$$\lim_{x\to+\infty}\left(1+\frac{1}{\frac{4-x}{6}}\right)^{\frac{4-x}{6}}=e,$$

since if $x\to+\infty$, $\frac{4-x}{6}\to+\infty$.

(c) This is a quotient of polynomials that tend to infinity. So, if we divide both polynomials by x^2 *we have:*

$$\lim_{x\to+\infty}\frac{7x-3}{4x^2+2}=\lim_{x\to+\infty}\frac{\frac{7x}{x^2}-\frac{3}{x^2}}{\frac{4x^2}{x^2}+\frac{2}{x^2}}=\lim_{x\to+\infty}\frac{\frac{7}{x}}{4}=\frac{0}{4}=0.$$

(d) *In this case, we can multiply and divide $\sqrt{x^2+1}-x$ by its conjugate:*

$$\lim_{x\to+\infty} \underbrace{\sqrt{x^2+1}}_{+\infty} - \underbrace{x}_{+\infty} - 2 = «+\infty - \infty» \text{ (indeterminate form)}$$

$$= \lim_{x\to+\infty} \left(\sqrt{x^2+1}-x\right)\cdot\frac{\left(\sqrt{x^2+1}+x\right)}{\sqrt{x^2+1}+x} - 2$$

$$= \lim_{x\to+\infty} \frac{x^2+1-x^2}{\sqrt{x^2+1}+x} - 2$$

$$= \lim_{x\to+\infty} \frac{1}{\sqrt{x^2+1}+x} - 2 = -2.$$

(e) *This limit is an indetermination of the form «$1^{+\infty}$». Therefore, we will resolve it using the number e. Indeed:*

$$\lim_{x\to+\infty}\left(\frac{x+5}{x+2}\right)^{x+3} = \lim_{x\to+\infty}\left(\frac{x+2+3}{x+2}\right)^{x+3}$$

$$= \lim_{x\to+\infty}\left(1+\frac{3}{x+2}\right)^{x+3} = \lim_{x\to+\infty}\left(1+\frac{1}{\frac{x+2}{3}}\right)^{x+3}$$

$$= \lim_{x\to+\infty}\left(1+\frac{1}{\frac{x+2}{3}}\right)^{\frac{x+2}{3}\cdot\frac{3}{x+2}\cdot(x+3)} = \lim_{x\to+\infty}\left[\left(1+\frac{1}{\frac{x+2}{3}}\right)^{\frac{x+2}{3}}\right]^{\frac{3(x+3)}{x+2}}$$

$$= e^{\lim_{x\to+\infty}\frac{3(x+3)}{x+2}} = e^3$$

(f) *This limit is an indetermination of the form «$+\infty-\infty$» that we will resolve by simply summing both fractions. Indeed:*

$$\lim_{x\to+\infty}\frac{x^2+x+2}{2x+3}+\frac{1-x^3}{x^2+5} = \lim_{x\to+\infty}\frac{-x^4-2\,x^3+7\,x^2+7\,x+13}{2\,x^3+3\,x^2+10\,x+15} = -\infty$$

(g) *To resolve this limit, we multiply and divide:*

$$\sqrt{3x^2+5x-9}-\sqrt{3x^2-x+1}$$

by its conjugate:

$$\lim_{x\to+\infty}\sqrt{3x^2+5x-9}-\sqrt{3x^2-x+1} = «+\infty-\infty» \text{ (indeterminate form)}$$

$$= \lim_{x\to+\infty}\left(\sqrt{3x^2+5x-9}-\sqrt{3x^2-x+1}\right)\cdot\frac{\sqrt{3x^2+5x-9}+\sqrt{3x^2-x+1}}{\sqrt{3x^2+5x-9}+\sqrt{3x^2-x+1}}$$

$$= \lim_{x\to+\infty} \frac{3x^2+5x-9-(3x^2-x+1)}{\sqrt{3x^2+5x-9}+\sqrt{3x^2-x+1}}$$

$$= \lim_{x\to+\infty} \frac{6x-10}{\sqrt{3x^2+5x-9}+\sqrt{3x^2-x+1}}$$

$$= \lim_{x\to+\infty} \frac{\frac{6x}{x}-\frac{10}{x}}{\sqrt{\frac{3x^2}{x^2}+\frac{5x}{x^2}-\frac{9}{x^2}}+\sqrt{\frac{3x^2}{x^2}-\frac{x}{x^2}+\frac{1}{x^2}}}$$

$$= \frac{6-0}{\sqrt{3+0-0}+\sqrt{3-0+0}} = \frac{6}{\sqrt{3}+\sqrt{3}} = \frac{6}{2\sqrt{3}} = \frac{3}{\sqrt{3}} = \sqrt{3}.$$

Problem 3.21. Calculate the following limits:

(a) $\lim_{x\to+\infty} \frac{3x}{\sqrt{4x^2-3}}$

(b) $\lim_{x\to-\infty} \left(\frac{2x+2}{-1+2x}\right)^{\frac{x}{2}}$

Solution to the problem 3.21. *(a) If we divide the numerator and the denominator by x we have:*

$$\lim_{x\to+\infty} \frac{3x}{\sqrt{4x^2-3}} = \lim_{x\to+\infty} \frac{3x/x}{\sqrt{4x^2/x^2-3/x^2}}$$

$$= \lim_{x\to+\infty} \frac{3}{\sqrt{4-3/x^2}} = \frac{3}{\sqrt{4}} = \frac{3}{2}.$$

(b) To resolve the indetermination we will rewrite the limit to make the number e appear:

$$\lim_{x\to-\infty} \left(\frac{2x+2}{-1+2x}\right)^{\frac{x}{2}} = \lim_{x\to-\infty} \left(1+\frac{2x+2}{-1+2x}-1\right)^{\frac{x}{2}}$$

$$= \lim_{x\to-\infty} \left(1+\frac{2x+2+1-2x}{-1+2x}\right)^{\frac{x}{2}}$$

$$= \lim_{x\to-\infty} \left(1+\frac{3}{-1+2x}\right)^{\frac{x}{2}}$$

$$= \lim_{x\to-\infty} \left(1+\frac{1}{\frac{-1+2x}{3}}\right)^{\frac{-1+2x}{3}\cdot\frac{3}{-1+2x}\cdot\frac{x}{2}}$$

$$= e^{\lim_{x\to-\infty} \frac{3x}{4x-2}} = e^{\frac{3}{4}}$$

where we have used that:

$$\lim_{x\to+\infty} \left(1+\frac{1}{\frac{-1+2x}{3}}\right)^{\frac{-1+2x}{3}} = e,$$

since if $x \to -\infty$, $\frac{-1+2x}{3} \to -\infty$.

Problem 3.22. Calculate the limits of the following rational functions where the denominator vanishes:

(a) $f(x) = \dfrac{x^3}{1-x^2}$

(b) $f(x) = \dfrac{1-x^3}{x^2-4x}$

(c) $f(x) = \dfrac{x^3+x}{3-2x^2}$

Solution to the problem 3.22. *(a) The roots of f are:*

$$1 - x^2 = 0 \iff x = \pm 1.$$

For each root we calculate the limit from the left and from the right:

$$\lim_{x\to 1} \frac{x^3}{1-x^2} = \frac{1}{0} = \begin{cases} \lim_{x\to 1^+} \frac{x^3}{1-x^2} = \frac{1}{0^-} = -\infty \\ \lim_{x\to 1^-} \frac{x^3}{1-x^2} = \frac{1}{0^+} = +\infty \end{cases}$$

$$\lim_{x\to -1} \frac{x^3}{1-x^2} = \frac{-1}{0} = \begin{cases} \lim_{x\to -1^+} \frac{x^3}{1-x^2} = \frac{-1}{0^+} = -\infty \\ \lim_{x\to -1^-} \frac{x^3}{1-x^2} = \frac{-1}{0^-} = +\infty. \end{cases}$$

Therefore, neither the limit of $f(x)$ when x tends to -1 nor when it tends to 1 exists.

(b) The roots of f are:

$$x^2 - 4x = 0 \iff x(x-4) = 0 \iff x = \{0, 4\}$$

For each root we calculate the limit from the left and from the right:

$$\lim_{x\to 0} \frac{1-x^3}{x^2-4x} = \frac{1}{0} = \begin{cases} \lim_{x\to 0^+} \frac{1-x^3}{x^2-4x} = \lim_{x\to 0^+} \frac{1-x^3}{x(x-4)} = \frac{1}{0^-} = -\infty \\ \lim_{x\to 0^-} \frac{1-x^3}{x^2-4x} = \lim_{x\to 0^+} \frac{1-x^3}{x(x-4)} = \frac{1}{0^+} = +\infty \end{cases}$$

$$\lim_{x\to 4} \frac{1-x^3}{x^2-4x} = \frac{-63}{0} = \begin{cases} \lim_{x\to 4^+} \frac{1-x^3}{x^2-4x} = \lim_{x\to 0^+} \frac{1-x^3}{x(x-4)} = \frac{-63}{0^+} = -\infty \\ \lim_{x\to 4^-} \frac{1-x^3}{x^2-4x} = \lim_{x\to 0^+} \frac{1-x^3}{x(x-4)} = \frac{-63}{0^-} = +\infty. \end{cases}$$

Therefore, neither the limit of $f(x)$ when x tends to 0 nor when it tends to 4 exists.

(c) The roots of f are:

$$3 - 2x^2 = 0 \implies x^2 = \frac{3}{2} \implies x = \pm\sqrt{\frac{3}{2}}$$

For each root we calculate the limit from the left and from the right:

$$\lim_{x\to\sqrt{\frac{3}{2}}}\frac{x^3+x}{3-2x^2}=\frac{\frac{5}{4}\sqrt{6}}{0}=\begin{cases}\lim_{x\to\sqrt{\frac{3}{2}}^+}\frac{x^3+x}{3-2x^2}=\frac{\frac{5}{4}\sqrt{6}}{0^-}=-\infty\\ \lim_{x\to\sqrt{\frac{3}{2}}^-}\frac{x^3+x}{3-2x^2}=\frac{\frac{5}{4}\sqrt{6}}{0^+}=+\infty\end{cases}$$

$$\lim_{x\to-\sqrt{\frac{3}{2}}}\frac{x^3+x}{3-2x^2}=\frac{-\frac{5}{4}\sqrt{6}}{0}=\begin{cases}\lim_{x\to-\sqrt{\frac{3}{2}}^+}\frac{x^3+x}{3-2x^2}=\frac{-\frac{5}{4}\sqrt{6}}{0^+}=-\infty\\ \lim_{x\to-\sqrt{\frac{3}{2}}^-}\frac{x^3+x}{3-2x^2}=\frac{-\frac{5}{4}\sqrt{6}}{0^-}=+\infty.\end{cases}$$

Therefore, neither the limit of $f(x)$ when x tends to $-\sqrt{\frac{3}{2}}$ nor when it tends to $\sqrt{\frac{3}{2}}$ exists.

Problem 3.23. Verify the following limits:

(a) $\lim_{x\to 0}\frac{x}{1+x}=0$

(b) $\lim_{x\to+\infty}\sin\left(\frac{1}{x^2}\right)=0$

(c) $\lim_{x\to+\infty}\frac{x+3}{8x+10}=\frac{1}{8}$

(d) $\lim_{x\to 0}|x|=0$

(e) $\nexists\lim_{x\to 2}\frac{8x}{x-2}$

(f) $\lim_{x\to+\infty}\frac{\sin(x)}{x}=0$

Solution to the problem 3.23. *(a) In this case, there is no indetermination and the limit can be calculated directly. Indeed:*

$$\lim_{x\to 0}\frac{x}{1+x}=\frac{0}{1+0}=0.$$

(b) In this case, there are also no indeterminations, so the limit can be calculated directly. Indeed:

$$\lim_{x\to\infty}\sin\left(\frac{1}{x^2}\right)=\sin\left(\lim_{x\to\infty}\frac{1}{x^2}\right)=\sin(0)=0.$$

(c) *This is a quotient of polynomials that tend to infinity. So, if we divide both polynomials by x we have:*

$$\lim_{x\to\infty}\frac{x+3}{8x+10}=\lim_{x\to\infty}\frac{\frac{x}{x}+\frac{3}{x}}{\frac{8x}{x}+\frac{10}{x}}=\lim_{x\to\infty}\frac{1+0}{8+0}=\frac{1}{8}.$$

(d) *To verify this limit we must check that the limit from the left is equal to the limit from the right. Indeed:*

$$\left.\begin{array}{l}\lim_{x\to 0^+}|x|=\lim_{x\to 0^+}x=0\\ \lim_{x\to 0^-}|x|=\lim_{x\to 0^-}-x=0\end{array}\right\}\Longrightarrow \lim_{x\to 0}|x|=0$$

(e) *In this case, we must also check if the limit from the left is equal to the limit from the right:*

$$\left.\begin{array}{l}\lim_{x\to 2^+}\frac{8x}{\underbrace{x-2}_{>0}}=\frac{16}{0^+}=+\infty\\ \lim_{x\to 2^-}\frac{8x}{\underbrace{x-2}_{<0}}=\frac{16}{0^-}=-\infty\end{array}\right\}\Longrightarrow \nexists\lim_{x\to 2}\frac{8x}{x-2}$$

(f) *This limit will be calculated considering the criterion of zero times bounded:*

$$\lim_{x\to+\infty}\frac{\sin(x)}{x}=\lim_{x\to+\infty}\frac{1}{x}\cdot\sin(x)=\text{«0·bounded»}=0.$$

Problem 3.24. Verify the following limits:

(a) $\nexists\lim_{x\to 0}e^{1/x}$

(b) $\lim_{x\to -2^+}e^{x/2}-6\ln|x+2|+3=+\infty$

Solution to the problem 3.24. (a) *To verify that this limit does not exist, we must check that the lateral limits are not equal. Indeed:*

$$\lim_{x\to 0^+}e^{1/x}=+\infty$$

since if $x\to 0^+$, *then* $\frac{1}{x}\to\frac{1}{0^+}=+\infty$ *and* $e^{+\infty}\to+\infty$. *On the other hand:*

$$\lim_{x\to 0^-}e^{1/x}=0$$

since if $x\to 0^-$, *then* $\frac{1}{x}\to\frac{1}{0^-}=-\infty$ *and* $e^{-\infty}\to 0$.

(*b*) *In this case, there is no indetermination and the limit can be calculated directly. Indeed:*

$$\lim_{x\to -2^+} \underbrace{e^{x/2}}_{\to e^{-1}} - \underbrace{6\ln|x+2|}_{\to -\infty} + 3 = -6(-\infty) = +\infty.$$

Problem 3.25. Calculate the following limits:

(a) $\displaystyle\lim_{x\to 8} \frac{x-8}{\sqrt[3]{x}-2}$

(b) $\displaystyle\lim_{x\to 0} \frac{6x-6\sin(x)}{x^3}$

(c) $\displaystyle\lim_{x\to +\infty} \left(\sqrt{\frac{x^2-x}{x^2-1}}\right)^{\frac{x+1}{3}}$

Help: $\sin(x) \sim_0 x$, $\sin(x) - x \sim_0 -\frac{1}{6}x^3$.

Solution to the problem 3.25. (*a*) *This limit is initially an indetermination of the form* «$\frac{0}{0}$»*. In this case, we will resolve it by making the change of variable* $x = y^3$*, which implies that instead of* $x \to 8$ *we will have* $y \to 2$*. Indeed:*

$$\lim_{x\to 8} \frac{x-8}{\sqrt[3]{x}-2} = \lim_{y\to 2} \frac{y^3-8}{y-2}.$$

Now, since it is a rational function, we can factor the numerator and simplify:

$$\lim_{y\to 2} \frac{y^3-8}{y-2} = \lim_{y\to 2} \frac{(y-2)(y^2+2y+4)}{y-2}$$

$$= \lim_{y\to 2} (y^2+2y+4) = 12.$$

(*b*) *This limit is an indetermination of the form* «$\frac{0}{0}$» *that we will resolve by applying equivalent infinitesimals. Indeed:*

$$\lim_{x\to 0} \frac{6x-6\sin(x)}{x^3} = \lim_{x\to 0} \frac{-6(\sin(x)-x)}{x^3}$$

$$= \lim_{x\to 0} \frac{-6(\sin(x)-x)}{x^3} \cdot \frac{-\frac{1}{6}x^3}{\sin(x)-x}$$

$$= \lim_{x\to 0} -6\frac{-1}{6} = 1.$$

(c) *This limit is an indetermination of the form «1^∞». We will resolve it by rewriting the limit in a form that makes the number e appear. Indeed:*

$$\lim_{x\to+\infty}\left(\sqrt{\frac{x^2-x}{x^2-1}}\right)^{\frac{x+1}{3}} = \lim_{x\to+\infty}\left(\left[\frac{x^2-x}{x^2-1}\right]^{\frac{1}{2}}\right)^{\frac{x+1}{3}} = \lim_{x\to+\infty}\left(\frac{x^2-x}{x^2-1}\right)^{\frac{x+1}{6}}$$

$$= \lim_{x\to+\infty}\left(1+\frac{x^2-x}{x^2-1}-1\right)^{\frac{x+1}{6}}$$

$$= \lim_{x\to+\infty}\left(1+\frac{x^2-x-x^2+1}{x^2-1}\right)^{\frac{x+1}{6}}$$

$$= \lim_{x\to+\infty}\left(1+\frac{-(x-1)}{x^2-1}\right)^{\frac{x+1}{6}}$$

$$= \lim_{x\to+\infty}\left(1+\frac{-(x-1)}{(x+1)(x-1)}\right)^{\frac{x+1}{6}}$$

$$= \lim_{x\to+\infty}\left(1+\frac{1}{-(x+1)}\right)^{\frac{x+1}{6}}$$

$$= \lim_{x\to+\infty}\left(1+\frac{1}{-(x+1)}\right)^{-(x+1)\frac{-1}{6}}$$

$$= e^{-\frac{1}{6}}.$$

where we have used that:

$$\lim_{x\to+\infty}\left(1+\frac{1}{-(x+1)}\right)^{-(x+1)} = e,$$

since if $x\to+\infty$, $-(x+1)\to-\infty$.

Problem 3.26. Knowing that $\lim_{x\to0}\frac{\sin(x)}{x}=1$, verify the following limits:

(a) $\lim_{x\to0}\frac{x}{\sin(x)}=1$

(b) $\lim_{x\to0}\frac{3x}{\sin(x)}=3$

Solution to the problem 3.26. *(a)* $\lim_{x\to0}\frac{x}{\sin(x)}=\lim_{x\to0}\frac{x}{\sin(x)}\cdot\underbrace{\frac{\sin(x)}{x}}_{\to1}=\lim_{x\to0}1=1.$

(b) $\lim_{x\to0}\frac{3x}{\sin(x)}=3\lim_{x\to0}\frac{x}{\sin(x)}=3\cdot1=3.$

Problem 3.27. Verify that:

(a) $\lim_{x\to 0} \frac{\sin(x) - x}{x^2} = 0$

(b) $\lim_{x\to 0} \frac{\sin(x) - x}{x^3} = -\frac{1}{6}$

Hint: $\sin(x) \sim_0 x$, $\sin(x) - x \sim_0 -\frac{1}{6}x^3$.

Solution to the problem 3.27. *To solve both limits, we will use the approximation* $\sin(x) - x \sim_0 -\frac{1}{6}x^3$, *i.e.,*

$$\lim_{x\to 0} \frac{\sin(x) - x}{-\frac{1}{6}x^3} = 1.$$

(a) $\lim_{x\to 0} \frac{\sin(x) - x}{x^2} = \lim_{x\to 0} \frac{\sin(x) - x}{-\frac{1}{6}x^3} \cdot \left(-\frac{1}{6}x\right) = \lim_{x\to 0} 1 \cdot \left(\frac{-1}{6}x\right) = 0.$

(b) $\lim_{x\to 0} \frac{\sin(x) - x}{x^3} = \lim_{x\to 0} \frac{\sin(x) - x}{-\frac{1}{6}x^3} \left(-\frac{1}{6}\right) = -\frac{1}{6}.$

Problem 3.28. Calculate the following limits:

(a) $\lim_{x\to +\infty} \frac{3x}{\sqrt{2x^2 - 3}}$

(b) $\lim_{x\to -\infty} \frac{|4x|}{2x - 7}$

(c) $\lim_{x\to 0} \frac{x - \ln(1+x)}{x^2}$

Hint: $\ln(1+x) \sim_0 x$, $\quad 2x - 2\ln(1+x) \sim_0 x^2$.

Solution to the problem 3.28. *(a) If we divide the numerator and the denominator by x we get:*

$$\lim_{x\to +\infty} \frac{3x}{\sqrt{2x^2 - 3}} = \lim_{x\to +\infty} \frac{3x/x}{\sqrt{2x^2/x^2 - 3/x^2}} = \lim_{x\to +\infty} \frac{3}{\sqrt{2 - 3/x^2}} = \frac{3}{\sqrt{2}}.$$

(b) Again, by dividing the numerator and the denominator by x we get:

$$\lim_{x\to -\infty} \frac{|4x|}{2x - 7} = \lim_{x\to -\infty} \frac{-4x}{2x - 7} = \lim_{x\to -\infty} \frac{-4x/x}{2x/x - 7/x} = \lim_{x\to -\infty} \frac{-4}{2 - 7/x} = -2.$$

(c) To solve the indeterminate form we will use the approximation $2x - 2\ln(1+x) \sim_0 x^2$, i.e.,

$$\lim_{x\to 0} \frac{2x - 2\ln(1+x)}{x^2} = 1.$$

Thus, we get:

$$\lim_{x\to 0} \frac{x - \ln(1+x)}{x^2} = \lim_{x\to 0} \underbrace{\frac{2x - 2\ln(1+x)}{x^2}}_{\to 1} \cdot \frac{1}{2} \cdot = \frac{1}{2}.$$

Problem 3.29. Calculate the following limits:

(a) $\lim_{x\to 0} \dfrac{1-x^2}{1-x^5}$

(b) $\lim_{x\to -1} \dfrac{1+x}{\sqrt{3+2x}-\sqrt{-1-2x}}$

(c) $\lim_{x\to -1} \dfrac{\sqrt{3+2x}-\sqrt{-1-2x}}{-x^2+5x-4+e}$

Solution to the problem 3.29. *(a) It is a quotient of polynomials. Therefore:*

$$\lim_{x\to 0} \frac{1-x^2}{1-x^5} = \frac{1}{1} = 1.$$

(b) In this case, both the numerator and the denominator are zero at $x=-1$, but the denominator is not a polynomial. Therefore, we multiply and divide by the conjugate of the denominator:

$$\begin{aligned}
\lim_{x\to -1} \frac{1+x}{\sqrt{3+2x}-\sqrt{-1-2x}} &= \lim_{x\to -1} \frac{1+x}{\sqrt{3+2x}-\sqrt{-1-2x}} \cdot \frac{\sqrt{3+2x}+\sqrt{-1-2x}}{\sqrt{3+2x}+\sqrt{-1-2x}} \\
&= \lim_{x\to -1} \frac{(1+x)(\sqrt{3+2x}+\sqrt{-1-2x})}{3+2x-(-1-2x)} \\
&= \lim_{x\to -1} \frac{(1+x)(\sqrt{3+2x}+\sqrt{-1-2x})}{4+4x} \\
&= \lim_{x\to -1} \frac{(1+x)(\sqrt{3+2x}+\sqrt{-1-2x})}{4(1+x)} \\
&= \lim_{x\to -1} \frac{\sqrt{3+2x}+\sqrt{-1-2x}}{4} \\
&= \frac{\sqrt{3+2(-1)}+\sqrt{-1-2(-1)}}{4} \\
&= \frac{\sqrt{1}+\sqrt{1}}{4} = \frac{2}{4} = \frac{1}{2}.
\end{aligned}$$

(c) In this case, there is no indeterminate form and the limit can be calculated directly:

$$\lim_{x \to -1} \frac{\sqrt{3+2x} - \sqrt{-1-2x}}{-x^2+5x-4+e} = \frac{0}{e-10} = 0.$$

Problem 3.30. Solve the following parts:

(a) Calculate a_1 and a_2 ($a_1, a_2 \in \mathbb{R}$) such that $\displaystyle\lim_{x \to +\infty} \left(a_1 x + a_2 - \frac{2x^3 - x^2 + 1}{3x^2 - 4} \right) = 1$

(b) Calculate $b \in \mathbb{R}$ such that $\displaystyle\lim_{x \to +\infty} \left(\frac{b+x}{x} \right)^{\frac{x}{2}} = 6$

(c) Calculate $c \in \mathbb{R}$ such that $\displaystyle\lim_{x \to 0} \frac{cx^3 \sin(x)}{(1-\cos(x))^2} = \ln(2)$

(d) Calculate the limit $L_1 = \displaystyle\lim_{x \to +\infty} \frac{3^x + 4^{x+1}}{4^{x-1}}$.

Hint: $\sin(x) \sim_0 x, \quad 1 - \cos(x) \sim_0 \frac{x^2}{2}$.

Solution to the problem 3.30. *(a) We incorporate a_1 and a_2 into the fraction:*

$$\lim_{x \to +\infty} \left(a_1 x + a_2 - \frac{2x^3 - x^2 + 1}{3x^2 - 4} \right) =$$

$$= \lim_{x \to +\infty} \left(\frac{(a_1 x + a_2)(3x^2 - 4) - (2x^3 - x^2 + 1)}{3x^2 - 4} \right)$$

$$= \lim_{x \to +\infty} \left(\frac{3a_1 x^3 - 4a_1 x + 3a_2 x^2 - 4a_2 - 2x^3 + x^2 - 1}{3x^2 - 4} \right)$$

$$= \lim_{x \to +\infty} \left(\frac{(3a_1 - 2)x^3 + (3a_2 + 1)x^2 - 4a_1 x - 4a_2 - 1}{3x^2 - 4} \right) = 1.$$

The coefficient of the cubic term in the numerator must be 0 and the coefficient of the quadratic term must be equal to the coefficient of the quadratic term in the denominator. Therefore, we have a linear system of equations:

$$3a_1 - 2 = 0$$
$$3a_2 + 1 = 3$$

whose solution is:

$$a_1 = 2/3, \qquad a_2 = 2/3.$$

(b) *We develop the limit to make the number e appear:*

$$\lim_{x\to+\infty}\left(\frac{b+x}{x}\right)^{\frac{x}{2}} = \lim_{x\to+\infty}\left(1+\frac{b}{x}\right)^{\frac{x}{2}} = \lim_{x\to+\infty}\left(1+\frac{1}{\frac{x}{b}}\right)^{\frac{x}{2}}$$

$$= \lim_{x\to+\infty}\left(1+\frac{1}{\frac{x}{b}}\right)^{\frac{x}{b}\frac{b}{x}\frac{x}{2}} = e^{\lim\limits_{x\to+\infty}\frac{b}{x}\frac{x}{2}}$$

$$= e^{\lim\limits_{x\to+\infty}\frac{b}{2}} = e^{\frac{b}{2}} = 6$$

where we have used that:

$$\lim_{x\to+\infty}\left(1+\frac{1}{\frac{x}{b}}\right)^{\frac{x}{b}} = e,$$

since if $x\to+\infty$, $\frac{x}{b}\to+\infty$. *Therefore:*

$$e^{\frac{b}{2}} = 6 \quad\Leftrightarrow\quad \frac{b}{2} = \ln(6) \quad\Leftrightarrow\quad b = 2\ln(6).$$

(c) *To develop this limit, we use equivalent infinitesimals:*

$$\lim_{x\to0}\frac{cx^3\sin(x)}{(1-\cos(x))^2} = \lim_{x\to0}\frac{cx^3\sin(x)}{(1-\cos(x))^2}\cdot\frac{x}{\sin(x)}\cdot\frac{(1-\cos(x))^2}{\left(\frac{x^2}{2}\right)^2}$$

$$= \lim_{x\to0}\frac{4cx^3x}{x^4} = \lim_{x\to0}4c$$

$$= 4c = \ln(2).$$

Therefore:

$$c = \frac{1}{4}\ln(2).$$

(d) *We rewrite this limit so that the exponents tend to* 1 *or* 0*:*

$$L_1 = \lim_{x\to+\infty}\frac{3^x+4^{x+1}}{4^{x-1}} = \lim_{x\to+\infty}\frac{3^x+4\cdot4^x}{4^{-1}4^x}\cdot\frac{\left(\frac{1}{4}\right)^x}{\left(\frac{1}{4}\right)^x}$$

$$= \lim_{x\to+\infty}\frac{\left(\frac{3}{4}\right)^x+4\left(\frac{4}{4}\right)^x}{4^{-1}\left(\frac{4}{4}\right)^x} = \frac{0+4\cdot1}{4^{-1}\cdot1} = 16.$$

Problem 3.31. Consider the function:

$$\text{sign}(x) = \begin{cases} 1, & \text{if } x > 0 \\ 0, & \text{if } x = 0 \\ -1, & \text{if } x < 0. \end{cases}$$

(a) Calculate $\text{sign}(f(x))$ when:

(1) $f(x) = x^2 - x - 1$

(2) $f(x) = \sin(x)$

(b) Prove that $\text{sign}(x) = \lim_{n\to\infty} \frac{2}{\pi} \arctan(nx)$

Hint: $\lim_{x\to+\infty} \arctan(x) = \pi/2 \quad \lim_{x\to-\infty} \arctan(x) = -\pi/2.$

Solution to the problem 3.31. *(a) (1) We need to calculate the function sign($f(x)$):*

$$sign(f(x)) = \begin{cases} 1 & \text{if } f(x) > 0 \\ 0 & \text{if } f(x) = 0 \\ -1 & \text{if } f(x) < 0 \end{cases}$$

therefore, we need to study the sign of the function $f(x)$. The changes of sign are given by:

$$f(x) = 0 \iff x^2 - x - 1 = 0 \iff x = \frac{1 \pm \sqrt{(-1)^2 - 4 \cdot (-1)}}{2} = \frac{1 \pm \sqrt{5}}{2}.$$

Moreover, since $\lim_{x\to\pm\infty} f(x) = +\infty$, we know that:

$$\begin{cases} \text{if } x \in \left(-\infty, \frac{1-\sqrt{5}}{2}\right) \cup \left(\frac{1+\sqrt{5}}{2}, \infty\right) & \implies f(x) > 0 \\ \text{if } x = \frac{1\pm\sqrt{5}}{2} \implies f(x) = 0 \\ \text{if } x \in \left(\frac{1-\sqrt{5}}{2}, \frac{1+\sqrt{5}}{2}\right) \implies f(x) < 0. \end{cases}$$

Therefore:

$$sign(f(x)) = \begin{cases} 1 & \text{if } x \in \left(-\infty, \frac{1-\sqrt{5}}{2}\right) \cup \left(\frac{1+\sqrt{5}}{2}, \infty\right) \\ 0 & \text{if } x = \frac{1\pm\sqrt{5}}{2} \\ -1 & \text{if } x \in \left(\frac{1-\sqrt{5}}{2}, \frac{1+\sqrt{5}}{2}\right). \end{cases}$$

The graph of this function can be seen in Figure 3.59 (left).

(2) We need to calculate the function sign($f(x)$) where $f(x) = \sin(x)$, $x \in [0, 2\pi)$.

$$\sin(x) = 0 \iff x \in \{k\pi, k \in \mathbb{Z}\},$$

therefore, $f(x) = 0$ *at* $x = 0, \pi$. *Furthermore:*

$$f(x) > 0 \text{ if } x \in (0, \pi) \quad \text{and} \quad f(x) < 0 \text{ if } x \in (\pi, 2\pi).$$

Therefore:

$$sign(f(x)) = \begin{cases} 1 & \text{if } x \in (0, \pi) \\ 0 & \text{if } x \in \{0, \pi\} \\ -1 & \text{if } x \in (\pi, 2\pi). \end{cases}$$

The graph of this function can be seen in Figure 3.59 (right).

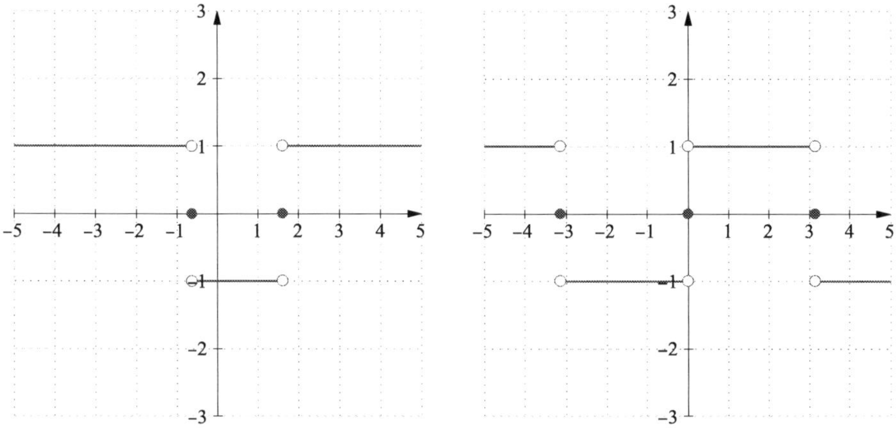

FIGURE 3.59
Graphs of the functions $\text{sign}(x^2 - x - 1)$ (left) and $\text{sign}(\sin(x))$ (right) (Problem 3.31).

(b) *Let's study the function* $f(x) = \lim_{n\to\infty} \frac{2}{\pi}\arctan(nx)$.

- *For* $x < 0$:

$$f(x) = \lim_{n\to\infty} \frac{2}{\pi}\arctan(\underbrace{n}_{\to\infty}\underbrace{x}_{<0}) = «\frac{2}{\pi}\arctan(-\infty)» = \frac{2}{\pi}\cdot\left(-\frac{\pi}{2}\right) = -1.$$

- *For* $x = 0$:

$$f(x) = \lim_{n\to\infty} \frac{2}{\pi}\arctan(\underbrace{n}_{\to\infty}\underbrace{x}_{0}) = \frac{2}{\pi}\arctan(0) = \frac{2}{\pi}\cdot 0 = 0.$$

- *For* $x > 0$:

$$f(x) = \lim_{n\to\infty} \frac{2}{\pi}\arctan(\underbrace{n}_{\to\infty}\underbrace{x}_{>0}) = «\frac{2}{\pi}\arctan(\infty)» = \frac{2}{\pi}\cdot\frac{\pi}{2} = 1.$$

Thus:

$$f(x) = \begin{cases} -1, & x < 0 \\ 0, & x = 0 \\ 1, & x > 0. \end{cases}$$

Therefore, $f(x) = \lim_{n\to\infty} \frac{2}{\pi}\arctan(nx) = sign(x)$, *as we wanted to prove.*

3.17 Solved Problems (Continuity)

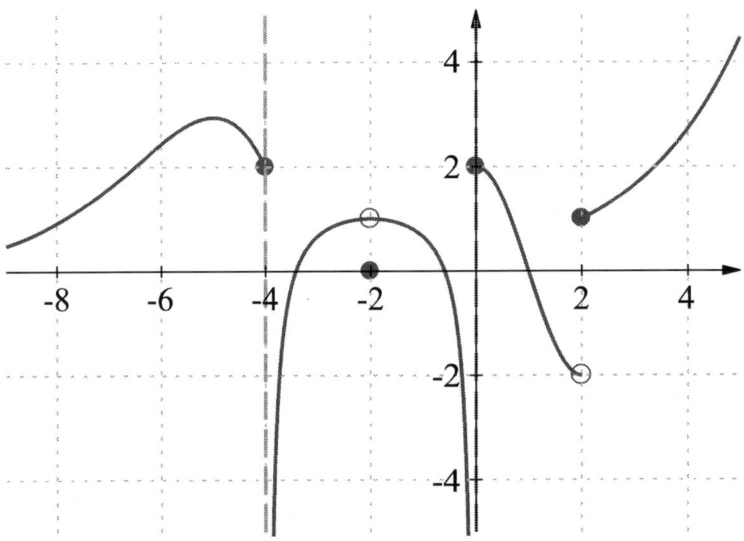

FIGURE 3.60
Graphical representation of a function and its discontinuities (Problem 3.32).

Problem 3.32. Justify and classify the discontinuities of the function represented graphically in Figure 3.60.

Solution to the problem 3.32.

- *At $x = -4$ the function has an essential discontinuity because:*

$$\lim_{x \to -4^+} f(x) = -\infty.$$

- *At $x = -2$ the function has a removable discontinuity because:*

$$\lim_{x \to -2} f(x) = 1 \neq 0 = f(-2).$$

- *At $x = 0$ the function has an essential discontinuity because:*

$$\lim_{x \to 0^-} f(x) = -\infty.$$

- *At $x = 2$ the function has a jump discontinuity because:*

$$\lim_{x \to 2^-} f(x) = -2 \neq 1 = \lim_{x \to 2^+} f(x) = f(2).$$

Problem 3.33. Let $f : \mathbb{R} \to \mathbb{R}$ be defined by:

$$f(x) = \begin{cases} e^x & x < 0 \\ x + a & x \geq 0. \end{cases}$$

How should a be chosen so that the function is continuous?

Solution to the problem 3.33. *The function $f(x)$ is continuous if $x \neq 0$ because it is polynomial and exponential. To also be continuous at $x = 0$, it is necessary that:*

$$f(0) = \lim_{x \to 0} f(x)$$

- $f(0) = 0 + a = a$
- $\lim_{x \to 0^-} f(x) = \lim_{x \to 0^-} e^x = e^0 = 1$
- $\lim_{x \to 0^+} f(x) = \lim_{x \to 0^+} x + a = a$

therefore, $a = 1$.

The graph of this function for the values $a = 1$ and $a = 2$ can be found in Figure 3.61.

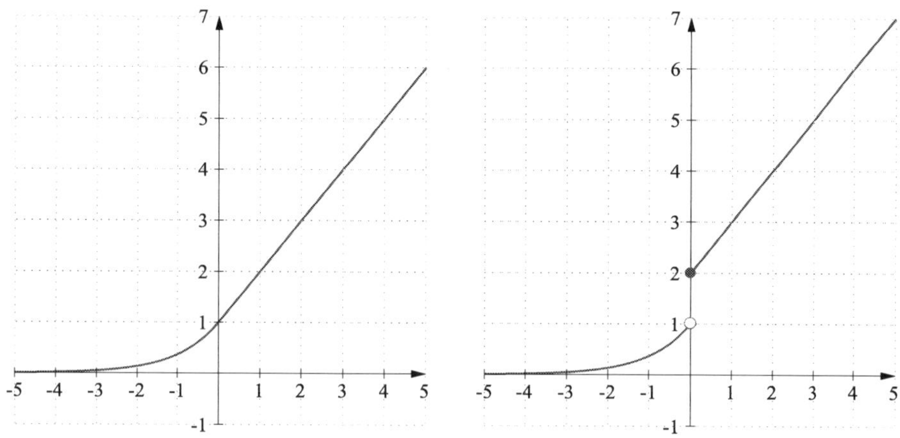

FIGURE 3.61
Graph of the function $f(x)$ for the values $a = 1$ (left) and $a = 2$ (right). It can be seen that only the function on the left is continuous. The function on the right has a jump discontinuity at $x = 0$ (Problem 3.33).

Problem 3.34. Let a, b, and c be three real parameters. Consider the function

$$f(x) = \begin{cases} \dfrac{x+a}{x^2+1} & , \quad x \leq 0 \\ \dfrac{x+c}{x^2-1} & , \quad x > 0, x \neq 1 \\ b & , \quad x = 1. \end{cases}$$

Calculate the values of a, b, and c that make $f \in C^0(\mathbb{R})$.

Solution to the problem 3.34. *The function may have discontinuities at points $x = 0$ and $x = 1$, as there is a change in the definition of the function at these points.*

Elsewhere, the function is continuous, as if $x \leq 0$, we have a quotient of polynomials where the denominator does not vanish, and if $x > 0, x \neq 1$ also, as $x^2 - 1 = 0$ at $x = \pm 1$ and $x = -1 \not> 0$.

Therefore, for $f \in \mathcal{C}^0(\mathbb{R})$, *i.e., for the function to be continuous over the entire* $\mathbb{R}$, *we must impose:*

$$f(0) = \lim_{x \to 0} f(x) \qquad \textit{and} \qquad f(1) = \lim_{x \to 1} f(x).$$

If we impose that the function be continuous at $x = 0$, *we get:*

$$\left.\begin{array}{l} f(0) = \dfrac{0+a}{0+1} = a \\ \displaystyle\lim_{x \to 0^-} f(x) = \lim_{x \to 0^-} \frac{x+a}{x^2+1} = a \\ \displaystyle\lim_{x \to 0^+} f(x) = \lim_{x \to 0^+} \frac{x+c}{x^2-1} = -c \end{array}\right\} \Longrightarrow \quad a - -c.$$

If we study the continuity at $x = 1$, *we get that* $f(1) = b$ *and:*

$$\lim_{x \to 1} f(x) = \lim_{x \to 1} \frac{x+c}{x^2-1} = \lim_{x \to 1} \frac{x+c}{(x-1)(x+1)}$$

$$= \begin{cases} \nexists & \textit{if } c \neq -1 \\ \displaystyle\lim_{x \to 1} \frac{x-1}{(x-1)(x+1)} = \lim_{x \to 1} \frac{1}{x+1} = \frac{1}{2} & \textit{if } c = -1. \end{cases}$$

Therefore, for the function to be continuous at $x = 1$ *we must have* $c = -1$ *and* $b = 1/2$. *If we also consider the condition we obtained for* $x = 0$, *we get:* $a = 1, b = 1/2$ *and* $c = -1$. *The graph of this function for different values of* a, b, *and* c *can be found in Figure 3.62.*

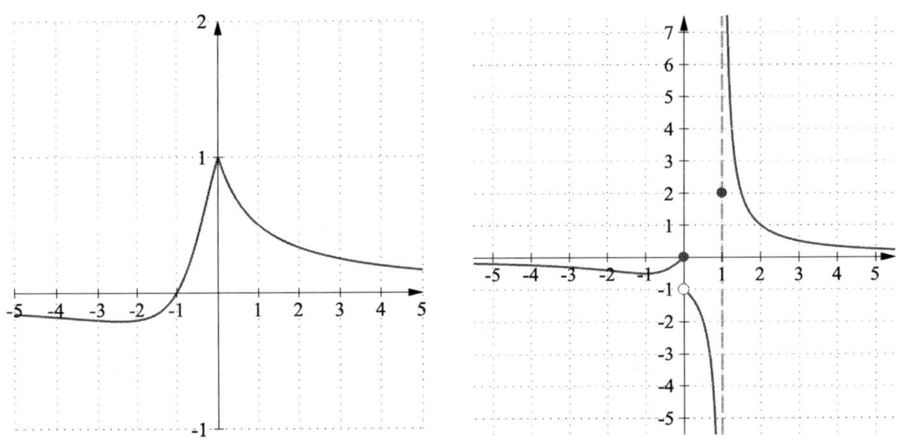

FIGURE 3.62
Graph of the function $f(x)$ for the values $a = 1, b = 1/2$ and $c = -1$ (left) and $a = 0, b = 2$ and $c = 1$ (right). It can be seen that only the function on the left is continuous. The function on the right has a jump discontinuity at $x = 0$ and an essential discontinuity at $x = 1$ (Problem 3.34).

Problem 3.35. Given the function:

$$f(x) = \begin{cases} a\left(\dfrac{\sqrt{x+3}-\sqrt{3}}{x}\right) & \text{if } x < 0 \\ b & \text{if } x = 0 \\ \left(\dfrac{\frac{1}{x}+1}{\frac{1}{x}}\right)^{\frac{1}{x}} & \text{if } 0 < x \leq 2 \\ c\left(\dfrac{x^2-x-2}{x^2+x-6}\right) & \text{if } x > 2 \end{cases}$$

find a, b and c $(a, b, c \in \mathbb{R})$ such that the function is continuous.

Solution to the problem 3.35. *For $x < 0$, the function $f(x)$ is defined as:*

$$a\left(\frac{\sqrt{x+3}-\sqrt{3}}{x}\right)$$

and, therefore, is a continuous function, as it is the quotient of continuous functions and the denominator does not vanish at any point.

For $0 < x < 2$, the function $f(x)$ is defined as:

$$\left(\frac{\frac{1}{x}+1}{\frac{1}{x}}\right)^{\frac{1}{x}}$$

and, therefore, is a continuous function as it is a composition of continuous functions and the denominator does not vanish at any point.

For $x > 2$, the function $f(x)$ is defined as:

$$c\left(\frac{x^2-x-2}{x^2+x-6}\right)$$

and, therefore, is a continuous function, as it is the quotient of continuous functions and the denominator does not vanish at any point. Note that the roots of $x^2 + x - 6$ are:

$$x = \frac{-1 \pm \sqrt{1+24}}{2} = \{-3, 2\}.$$

Therefore, continuity must only be imposed at $x = 0$ and $x = 2$.

The function will be continuous at $x = 0$ if:

$$f(0) = \lim_{x \to 0^-} f(x) = \lim_{x \to 0^+} f(x)$$

$$\begin{cases} f(0) = b \\ \lim_{x\to 0^-} f(x) = \lim_{x\to 0^-} a\left(\dfrac{\sqrt{x+3}-\sqrt{3}}{x}\right) \\ \qquad = \lim_{x\to 0^-} a\left(\dfrac{\sqrt{x+3}-\sqrt{3}}{x}\right)\left(\dfrac{\sqrt{x+3}+\sqrt{3}}{\sqrt{x+3}+\sqrt{3}}\right) \\ \qquad = \lim_{x\to 0^-} a\left(\dfrac{x+3-3}{x(\sqrt{x+3}+\sqrt{3})}\right) = \dfrac{a}{2\sqrt{3}} \\ \lim_{x\to 0^+} f(x) = \lim_{x\to 0^+} \left(\dfrac{\frac{1}{x}+1}{\frac{1}{x}}\right)^{\frac{1}{x}} = \lim_{x\to 0^+}\left(1+\dfrac{1}{\frac{1}{x}}\right)^{\frac{1}{x}} = e. \end{cases}$$

Therefore, the function will be continuous at $x = 0$ if $a = 2\sqrt{3}e$ and $b = e$.

Finally, the function is continuous at $x = 2$ if:

$$f(2) = \lim_{x\to 2^-} f(x) = \lim_{x\to 2^+} f(x)$$

$$\begin{cases} f(2) = \left(\dfrac{\frac{1}{2}+1}{\frac{1}{2}}\right)^{\frac{1}{2}} = \sqrt{3} \\ \lim_{x\to 2^-} f(x) = \lim_{x\to 2^-} \left(\dfrac{\frac{1}{x}+1}{\frac{1}{x}}\right)^{\frac{1}{x}} = \sqrt{3} \\ \lim_{x\to 2^+} f(x) = \lim_{x\to 2^+} c\left(\dfrac{x^2-x-2}{x^2+x-6}\right) = \lim_{x\to 2^+} c\dfrac{(x+1)(x-2)}{(x+3)(x-2)} = \lim_{x\to 2^+} c\dfrac{(x+1)}{(x+3)} = \dfrac{3c}{5}. \end{cases}$$

Therefore, the function will be continuous at $x = 2$ if:

$$c = \frac{5\sqrt{3}}{3}.$$

The graphical representation of this function for different values of a, b and c can be found in Figure 3.63.

Problem 3.36. Knowing that $\lim_{t\to 0} \dfrac{\ln(1+t)}{t} = 1$ define $f(1)$ so that the function:

$$f(x) = \frac{\ln(3x-2)}{x^3-1}$$

is continuous at $x = 1$.

Solution to the problem 3.36.

$$\lim_{x\to 1} f(x) = \lim_{x\to 1} \frac{\ln(3x-2)}{x^3-1} = «\frac{0}{0}» \text{ (indeterminacy)}.$$

To resolve the indeterminacy we will make the variable change:

$$3x-2 = 1+t \Longleftrightarrow t = 3x-3 \text{ and } x = \frac{t+3}{3}.$$

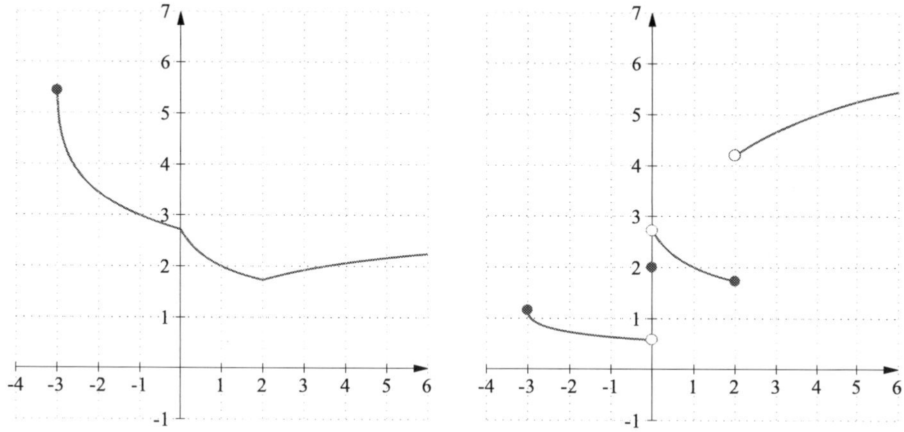

FIGURE 3.63
Graph of the function $f(x)$ for the values $a = 2\sqrt{3}e, b = e$ and $c = 5\sqrt{3}/3$ (left) and $a = 2, b = 2$ and $c = 7$ (right). It can be observed that only the function on the left is continuous. The function on the right presents a jump discontinuity at $x = 0$ and $x = 2$ (Problem 3.35).

Moreover $x \to 1 \Longrightarrow t \to 0$. *Then:*

$$\lim_{x\to 1} f(x) = \lim_{x\to 1} \frac{\ln(3x-2)}{x^3-1} = \lim_{t\to 0} \frac{\ln(1+t)}{\left(\frac{t+3}{3}\right)^3 - 1}$$

$$= \lim_{t\to 0} \frac{\ln(1+t)}{t} \cdot \frac{t}{\frac{1}{27}(t+3)^3 - 1}$$

$$= \lim_{t\to 0} 1 \cdot \frac{t}{\frac{1}{27}(t+3)^3 - 1} = \lim_{t\to 0} \frac{t}{\frac{1}{27}(t^3+9t^2+27t+27) - 1}$$

$$= \lim_{t\to 0} \frac{t}{\frac{1}{27}(t^3+9t^2+27t)} = \lim_{t\to 0} \frac{1}{\frac{1}{27}(t^2+9t+27)} = 1.$$

Therefore, since $\lim_{x\to 1} f(x) = 1$, f *will be continuous if* $f(1) = 1$.

Problem 3.37. Knowing that:

$$\lim_{x\to 0} \frac{\sin(x)}{x} = 1,$$

calculate the value of $k \in \mathbb{R}$ so that the following function is continuous throughout $\mathbb{R}$:

$$f(x) = \begin{cases} \dfrac{\sin(x-1)}{1-x} + k^2 x & x \neq 1 \\ k & x = 1. \end{cases}$$

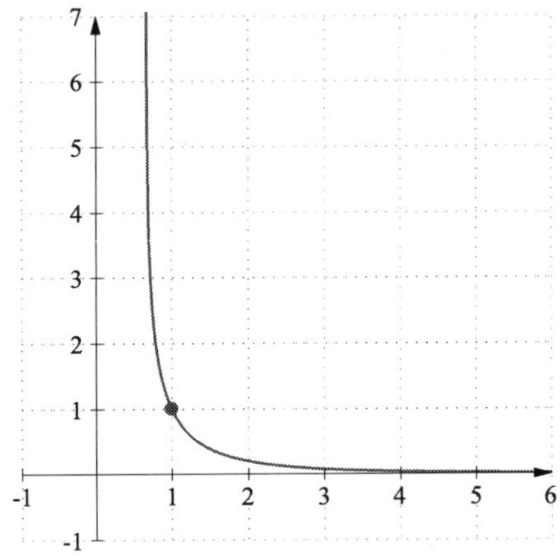

FIGURE 3.64
Graph of the function $f(x) = \frac{\ln(3x-2)}{x^3-1}$ redefining the value $f(1) = 1$ (Problem 3.36).

Solution to the problem 3.37. *If $x \neq 1$, the function f is continuous as it is the sum and quotient of continuous functions: the function k^2x is continuous as it is polynomial and the function:*

$$\frac{\sin(x-1)}{1-x}$$

is continuous as it is the quotient of two continuous functions where the denominator does not vanish.

Let's see if f is continuous at $x = 1$. f will be continuous at $x = 1$ if:

$$f(1) = \lim_{x\to 1} f(x).$$

Since $f(1) = k$, f will be continuous if $\lim_{x\to 1} f(x) = k$:

$$\begin{aligned}\lim_{x\to 1} f(x) &= \lim_{x\to 1} \frac{\sin(x-1)}{1-x} + k^2x \\ &= \lim_{y\to 0} \frac{\sin y}{-y} + k^2(y+1) \\ &= -\lim_{y\to 0} \frac{\sin y}{y} + \lim_{y\to 0} k^2(y+1) \\ &= -1 + k^2,\end{aligned}$$

where we have used the variable change $y = x - 1$, which implies that if before $x \to 1$ now $y \to 0$.

Therefore, for f to be continuous:

$$k = k^2 - 1 \iff k^2 - k - 1 = 0 \iff k = \frac{1}{2} \pm \frac{1}{2}\sqrt{5}.$$

The graphical representation of this function for different values of k can be found in Figure 3.65.

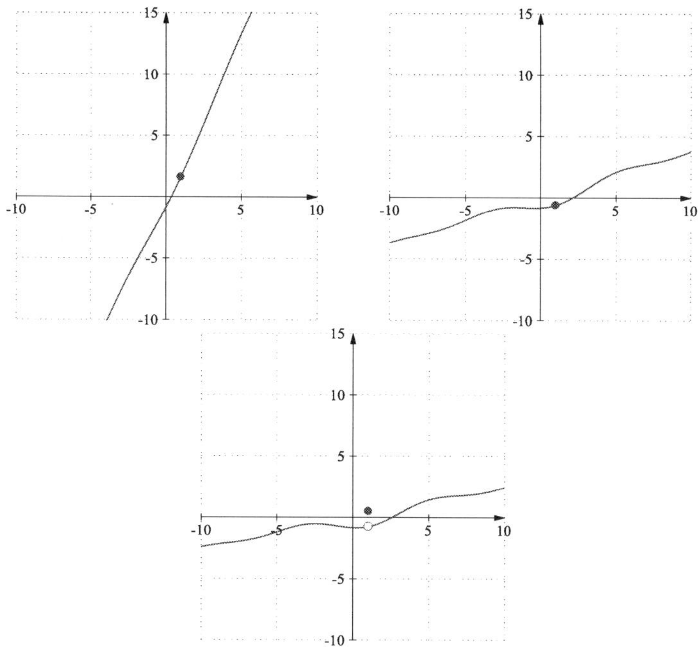

FIGURE 3.65
Graph of the function $f(x)$ for the values $k = 1/2 + 1/2\sqrt{5}$ (left), $k = 1/2 - 1/2\sqrt{5}$ (center) and $k = 0.5$ (right). The function on the right has a removable discontinuity at $x = 1$ (Problem 3.37).

Problem 3.38. In the function

$$f(x) = \begin{cases} \dfrac{px+1}{x-4} & x \leq 1 \\ \dfrac{x-1}{x^2-x} & x > 1 \end{cases}$$

(a) Find the value of p so that it is continuous at $x = 1$.

(b) Is there any other point where the function is discontinuous? Indicate the type of discontinuity.

Solution to the problem 3.38. *(a)*

$$\left.\begin{aligned} \lim_{x\to 1^-} f(x) &= \lim_{x\to 1^-} \frac{px+1}{x-4} = \frac{p+1}{-3} = -\frac{p+1}{3} \\ \lim_{x\to 1^+} f(x) &= \lim_{x\to 1^+} \frac{x-1}{x^2-x} = «\frac{0}{0}» \text{ (indeterminacy)} \\ &= \lim_{x\to 1^+} \frac{x-1}{x(x-1)} = \lim_{x\to 1^+} \frac{1}{x} = 1 \\ f(1) &= -\frac{p+1}{3}. \end{aligned}\right\}$$

To make f continuous:

$$-\frac{p+1}{3} = 1 \implies -p-1 = 3 \implies p = -4.$$

The graphical representation of this function for different values of p can be found in Figure 3.66.

(b) *The function is continuous at all other points because:*

$$\frac{px+1}{x-4}$$

has problems at $x = 4 \nleq 1$.

Similarly, the function:

$$\frac{x-1}{x^2-x} = \frac{x-1}{x(x-1)}$$

has problems at $x = 0 \ngtr 1$ and at $x = 1$, which has already been studied in section (a).

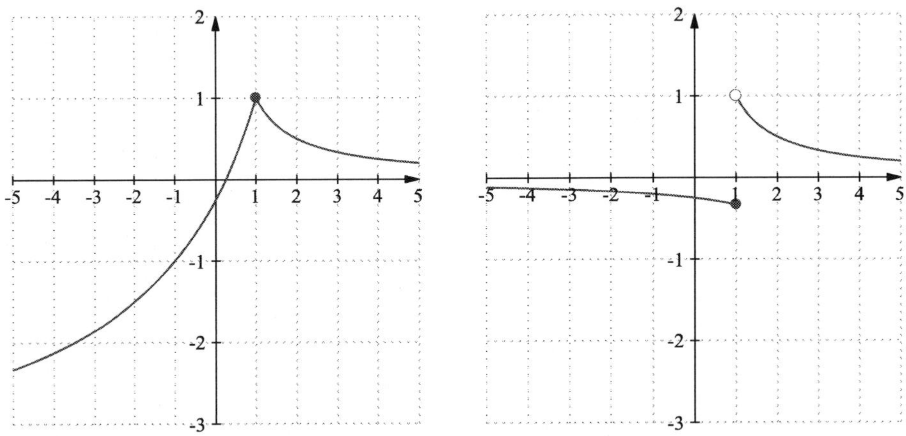

FIGURE 3.66
Graph of the function $f(x)$ from Problem 3.38 for the values $p = -4$ (left) and $p = 0$ (right). The function on the right has a jump discontinuity at $x = 1$.

Problem 3.39. Study the continuity of the following function:

$$f(x) = \begin{cases} |x-a| & x \leq a \\ 2|a| & x > a. \end{cases}$$

Solution to the problem 3.39. *If:*

$$x \leq a \implies x - a \leq 0 \implies |x-a| = a - x$$

therefore:

$$f(x) = \begin{cases} a - x & \text{if } x \leq a \\ 2|a| & \text{if } x > a. \end{cases}$$

f is continuous at $x \neq a$ because it is polynomial and constant. Let's study what happens at $x = a$:

$$\left.\begin{array}{l} \lim\limits_{x\to a^+} f(x) = \lim\limits_{x\to a^+} 2|a| = 2|a| \\ \lim\limits_{x\to a^-} f(x) = \lim\limits_{x\to a^-} a - x = 0 \end{array}\right\} \quad \textit{and } f(a) = 0.$$

Therefore, f is continuous at the point $x = a \iff a = 0$. Otherwise, f has a jump discontinuity (it is continuous on the left but not on the right).

Problem 3.40. Consider the function $f(x) = \dfrac{2^{\frac{1}{x-3}} - 1}{2^{\frac{1}{x-3}} + 1}$.

(a) Find the domain of the function.

(b) Study the continuity of the function and classify the discontinuities, if any.

(c) Is the function right-continuous at all points? Can the function be made right-continuous at all points? How?

Solution to the problem 3.40. *(a) If we look at the function, we can see that the point $x = 3$ is not in the domain of the function, as it vanishes the denominator. On the other hand, the denominator is always strictly positive and, therefore, never vanishes. Consequently, the domain of the function is:*

$$Dom(f) = \mathbb{R} - \{3\}.$$

(b) The function is continuous at all points except at $x = 3$, which must be studied. Indeed, if we calculate the lateral limits we have:

$$\lim_{x\to 3^-} \frac{2^{\frac{1}{x-3}} - 1}{2^{\frac{1}{x-3}} + 1} = \lim_{x\to 3^-} \frac{2^{\frac{1}{x-3}} - 1}{2^{\frac{1}{x-3}} + 1} \cdot \frac{2^{-\frac{1}{x-3}}}{2^{-\frac{1}{x-3}}}$$

$$= \text{«}\frac{2^{\frac{1}{0^-}} - 1}{2^{\frac{1}{0^-}} + 1}\text{»} = \text{«}\frac{2^{-\infty} - 1}{2^{-\infty} + 1}\text{»} = \text{«}\frac{0 - 1}{0 + 1}\text{»} = -1$$

$$\lim_{x\to 3^+} \frac{2^{\frac{1}{x-3}} - 1}{2^{\frac{1}{x-3}} + 1} = \lim_{x\to 3^-} \frac{2^{\frac{1}{x-3}} - 1}{2^{\frac{1}{x-3}} + 1} \cdot \frac{2^{-\frac{1}{x-3}}}{2^{-\frac{1}{x-3}}} = \lim_{x\to 3^+} \frac{1 - 2^{-\frac{1}{x-3}}}{1 + 2^{-\frac{1}{x-3}}}$$

$$= \text{«}\frac{1 - 2^{-\frac{1}{0^+}}}{1 + 2^{-\frac{1}{0^+}}}\text{»} = \text{«}\frac{1 - 2^{-\infty}}{1 + 2^{-\infty}}\text{»} = \frac{1 - 0}{1 + 0} = 1.$$

Therefore, the lateral limits exist and are different. This implies that at $x = 3$ the function has a jump discontinuity.

The graphical representation of this function can be found in Figure 3.67.

(c) The function is right-continuous at all points except at $x = 3$. We can make the function also right-continuous at $x = 3$ if we redefine the function as follows:

$$f(3) = \lim_{x\to 3^+} \frac{2^{\frac{1}{x-3}} - 1}{2^{\frac{1}{x-3}} + 1} = 1,$$

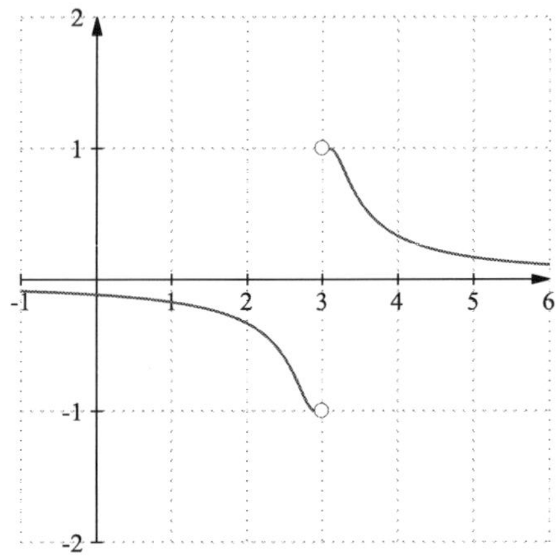

FIGURE 3.67
Graph of the function $f(x)$ from Problem 3.40. The function has a jump discontinuity at $x = 3$.

that is, if the function is redefined as:

$$f(x) = \begin{cases} \dfrac{2^{\frac{1}{x-3}} - 1}{2^{\frac{1}{x-3}} + 1}, & x \neq 3 \\ 1, & x = 3. \end{cases}$$

Problem 3.41. Study the continuity and classify the points of discontinuity for the functions:

(a) $f(x) = \begin{cases} \sqrt{|x|}, & x \neq 0 \\ 1, & x = 0 \end{cases}$

(b) $g(x) = \begin{cases} x - E(x), & x \geq 0 \\ -x - E(-x), & x < 0 \end{cases}$

(c) $h(x) = \begin{cases} \sqrt{x}\,\sin(\frac{1}{x}), & x > 0 \\ 0, & x = 0 \end{cases}$

Solution to the problem 3.41. *(a) The function $f(x)$ can be rewritten as:*

$$f(x) = \begin{cases} \sqrt{x} & \text{if } x > 0 \\ 1 & \text{if } x = 0 \\ \sqrt{-x} & \text{if } x < 0 \end{cases}$$

On the one hand, $f(x)$ *is continuous at all points* $x \neq 0$*. Let's study the continuity at* $x = 0$*:*

$$\lim_{x \to 0^+} \sqrt{x} = 0$$
$$\lim_{x \to 0^-} \sqrt{-x} = 0$$
$$f(0) = 1$$

Therefore, f *has a removable discontinuity at* $x = 0$*.*

The graphical representation of this function can be found in Figure 3.68.

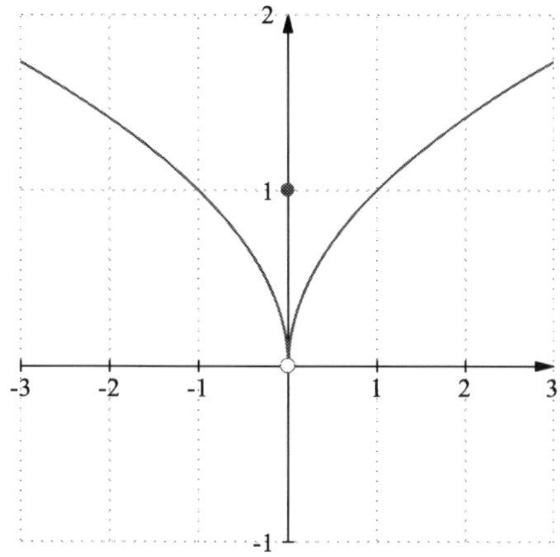

FIGURE 3.68
Graph of the function $f(x)$ from Problem 3.41 (a). The function has a removable discontinuity at $x = 0$.

(b) *We will start by studying the continuity of the function at* $x = 0$*:*

$$\lim_{x \to 0^+} g(x) = \lim_{x \to 0^+} x - E(x) = 0 - 0 = 0$$
$$\lim_{x \to 0^-} g(x) = \lim_{x \to 0^-} -x - E(-x) = 0 - 0 = 0$$
$$g(0) = 0 - 0 = 0$$

Therefore, the function is continuous at $x = 0$*. On the other hand, the function is discontinuous at any other integer, as can be seen below. Let's suppose* $z \in \mathbb{Z}, z > 0$*. Then:*

$$\lim_{x \to z^+} x - E(x) = \lim_{x \to 0^+} z - z = 0$$
$$\lim_{x \to z^-} x - E(x) = \lim_{x \to 0^-} z - (z - 1) = 1$$
$$g(z) = z - E(z) = 0$$

which implies that we have a jump discontinuity.

On the other hand, if $z \in \mathbb{Z}, z < 0$, then:

$$\lim_{x \to z^+} -x - E(-x) = \lim_{x \to 0^+} -z - (-z-1) = 1$$

$$\lim_{x \to z^-} -x - E(-x) = \lim_{x \to 0^-} -z - (-z) = 0$$

$$g(z) = -z - (-z) = 0$$

which implies that we also have a jump discontinuity.

At all other points, the function is continuous. Therefore, the function $g(x)$ is continuous in $\mathbb{R} - \{\mathbb{Z} - \{0\}\}$.

(c) *The function h is continuous at $x > 0$. Let's study what happens at zero:*

$$\lim_{x \to 0^+} \underbrace{\sqrt{x}}_{0} \underbrace{\sin\left(\frac{1}{x}\right)}_{bounded} = 0$$

$$h(0) = 0$$

Therefore, $h(x)$ is right-continuous at zero.

The graphical representation of this function can be found in Figure 3.69.

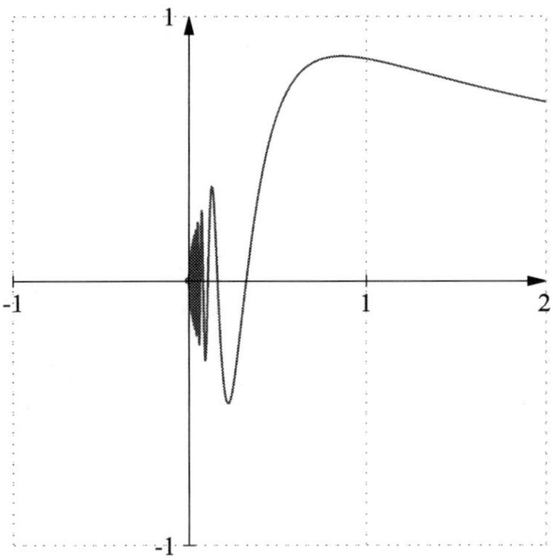

FIGURE 3.69
Graph of the function $h(x)$ from Problem 3.41(c). The function is right-continuous at $x = 0$.

Problem 3.42. Study the continuity of the following function:

$$f(x) = \begin{cases} e^{-\dfrac{1}{x^2}}, & x \leq 1 \\ e^{\dfrac{-1}{x(x-2)}}, & x > 1. \end{cases}$$

Solution to the problem 3.42. *The points $x = 0$ and $x = 2$ are not part of the domain of the function, as they vanish some of the denominators that appear in its definition. Thus, we must study the type of discontinuity that appears at these points. On the other hand, at $x = 1$ there is a change in the definition of the function and, therefore, we must also study the continuity at this point.*

- $\boxed{x = 0}$

$$\left. \begin{array}{l} \lim\limits_{x\to 0} e^{-\frac{1}{x^2}} = e^{\lim\limits_{x\to 0} -\frac{1}{x^2}} = "e^{-\infty}" = 0 \\ \nexists f(0) \end{array} \right\} \textit{removable discontinuity}$$

- $\boxed{x = 1}$

$$\left. \begin{array}{l} \lim\limits_{x\to 1^-} f(x) = \lim\limits_{x\to 1^-} e^{-\frac{1}{x^2}} = e^{-1} \\ \lim\limits_{x\to 1^+} f(x) = \lim\limits_{x\to 1^+} e^{\frac{-1}{x(x-2)}} = e^{\frac{-1}{1(-1)}} = e^{1} \end{array} \right\} \textit{jump discontinuity}$$

- $\boxed{x = 2}$

$$\lim_{x\to 2} e^{\frac{-1}{2(x-2)}} = \lim_{x\to 2} e^{\frac{1}{2(2-x)}};$$

if $x \to 2^+$ $\quad 2 - x < 0 \implies \frac{1}{2(2-x)} \to -\infty$ *and* $e^{-\infty} \to 0$;

if $x \to 2^-$ $\quad 2 - x > 0 \implies \frac{1}{2(2-x)} \to +\infty$ *and* $e^{+\infty} \to +\infty$.

Therefore:

$$\left. \begin{array}{l} \lim\limits_{x\to 2^+} f(x) = 0 \\ \lim\limits_{x\to 2^-} f(x) = +\infty \end{array} \right\} \textit{f has an essential discontinuity at } x = 2.$$

At other points, the function is continuous.
The graphical representation of this function can be found in Figure 3.70.

Problem 3.43. Let $f(x)$ be the function defined by

$$f(x) = \begin{cases} x - e^{-x}, & x \leq 1 \\ \dfrac{1}{4x - x^2 - 5} + a, & x > 1. \end{cases}$$

(a) Calculate the value of a that makes the function continuous.

(b) Prove that the function $f(x)$ has a zero in the interval $[0, 1]$.

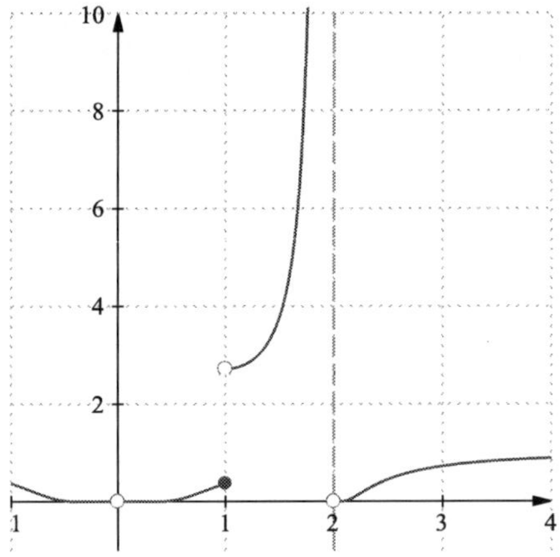

FIGURE 3.70
Graph of the function $f(x)$ from Problem 3.42.The function has a removable discontinuity at $x = 0$, a jump discontinuity at $x = 1$ and an asymptotic discontinuity at $x = 2$.

Solution to the problem 3.43. *(a) We need to calculate the value of a that makes the function continuous.*

- *If $x < 1$, the function is $x - e^{-x}$ in a neighborhood of the point and therefore is continuous as it is the sum of a polynomial and an exponential.*
- *If $x > 1$, the function is:*

$$\frac{1}{4x - x^2 - 5} + a$$

in a neighborhood of the point and therefore is continuous, as in the quotient, the denominator does not vanish at any point. The roots of $4x - x^2 - 5$ are:

$$x = \frac{-4 \pm \sqrt{16 - 20}}{-2} = \frac{-4 \pm \sqrt{-4}}{-2} \notin \mathbb{R}.$$

- *The function will be continuous at $x = 1$ if:*

$$f(1) = \lim_{x\to 1^-} f(x) = \lim_{x\to 1^+} f(x)$$

$$\begin{cases} f(1) = 1 - e^{-1} \\ \lim_{x\to 1^-} f(x) = \lim_{x\to 1^-} x - e^{-x} = 1 - e^{-1} \\ \lim_{x\to 1^+} f(x) = \lim_{x\to 1^+} \dfrac{1}{4x - x^2 - 5} + a = \dfrac{1}{-2} + a = -\dfrac{1}{2} + a. \end{cases}$$

Therefore, the function will be continuous if:

$$-\frac{1}{2} + a = 1 - e^{-1} \quad \Longrightarrow \quad a = 1 - e^{-1} + \frac{1}{2} = \frac{3}{2} - e^{-1}.$$

(b) We need to prove that $f(x)$ has a zero in the interval $[0, 1]$.

The function $f(x)$ is continuous in $[0, 1]$ because it is continuous in all its domain $(0, +\infty)$ since it is the difference of continuous functions. Additionally:

$$f(0) = -1$$

and:

$$f(1) = 1 - e^{-1} \approx 0.63,$$

that is:

$$f(0) \cdot f(1) < 0.$$

Therefore, Bolzano's theorem guarantees that the function $f(x)$ has a zero in the interval $[0, 1]$, which is what we wanted to prove.

Problem 3.44. Consider the function:

$$f(x) = 2x^2 - \ln(x).$$

Prove that $f(x)$ equals 3 for some point within the interval $[1, e]$.
Hint: $e \approx 2.7$ and $e^2 \approx 7.4$.

Solution to the problem 3.44. *The function $f(x)$ is continuous in the interval $[1, e]$, as it is continuous in its entire domain $(0, +\infty)$ as it is the difference of continuous functions. Additionally:*

$$f(1) = 2 - \ln(1) = 2$$

and:

$$f(e) = 2e^2 - \ln(e) = 2e^2 - 1 \approx 13.8.$$

Then, by the intermediate value theorem, for every value $y \in (2, 13.8)$ there exists a point $c \in (1, e)$ such that $f(c) = y$.

In particular, if we take $y = 3 \in (2, 13.8)$, we can ensure that there is at least one point $c \in (1, e)$ such that $f(c) = 3$, as stated in the problem.

The graphical representation of this function can be found in Figure 3.71.

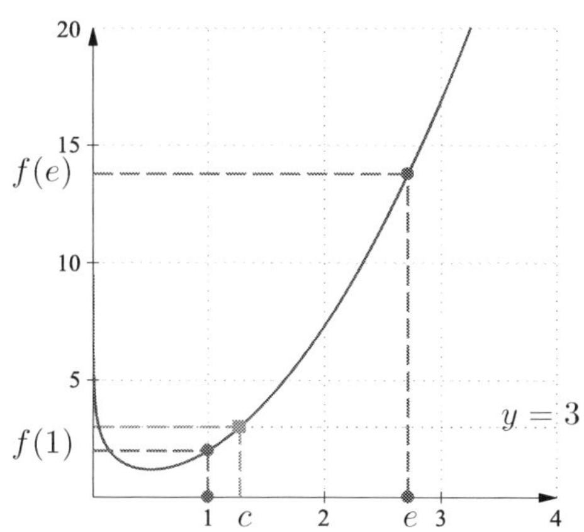

FIGURE 3.71
Graph of the function $f(x)$ from Problem 3.44.It can be observed that the function takes the value $y = 3$ at an undefined point c within the interval $(1, e)$.

Problem 3.45. Consider the function $f(x) = \sin^2(x)\cos(x)$. Prove that $f(x)$ has at least one crossing with the x-axis within the interval $(\frac{\pi}{4}, \frac{3\pi}{4})$.

Solution to the problem 3.45. *First, we note that the function $f(x)$ is continuous throughout $\mathbb{R}$ as it is the product of continuous functions (sine and cosine). That is, $f \in \mathcal{C}^0(\mathbb{R})$. We also observe that:*

$$f\left(\frac{\pi}{4}\right) = \frac{\sqrt{2}}{4} > 0$$
$$f\left(\frac{3\pi}{4}\right) = -\frac{\sqrt{2}}{4} < 0.$$

Then, by applying Bolzano's theorem, we can conclude that there exists a point $c \in (\frac{\pi}{4}, \frac{3\pi}{4})$ such that $f(c) = 0$.

Problem 3.46. We say that α is a fixed point of the function $f(x)$ if $f(\alpha) = \alpha$. Suppose $f : [a, b] \to \mathbb{R}$ is continuous and $f(a) < a$, $f(b) > b$. Prove that there is a fixed point of f within the interval (a, b).

Solution to the problem 3.46. *We know that f satisfies:*

$$f(a) < a \text{ and } f(b) > b,$$

and we want to find $\alpha \in (a, b)$ such that:

$$f(\alpha) = \alpha.$$

The above conditions can be rewritten as: we know that f satisfies:

$$f(a) - a < 0 \quad \text{and} \quad f(b) - b > 0,$$

and we we want to prove that: there is an $\alpha \in (a, b)$ such that:

$$f(\alpha) - \alpha = 0.$$

Therefore, if we consider the function $g(x) = f(x) - x$, we are in the hypotheses of Bolzano's theorem.

- *g is continuous in $[a, b]$ because it is the sum of two continuous functions in $[a, b]$*
- $g(a) = f(a) - a < 0$
- $g(b) = f(b) - b > 0.$

Therefore, we know that there is an $\alpha \in (a, b)$ such that:

$$g(\alpha) = 0 \iff f(\alpha) - \alpha = 0 \iff f(\alpha) = \alpha,$$

as we wanted to see.

Problem 3.47. Let $f : \mathbb{R} \to \mathbb{R}$ be a function defined as:

$$f(x) = x^{180} + \frac{260}{2 + x^2 + \sin^2(x)}.$$

Prove that there exists an $x_0 \in [0, 1]$ such that $f(x_0) = 100$.

Solution to the problem 3.47. *f is a continuous function in all $\mathbb{R}$, since it is the sum and quotient of continuous functions where the denominator does not vanish, since:*

$$2 + x^2 + \sin^2(x) \geq 2 \ \forall x \in \mathbb{R}.$$

Therefore, in particular it is continuous in $[0, 1]$.

Additionally:

$$f(0) = \frac{260}{2} = 130$$
$$f(1) = 1 + \frac{260}{2 + 1 + \sin^2(1)} = 1 + \frac{260}{3 + \sin^2(1)} \leq 1 + \frac{260}{3} = \frac{263}{3} \leq 100.$$

Then, since $f(1) < 100 < f(0)$, by the intermediate value theorem, there exists an $x_0 \in (0, 1)$ such that $f(x_0) = 100$, as we wanted to prove.

Problem 3.48. Prove that the equation $\ln(x) = x^2 - 4x$ has a real solution in the interval $[1, +\infty)$.

Solution to the problem 3.48. *x is a solution of the equation:*

$$\ln(x) = x^2 - 4x \iff \ln(x) - x^2 + 4x = 0.$$

Therefore, the problem is equivalent to proving that the function:

$$f(x) = \ln(x) - x^2 + 4x$$

has a zero in the interval $[1, +\infty]$. To do this we will use Bolzano's theorem.

- *f is continuous in any interval of the form $(a, +\infty)$ where $a > 0$. In particular, it is a continuous function for example in $(0.5, +\infty)$, and, therefore, it is also continuous for example in the interval $[1, 10]$.*
- $f(1) = \ln(1) - 1 + 4 = 3 > 0$
- $f(10) = \ln(10) - 100 + 40 = -60 + \ln(10) < 0.$

Therefore, by Bolzano's theorem we know that there is a $c \in (1, 10)$ such that:

$$f(c) = 0 \iff \ln(c) - c^2 + 4c = 0 \iff \ln(c) = c^2 - 4c,$$

and we have proved that the equation given has at least one real solution in $[1, +\infty)$.

4

Differentiation of Functions of a Real Variable

«God created everything by number, weight and measure.» (Isaac Newton, cited in the book *Symmetry in Plants*, by R. V. Jean and D. Barabé, 1998).

It was Newton and Leibniz who, at the end of the XVII century, developed the concept of derivative by studying the problem of finding the tangent line to a function at one of its points. In this chapter, this important concept is presented, as well as the techniques for its calculation, its relationship with the concept of continuity, and some of its applications, specifically: optimization of functions, approximation of functions through Taylor's theorem (with application in the calculation of numerical approximations of the value of a function at a point, in the characterization of relative extrema, and in the calculation of indeterminate limits at the origin), and L'Hôpital's rules for calculating certain types of indeterminate limits.

DOI: 10.1201/9781003399735-4

4.1 Derivative of a Function at a Point

4.1.1 Geometric Interpretation

Definition 4.1 (Derivative). The derivative of a function $f(x)$ at a point a is the slope of the tangent line to the function $f(x)$ at that point (Figure 4.1).

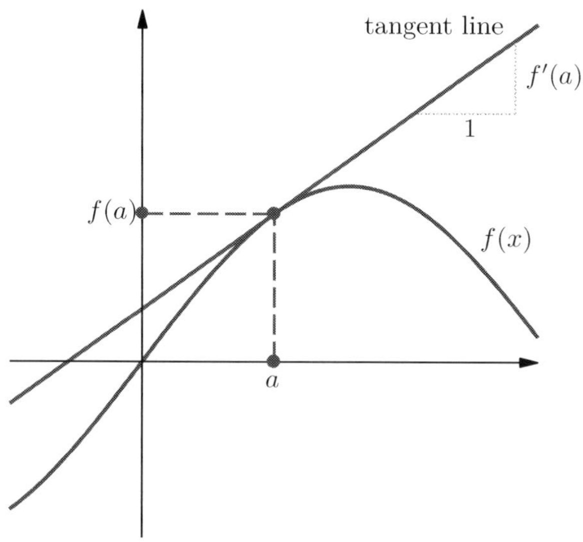

FIGURE 4.1
Geometric interpretation of the concept of the derivative of a function at a point a.

Thus, if we know how to determine the slope of the tangent line to $f(x)$ at a, we will have the value of the derivative at that point. But, how can we determine the slope? We know that the tangent line must pass through the point $(a, f(a))$ and, therefore, can be written as:

$$y = m(x - a) + f(a).$$

To determine the value of the slope, m, it would be very easy if we knew another point (x_1, y_1) on the tangent line (*see* Figure 4.2), since then:

$$y_1 = m(x_1 - a) + f(a)$$

and solving for m we would obtain:

$$m = \frac{y_1 - f(a)}{x_1 - a}.$$

The problem is that we do not know another point on the tangent line. What can we do? A very simple idea is to consider the point $a + h$ and calculate the *secant* line that passes through this point and the point a. *See* Figure 4.3.

The secant line, as it must pass through the point $(a, f(a))$, is of the form:

$$y = m_h(x - a) + f(a),$$

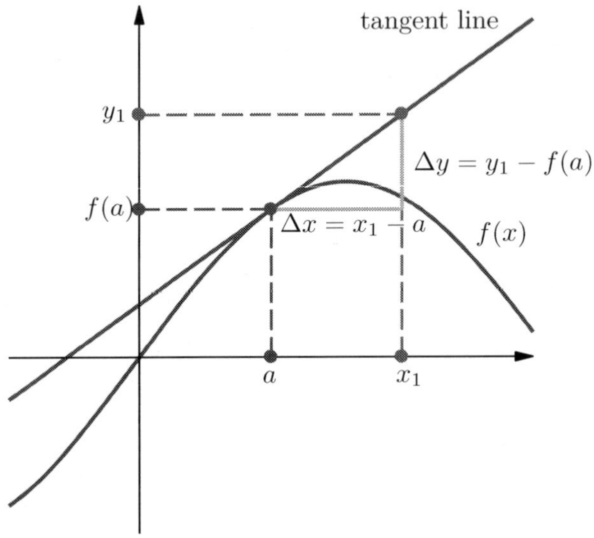

FIGURE 4.2
Determining the value of the slope would be easy if we knew two points on the tangent line.

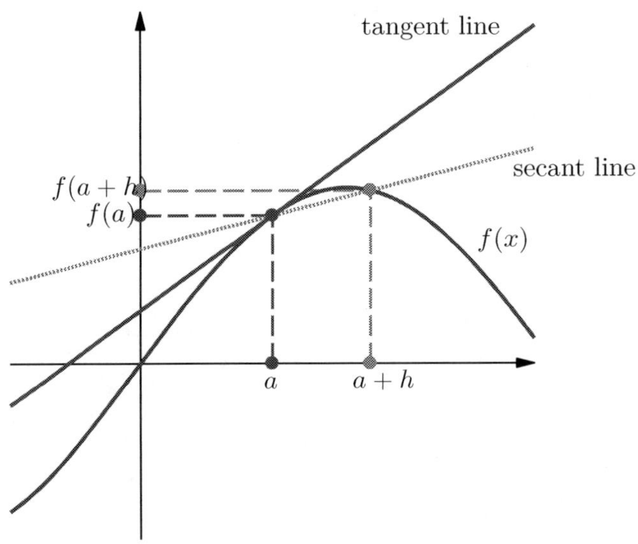

FIGURE 4.3
Secant line that passes through the points $(a, f(a))$ and $(a+h, f(a+h))$.

and imposing that it also passes through the point $(a+h, f(a+h))$:

$$f(a+h) = m_h(a+h-a) + f(a).$$

Finally, solving for m_h, we obtain that the slope of this secant line is:

$$m_h = \frac{f(a+h) - f(a)}{h}.$$

As illustrated in Figure 4.4, if we make h smaller and smaller, the point $a+h$ approaches the point a and, consequently, the secant line approaches the tangent line. That is, when h tends to zero, the slope of the secant line, m_h, approximates the slope of the tangent line, m:

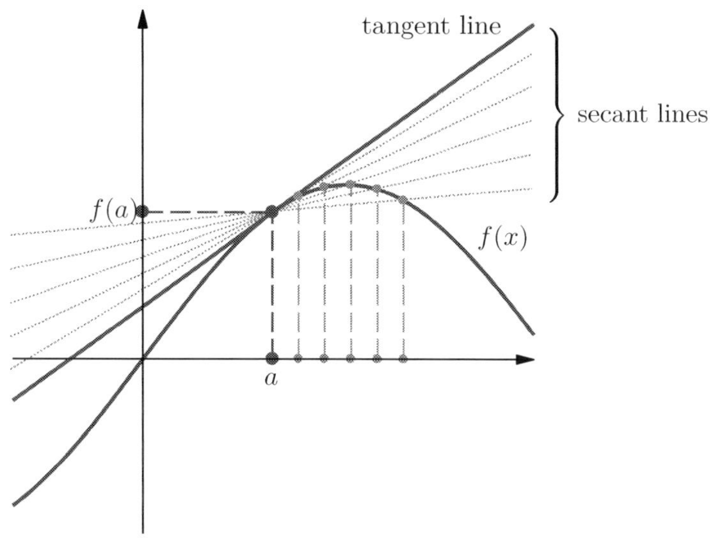

FIGURE 4.4
Secant lines that become the tangent line when h tends to 0.

$$m = \lim_{h \to 0} m_h = \lim_{h \to 0} \frac{f(a+h) - f(a)}{h}.$$

Finally, we can write the equation of the tangent line to the curve $y = f(x)$ at the point $(a, f(a))$ as:

$$y = f'(a)(x - a) + f(a),$$

or, equivalently, in the point-slope form:

$$y = \underbrace{f'(a)}_{m} x + \underbrace{f(a) - f'(a)a}_{n}.$$

where m is the slope of the line and n, the y-intercept.

4.1.2 Formal Definition of the Derivative

Definition 4.2 (Differentiable Function)**.** A function $f : A \subset \mathbb{R} \to \mathbb{R}$ is said to be differentiable at a point a if the limit exists:

$$\lim_{h \to 0} \frac{f(a+h) - f(a)}{h}.$$

If it exists, the above limit is called the derivative of $f(x)$ at the point a and is denoted $f'(a)$ (notation due to Newton), or also as $\frac{df}{dx}(a)$ (notation due to Leibniz).

Proposed Exercise 4.1. Prove that:

$$\lim_{h\to 0}\frac{f(a+h)-f(a)}{h}=\lim_{x\to a}\frac{f(x)-f(a)}{x-a},$$

using the substitution $h=x-a$.

Definition 4.3. An equivalent definition of the derivative of $f(x)$ at the point a is the following,

$$f'(a)=\lim_{x\to a}\frac{f(x)-f(a)}{x-a},$$

if the above limit exists.

Example 4.1. *Calculate the slope of the tangent line to $f(x)=x^2$ at $a=-1$.*

SOLUTION. We know, from the geometric interpretation of the derivative, that the slope of the tangent line at a point is given by the derivative of the function at that point. Thus, using the formal definition of the derivative (*see* Definition 4.2) we obtain:

$$\begin{aligned} m &= \lim_{h\to 0}\frac{f(-1+h)-f(-1)}{h}=\lim_{h\to 0}\frac{(-1+h)^2-(-1)^2}{h}\\ &= \lim_{h\to 0}\frac{1+h^2-2h-1}{h}=\lim_{h\to 0}\frac{h^2-2h}{h}=\lim_{h\to 0}\frac{h-2}{1}=-2. \end{aligned}$$

□

Example 4.2. *Calculate the slope of the tangent line to $f(x)=x^2$ at an arbitrary point a.*

SOLUTION.
As before, we have:

$$\begin{aligned} m &= \lim_{h\to 0}\frac{f(a+h)-f(a)}{h}=\lim_{h\to 0}\frac{(a+h)^2-a^2}{h}\\ &= \lim_{h\to 0}\frac{a^2+h^2+2ah-a^2}{h}=\lim_{h\to 0}\frac{h^2+2ah}{h}\\ &= \lim_{h\to 0}\frac{h+2a}{1}=2a. \end{aligned}$$

Thus, as illustrated in Figure 4.5, we can deduce that:

- if $a=-1$, the slope is $m=2\cdot(-1)=-2$ (as we obtained in the previous example);
- if $a=0$, the slope is $m=2\cdot(0)=0$;
- if $a=1$, the slope is $m=2\cdot 1=2$;
- if $a=2$, the slope is $m=2\cdot 2=4$;

and so on, for any other value. □

Example 4.3. *Calculate the derivative of the function $f(x)=|x|$ at the point $a=0$.*

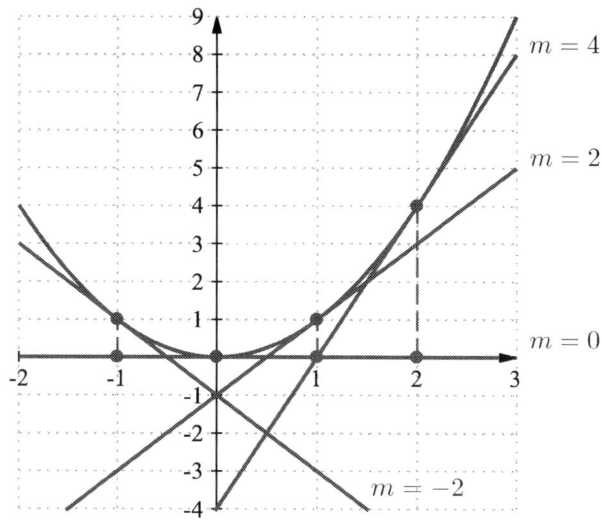

FIGURE 4.5
Tangent lines to the function $f(x) = x^2$ at different points.

SOLUTION. The function $f(x) = |x|$ can be written as:

$$f(x) = \begin{cases} -x, & x < 0 \\ x, & x \geq 0 \end{cases}$$

The calculation of the derivative is done by calculating the following limit:

$$f'(0) = \lim_{h \to 0} \frac{f(h) - f(0)}{h} = \lim_{h \to 0} \frac{f(h) - 0}{h}$$

But the value of $f(h)$ depends on whether $h < 0$ or $h \geq 0$, that is, whether we calculate the limit from the left or from the right. If we calculate the limit from the left:

$$\lim_{h \to 0^-} \frac{f(h) - 0}{h} = \lim_{h \to 0^-} \frac{-h - 0}{h} = \lim_{h \to 0^-} -1 = -1$$

whereas if we calculate it from the right, we obtain:

$$\lim_{h \to 0^+} \frac{f(h) - 0}{h} = \lim_{h \to 0^+} \frac{h - 0}{h} = \lim_{h \to 0^-} 1 = 1$$

Since the lateral limits are different, the limit does not exist. Consequently, the function $f(x) = |x|$ is not differentiable at $a = 0$. □

4.2 Domain of Differentiability

Definition 4.4 (Domain of Differentiability). Given a function $f : A \subset \mathbb{R} \to \mathbb{R}$, the domain of differentiability of f is the subset Ω of its domain ($\Omega \subset A$) consisting of all points where $f(x)$ is differentiable.

In Example 4.2, we saw that for the function $f(x) = x^2$, given a point a, the derivative at that point is $2a$. Therefore, the domain of differentiability of the function $f(x) = x^2$ is $\mathbb{R}$, since the derivative can be calculated at every point.

4.3 The Derivative Function

Definition 4.5 (Derivative Function). Given $f : A \subset \mathbb{R} \to \mathbb{R}$, a function with domain of differentiability $\Omega \subset A$, the derivative function of f is the function f' defined by:

$$\begin{array}{rcl} f' : \Omega & \longrightarrow & \mathbb{R} \\ x & \longmapsto & f'(x) \end{array}$$

That is, the derivative of the function $f(x)$ is denoted $f'(x)$ and is a function that at the point x gives us the value of the derivative at that point. In Example 4.2, we saw that for the function $f(x) = x^2$ and given a point a, the derivative at that point is $2a$. Therefore, the derivative function of the function $f(x) = x^2$ is the function $f'(x) = 2x$.

4.3.1 Calculating the Derivatives of Elementary Functions

Using the formal definition of the derivative, we will calculate the derivatives of elementary functions and study their domain of differentiability.

- Determining the derivative of the constant function: $f(x) = \lambda$, $\lambda \in \mathbb{R}$.

$$f'(x) = \lim_{h \to 0} \frac{f(x+h) - f(x)}{h} = \lim_{h \to 0} \frac{\lambda - \lambda}{h} = \lim_{h \to 0} \frac{0}{h} = 0.$$

  The domain of differentiability of constant functions is $\Omega = \mathbb{R}$ and the derivative function is $f'(x) = 0$. *See* an example in Figure 4.6.

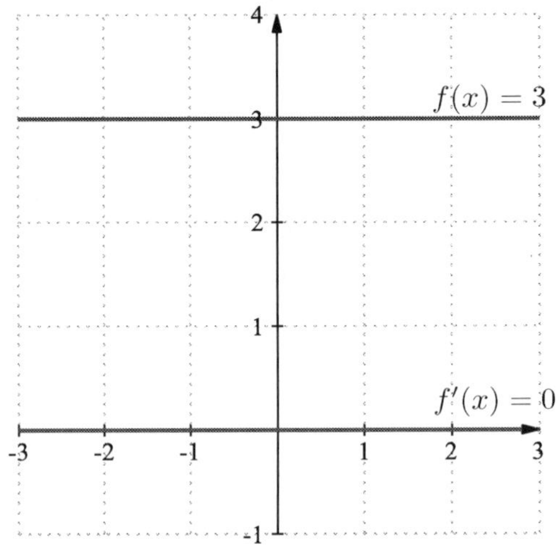

FIGURE 4.6
Graph of the constant function $f(x) = 3$ and its derivative $f'(x) = 0$.

The tangent line has a slope of 0, as it is always a horizontal line. In fact, for constant functions, the tangent line coincides with the function itself (which is a line).

- Determining the derivative of the identity function: $f(x) = x$.

$$f'(x) = \lim_{h\to 0} \frac{f(x+h) - f(x)}{h} = \lim_{h\to 0} \frac{x+h-x}{h} = \lim_{h\to 0} \frac{h}{h} = 1$$

The domain of differentiability of the identity function is $\Omega = \mathbb{R}$ and the derivative function is $f'(x) = 1$. *See* Figure 4.7. Since the identity function is a line, the tangent

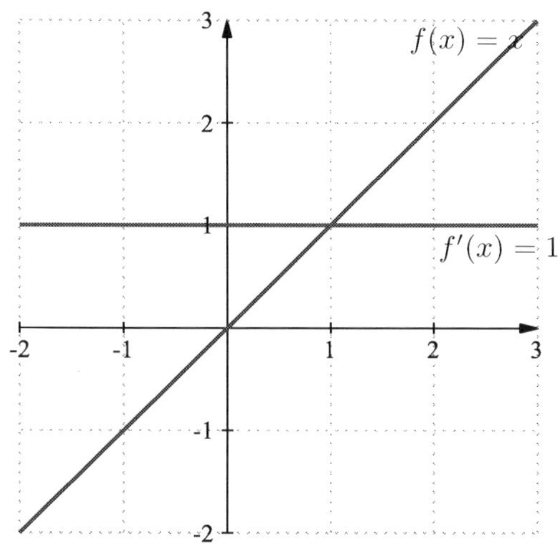

FIGURE 4.7
Graph of the identity function $f(x) = x$ and its derivative $f'(x) = 1$.

line at every point coincides with the function itself and thus has a slope of 1 at every point.

- Determining the derivative of a polynomial of any degree: $f(x) = x^n$, $n \in \mathbb{N}$.

$$f'(x) = \lim_{h\to 0} \frac{f(x+h) - f(x)}{h} = \lim_{h\to 0} \frac{(x+h)^n - x^n}{h}$$

Recall that the binomial theorem tells us that:

$$(x+h)^n = \sum_{k=0}^{n} \binom{n}{k} x^{n-k} h^k = \sum_{k=0}^{n} \frac{n!}{k!(n-k)!} x^{n-k} h^k$$

$$= x^n + nx^{n-1}h + \frac{n(n-1)}{2} x^{n-2}h^2 + \cdots + nxh^{n-1} + h^n.$$

Therefore:

$$f'(x) = \lim_{h\to 0} \frac{\cancel{x^n} + nx^{n-1}h + \frac{n(n-1)}{2}x^{n-2}h^2 + \cdots + nxh^{n-1} + h^n - \cancel{x^n}}{h}$$

$$= \lim_{h\to 0} \left(nx^{n-1} + \frac{n(n-1)}{2} x^{n-2}h + \cdots + nxh^{n-2} + h^{n-1} \right) = nx^{n-1}$$

We have demonstrated that the power function, $f(x) = x^n$, is differentiable at every $x \in \mathbb{R}$ and $f'(x) = nx^{n-1}$ for any natural $n > 0$, since the binomial theorem is valid only for positive integer exponents. Later, in Example 4.17, we will see that the previous rule is true for any $n \in \mathbb{R}$.

- Determining the derivative of the square root function, $f(x) = \sqrt{x}$.

$$
\begin{aligned}
f'(x) &= \lim_{h\to 0} \frac{f(x+h)-f(x)}{h} = \lim_{h\to 0} \frac{\sqrt{x+h}-\sqrt{x}}{h} \\
&= \lim_{h\to 0} \frac{\sqrt{x+h}-\sqrt{x}}{h} \cdot \frac{\sqrt{x+h}+\sqrt{x}}{\sqrt{x+h}+\sqrt{x}} \\
&= \lim_{h\to 0} \frac{x+h-x}{h(\sqrt{x+h}+\sqrt{x})} = \lim_{h\to 0} \frac{1}{\sqrt{x+h}+\sqrt{x}} = \frac{1}{2\sqrt{x}}
\end{aligned}
$$

The domain of differentiability of the square root function is $\mathbb{R}^+ = (0, +\infty)$ and its derivative is the function $f'(x) = \frac{1}{2\sqrt{x}}$. *See* Figure 4.8.

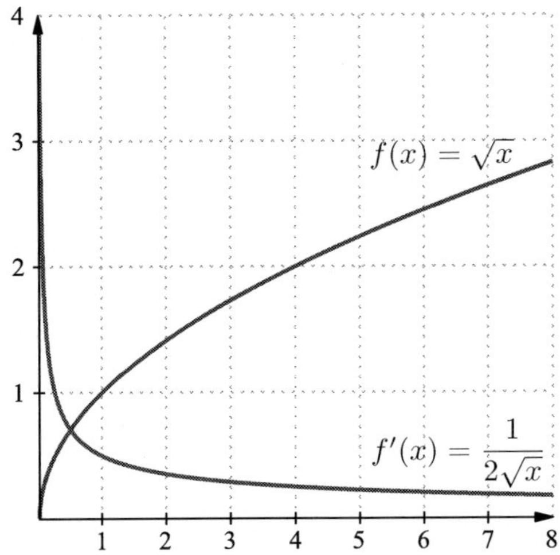

FIGURE 4.8
Graph of the square root function $f(x) = \sqrt{x}$ and its derivative $f'(x) = \frac{1}{2\sqrt{x}}$.

- Determining the derivative of the exponential function, $f(x) = e^x$.

$$
\begin{aligned}
f'(x) &= \lim_{h\to 0} \frac{f(x+h)-f(x)}{h} = \lim_{h\to 0} \frac{e^{x+h}-e^x}{h} \\
&= \lim_{h\to 0} \frac{e^x e^h - e^x}{h} = e^x \left(\lim_{h\to 0} \frac{e^h - 1}{h} \right)
\end{aligned}
$$

Consider the change $t = e^h - 1$; then, $h \to 0 \Leftrightarrow t \to 0$ and noting that $h = \ln(1+t)$ and that $\ln(1+t) \sim_0 t$:

$$
f'(x) = e^x \left(\lim_{h\to 0} \frac{e^h - 1}{h} \right) = e^x \underbrace{\lim_{t\to 0} \frac{t}{\ln(1+t)}}_{\to 1} = e^x.
$$

We will justify this last limit a bit more:

$$\lim_{t\to 0}\frac{t}{\ln(1+t)}=\lim_{t\to 0}\frac{1}{\frac{1}{t}\ln(1+t)}=\lim_{t\to 0}\frac{1}{\ln(1+t)^{1/t}}$$

Now, the limit of the quotient will exist if the limit of the denominator exists. In this case:

$$\lim_{t\to 0}\ln(1+t)^{1/t},$$

which is equivalent to:

$$\ln\left(\lim_{t\to 0}(1+t)^{1/t}\right).$$

Now, the limit:

$$\lim_{t\to 0}(1+t)^{1/t}$$

is immediate if we make the change of variable $t=\frac{1}{s}$, since:

$$\lim_{t\to 0}(1+t)^{1/t}=\lim_{s\to\infty}\left(1+\frac{1}{s}\right)^{s}=e.$$

Thus:

$$\ln\left(\lim_{t\to 0}(1+t)^{1/t}\right)=\ln(e)=1,$$

which is what we wanted to show.

The domain of differentiability of the function e^x is $\mathbb{R}$ and its derivative is itself, $f'(x)=f(x)=e^x$.

- Determining the derivative of the natural logarithm function, $f(x)=\ln(x)$.

$$f'(x)=\lim_{h\to 0}\frac{f(x+h)-f(x)}{h}=\lim_{h\to 0}\frac{\ln(x+h)-\ln(x)}{h}$$

$$=\lim_{h\to 0}\frac{\ln\left(\frac{x+h}{x}\right)}{h}=\lim_{h\to 0}\frac{\ln\left(1+\frac{h}{x}\right)}{h}$$

Consider the change $t=\frac{h}{x}$, then $h\to 0\Leftrightarrow t\to 0$ and noting that $h=xt$ and that $\ln(1+t)\sim_0 t$ (as we saw in the previous section):

$$f'(x)=\lim_{t\to 0}\frac{\ln(1+t)}{xt}=\frac{1}{x}.$$

The domain of differentiability of the function $\ln(x)$ is $\mathbb{R}^+$ and its derivative is the function $f'(x)=\frac{1}{x}$. *See* Figure 4.9.

- Determining the derivative of the sine function, $f(x)=\sin(x)$. For this demonstration, we will use that:

$$\lim_{h\to 0}\frac{\sin(h)}{h}=1.$$

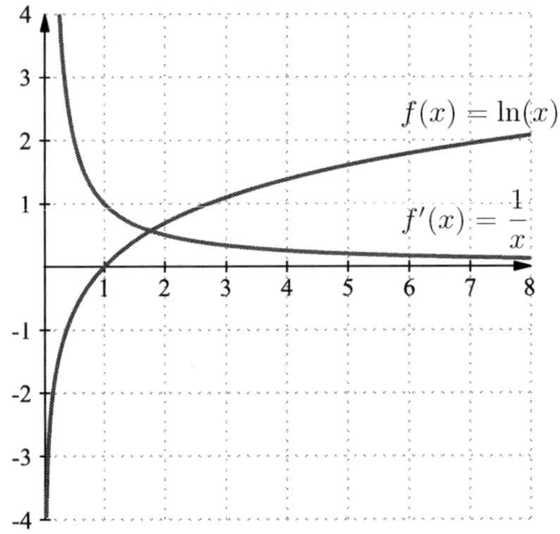

FIGURE 4.9
Graph of the natural logarithm function $f(x) = \ln(x)$ and its derivative $f'(x) = \frac{1}{x}$.

Indeed:

$$\begin{aligned}
f'(x) &= \lim_{h\to 0} \frac{\sin(x+h) - \sin(x)}{h} \\
&= \lim_{h\to 0} \frac{\sin(x)\cos(h) + \cos(x)\sin(h) - \sin(x)}{h} \\
&= \lim_{h\to 0} \frac{\sin(x)(\cos(h) - 1) + \cos(x)\sin(h)}{h} \\
&= \lim_{h\to 0} \sin(x)\frac{\cos(h) - 1}{h} + \lim_{h\to 0} \cos(x)\frac{\sin(h)}{h} \\
&= \sin(x) \lim_{h\to 0} \frac{\cos(h) - 1}{h}\frac{\cos(h) + 1}{\cos(h) + 1} + \cos(x) \lim_{h\to 0} \frac{\sin(h)}{h} \\
&= \sin(x) \lim_{h\to 0} \frac{\cos^2(h) - 1}{h(\cos(h) + 1)} + \cos(x) \lim_{h\to 0} \frac{\sin(h)}{h} \\
&= \sin(x) \lim_{h\to 0} \frac{-\sin^2(h)}{h(\cos(h) + 1)} + \cos(x) \lim_{h\to 0} \frac{\sin(h)}{h} \\
&= \sin(x) \lim_{h\to 0} \underbrace{\frac{\sin(h)}{h}}_{\to 1} \underbrace{\frac{-\sin(h)}{\cos(h) + 1}}_{\to 0} + \cos(x) \lim_{h\to 0} \underbrace{\frac{\sin(h)}{h}}_{\to 1} = \cos(x)
\end{aligned}$$

The domain of differentiability of the function $\sin(x)$ is $\mathbb{R}$ and its derivative is the function $f'(x) = \cos(x)$. *See* Figure 4.10.

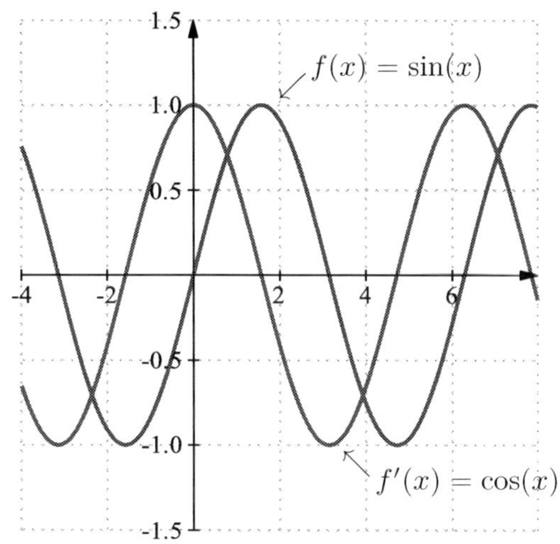

FIGURE 4.10
Graph of the function $f(x) = \sin(x)$ and its derivative $f'(x) = \cos(x)$.

- Determining the derivative of the cosine function, $f(x) = \cos(x)$.

$$\begin{aligned} f'(x) &= \lim_{h\to 0} \frac{\cos(x+h) - \cos(x)}{h} \\ &= \lim_{h\to 0} \frac{\cos(x)\cos(h) - \sin(x)\sin(h) - \cos(x)}{h} \\ &= \lim_{h\to 0} \cos(x) \underbrace{\frac{\cos(h)-1}{h}}_{\to 0} - \sin(x) \underbrace{\frac{\sin(h)}{h}}_{\to 1} = -\sin(x) \end{aligned}$$

The domain of differentiability of the function $\cos(x)$ is $\mathbb{R}$ and its derivative is the function $f'(x) = -\sin(x)$.

- Determining the derivative of the absolute value function, $f(x) = |x|$.

The function can be rewritten as:

$$f(x) = \begin{cases} x, & x \geq 0 \\ -x, & x < 0 \end{cases}$$

Since it is a piecewise-defined function, we will study its derivative according to the following cases:

- If $x > 0$, the derivative of the function is $f'(x) = 1$.
- If $x < 0$, the derivative is $f'(x) = -1$.
- If $x = 0$, we are at a point where the definition of the function changes, and therefore we must study the lateral limits. Note that:
  - $\lim_{h\to 0^+} \frac{f(0+h) - f(0)}{h} = \lim_{h\to 0^+} \frac{h}{h} = 1$
  - $\lim_{h\to 0^-} \frac{f(0+h) - f(0)}{h} = \lim_{h\to 0^-} \frac{-h}{h} = -1$

Since the limit:

$$\lim_{h \to 0} \frac{f(0+h) - f(0)}{h}$$

does not exist, the function is not differentiable at $x = 0$. If we observe Figure 4.11, we see that at the origin the function is continuous but presents a sharp point, that is, a tangent line cannot be drawn to the function at that point. Informally, if the graph of a function has sharp points, then it is not differentiable at those points.

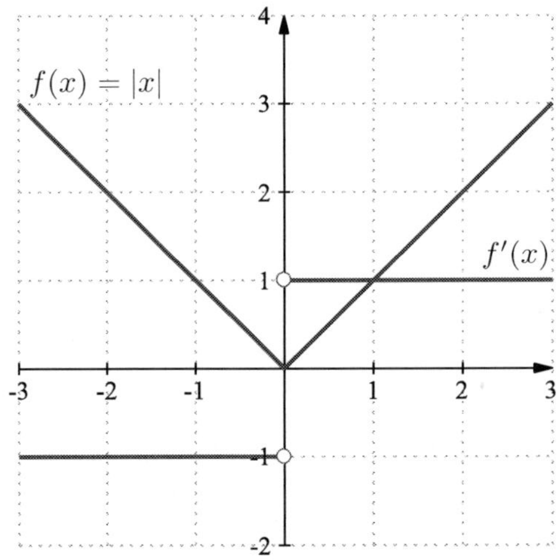

FIGURE 4.11
Graph of the absolute value function $f(x) = |x|$ and its derivative.

The domain of differentiability of the absolute value function is $\mathbb{R} - \{0\}$ and its derivative is the function:

$$f'(x) = \begin{cases} 1, & x > 0 \\ -1, & x < 0 \end{cases}$$

In summary, in Table 4.1, we have the derivatives that we have demonstrated so far. In the section *Techniques for Calculating Derivatives*, where we will be able to calculate any derivative, a more complete table will be provided.

TABLE 4.1
Table of derivatives of elementary functions.

$f(x)$	$f'(x)$
λ	0
x	1
x^n	nx^{n-1}
$\sqrt{x}$	$\frac{1}{2\sqrt{x}}$
$\ln(x)$	$\frac{1}{x}$
e^x	e^x
$\sin(x)$	$\cos(x)$
$\cos(x)$	$-\sin(x)$

4.4 Differentiability and Continuity

Next, we will see the relationship between two very important properties of functions: continuity and differentiability.

Proposition 4.1. If $f : A \subset \mathbb{R} \to \mathbb{R}$ is differentiable at a, then f is continuous at a.

Proof. Let f be differentiable at a, that is:

$$\lim_{x \to a} \frac{f(x) - f(a)}{x - a} = f'(a) \in \mathbb{R}.$$

We we want to prove that f is continuous at a, that is, that:

$$\lim_{x \to a} f(x) = f(a),$$

which is equivalent to checking that:

$$\lim_{x \to a} (f(x) - f(a)) = 0.$$

Let's calculate this limit and verify that it is zero using the hypothesis that f is differentiable at a:

$$\lim_{x \to a} (f(x) - f(a)) = \lim_{x \to a} \frac{f(x) - f(a)}{x - a} \cdot (x - a) = f'(a) \cdot 0 = 0$$

□

It is important to note that the converse is not true. That is, there are continuous functions that are not differentiable, such as the absolute value function. Recall that this function, although continuous, is not differentiable at the origin.

4.5 Lateral Derivatives

When analyzing the differentiability of a function $f(x)$ at a point a, it may occur that the limit that establishes the definition of the derivative does not exist, but the corresponding lateral limits do. This is the case, as we saw in the previous section, with the absolute value function at the origin. This motivates the following definitions.

Definition 4.6 (Right-hand lateral derivative). The right-hand **lateral** derivative of the function f at the point a is defined as:

$$f'_+(a) = \lim_{h \to 0^+} \frac{f(a + h) - f(a)}{h}.$$

Definition 4.7 (Left-hand lateral derivative)**.** The left-hand **lateral** derivative of the function f at the point a is defined as:

$$f'_{-}(a) = \lim_{h \to 0^-} \frac{f(a+h) - f(a)}{h}.$$

Proposition 4.2. A function f is differentiable at a if and only if the two lateral derivatives exist and are equal.

In the case where the lateral derivatives exist but are not equal, this can be interpreted geometrically (Figure 4.12) as the function having two distinct tangent half-lines at the point in question, as the limit of the secant lines from the right and left.

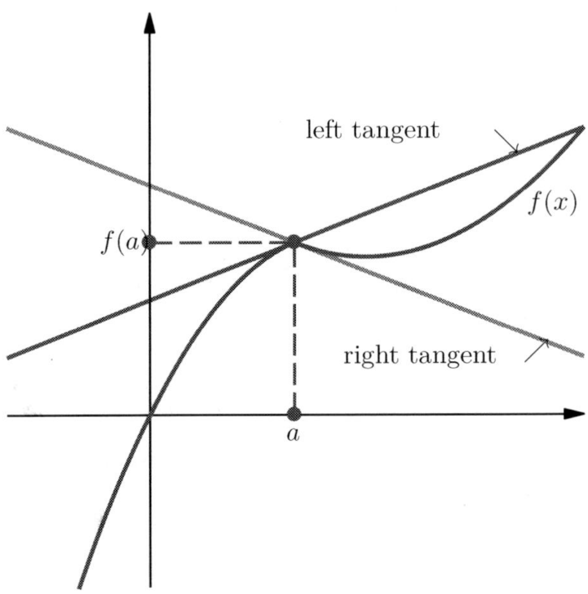

FIGURE 4.12
Geometric interpretation of the concept of lateral derivatives at a point a.

Example 4.4. *Calculate the derivative of the following function:*

$$f(x) = \begin{cases} 0, & \textit{if } x < 0 \\ 6x, & \textit{if } x \geq 0 \end{cases}$$

at the points where it is possible.

SOLUTION. At the points $x \neq 0$, we have:

$$f'(x) = \begin{cases} 0, & \text{if } x < 0 \\ 6, & \text{if } x > 0. \end{cases}$$

On the other hand, to study the differentiability at $x = 0$, given a change in the definition of the function, we must resort to the lateral derivatives. We observe that the left-hand lateral derivative is:

$$f'_{-}(0) = \lim_{h \to 0^-} \frac{f(0+h) - f(0)}{h} = \lim_{h \to 0^-} \frac{0}{h} = 0,$$

and from the right:

$$f'_+(0) = \lim_{h \to 0^+} \frac{f(0+h) - f(0)}{h} = \lim_{h \to 0^+} \frac{6h}{h} = 6.$$

The lateral derivatives exist but are different; therefore, the derivative does not exist at $x = 0$. As seen in Figure 4.13, at $x = 0$ the function $f(x)$ is continuous but has a cusp. □

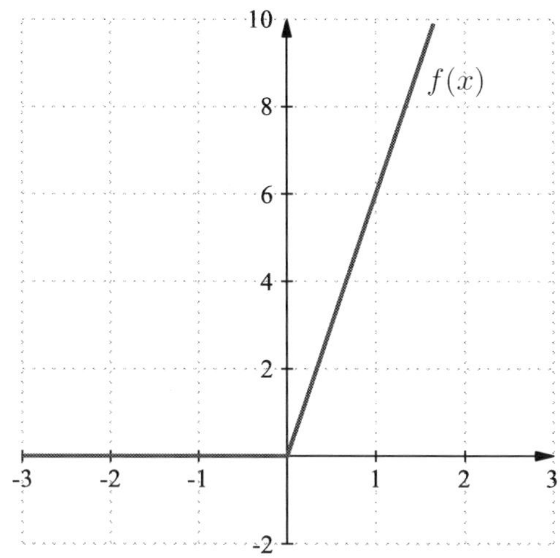

FIGURE 4.13
Graph of the function from Example 4.4.

4.6 Infinite Derivative

We know that a function may not be differentiable at a point when the two lateral derivatives are different. But what happens in the graphs in Figure 4.14? What is the slope of the tangent line at zero? This motivates the following definitions.

Definition 4.8 (Infinite derivative). $f(x)$ is said to have a derivative of $+\infty$ at the point a if:

$$f'_+(a) = f'_-(a) = +\infty$$

Definition 4.9 (Infinite derivative). $f(x)$ is said to have a derivative of $-\infty$ at the point a if:

$$f'_+(a) = f'_-(a) = -\infty$$

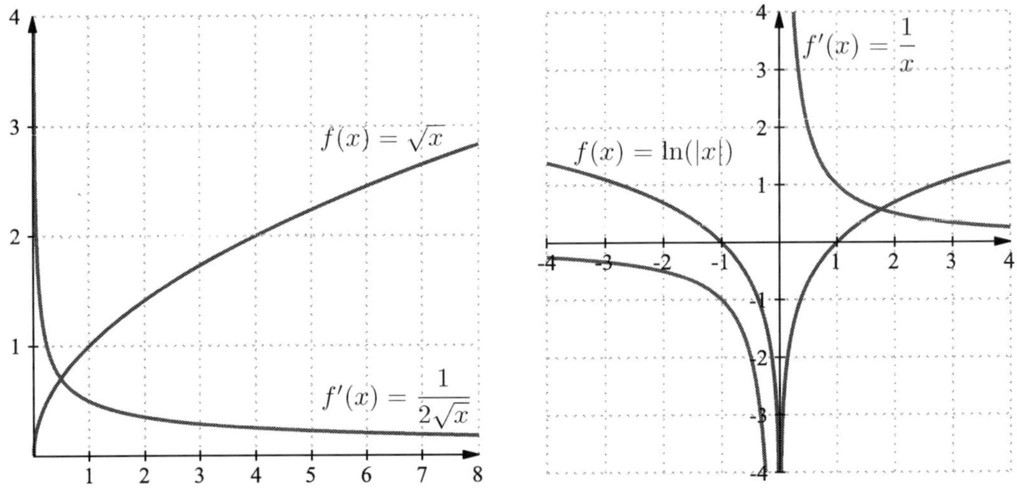

FIGURE 4.14
Geometric interpretation of the concept of infinite derivative.

Definition 4.10 (Infinite derivative). $f(x)$ is said to have a derivative of ∞ at the point a if:

$$f'_+(a) = \pm\infty$$
$$f'_-(a) = \mp\infty$$

4.7 Derivative and Operations with Functions

Let $f,\ g : A \subset \mathbb{R} \to \mathbb{R}$ be differentiable at a point a, then:

(a) $(f+g)'(a) = f'(a) + g'(a)$

(b) $(\lambda f)'(a) = \lambda f'(a),\ \lambda \in \mathbb{R}$

(c) $(f \cdot g)'(a) = f'(a) \cdot g(a) + f(a) \cdot g'(a)$

(d) $\left(\frac{f}{g}\right)'(a) = \frac{f'(a)g(a) - f(a)g'(a)}{g(a)^2}$

Proof.

(a) Let us show that $(f+g)'(a) = f'(a) + g'(a)$.

$$\begin{aligned}(f+g)'(a) &= \lim_{x\to a}\frac{(f+g)(x)-(f+g)(a)}{x-a}\\ &= \lim_{x\to a}\frac{f(x)+g(x)-f(a)-g(a)}{x-a}\\ &= \lim_{x\to a}\frac{f(x)-f(a)}{x-a}+\frac{g(x)-g(a)}{x-a}\\ &= \lim_{x\to a}\frac{f(x)-f(a)}{x-a}+\lim_{x\to a}\frac{g(x)-g(a)}{x-a}=f'(a)+g'(a)\end{aligned}$$

where in the last equality we have used the hypothesis that f and g are differentiable at point a.

(b) Let us show that $(\lambda f)'(a) = \lambda f'(a)$, $\lambda \in \mathbb{R}$.

$$\begin{aligned}(\lambda f)'(a) &= \lim_{x\to a}\frac{(\lambda f)(x)-(\lambda f)(a)}{x-a}=\lim_{x\to a}\frac{\lambda f(x)-\lambda f(a)}{x-a}\\ &= \lim_{x\to a}\lambda\frac{f(x)-f(a)}{x-a}=\lambda\left(\lim_{x\to a}\frac{f(x)-f(a)}{x-a}\right)=\lambda f'(a)\end{aligned}$$

(c) Let us show that $(f\cdot g)'(a) = f'(a)\cdot g(a) + f(a)\cdot g'(a)$.

$$\begin{aligned}(f\cdot g)'(a) &= \lim_{x\to a}\frac{(f\cdot g)(x)-(f\cdot g)(a)}{x-a}\\ &= \lim_{x\to a}\frac{f(x)g(x)-f(a)g(a)}{x-a}\\ &= \lim_{x\to a}\frac{f(x)g(x)-f(a)g(x)+f(a)g(x)-f(a)g(a)}{x-a}\\ &= \lim_{x\to a}\underbrace{\frac{f(x)-f(a)}{x-a}}_{\to f'(a)}g(x)+f(a)\underbrace{\frac{g(x)-g(a)}{x-a}}_{\to g'(a)}\end{aligned}$$

since f and g are differentiable at point a. Therefore:

$$\begin{aligned}(f\cdot g)'(a) &= \lim_{x\to a} f'(a)\,g(x)+f(a)\,g'(a)\\ &= f'(a)\underbrace{\left(\lim_{x\to a} g(x)\right)}_{\to g(a)}+f(a)\,g'(a)\end{aligned}$$

To complete the proof, it is necessary to justify that:

$$\lim_{x\to a} g(x) = g(a).$$

Since g is differentiable at point a this implies that it is continuous at a (*see* Proposition 4.1) and, therefore, it is proven that $\lim_{x\to a} g(x) = g(a)$.

(d) Let us show that:

$$\left(\frac{f}{g}\right)'(a) = \frac{f'(a)g(a) - f(a)g'(a)}{g(a)^2}.$$

Since $f/g = f \cdot (1/g)$, using the product rule, we have:

$$\left(\frac{f}{g}\right)'(a) = \left(f \cdot \frac{1}{g}\right)'(a) = f'(a) \cdot \frac{1}{g(a)} + f(a)\left(\frac{1}{g}\right)'(a).$$

Therefore, to calculate the derivative of a quotient we only need to calculate the derivative of the function $1/g$, i.e.:

$$\left(\frac{1}{g}\right)'(a).$$

$$\begin{aligned}
\left(\frac{1}{g}\right)'(a) &= \lim_{x\to a} \frac{\left(\frac{1}{g}\right)(x) - \left(\frac{1}{g}\right)(a)}{x-a} = \lim_{x\to a} \frac{\frac{1}{g(x)} - \frac{1}{g(a)}}{x-a} \\
&= \lim_{x\to a} \frac{\frac{1}{g(x)} - \frac{1}{g(a)}}{x-a} \cdot \frac{g(x)g(a)}{g(x)g(a)} = \lim_{x\to a} \frac{g(a)-g(x)}{g(x)g(a)(x-a)} \\
&= \lim_{x\to a} \frac{-1}{g(x)g(a)} \cdot \frac{g(x)-g(a)}{x-a} = \frac{-1}{g(a)^2}\, g'(a) = \frac{-g'(a)}{g(a)^2}
\end{aligned}$$

Thus:

$$\begin{aligned}
\left(\frac{f}{g}\right)'(a) &= \frac{f'(a)}{g(a)} + f(a)\left(\frac{1}{g}\right)'(a) = \frac{f'(a)}{g(a)} - f(a)\frac{g'(a)}{g(a)^2} \\
&= \frac{f'(a)g(a) - f(a)g'(a)}{g(a)^2}.
\end{aligned}$$

□

Example 4.5. *Let $h(x) = x^2 + \sin(x)$. Calculate $h'(x)$.*

SOLUTION. Consider $f(x) = x^2$ and $g(x) = \sin(x)$. Then, $h(x) = f(x) + g(x)$ and since, by property (a), the derivative of the sum is the sum of the derivatives:

$$h'(a) = (f+g)'(a) = f'(a) + g'(a) = 2a + \cos(a).$$

Example 4.6. *Let $h(x) = 4e^x$. Calculate $h'(x)$.*

SOLUTION.
Consider $f(x) = e^x$. Then $h(x) = 4f(x)$ and using property (b):

$$h'(a) = (4f)'(a) = 4f'(a) = 4e^a.$$

□

Example 4.7. *Let $h(x) = x^2\cos(x)$. Calculate $h'(x)$.*

SOLUTION. Take $f(x) = x^2$ and $g(x) = \cos(x)$. Then $h(x) = f(x)g(x)$ and using property (c):

$$\begin{aligned} h'(a) &= (f \cdot g)'(a) = f'(a)g(a) + f(a)g'(a) \\ &= 2a\cos(a) - a^2\sin(a). \end{aligned}$$

□

Example 4.8. *Calculate the derivative of the function* $h(x) = \tan(x)$.

SOLUTION. Take $f(x) = \sin(x)$ and $g(x) = \cos(x)$. Then:

$$h(x) = \frac{f(x)}{g(x)}$$

and using property (d):

$$\begin{aligned} h'(a) &= \left(\frac{f}{g}\right)'(a) = \frac{f'(a)g(a) - f(a)g'(a)}{g(a)^2} \\ &= \frac{\cos(a)\cos(a) - \sin(a)(-\sin(a))}{\cos^2(a)} = \frac{1}{\cos^2(a)} \end{aligned}$$

Observation 4.1. The derivative of the tangent function can be expressed in three different ways, considering that:

$$\sec^2(a) = \frac{1}{\cos^2(a)} = \frac{\sin^2(a) + \cos^2(a)}{\cos^2(a)} = 1 + \tan^2(a)$$

The two most commonly used options are:

$$\tan'(a) = \frac{1}{\cos^2(a)} = 1 + \tan^2(a)$$

□

Example 4.9. *Calculate the derivative of the function* $f(x) = \cot(x)$.

SOLUTION. Since:

$$f(x) = \cot(x) = \frac{1}{\tan(x)}$$

then, using property (d):

$$f'(x) = -\frac{(\tan(x))'}{\tan(x)^2} = -\frac{1}{\cos^2(x)}\frac{\cos^2(x)}{\sin(x)^2} = -\frac{1}{\sin^2(x)}$$

□

4.8 Chain Rule

In this section, we will see a rule that tells us how to calculate the derivative of the composition of two functions. This rule is known as the *chain rule*.

Proposition 4.3 (Chain Rule)**.** If we have functions f and g, such that:

$$\begin{array}{rcccc} g\circ f: & A\subset\mathbb{R} & \overset{f}{\longrightarrow} & B & \overset{g}{\longrightarrow} & \mathbb{R} \\ & a & \longmapsto & f(a) & \longmapsto & g(f(a)) \end{array}$$

where f is differentiable at a, and g is differentiable at $f(a)$, then the composite function $(g \circ f)$ is differentiable at a and satisfies:

$$(g\circ f)'(a) = g'(f(a))\cdot f'(a).$$

Proof. We want to prove that:

$$(g\circ f)'(a) = g'(f(a))\cdot f'(a).$$

Then:

$$\begin{aligned} (g\circ f)'(a) &= \lim_{x\to a}\frac{(g\circ f)(x)-(g\circ f)(a)}{x-a} \\ &= \lim_{x\to a}\frac{g(f(x))-g(f(a))}{x-a} \\ &= \lim_{x\to a}\frac{g(f(x))-g(f(a))}{x-a}\cdot\frac{f(x)-f(a)}{f(x)-f(a)} \\ &= \lim_{x\to a}\underbrace{\frac{g(f(x))-g(f(a))}{f(x)-f(a)}}_{\to g'(f(a))}\cdot\underbrace{\frac{f(x)-f(a)}{x-a}}_{\to f'(a)} \\ &= g'(f(a))f'(a) \end{aligned}$$

where in the last equality we have used that f is differentiable at a and, therefore, also continuous at a and that g is differentiable at $f(a)$.

To formally prove that:

$$\lim_{x\to a}\frac{g(f(x))-g(f(a))}{f(x)-f(a)} = g'(f(a))$$

we need to perform a change of variable $y = f(x)$. Here, it is essential to ensure that any variable changes made when taking a limit are defined by a continuous function. In this case, considering the change $y = f(x)$ at point a is possible, since the function $f(x)$ is continuous at point a.

Then, if $x \to a$ as f is a continuous function at a, $y = f(x) \to f(a)$, since:

$$\lim_{x\to a} f(x) = f(a).$$

Finally:

$$\lim_{x\to a}\frac{g(f(x))-g(f(a))}{f(x)-f(a)} = \lim_{y\to f(a)}\frac{g(y)-g(f(a))}{y-f(a)} = g'(f(a)).$$

□

Example 4.10. *Calculate the derivative of* $h(x) = (x^4 + 3x)^5$.

SOLUTION. Take $f(x) = x^4 + 3x$ and $g(x) = x^5$, then $h(x) = (g \circ f)(x)$. Therefore:

$$h'(a) = (g \circ f)'(a) = g'(f(a))f'(a).$$

Since $f'(x) = 4x^3 + 3$ and $g'(x) = 5x^4$, we obtain:

$$h'(a) = 5(f(a))^4(4a^3 + 3) = 5(a^4 + 3a)^4(4a^3 + 3)$$

□

Example 4.11. *Calculate the derivative of* $h(x) = \sin^5(x)$.

SOLUTION. Consider $f(x) = \sin(x)$ and $g(x) = x^5$, where $f'(x) = \cos(x)$ and $g'(x) = 5x^4$; then:

$$h'(a) = (g \circ f)'(a) = 5(f(a))^4 \cos(a) = 5\sin^4(a)\cos(a)$$

□

Proposed Exercise 4.2. Calculate the derivative of the following functions:

(a) $h(x) = \sqrt{x^2 + 3}$

(b) $h(x) = \ln^3(x)$

(c) $h(x) = \sin(x^2 + 1)$

(d) $h(x) = \cos(\sqrt{x})$

(e) $h(x) = \cos\left(\frac{1}{x}\right)$

(f) $h(x) = \sin(\ln(x) + 1)$

(g) $h(x) = \sin^3(3x^2 + 1)$

(h) $h(x) = \ln(\cos(2x + 3))$

(i) $h(x) = \sqrt{1 + \ln(x^2 + 1)}$

(j) $h(x) = \sin(\cos(\ln(x)))$

4.8.1 Derivative of the Inverse Function

One of the multiple applications of the chain rule is the calculation of the derivative of the inverse function.

Let $f(x)$ be an invertible function, then the derivative of the inverse function, $f^{-1}(x)$, can be calculated by differentiating the following identity:

$$(f \circ f^{-1})(x) = x,$$

where the chain rule is applied to calculate $(f \circ f^{-1})'(x)$.

Specifically, suppose we know $f(x)$, its derivative $f'(x)$, and its inverse function $f^{-1}(x)$. Then, if we differentiate the identity:

$$(f \circ f^{-1})(x) = x$$

using the chain rule we obtain:

$$f(f^{-1}(x)) = x \implies f'(f^{-1}(x)) \cdot (f^{-1})'(x) = 1$$

$$\implies (f^{-1})'(x) = \frac{1}{f'(f^{-1}(x))}.$$

Example 4.12. *Find the derivative of the function* $g(x) = \ln(x)$, *assuming the derivative of the exponential function is known.*

SOLUTION. Since the logarithm and the exponential are inverse functions:

$$e^{\ln(x)} = x \quad \forall x > 0.$$

Then, if we differentiate with respect to x:

$$e^{\ln(x)}(\ln(x))' = 1 \quad \Longrightarrow \quad g'(x) = (\ln(x))' = \frac{1}{e^{\ln(x)}} = \frac{1}{x},$$

and, therefore, we see that the derivative of the logarithm function is $1/x$ whenever $x > 0$.□

Example 4.13. *Using that if* $f(x) = \cos(x)$, $f'(x) = -\sin(x)$, *prove that if* $g(x) = \arccos(x)$, $g'(x) = -\frac{1}{\sqrt{1-x^2}}$.

SOLUTION. We know that the function $f(x) = \cos(x)$ restricted to the interval $[0, \pi]$ is invertible and its inverse is the function $g(x) = \arccos(x)$. Since f and g are inverse functions:

$$\cos(\arccos(x)) = x.$$

Then, by differentiating and applying the chain rule:

$$-\sin(\arccos(x))(\arccos(x))' = 1$$

$$g'(x) = (\arccos(x))' = -\frac{1}{\sin(\arccos(x))}.$$

Using that $\sin^2(x) + \cos^2(x) = 1$, we obtain:

$$\sin(x) = \begin{cases} \sqrt{1-\cos^2(x)} & x \in [0, \pi] \\ -\sqrt{1-\cos^2(x)} & x \in [\pi, 2\pi] \end{cases}$$

Then, since $\arccos(x) \in [0, \pi]$:

$$\sin(\arccos(x)) = \sqrt{1-\cos^2(\arccos(x))} = \sqrt{1-x^2},$$

and, therefore:

$$(\arccos(x))' = -\frac{1}{\sqrt{1-x^2}}.$$

□

Example 4.14. *Using the chain rule, prove that:*

$$f(x) = \frac{1}{g(x)} \quad \Longrightarrow \quad f'(x) = -\frac{g'(x)}{g(x)^2}.$$

SOLUTION. Taking logarithms in the initial expression:

$$\ln(f(x)) = \ln\left(\frac{1}{g(x)}\right) = -\ln(g(x)).$$

Differentiating and applying the chain rule:

$$(\ln(f(x)))' = -(\ln(g(x)))' \quad \Longrightarrow \quad \frac{1}{f(x)}f'(x) = -\frac{1}{g(x)}g'(x).$$

Thus:

$$f'(x) = -\frac{f(x)g'(x)}{g(x)} = -\frac{g'(x)}{g(x)^2}.$$

□

4.9 Techniques for Calculating Derivatives

This section aims to compile all the techniques for calculating derivatives seen so far. With them, we should be able to calculate the derivative of any given function, whether by the formal definition via limits, with the table of elementary derivatives (which we have proved from the definition), with differentiation properties and operations, with the chain rule, or with the derivative of the inverse function. Let's see some examples of function derivative calculations here.

Example 4.15. *Calculate the derivative of the function* $f(x) = a^x$, *with* $a \in \mathbb{R}^+$.

SOLUTION. Let's take the function $f(x) = a^x$ and apply logarithms:

$$\ln(f(x)) = \ln(a^x).$$

By the properties of the logarithm, we know that $\ln(a^x) = x\ln(a)$. Therefore, we have:

$$\ln(f(x)) = x\ln(a),$$

where, by differentiating this equality (applying the chain rule), we obtain:

$$\frac{f'(x)}{f(x)} = \ln(a),$$

and, finally, solving for $f'(x)$:

$$f'(x) = a^x\ln(a).$$

□

Example 4.16. *Calculate the derivative of the function* $\arctan(x)$.

SOLUTION. We know that the inverse function of $\tan(x)$ restricted to the domain $[-\frac{\pi}{2}, \frac{\pi}{2}]$ is precisely the function $\arctan(x)$. Therefore, we will use the derivative of the inverse function to calculate its derivative. Being inverse functions, we have:

$$\tan(\arctan(x)) = x.$$

Differentiating the previous identity (applying the chain rule):

$$(1 + \tan^2(\arctan(x)))(\arctan(x))' = 1,$$

and, therefore:

$$(\arctan(x))' = \frac{1}{1+x^2}.$$

□

Proposed Exercise 4.3. Prove that the derivative of $\ln(bx)$ is:

$$\frac{1}{x},$$

for any constant $b \in \mathbb{R}$, as long as $bx > 0$.

Example 4.17. *Calculate the derivative of the function* $f(x) = x^n$, *with* $n \in \mathbb{R}$.

SOLUTION. Let's first assume the case $x > 0$. We can write:

$$x^n = e^{n \ln(x)}.$$

We differentiate the previous equality using the chain rule and obtain:

$$(x^n)' = e^{n \ln(x)} \frac{n}{x},$$

that is:

$$(x^n)' = nx^{n-1}.$$

Now let's assume that $x < 0$. We can write:

$$\ln|x^n| = n \ln|x|.$$

Using the property proved in the Proposed Exercise 4.3, taking $b = -1$, we have that the derivative of $\ln|x|$ is $\frac{1}{x}$. Thus, by differentiating the previous equality:

$$\frac{(x^n)'}{x^n} = \frac{n}{x},$$

hence solving:

$$(x^n)' = nx^{n-1}.$$

□

Proposed Exercise 4.4. Calculate the derivative of the following functions:

(a) $f(x) = \arcsin(x)$

(b) $f(x) = \tan(5 - \sin(2x))$

(c) $f(x) = \arctan(x)$

(d) $f(x) = 3^{2x^2}\sqrt{x} + (x^2 - 2)^4$

(e) $f(x) = \sqrt[3]{x^2} + \sqrt{x} + \dfrac{5}{x^5}$

(f) $f(x) = 10^{\sqrt{x}} + e^{3-x^2}$

(g) $f(x) = \sqrt[3]{\dfrac{x^2+1}{x^2-1}}$

(h) $f(x) = x^x + \ln\left(\dfrac{e^x+1}{e^x-1}\right)$

Proposed Exercise 4.5. Calculate the derivatives of the following functions:

(a) $f(x) = \arctan(\arctan(x))$

(b) $f(x) = \sqrt{(\sin(x))^{\cos(x)}}$

(c) $f(x) = 2\sqrt{ax} + \sqrt[5]{x^2} + \dfrac{1}{x} + \dfrac{3}{x^5}$

(d) $f(x) = e^{-3x}\sqrt{x^2+c} + x^{x^3+1}$

(e) $f(x) = \ln\left(\sqrt{\dfrac{1+\sin(x)}{\tan x}}\right)$

(f) $f(x) = \tan(2x) + \tan(x^3) + \tan^3(x)$

(g) $f(x) = \cos(x^b + ax)$

(h) $f(x) = (\sin(x^2))^2$

(i) $f(x) = x^{\sin(x)}$

(j) $f(x) = e^{a(x-1)} + \ln(bx)$

(k) $f(x) = \dfrac{1}{e^{-2x} - 1}$

(l) $f(x) = \dfrac{x^2-1}{cx^4}$

(m) $f(x) = \sqrt{\ln(\tan(x))}$

(n) $f(x) = (\sin(x))^{\tan(x)}$

(o) $f(x) = \dfrac{4\tan^3(x) - 3^{2x}}{\cos(5x - x^3)}$

(p) $f(x) = (\ln(x^4 - 5x))^5$

(q) $f(x) = \dfrac{2a^2x^2\sqrt{a^2-x^2}}{x^3}$

(r) $f(x) = \arctan\left(\dfrac{\tan(x)}{\sqrt{2}}\right)$

(s) $f(x) = \dfrac{1}{x^4} + \dfrac{5}{2a^2x^2} - \dfrac{1}{\sqrt{a^2-x^2}}$

(t) $f(x) = \dfrac{1}{b^2+x}\left(2\sqrt{x^3} + 3b^2\sqrt{x}\right)$

(u) $f(x) = 3b\arctan\left(\dfrac{\sqrt{x}}{b}\right)$

(v) $f(x) = \dfrac{\sqrt{x^2+2}}{\sqrt{cx}}$

We include in Table 4.2 a complete table of derivatives, most of which have already been proved.

4.10 Extreme Values

The maximum and minimum values of a function, known as *extremes*, are the largest (maximum) or smallest (minimum) values taken by a function, either within a closed interval or in the domain of the function as a whole. The process of finding the extremes of a function is called *optimization*, and we will see that it is one of the most important applications of differentiation.

TABLE 4.2
Table of derivatives.

$f(x)$	$f'(x)$
λ	0
x	1
x^n	nx^{n-1}
$\sqrt{x}$	$\frac{1}{2\sqrt{x}}$
$\ln(x)$	$\frac{1}{x}$
e^x	e^x
$\log_a(x)$	$\frac{1}{x\ln(a)}$
a^x	$a^x \ln(a)$
$\sin(x)$	$\cos(x)$
$\cos(x)$	$-\sin(x)$
$\tan(x)$	$\sec^2(x)$
$\sec(x)$	$\sec(x)\tan(x)$
$\cot(x)$	$-\csc^2(x)$
$\csc(x)$	$-\csc(x)\cot(x)$
$\arcsin(x)$	$\frac{1}{\sqrt{1-x^2}}$
$\arccos(x)$	$\frac{-1}{\sqrt{1-x^2}}$
$\arctan(x)$	$\frac{1}{1+x^2}$
$\sinh(x)$	$\cosh(x)$
$\cosh(x)$	$\sinh(x)$

Definition 4.11 (Absolute (or global) extremes on a closed interval). Given $f : [a, b] \to \mathbb{R}$ it is called:

- *absolute maximum* of f on $[a, b]$ the point $x_{max} \in [a, b]$ such that:

$$f(x_{max}) \geq f(x), \quad \forall x \in [a, b];$$

- *absolute minimum* of f on $[a, b]$ the point $x_{min} \in [a, b]$ such that:

$$f(x_{min}) \leq f(x), \quad \forall x \in [a, b].$$

In Figure 4.15, it is shown, on the graph of two given functions, what their absolute maximum and minimum values are on a given interval.

Remark 4.1. By Weierstrass' Theorem 3.2, we know that if f is continuous on $[a, b]$, then there exist x_{min} and x_{max}.

Definition 4.12 (Relative (or local) extremes on a closed interval). Given $f : [a, b] \to \mathbb{R}$, it is said that the function has a:

- *relative maximum* at $x_0 \in [a, b]$ if $f(x_0) \geq f(x)$ for all x in some open interval containing x_0;

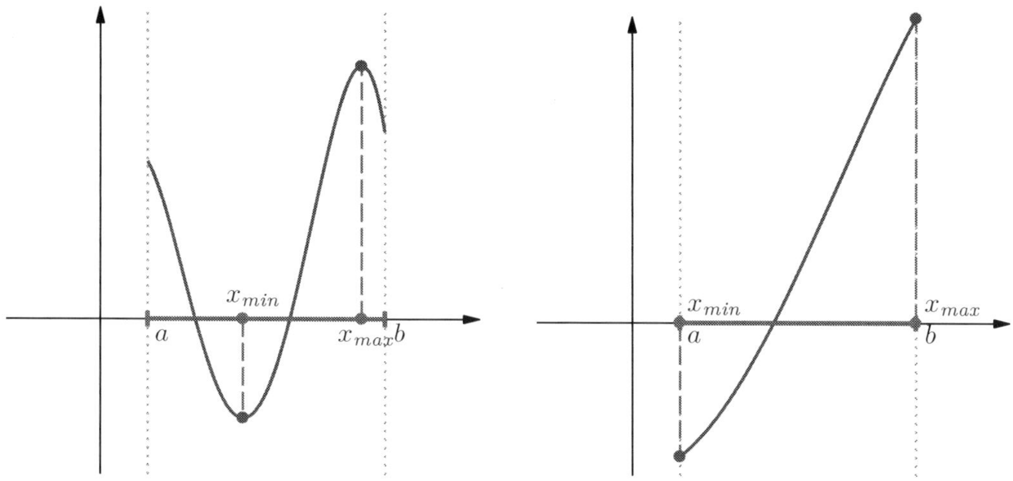

FIGURE 4.15
Graphical interpretation of the concept of absolute maximum and minimum.

- *relative minimum* at $x_0 \in [a,b]$ if $f(x_0) \leq f(x)$ for all x in some open interval containing x_0.

Remark 4.2 (Clarification about relative extremes). In Definition 4.12, the ends of the interval a and b can also be considered as relative extremes. In this sense, although any open interval containing a or b —for example, $(a-\delta, a+\delta)$ or $(b-\delta, b+\delta)$, for some $\delta > 0$, respectively— will clearly be outside the interval $[a,b]$, what we will do is consider what we call the *induced* interval, that is, the interval

$$(a-\delta, a+\delta) \cap [a,b] = [a, a+\delta)$$

for the point $x = a$ and the interval

$$(b-\delta, b+\delta) \cap [a,b] = (b-\delta, b]$$

for the point $x = b$.

In Figure 4.16, it is shown, on the graph of two given functions, what their relative maximum and minimum values are on a given interval.

Observation 4.2. Given $f : [a,b] \to \mathbb{R}$, if x_0 is an absolute maximum, then x_0 is also a relative maximum.

Definition 4.13 (Critical point). A point x_0 of the domain of a function is called a *critical point* if $f'(x_0) = 0$ or if $f'(x_0)$ does not exist.

In Figure 4.17, it is shown, on the graph of two given functions, what their critical points are on a given interval.

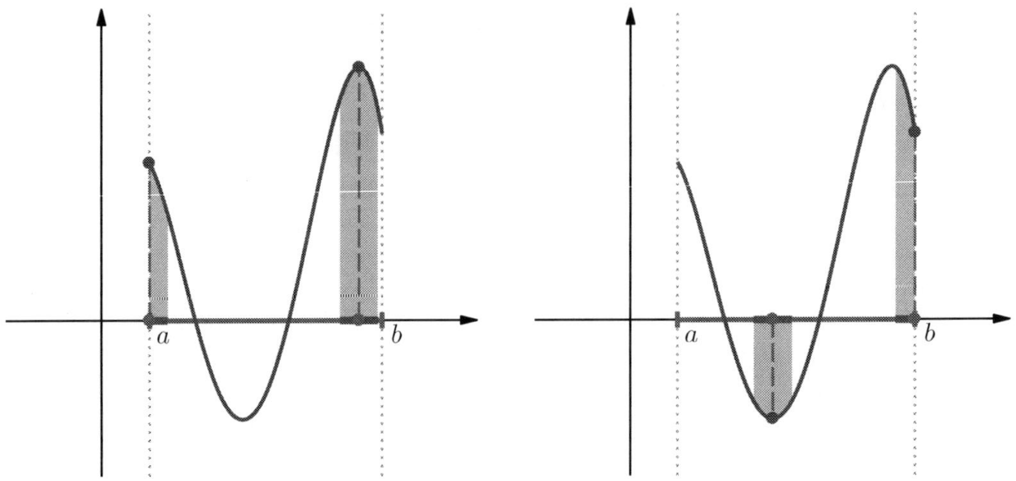

FIGURE 4.16
Relative maxima (left) and minima (right) of a function.

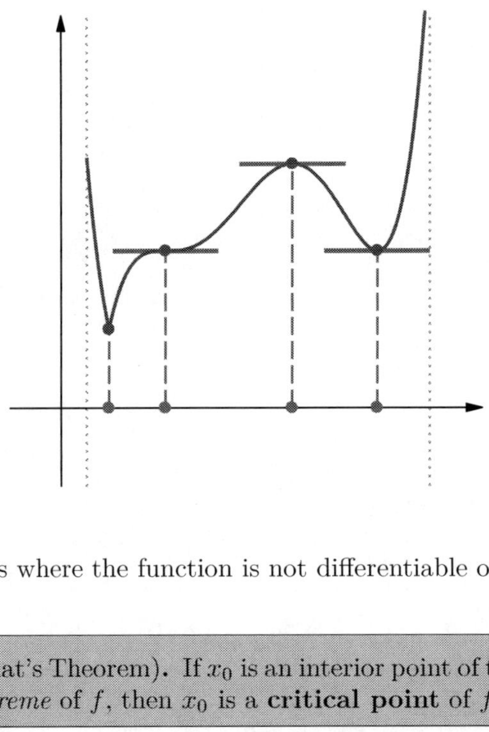

FIGURE 4.17
Critical points are points where the function is not differentiable or the derivative is zero.

Proposition 4.4 (Fermat's Theorem)**.** If x_0 is an interior point of the domain of a function f and is also a *local extreme* of f, then x_0 is a **critical point** of f.

Proof. Let's suppose that $f(x_0)$ is a local minimum (for a local maximum, the procedure would be similar). If $f'(x_0)$ does not exist, then x_0 is a critical point, as we wanted to prove. Suppose, then, that $f'(x_0)$ exists. We now need to prove that $f'(x_0) = 0$.

Since $f(x_0)$ is a local minimum, it holds that $f(x_0) \leq f(x_0 + h)$ for any $h \neq 0$ sufficiently small. This is equivalent to:

$$f(x_0 + h) - f(x_0) \geq 0.$$

We divide this inequality by h and get:

$$\frac{f(x_0+h)-f(x_0)}{h} \geq 0 \quad if \quad h>0,$$

$$\frac{f(x_0+h)-f(x_0)}{h} \leq 0 \quad if \quad h<0.$$

Since the derivative of f exists at x_0, we know that the lateral derivatives coincide with this derivative. Therefore, using the previous inequalities, we can bound the derivative of f at point x_0 as follows:

$$f'(x_0) = f'_+(x_0) = \lim_{h\to 0+} \frac{f(x_0+h)-f(x_0)}{h} \geq \lim_{h\to 0+} 0 = 0,$$

$$f'(x_0) = f'_-(x_0) = \lim_{h\to 0-} \frac{f(x_0+h)-f(x_0)}{h} \leq \lim_{h\to 0-} 0 = 0.$$

Thus, $f'(x_0) \geq 0$ and $f'(x_0) \leq 0$. The only possibility is that $f'(x_0) = 0$, as we wanted to prove. □

Remark 4.3. The converse implication of Fermat's theorem is not true. For example, $x_0 = 0$ is a critical point of $f(x) = x^3$, but it is not a local extreme of that function but an inflection point.

From Fermat's theorem, it follows that, to determine the absolute extremes of a function on a closed interval, we must evaluate f at the points that are candidates to be so, which are:

(A) the points where f is not continuous,

(B) the critical points,

(C) the extreme points of the interval.

We will evaluate the function at all the candidate points and take the highest and lowest of such values to determine the absolute extremes.

Example 4.18. *Determine the extremes of the function:*

$$f(x) = \sqrt[3]{x^2(6-x)}$$

on the closed interval $[-6, 8]$.

Solution. Let's first see what the candidate points are:

(A) We observe that $f(x)$ is continuous on $[-6, 8]$. Therefore, there is no candidate point due to discontinuity.

(B) Since $f'(x) = \dfrac{4-x}{\sqrt[3]{x}\sqrt[3]{(6-x)^2}}$, the function is not differentiable at $x = 0$ and $x = 6$. Therefore, these are critical points. On the other hand, we observe that the derivative is zero if $x = 4$. This is, then, another critical point of the function.

(C) Finally, the extreme points of the closed interval where we want to optimize the function are $x = -6$ and $x = 8$.

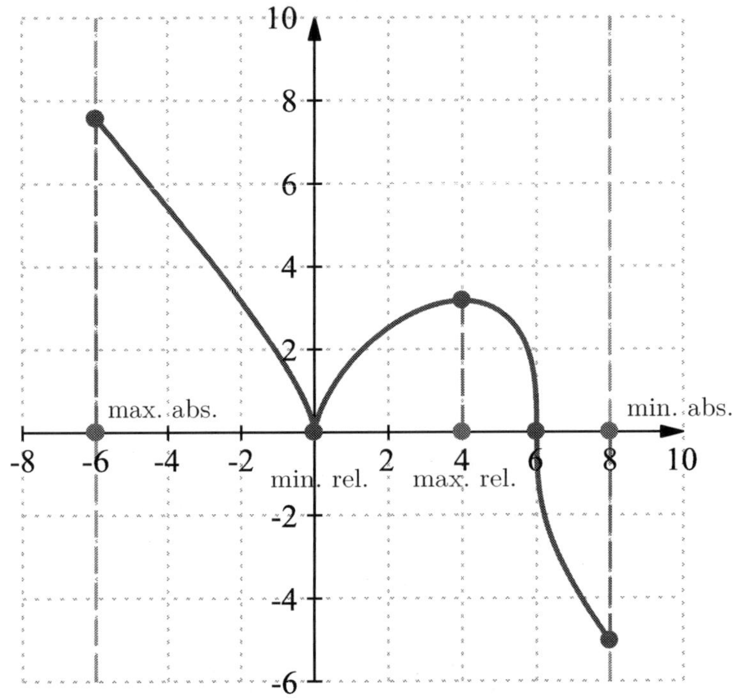

FIGURE 4.18
Graph of the function $f(x) = \sqrt[3]{x^2(6-x)}$ and location of its extremes.

Once we have the candidates, we evaluate the function at these points. *See* Figure 4.18.

$x = -6$	(interval end)	$f(-6) = 7.5595$	absolute maximum
$x = 0$	(critical point)	$f(0) = 0$	
$x = 4$	(critical point)	$f(4) = 3.1748$	
$x = 6$	(critical point)	$f(6) = 0$	
$x = 8$	(interval end)	$f(8) = -5.0397$	absolute minimum

□

For now, we only know how to determine absolute extremes. In the next section, we will see an important theorem that forms the basis for establishing the criterion that tells us whether the candidate points are or are not relative extremes and of what type.

4.11 Mean Value Theorem

The mean value theorem is one of the main theorems of calculus. It was first stated by the mathematician Joseph Louis Lagrange (1736-1813) and relates the values of a function to the values of its derivative. In this section, we establish this important theorem and see some of its applications.

Before considering the main result, let's see, below, a particular case: Rolle's theorem.

Theorem 4.1 (Rolle's). Let $f : [a,b] \to \mathbb{R}$ be a continuous function on $[a,b]$, differentiable on (a,b) and such that $f(a) = f(b)$. Then, there is at least one point $\xi \in (a,b)$, such that $f'(\xi) = 0$ (*see* Figure 4.19).

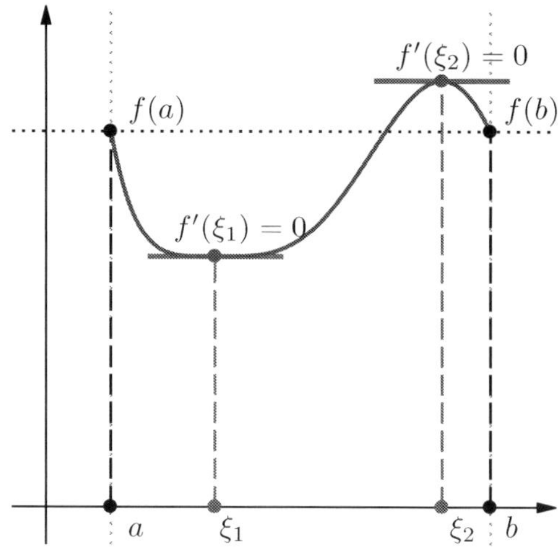

FIGURE 4.19
Geometric interpretation of Rolle's theorem.

Proof. If f is constant on the interval, then:

$$f'(x) = 0, \quad \forall x \in (a, b)$$

and, therefore, we can take any $\xi \in (a, b)$.

Suppose then that f is not constant. Since f is continuous on $[a, b]$, by Weierstrass' Theorem 3.2, there exist absolute maximum and minimum on $[a, b]$ which we will call x_{max} and x_{min} respectively. If x_{max} and x_{min} were a and b, since $f(a) = f(b)$, then f would have to be constant. Therefore, at least one of the extremes cannot be either a or b, that is, $\exists \xi \in (a, b)$ such that $\xi = x_{max}$ or $\xi = x_{min}$. Then, since f is differentiable on (a, b) and at point $\xi \in (a, b)$ there is an absolute maximum or minimum, by Fermat's theorem (*see* Proposition 4.4), $f'(\xi)$ must be zero, as we wanted to prove. □

Remark 4.4. The hypothesis of differentiability of the function f is essential as illustrated in Figure 4.20, where the function takes the same value at the ends of the interval, but there is no point with a zero derivative. Note, however, that there is a point where the function is not differentiable (cusp point) and that is why we cannot apply Rolle's theorem in this case.

Next, we generalize Rolle's theorem with one of the most significant results in calculus: the mean value theorem.

Theorem 4.2 (Mean Value)**.** Let $f : [a, b] \to \mathbb{R}$ be a continuous function on $[a, b]$ and **differentiable** on (a, b). Then there is at least one point $\xi \in (a, b)$ such that:

$$f'(\xi) = \frac{f(b) - f(a)}{b - a}.$$

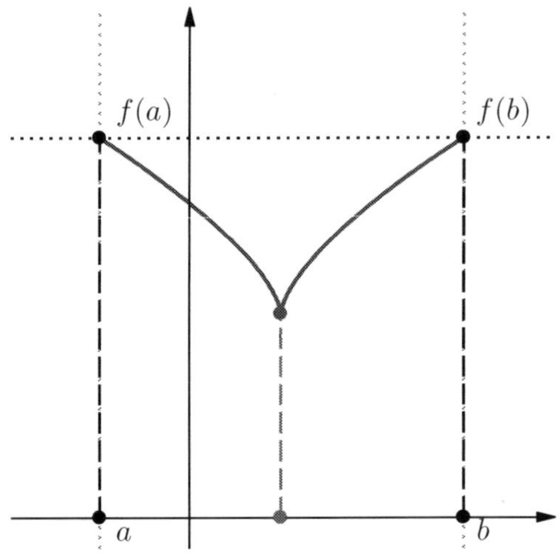

FIGURE 4.20
Importance of the hypothesis of differentiability of the function in Rolle's theorem.

Observation 4.3. A geometric interpretation of the mean value theorem (*see* Figure 4.21) is that there is at least one point $\xi \in (a, b)$, such that the tangent line to the graph of f at point $(\xi, f(\xi))$ is parallel to the line joining $(a, f(a))$ and $(b, f(b))$.

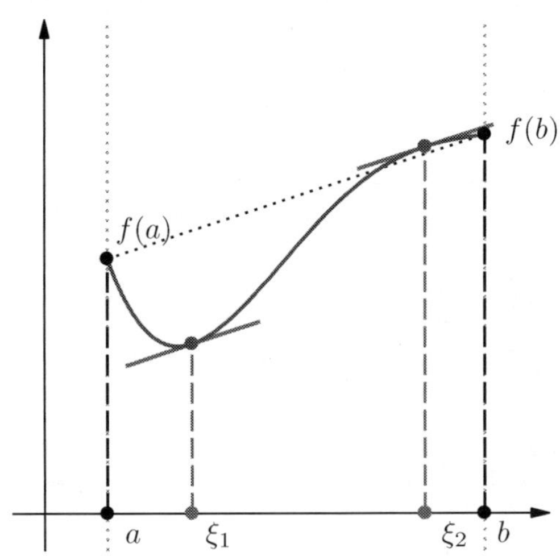

FIGURE 4.21
Geometric interpretation of the mean value theorem.

Proof. Consider the function:

$$\varphi(x) = (b - a)f(x) - (f(b) - f(a))x.$$

φ is continuous on $[a, b]$ and differentiable on (a, b) because the functions $f(x)$ and x are. Furthermore:

$$\varphi(a) = (b - a)f(a) - (f(b) - f(a))a = bf(a) - af(b)$$
$$\varphi(b) = (b - a)f(b) - (f(b) - f(a))b = -af(b) + bf(a)$$

that is, $\varphi(a) = \varphi(b)$. Then by Rolle's theorem, $\exists\ \xi \in (a, b),\ \varphi'(\xi) = 0$. If we differentiate $\varphi(x)$, we have:

$$\varphi'(x) = (b - a)f'(x) - (f(b) - f(a))$$

and evaluating at ξ:

$$0 = \varphi'(\xi) = (b - a)f'(\xi) - (f(b) - f(a)) \quad \Longrightarrow \quad f'(\xi) = \frac{f(b) - f(a)}{b - a}.$$

□

4.11.1 Monotone Functions and the First Derivative Test

Next, we will demonstrate that the sign of the derivative determines if a function $f(x)$ is increasing or decreasing. First, we need to define the concepts of increasing and decreasing in a formal manner.

Definition 4.14 (Monotone Functions)**.** Given $f : [a, b] \longrightarrow \mathbb{R}$, if for any $x_1, x_2 \in [a, b]$ with $x_1 < x_2$

- we have $f(x_1) \leq f(x_2)$, we say that f is *monotonically increasing* on $[a, b]$;
- we have $f(x_1) \geq f(x_2)$, we say that f is *monotonically decreasing* on $[a, b]$;
- we have $f(x_1) < f(x_2)$, we say that f is *strictly increasing* on $[a, b]$;
- we have $f(x_1) > f(x_2)$, we say that f is *strictly decreasing* on $[a, b]$.

Figure 4.22 shows some examples of monotonically increasing/decreasing and strictly increasing/decreasing functions.

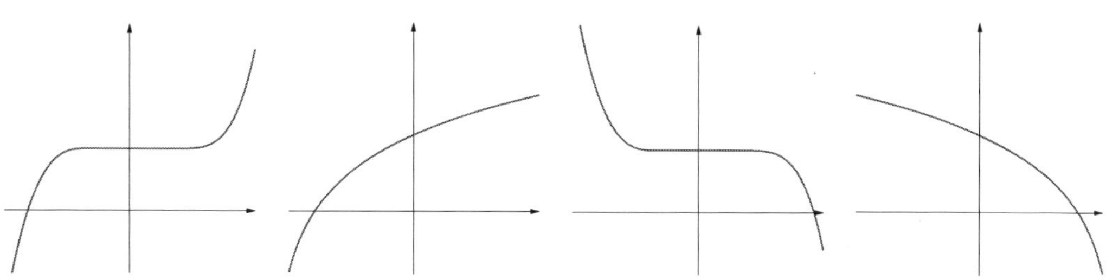

FIGURE 4.22
Examples of monotonically increasing/decreasing functions (first and third) and strictly increasing/decreasing functions (second and fourth).

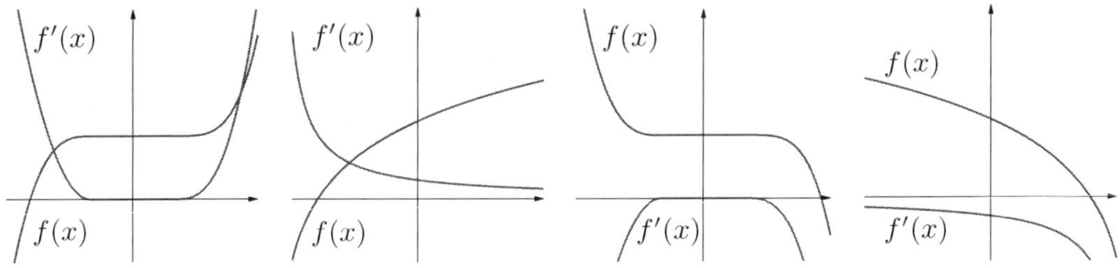

FIGURE 4.23
Examples of monotonically increasing/decreasing functions (first and third) and strictly increasing/decreasing functions (second and fourth) along with their derivatives.

Theorem 4.3. Let $f : [a,b] \subset \mathbb{R}$ be differentiable on (a,b).

(i) If $f'(x) \geq 0$ for all $x \in (a,b)$, then f is *monotonically increasing* on (a,b).

(ii) If $f'(x) \leq 0$ for all $x \in (a,b)$, then f is *monotonically decreasing* on (a,b).

(iii) If $f'(x) = 0$ for all $x \in (a,b)$, then f is *constant* on (a,b).

Figure 4.24 shows the relationship between the sign of the derivative and the increase/decrease of a function. We observe that when the function is strictly decreasing, the tangent lines have a negative slope; when the function is constant, the slope of the tangent line is zero; and, finally, when the function is increasing, the tangent lines have a positive slope.

Proof. (i) Suppose $f'(x) \geq 0$ for all $x \in (a,b)$. If x_1 and x_2 are two arbitrary points in (a,b) that satisfy $x_1 < x_2$, we can apply the mean value theorem (*see* Theorem 4.2) to f in the interval $[x_1, x_2]$ to obtain a point $\xi \in (x_1, x_2)$ such that:

$$f'(\xi) = \frac{f(x_2) - f(x_1)}{x_2 - x_1}.$$

Since $f'(\xi) \geq 0$ and $x_2 - x_1 > 0$, it follows that $f(x_2) - f(x_1) \geq 0$. Therefore, $f(x_1) \leq f(x_2)$ as desired.

(ii) The proof of this case is omitted, as it is analogous to the previous one.

(iii) Suppose $f'(x) = 0$ for all $x \in (a,b)$. If x_1 and x_2 are two arbitrary points in (a,b) that satisfy $x_1 < x_2$, we can apply the mean value theorem (*see* Theorem 4.2) to f in the interval $[x_1, x_2]$ to obtain a point $\xi \in (x_1, x_2)$ such that:

$$f'(\xi) = \frac{f(x_2) - f(x_1)}{x_2 - x_1}.$$

Since $f'(\xi) = 0$, it follows that $f(x_1) = f(x_2)$ as desired.

□

There is a very useful criterion to determine if a critical point of a function f is a maximum, a minimum, or neither. This criterion is often called the *first derivative test* and we will see it below.

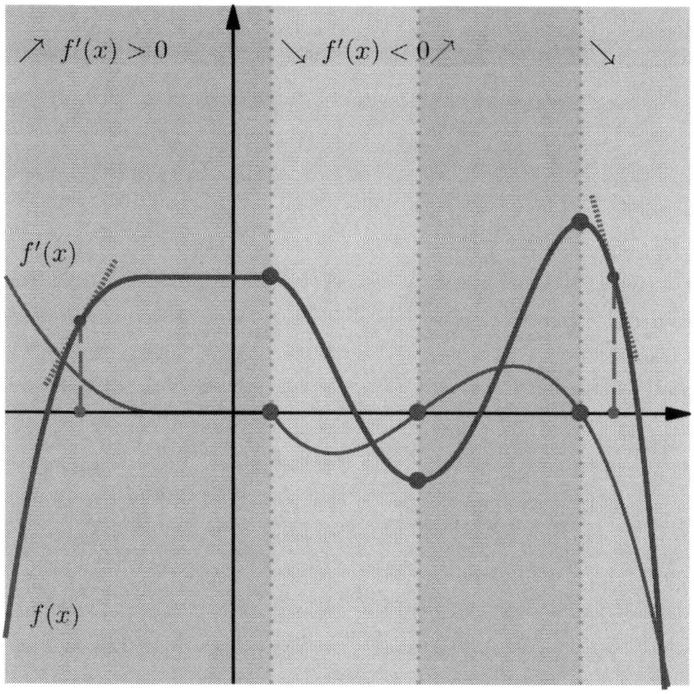

FIGURE 4.24
Relationship between the sign of the derivative and the increase/decrease of a function.

Theorem 4.4 (First Derivative Test for Characterizing Extrema). Given $f : [a, b] \subset \mathbb{R}$ continuous on $[a, b]$, and let x_0 be a critical point in the interior of $[a, b]$. Suppose f is differentiable in the open intervals (a, x_0) and (x_0, b). Then

- x_0 is a **relative maximum** if there exists $\delta > 0$ such that
  - $f'(x) > 0$, for all $x \in (x_0 - \delta, x_0)$
  - $f'(x) < 0$, for all $x \in (x_0, x_0 + \delta)$
- x_0 is a **relative minimum** if there exists $\delta > 0$ such that
  - $f'(x) < 0$, for all $x \in (x_0 - \delta, x_0)$
  - $f'(x) > 0$, for all $x \in (x_0, x_0 + \delta)$

Figure 4.25 illustrates the first derivative test for characterizing extrema.

To determine the extrema of a function on a closed interval, we must find the candidate points which, recall, are:

(A) points where f is not continuous,

(B) critical points,

(C) endpoints of the interval,

and then use the first derivative test (*see* Theorem 4.4) to determine, for each candidate, whether it is a relative maximum/minimum or not an extremum. Finally, we will evaluate

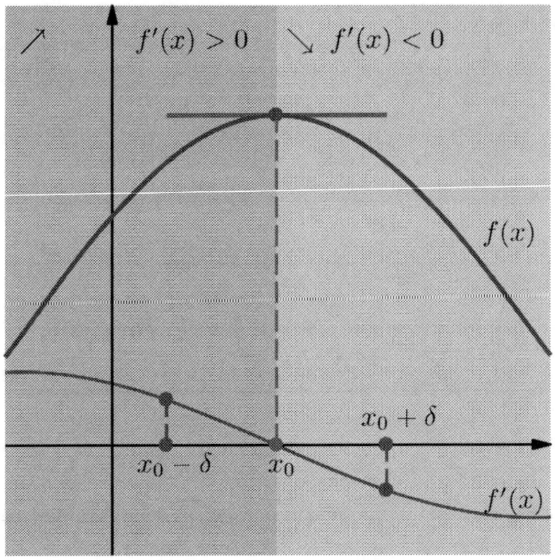

FIGURE 4.25
First derivative test for a relative maximum.

the function at all the extrema and take the highest and lowest of such values to determine the absolute extrema.

Example 4.19. *Determine the extrema of the function:*

$$f(x) = \frac{x^3}{(x+1)^2}$$

in $\mathbb{R}$.

SOLUTION. First, we look for the candidate points for extremum which are:

(A) Points where f is not continuous. In this example $f(x)$ is continuous in $\mathbb{R} - \{-1\}$ as it is the quotient of polynomials where the denominator does not vanish. At $x = -1$, the denominator vanishes and the function has a vertical asymptote:

$$\lim_{x \to -1} f(x) = \lim_{x \to -1} \frac{x^3}{(x+1)^2} = \frac{-1}{0^+} = -\infty.$$

(B) Critical points. We calculate the derivative of $f(x)$:

$$f'(x) = \frac{3x^2(x+1)^2 - x^3 2(x+1)}{(x+1)^4} = \frac{3x^2(x+1) - 2x^3}{(x+1)^3} = \frac{x^2(x+3)}{(x+1)^3}.$$

We see that $f(x)$ is not differentiable at $x = -1$ (we already knew this because f is not continuous at -1). At other points, the function is differentiable. On the other hand, the derivative function vanishes at $x = 0$ and $x = -3$.

(C) The endpoints of the interval. In this example, we are not asked to optimize in a closed interval but in the entire real line, $\mathbb{R}$. In this case, the equivalent is to study the behavior of the function as we tend to $\pm\infty$.

x_0	$f(x_0)$
-3	$f(-3) = -27/4 = -6.75$
-1	$\nexists f(-1)$ but $\lim_{x\to -1} f(x) = -\infty$
0	$f(0) = 0$
$-\infty$	$\lim_{x\to -\infty} f(x) = \lim_{x\to -\infty} \frac{x^3/x^2}{(x/x+1/x)^2} = \lim_{x\to -\infty} \frac{x}{(1+1/x)^2} = -\infty$
$+\infty$	$\lim_{x\to +\infty} f(x) = \lim_{x\to +\infty} \frac{x^3/x^2}{(x/x+1/x)^2} = \lim_{x\to +\infty} \frac{x}{(1+1/x)^2} = +\infty$

Once we have all the candidates, we make a table of values.

We observe that the function will not have absolute minimums since:

$$\lim_{x\to -1} f(x) = -\infty, \qquad \lim_{x\to -\infty} f(x) = -\infty.$$

On the other hand, it will not have absolute maximums either, since:

$$\lim_{x\to +\infty} f(x) = +\infty.$$

Finally, we aim to demonstrate if the candidate points are relative extrema or not using the first derivative test (*see* Theorem 4.4).

At $x = -3$, we have a relative maximum, since:

$$\begin{array}{lllll} \text{if } x < -3 & \Longrightarrow & f'(x) > 0 & \Longrightarrow & f(x) \text{ increasing} \\ \text{if } x > -3 & \Longrightarrow & f'(x) < 0 & \Longrightarrow & f(x) \text{ decreasing} \end{array}$$

At $x = -1$, there is no extremum, since $\lim_{x\to -1} f(x) = -\infty$.

At $x = 0$, there is no extremum, since:

$$\begin{array}{lllll} \text{if } -1 < x < 0 & \Longrightarrow & f'(x) > 0 & \Longrightarrow & f(x) \text{ increasing} \\ \text{if } x > 0 & \Longrightarrow & f'(x) > 0 & \Longrightarrow & f(x) \text{ increasing} \end{array}$$

If we observe Figure 4.26, we can see that, indeed, the only extremum the function has is at $x = -3$, where it has a relative maximum.

□

Proposed Exercise 4.6. Calculate the extrema of the function:

$$f(x) = \sqrt[3]{(x+1)^2} + \sqrt[3]{(x-1)^2},$$

whose graph is shown in Figure 4.27.

Example 4.20. *Calculate the absolute minimum of the function* $f(x, y) = x^2 + \sin(y) + 1$ *restricted to the points that satisfy:*

$$\{x - y^2 = 0\,,\ -1 \le x \le 1\,,\ -1 \le y \le 1\}$$

SOLUTION. We observe that this minimization problem can be written as:

$$\min \quad x^2 + \sin(y) + 1$$

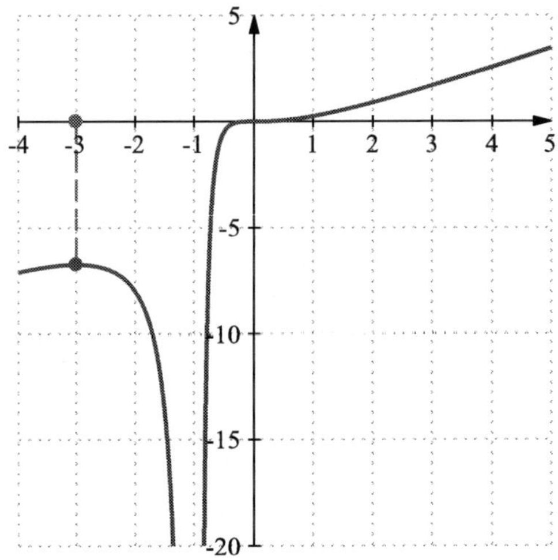

FIGURE 4.26
Graph of the function $f(x) = \dfrac{x^3}{(x+1)^2}$ and location of its extrema.

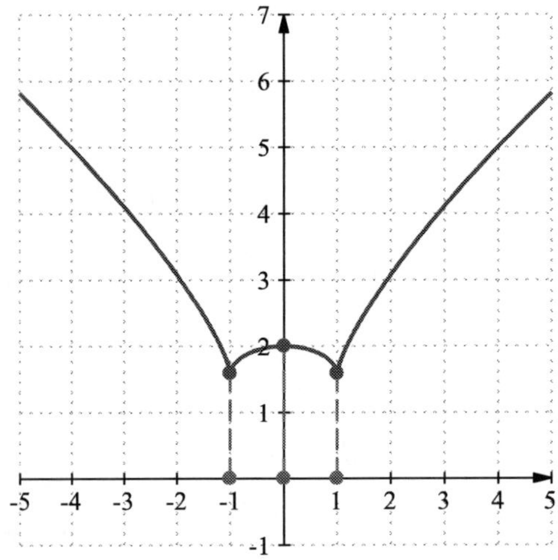

FIGURE 4.27
Graph of the function $f(x) = \sqrt[3]{(x+1)^2} + \sqrt[3]{(x-1)^2}$ and location of its extrema.

restricted to:

$$x - y^2 = 0$$
$$-1 \leq x \leq 1$$
$$-1 \leq y \leq 1$$

In Figure 4.28 we have represented the function to be optimized, as well as the constraints of the problem.

This is a problem of optimization in two variables and, therefore, we might think that it is beyond the scope of this book. However, there are some problems that can be easily transformed into a one-dimensional problem and, therefore, we can solve them using the tools we have seen previously. In this case, we will use that the constraint $x - y^2 = 0$ allows us to solve for x obtaining $x = y^2$. Thus, substituting x with y^2 we get that the

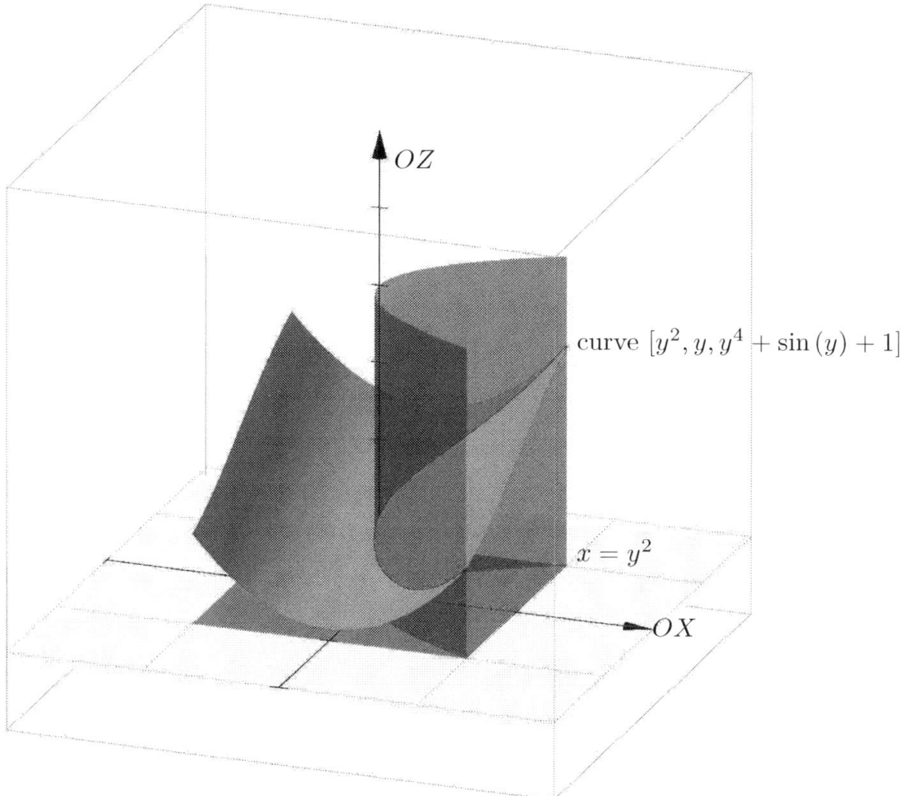

FIGURE 4.28
Graph of the function $f(x, y) = x^2 + \sin(y) + 1$ restricted to the points that satisfy $\{x - y^2 = 0\,,\ -1 \le x \le 1\,,\ -1 \le y \le 1\}$.

original two-dimensional problem is equivalent to solving the one-dimensional minimization problem:

$$\min_{\substack{-1 \le y^2 \le 1 \\ -1 \le y \le 1}} \quad y^4 + \sin(y) + 1$$

Notice, moreover, that the constraint $-1 \le y^2 \le 1$ is true if $y^2 \le 1 \Longrightarrow -1 \le y \le 1$ and, therefore, the above problem is equivalent to solving:

$$\min_{-1 \le y \le 1} \quad y^4 + \sin(y) + 1.$$

In Figure 4.29 we see represented the function $g(y) = y^4 + \sin(y) + 1$.

Let us study the extrema of the function $g(y) = y^4 + \sin(y) + 1$ in the closed interval $[-1, 1]$. First, we look for the candidate points for extremum which are:

(A) Points where g is not continuous. In this case, the function is continuous on $\mathbb{R}$.

(B) Critical points. The function g is differentiable on $\mathbb{R}$ being:

$$g'(y) = 4y^3 + \cos(y).$$

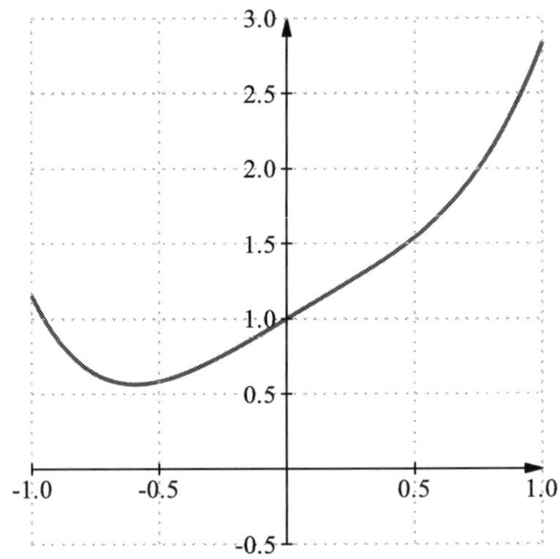

FIGURE 4.29
Graph of the function $g(y) = y^4 + \sin(y) + 1$.

We study the points where the derivative vanishes. In this case, we have a nonlinear equation and we resort to some algebraic computation system to obtain that $y \approx -0.59198$.

(C) The endpoints of the interval, which are $y = -1$ and $y = 1$.

Since we have a continuous function on the closed interval $[-1, 1]$, by the extreme value theorem we know that there is an absolute maximum and minimum. To determine them, we evaluate the function at the three candidates and construct the following table of values:

x_0	$g(x_0)$
-1	1.15853
-0.59198	0.56480
1	2.84147

We observe that the absolute maximum is reached at $x = 1$ and the absolute minimum at $x = -0.59198$. Thus, the solution to the problem is:

$$y \approx -0.59198 \quad \Longrightarrow \quad x = y^2 \approx 0.35045.$$

□

4.12 Higher Order Derivatives

Definition 4.15 (Second Derivative)**.** If $f : A \subset \mathbb{R} \to \mathbb{R}$ is differentiable in its domain of differentiability $\Omega \subset A$, since $f'(x)$ is a function defined on Ω we can consider its differentiability, that is, we consider whether for $a \in \Omega$ the limit exists:

$$\lim_{h \to 0} \frac{f'(a+h) - f'(a)}{h}.$$

If this limit exists and is finite, it is called the **second derivative** of f at point a and is denoted by $f''(a)$.

If the function $f' : \Omega \to \mathbb{R}$ is differentiable on $\Omega_1 \subset \Omega$ we can define the second derivative function:

$$\begin{aligned} f'' : \Omega_1 &\longrightarrow \mathbb{R} \\ x &\longmapsto f''(x) \end{aligned}$$

Example 4.21. *Consider the function $f(x) = \sin(x)$. Calculate $f'(x)$. If $f'(x)$ is differentiable, calculate $f''(x)$.*

SOLUTION.

$$f(x) = \sin(x) \Longrightarrow f'(x) = \cos(x)$$

$f'(x)$ is differentiable and, therefore, we can calculate the second derivative:

$$f''(x) = (\cos(x))' = -\sin(x)$$

□

Definition 4.16 (Higher Order Derivative)**.** If $k \in \mathbb{N}$, we define the k-th derivative of f at point a as:

$$f^{(k)}(a) = \lim_{h \to 0} \frac{f^{(k-1)}(a+h) - f^{(k-1)}(a)}{h}$$

if this limit exists and is finite. These derivatives, if they exist, are called **higher order derivatives** or successive derivatives of the function f.

Example 4.22. *Calculate the higher order derivatives of the function $f(x) = x^5$.*

SOLUTION.

- $f(x) = x^5$
- $f'(x) = 5x^4$
- $f''(x) = 20x^3$
- $f'''(x) = 60x^2$
- $f^{(4)}(x) = 120x$
- $f^{(5)}(x) = 120$
- $f^{(k)}(x) = 0, \quad \forall k \geq 6$

In Figure 4.30 we see represented the successive derivatives of the function $f(x) = x^5$.

□

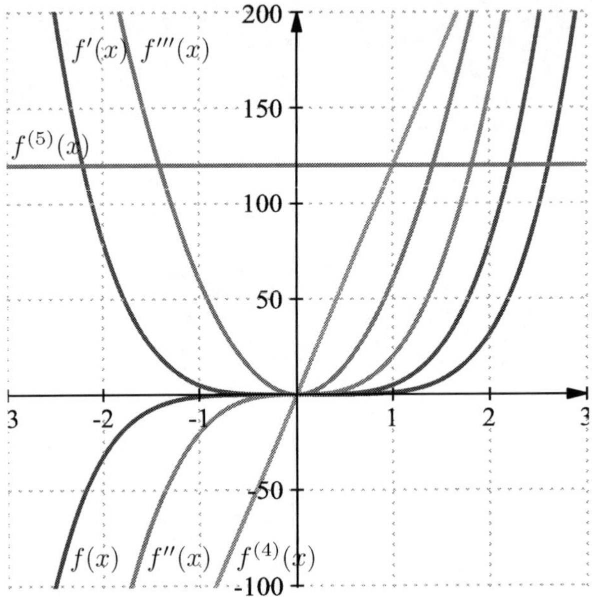

FIGURE 4.30
Successive derivatives of the function $f(x) = x^5$.

Example 4.23. *Calculate the higher order derivatives of the function* $g(x) = e^x$.

SOLUTION.

- $g(x) = e^x$
- $g'(x) = e^x$
- $g^{(k)}(x) = e^x, \quad \forall k \geq 2$

□

Example 4.24. *Calculate the higher order derivatives of the function* $h(x) = \sin(x)$.

SOLUTION.

- $h(x) = \sin(x)$
- $h'(x) = \cos(x)$
- $h''(x) = -\sin(x)$
- $h'''(x) = -\cos(x)$
- $h^{(4)}(x) = f(x) = \sin(x)$

- $h^{(k)}(x) = \begin{cases} \sin(x), & k = 4n,\ n \in \mathbb{N} \\ \cos(x), & k = 4n+1,\ n \in \mathbb{N} \\ -\sin(x), & k = 4n+2,\ n \in \mathbb{N} \\ -\cos(x), & k = 4n+3,\ n \in \mathbb{N} \end{cases}$

□

Definition 4.17 (Class of Differentiable Functions). If $f : A \subset \mathbb{R} \to \mathbb{R}$ is differentiable on $\Omega \subset A$ and its derivative f' is continuous on Ω we say that f belongs to the class of functions $\mathcal{C}^1(\Omega)$.

In general, we say that $f \in \mathcal{C}^k(\Omega)$ if we can differentiate it k times and the successive derivative functions are continuous on Ω, but the $k+1$ derivative function cannot be calculated or is not continuous.

If $f \in \mathcal{C}^k(\Omega)$ for all $k \in \mathbb{N}$, we say that f is an infinitely differentiable function and we represent it as $f \in \mathcal{C}^\infty(\Omega)$. That is, we say that f belongs to the set of infinitely differentiable functions on the set Ω.

Example 4.25. *Determine the value of k if* $f \in \mathcal{C}^k(\mathbb{R})$ *for the functions:*

(a) $f(x) = |x|$

(b) $g(x) = \begin{cases} x^3 + 3x^2 + 6x + 6, & x < 0 \\ 6e^x, & x \geq 0 \end{cases}$

(c) $h(x) = \begin{cases} -x^2, & x < 0 \\ x^3, & x \geq 0 \end{cases}$

SOLUTION.

(a) $f(x) = |x|$

$$f'(x) = \begin{cases} -1, & x < 0 \\ 1, & x > 0 \end{cases}$$

$f(x) = |x| \in \mathcal{C}^0(\Omega)$ because it is continuous, but not differentiable at $x = 0$. Note that the first derivative function is not defined for $x = 0$, as the lateral derivatives, in this case, exist but are different with $f'_+(0) = 1$ and $f'_-(0) = -1$. *See* Figure 4.31.

(b) $g(x) = \begin{cases} x^3 + 3x^2 + 6x + 6, & x < 0 \\ 6e^x, & x \geq 0 \end{cases}$

$$g'(x) = \begin{cases} 3x^2 + 6x + 6, & x < 0 \\ 6e^x, & x \geq 0 \end{cases}$$

$$g''(x) = \begin{cases} 6x + 6, & x < 0 \\ 6e^x, & x \geq 0 \end{cases}$$

$$g'''(x) = \begin{cases} 6, & x < 0 \\ 6e^x, & x \geq 0 \end{cases}$$

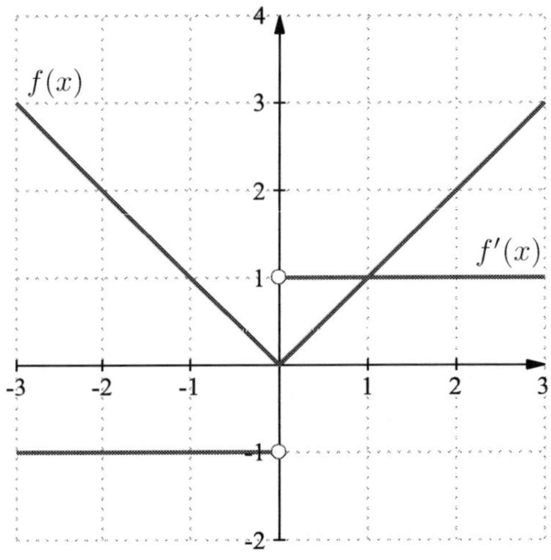

FIGURE 4.31
Graph of the absolute value function and its derivative.

$$g^{(4)}(x) = \begin{cases} 0, & x < 0 \\ 6e^x, & x > 0 \end{cases}$$

$g(x) \in \mathcal{C}^3(\mathbb{R})$ because the third derivative is continuous, but not differentiable at $x = 0$. Note that the fourth derivative function is not defined for $x = 0$, as the lateral derivatives, in this case, exist but are different with $g_+^{(4)}(0) = 6$ and $g_-^{(4)}(0) = 0$. *See* Figure 4.32.

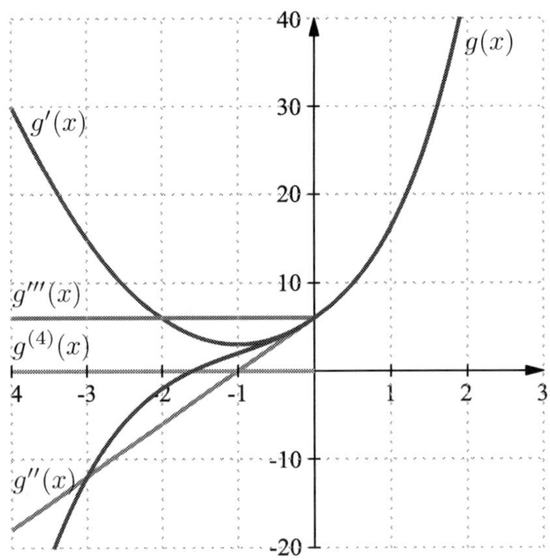

FIGURE 4.32
Graph of the function $g(x)$ and its derivatives.

(c) $h(x) = \begin{cases} -x^2, & x < 0 \\ x^3, & x \geq 0 \end{cases}$

smallskip

$$h'(x) = \begin{cases} -2x, & x < 0 \\ 3x^2, & x \geq 0 \end{cases}$$

$$h''(x) = \begin{cases} -2, & x < 0 \\ 6x, & x > 0 \end{cases}$$

$h(x) \in \mathcal{C}^1(\mathbb{R})$ because the first derivative function is continuous, but not differentiable at $x = 0$. Note that the second derivative function is not defined for $x = 0$, as the lateral derivatives, in this case, exist but are different, with $h''_+(0) = 0$ and $h''_-(0) = -2$. *See* Figure 4.33. □

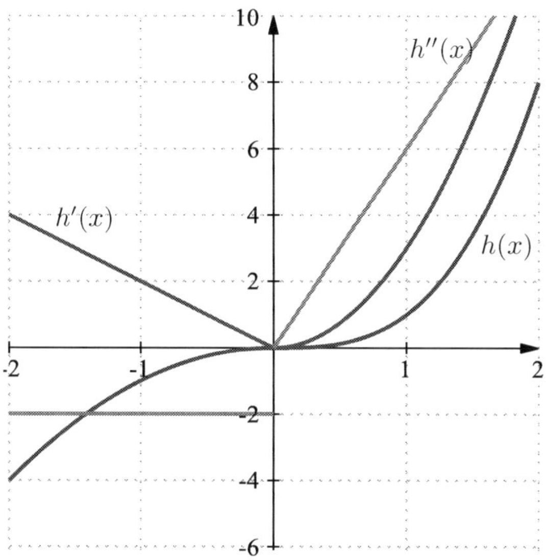

FIGURE 4.33
Graph of the function $h(x)$ and its derivatives.

Example 4.26. *Determine the value of k if $f \in \mathcal{C}^k(\mathbb{R})$ for:*

$$f(x) = \begin{cases} x^3 \sin\left(\frac{1}{x}\right), & \textit{if } x \neq 0 \\ 0, & \textit{if } x = 0. \end{cases}$$

Solution. Let us start by calculating the first derivative. It is evident that:

$$f'(x) = 3x^2 \sin\left(\frac{1}{x}\right) - x\cos\left(\frac{1}{x}\right), \quad \text{if } x \neq 0.$$

Now we must study what happens at $x = 0$. For this, we will use the lateral derivatives. Recall that the right-hand derivative of the function f at 0 is given by the limit:

$$f'_+(0) = \lim_{h\to 0^+} \frac{f(0+h) - f(0)}{h} = \lim_{h\to 0^+} \frac{h^3 \sin\left(\frac{1}{h}\right)}{h} = \text{«}0\cdot \text{ bounded»} = 0,$$

and the left-hand derivative:

$$f'_-(0) = \lim_{h\to 0^-} \frac{f(0+h) - f(0)}{h} = \lim_{h\to 0^-} \frac{h^3 \sin\left(\frac{1}{h}\right)}{h} = \text{«}0\cdot \text{ bounded»} = 0.$$

Thus, since the lateral derivatives exist and are equal, we can assert that:

$$f'(0) = 0.$$

We observe that the first derivative function is continuous because at $x = 0$ we have:

$$\lim_{x \to 0} f'(x) = f'(0),$$

and at other points, the function is continuous as it is the product and difference of continuous functions.

Now let us calculate the second derivative. On one hand:

$$f''(x) = 6x \sin\left(\frac{1}{x}\right) - 4\cos\left(\frac{1}{x}\right) - \frac{\sin\left(\frac{1}{x}\right)}{x}, \quad \text{if } x \neq 0.$$

Now we must study what happens at $x = 0$. For this, we will use the lateral derivatives. Recall that the right-hand derivative of the function f' at 0 is given by the limit:

$$\begin{aligned} f''_{+}(0) &= \lim_{h \to 0^+} \frac{f'(0+h) - f'(0)}{h} = \lim_{h \to 0^+} \frac{3h^2 \sin\left(\frac{1}{h}\right) - h\cos\left(\frac{1}{h}\right)}{h} \\ &= \underbrace{\lim_{h \to 0^+} 3h \sin\left(\frac{1}{h}\right)}_{\text{«0· bounded»} = 0} - \underbrace{\lim_{h \to 0^+} \cos\left(\frac{1}{h}\right)}_{\nexists} = \nexists \end{aligned}$$

As one of the lateral derivatives does not exist, we can assert that the second derivative does not exist at point $x = 0$. Thus, since the function f is differentiable only once and this first derivative is a continuous function, we can assert that $f \in \mathcal{C}^1(\mathbb{R})$. □

4.13 Taylor's Theorem

A highly useful technique in the study of real functions is the approximation of functions by means of polynomials. In this section, a fundamental theorem of this area, dating back to Brook Taylor (1685-1731), is demonstrated. Taylor's theorem has multiple applications, and in this book, by way of example, the following are illustrated: numerical approximation of the value of a function at a point and estimation of the error committed; characterization of relative extrema; and calculation of limits at the origin.

Taylor's theorem can be considered as an extension of the mean value theorem for higher-order derivatives. While the mean value theorem relates the values of a function to its derivative, Taylor's theorem provides a relationship between the values of a function and its higher-order derivatives.

If a function $f(x)$ has a derivative of order n at a point x_0, we can construct a polynomial $p_n(x)$ of degree n such that:

$$\begin{aligned} p_n(x_0) &= f(x_0) \\ p_n^{(k)}(x_0) &= f^{(k)}(x_0), \quad k = 1, 2, \ldots, n. \end{aligned}$$

In fact, the polynomial:

$$p_n(x) = f(x_0) + f'(x_0)(x - x_0) + \frac{f''(x_0)}{2!}(x - x_0)^2 + \cdots + \frac{f^{(n)}(x_0)}{n!}(x - x_0)^n,$$

has the property that it and its derivatives up to order n coincide with the function f and its derivatives up to order n at the specific point x_0. This polynomial p_n is called the *n-th Taylor polynomial of f at x_0*. It is natural to expect that this polynomial provides a reasonable approximation of f at points close to x_0. In order to measure the quality of the approximation, it is necessary to have information about the *residual* R_n which is defined as the difference between the function f and the polynomial p_n. The following theorem provides this information.

Theorem 4.5 (Taylor's). Let $n \in \mathbb{N}$, and let $f : [a, b] \subset \mathbb{R} \to \mathbb{R}$ be such that $f(x) \in C^n(a, b)$ and such that $f^{(n+1)}$ exists in the open interval (a, b). If $x_0 \in [a, b]$, then for any $x \in [a, b]$ there exists a point $\xi \in (a, b)$ such that:

$$f(x) = f(x_0) + f'(x_0)(x - x_0) + \frac{f''(x_0)}{2!}(x - x_0)^2 + \cdots$$

$$+ \frac{f^{(n)}(x_0)}{n!}(x - x_0)^n + \frac{f^{(n+1)}(\xi)}{(n+1)!}(x - x_0)^{n+1}.$$

Proof. Given x and x_0. Let J be the closed interval with endpoints x and x_0. We then define the function F on J by:

$$F(t) = f(x) - f(t) - (x - t)f'(t) - \cdots - \frac{(x - t)^n}{n!} f^{(n)}(t)$$

for $t \in J$. Then by differentiating we have:

$$F'(t) = -\frac{(x - t)^n}{n!} f^{(n+1)}(t).$$

We now define the function G on J by:

$$G(t) = F(t) - \left(\frac{x - t}{x - x_0}\right)^{n+1} F(x_0)$$

for $t \in J$. Then $G(x_0) = G(x) = 0$. Applying Rolle's theorem (*see* Theorem 4.1) to the function G, we obtain that there exists a point ξ between x and x_0 such that:

$$0 = G'(\xi) = F'(\xi) + (n + 1)\frac{(x - \xi)^n}{(x - x_0)^{n+1}} F(x_0).$$

From which we obtain:

$$F(x_0) = -\frac{1}{n + 1} \frac{(x - x_0)^{n+1}}{(x - \xi)^n} F'(\xi)$$

and substituting the expression for F':

$$F(x_0) = \frac{1}{n + 1} \frac{(x - x_0)^{n+1}}{(x - \xi)^n} \frac{(x - \xi)^n}{n!} f^{(n+1)}(\xi) = \frac{f^{(n+1)}(\xi)}{(n + 1)!}(x - x_0)^{n+1}.$$

which is the stated result. □

We will use the notation p_n for the n-th Taylor polynomial of f, and R_n for the *residual* (also called the complementary term). Thus, the conclusion of Taylor's theorem can be written as $f(x) = p_n(x) + R_n(x)$, where R_n is given by:

$$R_n(x) = \frac{f^{(n+1)}(\xi)}{(n + 1)!}(x - x_0)^{n+1}$$

for some point ξ between x and x_0. In other words, $R_n(x)$ is the error committed when approximating the function by its n-th degree Taylor polynomial in the neighborhood of the point x_0. In Figure 4.34, the function e^x and its Taylor polynomials of degrees 1, 2, and 3 are shown as an example, as well as the residual (or error) of each of them.

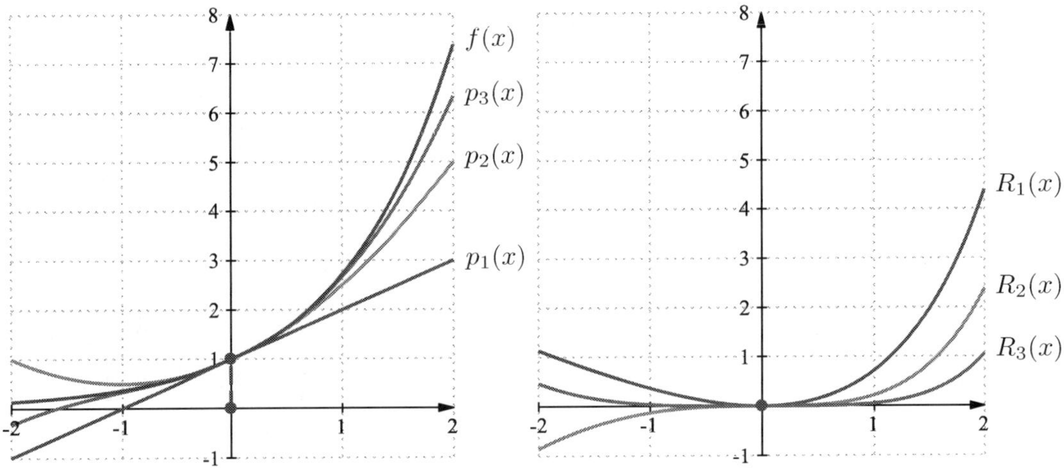

FIGURE 4.34
Graph of the function e^x and its Taylor polynomials centered at the origin $p_1(x) = 1 + x$, $p_2(x) = 1 + x + \dfrac{x^2}{2}$, and $p_3(x) = 1 + x + \dfrac{x^2}{2} + \dfrac{x^3}{6}$ (left). Graph of the error committed when approximating the function by each of the Taylor polynomials: $R_1(x) = f(x) - p_1(x)$, $R_2(x) = f(x) - p_2(x)$, $R_3(x) = f(x) - p_3(x)$ (right).

Observation 4.4. It is important to observe that the point ξ where the (n+1)-th derivative is evaluated is different for each point x where the error is to be evaluated. Therefore, if $f(x) \in C^n(a,b)$ and if $x_0 \in (a,b)$, for all $x \in (a,b)$ there exists a $\xi \in (a,b)$ such that:

$$f(x) = f(x_0) + f'(x_0)(x - x_0) + \frac{f''(x_0)}{2!}(x - x_0)^2 + \cdots$$

$$+ \frac{f^{(n)}(x_0)}{n!}(x - x_0)^n + \underbrace{\frac{f^{(n+1)}(\xi)}{(n+1)!}(x - x_0)^{n+1}}_{R_n(x)}$$

$$= \sum_{k=0}^{n} \frac{f^{(k)}(x_0)}{k!}(x - x_0)^k + R_n(x)$$

Observation 4.5. Note that:

- $p_n^{(k)}(x_0) = f^{(k)}(x_0), \quad k \leq n$
- $p_n^{(k)}(x_0) = 0, \quad k > n$

Observation 4.6. The first non-zero term of the polynomial p_n is called the *principal term.*

Observation 4.7. The Taylor series using the development at the point $x_0 = 0$ is often called the Maclaurin series.

Example 4.27. *Consider the function $f(x) = e^x$. Calculate the 5th degree Taylor polynomial at $x = 0$.*

SOLUTION. First, we must calculate the derivatives at $x = 0$:

$$f^{(i)}(x) = e^x, \ \forall i \in \mathbb{N} \quad \Longrightarrow \quad f^{(i)}(0) = e^0 = 1, \ \forall i \in \mathbb{N}$$

Thus, we can construct the Taylor polynomial as:

$$\begin{aligned} p_5(x) &= f(0) + f'(0)x + \frac{f''(0)}{2!}x^2 + \frac{f'''(0)}{3!}x^3 + \frac{f^{(4)}(0)}{4!}x^4 + \frac{f^{(5)}(0)}{5!}x^5 \\ &= 1 + x + \frac{1}{2}x^2 + \frac{1}{6}x^3 + \frac{1}{24}x^4 + \frac{1}{120}x^5 \end{aligned}$$

□

The Taylor polynomial allows us to have a polynomial approximation of a function around a point $x = x_0$. The larger the degree of the Taylor polynomial, the larger the neighborhood of the point x_0 where the function and the polynomial will resemble each other. This is illustrated in the following example.

Example 4.28. *Using the 5th degree Taylor polynomial at $x = 0$ of the function $f(x) = e^x$ calculated in Example 4.27, find an approximation of the values of $e^{i/10}$, $i = 0, 1, \ldots, 10$.*

SOLUTION.

TABLE 4.3
Approximations and associated error.

i	$x = e^{i/10}$	$p_5(x)$	e^x	$e^x - p_5(x)$
0	e^0	1	1	0
1	$e^{0.1}$	1.105170917	1.105170918	10^{-9}
2	$e^{0.2}$	1.221402667	1.221402758	$0.91 \cdot 10^{-7}$
3	$e^{0.3}$	1.349857750	1.349858808	$0.11 \cdot 10^{-5}$
4	$e^{0.4}$	1.491818667	1.491824698	$0.60 \cdot 10^{-5}$
5	$e^{0.5}$	1.648697917	1.648721271	$0.23 \cdot 10^{-4}$
6	$e^{0.6}$	1.822048000	1.822118800	$0.71 \cdot 10^{-4}$
7	$e^{0.7}$	2.013571417	2.013752707	$0.18 \cdot 10^{-3}$
8	$e^{0.8}$	2.225130667	2.225540928	$0.41 \cdot 10^{-3}$
9	$e^{0.9}$	2.458758250	2.459603111	$0.84 \cdot 10^{-3}$
10	e^1	2.716666667	2.718281828	$0.16 \cdot 10^{-2}$

As illustrated in Table 4.3 and in Figure 4.35, as we move away from the point $x = 0$, $p_5(x)$ approximates e^x worse. □

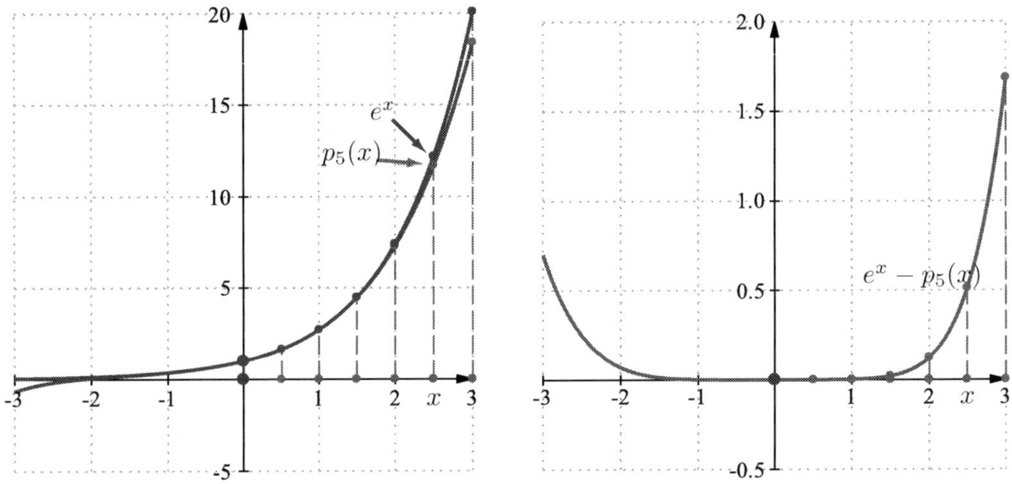

FIGURE 4.35
Graph of the function e^x and its 5th degree Taylor polynomial centered at the origin (left) and the residual (right).

Example 4.29. *Calculate the Taylor polynomials of degrees* 1, 2 *and* 3 *of the function* $f(x) = \sin(x)$ *at* $x = 1$.

SOLUTION.

$$
\begin{aligned}
p_1(x) &= \sin(1) + \cos(1)(x-1) \\
p_2(x) &= \sin(1) + \cos(1)(x-1) - \frac{1}{2}\sin(1)(x-1)^2 \\
p_3(x) &= \sin(1) + \cos(1)(x-1) - \frac{1}{2}\sin(1)(x-1)^2 - \frac{1}{6}\cos(1)(x-1)^3
\end{aligned}
$$

In Figure 4.36, the function is shown along with the three Taylor polynomials calculated in this example. It is observed that, as expected, the higher the degree of the Taylor polynomial, the better we approximate the function. □

Example 4.30. *Consider the function* $f(x) = x\sin(x)$. *Determine the Taylor polynomials of* f *up to degree* 10 *at the point* $x = 0$.

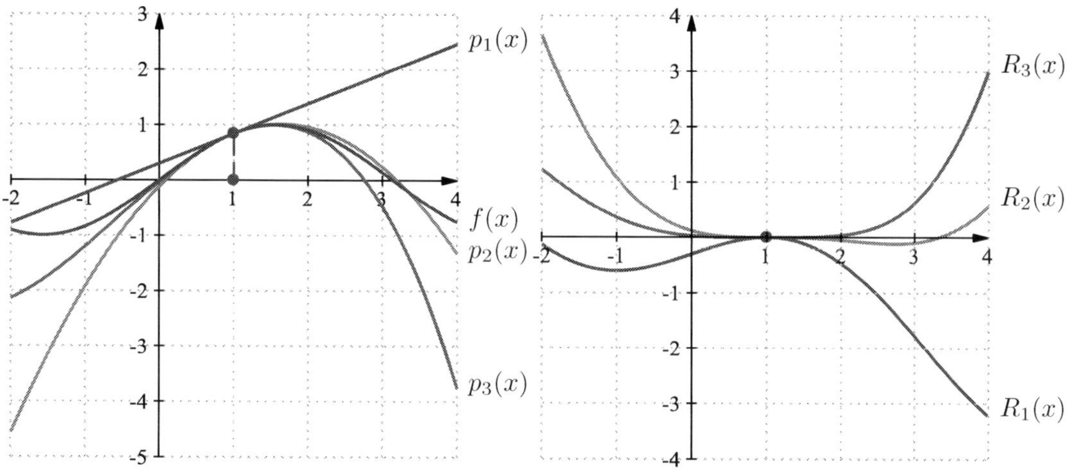

FIGURE 4.36
Graph of the function $\sin(x)$ and its 1st, 2nd, and 3rd degree Taylor polynomials centered at $x = 1$ (left) and the associated residuals (right).

SOLUTION. We first calculate the derivatives at $x = 0$ up to order 10.

$$\begin{aligned}
f(x) &= x\sin(x), & \Longrightarrow && f(0) &= 0\\
f'(x) &= \sin(x) + x\cos(x) & \Longrightarrow && f'(0) &= 0\\
f''(x) &= 2\cos(x) - x\sin(x) & \Longrightarrow && f''(0) &= 2\\
f^{(3)}(x) &= -3\sin(x) - x\cos(x) & \Longrightarrow && f^{(3)}(0) &= 0\\
f^{(4)}(x) &= -4\cos(x) + x\sin(x) & \Longrightarrow && f^{(4)}(0) &= -4\\
f^{(5)}(x) &= 5\sin(x) + x\cos(x) & \Longrightarrow && f^{(5)}(0) &= 0\\
f^{(6)}(x) &= 6\cos(x) - x\sin(x) & \Longrightarrow && f^{(6)}(0) &= 6\\
f^{(7)}(x) &= -7\sin(x) - x\cos(x) & \Longrightarrow && f^{(7)}(0) &= 0\\
f^{(8)}(x) &= -8\cos(x) + x\sin(x) & \Longrightarrow && f^{(8)}(0) &= -8\\
f^{(9)}(x) &= 9\sin(x) + x\cos(x) & \Longrightarrow && f^{(9)}(0) &= 0\\
f^{(10)}(x) &= 10\cos(x) - x\sin(x) & \Longrightarrow && f^{(10)}(0) &= 10
\end{aligned}$$

Therefore, we have:

$$\begin{aligned}
p_0(x) =& p_1(x) = 0 \\
p_2(x) =& p_3(x) = \frac{2}{2!}x^2 = x^2 \\
p_4(x) =& p_5(x) = x^2 + \frac{-4}{4!}x^4 = x^2 - \frac{1}{6}x^4 \\
p_6(x) =& p_7(x) = x^2 - \frac{1}{6}x^4 + \frac{6}{6!}x^6 = x^2 - \frac{1}{6}x^4 + \frac{1}{120}x^6 \\
p_8(x) =& p_9(x) = x^2 - \frac{1}{6}x^4 + \frac{1}{120}x^6 - \frac{1}{5040}x^8 \\
p_{10}(x) =& x^2 - \frac{1}{6}x^4 + \frac{1}{120}x^6 - \frac{1}{5040}x^8 + \frac{1}{362880}x^{10}
\end{aligned}$$

Figure 4.37 compares the function $f(x) = x\sin(x)$ with its respective calculated Taylor polynomials. As expected, the higher the degree of the polynomial, the better the approximation. □

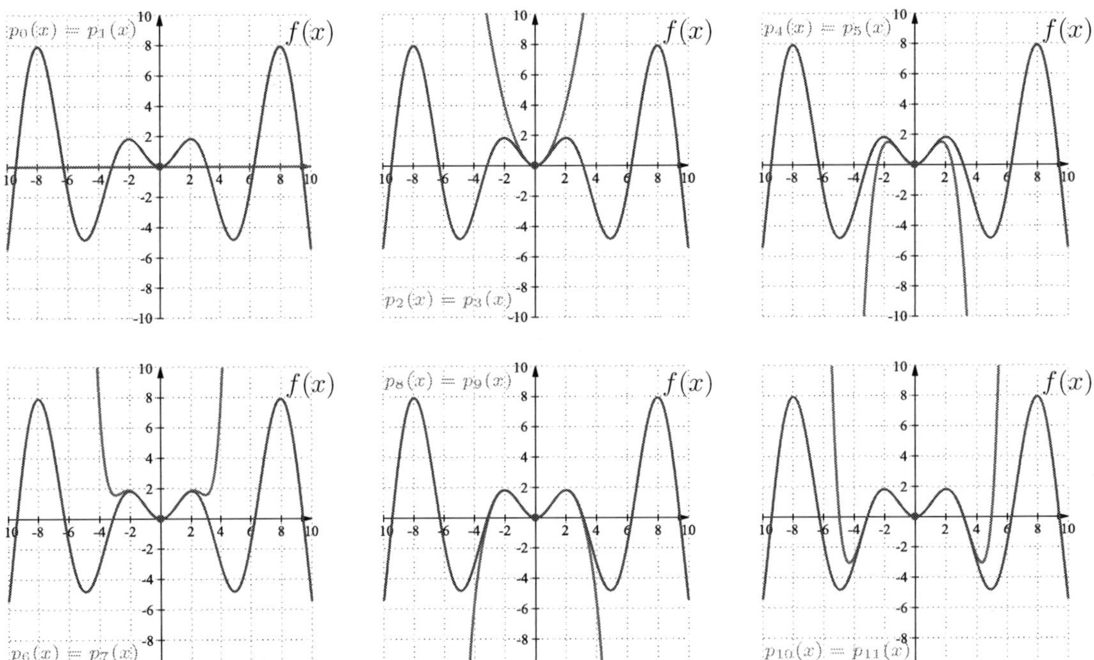

FIGURE 4.37
Graphs of the Taylor polynomials of the function $f(x) = x\sin(x)$ at $x = 0$ for different degrees of the polynomial.

4.13.1 Applications of the Taylor Series

The term corresponding to the residual R_n in Taylor's theorem can be used to estimate the error when approximating the value of a function at a certain point by its Taylor polynomial

p_n evaluated at that point. There are two possible cases. First, if the number n is fixed in advance, then the aim is to estimate the error made in the approximation. The second, if a certain desired precision in the approximation is specified, then the objective is to find the value of n that would provide such precision. In this section, we will see how to solve these two cases.

Another application of Taylor's theorem is the characterization of relative extrema. Recall that one way to determine if f has a maximum or minimum relative (or neither) at a point is to use the first derivative criterion (*see* Theorem 4.4). Here we will see that higher-order derivatives, if they exist, can also be used in this determination.

Finally, some of the most powerful limit calculation techniques we have studied are based on the properties of differentiable functions. Specifically, Taylor series allow us to know if two functions are locally equivalent at a point, and we can deduce the infinitesimals we have seen in Section 3.11.1. In this section, we will also see this useful application of Taylor series.

Applications of the Taylor series: residual estimation

Example 4.31. *Calculate the 6th-degree Taylor polynomial of the function* $f(x) = \cos(x)$ *at* $x = 0$ *and bound the remainder.*

SOLUTION. Given that:

$$\begin{aligned} f(0) &= \cos(0) = 1, & f'(0) &= -\sin(0) = 0 \\ f''(0) &= -\cos(0) = -1, & f'''(0) &= \sin(0) = 0 \\ f^{(4)}(0) &= f(0) = 1, & f^{(5)}(0) &= f'(0) = 0 \\ f^{(6)}(0) &= f''(0) = -1, & f^{(7)}(x) &= f'''(x) = \sin(x) \end{aligned}$$

Then:

$$p_6(x) = 1 - \frac{1}{2}x^2 + \frac{1}{24}x^4 - \frac{1}{720}x^6$$

Furthermore, the absolute value of the remainder is given by:

$$|R_6(x)| = |f(x) - p_6(x)| = \left| \frac{1}{5040} \sin(\xi) x^7 \right| \le \frac{|x|^7}{5040}$$

Figure 4.38 shows the function and its Taylor polynomial of degree 6, while the following table shows the error bound and the exact error for different values. □

x	$\cos(x)$	$p_6(x)$	$\lvert x\rvert^7/5040$	$\lvert R_6(x)\rvert$
0	1	1	0	0
1	0.54030	0.54028	0.00020	$0.24 \cdot 10^{-4}$
2	−0.41615	−0.42222	0.02540	0.00608

Example 4.32. *Calculate* $\cos(1)$ *with an error less than* 0.001 *using the Taylor formula.*

SOLUTION. To approximate $\cos(1)$ we use the Taylor polynomial of the function $\cos(x)$ at the point $x_0 = 0$. The degree of the polynomial n is determined by the required precision: in this case, we are asked for the error of the approximation to be less than 0.001, that is,

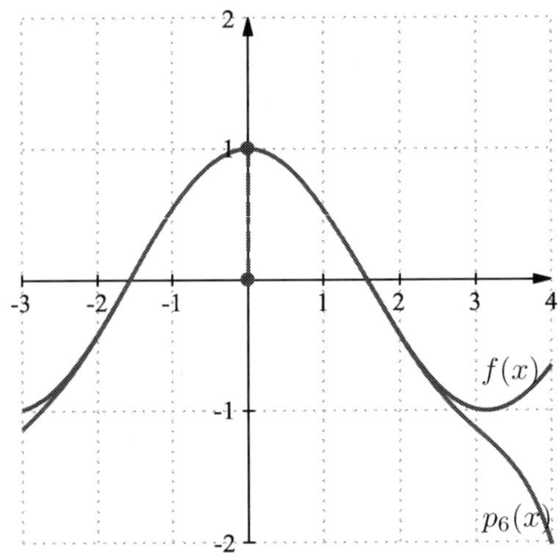

FIGURE 4.38
Graph of the function $f(x) = \cos(x)$ and its Taylor polynomial at $x = 0$ of degree 6.

0.001 must be a strict upper bound of the remainder associated with the polynomial of degree n evaluated at 1:

$$|R_n(1)| = \left|\tfrac{f^{(n+1)}(\xi)}{(n+1)!}1^{n+1}\right| < 0.001, \quad \xi \in (0,1)$$

We know that the $(n+1)$-th derivative of the cosine function can be either $\pm\sin$ or $\pm\cos$, depending on the value of n. Given that $\xi \in (0,1)$, we know that:

$$|f^{(n+1)}(\xi)| < 1$$

and, therefore:

$$|R_n(1)| < \tfrac{1}{(n+1)!} < 0.001 \quad \Leftrightarrow \quad (n+1)! > 1000$$

Thus, taking $n = 6$ we have $(n+1)! = 7! = 5040 > 1000$. We must then construct the Taylor polynomial of degree 6 of $f(x) = \cos(x)$ at the point $x_0 = 0$, that is:

$$\begin{aligned} p_6(x) &= f(0) + f'(0)x + \frac{f''(0)}{2}x^2 + \frac{f^{(3)}(0)}{3!}x^3 + \\ &\quad \frac{f^{(4)}(0)}{4!}x^4 + \frac{f^{(5)}(0)}{5!} + \frac{f^{(6)}(0)}{6!}x^6. \end{aligned}$$

We have:

$$\begin{array}{ll} f(0) = \cos(0) = 1 & f^{(4)} = \cos(0) = 1 \\ f'(0) = -\sin(0) = 0 & f^{(5)} = -\sin(0) = 0 \\ f''(0) = -\cos(0) = -1 & f^{(6)} = -\cos(0) = -1 \\ f^{(3)}(0) = \sin(0) = 0 & \end{array}$$

and thus:

$$p_6(x) = 1 - \frac{x^2}{2} + \frac{x^4}{24} - \frac{x^6}{720}$$

and, therefore:

$$\cos(1) \approx p_6(1) = 1 - \frac{1}{2} + \frac{1}{24} - \frac{1}{720} = \frac{389}{720} \approx 0.5402777778.$$

□

Applications of the Taylor Series: Characterization of Relative Extrema

Another application of Taylor's theorem is the characterization of relative extrema. We know that if a function $f : [a,b] \subset \mathbb{R} \to \mathbb{R}$ is differentiable at an interior point c of the interval $[a,b]$. Then, a necessary condition for f to have a relative extremum at c is that $f'(c) = 0$. Recall that one way to determine if f has a relative maximum or minimum (or neither) at c is to use the first derivative test (*see* Theorem 4.4). Here we will see that higher order derivatives, if they exist, can also be used in this determination.

Theorem 4.6 (Higher Order Derivative Test). Let $f : [a,b] \subset \mathbb{R} \to \mathbb{R}$ be a function such that $f(x) \in \mathcal{C}^n(a,b)$ with $n \geq 2$ and let $x_0 \in (a,b)$ such that $f'(x_0) = \cdots = f^{n-1}(x_0) = 0$, but $f^{(n)}(x_0) \neq 0$.

(a) If n is even and $f^{(n)}(x_0) > 0$, then f has a relative minimum at x_0.

(b) If n is even and $f^{(n)}(x_0) < 0$, then f has a relative maximum at x_0.

(c) If n is odd, then f has neither a relative minimum nor a relative maximum at x_0.

Proof. We use Taylor's theorem at x_0 and obtain that for $x \in [a,b]$:

$$f(x) = p_{n-1}(x) + R_{n-1}(x) = f(x_0) + \frac{f^{(n)}(\xi)}{n!}(x - x_0)^n,$$

where ξ is some point between x_0 and x. Since $f^{(n)}$ is continuous, if $f^{(n)}(x_0) \neq 0$, then there exists an interval J containing x_0 such that $f^{(n)}(x)$ will have the same sign as $f^{(n)}(x_0)$ for $x \in J$. If we take $x \in J$ then the point ξ also belongs to J and, therefore, $f^{(n)}(\xi)$ and $f^{(n)}(x_0)$ will have the same sign. Then:

(a) If n is even and $f^{(n)}(x_0) > 0$, then for $x \in J$ we have $f^{(n)}(\xi) > 0$ and $(x - x_0)^n \geq 0$ so that $R_{n-1}(x) \geq 0$. Therefore, $f(x) \geq f(x_0)$ for $x \in J$ and, consequently, f has a relative minimum at x_0.

(b) If n is even and $f^{(n)}(x_0) < 0$, then similarly to the previous case we obtain that $R_{n-1}(x) \leq 0$. Therefore, $f(x) \leq f(x_0)$ for $x \in J$ and, consequently, f has a relative maximum at x_0.

(c) If n is odd, then $(x - x_0)^n$ is strictly positive if $x > x_0$ but strictly negative if $x < x_0$. Therefore, if $x \in J$ then $R_{n-1}(x)$ will have opposite signs to the left and right of x_0. Consequently, f has neither a relative minimum nor a relative maximum at x_0.

□

Example 4.33. *Let $f(x) = 2x^3 - 3x^2 - 12x + 5$. Determine the relative maxima and minima.*

SOLUTION. First, we observe that $f(x) \in \mathcal{C}^\infty(\mathbb{R})$ since it is a polynomial. We have:

$$f'(x) = 6x^2 - 6x - 12 = 6(x^2 - x - 2) = 6(x - 2)(x + 1)$$

and:

$$f''(x) = 12x - 6.$$

The critical points are -1 and 2. Since $f''(-1) = -18 < 0$ and $f''(2) = 18 > 0$ we can conclude, by the higher order derivative test, that there is a local maximum at $x = -1$ and a local minimum at $x = 2$. □

Observation 4.8. It is important to note that the first derivative test is more general than the higher order derivative test. For example, we can apply the first derivative test even at critical points where the function is not differentiable (assuming, of course, that it is continuous).

Applications of the Taylor Series: Infinitesimals and the Taylor Series

To determine how quickly a function grows or decreases, Edmund Landau (1877–1938) introduced the notation of orders of magnitude that bears his name (or colloquially called Big O notation) which will be very useful in this section.

Definition 4.18 (Landau's Notation). If $a \in \mathbb{R}$, we write:

$$f(x) = O(g(x)), \quad x \to a,$$

if and only if, there exist two constants d and C such that:

$$|f(x)| \leq C|g(x)|, \quad |x - a| < d.$$

Example 4.34. *For example, using Landau's notation, the Taylor expansion of the exponential function centered at the origin can be written as:*

$$e^x = 1 + x + \frac{x^2}{2} + O(x^3), \quad x \to 0,$$

where the last term means that the Taylor error term is less, in absolute value, than a constant times x^3, provided that x is near 0.

Now let's see how, using Taylor series and Landau's notation, we can easily demonstrate that $\sin(x) \sim_0 x$. That is, we can prove that:

$$\lim_{x \to 0} \frac{\sin(x)}{x} = 1.$$

Since we want to calculate the limit of the function at 0, we are interested in knowing the behavior of the function $\sin(x)$ around the origin. To do this, we calculate the Taylor series around this point. If we take $f(x) = \sin(x) \in \mathcal{C}^\infty(\mathbb{R})$:

$$\begin{aligned} f(x) &= f(0) + f'(0)x + \frac{f''(0)}{2!}x^2 + \frac{f'''(0)}{3!}x^3 + \cdots \\ &= \sin(0) + \cos(0)x - \frac{\sin(0)}{2!}x^2 - \frac{\cos(0)}{3!}x^3 + \cdots \\ &= x - \frac{x^3}{6} + \frac{x^5}{120} - \frac{x^7}{5040} + \cdots = x + O(x^3), \quad x \to 0, \end{aligned}$$

So:

$$\lim_{x \to 0} \frac{\sin(x)}{x} = \lim_{x \to 0} \frac{x + O(x^3)}{x} = \lim_{x \to 0} \left(1 + O(x^2)\right) = 1 + 0 = 1,$$

just as we wanted to prove.

Now let's see that using Taylor series and Landau's notation we can solve most limits at $x = 0$ that present indeterminacy.

Example 4.35. *Calculate:*

$$\lim_{x\to 0} \frac{\sin(x) - x}{x^3}$$

SOLUTION. Since we want to calculate the limit at $x = 0$, we are interested in knowing the behavior of the function $f(x) = \sin(x) - x$ near this point. To do this, we calculate the Taylor series centered at the origin:

$$\begin{array}{ll} f(x) = \sin(x) - x & f(0) = \sin 0 - 0 = 0 \\ f'(x) = \cos(x) - 1 & f'(0) = \cos 0 - 1 = 1 - 1 = 0 \\ f''(x) = -\sin(x) & f''(0) = -\sin 0 = 0 \\ f'''(x) = -\cos(x) & f'''(0) = -\cos 0 = -1 \end{array}$$

Therefore:

$$f(x) = 0 + 0x + \frac{0}{2!}x^2 - \frac{1}{3!}x^3 + O(x^4) = -\frac{x^3}{6} + O(x^4), \quad x \to 0,$$

and then:

$$\lim_{x\to 0} \frac{\sin(x) - x}{x^3} = \lim_{x\to 0} \frac{-\frac{x^3}{6} + O(x^4)}{x^3} = \lim_{x\to 0}\left(-\frac{1}{6} + O(x)\right) = -\frac{1}{6}.$$

□

Example 4.36. *Calculate:*

$$\lim_{x\to 0} \frac{e^x - e^{-3x}}{\ln(x+1)}$$

SOLUTION. First, we calculate the Taylor series of the functions $f(x) = e^x - e^{-3x}$ and $g(x) = \ln(x+1)$ around the point $x = 0$:

$$f(x) = e^x - e^{-3x} \quad \Rightarrow \quad f(0) = 0, \qquad g(x) = \ln(x+1) \quad \Rightarrow \quad g(0) = 0$$

$$f'(x) = e^x + 3e^{-3x} \quad \Rightarrow \quad f'(0) = 4, \qquad g'(x) = \tfrac{1}{x+1} \quad \Rightarrow \quad g'(0) = 1$$

Therefore:

$$f(x) = 4x + O(x^2), \quad g(x) = x + O(x^2), \quad x \to 0,$$

and then:

$$\lim_{x\to 0} \frac{e^x - e^{-3x}}{\ln(x+1)} = \lim_{x\to 0} \frac{4x + O(x^2)}{x + O(x^2)} = \lim_{x\to 0} \frac{4 + O(x)}{1 + O(x)} = \frac{4+0}{1+0} = 4$$

□

4.14 L'Hôpital's Rules

The French mathematician Guillaume François L'Hôpital (1661 - 1704) was the author of the first text written on differential calculus, *Analyse des infiniment petits* (1696). The theorem on limits, which came to be known as L'Hôpital's rule, first appeared in this book, although it is known that the rule is more due to his teacher, the Swiss mathematician Johann Bernoulli (1667-1748) who developed and proved it. The original theorem was refined and extended, and the different results are collectively known as L'Hôpital's rules, which we will see in this section. These rules use derivatives to evaluate certain limits of functions that are in indeterminate form and will allow us to prove the orders of infinitude seen in the chapter on limits, functions, and continuity.

Theorem 4.7. Cauchy's Mean Value Theorem Let $f, g : [a, b] \subset \mathbb{R} \to \mathbb{R}$ be two functions continuous on $[a, b]$ and differentiable on the open interval (a, b). Then, there exists $\xi \in [a, b]$ such that:

$$(f(b) - f(a))\, g'(\xi) = (g(b) - g(a))\, f'(\xi).$$

Proof. Consider the function $h : [a, b] \subset \mathbb{R} \to \mathbb{R}$ defined as:

$$h(x) = (f(b) - f(a))\, g(x) - (g(b) - g(a))\, f(x) + f(a)g(b) - f(b)g(a).$$

Since f and g are continuous on $[a, b]$ and differentiable on the open interval (a, b), so is the function h with:

$$h'(x) = (f(b) - f(a))\, g'(x) - (g(b) - g(a))\, f'(x).$$

Observing the function $h(x)$ it is evident that $h(a) = h(b) = 0$ and, therefore, by applying Rolle's theorem (*see* Theorem 4.1), there exists $\xi \in (a, b)$ such that $h'(\xi) = 0$, which is what we wanted to prove. □

Theorem 4.8 (L'Hôpital's Rule: Case 0/0). Let I be a closed interval, $a \in I$ and $f, g : I\backslash\{a\} \to \mathbb{R}$ functions verifying:

(a) f and g are differentiable on $I\backslash\{a\}$

(b) $g'(x) \neq 0$ for all $x \in I\backslash\{a\}$

(c) $\lim_{x\to a} f(x) = \lim_{x\to a} g(x) = 0$

Then it holds that $g(x) \neq 0$ for all $x \in I\backslash\{a\}$, so the functions f/g and f'/g' are defined on $I\backslash\{a\}$ and it is verified that:

(i) If $\lim_{x\to a} \dfrac{f'(x)}{g'(x)} = L \in \mathbb{R}$ then $\lim_{x\to a} \dfrac{f(x)}{g(x)} = L$

(ii) If $\lim_{x\to a} \dfrac{f'(x)}{g'(x)} = \pm\infty$ then $\lim_{x\to a} \dfrac{f(x)}{g(x)} = \pm\infty$

Proof. Note that hypothesis (c) allows us to extend the functions f and g, defining $f(a) = g(a) = 0$, to obtain continuous functions at point a. We can continue denoting by f and g the extensions thus obtained, since the local nature of the derivative concept means that hypotheses (a) and (b) continue to be verified, and the thesis being sought does not involve at all the values we have given to our functions at point a. In summary, we can assume that $f, g : I \to \mathbb{R}$ are continuous functions on I with $f(a) = g(a) = 0$, and differentiable on $I\backslash\{a\}$ with $g'(x) \neq 0$ for all $x \in I\backslash\{a\}$.

To obtain the first statement of the theorem, let $x \in I\backslash\{a\}$. If $a < x$, since g is continuous on the interval $[a, x] \subset I$ and differentiable on $(a, x) \subset I\backslash\{a\}$, the Mean Value Theorem provides us with a $\xi \in (a, x)$ verifying that $g(x) = g(x) - g(a) = g'(c)(x - a)$, but by hypothesis we have $g'(c) \neq 0$, so $g(x) \neq 0$. In the case $x < a$, reasoning with the interval $[x, a]$ we reach the same conclusion. Thus we have demonstrated that under the hypotheses of the theorem it holds that $g(x) \neq 0$ for, all $x \in I\backslash\{a\}$.

Now let us prove that (i) and (ii) are verified. Again for $x \in I$ with $a < x$, the functions f and g are continuous on $[a, x]$ and differentiable on (a, x), then by Cauchy's Mean Value Theorem (*see* Theorem 4.7) there exists $\xi_x \in (a, x)$ such that:

$$(f(x) - f(a))\, g'(\xi_x) = (g(x) - g(a))\, f'(\xi_x)$$

that is:

$$\frac{f(x)}{g(x)} = \frac{f'(\xi_x)}{g'(\xi_x)}.$$

In the case $x < a$ we would reason with the interval $[x, a]$ obtaining the same conclusion, only now $\xi_x \in (x, a)$. In both cases we have:

$$0 < |\xi_x - a| < |x - a|$$

and we have proven:

$$\text{for all } x \in I \backslash \{a\} \text{ exists } \xi_x \in I \text{ such that } 0 < |\xi_x - a| < |x - a| \text{ y } \frac{f(x)}{g(x)} = \frac{f'(\xi_x)}{g'(\xi_x)}. \tag{4.1}$$

From here the rest of the proof is quite immediate, but must be done considering separately the two cases to be demonstrated.

(i) If:

$$\lim_{x \to a} \frac{f'(x)}{g'(x)} = L \in \mathbb{R}$$

then, by the formal definition of limit, we know that given $\varepsilon > 0$ we can find $\delta > 0$ such that:

$$y \in I, \quad 0 < |y - a| < \delta \quad \Longrightarrow \quad \left| \frac{f'(y)}{g'(y)} - L \right| < \varepsilon. \tag{4.2}$$

Then, for $x \in I$ with $0 < |x - a| < \delta$, by applying equation (4.1) we have:

$$0 < |\xi_x - a| < \delta,$$

so we can use equation (4.2) with $y = \xi_x$ to obtain:

$$\left| \frac{f(x)}{g(x)} - L \right| = \left| \frac{f'(\xi_x)}{g'(\xi_x)} - L \right| < \varepsilon,$$

just as we wanted to prove.

(ii) If:

$$\lim_{x \to a} \frac{f'(x)}{g'(x)} = +\infty$$

then, by the formal definition of limit, we know that given $K \in \mathbb{R}$ we have $\delta > 0$ such that:

$$y \in I, \quad 0 < |y - a| < \delta \quad \Longrightarrow \quad \frac{f'(y)}{g'(y)} > K. \tag{4.3}$$

Then, for $x \in I$ with $0 < |x - a| < \delta$ applying equations (4.1) and (4.3) we conclude that:

$$\frac{f(x)}{g(x)} = \frac{f'(\xi_x)}{g'(\xi_x)} > K$$

just as we wanted to prove. In the case that:

$$\lim_{x \to a} \frac{f'(x)}{g'(x)} = -\infty$$

we apply the same proof as above but changing f for $-f$.

□

Example 4.37. *Calculate the limit:*

$$\lim_{x\to 0} \frac{\sin(x)}{x}.$$

SOLUTION. Given that the functions $f(x) = \sin(x)$ and $g(x) = x$ are continuous and differentiable on $\mathbb{R}$ and also $g'(x) = 1 \neq 0$, by applying L'Hôpital's rule:

$$\lim_{x\to 0} \frac{\sin(x)}{x} = «\frac{0}{0}» \text{ (indeterminate)} = \lim_{x\to 0} \frac{(\sin(x))'}{(x)'} = \lim_{x\to 0} \frac{\cos(x)}{1} = 1$$

□

It is possible to extend L'Hôpital's rule to the case of indeterminacies of the type ∞/∞ as seen in the following theorem whose proof is omitted.

Theorem 4.9 (L'Hôpital's Rule: Case ∞/∞). Let I be a closed interval, $a \in I$ and $f, g : I\backslash\{a\} \to \mathbb{R}$ functions verifying:

(a) f and g are differentiable on $I\backslash\{a\}$

(b) $g'(x) \neq 0$ for all $x \in I\backslash\{a\}$

(c) $\lim_{x\to a} g(x) = \pm\infty$

Then there exists $\rho > 0$ such that, for $x \in I$ with $0 < |x - a| < \rho$, we have $g(x) \neq 0$. Furthermore, it is verified that:

(i) If $\lim_{x\to a} \frac{f'(x)}{g'(x)} = L \in \mathbb{R}$ then $\lim_{x\to a} \frac{f(x)}{g(x)} = L$

(ii) If $\lim_{x\to a} \frac{f'(x)}{g'(x)} = \pm\infty$ then $\lim_{x\to a} \frac{f(x)}{g(x)} = \pm\infty$

To complete all the cases, we will analyze the adaptation of the L'Hôpital's rules seen to the case where the limit is not at a point $a \in R$ but at $+\infty$ or at $-\infty$.

Theorem 4.10 (L'Hôpital's Rule at ∞). Let I be a closed interval and $f, g : I \to \mathbb{R}$ functions verifying:

(a) f and g are differentiable on I

(b) $g'(x) \neq 0$ for all $x \in I$

Let us also assume that one of the following conditions is met:

(c1) $\lim_{x\to+\infty} f(x) = \lim_{x\to+\infty} g(x) = 0$

(c2) $\lim_{x\to+\infty} g(x) = \pm\infty$

Then there exists $M > 0$ such that, for $x \in I$ with $x > M$, we have $g(x) \neq 0$. Furthermore, it is verified that:

(i) If $\lim_{x\to+\infty} \frac{f'(x)}{g'(x)} = L \in \mathbb{R}$ then $\lim_{x\to+\infty} \frac{f(x)}{g(x)} = L$

(ii) If $\lim_{x\to+\infty} \frac{f'(x)}{g'(x)} = \pm\infty$ then $\lim_{x\to+\infty} \frac{f(x)}{g(x)} = \pm\infty$

An analogous statement can be written for the behavior at $-\infty$.

Proof. First, hypothesis (b) implies that g is strictly monotonic, so it can cancel at most once in I, which guarantees the existence of M. We now use a procedure to reduce the study of the behavior of functions at $+\infty$ to the behavior at 0 of suitable functions. The case $-\infty$ would have an analogous treatment.

We fix $\alpha \in I$, with $\alpha > M$, we take $J = [0, 1/\alpha)$ and define the functions:

$$\phi, \psi : J\backslash\{0\} \to \mathbb{R}$$

as:

$$\phi(x) = f(1/x), \qquad \psi(x) = g(1/x), \qquad \forall x \in J\backslash\{0\}$$

By the definition given of these functions, we have that the behavior at $+\infty$ of the quotient f/g will be the same as that of ϕ/ψ at 0.

By the chain rule, ϕ and ψ are differentiable functions on $J\backslash\{0\}$ with:

$$\phi'(x) = -\frac{1}{x^2} f'(1/x), \qquad \psi'(x) = -\frac{1}{x^2} g'(1/x) \neq 0, \qquad \forall x \in J\backslash\{0\}$$

If (c1) is verified we have that:

$$\lim_{x\to 0} \phi(x) = \lim_{x\to 0} \psi(x) = 0,$$

so the functions ϕ and ψ verify the hypotheses of L'Hôpital's rule (case 0/0) (*see* Theorem 4.8). If (c2) is verified, then:

$$\lim_{x\to 0} \psi(x) = \pm\infty$$

and the hypotheses of L'Hôpital's rule (case ∞/∞) (*see* Theorem 4.9) are met. Therefore, we can apply the appropriate rule to the functions ϕ and ψ using the interval J, to obtain the behavior of the quotient ϕ/ψ at 0. Taking into account that:

$$\frac{\phi'(x)}{\psi'(x)} = \frac{f'(1/x)}{g'(1/x)}, \qquad \forall x \in J\backslash\{0\}$$

the behavior at 0 of the quotient ϕ'/ψ' is the same as that of f'/g' at $+\infty$. Therefore:

$$\lim_{x\to+\infty} \frac{f'(x)}{g'(x)} = L \quad \Leftrightarrow \quad \lim_{x\to 0} \frac{\phi'(x)}{\psi'(x)} = L \quad \Rightarrow \quad \lim_{x\to 0} \frac{\phi(x)}{\psi(x)} = L \quad \Leftrightarrow \quad \lim_{x\to+\infty} \frac{f(x)}{g(x)} = L$$

and an analogous reasoning can be made in case (ii). □

Let's see that L'Hôpital's rules allow us to demonstrate the orders of infinitude seen in the chapter on functions, limits, and continuity.

Example 4.38. *Calculate the limit:*

$$\lim_{x\to+\infty} \frac{\ln(x)}{x}.$$

Solution. Consider the interval $(1, +\infty)$ and the functions $f(x) = \ln(x)$ and $g(x) = x$, then the hypotheses of L'Hôpital's rule are met and we have:

$$\begin{aligned}\lim_{x\to+\infty} \frac{\ln(x)}{x} &= «\frac{+\infty}{+\infty}» \text{ (indeterminate)}\\ &= \lim_{x\to+\infty} \frac{(\ln(x))'}{(x)'} = \lim_{x\to+\infty} \frac{1/x}{1} = \lim_{x\to+\infty} \frac{1}{x} = 0\end{aligned}$$

That is, we have just demonstrated that $g(x) = x$ is an infinity of higher order than $f(x) - \ln(x)$ at infinity. □

Example 4.39. *Calculate the limit:*

$$\lim_{x\to+\infty} \frac{x^2}{e^x}.$$

Solution. Consider the real line $\mathbb{R}$ and the functions $f(x) = x^2$ and $g(x) = e^x$; then the hypotheses of L'Hôpital's rule are met and we have:

$$\lim_{x\to+\infty} \frac{x^2}{e^x} = \lim_{x\to+\infty} \frac{(x^2)'}{(e^x)'} = \lim_{x\to+\infty} \frac{2x}{e^x}$$

We observe that the functions $2x$ and e^x meet the hypotheses of L'Hôpital's rule and, therefore:

$$\lim_{x\to+\infty} \frac{2x}{e^x} = \lim_{x\to+\infty} \frac{(2x)'}{(e^x)'} = \lim_{x\to+\infty} \frac{2}{e^x} = 0$$

That is, we have just demonstrated that $g(x) = e^x$ is an infinity of higher order than $f(x) = x^2$ at infinity. □

Example 4.40. *Calculate the limit:*

$$\lim_{x\to+\infty} \frac{x^2}{2^x}.$$

Solution. Consider the real line $\mathbb{R}$ and the functions $f(x) = x^2$ and $g(x) = 2^x$, then the hypotheses of L'Hôpital's rule are met and we have:

$$\lim_{x\to+\infty} \frac{x^2}{2^x} = \lim_{x\to+\infty} \frac{(x^2)'}{(2^x)'} = \lim_{x\to+\infty} \frac{2x}{2^x \ln(2)}$$

We observe that the functions $2x$ and $2^x \ln(2)$ meet the hypotheses of L'Hôpital's rule and, therefore:

$$\lim_{x\to+\infty} \frac{2x}{2^x \ln(2)} = \lim_{x\to+\infty} \frac{(2x)'}{(2^x \ln(2))'} = \lim_{x\to+\infty} \frac{2}{2^x(\ln(2))^2} = 0$$

That is, we have just demonstrated that $g(x) = 2^x$ is an infinity of higher order than $f(x) = x^2$ at infinity. □

Example 4.41. *Let's calculate different limits using L'Hôpital's rule:*

- $\displaystyle\lim_{x\to 0} \frac{x - \sin(x)}{x} = «\frac{0}{0}» \text{ (indeterminate)} = \lim_{x\to 0} \frac{1 - \cos(x)}{1} = 0$
- $\displaystyle\lim_{x\to 3} \frac{x^2 - 9}{x^2 - 5x + 6} = «\frac{0}{0}» \text{ (indeterminate)} = \lim_{x\to 3} \frac{2x}{2x - 5} = \frac{6}{1} = 6$

- $\lim_{x\to 0} \frac{\sin(x)}{x} = «\frac{0}{0}»$ (indeterminate) $= \lim_{x\to 0} \frac{\cos(x)}{1} = 1$
- $\lim_{x\to 0} \frac{1-\cos(x)}{\tan(x)} = «\frac{0}{0}»$ (indeterminate) $= \lim_{x\to 0} \frac{\sin(x)}{\frac{1}{\cos^2(x)}} = \lim_{x\to 0} \sin(x)\cos^2(x) = 0$
- $\lim_{x\to 0} \frac{e^x - e^{-3x}}{\ln(x+1)} = «\frac{0}{0}»$ (indeterminate) $== \lim_{x\to 0} \frac{e^x + 3e^{-3x}}{\frac{1}{x+1}} = \frac{4}{1} = 4$
- $\lim_{x\to 0} \frac{\sin(2x)}{3x - x^2} = «\frac{0}{0}»$ (indeterminate) $= \lim_{x\to 0} \frac{2\cos(2x)}{3-2x} = \frac{2}{3}$
- $\lim_{x\to +\infty} \frac{\ln(x)}{\sqrt{x}} = «\frac{+\infty}{+\infty}»$ (indeterminate) $= \lim_{x\to +\infty} \frac{\frac{1}{x}}{\frac{1}{2\sqrt{x}}} = \lim_{x\to +\infty} \frac{2\sqrt{x}}{x} = 0$

□

Example 4.42. *Calculate:*

$$\lim_{x\to +\infty} \frac{x+\sin(x)}{x-\sin(x)}.$$

SOLUTION.

$$\begin{aligned} \lim_{x\to +\infty} \frac{x+\sin(x)}{x-\sin(x)} &= «\frac{+\infty}{+\infty}» \text{ (indeterminate)} \\ &= \lim_{x\to +\infty} \frac{\frac{x}{x} + \frac{\sin(x)}{x}}{\frac{x}{x} - \frac{\sin(x)}{x}} = \frac{1+0}{1-0} = 1 \end{aligned}$$

In this case, we cannot apply L'Hôpital's rule because the derivative of the denominator, $1 - \cos(x)$, cancels as we approach infinity.

If we had used L'Hôpital's rule we would have obtained an erroneous result:

$$\lim_{x\to +\infty} \frac{x+\sin(x)}{x-\sin(x)} = \lim_{x\to +\infty} \frac{1+\cos(x)}{1-\cos(x)} = \nexists$$

□

Example 4.43. *Calculate:*

$$\lim_{x\to 0} \frac{x^2 \sin(\frac{1}{x})}{\sin(x)}.$$

SOLUTION.

$$\lim_{x\to 0} \frac{x^2 \sin(\frac{1}{x})}{\sin(x)} = «\frac{0}{0}» \text{ (indeterminate)}$$

In this case, we cannot apply L'Hôpital's rule because the numerator is not differentiable at 0:

$$\lim_{x\to 0} \frac{x^2 \sin\left(\frac{1}{x}\right)}{\sin(x)} = \lim_{x\to 0} x \underbrace{\frac{x}{\sin(x)}}_{\to 1} \sin\left(\frac{1}{x}\right) = \lim_{x\to 0} x \underbrace{\sin(\frac{1}{x})}_{\text{bounded}} = «0\cdot \text{ bounded}» = 0$$

□

Example 4.44. *L'Hôpital's rules must be applied with caution, as they only hold in the case where $g'(x)$ does not change sign infinitely many times in a neighborhood of ∞. For example, consider the limit $f(x)/g(x)$ with:*

$$f(x) = x + \cos(x)\sin(x)$$
$$g(x) = e^{\sin(x)}(x + \cos(x)\sin(x))$$

when $x \to \infty$. While $f(x)$ and $g(x)$ tend to ∞ as $x \to \infty$, the limit of their quotient does not exist although it is bounded in the interval $[1/e, e]$. On the other hand, the limit of $f'(x)/g'(x)$ tends to 0. This is because the hypotheses of L'Hôpital's rules are not satisfied. See Figure 4.39.

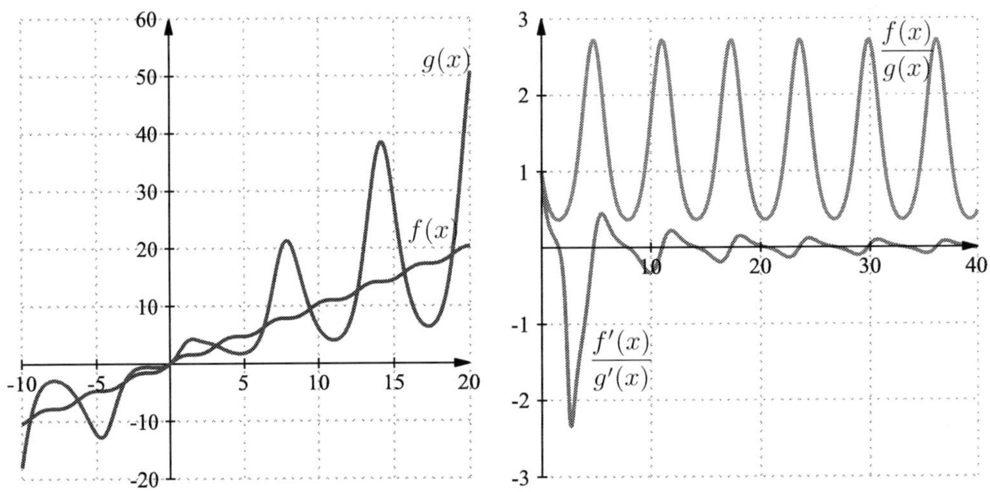

FIGURE 4.39
Graphs of $f(x) = x + \cos(x)\sin(x)$ and $g(x) = e^{\sin(x)}(x + \cos(x)\sin(x))$ (left) and their quotient and the quotient of their derivatives (right).

Example 4.45. *L'Hôpital's rules must be applied with caution, as they only hold in the case where $g'(x)$ does not change sign infinitely many times in a neighborhood of 0. For example, consider the limit $f(x)/g(x)$ with:*

$$f(x) = x\sin(x^{-4})e^{(-1/x^2)}$$
$$g(x) = e^{(-1/x^2)}$$

when $x \to 0$. While $f(x)$ and $g(x)$ tend to 0 as $x \to 0$, the limit of their quotient is 0, while the limit of $f'(x)/g'(x)$ is not bounded on the real line. See Figure 4.40.

The indeterminate forms $0 \cdot \infty$, $\infty - \infty$, ∞^0, 0^0, 1^∞ can be reduced to the cases we have seen above by means of algebraic manipulations and the use of logarithmic and exponential functions. The relevant techniques for each type of indeterminate form are illustrated below.

(1) **Indeterminate forms** $0 \cdot \infty$.

If we want to calculate $\lim\limits_{x\to a} f(x)g(x)$ with:

$$\lim_{x\to a} f(x) = 0 \text{ and } \lim_{x\to a} g(x) = \infty$$

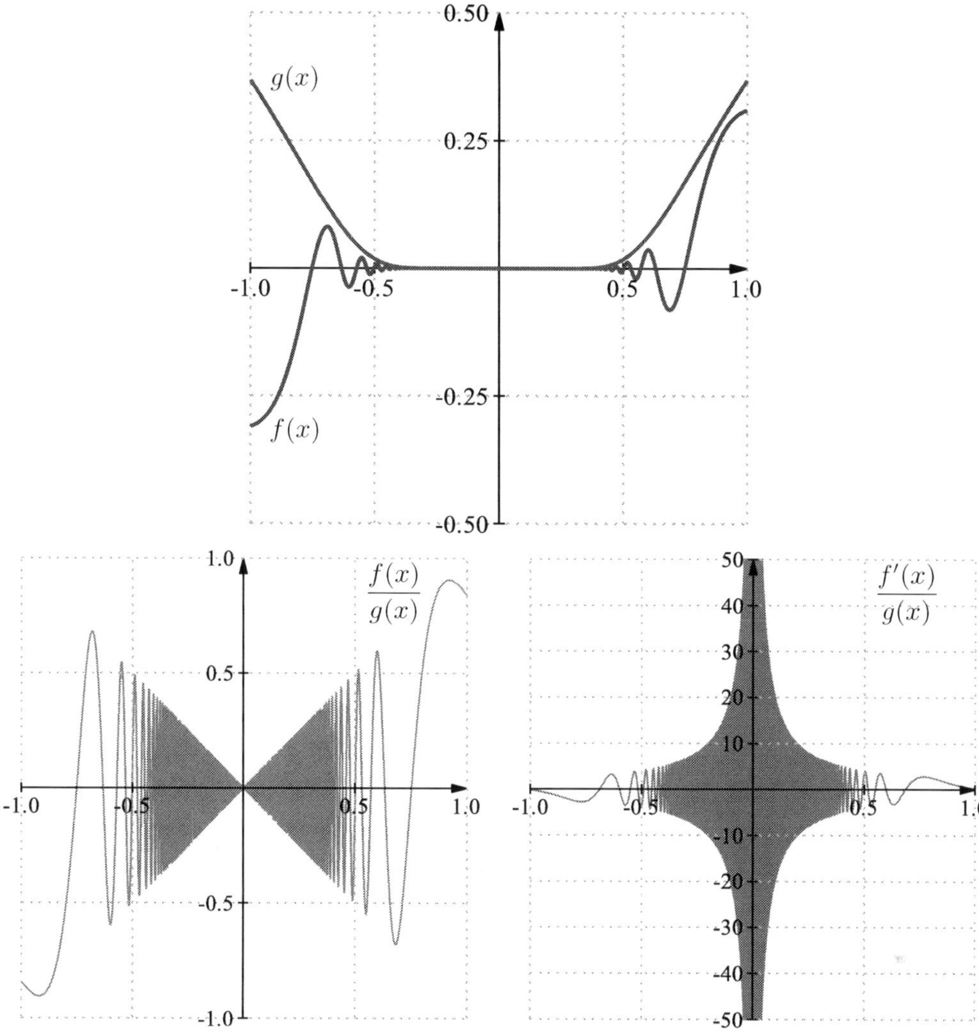

FIGURE 4.40
Graphs of the functions $f(x) = x\sin(x^{-4})e^{(-1/x^2)}$ and $g(x) = e^{(-1/x^2)}$ (above) as well as the functions $\frac{f(x)}{g(x)}$ (below left) and $\frac{f'(x)}{g'(x)}$ (below right).

then we can write:

$$\lim_{x\to a} \frac{f(x)}{\frac{1}{g(x)}} = \frac{0}{0} \quad \text{or} \quad \lim_{x\to a} \frac{g(x)}{\frac{1}{f(x)}} = \frac{\infty}{\infty}$$

(2) **Indeterminate forms** $\infty - \infty$.

If we want to calculate $\lim_{x\to a} (f(x) - g(x))$ with:

$$\lim_{x\to a} f(x) = \infty \quad \text{and} \quad \lim_{x\to a} g(x) = \infty$$

then we can write:

$$\lim_{x\to a} f(x) - g(x) = \lim_{x\to a} \frac{\frac{1}{g(x)} - \frac{1}{f(x)}}{\frac{1}{f(x)g(x)}} = \frac{0}{0}$$

(3) **Indeterminate forms** $\infty^0, 0^0, 1^\infty$.

If we want to calculate $\lim_{x\to a} f(x)^{g(x)}$ with:

$$\lim_{x\to a} f(x) = \infty \text{ and } \lim_{x\to a} g(x) = 0$$

or:

$$\lim_{x\to a} f(x) = 0 \text{ and } \lim_{x\to a} g(x) = 0$$

or:

$$\lim_{x\to a} f(x) = 1 \text{ and } \lim_{x\to a} g(x) = 0$$

we can write:

$$f(x)^{g(x)} = e^{\ln(f(x)^{g(x)})} = e^{g(x)\ln(f(x))},$$

and then:

$$\lim_{x\to a} f(x)^{g(x)} = \lim_{x\to a} e^{g(x)\ln(f(x))} = e^{\lim_{x\to a} g(x)\ln(f(x))}$$

where $\lim_{x\to a} g(x)\ln(f(x))$ is an indeterminacy $0 \cdot \infty$.

Proposed Exercise 4.7. Check that:

(a) The limit $\lim_{x\to 0} x\ln^2(x)$ is resolved using (1).

(b) The limit $\lim_{x\to 1^-} \ln(x)\ln(1-x)$ is resolved using (1).

(c) The limit $\lim_{x\to \frac{3}{4}} (3-4x)\tan\left(\frac{2\pi}{3}x\right)$ is resolved using (1).

(d) The limit $\lim_{x\to 0} \left(\frac{1}{\sin(x)} - \frac{1}{x}\right)$ is resolved using (2).

(e) The limit $\lim_{x\to 0^+} x^x$ is resolved using (3).

(f) The limit $\lim_{x\to 1} x^{\frac{1}{1-x}}$ is resolved using (3).

4.15 Solved Problems (Differentiability)

Problem 4.1. Calculate the second derivative of the functions:

(a) $f(x) = e^{x^2}$

(b) $g(x) = (1 + x^2)\arctan(x)$

(c) $h(x) = \ln\left(\sqrt[3]{1 + x^2}\right)$

(d) $i(x) = a\cosh\left(\frac{x}{a}\right)$

Solution to the problem 4.1. (a) $f'(x) = 2xe^{x^2}$, $f''(x) = 2e^{x^2} + 4x^2e^{x^2}$

(b) $g'(x) = 2x\arctan(x) + 1$, $g''(x) = 2\arctan(x) + \dfrac{2x}{1+x^2}$

(c) $h'(x) = \dfrac{2x}{3(1+x^2)}$, $h''(x) = \dfrac{2}{3(1+x^2)} - \dfrac{4x^2}{3(1+x^2)^2}$

(d) $i'(x) = \sinh\left(\dfrac{x}{a}\right)$, $i''(x) = \dfrac{1}{a}\cosh\left(\dfrac{x}{a}\right)$

Problem 4.2. If $f : \mathbb{R} \longrightarrow \mathbb{R}$ is a differentiable function, calculate the derivatives of the following functions:

(a) $g(x) = f(x^2 + 1)$

(b) $g(x) = \dfrac{f(x) + 1}{x^2 + 1}$

(c) $g(x) = f(\sin^2(x)) + \cos(f(x))$

(d) $g(x) = e^{f(x)} + x^2 f(x) + f(f(x))$

Solution to the problem 4.2. *Using the chain rule, we have:*

(a) $g'(x) = 2xf'(x^2 + 1)$

(b) $g'(x) = \dfrac{f'(x)(x^2+1) - 2x(f(x)+1)}{(x^2+1)^2}$

(c) $g'(x) = 2\sin(x)\cos(x)f'(\sin^2(x)) - \sin(f(x))f'(x)$

(d) $g'(x) = f'(x)e^{f(x)} + 2xf(x) + x^2f'(x) + f'(f(x))f'(x)$

Problem 4.3. Consider the function $f : \mathbb{R} \longrightarrow \mathbb{R}$ defined by:

$$f(x) = \begin{cases} x^2, & \text{if } x \geq 0 \\ -x^2, & \text{if } x < 0 \end{cases}$$

(a) Study the continuity of the function on $\mathbb{R}$.

(b) Study the differentiability of the function on $\mathbb{R}$. Calculate the derivative function f'.

(c) Using the definition of the derivative of a function at a point, calculate $f'(-3)$.

(d) Study the existence of the second derivative of f.

Solution to the problem 4.3. *(a) f is continuous at $x = 0$ since:*

$$\lim_{x\to 0^-} f(x) = \lim_{x\to 0^-} -x^2 = 0 = \lim_{x\to 0^+} f(x) = \lim_{x\to 0^+} x^2 = f(0).$$

The function f is also continuous at other points of $\mathbb{R}$, as it is defined by polynomials. Therefore, $f \in \mathcal{C}^0(\mathbb{R})$.

(b) The function is differentiable at all points $x \neq 0$ as it is defined by polynomials. f is also differentiable at $x = 0$ because:

$$\lim_{h\to 0^+} \frac{f(h) - f(0)}{h} = \lim_{h\to 0^+} \frac{h^2}{h} = 0$$
$$\lim_{h\to 0^-} \frac{f(h) - f(0)}{h} = \lim_{h\to 0^-} \frac{-h^2}{h} = 0$$

therefore $f'(0) = 0$. Finally, we can write:

$$f'(x) = \begin{cases} 2x, & \text{if } x \geq 0 \\ -2x, & \text{if } x < 0 \end{cases}$$

(c) Using the definition of the derivative of a function at a point, we have:

$$f'(-3) = \lim_{h\to 0} \frac{f(-3+h) - f(-3)}{h} = \lim_{h\to 0} \frac{-(-3+h)^2 + (-3)^2}{h} = 6.$$

(d) It is evident that f' is differentiable for $x < 0$ and $x > 0$. Let's study the differentiability of f' at $x = 0$:

$$\lim_{h\to 0^+} \frac{f'(h) - f'(0)}{h} = \lim_{h\to 0^+} \frac{2h}{h} = 2$$
$$\lim_{h\to 0^-} \frac{f'(h) - f'(0)}{h} = \lim_{h\to 0^-} \frac{-2h}{h} = -2$$

Since the two limits are different, $f''(0)$ does not exist. Therefore, the second derivative of f exists on $\mathbb{R} - \{0\}$ and its value is:

$$f''(x) = \begin{cases} 2, & \text{if } x > 0 \\ -2, & \text{if } x < 0 \end{cases}$$

In Figure 4.41, the function $f(x)$ and its first and second derivatives are plotted. We can verify the results obtained in the different sections.

Problem 4.4. Calculate the lateral derivatives of the following functions at the indicated points:

(a) $f(x) = \begin{cases} \dfrac{x}{1+e^{1/x}}, & \text{if } x \neq 0 \\ 0, & \text{if } x = 0 \end{cases}$ at the point $a = 0$

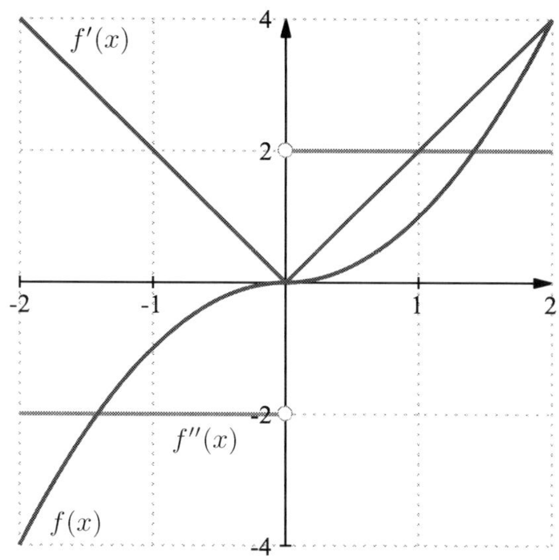

FIGURE 4.41
Graph of the function from Problem 4.3 and its first and second derivatives.

(b) $f(x) = \begin{cases} (x-2)\arctan\left(\dfrac{1}{x-2}\right), & \text{if } x \neq 2 \\ 0, & \text{if } x = 2 \end{cases}$ at $a = 2$.

Solution to the problem 4.4. (a) *Using the definition of the derivative:*

$$f'(0) = \lim_{x\to 0} \frac{f(x) - f(0)}{x - 0} = \lim_{x\to 0} \frac{\frac{x}{1+e^{1/x}}}{x} = \lim_{x\to 0} \frac{1}{1+e^{1/x}}$$

$$= \begin{cases} \displaystyle\lim_{x\to 0^+} \frac{1}{1+e^{1/x}} = \frac{1}{1+e^{1/0^+}} = \frac{1}{1+e^{+\infty}} = \frac{1}{+\infty} = 0 \\ \\ \displaystyle\lim_{x\to 0^-} \frac{1}{1+e^{1/x}} = \frac{1}{1+e^{1/0^-}} = \frac{1}{1+e^{-\infty}} = \frac{1}{1+0} = 1. \end{cases}$$

Therefore, $f'_+(0) = 0$ and $f'_-(0) = 1$, and since the lateral derivatives do not match, the function f is not differentiable at zero.

(b) *Using the definition of the derivative:*

$$f'(2) = \lim_{x\to 2} \frac{f(x) - f(2)}{x - 2} = \lim_{x\to 2} \frac{(x-2)\arctan\left(\frac{1}{x-2}\right)}{x-2}$$

$$= \lim_{x\to 2} \arctan\left(\frac{1}{x-2}\right).$$

Thus:

$$f'_+(2) = \lim_{x\to 2^+} \arctan\left(\frac{1}{x-2}\right) = \lim_{y\to +\infty} \arctan(y) = \frac{\pi}{2},$$

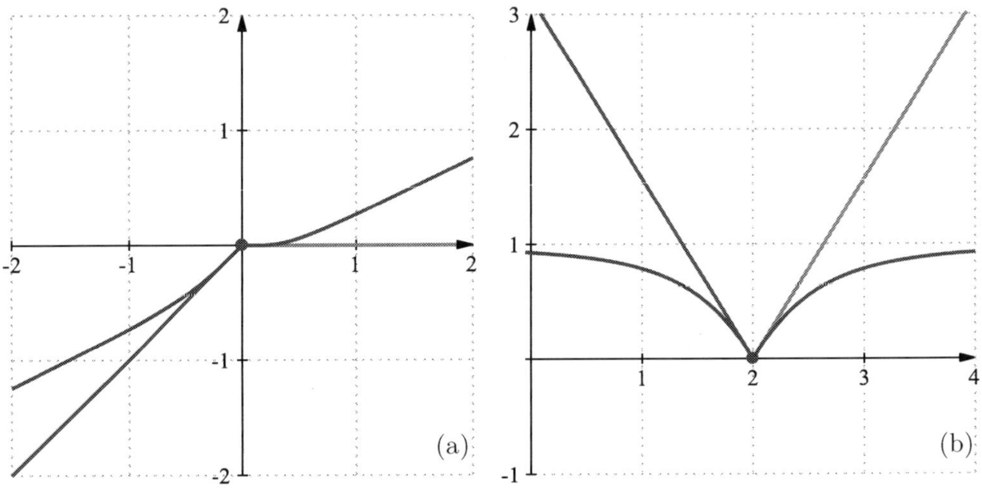

FIGURE 4.42
Graphs of the functions from Problem 4.4 and lateral tangent lines at the studied points.

and:

$$f'_{+}(2) = \lim_{x \to 2^-} \arctan\left(\frac{1}{x-2}\right) = \lim_{y \to -\infty} \arctan(y) = -\frac{\pi}{2},$$

where in the two previous limits the change of variable was made:

$$y = \frac{1}{x-2}$$

Therefore, $f'_{+}(2) = \frac{\pi}{2}$ and $f'_{-}(0) = -\frac{\pi}{2}$, and since the lateral derivatives do not match, the function f is not differentiable at the point $x = 2$.

In Figure 4.42, the graphs of the functions and the lateral tangent lines at the studied points are shown.

Problem 4.5. Consider the function $f : \mathbb{R} \longrightarrow \mathbb{R}$ defined by:

$$f(x) = \begin{cases} 2ax + 3, & \text{if } x < 1 \\ 3, & \text{if } x = 1 \\ \dfrac{x^2 - bx}{x+5}, & \text{if } x > 1 \end{cases}$$

Study the continuity and differentiability of this function based on the values of the parameters $a, b \in \mathbb{R}$.

Solution to the problem 4.5. *Let's first study the interior of the intervals:*

- *in $(-\infty, 1)$, f is continuous and differentiable $\forall a, b \in \mathbb{R}$.*
- *in $(1, \infty)$, f is continuous and differentiable $\forall a, b \in \mathbb{R}$ because the denominator does not vanish in this interval.*

Now let's study the point $x = 1$:

- *Continuity:*

$$f(1) = 3$$

$$\lim_{x\to 1^-} f(x) = \lim_{x\to 1^-} (2ax + 3) = 2a + 3$$

$$\lim_{x\to 1^+} f(x) = \lim_{x\to 1^+} \frac{x^2 - bx}{x + 5} = \frac{1-b}{6}$$

and thus f is continuous at $x = 1$ if and only if:

$$3 = 2a + 3 = \frac{1-b}{6} \quad \Leftrightarrow \quad \left\{ \begin{array}{l} 3 = 2a + 3 \\ 3 = \dfrac{1-b}{6} \end{array} \right. \quad \Leftrightarrow \quad \left\{ \begin{array}{l} a = 0 \\ b = -17 \end{array} \right.$$

Therefore, f is continuous on $\mathbb{R}$ if and only if, $a = 0$ and $b = -17$.

- *Differentiability: We need to study only the case where f is continuous, i.e., $a = 0$ and $b = -17$.*

$$\lim_{h\to 0^+} \frac{f(1+h) - f(1)}{h} = \lim_{h\to 0^+} \frac{\frac{1}{6}(1+h)^2 + \frac{17}{6}(1+h) - 3}{h} = \frac{19}{6}$$

$$\lim_{h\to 0^-} \frac{f(1+h) - f(1)}{h} = \lim_{h\to 0^-} \frac{3-3}{h} = 0$$

and, therefore, f is not differentiable at $x = 1$ for any value of $a, b \in \mathbb{R}$.

Problem 4.6. Study the differentiability of the indicated functions, calculating their domain of differentiability. Write the expression of the corresponding derivative function, if it exists.

(a) $f(x) = \sin(x) \cdot \lfloor x \rfloor$

(b) $f(x) = x|x|$

(c) $f(x) = \left\{ \begin{array}{ll} 1 - x, & \text{if } x \leq 0 \\ e^{-x}, & \text{if } x > 0 \end{array} \right.$

(d) $f(x) = x^2 - \lfloor x^2 \rfloor$

Solution to the problem 4.6. (a) *Let's study the continuity of the function $f(x)$ at the points $k \in \mathbb{Z}$. First, we calculate the lateral limits:*

$$\lim_{x\to k^-} \sin(x) \cdot \lfloor x \rfloor = \sin(k) \cdot (k-1)$$

$$\lim_{x\to k^+} \sin(x) \cdot \lfloor x \rfloor = \sin(k) \cdot (k)$$

We observe that except for the case $k = 0$, where both limits give zero, the above limits are different. Thus, the function has jump discontinuities at all integers except *at the point $x = 0$, where it is continuous. Now let's study the differentiability of the function at this point. The lateral derivatives are given by:*

$$\lim_{h\to 0^-} \frac{f(h)-f(0)}{h} = \lim_{h\to 0^-} \frac{\sin(h)\cdot\lfloor h\rfloor - \sin(0)\cdot\lfloor 0\rfloor}{h} = \lim_{h\to 0^-} \frac{\sin(h)\cdot(-1)}{h} = -1,$$
$$\lim_{h\to 0^+} \frac{f(h)-f(0)}{h} = \lim_{h\to 0^+} \frac{\sin(h)\cdot\lfloor h\rfloor - \sin(0)\cdot\lfloor 0\rfloor}{h} = \lim_{h\to 0^+} \frac{\sin(h)\cdot(0)}{h} = 0.$$

Since the lateral derivatives at $x = 0$ are different, we can assure that the function is not differentiable at this point.

Now let's study the non-integer real numbers. Let $a \in \mathbb{R} - \mathbb{Z}$. Then:

$$\begin{aligned}\lim_{h\to 0} \frac{f(a+h)-f(a)}{h} &= \lim_{h\to 0} \frac{\sin(a+h)\lfloor a+h\rfloor - \sin(a)\lfloor a\rfloor}{h}\\ &= \lim_{h\to 0} \frac{\sin(a+h)\lfloor a\rfloor - \sin(a)\lfloor a\rfloor}{h}\\ &= \lim_{h\to 0} \frac{\sin(a+h)-\sin(a)}{h}\lfloor a\rfloor\\ &= \underbrace{\lim_{h\to 0} \frac{\sin(a+h)-\sin(a)}{h}}_{\to\cos(a)}\lfloor a\rfloor = \cos(a)\lfloor a\rfloor\end{aligned}$$

Therefore, f is differentiable on $\mathbb{R} - \mathbb{Z}$ and its derivative is:

$$f'(x) = \cos(x)\lfloor x\rfloor.$$

In Figure 4.43 (top left), the function of this section is plotted and the obtained results can be verified.

(b) *The function $f(x) = x|x|$ can be expressed as:*

$$f(x) = x|x| = \begin{cases} x^2\ , & x \geq 0\\ -x^2\ , & x < 0\end{cases}$$

The differentiability of this function is studied in Problem 4.3.

In Figure 4.43 (top right), the function of this section is plotted.

(c) *It is evident that f is differentiable for $x < 0$ and $x > 0$. Let's study the differentiability at $x = 0$:*

$$\lim_{h\to 0^+} \frac{f(h)-f(0)}{h} = \lim_{h\to 0^+} \frac{e^{-h}-1}{h} \overset{(*)}{=} -1$$
$$\lim_{h\to 0^-} \frac{f(h)-f(0)}{h} = \lim_{h\to 0^-} \frac{\not{1}-h-\not{1}}{h} = -1$$

To calculate the first of the limits, consider the change $t = e^{-h} - 1$, then $h \to 0^+ \Leftrightarrow t \to 0^-$ and $h = -\ln(1+t)$. Using that $\ln(1+t) \sim_0 t$:

$$\lim_{h\to 0^+} \frac{e^{-h}-1}{h} = \lim_{t\to 0^-} \frac{t}{-\ln(1+t)} = -1.$$

Therefore, f is differentiable on $\mathbb{R}$ and $f'(x) = \begin{cases} -1\ , & x \leq 0\\ -e^{-x}\ , & x > 0\end{cases}$

In Figure 4.43 (bottom left), the function of this section is plotted and the obtained results can be verified.

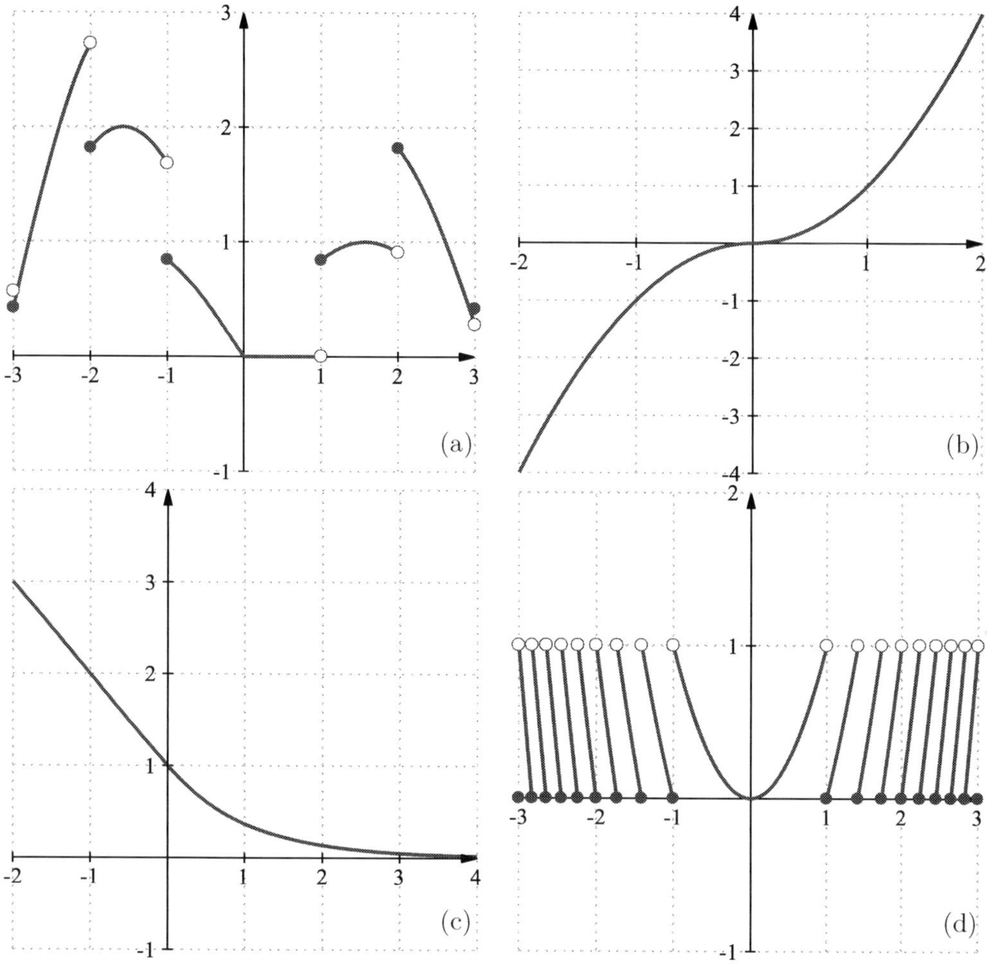

FIGURE 4.43
Graphs of the functions from Problem 4.6.

(d) *Note that if* $x \in (-1, 1)$ *then* $x^2 \in (0, 1)$ *and, therefore:*

$$\lfloor x^2 \rfloor = 0 \quad \textit{if} \quad x \in (-1, 1).$$

Therefore:

$$f(x) = x^2 - \lfloor x^2 \rfloor = x^2 \quad \textit{if} \quad x \in (-1, 1).$$

That is, the function $f(x)$ *in the interval* $(-1, 1)$ *coincides with the parabola* x^2*. Thus, we can assure that the function* $f(x)$ *is continuous and differentiable in this interval. The function* $\lfloor x^2 \rfloor$ *has jump discontinuities at* $\pm\sqrt{|k|}$, $\forall k \in \mathbb{Z} - \{0\}$ *and consequently is not differentiable at these points.*

What happens at the other points? Consider a real number a that is not the square root of any natural number. Then:

$$\begin{aligned}\lim_{h\to 0}\frac{f(a+h)-f(a)}{h} &= \lim_{h\to 0}\frac{(a+h)^2-\lfloor (a+h)^2\rfloor - a^2+\lfloor a^2\rfloor}{h}\\ &= \lim_{h\to 0}\frac{a^2+2ah+h^2-a^2+\lfloor a^2\rfloor-\lfloor (a+h)^2\rfloor}{h}\\ &= \lim_{h\to 0}\frac{h(2a+h)}{h}+\lim_{h\to 0}\frac{\lfloor a^2\rfloor-\lfloor (a+h)^2\rfloor}{h}\\ &= \lim_{h\to 0}\frac{h(2a+h)}{h}+\lim_{h\to 0}\frac{\lfloor a^2\rfloor-\lfloor a^2\rfloor}{h}\\ &= 2a+0=2a\end{aligned}$$

Therefore, f is differentiable on:

$$\mathbb{R}-\{x\in\mathbb{R}|x=\pm\sqrt{|k|},\ k\in\mathbb{Z}-\{0\}\}$$

and:

$$f'(x)=2x,\quad x\neq\pm\sqrt{|k|},\quad k\in\mathbb{Z}-\{0\}.$$

In Figure 4.43 (bottom right), the function of this section is plotted and the obtained results can be verified.

Problem 4.7. Study the continuity and differentiability of the functions $f_n:\mathbb{R}\longrightarrow\mathbb{R}$, defined by:

$$f_n(x)=\begin{cases} x^n\sin\left(\dfrac{1}{x}\right), & \text{if } x\neq 0\\ 0, & \text{if } x=0\end{cases}$$

based on the values of $n\in\mathbb{N}$.

Solution to the problem 4.7. *The functions f_n are continuous for all $x\neq 0$. The functions f_n are also continuous at $x=0$, because:*

$$\lim_{x\to 0}f_n(x)=\lim_{x\to 0}x^n\sin\left(\frac{1}{x}\right)=\text{«}0\cdot\ \textit{bounded}\text{»}=0=f_n(0).$$

In the above limit, we have used that when taking the limit as x tends to 0, we evaluate the function $f_n(x)$ at points $x\neq 0$ close to zero. That is why we wrote:

$$f_n(x)=x^n\sin\left(\frac{1}{x}\right).$$

We also used that for $n\geq 1$, if $x\to 0$, then $x^n\to 0$.

Now let's study the differentiability.

- *If $x\neq 0$:*

$$f_n'(x)=nx^{n-1}\sin\left(\frac{1}{x}\right)+x^n\cos\left(\frac{1}{x}\right)\left(-\frac{1}{x^2}\right)=nx^{n-1}\sin\left(\frac{1}{x}\right)-x^{n-2}\cos\left(\frac{1}{x}\right).$$

- *If $x = 0$, we use the definition of the derivative:*

$$
\begin{aligned}
f_n'(0) &= \lim_{x\to 0}\frac{f_n(x)-f_n(0)}{x-0}\\
&= \lim_{x\to 0}\frac{x^n\sin\left(\frac{1}{x}\right)}{x}\\
&= \lim_{x\to 0} x^{n-1}\sin\left(\frac{1}{x}\right)\\
&= \begin{cases} \text{«}0\cdot \textit{bounded}\text{»} = 0, & \textit{if } n>1\\ \nexists, & \textit{if } n=1. \end{cases}
\end{aligned}
$$

Therefore, the functions $f_n(x)$ are continuous on $\mathbb{R}$, $n \in \mathbb{N}$. They are also differentiable on $\mathbb{R}$, except the function $f_1(x)$ which is differentiable on $\mathbb{R} - \{0\}$.

Problem 4.8. Study the differentiability and calculate the derivative of the following functions:

(a) $f(x) = |x| + x$

(b) $f(x) = e^{-x^2} + \sqrt{x}$

(c) $f(x) = \dfrac{x+1}{x-1}$

(d) $f(x) = \dfrac{\sqrt{x^2+1}}{x}$

(e) $f(x) = \cos^2(x) + \sin(2x) + \arctan(x)$

(f) $f(x) = \dfrac{1}{x} + \sin\left(\dfrac{1}{x^2}\right) + \ln\left(x^2+1\right)$

(g) $f(t) = \begin{cases} |t|, & t \le 1\\ -(t-1)^2, & t>1 \end{cases}$

(h) $f(t) = \begin{cases} e^t, & t \le 0\\ 1, & 0<t<1\\ \ln(t), & t \ge 1 \end{cases}$

Solution to the problem 4.8. (a) *The function can be rewritten as:*

$$f(x) = \begin{cases} 0, & x<0\\ 2x, & x \ge 0 \end{cases}$$

The function is continuous on $\mathbb{R}$, but not differentiable at $x = 0$, since $f'_-(0) = 0$ and $f'_+(0) = 2$. The derivative function is:

$$f'(x) = \begin{cases} 0, & x<0\\ 2, & x>0 \end{cases}$$

In Figure 4.44, the results obtained can be verified by observing the graph of the function and its derivative.

(b) *The function is defined on $[0, +\infty)$ and is continuous in its domain of definition. The function is differentiable on $(0, +\infty)$ and its derivative is:*

$$f'(x) = -2xe^{-x^2} + \frac{1}{2\sqrt{x}}$$

In Figure 4.44, the results obtained can be verified by observing the graph of the function and its derivative.

(c) *The function is a quotient of polynomials. Therefore, it is continuous except at the points where the denominator vanishes. That is, f is continuous on $\mathbb{R} - \{1\}$. The function is also differentiable on this set and its derivative is given by:*

$$f'(x) = \frac{-2}{(x-1)^2}$$

In Figure 4.44, the results obtained can be verified by observing the graph of the function and its derivative.

(d) *The function is a quotient of a radical (whose radicand is always positive) and a polynomial. Therefore, it is continuous except at the points where the denominator vanishes. That is, f is continuous on $\mathbb{R} - \{0\}$. The function is also differentiable on this set with derivative:*

$$f'(x) = \frac{-1}{x^2\sqrt{x+1}}$$

In Figure 4.44, the results obtained can be verified by observing the graph of the function and its derivative.

(e) *The function is the sum of three continuous and infinitely differentiable functions on $\mathbb{R}$. Therefore, it is differentiable on $\mathbb{R}$ and its derivative is:*

$$f'(x) = -2\sin(x)\cos(x) + 2\cos(2x) + \frac{1}{1+x^2}$$

In Figure 4.44, the results obtained can be verified by observing the graph of the function and its derivative.

(f) *The function is continuous and differentiable on $\mathbb{R} - \{0\}$ (being the sum of continuous and differentiable functions, except at $x = 0$) with derivative:*

$$f'(x) = -\frac{1}{x^2} - \frac{2}{x^3}\cos\left(\frac{1}{x^2}\right) + \frac{2x}{1+x^2}$$

In Figure 4.44, the results obtained can be verified by observing the graph of the function and its derivative.

(g) *The function can be re-expressed as:*

$$f(t) = \begin{cases} -t, & t \le 0 \\ t, & 0 < t \le 1 \\ -(t-1)^2, & t > 1 \end{cases}$$

This function is continuous on $\mathbb{R} - \{1\}$. The function is differentiable for $t < 0$, $0 < t < 1$ and $t > 1$. At $t = 1$ the function cannot be differentiable (as it is not continuous). Let's see that at $t = 0$ the function is also not differentiable:

$$f'_-(0) = -1 \neq 1 = f'_+(0)$$

Therefore, f is differentiable on $\mathbb{R} - \{0, 1\}$ and the derivative is:

$$f'(t) = \begin{cases} -1, & t < 0 \\ 1, & 0 < t < 1 \\ -2(t-1), & t > 1 \end{cases}$$

In Figure 4.44, the results obtained can be verified by observing the graph of the function and its derivative.

(h) *The function is continuous at $t = 0$, but not at $t = 1$. The function cannot be differentiable at $t = 1$. At $t = 0$ it is also not differentiable, because:*

$$f'_{-}(0) = 1 \neq 0 = f'_{+}(0)$$

Therefore, f is differentiable on $\mathbb{R} - \{0, 1\}$ and the derivative is:

$$f'(t) = \begin{cases} e^t, & t < 0 \\ 0, & 0 < t < 1 \\ \frac{1}{t}, & t > 1 \end{cases}$$

In Figure 4.44, the results obtained can be verified by observing the graph of the function and its derivative.

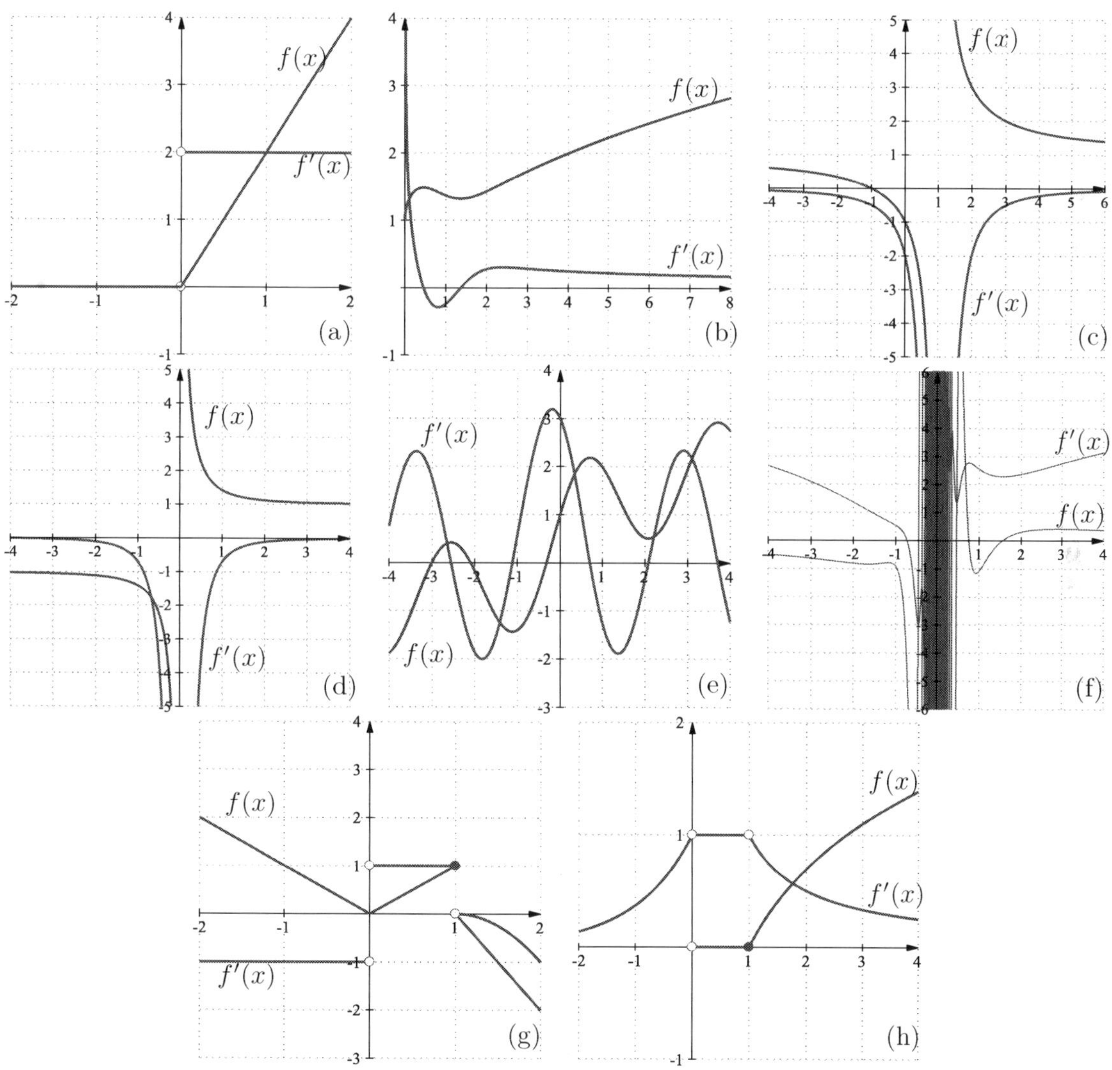

FIGURE 4.44
Graphs of the functions from Problem 4.8 and their respective derivatives.

Problem 4.9. At what points is the slope of the tangent line to $f(x) = x^3 - 6x$ parallel to the segment that joins $P_1 = (0, 0)$ and $P_2 = (2, -4)$?

Solution to the problem 4.9. *The slope of the given segment is:*

$$m_s = \frac{y_2 - y_1}{x_2 - x_1} = \frac{-4 - 0}{2 - 0} = -2.$$

We are looking for points where the tangent has a slope of -2*, that is:*

$$f'(x) = 3x^2 - 6 = -2 \quad \Leftrightarrow \quad x^2 = \frac{4}{3} \quad \Leftrightarrow \quad x = \pm\frac{2}{\sqrt{3}}$$

The points we are looking for are then:

$$\left(\frac{2}{\sqrt{3}}, -\frac{28}{3\sqrt{3}}\right), \quad \left(-\frac{2}{\sqrt{3}}, \frac{28}{3\sqrt{3}}\right).$$

In Figure 4.45, the function and the points found are represented, and it is possible to verify at first glance that at these points the slope of the tangent line to $f(x) = x^3 - 6x$ *is parallel to the segment that joins* $P_1 = (0, 0)$ *and* $P_2 = (2, -4)$.

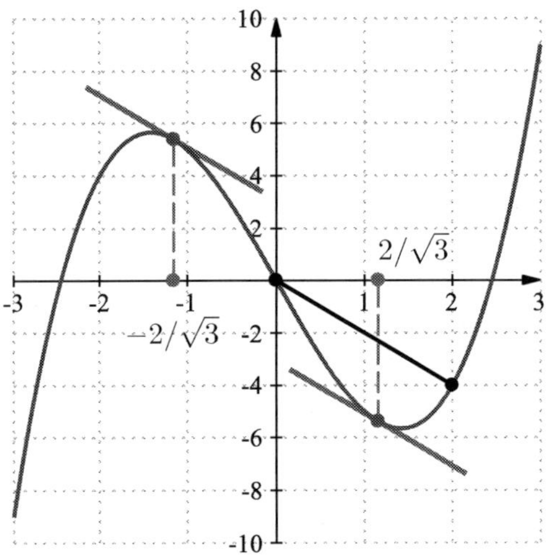

FIGURE 4.45
Points where the slope of the tangent line to $f(x) = x^3 - 6x$ is parallel to the segment that joins $P_1 = (0, 0)$ and $P_2 = (2, -4)$.

Problem 4.10. Find the points where the tangent line to the curve $y = x - \frac{1}{x}$ is parallel to the line $2x - y = 5$.

Solution to the problem 4.10. *The line* $2x - y = 5$ *has slope* $m = 2$*. We need to find the points* $(x, f(x))$*, with:*

$$f(x) = x - \frac{1}{x},$$

such that $f'(x) = 2$. We have:

$$f'(x) = 1 + \frac{1}{x^2} = 2 \quad \Leftrightarrow \quad x^2 = 1 \quad \Leftrightarrow \quad x = \pm 1,$$

and since $f(1) = 0$ and $f(-1) = 0$, the points we are looking for are $(-1, 0)$ and $(1, 0)$.

In Figure 4.46, the function and the points found are represented, and it can be visually verified that at these points the tangent line to $y = x - \frac{1}{x}$ is parallel to the line $2x - y = 5$.

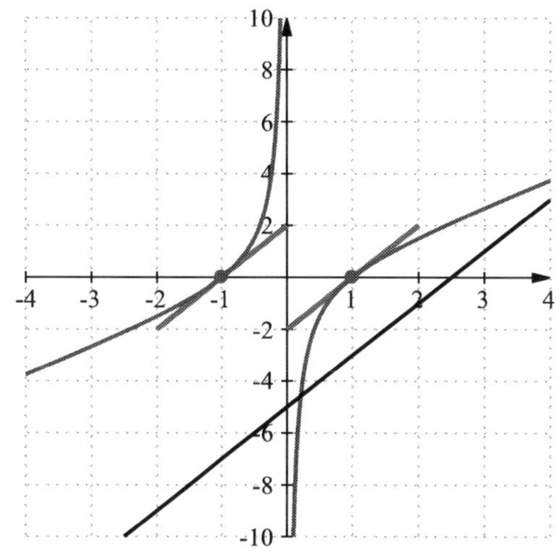

FIGURE 4.46

Points where the tangent line to the curve $y = x - \dfrac{1}{x}$ is parallel to the line $2x - y = 5$.

Problem 4.11. Find the points where the tangent line to the curve:

$$y = 3x^4 + 4x^3 - 12x^2 + 20$$

is parallel to the x-axis.

Solution to the problem 4.11. *To be parallel to the x-axis, the tangent line $f(x) = 3x^4 + 4x^3 - 12x^2 + 20$ must have zero slope, that is:*

$$\begin{aligned} f'(x) &= 12x^3 + 12x^2 - 24x \\ &= 12x(x-1)(x+2) = 0 \quad \Leftrightarrow \quad \begin{cases} x = 0 & \to \quad f(0) = 20 \\ x = 1 & \to \quad f(1) = 15 \\ x = -2 & \to \quad f(-2) = -12 \end{cases} \end{aligned}$$

and, therefore, the points of horizontal tangent are $(0, 20)$, $(1, 15)$ and $(-2, -12)$. In Figure 4.47, the function and the points found are represented, and it can be visually verified that at these points the tangent line to the curve $y = 3x^4 + 4x^3 - 12x^2 + 20$ is parallel to the x-axis.

Problem 4.12. Calculate the slope of the tangents to the parabola $y = -x^2 - 4x + 1$ at its points of intersection with the x-axis.

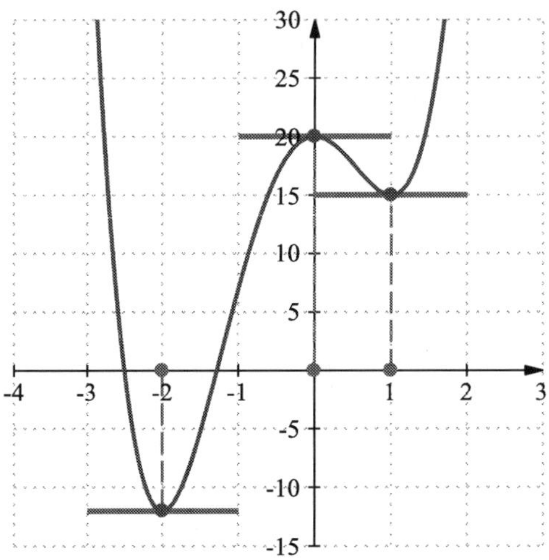

FIGURE 4.47
Points where the tangent line to the curve $y = 3x^4 + 4x^3 - 12x^2 + 20$ is parallel to the x-axis.

Solution to the problem 4.12. *The parabola intersects the x-axis at points such that:*

$$f(x) = -x^2 - 4x + 1 = 0 \quad \Leftrightarrow \quad x = -2 \pm \sqrt{5}$$

The slopes of the tangents to f at these points are obtained by evaluating the derivative $f'(x) = -2x - 4$ and are:

$$f'(-2 \pm \sqrt{5}) = \mp 2\sqrt{5}.$$

Problem 4.13. Prove that in the parabola with equation $y = Ax^2 + Bx + C$, the chord that joins the points with abscissa $x = a$ and $x = b$ is parallel to the tangent line to the parabola at the point with abscissa $x = \frac{a+b}{2}$.

Solution to the problem 4.13. *To see that the chord and the tangent are parallel, we must see that they have the same slope. Let f be the function that defines the parabola, the slope of the chord is:*

$$\begin{aligned} m_c &= \frac{f(b) - f(a)}{b - a} = \frac{Ab^2 + Bb + \not{C} - Aa^2 - Ba - \not{C}}{b - a} \\ &= \frac{A(b^2 - a^2) + B(b - a)}{b - a} = A(b + a) + B \end{aligned}$$

and on the other hand, the slope of the tangent to the parabola at $x = (a + b)/2$ is:

$$m_t = f'\left(\frac{a+b}{2}\right) = \not{2}A\frac{a+b}{\not{2}} + B$$

and, therefore, indeed, $m_c = m_t$.

Problem 4.14. Let f be the real function defined by:

$$f(x) = 5 - (x^2+1) \cdot \ln(x^2+1) - x^2, \ \forall x \in [0, e]$$

Determine, if possible, the number of roots in $[0, e]$.

Solution to the problem 4.14. *We need to calculate the number of roots (or zeros) of the function:*

$$f(x) = 5 - (x^2+1) \cdot \ln(x^2+1) - x^2.$$

This function is continuous in $[0, e]$ and differentiable in $(0, e)$, since it belongs to the class $\mathcal{C}^\infty(\mathbb{R})$. Given that:

$$f(0) = 5 > 0, \quad f(e) = -20.23 < 0,$$

by Bolzano's theorem, we know that there is at least one zero in this interval. To determine how many zeros there are, we need to study the growth and decrease of the function. Since the function is differentiable, we know that at points where $f'(x) > 0$, the function will be increasing and where $f'(x) < 0$, the function will be decreasing. To calculate the sign of $f'(x)$, we need to find the points where $f'(x)$ is zero.

$$\begin{aligned} f'(x) &= -2x\ln(x^2+1) - (x^2+1)\frac{2x}{x^2+1} - 2x \\ &= -2x\ln(x^2+1) - 2x - 2x \\ &= -2x(\ln(x^2+1) + 2) \end{aligned}$$

Therefore, $f'(x) = 0$ if $x = 0$ or if:

$$\ln(x^2+1) = -2 \quad \Longrightarrow \quad x^2+1 = e^{-2} \quad \Longrightarrow \quad x^2 = e^{-2} - 1 < 0.$$

Consequently, the only critical point in $[0, e]$ is $x = 0$.

Furthermore, given that:

$$f'(e) = -22.44 < 0,$$

the function f is strictly decreasing in the interval $[0, e]$, and, therefore, we have only one zero in the interval $[0, e]$. We can corroborate the result obtained by observing Figure 4.48.

Problem 4.15. Demonstrate using the definition of the derivative that if $f(x) = x^3$, then $f'(x) = 3x^2$.

Solution to the problem 4.15. *Applying the definition of the derivative of a function at a point we have that:*

$$\begin{aligned} f'(a) &= \lim_{h \to 0} \frac{f(a+h) - f(a)}{h} = \lim_{h \to 0} \frac{(a+h)^3 - a^3}{h} = \lim_{h \to 0} \frac{a^3 + 3a^2h + 3ah^2 + h^3 - a^3}{h} \\ &= \lim_{h \to 0} \frac{3a^2h + 3ah^2 + h^3}{h} = \lim_{h \to 0} 3a^2 + 3ah + h^2 = 3a^2 \end{aligned}$$

as we wanted to demonstrate.

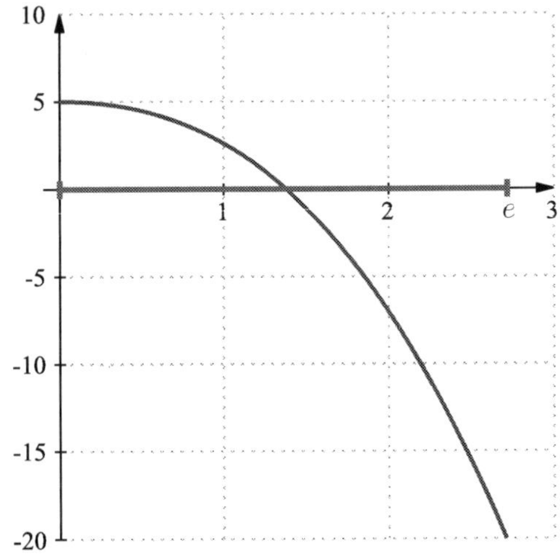

FIGURE 4.48
Graph of the function $f(x) = 5 - (x^2 + 1) \cdot \ln(x^2 + 1) - x^2$ in the interval $[0, e]$.

Problem 4.16. Calculate the derivative of the following functions:

(a) $f(x) = \ln\left(\frac{5+x}{5-x}\right)$

(b) $f(x) = \frac{3\cos(x)}{e^{x^2}}$

(c) $f(x) = \ln\left(\sqrt{1+\sin^2(x)}\right)$

Solution to the problem 4.16. (a) *We must calculate the derivative of the function:*

$$f(x) = \ln\left(\frac{5+x}{5-x}\right).$$

Applying the chain rule and the derivative of a quotient:

$$\begin{aligned} f'(x) &= \frac{1}{\frac{5+x}{5-x}} \cdot \left(\frac{5+x}{5-x}\right)' = \frac{5-x}{5+x} \cdot \frac{(5-x)-(5+x)(-1)}{(5-x)^2} = \frac{5-x}{5+x} \cdot \frac{10}{(5-x)^2} \\ &= \frac{10}{(5+x)(5-x)} = \frac{10}{25-x^2}. \end{aligned}$$

(b) *We must calculate the derivative of the function:*

$$f(x) = \frac{3\cos(x)}{e^{x^2}}.$$

Applying the derivative of a quotient:

$$f'(x) = \frac{-3\sin(x)e^{x^2} - 3\cos(x)e^{x^2}2x}{(e^{x^2})^2} = \frac{-3\sin(x) - 6x\cos(x)}{e^{x^2}}.$$

(c) *We must calculate the derivative of the function:*

$$f(x) = \ln\left(\sqrt{1+\sin^2(x)}\right).$$

Applying the chain rule:

$$f'(x) = \frac{1}{\sqrt{1+\sin^2(x)}} \cdot \frac{1}{2\sqrt{1+\sin^2(x)}} \cdot 2\sin(x)\cos(x) = \frac{\sin(x)\cos(x)}{1+\sin^2(x)}.$$

Problem 4.17. Calculate the derivative of the following functions:

(a) $f(x) = \sin(x)\sin(x^2)$

(b) $f(x) = \dfrac{x}{\ln\left(\frac{1}{x}\right)}$

(c) $f(x) = \ln\left(\sqrt{1+\cos^2(x)}\right)$

(d) $f(x) = \dfrac{2\sin(x^2)}{e^{x^2}}$

Solution to the problem 4.17. (a) *We must calculate the derivative of the function* $f(x) = \sin(x)\sin(x^2)$. *Applying the derivative of a product and the chain rule:*

$$\begin{aligned} f'(x) &= \cos(x)\sin(x^2) + \sin(x)\cos(x^2)2x \\ &= \sin(x^2)\cos(x) + 2x\sin(x)\cos(x^2). \end{aligned}$$

(b) *We must calculate the derivative of the function:*

$$f(x) = \frac{x}{\ln\left(\frac{1}{x}\right)}.$$

Rewriting the logarithm as:

$$f(x) = \frac{x}{\ln\left(\frac{1}{x}\right)} = \frac{x}{-\ln(x)}.$$

and applying the derivative of a quotient:

$$f'(x) = \frac{-\ln(x) - x\frac{-1}{x}}{\ln^2(x)} = \frac{1-\ln(x)}{\ln^2(x)}.$$

(c) *We must calculate the derivative of the function* $f(x) = \ln\left(\sqrt{1+\cos^2(x)}\right)$. *Applying the chain rule:*

$$\begin{aligned} f'(x) &= \frac{1}{\sqrt{1+\cos^2(x)}} \cdot \frac{1}{2}\left(1+\cos^2(x)\right)^{-1/2} \cdot 2\cos(x)(-\sin(x)) \\ &= -\frac{1}{2}\frac{1}{1+\cos^2(x)} \cdot 2\sin(x)\cos(x) \\ &= \frac{-\sin(x)\cos(x)}{1+\cos^2(x)}. \end{aligned}$$

(d) *We must calculate the derivative of the function:*

$$f(x) = \frac{2\sin(x^2)}{e^{x^2}}.$$

Applying the derivative of a quotient and the chain rule:

$$\begin{aligned} g'(x) &= \frac{2\cos(x^2)\cdot 2xe^{x^2} - 2\sin(x^2)e^{x^2}\cdot 2x}{\left(e^{x^2}\right)^2} \\ &= \frac{4x\left(\cos(x^2) - \sin(x^2)\right)}{e^{x^2}} \end{aligned}$$

Problem 4.18. Prove that the derivative of the function $f(x) = 2^x$ is $f'(x) = 2^x\ln(2)$. Help:

(1) Take logarithms in the equation $f(x) = 2^x$ and simplify the result.

(2) Differentiate the equation obtained in step (1) using the chain rule. You will obtain an equation where $f'(x)$ will appear.

(3) Isolate the derivative $f'(x)$ to obtain the desired result.

Solution to the problem 4.18. *We know that* $f(x) = 2^x$*, therefore:*

$$\ln(f(x)) = \ln(2^x) = x\ln(2).$$

If we differentiate this expression, applying the chain rule, we obtain:

$$\begin{aligned} (\ln(f(x)))' = (x\ln(2))' \quad &\Longrightarrow \quad \frac{1}{f(x)}\cdot f'(x) = \ln(2) \quad \Longrightarrow \\ &\Longrightarrow \quad f'(x) = \ln(2)f(x) = \ln(2)2^x = 2^x\ln(2) \end{aligned}$$

as we wanted to show.

Problem 4.19. Deduce the derivative of the function $g(x) = \sqrt{x}$ knowing that $f(x) = x^2$ and $g(x)$ are inverse functions with respect to composition, that is:

$$(f\circ g)(x) = (g\circ f)(x) = x.$$

Help: (1) Note that $f'(x) = 2x$.
(2) Apply the chain rule appropriately.

Solution to the problem 4.19. *Applying the chain rule to the derivative of* $(f\circ g)(x) = x$ *we obtain:*

$$\begin{aligned} (f\circ g)'(x) = x' \quad &\Longleftrightarrow \quad f'(g(x))\cdot g'(x) = 1 \\ &\Longleftrightarrow \quad g'(x) = \frac{1}{f'(g(x))} \\ &\Longleftrightarrow \quad g'(x) = \frac{1}{2\sqrt{x}} \end{aligned}$$

since $f'(x) = 2x$ *and* $g(x) = \sqrt{x}$*, and therefore* $f'(g(x)) = 2\sqrt{x}$*.*

Problem 4.20. Prove, by calculating the derivative of the inverse function, that if:

$$g(x) = \arcsin(x)$$

then:

$$g'(x) = \frac{1}{\sqrt{1-x^2}}.$$

Help: express $\cos(x)$ in terms of $\sin(x)$ using the relationship:

$$\cos^2(x) + \sin^2(x) = 1.$$

Remember that the sine function, to be invertible, must be restricted to the interval $[-\frac{\pi}{2}, \frac{\pi}{2}]$.

Solution to the problem 4.20. *The function* $\arcsin(x)$ *is the inverse of the function* $\sin(x)$ *when restricted to the interval* $[-\frac{\pi}{2}, \frac{\pi}{2}]$. *Recall that one way to calculate the derivative of the inverse of a function is:*

$$\left(f^{-1}(x)\right)' = \frac{1}{f'(f^{-1}(x))} = \frac{1}{\cos(\arcsin(x))}.$$

Following the given suggestion, recall that $\sin^2(x) + \cos^2(x) = 1$, *and, therefore:*

$$\cos(x) = \pm\sqrt{1-\sin^2(x)}.$$

Thus:

$$\cos(x) = \begin{cases} +\sqrt{1-\sin^2(x)} & , x \in [-\frac{\pi}{2}, \frac{\pi}{2}] \\ -\sqrt{1-\sin^2(x)} & , x \in [\frac{\pi}{2}, \frac{3\pi}{2}] \end{cases}$$

Since the values of the function $\arcsin(x)$ *always belong to the interval* $[-\frac{\pi}{2}, \frac{\pi}{2}]$:

$$\cos(\arcsin(x)) = +\sqrt{1-\sin^2(\arcsin(x))} = +\sqrt{1-x^2},$$

therefore:

$$\left(f^{-1}(x)\right)' = \frac{1}{\sqrt{1-x^2}}.$$

Problem 4.21. Let $f(x)$ and $g(x)$ be two differentiable functions. Consider the function:

$$h(x) = f(x)^{g(x)}. \tag{4.4}$$

(a) Calculate $h'(x)$ following these steps:

  (i) Apply the natural logarithm to both sides of equation (4.4).

  (ii) Differentiate the result from step (i) using the chain rule.

  (iii) The result from step (ii) is a function of $h'(x)$. Solve for $h'(x)$ (i.e., isolate $h'(x)$).

(b) Calculate the derivative of the function $h(x) = (2x)^{\cos(x)}$ using the previous result.

(c) Given the point $P = (\frac{1}{2}, 1)$, find the equation of the tangent line to the graph of the function $h(x) = (2x)^{\cos(x)}$ at this point.

Solution to the problem 4.21. *(a) Following the suggested steps:*

(i) Apply the natural logarithm to both sides of the equation:

$$\ln(h(x)) = g(x)\ln(f(x)).$$

(ii) Differentiate the result obtained in the previous step:

$$\frac{h'(x)}{h(x)} = g'(x)\ln(f(x)) + g(x)\frac{f'(x)}{f(x)}.$$

(iii) Isolate $h'(x)$ in the previous expressions:

$$h'(x) = h(x)\left(g'(x)\ln(f(x)) + g(x)\frac{f'(x)}{f(x)}\right)$$

(b) Using the previous result, calculate the derivative of the function:

$$h(x) = (2x)^{\cos(x)}.$$

In this case, $f(x) = 2x$ and $g(x) = \cos(x)$, therefore, substituting into the expression obtained in step (iii) we obtain:

$$\begin{aligned} h'(x) &= h(x)\left(g'(x)\ln(f(x)) + g(x)\frac{f'(x)}{f(x)}\right) \\ &= (2x)^{\cos(x)}\left(-\sin(x)\ln(2x) + \cos(x)\frac{2}{2x}\right) \end{aligned}$$

which can be simplified as:

$$h'(x) = (2x)^{\cos(x)}\left(-\sin(x)\ln(2x) + \frac{\cos(x)}{x}\right).$$

(c) We need to find the equation of the tangent line to the graph of the function $h(x) = (2x)^{\cos(x)}$ at the point $P = (\frac{1}{2}, 1)$. At the point $x = \frac{1}{2}$, the slope of the tangent line is $h'(\frac{1}{2}) = 2\cos\frac{1}{2}$. The tangent line, $r(x)$, at the point $x = \frac{1}{2}$ can be written as:

$$r(x) = h'\left(\frac{1}{2}\right)(x - \frac{1}{2}) + h\left(\frac{1}{2}\right)$$

where substituting the values of $h'\left(\frac{1}{2}\right)$ and $h\left(\frac{1}{2}\right)$ gives:

$$r(x) = 2\cos\left(\frac{1}{2}\right)(x - \frac{1}{2}) + 1.$$

Problem 4.22. Consider the function $f(x) = 2x^2 - \ln(x)$.

(a) Determine the equation of the tangent line to the graph of $f(x)$ at the point $(1, 2)$.

(b) Prove that $f(x)$ is equal to 3 for some point within the interval $[1, e]$.
Help: $e \approx 2.7$ and $e^2 \approx 7.4$.

Solution to the problem 4.22. (a) *The equation of the tangent line at the point $x_0 = 1$ with $f(x_0) = 2$ is:*

$$y = f'(x_0)(x - x_0) + f(x_0).$$

Therefore, considering that $f'(x) = 4x - \frac{1}{x}$ and, consequently, that $f'(1) = 3$ we obtain the tangent line:

$$y = 3(x - 1) + 2 = 3x - 1.$$

In Figure 4.49 (left) the function and the obtained tangent line at the point $(1, 2)$ are plotted.

(b) *The function $f(x)$ is continuous in the interval $[1, e]$, as it is continuous in its entire domain $(0, +\infty)$ being the difference of continuous functions. Additionally:*

$$f(1) = 2 - \ln(1) = 2, \quad f(e) = 2e^2 - \ln(e) = 2e^2 - 1 \approx 13.8.$$

Thus, by the intermediate value theorem, for any value $y \in (2, 13.8)$ there exists a point $c \in (1, e)$ such that $f(c) = y$. In particular, if we take:

$$y = 3 \in (2, 13.8),$$

we can assure that there is at least one point $c \in (1, e)$ such that $f(c) = 3$, as stated in the problem.

In Figure 4.49 (right) the function is plotted, and the existence of a point in the interval $c \in (1, e)$ such that $f(c) = 3$ can be visually verified.

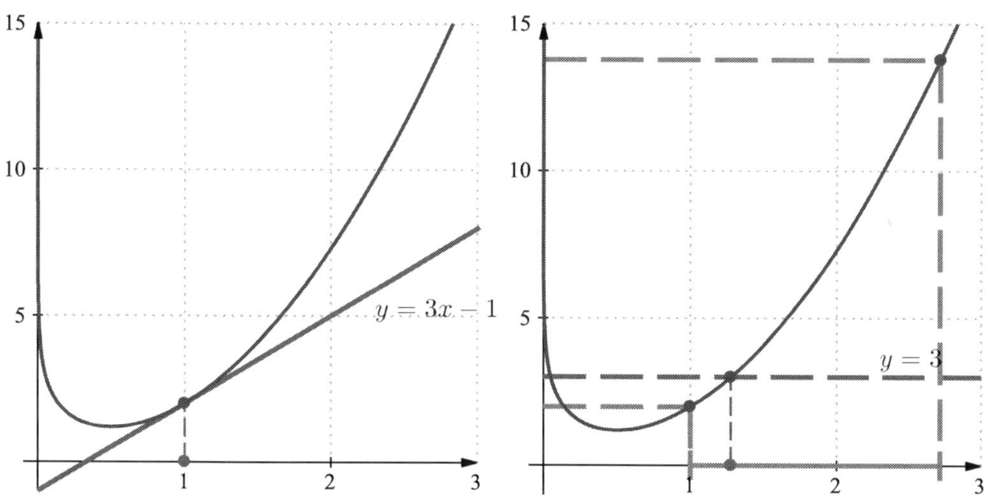

FIGURE 4.49
Graph of the function in Problem 4.22 and its tangent line at the point $(1, 2)$.

Problem 4.23. Consider the function $f(x) = \sin^2(x)\cos(x)$. Determine the equation of the tangent line to the graph of $f(x)$ at the point $(\frac{\pi}{2}, 0)$.

Solution to the problem 4.23. *The equation of the tangent line to the function $f(x)$ passing through the point $(\frac{\pi}{2}, 0)$ is:*

$$y = 0 + f'\left(\frac{\pi}{2}\right)\left(x - \frac{\pi}{2}\right).$$

It is necessary, then, to calculate the derivative of the function $f(x)$ and evaluate it at $x = \frac{\pi}{2}$. Applying the formula for the derivative of a product and the chain rule:

$$\begin{aligned} f'(x) &= [\sin^2(x)]'\cos(x) + \sin^2(x)[\cos(x)]' \\ &= 2\sin(x)\cos(x)\cos(x) + \sin^2(x)(-\sin(x)) \\ &= 2\sin(x)\cos^2(x) - \sin^3(x). \end{aligned}$$

Evaluating $f'(x)$ at $x = \frac{\pi}{2}$, we obtain $f'\left(\frac{\pi}{2}\right) = -1$.

Therefore, the equation of the line is:

$$y = \frac{\pi}{2} - x.$$

In Figure 4.50 the function and the tangent line obtained at the point $(\frac{\pi}{2}, 0)$ are plotted.

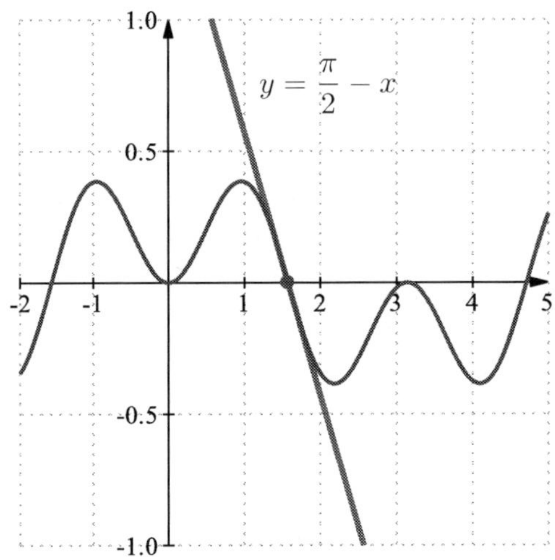

FIGURE 4.50
Graph of the function $f(x) = \sin^2(x)\cos(x)$ and its tangent line at the point $(\frac{\pi}{2}, 0)$.

Problem 4.24. Consider the function $f(x) = x^{\sin(x)}$, $x > 0$.

(a) Determine the equation of the tangent line to the graph of $f(x)$ at the point $(1, 1)$.

(b) Prove that $f(x)$ has at least one critical point within the interval $[1, e]$. Use, if necessary, that $\sin(1) > 0$, $\cos(1) > 0$, $\cos(e) < 0$, $\sin(e) > 0$ and $e\cos(e) < -\sin(e)$.

Solution to the problem 4.24. (a) *To write the equation of the tangent line, it is necessary to calculate the derivative of the function $f(x) = x^{\sin(x)}$. We will calculate* *this derivative by applying logarithms to both sides of the equality and differentiating:*

$$\begin{aligned} f(x) = x^{\sin(x)} &\iff \ln(f(x)) = \sin(x)\ln(x) \\ &\iff \frac{f'(x)}{f(x)} = \cos(x)\ln(x) + \frac{\sin(x)}{x} \\ &\iff f'(x) = f(x)\left(\cos(x)\ln(x) + \frac{\sin(x)}{x}\right) \\ &\iff f'(x) = x^{\sin(x)}\left(\cos(x)\ln(x) + \frac{\sin(x)}{x}\right). \end{aligned}$$

The equation of the tangent line at the point $(1,1)$ *is* $y-1 = f'(1)(x-1)$. *The slope of the tangent line is* $f'(1)$, *which is:*

$$f'(1) = 1^{\sin(1)}\left(\cos(1)\ln(1) + \frac{\sin(1)}{1}\right) = \sin(1).$$

Therefore, the equation of the tangent line is:

$$y - 1 = \sin(1)(x-1).$$

In Figure 4.51 the function $f(x)$, *as well as the tangent line at the point* $(1,1)$, *are plotted.*

(b) *The derivative of the function* $f(x)$ *is continuous in the interval* $[1, e]$ *because it is formed by continuous functions. Then, evaluating* $f'(x)$ *at the endpoints of the interval we get:*

$$\begin{aligned} f'(1) &= \sin(1) > 0 \\ f'(e) &= e^{\sin(e)}\left(\cos(e)\ln(e) + \frac{\sin(e)}{e}\right) \\ &= \underbrace{e^{\sin(e)}}_{>0}\underbrace{\left(\cos(e) + \frac{\sin(e)}{e}\right)}_{<0,\ \text{as } e\cos(e) < -\sin(e)} < 0. \end{aligned}$$

Therefore, since there is a sign change at the endpoints, by Bolzano's theorem, we can guarantee the existence of a point $c \in (1, e)$ *such that* $f'(c) = 0$, *i.e., we can guarantee the existence of a critical point.*

In Figure 4.51 these results can be verified.

Problem 4.25. Given the function:

$$f(x) = e^{ix},$$

where i is the imaginary unit, calculate $f''(x) + f(x)$.

Solution to the problem 4.25. *Given:*

$$f(x) = e^{ix}$$

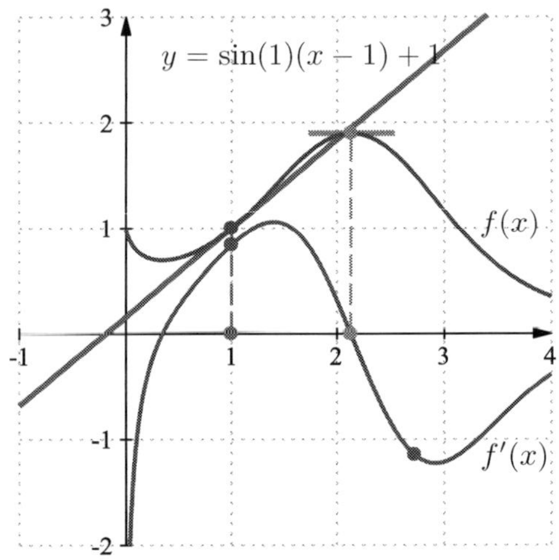

FIGURE 4.51
Graph of the function $f(x) = x^{\sin(x)}$, $x > 0$, its derivative, and the tangent line at the point $(1, 1)$.

We need to calculate $f''(x) + f(x)$. *We calculate the derivatives of the function:*

$$\begin{aligned} f(x) &= e^{ix} \\ f'(x) &= ie^{ix} \\ f''(x) &= i \cdot ie^{ix} = i^2 e^{ix} = -e^{ix} \quad \text{because} \quad i^2 = -1. \end{aligned}$$

Therefore:

$$f''(x) + f(x) = -e^{ix} + e^{ix} = 0.$$

4.16 Solved Problems (Extremes)

Problem 4.26. Find the absolute extremes of the function $f(x) = x^3 - 3x$ in the interval $[0, 2]$.

Solution to the problem 4.26. *First, we calculate the candidates for extremum:*

(A) ***Points where the function is not continuous.*** *Since* $f \in \mathcal{C}^\infty(\mathbb{R})$, *we know that* f *is continuous in* $[0, 2]$. *We do not have any candidate for an extremum.*

(B) ***Critical points.*** *Since* $f \in \mathcal{C}^\infty(\mathbb{R})$, *there are no points where the function is not differentiable. We study the points where the derivative is zero:*

$$f'(x) = 3x^2 - 3 = 0 \quad \Leftrightarrow \quad x^2 = 1 \Leftrightarrow x = \pm 1.$$

We discard the point $x = -1$ *as it lies outside the interval. Therefore, we have* $x = 1$ *as the only candidate for an extremum.*

(C) ***Endpoints of the interval.*** *The endpoints of the interval are* $x = 0$ *and* $x = 2$.

Thus, the candidates for extremum are the points $x = 0$, $x = 1$, and $x = 2$. We study the growth of the function in the intervals generated by these points.

- *Interval $(0, 1)$. If we consider, for example, the point $x = 1/2$, the value of $f'(1/2) = -9/4$ is negative, so the function is decreasing.*
- *Interval $(1, 2)$. If we consider, for example, the point $x = 3/2$, the value of $f'(3/2) = 15/4$ is positive, so the function is increasing.*

Therefore, $x = 0$ is a maximum, $x = 1$ is a minimum, and $x = 2$ is a maximum.
If we also evaluate the function at these points:

$$f(0) = 0$$
$$f(1) = -2$$
$$f(2) = 2$$

we can conclude that:

- *$x = 0$ is a relative maximum,*
- *$x = 1$ is a global minimum, and*
- *$x = 2$ is a global maximum.*

We can verify the results obtained by observing Figure 4.52.

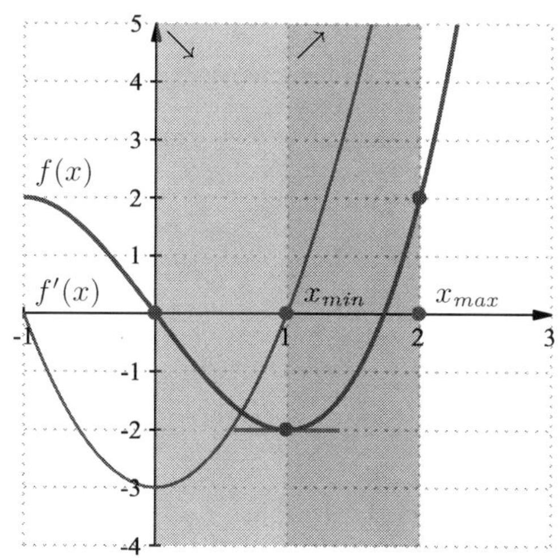

FIGURE 4.52
Graph of the function $f(x) = x^3 - 3x$ in the interval $[0, 2]$.

Problem 4.27. Find the relative extremes of $f(x) = x + \sqrt{x^3}$.

Solution to the problem 4.27. *The function f is defined in $[0, \infty)$, as we cannot calculate the square root of a negative number. First, we calculate the candidates for extremum:*

(A) Points where the function is not continuous. *The function f is continuous in $[0, \infty)$. We do not have any candidate for an extremum.*

(B) ***Critical points***. *The function is differentiable in* $(0, +\infty)$. *We study the points where the derivative is zero:*

$$f'(x) = 1 + \frac{3}{2}\sqrt{x} > 0, \ \forall x \in (0, +\infty)$$

Therefore, there are no critical points.

(C) ***Endpoints of the interval***. *The only endpoint of the interval is* $x = 0$.

The only candidate for an extremum is the point $x = 0$. *We study the growth of the function in the interval* $(0, +\infty)$.

- *Interval* $(0, +\infty)$. *If we consider, for example, the point* $x = 1$, *the value of* $f'(1) = 5/2$ *is positive, so the function is increasing.*

Therefore, $x = 0$ *is a minimum.*
If we also evaluate the function at this point:

$$f(0) = 0$$

and calculate the limit of the function as x *tends to* $+\infty$:

$$\lim_{x \to +\infty} f(x) = +\infty$$

we can conclude that:

- $x = 0$ *is an absolute minimum.*

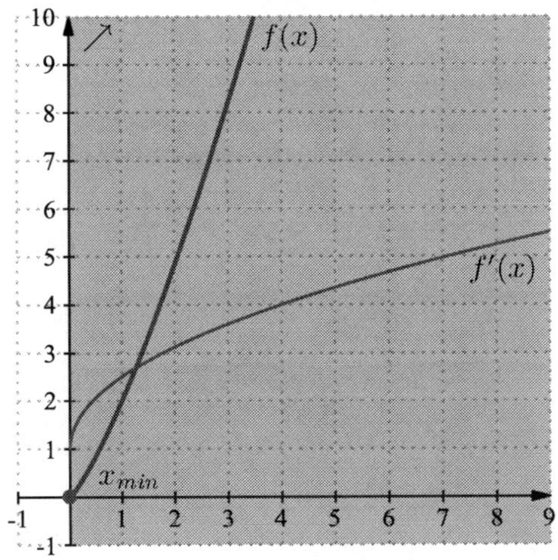

FIGURE 4.53
Graph of the function $f(x) = x + \sqrt{x^3}$.

Problem 4.28. Find the extremes of the function $f(x) = x \ln(x)$ in the interval $[0.1, 3]$.

Solution to the problem 4.28. *The function* f *is defined over the entire interval* $[0.1, 3]$. *First, we calculate the candidates for extremum:*

(A) Points where the function is not continuous. *The function f is continuous in $[0.1, 3]$. We do not have any candidate for an extremum.*

(B) Critical points. *The function is differentiable in $[0.1, 3]$. We study the points where the derivative is zero:*

$$f'(x) = 1 + \ln(x) = 0 \Leftrightarrow \ln(x) = -1 \Leftrightarrow x = e^{-1}$$

Since $x = e^{-1}$ lies within the interval $[0.1, 3]$, this is a candidate for an extremum.

(C) Endpoints of the interval. *The endpoints of the interval are $x = 0.1$ and $x = 3$.*

Thus, the candidates for extremum are the points $x = 0.1$, $x = e^{-1}$, and $x = 3$. We study the growth of
the function in the intervals generated by these points.

- *Interval $(0.1, e^{-1})$. If we consider, for example, the point $x = 1/5$, the value of $f'(1/5) = 1 - \ln(5)$ is negative, so the function is decreasing.*
- *Interval $(e^{-1}, 3)$. If we consider, for example, the point $x = 2$, the value of $f'(2) = 1 + \ln(2)$ is positive, so the function is increasing.*

Therefore, $x = 0.1$ is a maximum, $x = e^{-1}$ is a minimum, and $x = 3$ is a maximum. If we also evaluate the function at these points:

$$f(0.1) = -0.2303$$
$$f(e^{-1}) = -0.3679$$
$$f(3) = 3.2958$$

we can conclude that:

- *$x = 0.1$ is a relative maximum,*
- *$x = e^{-1}$ is an absolute minimum, and*
- *$x = 3$ is an absolute maximum.*

Figure 4.54 shows the graph of the function, and the extremes found can be observed.

Problem 4.29. Find the extremes of the function $f(x) = x\sin(x)$ in the interval $[0, 6]$.

Solution to the problem 4.29. *First, we calculate the candidates for extremum:*

(A) Points where the function is not continuous. *The function f is continuous in $[0, 6]$. We do not have any candidate for an extremum.*

(B) Critical points. *The function is differentiable in $[0, 6]$. We study the points where the derivative is zero:*

$$f'(x) = \sin(x) + x\cos(x) = 0 \Leftrightarrow -x = \tan(x)$$

This equation cannot be solved analytically; therefore, we will use Bolzano's Theorem. To do this, we must first identify in which intervals we will find the zeros. In this case, the critical points are given by the intersection of the line $y = -x$ and the curve $y = \tan(x)$. As shown in Figure 4.55,

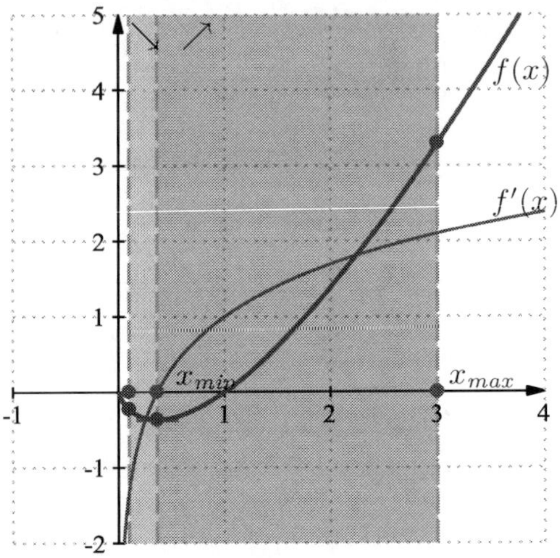

FIGURE 4.54
Graph of the function $f(x) = x\ln(x)$ in the interval $[0.1, 3]$.

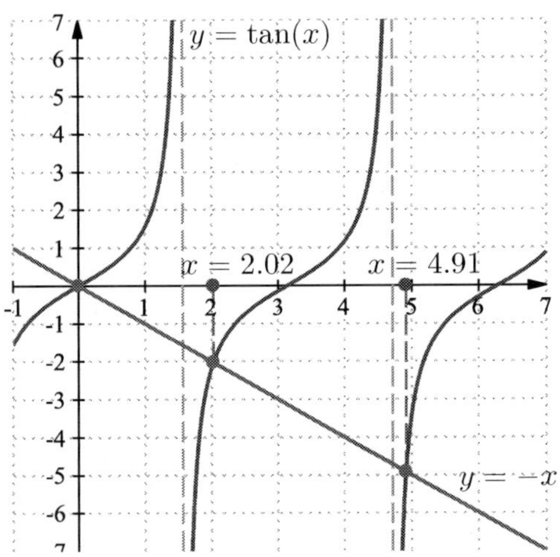

FIGURE 4.55
Graph of the functions $y = -x$ and $y = \tan(x)$.

these functions intersect at three points in the interval $[0, 6]$. *The first intersection point is* $x = 0$, *the second lies in the interval* $(\frac{\pi}{2}, \pi]$, *and the third in the interval* $(\frac{3\pi}{2}, 2\pi]$. *To determine the last two intersection points, we will use the bisection method applied to the function* $f'(x) = \sin(x) + x\cos(x)$.

First, consider the interval $[\frac{\pi}{2}, \pi]$. *In Table 4.4, we observe that the sought intersection point with two correct decimal places is* $x = 2.02$.

Next, consider the interval $[\frac{3\pi}{2}, 2\pi]$. *In Table 4.5, we observe that the sought intersection point with two correct decimal places is* $x = 4.91$.

Thus, the critical points are $x = 2.02$ *and* $x = 4.91$.

TABLE 4.4
Bisection method applied to the function $\sin(x) + x\cos(x)$ in the interval $[\frac{\pi}{2}, \pi]$.

x_e	x_d	$f'(x_e)$	$f'(x_d)$	x_m	$f'(x_m)$
1.5708	3.1416	0.99999	−3.1416	2.3562	−0.95900
1.5708	2.3562	0.99999	−0.95900	1.9635	0.17247
1.9635	2.3562	0.17247	−0.95900	2.1598	−0.36831
1.9635	2.1598	0.17247	−0.36831	2.0616	−0.08974
1.9635	2.0616	0.17247	−0.08974	2.0126	0.04345
2.0126	2.0616	0.04345	−0.08974	2.0371	−0.02262
2.0126	2.0371	0.04345	−0.02262	2.0249	0.01041
2.0249	2.0371	0.01041	−0.02262	2.0310	−0.00607
2.0249	2.0310	0.01041	−0.00607	2.0279	0.00232
2.0279	2.0310	0.00232	−0.00607	2.0295	−0.00202
2.0279	2.0295	0.00232	−0.00201	2.0288	−0.00011

TABLE 4.5
Bisection method applied to the function $\sin(x) + x\cos(x)$ in the interval $[\frac{3\pi}{2}, 2\pi]$.

x_e	x_d	$f'(x_e)$	$f'(x_d)$	x_m	$f'(x_m)$
4.7124	6.2832	−0.99995	6.2832	5.4978	3.1805
4.7124	5.4978	−0.99995	3.1805	5.1051	1.0298
4.7124	5.1051	−0.99995	1.0298	4.9088	−0.02282
4.9088	5.1051	−0.02282	1.0298	5.0070	0.49698
4.9088	5.0070	−0.02282	0.49698	4.9579	0.23499
4.9088	4.9579	−0.02282	0.23499	4.9334	0.10582
4.9088	4.9334	−0.02282	0.10582	4.9211	0.04140
4.9088	4.9211	−0.02282	0.04140	4.9150	0.00951
4.9088	4.9150	−0.02282	0.00951	4.9119	−0.00667
4.9119	4.9150	−0.00667	0.00951	4.9135	0.00167

(C) Endpoints of the interval. *The endpoints of the interval are* $x = 0$ *and* $x = 6$.

Thus, the candidates for extremum are the points $x = 0$, $x = 2.02$, $x = 4.91$, *and* $x = 6$. *We study the growth of the function in the intervals generated by these points.*

- *Interval* $(0, 2.02)$. *If we consider, for example, the point* $x = 1$, *the value of* $f'(1) = \sin(1) + \cos(1)$ *is positive, so the function is increasing.*
- *Interval* $(2.02, 4.91)$. *If we consider, for example, the point* $x = 3$, *the value of* $f'(3) = \sin(3) + 3\cos(3)$ *is negative, so the function is decreasing.*
- *Interval* $(4.91, 6)$. *If we consider, for example, the point* $x = 5$, *the value of* $f'(5) = \sin(5) + 5\cos(5)$ *is positive, so the function is increasing.*

Figure 4.56 schematically represents the growth/decrease of the function in each of the studied intervals.

Therefore, $x = 0$ *is a minimum,* $x = 2.02$ *is a maximum,* $x = 4.91$ *is a minimum, and* $x = 6$ *is a maximum.*

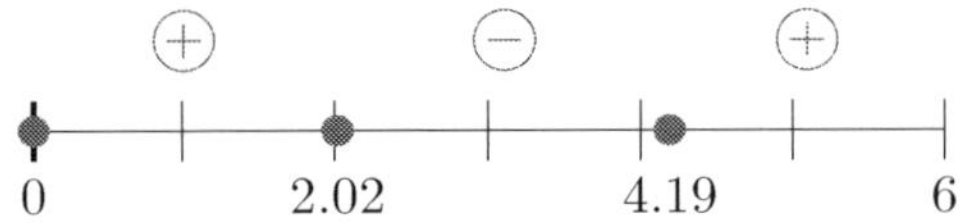

FIGURE 4.56
Schematic representation of the growth/decrease of the function.

If we also evaluate the function at these points:

$$f(0) = 0$$
$$f(2.02) = 1.8196$$
$$f(4.91) = -4.8144$$
$$f(6) = -1.6764$$

we can conclude that:

- $x = 0$ *is a relative minimum.*
- $x = 2.02$ *is an absolute maximum.*
- $x = 4.91$ *is an absolute minimum.*
- $x = 6$ *is a relative maximum.*

Figure 4.57 shows the graph of the studied function, and we can verify the obtained results.

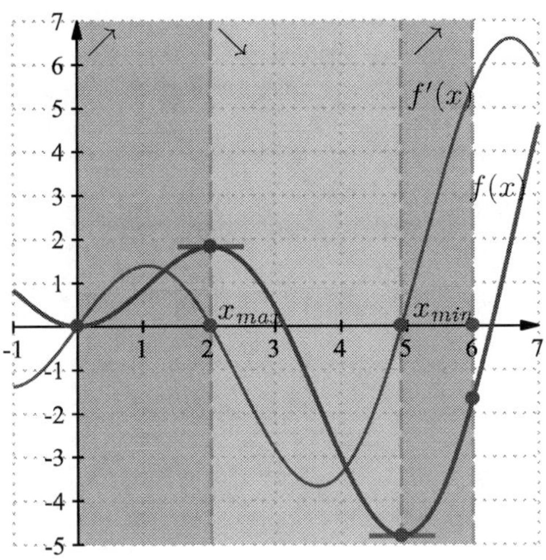

FIGURE 4.57
Graph of the function $f(x) = x\sin(x)$ in the interval $[0, 6]$.

Problem 4.30. Find the local and global extremes of the function:

$$f(x) = \begin{cases} 3x^4 - 4x^3, & x \leq 2 \\ 20 - x^2, & x > 2 \end{cases}$$

in the interval $[-1, 5]$.

Solution to the problem 4.30. *First, we calculate the candidates for extremum:*

(A) points where the function is not continuous.

(B) critical points.

(C) endpoints of the interval.

We will study each of the previous points.

(A) *For $x \neq 2$, the function is continuous, since it is polynomial. Thus, we only need to study the continuity of the function at the point $x = 2$.*

The function will be continuous at $x = 2$ if:

$$f(2) = \lim_{x \to 2^+} f(x) = \lim_{x \to 2^-} f(x).$$

$$\left.\begin{aligned} f(2) &= 3 \cdot 2^4 - 4 \cdot 2^3 = 2^3(6-4) = 2^4 = 16 \\ \lim_{x \to 2^+} f(x) &= \lim_{x \to 2^+} 20 - x^2 = 20 - 2^2 = 16 \\ \lim_{x \to 2^-} f(x) &= \lim_{x \to 2^-} 3x^4 - 4x^3 = 3 \cdot 2^4 - 4 \cdot 2^3 = 16 \end{aligned}\right\}$$

Therefore, the function is continuous in $\mathbb{R}$, that is, $f(x) \in \mathcal{C}^0(\mathbb{R})$.

(B) *For $x \neq 2$, the derivative is:*

$$f'(x) = \begin{cases} 12x^3 - 12x^2 & x < 2 \\ -2x & x > 2 \end{cases}$$

We calculate the derivative at the point $x = 2$ using the lateral derivatives:

$$\begin{aligned} f'_-(2) &= \lim_{x \to 2^-} f'(x) = \lim_{x \to 2^-} 12x^3 - 12x^2 = 12 \cdot 2^3 - 12 \cdot 2^2 \\ &= 12 \cdot 2^2 \cdot (2-1) = 48 \\ f'_+(2) &= \lim_{x \to 2^+} f'(x) = \lim_{x \to 2^-} -2x = -2 \cdot 2 = -4 \end{aligned}$$

Therefore, since $f'_-(2) \neq f'_+(2)$, the function is not differentiable at $x = 2$ and, therefore, $x = 2$ is a candidate for an extremum. Regarding the rest of the critical points, we must find the points such that $f'(x) = 0$. Given that the derivative has a different definition if $x > 2$ or $x < 2$, we must solve $f'(x) = 0$ in both intervals separately.

- *If $x < 2$ $f'(x) = 0 \Longleftrightarrow 12x^3 - 12x^2 = 0 \Longleftrightarrow 12x^2(x-1) = 0 \Longleftrightarrow x = 0, 1$*
- *If $x > 2$ $f'(x) = 0 \Longleftrightarrow -2x = 0 \Longleftrightarrow x = 0$ but since $0 < 2$, there are no critical points for $x > 2$.*

Therefore, the points $x = 0$ and $x = 1$ are candidates for an extremum.

(C) The endpoints of the interval $x = -1$ and $x = 5$.

Therefore, the candidates for an extremum are $x = -1, 0, 1, 2$, and 5. With these points and their images, we can construct a table of values (see Table 4.6). Since the function is continuous in $[-1, 5]$, by the maximum and minimum theorem (Weierstrass), we know that the absolute maximum and minimum exist in that interval, and according to the Table $x = 2$ is the absolute maximum and $x = 5$ is the absolute minimum.

TABLE 4.6
Images of the candidates for the extremum of Problem 4.30.

x	$f(x)$
-1	7
0	0
1	-1
2	16
5	-5

To classify the other points, we need to study the growth/decrease of the function, that is, the sign of the derivative. Since we need to study the behavior of the function around the points $x = -1, 0$, and 1, that is, at points less than 2, we need to study the sign of the function:

$$12x^3 - 12x^2 = 12x^2(x - 1).$$

We observe that the sign of the derivative matches the sign of the factor $x - 1$. Specifically,

- *If $x < 1$, the $f'(x) = \oplus \cdot \ominus < 0$ and, therefore, the function is decreasing.*
- *If $x > 1$, the $f'(x) = \oplus \cdot \oplus > 0$ and, therefore, the function is increasing.*

That is, the function is decreasing to the right of $x = -1$, so at this point, we have a relative maximum. The function is decreasing both to the left and right of the point $x = 0$; therefore, there is no extremum at this point. Finally, the function $f(x)$ decreases to the left of the point $x = 1$ and increases to the right, so at $x = 1$ we have a relative minimum.

In summary:

- *$x = -1$ is a relative maximum,*
- *$x = 0$ is not an extremum,*
- *$x = 1$ is a relative minimum,*
- *$x = 2$ is an absolute maximum,*
- *$x = 5$ is an absolute minimum.*

Finally, in Figure 4.58, we can verify the obtained results.

Problem 4.31. Let $f(x)$ be the function defined by:

$$f(x) = \begin{cases} x - e^{-x} & x \leq 1 \\ \dfrac{1}{4x - x^2 - 5} + a & x > 1 \end{cases}$$

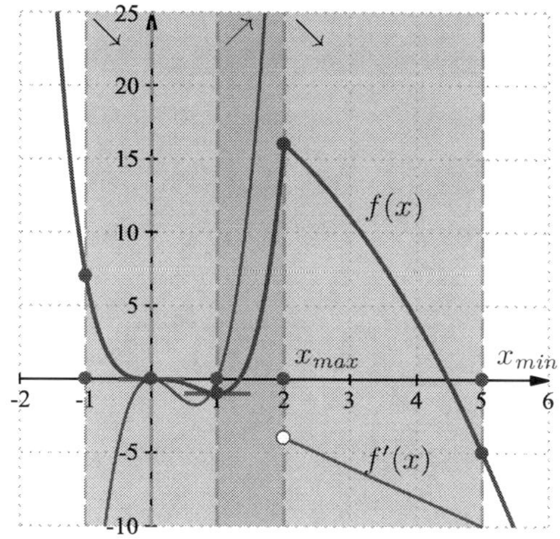

FIGURE 4.58
Graph of the function of Problem 4.30 in the interval $[-1, 5]$.

(a) Calculate the value of a that makes the function continuous.
In the following parts, take the value of a obtained in part (a).

(b) Calculate $f'(x)$ for all values of $x \in \mathbb{R}$ knowing that $f'_-(1) = 1 + e^{-1}$ and that $f'_+(1) = -1/2$.

(c) Find the extrema of the function in the interval $[0, 5]$.

(d) Prove that the function $f(x)$ has a zero in the interval $[0, 1]$.

Hint: $e^{-1} \approx 0.37$.

Solution to the problem 4.31. (a) *We must calculate the value of a that makes the function continuous.*

- *If $x < 1$, the function is $x - e^{-x}$ in a neighborhood of the point and, therefore, is continuous as it is the sum of a polynomial and an exponential function.*
- *If $x > 1$, the function is:*

$$\frac{1}{4x - x^2 - 5} + a$$

in a neighborhood of the point and, therefore, is continuous, as in the quotient, the denominator does not vanish at any point. The roots of $4x - x^2 - 5$ are:

$$x = \frac{-4 \pm \sqrt{16 - 20}}{-2} = \frac{-4 \pm \sqrt{-4}}{-2} \notin \mathbb{R}.$$

- *The function will be continuous at $x = 1$ if:*

$$f(1) = \lim_{x \to 1^-} f(x) = \lim_{x \to 1^+} f(x)$$

$$\begin{cases} f(1) = 1 - e^{-1} \\ \lim\limits_{x \to 1^-} f(x) = \lim\limits_{x \to 1^-} x - e^{-x} = 1 - e^{-1} \\ \lim\limits_{x \to 1^+} f(x) = \lim\limits_{x \to 1^+} \frac{1}{4x - x^2 - 5} + a = \frac{1}{-2} + a = -\frac{1}{2} + a \end{cases}$$

Thus, the function will be continuous if:

$$-\frac{1}{2} + a = 1 - e^{-1} \iff a = 1 - e^{-1} + \frac{1}{2} = \frac{3}{2} - e^{-1}.$$

(b) *We must calculate $f'(x)$ when:*

$$a = \frac{3}{2} - e^{-1}.$$

- *If $x < 1$, the function is $x - e^{-x}$ in a neighborhood of the point and, therefore, is differentiable, with its derivative $1 + e^{-x}$.*
- *If $x > 1$, the function is:*

$$\frac{1}{4x - x^2 - 5} + a$$

in a neighborhood of the point and, therefore, its derivative is:

$$-\frac{4 - 2x}{(4x - x^2 - 5)^2} = \frac{2x - 4}{(4x - x^2 - 5)^2} = \frac{2(x - 2)}{(4x - x^2 - 5)^2}.$$

- *It is given that $f'_-(1) = 1 + e^{-1}$ and $f'_+(1) = -\frac{1}{2}$, therefore, the function is not differentiable at $x = 1$.*

The derivative function is therefore:

$$f'(x) = \begin{cases} 1 + e^{-x} & x < 1 \\ \dfrac{2(x - 2)}{(4x - x^2 - 5)^2} & x > 1 \end{cases}$$

(c) *We must find the extrema of the function in the interval $[0, 5]$.*

We will calculate the candidates for extrema. From part (a) we know that the function is continuous and from part (b) we know that it is differentiable at all points except at $x = 1$.

We calculate the points of zero derivative:

If $x < 1$:

$$f'(x) = 1 + e^{-x} > 0.$$

Therefore, there are no points of zero derivative in $[0, 1]$.

If $x > 1$:

$$f'(x) = \frac{2(x - 2)}{(4x - x^2 - 5)^2},$$

which vanishes at the point $x = 2$.

In summary:

(A) Points where f is not continuous: $\emptyset$.

(B) Critical points: $x = 1,\ x = 2$.

(C) Endpoints of the interval: $x = 0,\ x = 5$.

TABLE 4.7
Values of the function.

x	$f(x)$
0	-1
1	$1 - e^{-1} \approx 0.63$
2	$1/2 - e^{-1} \approx 0.13$
5	$7/5 - e^{-1} \approx 1.033$

TABLE 4.8
Study of the growth/decay of the function.

Interval	Sign of $f'(x)$	Behavior
$0 < x < 1$	$f'(x) = 1 + e^{-x} > 0$	Increasing
$1 < x < 2$	$f'(x) = \dfrac{2(x-2)}{(4x - x^2 - 5)^2} < 0$	Decreasing
$2 < x < 5$	$f'(x) = \dfrac{2(x-2)}{(4x - x^2 - 5)^2} > 0$	Increasing

Let's now classify which are extrema and of what type.

Given that we have a continuous function on a closed interval, we know that absolute maxima and minima exist, and looking at Table 4.7, we see that they are $x = 5$ *and* $x = 0$*, respectively.*

In Table 4.8 we conduct a study of the growth/decay of the function around the points $x = 1$ *and* $x = 2$*. We observe that at* $x = 1$ *and* $x = 2$ *a relative maximum and minimum are reached, respectively.*

In summary, the extrema of the function are:

- $x = 0$ *is an absolute minimum,*
- $x = 1$ *is a relative maximum,*
- $x = 2$ *is a relative minimum,*
- $x = 5$ *is an absolute maximum.*

In Figure 4.59 the function is plotted and the found extrema can be checked.

(d) *We have to prove that* $f(x)$ *has a zero in the interval* $[0, 1]$*.*

The function $f(x)$ *is continuous in* $[0, 1]$ *and also:*

$$f(0) = -1, \quad f(1) = 1 - e^{-1} \approx 0.63,$$

that is, $f(0) \cdot f(1) < 0$*.*

Therefore, Bolzano's theorem guarantees that the function $f(x)$ *has a zero in the interval* $[0, 1]$*, which is what we wanted to prove.*

4.17 Solved Problems (Optimization)

Problem 4.32. Find two positive numbers that add up to 30 and minimize the sum of their squares.

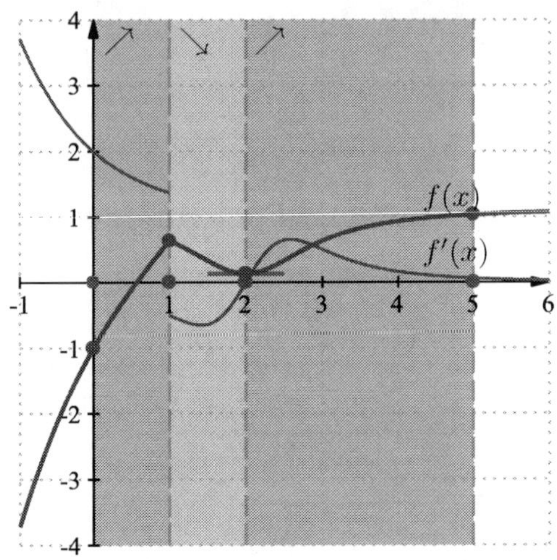

FIGURE 4.59
Graph of the function for Problem 4.31.

Solution to the problem 4.32. *Let $p, q \in \mathbb{R}$, $p, q > 0$, such that:*

$$p + q = 30 \Leftrightarrow q = 30 - p.$$

The sum of squares is:

$$S(p, q) = p^2 + q^2 = p^2 + (30 - p)^2 = S(p).$$

The critical point of this function is:

$$S'(p) = 2p - 2(30 - p) = 4p - 60 = 0 \Leftrightarrow p = 15$$

Since $S(p)$ is a parabola with a positive leading coefficient, $p = q = 15$ is an absolute minimum in $\mathbb{R}^+$.

Problem 4.33. We have a wire of 1 m in length. How should it be divided to construct a circle and a square so that the sum of the areas of the circle and the square is minimized?

Solution to the problem 4.33. *Let x be the piece of wire used to construct the circle and let y be the piece of wire used to construct the square. We have the condition that $x + y = 1$, that is, $y = 1 - x$.*

If x is the circumference of the circle, the radius of the circle will be:

$$r = \frac{x}{2\pi}.$$

Similarly, if $y = 1 - x$ is the perimeter of the square, the side will be:

$$l = \frac{y}{4} = \frac{1 - x}{4}.$$

The area of the circle is:

$$A_\circ = \pi r^2 = \frac{x^2}{4\pi}$$

while the area of the square is:

$$A_{\square} = \left(\frac{1-x}{4}\right)^2.$$

Therefore, the function we want to minimize is:

$$S(x) = \frac{x^2}{4\pi} + \left(\frac{1-x}{4}\right)^2$$

This function is differentiable in (0,1) and, therefore, its absolute minimum is found either at a critical point or at the endpoints of the interval.

Let's study the critical points:

$$S'(x) = \frac{x}{2\pi} + \frac{x-1}{8} = 0 \Leftrightarrow x = \frac{\pi}{4+\pi}$$

The point $x = \frac{\pi}{4+\pi}$*, along with the endpoints of the interval* $x = 0$ *and* $x = 1$*, are the candidates for the absolute minimum.*

Let's study the growth in the intervals generated by these points.

- *Interval* $\left(0, \frac{\pi}{4+\pi}\right)$*. If we consider, for example, the point* $x = 1/5$*, the value of* $f'(1/5) = \frac{1-\pi}{10\pi}$ *is negative, so the function is decreasing.*
- *Interval* $\left(\frac{\pi}{4+\pi}, 1\right)$*. If we consider, for example, the point* $x = 1/2$*, the value of* $f'(1/2) = \frac{4-\pi}{16\pi}$ *is positive, so the function is increasing.*

Therefore, $x = 0$ *is a maximum,* $x = \frac{\pi}{4+\pi}$ *is a minimum, and* $x = 1$ *is a maximum.*

If we also evaluate the function at these points:

$$f(0) = 1/16 \approx 0.0625$$

$$f\left(\frac{\pi}{4+\pi}\right) = \frac{1}{4(4+\pi)} \approx 0.0350$$

$$f(1) = \frac{1}{4\pi} \approx 0.0796$$

we can conclude that

- $x = 0$ *is a relative maximum,*
- $x = \frac{\pi}{4+\pi}$ *is an absolute minimum, and*
- $x = 1$ *is an absolute maximum.*

Therefore, as illustrated in Figure 4.60, we should allocate $x = \frac{\pi}{4+\pi}$ *m of wire for the circle and* $y = 1 - x = \frac{4}{4+\pi}$ *m of wire for the square. We will obtain a sum of areas equal to* $\frac{1}{4(4+\pi)}$ m^2*.*

Problem 4.34. We want to construct a cylindrical container with a lid, with a volume of $100m^3$. What should its dimensions be to use the minimum amount of material?

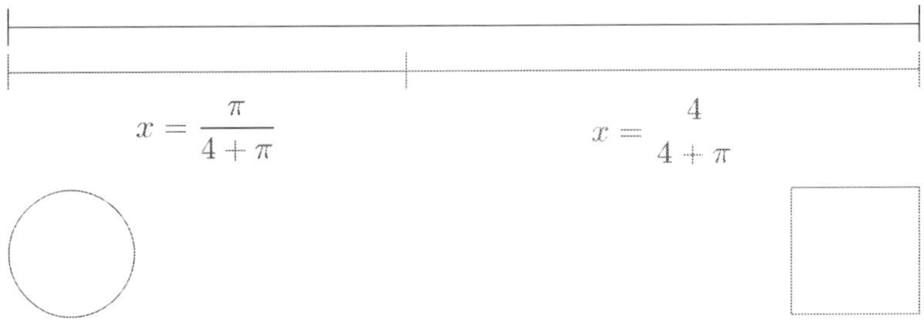

FIGURE 4.60
How to divide 1 m of wire to construct a circle and a square so that the sum of the areas is minimized.

Solution to the problem 4.34. *Let R and H be the radius and height of the container, respectively. We know that its volume is $V = \pi R^2 H = 100\ m^3$. Expressing the height H as a function of the radius, we obtain:*

$$H = \frac{V}{\pi R^2}$$

We seek the values of R and H that minimize the total area of the container, that is, the absolute minimum of the function:

$$A(R) = 2\pi R^2 + 2\pi RH = 2\pi R\left(R + \frac{V}{\pi R^2}\right)$$

We look for the critical points of $A(R)$:

$$A'(R) = \frac{2\left(2\pi R^3 - V\right)}{R^2} = 0 \quad \Leftrightarrow \quad R = \sqrt[3]{\frac{V}{2\pi}} \approx 2.5153$$

The value we obtained is a minimum since before this value the function is decreasing ($f'(2) < 0$) and after it, the function is increasing ($f'(3) > 0$). Additionally, it is a global minimum, since:

$$\lim_{R\to 0^+} A(R) = \lim_{R\to +\infty} A(R) = +\infty.$$

Problem 4.35. Calculate the points on the graph of the function $f(x) = \frac{1}{1+x^2}$ where the tangent has the maximum slope.

Solution to the problem 4.35. *Given a point $(x, f(x))$, the slope of the tangent line is given by $f'(x)$. Therefore, we are asked to solve the following problem:*

$$\max_{x\in\mathbb{R}} f'(x) = \max_{x\in\mathbb{R}} \quad -\frac{2x}{(1+x^2)^2}.$$

Since $f' \in \mathcal{C}^\infty(\mathbb{R})$, the candidates for the maximum will be critical points of the function $f'(x)$, that is, we need to solve the equation:

$$f''(x) = 0.$$

This is:

$$f''(x) = \frac{2(3x^2 - 1)}{(1+x^2)^3}$$

So:

$$f''(x) = 0 \quad \Leftrightarrow \quad 2(3x^2 - 1) = 0,$$

therefore:

$$2(3x^2 - 1) = 0 \quad \Leftrightarrow \quad 3x^2 = 1 \quad \Leftrightarrow \quad x = \pm\frac{1}{\sqrt{3}}$$

The candidates for extrema are $x = +\frac{1}{\sqrt{3}}$ *and* $x = -\frac{1}{\sqrt{3}}$. *Let's study the behavior of the function* f' *in the intervals that limit these points.*

- *Interval* $\left(-\infty, -\frac{1}{\sqrt{3}}\right)$. *If we consider, for example, the point* $x = -1$, *the value of* $f''(-1) = 1/2$ *is positive, so the function is increasing.*
- *Interval* $\left(-\frac{1}{\sqrt{3}}, \frac{1}{\sqrt{3}}\right)$. *If we consider, for example, the point* $x = 0$, *the value of* $f''(0) = -2$ *is negative, so the function is decreasing.*
- *Interval* $\left(\frac{1}{\sqrt{3}}, +\infty\right)$. *If we consider, for example, the point* $x = 1$, *the value of* $f''(1) = 1/2$ *is positive, so the function is increasing.*

Therefore, $x = -\frac{1}{\sqrt{3}}$ *is a maximum and* $x = \frac{1}{\sqrt{3}}$ *is a minimum. If we evaluate the function at these points:*

$$f'\left(-\frac{1}{\sqrt{3}}\right) = \frac{3\sqrt{3}}{8} \approx 0.6495$$

$$f'\left(\frac{1}{\sqrt{3}}\right) = -\frac{3\sqrt{3}}{8} \approx -0.6495$$

and we also have that:

$$\lim_{x \to \pm\infty} f'(x) = 0.$$

Thus, we can conclude that:

- $x = -\frac{1}{\sqrt{3}}$ *is an absolute maximum and*
- $x = \frac{1}{\sqrt{3}}$ *is an absolute minimum.*

In Figure 4.61 the function $f(x)$ *and its derivative* $f'(x)$ *are plotted. We observe that at* $x = -\frac{1}{\sqrt{3}}$ *the derivative has a maximum and the tangent line to* f *at this point has the maximum slope. Similarly, at* $x = \frac{1}{\sqrt{3}}$ *the derivative has a minimum, and the tangent line to* f *at this point has the minimum slope.*

Problem 4.36. From a rectangular mirror 2 m long and 1 m high, a right triangle with a base of 30 cm and a height of 20 cm has broken off from one of the corners. How should another mirror be cut with sides parallel to the initial mirror so that the area of this new mirror is maximized?

Solution to the problem 4.36. *Figure 4.62 schematically shows the problem setup.*

We need to determine the point (x, y) *that maximizes the area* $A(x) = (200 - x)(100 - y)$, *where* $0 \leq x \leq 30$ *and* $0 \leq y \leq 20$.

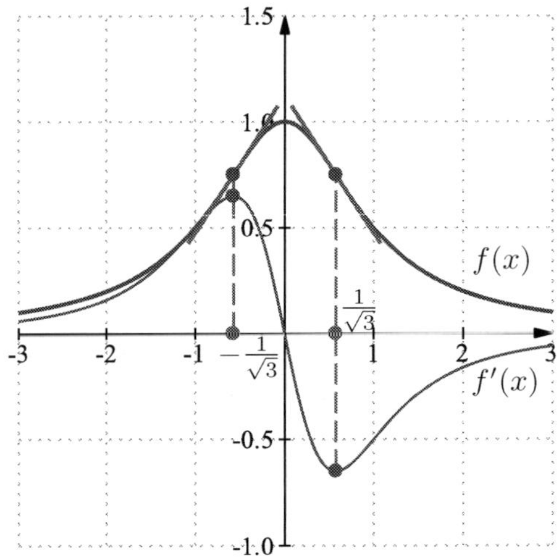

FIGURE 4.61
Graph of the function $f(x) = \dfrac{1}{1+x^2}$ and its derivative.

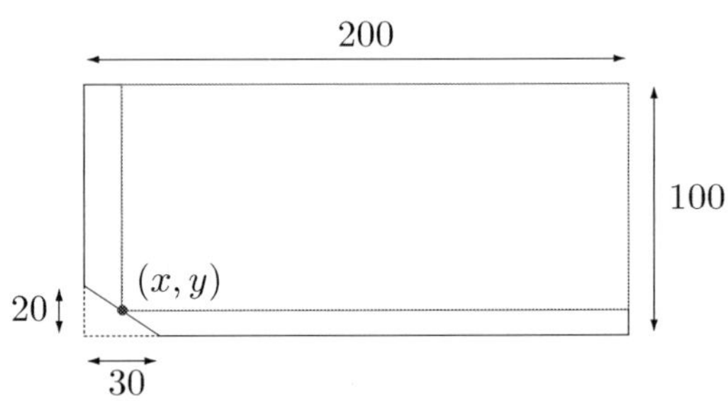

FIGURE 4.62
Problem setup for Problem 4.36.

By similarity of triangles, we have:

$$\frac{30-x}{y} = \frac{30}{20} \Rightarrow y = \frac{2}{3}(30-x)$$

and then:

$$A(x) = (200-x)(100-y) = \frac{2}{3}(200-x)(120+x) = 16000 + \frac{160}{3}x - \frac{2}{3}x^2$$

We seek the absolute maximum of this function in the interval [0,30]. Since the function is differentiable in (0,30), we know that the maximum can be either a critical point or one of the endpoints of the interval. Let's first study the critical points:

$$A'(x) = \frac{160}{3} - \frac{4}{3}x = 0 \quad \Leftrightarrow \quad x = 40 \notin (0,30)$$

Thus, the absolute maximum is at one of the endpoints of the interval:

$$A(0) = \frac{2}{3}(200-0)(120+0) = 16000,$$
$$A(30) = \frac{2}{3}(200-30)(120+30) = 17000,$$

and, therefore, we should cut the mirror at the point $(30,0)$.

Problem 4.37. We need to construct a flowerbed in the shape of a circular sector with a perimeter of 20 m. Calculate the radius of the sector to obtain the maximum area.

Solution to the problem 4.37. *Figure 4.63 schematically shows the problem setup. If we denote by α the angle of the circular sector and by r the radius, we must determine r and α such that, knowing that the perimeter is 20, the area is maximized. We also know that $r \geq 0$ and that $\alpha \in [0, 2\pi)$.*

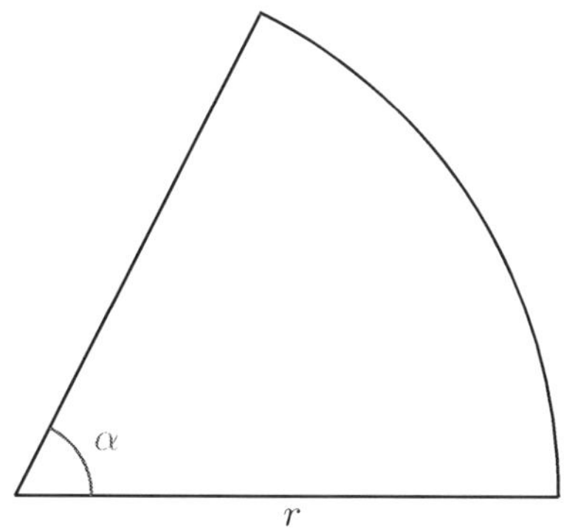

FIGURE 4.63
Problem setup for Problem 4.37.

Since the perimeter of a circular sector is $P = \alpha \cdot r + 2r$, we know that $\alpha \cdot r + 2r = 20$ or, equivalently:

$$r = \frac{20}{\alpha + 2}.$$

Moreover, the area is:

$$A(\alpha) = \frac{\alpha}{2} r^2 = \frac{\alpha}{2}\left(\frac{20}{\alpha+2}\right)^2 = \frac{200\alpha}{(\alpha+2)^2}.$$

We seek the absolute maximum of this function:

$$A'(\alpha) = \frac{200(2+\alpha)^2 - 2(2+\alpha)200\alpha}{(2+\alpha)^4} = \frac{200(2+\alpha) - 400\alpha}{(2+\alpha)^3}$$

$$= \frac{400 - 200\alpha}{(2+\alpha)^3} = 0 \quad \Leftrightarrow \quad 400 - 200\alpha = 0 \quad \Leftrightarrow \quad \alpha = 2.$$

Therefore, the critical point is $\alpha = 2$ which implies:

$$r = \frac{20}{2+2} = 5.$$

For $\alpha = 2$ and $r = 5$, the area is maximized and is $A = r^2 = 25$.

Problem 4.38. Find the points on the graph of the function $y^2 = 4x$ such that the distance to the point $(4, 0)$ is minimized. Calculate this distance.

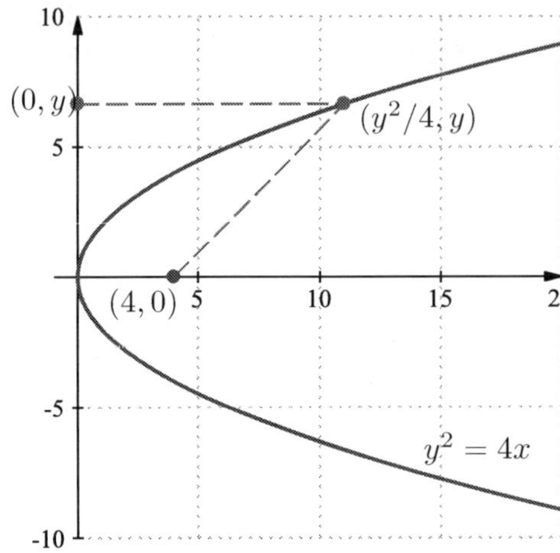

FIGURE 4.64
Setup of Problem 4.38.

Solution to the problem 4.38. *Figure 4.64 schematically shows the problem's setup. Note that the function* $y^2 = 4x$ *is a parabola with interchanged axes. Therefore, we seek the points on the parabola that are at the shortest distance from the point* $(4, 0)$.

First, note that minimizing the distance to the point is equivalent to minimizing the squared distance. Thus, given a point $(x, y) \in \mathbb{R}^2$, *the squared distance from this point to the point* $(4, 0)$ *is given by the function:*

$$D(x, y) = (x - 4)^2 + y^2.$$

If we impose that the point $(x, y) \in \mathbb{R}^2$ *is on the parabola (i.e., it satisfies the relation* $y^2 = 4x$*) we obtain:*

$$D(x) = (x - 4)^2 + 4x.$$

Our goal is to find the absolute minimum of this function. Since $D(x)$ *is of class* $\mathcal{C}^\infty(\mathbb{R})$, *we seek the critical points. First, we differentiate:*

$$D'(x) = 2(x - 4) + 4 = 2x - 4$$

and setting it to zero we obtain:

$$D'(x) = 0 \iff x = 2.$$

Since the function $D(x)$ *is of class* $\mathcal{C}^\infty(\mathbb{R})$, *the minimum will be a critical point or the extreme* $x = 0$. *We evaluate the function at these points:*

$$D(0) = 16$$

$$D(2) = 12$$

Thus, at $x = 0$ *there is a local maximum and at* $x = 2$ *the sought global minimum. Moreover, if* $x = 2$, $y = \pm\sqrt{4x} = \pm\sqrt{8} = \pm 2\sqrt{2}$. *Therefore, the points that minimize the distance are:*

$$(2, -2\sqrt{2}) \qquad \textit{and} \qquad (2, 2\sqrt{2}).$$

In Figures 4.65, the obtained results can be verified.

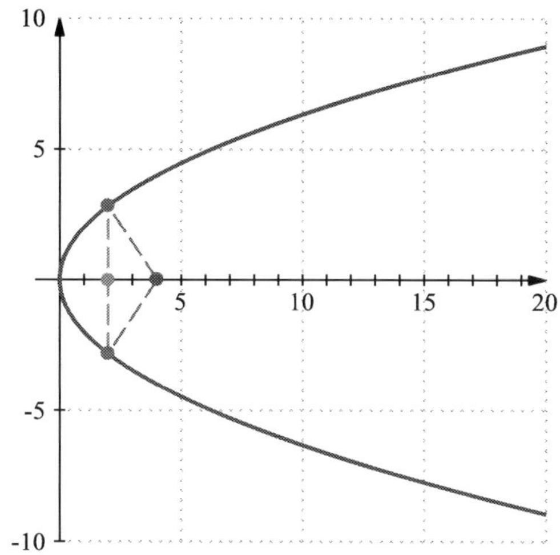

FIGURE 4.65
Points on the graph of the function $y^2 = 4x$ that minimize the distance to the point $(4, 0)$.

Problem 4.39. Find the point on the parabola $y = 2x^2$ that is closest to the point $(9, 0)$.

Solution to the problem 4.39. *Figure 4.66 schematically shows the problem's setup. As in the previous problem, instead of minimizing the distance from a point on the parabola to the point $(9, 0)$, we will minimize the squared distance.*

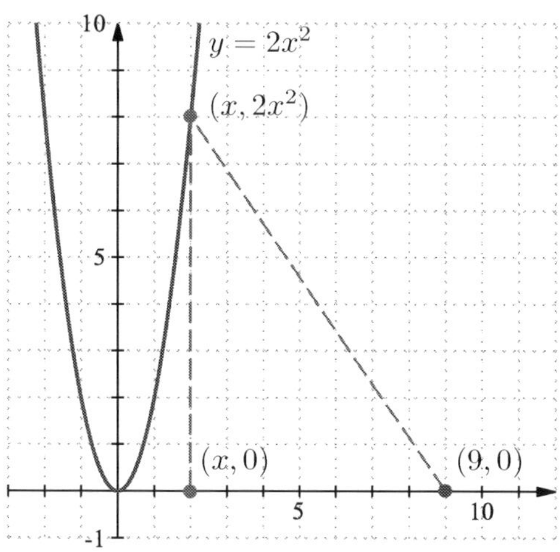

FIGURE 4.66
Setup of Problem 4.39.

If we consider a point $(x, y) \in \mathbb{R}^2$, to minimize the distance from this point to $(9, 0)$, we will minimize the squared distance (which is equivalent), where:

$$d^2 = (x - 9)^2 + y^2.$$

Additionally, we must impose that the point is on the parabola.

$$\min_{x \in \mathbb{R}} \quad (x - 9)^2 + (2x^2)^2 = \min_{x \in \mathbb{R}} \quad x^2 - 18x + 81 + 4x^4.$$

Thus, we must find the minimum of the function:

$$f(x) = 4x^4 + x^2 - 18x + 81.$$

Since $f \in \mathcal{C}^\infty(\mathbb{R})$, *the candidates for the minimum are the critical points.*

$$f'(x) = 16x^3 + 2x - 18 = 0 \quad \Longleftrightarrow \quad 8x^3 + x - 9 = 0.$$

We have a cubic equation of which $x = 1$ *is one of the solutions (it can be verified by substitution). Therefore, by Ruffini:*

$$\begin{array}{c|cccc} & 8 & 0 & 1 & -9 \\ 1 & & 8 & 8 & 9 \\ \hline & 8 & 8 & 9 & 0 \end{array}$$

we obtain $f'(x) = (x-1)(8x^2 + 8x + 9)$. *If we now look for the solutions of the equation* $8x^2 + 8x + 9 = 0$, *using the quadratic formula we have:*

$$x = \frac{-8 \pm \sqrt{8^2 - 4 \cdot 8 \cdot 9}}{16} \notin \mathbb{R}.$$

Therefore, the only candidate is $x = 1$. *Note that* $x = 1 \Longrightarrow y = 2$. *Since* $f'(x) > 0$ *if* $x < 1$ *and* $f'(x) < 0$ *if* $x > 1$, *at* $x = 1$ *we have a minimum. Thus, the point that minimizes the distance is the point* $(1, 2)$ *as observed in Figure 4.67.*

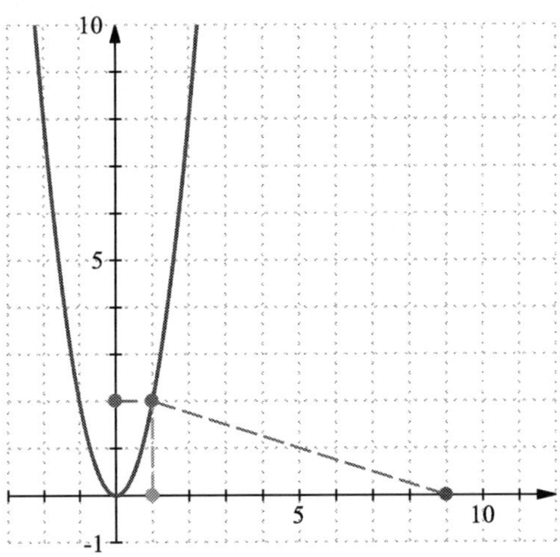

FIGURE 4.67
Point on the parabola $y = 2x^2$ that is closest to the point $(9, 0)$.

Problem 4.40. Determine the height of the right circular cylinder of maximum volume that can be inscribed in a right circular cone of height H meters.

Solution to the problem 4.40. *Let* R *be the radius of the cone and* H *its height. Let* r *be the radius and* h *the height of the cylinder. Consider the main section of the cone, as illustrated in Figure 4.68*

By similarity of triangles, we have:

$$\frac{R}{H} = \frac{r}{H-h} \quad \Leftrightarrow \quad r = \frac{R(H-h)}{H}$$

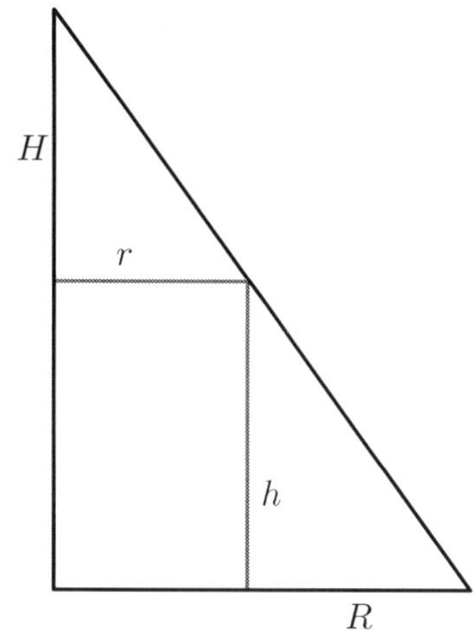

FIGURE 4.68
Section of the cone with radius R and height H as well as the inscribed right circular cylinder with radius r and height h.

Using that the volume of the cylinder is $V = \pi r^2 h$, *we can write:*

$$V(h) = \pi \frac{R^2(H-h)^2}{H^2} h = \frac{\pi R^2}{H^2}(H-h)^2 h$$

We seek an absolute maximum of this function. First, we study the critical points:

$$V'(h) = \frac{\pi R^2}{H^2}(H-h)(H-3h) = 0 \quad \Leftrightarrow \quad h = H \quad or \quad h = \frac{H}{3}$$

The solution $h = H$ *does not make sense, as it corresponds to a cylinder with radius 0 and, consequently, volume 0. On the other hand, we check that:*

$$V''(h) = \frac{2\pi R^2}{H^2}(3h - 2H); \quad V''\left(\frac{H}{3}\right) = -2\pi \frac{R^2}{H} < 0$$

and, therefore, the solution $h = \dfrac{H}{3}$ *corresponds to a maximum of* $V(h)$.

Problem 4.41. A person carries a very thin glass through an L-shaped street, where one part of the street is 4 m wide and the other 3 m wide. What will be the maximum length that the glass can have to be able to pass through? Hint: assume the glass has negligible thickness.

Solution to the problem 4.41. *The situation described is illustrated in Figure 4.69:*

Note that if one end of the glass is at the point $(-3, s)$, *then the other end is at the point* $(-\frac{12}{s}, 4)$. *This is because the line passing through the points* $(-3, s)$ *and* $(0, 0)$ *is* $y = -\frac{s}{3}x$, *and thus at* $y = 4$ *it takes the value* $x = -\frac{12}{s}$.

We calculate the minimum distance between the points $(-3, s)$ *and* $(-\frac{12}{s}, 4)$. *Then, this will be the maximum length that the glass can have to pass through this street.*

Thus, we pose the problem:

$$\min_{s \leq 4} d^2,$$

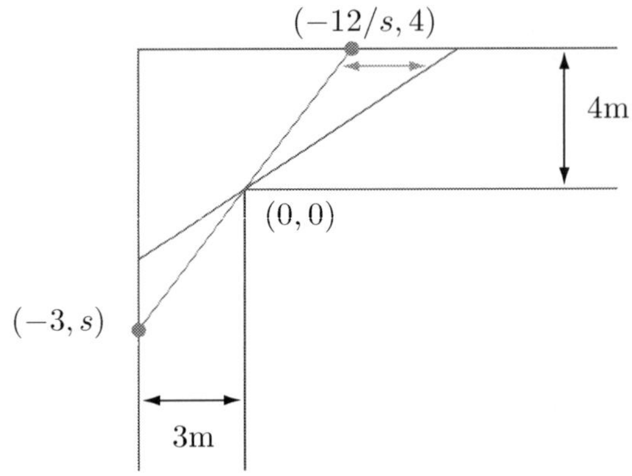

FIGURE 4.69
Setup of Problem 4.41.

where d is the distance between the endpoints of the glass. That is:

$$d^2 = \left(-\frac{12}{s} - (-3)\right)^2 + (4 - s)^2$$

Therefore, if we consider the function:

$$f(s) = \left(-\frac{12}{s} - (-3)\right)^2 + (4 - s)^2$$

we must find its minimum in the open interval $(-\infty, 0)$. *Since the function is continuous and differentiable in* $(-\infty, 0)$ *as it is polynomial, we only need to study the critical points.*

$$f(s) = \frac{12^2}{s^2} + 9 - \frac{3 \cdot 24}{s} + 16 + s^2 - 8s = \frac{144}{s^2} - \frac{72}{s} - 8s + s^2 + 25.$$

and:

$$f'(s) = -\frac{288}{s^3} + \frac{72}{s^2} - 8 + 2s = 0 \iff -144 + 36s - 4s^3 + s^4 = 0$$

If we solve the above equation using some algebraic manipulation software, we get two imaginary solutions and two real solutions which are $s = 4$ *and* -3.301927249. *Of these two solutions, we keep the negative one since we are looking for* $s \in (-\infty, 0)$. *Therefore, we denote* $s_0 = -3.301927249$. *Note that:*

$$f'(s) < 0 \quad \text{if} \quad s < s_0,$$

$$f'(s) > 0 \quad \text{if} \quad s > s_0.$$

Thus, at $s = s_0$ *we have a minimum.*
In this case, the squared distance we have for the glass is:

$$f(s_0) = (-\frac{12}{s_0} - (-3))^2 + (4 - s_0)^2 = 97.33129765,$$

Finally, the distance is obtained by calculating the square root:

$$d = \sqrt{97.33129765} = 9.865662555.$$

Problem 4.42. A collector has 50 items, consisting of stamps and coins. If another collector gives him three stamps in exchange for one coin, the product of the number of remaining coins and stamps is maximized. How many stamps and coins did he initially have?

Solution to the problem 4.42. *Let $s, m \in \mathbb{N}$ be the initial numbers of stamps and coins respectively. Then:*

$$s + m = 50 \Leftrightarrow m = 50 - s.$$

After the exchange, the product:

$$p(s) = (s+3)(m-1) = (s+3)(49-s)$$

is an absolute maximum. If the maximum were a relative extremum, we would have:

$$p'(s) = 49 - s - s - 3 = 46 - 2s = 0 \quad \Leftrightarrow \quad s = 23 \quad \to \quad p(23) = 26 \cdot 26 = 676$$

On the other hand, at the interval's extremes, in this case, $s = 0$ or $s = 49$ (since $m \geq 1$ to make the exchange possible as stated in the problem), we have $p(0) = p(49) = 0$. Therefore, $s = 23$ and $m = 27$.

Problem 4.43. The trajectory of a projectile fired by an artillery cannon located at the origin of coordinates is the parabola:

$$f(x) = -k(1 + \tan^2(\alpha))x^2 + \tan(\alpha)x,$$

where k is a positive constant that depends on the characteristics of the cannon and α is the angle formed by the positive semi-axis of the abscissas and the cannon. The angle α is assumed to be between 0 and $\frac{\pi}{2}$. Calculate the angle α for which the above parabola cuts the OX axis as far as possible from the origin.

Solution to the problem 4.43. *Figure 4.70 outlines the problem statement.*

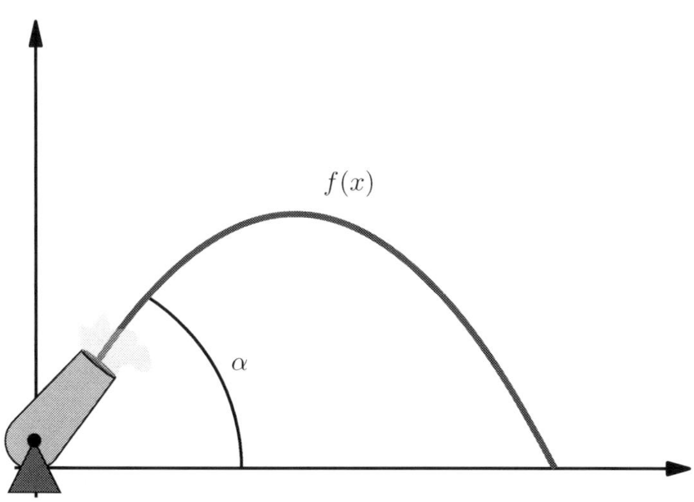

FIGURE 4.70
Problem Statement 4.43.

The parabola:

$$f(x) = -k(1+\tan^2(\alpha))x^2 + (\tan(\alpha))x$$

cuts the OX axis when $f(x) = 0$:

$$\begin{aligned} f(x)=0 \quad &\Longleftrightarrow -k(1+\tan^2(\alpha))x^2 + (\tan(\alpha))x = 0 \\ &\Longleftrightarrow (-k(1+\tan^2(\alpha))x + (\tan(\alpha)))x = 0 \\ &\Longleftrightarrow \begin{cases} x = 0 \\ x = \dfrac{(\tan(\alpha))}{k(1+\tan^2(\alpha))} \end{cases} \end{aligned}$$

The point $x = 0$ is where the projectile is fired from, and the point $x = \frac{(\tan(\alpha))}{k(1+\tan^2(\alpha))}$ is where it lands. We need to calculate the angle that makes the parabola cut as far as possible; therefore, we must solve the following problem:

$$\max_{\alpha\in[0,\frac{\pi}{2}]} \quad \frac{(\tan(\alpha))}{k(1+\tan^2(\alpha))}.$$

Consider the function:

$$g(\alpha) = \frac{(\tan(\alpha))}{k(1+\tan^2(\alpha))}.$$

Note that this function takes the value 0 *at $\alpha = 0$ and is not defined at $\alpha = \pi/2$. If we study the limit at $\pi/2$:*

$$\lim_{\alpha\to\frac{\pi}{2}} \frac{(\tan(\alpha))}{k(1+\tan^2(\alpha))}$$

it is an indeterminate form of the type $\frac{\infty}{\infty}$. Using L'Hôpital's rule:

$$\lim_{\alpha\to\frac{\pi}{2}} \frac{(\tan(\alpha))}{k(1+\tan^2(\alpha))} = \lim_{\alpha\to\frac{\pi}{2}} \frac{1}{2\tan\alpha} = 0.$$

Thus, at the extremes of the interval to be studied, $[0, \pi/2]$, the function takes the value (or tends to) 0*. Therefore, the maximum of the function is not reached. Since $g(\alpha)$ is continuous in $\alpha \in (0, \frac{\pi}{2})$ and differentiable in $(0, \frac{\pi}{2})$, we only need to study the critical points.*

$$\begin{aligned} g'(\alpha) &= \frac{(1+\tan^2(\alpha))k(1+\tan^2(\alpha)) - (\tan(\alpha))\, 2k(\tan(\alpha))(1+\tan^2(\alpha))}{k^2(1+\tan^2(\alpha))^2} \\ &= \frac{k(1+\tan^2(\alpha)) - 2k\tan^2(\alpha)}{k^2(1+\tan^2(\alpha))} = \frac{k(1-\tan^2(\alpha))}{k(1+\tan^2(\alpha))} \end{aligned}$$

Therefore:

$$g'(\alpha) = 0 \quad \Longleftrightarrow \quad 1-\tan^2(\alpha) = 0 \quad \Longleftrightarrow \quad \tan^2(\alpha) = 1 \quad \Longleftrightarrow \quad (\tan(\alpha)) = \pm 1.$$

The only angle, within the interval $\alpha \in [0, \frac{\pi}{2}]$, that verifies $(\tan(\alpha)) = \pm 1$ is $\alpha = \frac{\pi}{4}$. Moreover, since:

- $g'(\alpha) > 0$ *if* $\alpha < \frac{\pi}{4}$,
- $g'(\alpha) < 0$ *if* $\alpha > \frac{\pi}{4}$,

for this angle we have a maximum. Therefore, the sought angle is $\alpha = \frac{\pi}{4}$.

In Figure 4.71, the different trajectories are drawn when $k = 2$ for the optimal angle, $\alpha = \frac{\pi}{4}$, and for larger and smaller angles.

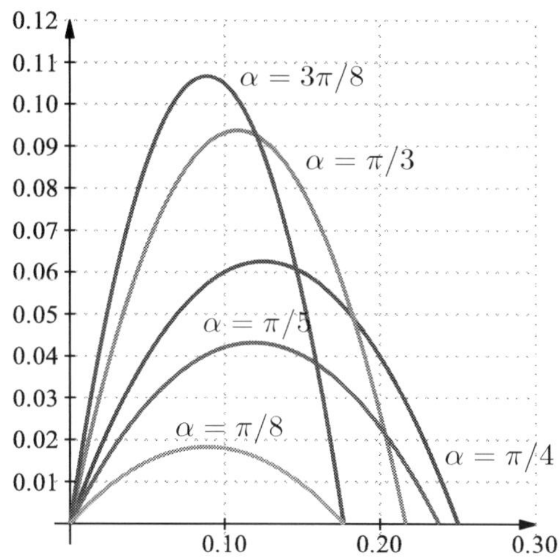

FIGURE 4.71
Projectile trajectories for different angles.

Problem 4.44. A bakery specializes in two types of bread: cereal bread and olive bread. For each loaf of cereal bread, the profit is 1 euro, and for each olive bread baguette, the profit is 2 euros (net profit). The goal is to determine the optimal production quantity to maximize daily profit, knowing that there are budget constraints: the daily investment cannot exceed 300 euros. The cost of producing a piece of cereal bread is 0.25 euros, and the cost of producing a piece of olive bread is 0.75 euros. Calculate the number of cereal and olive loaves that need to be produced to maximize the profits.

Solution to the problem 4.44. *Let n_c and n_o denote the number of cereal bread loaves and olive bread loaves produced in a day, respectively. The profit obtained by producing n_c loaves of cereal bread and n_o loaves of olive bread is:*

$$B = n_c + 2n_o.$$

We also know that the production cost is $0.25n_c + 0.75n_o$ and that this cost cannot exceed 300 euros.

Therefore, the problem we want to solve is to maximize the function:

$$B(n_c, n_o) = n_c + 2n_o$$

with the constraints $n_c \geq 0, n_o \geq 0$ and $0.25n_c + 0.75n_o = 300$.

Using $0.25n_c + 0.75n_o = 300$, we can re-express $n_c = 1200 - 3n_o$ so the original problem is equivalent to:

$$B(n_o) = 1200 - n_o$$

with the constraint $0 \leq n_o \leq 400$, since the daily investment is limited to 300 euros (producing 400 olive breads at a cost of 0.75 each reaches the budget limit). Therefore, we must find the maximum of the function $1200 - n_o$ in the interval $[0, 400]$. Since the function is a straight line, the maximum will be reached at one of the two extremes (evidently, being a straight line, we have no discontinuity points, no critical points, and we also know that the maximum exists because it is a continuous function in a closed interval). Let's evaluate the function at the extremes:

- $n_o = 0 \implies B(n_o) = 1200$ *euros*
- $n_o = 400 \implies B(n_o) = 800$ *euros*

Therefore, the optimal production is: $n_o = 0$ *and* $n_c = 1200$ *with a profit of 1200 euros daily.*

4.18 Solved Problems (Taylor)

Problem 4.45. Find the equation of the parabola that best approximates the curve $f(x) = e^x \ln(x+1)$ at the point $(0,0)$.

Solution to the problem 4.45. *The parabola that best approximates the function* $f(x)$ *at the point* $(0,0)$ *is its second-degree Taylor polynomial centered at the origin, that is:*

$$p(x) = f(0) + f'(0)x + \frac{f''(0)}{2}x^2.$$

$$f(x) = e^x \ln(x+1) \implies f(0) = e^0 \ln(1) = 1 \cdot 0 = 0$$

$$f'(x) = e^x \ln(x+1) + e^x \frac{1}{x+1} = f(x) + \frac{e^x}{x+1} \implies f'(0) = 0 + \frac{1}{1} = 1$$

$$f''(x) = f'(x) + \frac{e^x(x+1) - e^x}{(x+1)^2} \implies f''(0) = 1 + \frac{1-1}{1} = 1.$$

Therefore:

$$p(x) = 0 + x + \frac{1}{2}x^2 = x + \frac{1}{2}x^2.$$

Figure 4.72 shows the graph of the function $f(x)$*, as well as the polynomial found.*

Problem 4.46. (a) State the Taylor theorem for the exponential function at $x_0 = 0$.
(b) Calculate e with an error less than 10^{-5}.

Solution to the problem 4.46. (a) *The* n*-th degree Taylor polynomial of* $f(x)$ *around the point* $x = 0$ *is:*

$$p_n(x) = f(0) + f'(0)x + \frac{1}{2}f''(0)x^2 + \frac{1}{3!}f'''(x)x^3 + \cdots + \frac{1}{n!}f^n(0)x^n.$$

In the case where $f(x) = e^x$*, since* $f^k(x) = f(x) = e^x$*, for all* $k \in \mathbb{N}$*, we have that* $f^k(0) = e^0 = 1, \forall k \in \mathbb{N}$*. Therefore, the Taylor polynomial is:*

$$p_n(x) = 1 + x + \frac{1}{2}x^2 + \frac{1}{6}x^3 + \frac{1}{24}x^4 + \cdots + \frac{1}{n!}x^n.$$

Additionally, we know that, given a point $x_0 \in (a,b)$*, the difference between the value of the function* f *and the polynomial* p_n *at this point can be expressed as:*

$$f(x_0) - p_n(x_0) = \frac{1}{(n+1)!}f^{n+1}(\xi)x^{n+1},$$

for some $\xi \in (a,b)$*.*

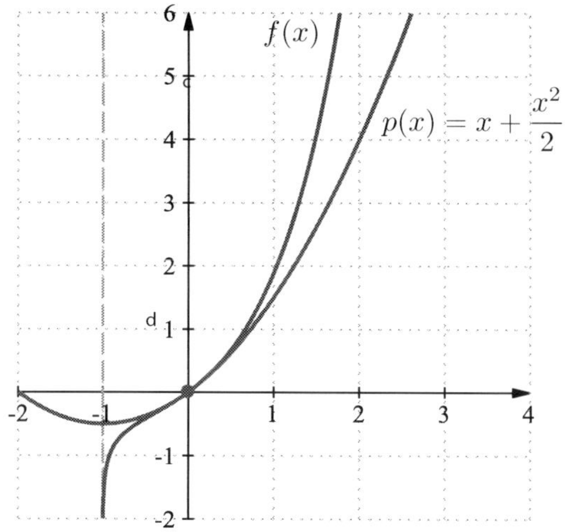

FIGURE 4.72
Graph of the function $f(x) = e^x \ln(x+1)$ and its second-degree Taylor polynomial centered at the origin.

(b) *We will consider the function $f(x) = e^x$ and apply Taylor's theorem with $x_0 = 0$ and $x = 1$. We want to determine n such that $|R_n(1)| < 10^{-5}$. Using the formula for the remainder, we know that:*

$$R_n(1) = \frac{e^\xi}{(n+1)!},$$

for some $\xi \in (0,1)$. Here we will consider that the exponential function is increasing and that $e^1 < 3$. That is, we know that for all $\xi \in (0,1)$ it is satisfied that $e^\xi < 3$. Thus, we look for a value of n such that:

$$R_n(1) \leq \frac{3}{(n+1)!} < 10^{-5},$$

that is, solving the above inequality, we look for n such that:

$$(n+1)! > 3 \cdot 10^5$$

We observe that $n = 8$ provides the desired precision, since:

$$(8+1)! = 362880 > 3 \cdot 10^5.$$

A lower value of n would not be sufficient. Finally:

$$e \approx p_8(1) = 1 + 1 + \frac{1}{2!} + \cdots \frac{1}{8!} = 2.71828$$

with an error less than 10^{-5}.

Problem 4.47. Consider the function $f(x) = \arctan(x) - \ln(x) - \frac{\pi}{4}$.

(a) Calculate the tangent line to the function $f(x)$ at the point $x = 1$.
Hint: recall that $\arctan(1) = \pi/4$ and that the derivative of the function $\arctan(x)$ is $1/(1+x^2)$.

(b) Calculate the second-degree Taylor polynomial of the function $f(x)$ around the point $x = 1$.

(c) Calculate the following limit:

$$\lim_{x \to 1} \frac{\arctan(x) - \ln(x) - \frac{\pi}{4}}{1 - x}.$$

Solution to the problem 4.47. (a) *Consider the function* $f(x) = \arctan(x) - \ln(x) - \frac{\pi}{4}$*. We need to calculate the tangent line to* $f(x)$ *at the point* $x = 1$*, which is equivalent to calculating the first-degree Taylor polynomial of the function.*

$$p_1(x) = f(1) + f'(1)(x - 1)$$

$$f(x) = \arctan(x) - \ln(x) - \tfrac{\pi}{4} \implies f(1) = \arctan(1) - \ln(1) - \frac{\pi}{4} = \frac{\pi}{4} - 0 - \frac{\pi}{4} = 0$$

$$f'(x) = \frac{1}{1 + x^2} - \frac{1}{x} \implies f'(1) = \frac{1}{2} - 1 = -\frac{1}{2}$$

Therefore, the tangent line is:

$$p_1(x) = 0 - \frac{1}{2}(x - 1) = \frac{1}{2} - \frac{1}{2}x.$$

(b) *We need to calculate the second-degree Taylor polynomial of the function. Since we already have the tangent line, we only need to add the second-degree term:*

$$p_2(x) = f(1) + f'(1)(x - 1) + \frac{f''(1)}{2}(x - 1)^2 = p_1(x) + \frac{f''(1)}{2}(x - 1)^2.$$

Given that:

$$f'(x) = \frac{1}{1 + x^2} - \frac{1}{x}$$

$$f''(x) = -\frac{2x}{(1 + x^2)^2} + \frac{1}{x^2} \implies f''(1) = -\frac{2}{4} + \frac{1}{1} = \frac{1}{2}.$$

Therefore, the second-degree Taylor polynomial is:

$$p_2(x) = 0 - \frac{1}{2}(x - 1) + \frac{1}{2} \cdot \frac{1}{2}(x - 1)^2 = -\frac{1}{2}(x - 1) + \frac{1}{4}(x - 1)^2.$$

Figure 4.73 shows the graph of the function $f(x)$ *as well as its first- and second-degree Taylor polynomials centered at* $x = 1$*.*

(c) *To calculate the given limit, we will use the first-degree Taylor polynomial to write:*

$$f(x) = p_1(x) + O\left((x - 1)^2\right) = -\frac{1}{2}(x - 1) + O\left((x - 1)^2\right).$$

Therefore:

$$\begin{aligned}
\lim_{x \to 1} \frac{\arctan(x) - \ln(x) - \frac{\pi}{4}}{1 - x} &= \lim_{x \to 1} \frac{-\frac{1}{2}(x - 1) + O\left((x - 1)^2\right)}{1 - x} \\
&= \lim_{x \to 1} -\frac{-\frac{1}{2}(x - 1) + O\left((x - 1)^2\right)}{x - 1} \\
&= \lim_{x \to 1} -\frac{-\frac{1}{2} + O\left((x - 1)\right)}{1} \\
&= \lim_{x \to 1} \frac{1}{2} - O\left((x - 1)\right) = \frac{1}{2} + 0 = \frac{1}{2}.
\end{aligned}$$

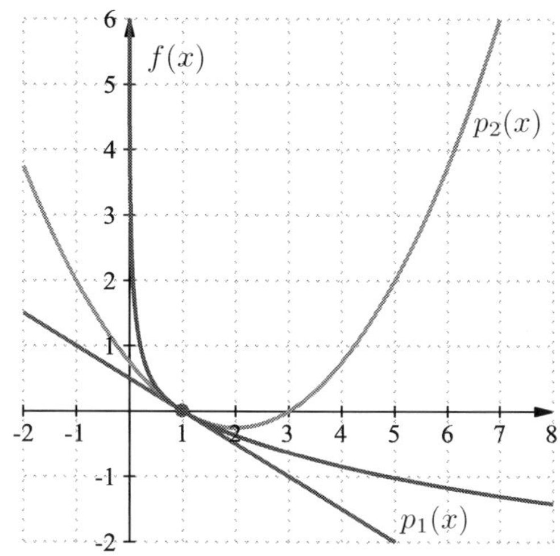

FIGURE 4.73
Graph of the function $f(x) = \arctan(x) - \ln(x) - \frac{\pi}{4}$ as well as its first- and second-degree Taylor polynomials centered at $x = 1$.

Problem 4.48. Calculate the following limits:

(a) $\lim_{x\to 0} \dfrac{e^x - 1 - x}{x^2}$

(b) $\lim_{h\to 0} \dfrac{\ln(2+h) - \ln(2)}{h}$

Solution to the problem 4.48. (a) *For the given limit, we have:*

$$\lim_{x\to 0} \frac{e^x - 1 - x}{x^2} = «\frac{0}{0}» \text{ (indeterminate form).}$$

To solve this limit, we will use the Taylor polynomial of the function $f(x) = e^x - 1 - x$ *around the point* $x = 0$.

$$\begin{aligned} f(x) &= e^x - 1 - x &&\Longrightarrow \quad f(0) = e^0 - 1 - 0 = 0 \\ f'(x) &= e^x - 1 &&\Longrightarrow \quad f'(0) = e^0 - 1 = 0 \\ f''(x) &= e^x &&\Longrightarrow \quad f''(0) = 1 \end{aligned}$$

Therefore, we know that:

$$f(x) = f(0) + f'(0)x + \frac{f''(0)}{2}x^2 + O(x^3) = \frac{1}{2}x^2 + O(x^3).$$

Substituting into the limit:

$$\lim_{x\to 0} \frac{e^x - 1 - x}{x^2} = \lim_{x\to 0} \frac{\frac{1}{2}x^2 + O(x^3)}{x^2} = \lim_{x\to 0} \frac{1}{2} + O(x) = \frac{1}{2} + 0 = \frac{1}{2}.$$

(b) *For the given limit, we have:*

$$\lim_{h\to 0}\frac{\ln(2+h)-\ln(2)}{h} = «\frac{0}{0}» \text{ (indeterminate form)}.$$

Note that if we consider the function $f(x) = \ln(x)$ *and use the definition of the derivative of the function at the point* $x = 2$*:*

$$f'(2) = \lim_{h\to 0}\frac{f(2+h)-f(2)}{h} = \lim_{h\to 0}\frac{\ln(2+h)-\ln(2)}{h}.$$

Therefore, the value of the derivative of the function $f(x) = \ln(x)$ *at the point* $x = 2$ *is:*

$$f'(x) = \frac{1}{x} \quad\Longrightarrow\quad f'(2) = \frac{1}{2} \quad\Longrightarrow\quad L_2 = \frac{1}{2}.$$

Another alternative to calculate the above limit is to use the Taylor polynomial of the function $g(h) = \ln(2+h) - \ln(2)$ *at the point* $h = 0$.

$$\begin{aligned} g(h) &= \ln(2+h)-\ln(2) &&\Longrightarrow\quad g(0) = \ln(2)-\ln(2) = 0\\ g'(h) &= \tfrac{1}{2+h} &&\Longrightarrow\quad g'(0) = \tfrac{1}{2} \end{aligned}$$

Therefore, we know that:

$$g(h) = g(0) + g'(0)h + O(h^2) = \frac{1}{2}h + O(h^2).$$

Substituting into the limit:

$$\lim_{h\to 0}\frac{\ln(2+h)-\ln(2)}{h} = \lim_{h\to 0}\frac{\frac{1}{2}h + O(h^2)}{h} = \lim_{h\to 0}\frac{1}{2} + O(h) = \frac{1}{2} + 0 = \frac{1}{2}.$$

4.19 Solved Problems (Limits)

Problem 4.49. Study the limits:

(a) $\displaystyle\lim_{x\to 0}\frac{\arctan(ax)}{\ln(1+bx)}$ $(b \neq 0)$

(b) $\displaystyle\lim_{x\to 0}\frac{\sin(x)+\sin(2x)}{2x}$

(c) $\displaystyle\lim_{x\to 2}\frac{x^2-4}{x^2-3x+2}$

(d) $\displaystyle\lim_{x\to 1}\frac{x^m-1}{x-1}$

(e) $\displaystyle\lim_{x\to +\infty}\frac{\ln(1+e^x)}{x}$

(f) $\displaystyle\lim_{x\to -\infty}\frac{\ln(1+e^x)}{x}$

(g) $\displaystyle\lim_{x\to +\infty}\frac{x+\sin(x)}{x-\sin(x)}$

(h) $\displaystyle\lim_{x\to +\infty}\frac{\ln(x)}{\sqrt{x}}$

(i) $\displaystyle\lim_{x\to 1}\frac{\tan(2x-2)-2(x-1)}{(x-1)^3}$

Solution to the problem 4.49. (a) $\lim\limits_{x\to 0} \dfrac{\arctan(ax)}{\ln(1+bx)} = «\dfrac{0}{0}»$ (indeterminate).

To solve the indeterminate form, we consider the Taylor polynomials of the functions in the numerator and denominator around the point where we are calculating the limit, $x = 0$. It will be sufficient to calculate the Taylor polynomial up to the main term.

First, consider the function $f(x) = \arctan(ax)$, then:

$$f(x) = \arctan(ax) \quad f(0) = 0$$
$$f'(x) = \frac{a}{1+a^2x^2} \quad f'(0) = a.$$

Therefore, the first-degree Taylor polynomial of f is: $p_1^f(x) = 0 + ax$, and additionally, we can write that in a neighborhood of the point $x = 0$, $f(x) = ax + O(x^2)$.

Similarly, consider the function $g(x) = \ln(1+bx)$, where:

$$g(x) = \ln(1+bx) \quad g(0) = 0$$
$$g'(x) = \tfrac{b}{1+bx} \quad g'(0) = b.$$

Therefore: $p_1^g(x) = bx$ and $g(x) = bx + O(x^2)$.

Then we can solve the limit as:

$$\lim_{x\to 0} \frac{\arctan(ax)}{\ln(1+bx)} = \lim_{x\to 0} \frac{ax + O(x^2)}{bx + O(x^2)} = \lim_{x\to 0} \frac{a + O(x)}{b + O(x)} = \frac{a+0}{b+0} = \frac{a}{b},$$

where we have used that $b \neq 0$.

Furthermore, we can deduce from this problem that if $a \neq 0$ and $b \neq 0$, then $\arctan(ax) \sim_0 ax$ and $\ln(1+bx) \sim_0 bx$.

Another way to solve this problem is to apply L'Hôpital's rule:

$$\lim_{x\to 0} \frac{\arctan(ax)}{\ln(1+bx)} = «\frac{0}{0}» \text{ (indeterminate).}$$

The functions $\arctan(ax)$ and $\ln(1+bx)$ are continuous and differentiable in a neighborhood of the point $x = 0$ and furthermore, $(\ln(1+bx))' = \frac{b}{1+bx}$ if $b \neq 0$ does not vanish in a neighborhood of the point $x = 0$. Therefore:

$$\lim_{x\to 0} \frac{\arctan(ax)}{\ln(1+bx)} = \lim_{x\to 0} \frac{\frac{a}{1+a^2x^2}}{\frac{b}{1+bx}} = \frac{a}{b}.$$

Note that to know that the indeterminate form was of the type $0/0$, we had to evaluate the functions $f(x) = \arctan(ax)$ and $g(x) = \ln(1+bx)$ at the point $x = 0$. Additionally, we had to calculate the derivatives $f'(x)$ and $g'(x)$ and also evaluate them at $x = 0$. That is, we needed the same information as for making the respective Taylor polynomials.

(b) $\lim\limits_{x\to 0} \dfrac{\sin(x) + \sin(2x)}{2x} = «\dfrac{0}{0}»$ (indeterminate).

We know that $\sin(2x) = 2\sin(x)\cos(x)$. *Therefore:*

$$\lim_{x\to 0} \frac{\sin(x)+\sin(2x)}{2x} = \lim_{x\to 0} \frac{\sin(x)(1+2\cos(x))}{2x}$$
$$= \lim_{x\to 0} \frac{\sin(x)}{x}\frac{1+2\cos(x)}{2} = \lim_{x\to 0} \frac{1+2\cos(x)}{2} = \frac{3}{2},$$

where we have used that $\sin(x) \sim_0 x$. *Let's prove it using the Taylor series of the function* $f(x) = \sin(x)$ *around zero:*

$$f(x) = \sin(x) \quad f(0) = 0$$
$$f'(x) = \cos(x) \quad f'(0) = 1.$$

Therefore, $f(x) = x + O(x^2)$ *around zero and:*

$$\lim_{x=0} \frac{\sin(x)}{x} = \lim_{x=0} \frac{x+O(x^2)}{x} = \lim_{x=0} \frac{1+O(x)}{1} = 1,$$

as we wanted to demonstrate.

If we had done this limit using L'Hôpital's rule, we would have obtained:

$$\lim_{x\to 0} \frac{\sin(x)+\sin(2x)}{2x} = \frac{0}{0} = \lim_{x\to 0} \frac{\cos(x)+2\cos(x)}{2} = \frac{3}{2}.$$

Note that we could use L'Hôpital's rule because the functions in the numerator and denominator are $\mathcal{C}^\infty(\mathbb{R})$ *and furthermore, the derivative of the denominator is* $2 \neq 0$ *in a neighborhood of the point* $x = 0$.

(c) $\lim_{x\to 2} \frac{x^2-4}{x^2-3x+2} = «\frac{0}{0}»$ (indeterminate).

In this case, we will use the Taylor series up to the second order around the point $x = 2$ *to factorize the polynomials that appear in the numerator and denominator:*

$$f(x) = x^2 - 4 \quad f(2) = 0$$
$$f'(x) = 2x \quad f'(2) = 4$$
$$f''(x) = 2 \quad f''(2) = 2,$$

and:

$$g(x) = x^2 - 3x + 2 \quad g(2) = 0$$
$$g'(x) = 2x - 3 \quad g'(2) = 1$$
$$g''(x) = 2 \quad g''(2) = 2,$$

Therefore:

$$f(x) = 4(x-2) + \frac{2}{2}(x-2)^2 = 4(x-2) + (x-2)^2$$

and:

$$g(x) = (x-2) + (x-2)^2.$$

Thus:

$$\lim_{x\to 2} \frac{x^2-4}{x^2-3x+2} = \lim_{x\to 2} \frac{4(x-2)+(x-2)^2}{(x-2)+(x-2)^2} = \lim_{x\to 2} \frac{4+(x-2)}{1+(x-2)} = 4.$$

Another way would be to factorize numerator and denominator as:

$$\lim_{x\to 2}\frac{x^2-4}{x^2-3x+2}=\lim_{x\to 2}\frac{(x-2)(x+2)}{(x-2)(x-1)}=\lim_{x\to 2}\frac{x+2}{x-1}=4.$$

Finally, if we had used L'Hôpital's rule we would have obtained:

$$\lim_{x\to 2}\frac{x^2-4}{x^2-3x+2}=\frac{0}{0}=\lim_{x\to 2}\frac{2x}{2x-3}=\frac{4}{1}=4.$$

(d) $\lim_{x\to 1}\dfrac{x^m-1}{x-1}=«\dfrac{0}{0}»$ (indeterminate).

The first way to solve this problem is by using the factorization of x^m-1*:*

$$x^m-1=(x-1)(1+x+x^2+x^3+\cdots+x^{m-1}),$$

therefore:

$$\begin{aligned}\lim_{x\to 1}\frac{x^m-1}{x-1}&=\lim_{x\to 1}\frac{(x-1)(1+x+x^2+x^3+\cdots+x^{m-1})}{x-1}\\&=\lim_{x\to 1}\left(1+x+x^2+x^3+\cdots+x^{m-1}\right)=m.\end{aligned}$$

Another possibility is to consider the second-degree Taylor polynomial of the function $f(x)=x^m-1$ *around the point* $x=1$*.*

$$\begin{array}{ll}f(x)=x^m-1 & f(1)=0\\ f'(x)=mx^{m-1} & f'(1)=m,\end{array}$$

therefore:

$$\begin{aligned}\lim_{x\to 1}\frac{x^m-1}{x-1}&=\lim_{x\to 1}\frac{m(x-1)+O((x-1)^2)}{x-1}\\&=\lim_{x\to 1}m+O((x-1))=m.\end{aligned}$$

Finally, using L'Hôpital's rule we obtain:

$$\lim_{x\to 1}\frac{x^m-1}{x-1}=\frac{0}{0}=\lim_{x\to 1}\frac{mx^{m-1}}{1}=m.$$

(e) $\lim_{x\to+\infty}\dfrac{\ln(1+e^x)}{x}=«\dfrac{+\infty}{+\infty}»$ (indeterminate).

First, consider the change of variable $t=\frac{1}{x}$*. Note that if* $x\to+\infty$*, then* $t\to 0^+$*; thus, we obtain:*

$$\lim_{x\to+\infty}\frac{\ln(1+e^x)}{x}=\lim_{t\to 0^+}\frac{\ln(1+e^{\frac{1}{t}})}{\frac{1}{t}}=\lim_{t\to 0^+}t\ln(1+e^{\frac{1}{t}}).$$

Let's calculate this limit using L'Hôpital's rule:

$$\begin{aligned}\lim_{x\to+\infty}\frac{\ln(1+e^x)}{x}&=«\frac{+\infty}{+\infty}»\text{ (indeterminate)}\\&=\lim_{x\to+\infty}\frac{\frac{e^x}{1+e^x}}{1}=\lim_{x\to+\infty}\frac{e^x}{1+e^x}\\&=\lim_{x\to+\infty}\frac{1}{\frac{1}{e^x}+1}=\frac{1}{1+0}=1.\end{aligned}$$

(f) $\lim_{x\to-\infty} \dfrac{\ln(1+e^x)}{x} = \dfrac{0}{-\infty} = 0 \cdot \dfrac{1}{-\infty} = 0 \cdot 0 = 0$

(g) $\lim_{x\to+\infty} \dfrac{x+\sin(x)}{x-\sin(x)} = «\dfrac{+\infty}{+\infty}»$ (indeterminate).

This indeterminate form can be resolved by dividing the numerator and denominator by x:

$$\lim_{x\to+\infty} \frac{x+\sin(x)}{x-\sin(x)} = \lim_{x\to+\infty} \frac{\frac{x}{x}+\frac{\sin(x)}{x}}{\frac{x}{x}-\frac{\sin(x)}{x}} = \frac{1+0}{1+0} = 1.$$

In this case, we cannot apply L'Hôpital's rule, because if we consider the derivative of the denominator $(x-\sin(x))' = 1-\cos(x)$, we see that we cannot assert that starting from some number, a, $(x-\sin(x))' \neq 0$, for all $x \geq a$. In fact, no matter what value a is, we can always find an $x > a$ for which $1-\cos(x) = 0$.

(h) $\lim_{x\to+\infty} \dfrac{\ln(x)}{\sqrt{x}} = «\dfrac{+\infty}{+\infty}»$ (indeterminate).

By orders of infinity, we know that: $\sqrt{x} << \ln(x)$ and, therefore:

$$\lim_{x\to+\infty} \frac{\ln(x)}{\sqrt{x}} = 0.$$

If we calculate the limit using L'Hôpital's rule we obtain:

$$\lim_{x\to+\infty} \frac{\ln(x)}{\sqrt{x}} = \frac{+\infty}{+\infty} = \lim_{x\to+\infty} \frac{\frac{1}{x}}{\frac{1}{2\sqrt{x}}} = \lim_{x\to+\infty} 2\frac{1}{\sqrt{x}} = 0.$$

(i) $\lim_{x\to1} \dfrac{\tan(2x-2)-2(x-1)}{(x-1)^3} = «\dfrac{0}{0}»$ (indeterminate).

Let's solve this limit first by calculating the Taylor series of the numerator function $f(x) = \tan(2x-2)-2(x-1)$.

$$\begin{aligned}
f(x) &= \tan(2x-2)-2(x-1)\\
f'(x) &= 2(1+\tan^2(2x-2))-2 = 2\tan^2(2x-2)\\
f''(x) &= 8\tan(2x-2)(1+\tan^2(2x-2)) = 8\tan(2x-2)+8\tan^3(2x-2)\\
f'''(x) &= 16(1+\tan^2(2x-2))+48\tan^2(2x-2)(1+\tan^2(2x-2))\\
&= 16+64\tan^2(2x-2)+48\tan^4(2x-2)
\end{aligned}$$

Therefore:

$$f(1) = f'(1) = f''(1) = 0 \quad y \quad f'''(1) = 16,$$

and we can write that in a neighborhood of the point $x = 1$:

$$f(x) = \frac{16}{3!}(x-1)^3 + O((x-1)^4).$$

Thus:

$$\lim_{x\to 1} \frac{\tan(2x-2)-2(x-1)}{(x-1)^3} = \lim_{x\to 1} \frac{\frac{8}{3}(x-1)^3 + O((x-1)^4)}{(x-1)^3}$$
$$= \lim_{x\to 1} \frac{\frac{8}{3} + O((x-1))}{1} = \frac{8}{3}.$$

This limit can also be solved by applying L'Hôpital's rule three times consecutively:

$$\lim_{x\to 1} \frac{\tan(2x-2)-2(x-1)}{(x-1)^3} = «\frac{0}{0}» \text{ (indeterminate)}$$
$$= \lim_{x\to 1} \frac{f(x)}{(x-1)^3} = \lim_{x\to 1} \frac{f'(x)}{3(x-1)^2} = «\frac{0}{0}» \text{ (indeterminate)}$$
$$= \lim_{x\to 1} \frac{f''(x)}{6(x-1)} = «\frac{0}{0}» \text{ (indeterminate)}$$
$$= \lim_{x\to 1} \frac{f'''(x)}{6} = \frac{16}{6} = \frac{8}{3}.$$

Part IV

Integral Calculus

5

Integration

«Nature laughs at the difficulties of integration». (Pierre-Simon Laplace, cited in the book *The Armchair Science Reader*, by I. Gordon and S. Sorkin, New York, 1959).

This chapter has two distinct parts. The first part introduces the concept of antiderivative (or indefinite integral), which is the inverse operator of the derivative. The second part introduces the concept of the Riemann integral (or definite integral) whose goal is the calculation of the area under a function. Subsequently, the fundamental theorem of calculus is presented, which relates both parts in one of the most important results of calculus. Before the appearance of this theorem, the approximate calculation of areas, which had been worked on since Archimedes, was a branch of mathematics separate from the differential calculus developed by Newton, Barrow, and Leibniz in the 18th century, which gave rise to concepts such as the derivative. With the statement of the fundamental theorem of calculus, both branches converged as it was demonstrated that the study of the area under a function was intimately related to differential calculus. Finally, in this chapter, the concept of the definite integral will be extended to unbounded functions and/or with an unbounded interval of integration, which are called improper integrals.

DOI: 10.1201/9781003399735-5

5.1 Primitive of a Function

Definition 5.1. Given a real function of a real variable defined over a closed interval $f : [a, b] \to \mathbb{R}$, any function $F : [a, b] \to \mathbb{R}$ such that its derivative is equal to $f(x)$ is called a **primitive** of $f(x)$, that is:

$$F'(x) = f(x), \quad \forall x \in [a, b].$$

To denote that the function $F(x)$ is a primitive of $f(x)$, the most common notation is:

$$F = \int f(x)\,\mathrm{d}x.$$

Observation 5.1. If F is a primitive of f, then the function $F + C$, where $C \in \mathbb{R}$, is also a primitive of f, since:

$$(F + C)' = F' = f.$$

That is, every function has infinitely many primitives that differ from each other by an additive constant called the integration constant.

This last observation can be presented through the following proposition.

Proposition 5.1. Let F_1 and F_2 be two primitives of the function $f : [a, b] \to \mathbb{R}$. Then, there exists a constant $C \in \mathbb{R}$ such that:

$$F_1 - F_2 = C$$

for all $x \in [a, b]$.

Proof. We know that F_1 and F_2 are primitives of the function $f : [a, b] \to \mathbb{R}$, that is:

$$F_1' = f \quad F_2' = f.$$

Therefore, if we calculate the derivative of the difference between F_1 and F_2 we obtain:

$$(F_1 - F_2)' = F_1' - F_2' = 0.$$

If the derivative of $(F_1 - F_2)$ is the constant function 0, then there exists a constant $C \in \mathbb{R}$ such that:

$$F_1 - F_2 = C,$$

as we wanted to demonstrate. □

Example 5.1. *Verify that:*

$$F(x) = \sin(x) - x\cos(x) + 2$$

is a primitive of the function $f(x) = x\sin(x)$.

SOLUTION.
$F(x)$ will be a primitive if it verifies that $F'(x) = f(x)$. Indeed:

$$F'(x) = \cos(x) - \cos(x) + x\sin(x) = x\sin(x).$$

Therefore, we can write that:

$$\int x\sin(x)\,\mathrm{d}x = \sin(x) - x\cos(x) + 2.$$

□

5.2 Indefinite Integral of a Function

Definition 5.2. Given a real function of a real variable $f : [a,b] \to \mathbb{R}$, the **indefinite integral** of f is the set of all its primitives. The indefinite integral of f is denoted by:

$$\int f(x)\,\mathrm{d}x$$

The notation for denoting a primitive or the indefinite integral is the same, but they are distinguished depending on whether it is a single function or a set of functions.

For example:

$$\int x\sin(x)\,\mathrm{d}x = \sin(x) - x\cos(x) + 2,$$

denotes a primitive of $x\sin(x)$, while:

$$\int x\sin(x)\,\mathrm{d}x = \sin(x) - x\cos(x) + C, \quad C \in \mathbb{R},$$

is the indefinite integral of the function $x\sin(x)$.

Property 5.1. The indefinite integral has the property of linearity, that is:

- $\displaystyle\int (f(x) + g(x))\,\mathrm{d}x = \int f(x)\,\mathrm{d}x + \int g(x)\,\mathrm{d}x$
- $\displaystyle\int kf(x)\,\mathrm{d}x = k\int f(x)\,\mathrm{d}x, \quad \forall k \in \mathbb{R}$

Proof. The proof is an immediate consequence of the linearity of the derivative, that is, the fact that:

$$[f(x) + g(x)]' = f'(x) + g'(x)$$
$$[k \cdot f(x)]' = k \cdot f'(x)$$

□

5.3 Calculus of Primitives

The process for finding the primitive of a function is the inverse process of differentiation. Certain primitives can be easily calculated, called immediate or almost-immediate primitives. In other cases, to calculate them, it will be necessary to use the methods of integration by parts, variable change, or decomposition into simple fractions.

5.3.1 Immediate Primitives

Immediate primitives are those that can be calculated simply by reading backwards a table of derivatives. That is, when the functions we need to integrate are directly the derivative of an elementary function. In this case, the calculation of the primitive is immediate by looking at a table of integrals like Table 5.1.

TABLE 5.1
Table of Immediate Integrals.

Integrals of simple functions
$\int x^n \, dx = \dfrac{1}{n+1} x^{n+1} + C, \; n \neq -1$
$\int \dfrac{1}{x} \, dx = \ln\lvert x\rvert + C$
$\int e^x \, dx = e^x + C$
$\int \cos(x) \, dx = \sin(x) + C$
$\int -\sin(x) \, dx = \cos(x) + C$
$\int \dfrac{1}{1+x^2} \, dx = \arctan(x) + C$
$\int \dfrac{1}{\cos^2(x)} \, dx = \tan(x) + C$
$\int \dfrac{1}{\sqrt{1-x^2}} \, dx = \arcsin(x) + C$
$\int \dfrac{-1}{\sqrt{1-x^2}} \, dx = \arccos(x) + C$
$\int a^x \, dx = \dfrac{a^x}{\ln(a)} + C, \; a > 0, \, a \neq 1$

Example 5.2. *Calculate a primitive of the function x^2.*

SOLUTION.

$$\int x^2 \, dx = \frac{x^3}{3} \quad \text{since} \quad \left(\frac{x^3}{3}\right)' = \frac{3x^2}{3} = x^2.$$

□

Proposed Exercise 5.1. Calculate a primitive of the functions:

(a) e^x

(b) $\dfrac{1}{x}$

(c) x^2

(d) $\cos(x)$

(e) $\dfrac{1}{\sqrt{x}}$

(f) $\dfrac{1}{1+x^2}$

Observation 5.2. Remember that to check if we have calculated the primitive correctly, we just have to calculate the derivative of the result and verify that we get the initial function.

Proposition 5.2. A primitive of $\frac{1}{x+a}$ is the function $\ln|x+a|$. That is:

$$\int \frac{1}{x+a}\,\mathrm{d}x = \ln|x+a|.$$

Proof. To prove this proposition, we will differentiate $\ln|x+a|$ and check that we obtain the function $\frac{1}{x+a}$.

The function $\ln|x+a|$, which is not defined at $x=-a$, can be written as:

$$\ln|x+a| = \begin{cases} \ln(x+a), & x > -a \\ \ln(-x-a), & x < -a \end{cases}$$

where by differentiating:

$$(\ln|x+a|)' = \begin{cases} \dfrac{1}{x+a}, & x > -a \\ \dfrac{1}{-x-a}\cdot(-1), & x < -a \end{cases}$$

which coincides with the function $\frac{1}{x+a}$, since at $x=-a$ neither function is defined. □

Observation 5.3. The function $\frac{1}{x+a}$ is defined for all $x \neq -a$, both positive and negative; therefore, the primitive must also be defined for all $x \neq -a$. The logarithm function is not defined for negative numbers; that is why the absolute value appears.

5.3.2 Almost-immediate Primitives

Almost-immediate primitives are those that can be calculated simply using the chain rule and the Table 5.1 of immediate integrals. This becomes clearer in the following proposition.

Proposition 5.3. If $F(x)$ is a primitive of the function $f(x)$, given another function $g(x)$ we have that:

$$\int f(g(x))g'(x) = F(g(x)).$$

These primitives are called almost-immediate.

Proof. Since $F(x)$ is a primitive of $f(x)$, then $F'(x) = f(x)$. Therefore, if we differentiate applying the chain rule:

$$(F(g(x))' = F'(g(x))g'(x) = f(g(x))g'(x),$$

as we wanted to demonstrate. □

Example 5.3. *Calculate the primitive of the function:*

$$\frac{\cos(x)}{\sin(x)}.$$

SOLUTION. If we consider the functions $f(x) = \frac{1}{x}$, $g(x) = \sin(x)$, since $g'(x) = \cos(x)$ we have that:

$$\frac{\cos(x)}{\sin(x)} = \frac{1}{\sin(x)}\cos(x) = f(g(x))g'(x).$$

Therefore, to calculate the primitive we only need to calculate a primitive of f; in this case, $F(x) = \ln|x|$. Then:

$$\int \frac{\cos(x)}{\sin(x)}\,\mathrm{d}x = \int f(g(x))g'(x) = F(g(x)) = \ln|\sin(x)|.$$

Indeed:

$$\Big(\ln|\sin(x)|\Big)' = \frac{1}{\sin(x)}(\sin(x))' = \frac{\cos(x)}{\sin(x)}.$$

□

Proposed Exercise 5.2. Calculate the following primitives:

(a) $\displaystyle\int \frac{3x^2}{\cos^2(x^3)}\,\mathrm{d}x$

(b) $\displaystyle\int 2xe^{x^2}\,\mathrm{d}x$

(c) $\displaystyle\int \frac{-\sin(x)}{1+\cos^2(x)}\,\mathrm{d}x$

5.3.3 Method of Integration by Parts

The method of integration by parts allows calculating the integral of a product of two functions using a formula that leads to a simpler integral than the initial one.

Proposition 5.4. Let $u(x)$ and $v(x)$ be differentiable functions. Then:

$$\int u(x)v'(x)\,\mathrm{d}x = u(x)v(x) - \int v(x)u'(x)\,\mathrm{d}x.$$

Proof. Since $u(x)$ and $v(x)$ are differentiable, so is the product function $u(x) \cdot v(x)$. If we differentiate this function we obtain:

$$(u(x)v(x))' = u'(x)v(x) + u(x)v'(x),$$

and if we integrate each of the terms we obtain:

$$\underbrace{\int (u(x)v(x))'\,\mathrm{d}x}_{u(x)v(x)} = \int u'(x)v(x)\,\mathrm{d}x + \int u(x)v'(x)\,\mathrm{d}x.$$

If we now rearrange the terms of the previous expression, we obtain:

$$\int u(x)v'(x)\ \mathrm{d}x = u(x)v(x) - \int v(x)u'(x)\ \mathrm{d}x,$$

as we wanted to demonstrate. □

As deduced from the formula, the method of integration by parts is applicable when the integrand can be expressed as a product of two functions, one of which, $v'(x)$, has an immediate integral and the other, $u(x)$, when differentiated, leads to a function, $u'(x)$, such that the new integrand $v(x)u'(x)$ is simpler. It is important to skillfully choose the functions $u(x)$ and $v'(x)$ so that when applying the integration by parts formula we arrive at a simpler integral than the initial one.

Example 5.4. *Calculate the indefinite integral of $f(x) = \ln(x)$.*

SOLUTION.

$$\int \ln(x)\,\mathrm{d}x = \int \underbrace{\ln(x)}_{u(x)}\ \underbrace{1\,\mathrm{d}x}_{v'(x)\,\mathrm{d}x} \overset{(*)}{=} \ln(x)x - \int x\frac{1}{x}\,\mathrm{d}x$$

$$(*)\left[\begin{array}{lll} u(x) = \ln(x) & \Rightarrow & u'(x)\,\mathrm{d}x = \frac{1}{x}\,\mathrm{d}x \\ v'(x)\,\mathrm{d}x = 1\,\mathrm{d}x & \Rightarrow & v(x) = x \end{array}\right]$$

$$= x\ln(x) - \int 1\,\mathrm{d}x = x\ln(x) - x + C, \quad C \in \mathbb{R}.$$

□

Observation 5.4. Note that, sometimes, as in Example 5.4, one of the parts remains *hidden* in the initial integral, by considering $v'(x) = 1$.

Example 5.5. *Calculate the indefinite integral of $f(x) = xe^x$.*

SOLUTION.

$$\int xe^x\,\mathrm{d}x = \int \underbrace{x}_{u(x)}\underbrace{e^x\,\mathrm{d}x}_{v'(x)\,\mathrm{d}x} \stackrel{(*)}{=} xe^x - \int e^x\,\mathrm{d}x$$

$$(*)\left[\begin{array}{lll} u(x) = x & \Rightarrow & u'(x)\,\mathrm{d}x = 1\,\mathrm{d}x \\ v'(x)\,\mathrm{d}x = e^x\,\mathrm{d}x & \Rightarrow & v(x) = e^x \end{array}\right]$$

$$= xe^x - e^x + C, \quad C \subset \mathbb{R}.$$

□

In the following example, it is necessary to apply the method of integration by parts twice.

Example 5.6. *Calculate the indefinite integral of $f(x) = x^2\sin(2x)$.*

SOLUTION.

$$\int x^2\sin(2x)\,\mathrm{d}x = \int \underbrace{x^2}_{u(x)}\underbrace{\sin(2x)\,\mathrm{d}x}_{v'(x)\,\mathrm{d}x} \stackrel{(*)}{=}$$

$$(*)\left[\begin{array}{lll} u(x) = x^2 & \Rightarrow & u'(x)\,\mathrm{d}x = 2x\,\mathrm{d}x \\ v'(x)\,\mathrm{d}x = \sin(2x)\,\mathrm{d}x & \Rightarrow & v(x) = -\frac{1}{2}\cos(2x) \end{array}\right]$$

$$= -\frac{1}{2}x^2\cos(2x) + \underbrace{\int x\cos(2x)\,\mathrm{d}x}_{\text{integration by parts}}$$

$$\int x\cos(2x)\,\mathrm{d}x = \int \underbrace{x}_{u(x)}\underbrace{\cos(2x)\,\mathrm{d}x}_{v'(x)\,\mathrm{d}x} \stackrel{(*)}{=}$$

$$(*)\left[\begin{array}{lll} u(x) = x & \Rightarrow & u'(x)\,\mathrm{d}x = 1\,\mathrm{d}x \\ v'(x)\,\mathrm{d}x = \cos(2x)\,\mathrm{d}x & \Rightarrow & v(x) = \frac{1}{2}\sin(2x) \end{array}\right]$$

$$= \frac{1}{2}x\sin(2x) - \int \frac{1}{2}\sin(2x)\,\mathrm{d}x$$
$$= \frac{1}{2}x\sin(2x) + \frac{1}{4}\cos(2x) + C, C \in \mathbb{R}.$$

Therefore, we have:

$$\int x^2\sin(2x)\,\mathrm{d}x = -\frac{1}{2}x^2\cos(2x) + \frac{1}{2}x\sin(2x) + \frac{1}{4}\cos(2x) + C.$$

□

Proposed Exercise 5.3. Calculate the primitives of the following functions using integration by parts:

(a) $\int xe^{-2x}\,\mathrm{d}x$

(b) $\int x\cos(x)\,\mathrm{d}x$

(c) $\int x\ln(x)\,\mathrm{d}x$

(d) $\int x^2\cos(x)\,\mathrm{d}x$

(e) $\int x^2\mathrm{e}^x\,\mathrm{d}x$

(f) $\int \mathrm{e}^x\cos(x)\,\mathrm{d}x$

5.3.4 Variable Change Method

This method consists of using a change of variable with the aim of rewriting the initial integral into a simpler one. The difficulty of this method lies in choosing the appropriate variable change.

Proposition 5.5. If $t = g(x)$ is a differentiable function whose image is an interval I, and f is continuous in I, then:

$$\int f(g(x))g'(x)\,\mathrm{d}x = \int f(t)\,\mathrm{d}t.$$

Proof. Let $F(t)$ be an antiderivative of the function $f(t)$; then, by the chain rule we have:

$$(F(g(x)))' = F'(g(x))g'(x) = f(g(x))g'(x).$$

That is, $F(g(x))$ is an antiderivative of the function $f(g(x))g'(x)$, which we can write as:

$$\int f(g(x))g'(x)\,\mathrm{d}x = F(g(x)) + C,\ C \in \mathbb{R}.$$

Now we use the variable change $t = g(x)$ to obtain:

$$\int f(g(x))g'(x)\,\mathrm{d}x = F(t) + C,\, C \in \mathbb{R}.$$

Finally, recalling that $F(t)$ is an antiderivative of the function $f(t)$, we get:

$$\int f(g(x))g'(x)\,\mathrm{d}x = \int f(t)\,\mathrm{d}t$$

just as we wanted to prove. □

It is common, and we will do so in this book, to apply the result of the previous proposition using the following notation:

$$\int f(g(x))g'(x)\,\mathrm{d}x \overset{(*)}{=} \int f(t)\,\mathrm{d}t = F(t) = F(g(x)).$$

$$(*)\left[\begin{array}{rcl} t & = & g(x) \\ \mathrm{d}t & = & g'(x)\,\mathrm{d}x \end{array}\right]$$

It is important to note that the last step consists of undoing the variable change, i.e., replacing t by $g(x)$.

Observation 5.5. Almost-immediate integrals, $\int f(g(x))g'(x)\,\mathrm{d}x$, can always be solved by the variable change method (using the change $t = g(x)$).

Example 5.7. *Calculate the indefinite integral of the function* $f(x) = x\sqrt{1+x^2}$.

SOLUTION.

$$\int x\sqrt{1+x^2}\,\mathrm{d}x \overset{(*)}{=} \int \sqrt{t}\frac{\mathrm{d}t}{2} = \frac{1}{2}\int t^{1/2}\,\mathrm{d}t = \frac{1}{2}\frac{2}{3}t^{3/2} + C$$

$$(*)\left[\begin{array}{rcl} t & = & 1+x^2 \\ \mathrm{d}t & = & 2x\,\mathrm{d}x \Rightarrow x\,\mathrm{d}x = \mathrm{d}t/2 \end{array}\right]$$

$$= \frac{1}{3}t\sqrt{t} + C = \frac{1}{3}(1+x^2)\sqrt{1+x^2} + C.$$

□

Example 5.8. *Calculate an antiderivative of the function* $f(x) = \dfrac{1}{1+\sqrt{x}}$ *using the change* $x = t^2$.

SOLUTION.

$$\int \frac{1}{1+\sqrt{x}}\,\mathrm{d}x \overset{(*)}{=} \int \frac{1}{1+\sqrt{t^2}}\,2t\,\mathrm{d}t = \int \frac{2t}{1+t}\,\mathrm{d}t =$$

$$(*)\left[\begin{array}{rcl} x & = & t^2 \\ \mathrm{d}x & = & 2t\,\mathrm{d}t \end{array}\right] \quad \sqrt{t^2} = t \text{ because } t \geq 0$$

$$= \int \frac{2(1+t)-2}{1+t}\,\mathrm{d}t = \int \left(2 - \frac{2}{1+t}\right)\,\mathrm{d}t$$

$$= 2t - 2\ln|1+t| + C$$

$$= 2\sqrt{x} - 2\ln|1+\sqrt{x}| = 2\sqrt{x} - 2\ln(1+\sqrt{x}) + C.$$

□

Proposed Exercise 5.4. Calculate the antiderivatives of the following functions using the variable change method.

(a) $\displaystyle\int x\sqrt{1-x^2}\,\mathrm{d}x$

(b) $\displaystyle\int x^2(1+2x^3)^3\,\mathrm{d}x$

(c) $\displaystyle\int \frac{e^{\sqrt{x}}}{\sqrt{x}}\,\mathrm{d}x$

(d) $\displaystyle\int \frac{x}{\sqrt{1-x^4}}\,\mathrm{d}x$

5.3.5 Antiderivatives of Rational Functions

In this section, we will see how to calculate antiderivatives of rational functions $\frac{P(x)}{Q(x)}$, where $P(x)$ and $Q(x)$ are polynomials with $Q(x) \neq 0$. The procedure consists of three steps:

Step (1) Reduction to a fraction where the degree of the numerator is less than the degree of the denominator.

The first step is to ensure that the degree of the numerator polynomial is less than the degree of the denominator polynomial. Otherwise, we will perform the polynomial division of $P(x)$ by $Q(x)$ obtaining:

$$P(x) = Q(x)s(x) + r(x),$$

where $s(x)$ is the quotient and $r(x)$ the remainder. Thus:

$$\frac{P(x)}{Q(x)} = \frac{Q(x)s(x) + r(x)}{Q(x)} = s(x) + \frac{r(x)}{Q(x)}$$

and integrating:

$$\int \frac{P(x)}{Q(x)}\,\mathrm{d}x = \underbrace{\int s(x)\,\mathrm{d}x}_{\text{immediate}} + \int \frac{r(x)}{Q(x)}\,\mathrm{d}x$$

where instead of the initial integral, we have obtained an immediate integral and another with a rational function where, now yes, the degree of the numerator polynomial is less than the degree of the denominator polynomial:

$$\text{degree}(r(x)) < \text{degree}(Q(x)).$$

Example 5.9. *Let's see some examples of how to reduce a rational function to another where the degree of the numerator is less than the degree of the denominator.*

- $\displaystyle\int \frac{x^2+3}{x}\,\mathrm{d}x = \int \left(x + \frac{3}{x}\right)\mathrm{d}x = \frac{x^2}{2} + \int \frac{3}{x}\,\mathrm{d}x$

- $\displaystyle\int \frac{2x}{x+1}\,\mathrm{d}x = \int \frac{2(x+1)-2}{x+1}\,\mathrm{d}x = \int \left(2 - \frac{2}{x+1}\right)\mathrm{d}x = 2x - \int \frac{2}{x+1}\,\mathrm{d}x$

- $\displaystyle\int \frac{3x^2}{x+7} = \int \frac{(3x-21)(x+7)+147}{x+7}\,\mathrm{d}x = \int (3x-21) + \frac{147}{x+7}\,\mathrm{d}x =$

$$= \frac{(3x-21)^2}{6} + \int \frac{147}{x+7}\,\mathrm{d}x$$

- $\displaystyle\int \frac{x^3-4x^2+2x+5}{x^2-5x+6} = \int \frac{(x^2-5x+6)(x+1)+(x-1)}{x^2-5x+6}\,\mathrm{d}x =$

$$= \int (x+1) + \frac{x-1}{x^2-5x+6}\,\mathrm{d}x =$$

$$= \frac{(x+1)^2}{2} + \int \frac{x-1}{x^2-5x+6}\,\mathrm{d}x$$

Step (2) Decomposition into partial fractions.

The second step consists of calculating the decomposition into partial fractions of the function $\frac{P(x)}{Q(x)}$ where, recall, the degree of the numerator polynomial is less than that of the denominator. To obtain the decomposition, we must first factorize $Q(x)$ and then:

(i) For each factor that is a simple real root, $(x-a)$, we will introduce a term of the type:

$$\frac{A}{x-a},$$

where A is a real number.

(ii) For each factor that is a multiple real root, $(x-a)^m$, we will introduce the terms:

$$\sum_{k=1}^{m} \frac{A_k}{(x-a)^k} = \frac{A_1}{x-a} + \frac{A_2}{(x-a)^2} + \cdots + \frac{A_m}{(x-a)^m},$$

where $A_1, \ldots, A_m$ are real numbers.

(iii) For each factor that is a simple complex root, that is, an irreducible quadratic monomial, (x^2+px+q), we will introduce the term:

$$\frac{Mx+N}{x^2+px+q},$$

where M and N are real numbers.

(iv) For each factor that is a multiple complex root, $(x^2+px+q)^m$, we will introduce the terms:

$$\sum_{k=1}^{m} \frac{M_k x + N_k}{(x^2+px+q)^k},$$

where $M_1, N_1, \ldots, M_m, N_m$ are real numbers.

In the case that any of the roots are multiple complex, instead of using the decomposition into partial fractions, it is simpler to use the Hermite method, which decomposes the integrand into the following form:

$$\frac{P(x)}{Q(x)} = \left[\frac{P_1(x)}{Q_1(x)}\right]' + \frac{P_2(x)}{Q_2(x)},$$

where $Q_1(x)$ is the greatest common divisor of $Q(x)$ and $Q'(x)$; $Q_2(x) = \frac{Q(x)}{Q_1(x)}$ and has all its roots simple; and the polynomials with indeterminate coefficients $P_1(x)$ and $P_2(x)$ are of degrees less than $Q_1(x)$ and $Q_2(x)$, respectively. Note that through this method the initial integral is rewritten as:

$$\int \frac{P(x)}{Q(x)}\,dx = \frac{P_1(x)}{Q_1(x)} + \int \frac{P_2(x)}{Q_2(x)}\,dx,$$

and thus, it consists of a rational part already integrated and another rational part with all simple roots in the denominator.

The indeterminate coefficients of the partial fractions are obtained by summing these fractions and imposing that the result obtained be equal to the initial fraction $P(x)/Q(x)$.

Example 5.10. *Calculate the decomposition into partial fractions of the function:*

$$\frac{7x-11}{(x-3)(x+2)}.$$

SOLUTION.

The first step is to ensure that the degree of the numerator polynomial is less than the degree of the denominator polynomial. Next, we observe that the denominator is already factorized and only simple real roots appear. Therefore:

$$\frac{7x-11}{(x-3)(x+2)} = \frac{A}{x-3} + \frac{B}{x+2}.$$

To find the coefficients A and B we sum the partial fractions:

$$\frac{A(x+2)+B(x-3)}{(x-3)(x+2)} = \frac{(A+B)x+2A-3B}{(x-3)(x+2)}.$$

Evidently, the result obtained must match the initial fraction. As the denominators are equal, we only need to impose that the numerators be equal. Thus, we obtain two equations with two unknowns:

$$A+B=7$$

$$2A-3B=-11$$

Finally, solving the system we get $A=2$ and $B=5$. Therefore:

$$\frac{7x-11}{(x-3)(x+2)} = \frac{2}{x-3} + \frac{5}{x+2}.$$

□

Example 5.11. *Calculate the decomposition into partial fractions of the function:*

$$\frac{2x-1}{(x-3)^2(x+1)}.$$

SOLUTION.

The first step is to ensure that the degree of the numerator polynomial is less than the degree of the denominator polynomial. Next, we observe that the denominator is already factorized and contains a simple real root and a multiple real root. Therefore:

$$\begin{aligned}\frac{2x-1}{(x-3)^2(x+1)} &= \frac{A}{x-3} + \frac{B}{(x-3)^2} + \frac{C}{x+1} \\ &= \frac{A(x-3)(x+1)+B(x+1)+C(x-3)^2}{(x-3)^2(x+1)}.\end{aligned}$$

Evidently, the result obtained must match the initial fraction. As the denominators are equal, we only need to impose that the numerators be equal. Thus, for any value of x it must be true that:

$$2x-1 = A(x-3)(x+1)+B(x+1)+C(x-3)^2.$$

Specifically:

- If we substitute $x=-1$, we obtain $-3=16C \Rightarrow C=-3/16$.
- If we substitute $x=3$, we obtain $5=4B \Rightarrow B=5/4$.

– If we differentiate, we obtain:

$$2 = A(x+1) + A(x-3) + B + 2C(x-3)$$

and then substituting $x = 3$:

$$2 = 4A + B$$

where using $B = 5/4$, we obtain $A = 3/16$.

Therefore:

$$\frac{2x-1}{(x-3)^2(x+1)} = \frac{3/16}{x-3} + \frac{5/4}{(x-3)^2} + \frac{-3/16}{x+1}.$$

□

Example 5.12. *Calculate the decomposition into partial fractions of the function:*

$$\frac{2x^2+x+1}{x(x^2+1)}.$$

SOLUTION.

The first step is to ensure that the degree of the numerator polynomial is less than the degree of the denominator polynomial. Next, we observe that, in this case, the denominator is already factorized and contains a simple real root and a simple complex root. Therefore:

$$\begin{aligned}\frac{2x^2+x+1}{x(x^2+1)} &= \frac{A}{x} + \frac{Bx+C}{x^2+1} = \frac{A(x^2+1)+x(Bx+C)}{x(x^2+1)} \\ &= \frac{(A+B)x^2+Cx+A}{x(x^2+1)}\end{aligned}$$

Matching the coefficients:

$$2x^2+x+1 = (A+B)x^2+Cx+A$$

we obtain $A = B = C = 1$ and, therefore:

$$\frac{2x^2+x+1}{x(x^2+1)} = \frac{1}{x} + \frac{x+1}{x^2+1}.$$

□

Example 5.13. *Calculate the following integral:*

$$\int \frac{1}{(1+x^2)^3}\,\mathrm{d}x.$$

SOLUTION.

The function to be integrated is rational with the degree of the numerator polynomial less than that of the denominator. Then, we observe that, in this case, the denominator is already factorized and contains a complex root of multiplicity 3.

In this case, instead of decomposing into partial fractions and then integrating all the terms, it is more convenient to use the Hermite method. Recall that this method consists of decomposing the integrand into the following form:

$$\frac{P(x)}{Q(x)} = \left[\frac{P_1(x)}{Q_1(x)}\right]' + \frac{P_2(x)}{Q_2(x)}$$

where $Q_1(x)$ is the greatest common divisor of $Q(x)$ and $Q'(x)$; $Q_2(x) = \frac{Q(x)}{Q_1(x)}$; and the polynomials $P_1(x)$ and $P_2(x)$ are of degrees less than $Q_1(x)$ and $Q_2(x)$, respectively. Specifically, in this example:

$$Q(x) = (1 + x^2)^3,$$

therefore, its derivative is:

$$Q'(x) = 6x(1 + x^2)^2.$$

Thus:

$$Q_1(x) = \gcd(Q(x), Q'(x)) = (1 + x^2)^2,$$

$$Q_2(x) = \frac{Q(x)}{Q_1(x)} = \frac{(1 + x^2)^3}{(1 + x^2)^2} = 1 + x^2.$$

So the Hermite decomposition is in the form:

$$\frac{1}{(1 + x^2)^3} = \left[\frac{Ax^3 + Bx^2 + Cx + D}{(1 + x^2)^2}\right]' + \frac{Ex + F}{1 + x^2}.$$

To determine the coefficients, we calculate the indicated derivative and group terms obtaining:

$$\frac{1}{(1 + x^2)^3} = \frac{C + F + (2B - 4D + E)x + (3A - 3C + 2F)x^2}{(1 + x^2)^3} +$$

$$+ \frac{(-2B + 2E)x^3 + (-A + F)x^4 + Ex^5}{(1 + x^2)^3}$$

where, by matching coefficients, we get: $E = 0$, $B = 0$, $D = 0$, $A = F = \frac{3}{8}$, $C = \frac{5}{8}$. Finally, the initial integral can be easily calculated:

$$\int \frac{1}{(1 + x^2)^3}\,\mathrm{d}x = \frac{\frac{3}{8}x^3 + \frac{5}{8}x}{(1 + x^2)^2} + \int \frac{\frac{3}{8}}{1 + x^2}\,\mathrm{d}x = \frac{3x^3 + 5x}{8(1 + x^2)^2} + \frac{3}{8}\arctan(x) + C,$$

where $C \in \mathbb{R}$. □

Proposed Exercise 5.5. Decompose into partial fractions the quotient:

$$\frac{1}{(x - 2)(x - 1)^2(x^2 + 1)(x^2 + 4)^2}.$$

Step (3) Integration of each of the terms.

The third step is to calculate the antiderivative of each of the obtained simple fractions. Thus, we need to know how to integrate the following types of simple fractions.

(i) The antiderivatives of simple fractions of the type $\frac{A}{x-a}$ are immediate:

$$\int \frac{A}{x-a}\,\mathrm{d}x = A\ln|x-a| + C.$$

(ii) The antiderivatives of simple fractions of the type $\frac{A}{(x-a)^m}$ are immediate:

$$\int \frac{A}{(x-a)^m}\,\mathrm{d}x = \int A(x-a)^{-m}\,\mathrm{d}x = \frac{A}{-m+1}(x-a)^{-m+1}$$
$$= -\frac{A}{(m-1)(x-a)^{m-1}} + C.$$

(iii) The antiderivatives of simple fractions of the type $\frac{Mx+N}{x^2+px+q}$ require more algebra to be calculated. Specifically:

$$\int \frac{Mx+N}{x^2+px+q}\,\mathrm{d}x = \int \frac{\frac{M}{2}(2x+p) + N - \frac{M}{2}p}{x^2+px+q}\,\mathrm{d}x$$
$$= \int \frac{M}{2}\frac{\overbrace{2x+p}^{\text{derivative of the denominator}}}{x^2+px+q}\,\mathrm{d}x + \int \frac{N-\frac{M}{2}p}{x^2+px+q}\,\mathrm{d}x$$
$$= \frac{M}{2}\ln|x^2+px+q| + \int \frac{N-\frac{M}{2}p}{x^2+px+q}\,\mathrm{d}x$$
$$= \frac{M}{2}\ln|x^2+px+q| + \int \frac{N-\frac{M}{2}p}{(x-r)^2+s^2}\,\mathrm{d}x$$

where $r = -p/2,\ s = +\sqrt{q-p^2/4}$:

$$= \frac{M}{2}\ln|x^2+px+q| + \frac{N+Mr}{s}\arctan\Big(\frac{x-r}{s}\Big) + C$$

Example 5.14. *Let us calculate the following antiderivatives:*

(a) $\displaystyle\int \frac{x^3+3}{x}\,\mathrm{d}x = \int \Big(x^2 + \frac{3}{x}\Big)\,\mathrm{d}x = \frac{x^3}{3} + 3\ln|x| + C$

(b) $\displaystyle\int \frac{1}{x^2-6x+9}\,\mathrm{d}x = \int \frac{1}{(x-3)^2}\,\mathrm{d}x = \int (x-3)^{-2}\,\mathrm{d}x = -(x-3)^{-1} = -\frac{1}{x-3} + C$

(c) $\displaystyle\int \frac{2x-5}{x^2-5x+1}\,\mathrm{d}x = \ln|x^2-5x+1| + C$

(d) $\displaystyle\int \frac{x+3}{x^2-4}\,\mathrm{d}x = \int \frac{x+3}{(x-2)(x+2)}\,\mathrm{d}x = \int \Big(\frac{5}{4}\frac{1}{x-2} - \frac{1}{4}\frac{1}{x+2}\Big)\,\mathrm{d}x$

$$= \frac{5}{4}\ln|x-2| - \frac{1}{4}\ln|x+2| + C$$

(e) $\displaystyle\int \frac{2x-1}{(x-3)^2(x+1)}\,\mathrm{d}x = \int \Big(\frac{3/16}{x-3} + \frac{5/4}{(x-3)^2} - \frac{3/16}{x+1}\Big)\,\mathrm{d}x$

$$= \frac{3}{16}\ln|x-3| - \frac{5}{4}\frac{1}{x-3} - \frac{3}{16}\ln|x+1| + C$$

(f) $$\int \frac{x+1}{x^2+2x+5}\,\mathrm{d}x = \int \frac{1}{2}\frac{2x+2}{x^2+2x+5}\,\mathrm{d}x = \frac{1}{2}\ln|x^2+2x+5| + C$$

(g) $$\int \frac{4x+6}{x^2+2x+5}\,\mathrm{d}x = \int \frac{2(2x+2)+2}{x^2+2x+5}\,\mathrm{d}x = \int\left(2\frac{2x+2}{x^2+2x+5} + \frac{2}{x^2+2x+5}\right)\mathrm{d}x$$
$$= 2\ln|x^2+2x+5| + \int \frac{2}{(x+1)^2+4}\,\mathrm{d}x.$$

In this case, the pending integral needs more development:

$$2\ln|x^2+2x+5| + \int \frac{2}{(x+1)^2+4}\,\mathrm{d}x = 2\ln(x^2+2x+5) + \int \frac{1}{4}\frac{2}{(\frac{x+1}{2})^2+1}\,\mathrm{d}x$$
$$= 2\ln|x^2+2x+5| + \int \frac{1/2}{(\frac{x+1}{2})^2+1}\,\mathrm{d}x$$
$$= 2\ln|x^2+2x+5| + \arctan(\frac{x+1}{2}) + C$$

(h) $$\int \frac{1}{x^2+8x+25}\,\mathrm{d}x = \int \frac{1}{(x+4)^2+9}\,\mathrm{d}x = \int \frac{1}{9}\frac{1}{(\frac{x+4}{3})^2+1}\,\mathrm{d}x$$
$$= \frac{1}{3}\int \frac{1/3}{(\frac{x+4}{3})^2+1}\,\mathrm{d}x = \frac{1}{3}\arctan\left(\frac{x+4}{3}\right) + C$$

(i) $$\int \frac{10x+2}{x^5-x^4-x+1}\,\mathrm{d}x = \int \frac{10x+2}{(x+1)(x-1)^2(x^2+1)}\,\mathrm{d}x$$
$$= \int\left(\frac{A}{x+1} + \frac{B}{x-1} + \frac{C}{(x-1)^2} + \frac{\mathrm{d}x+E}{x^2+1}\right)\mathrm{d}x$$

To be able to calculate the integral, we must find the indeterminate coefficients of the decomposition into simple fractions:

$$\frac{A}{x+1} + \frac{B}{x-1} + \frac{C}{(x-1)^2} + \frac{\mathrm{d}x+E}{x^2+1} =$$
$$= \frac{A(x-1)^2(x^2+1) + B(x+1)(x-1)(x^2+1)}{x^5-x^4-x+1} +$$
$$+ \frac{C(x+1)(x^2+1) + (\mathrm{d}x+E)(x+1)(x-1)^2}{x^5-x^4-x+1}$$

Therefore, we must impose that:

$$10x+2 = A(x-1)^2(x^2+1) + B(x+1)(x-1)(x^2+1) +$$
$$+ C(x+1)(x^2+1) + (\mathrm{d}x+E)(x+1)(x-1)^2$$

Substituting for $x=-1$ *we get* $-8 = 8A \quad\Rightarrow\quad A=-1.$

Substituting for $x=1$ *we get* $12 = 4C \quad\Rightarrow\quad C=3.$

Differentiating and substituting for $x = 1$ we get:

$$10 = 2A(x-1)(x^2+1) + A(x-1)^2 2x + B(x-1)(x^2+1) +$$
$$+ B(x+1)(x^2+1) + B(x+1)(x-1)2x + C(x^2+1) + C(x+1)2x +$$
$$+ D(x+1)(x-1)^2 + (\,\mathrm{d}x + E)(x-1)^2 + (\,\mathrm{d}x + E)(x+1)2(x-1)$$

and, therefore, substituting in $x = 1$ we have:

$$10 = 4B + 2C + 4C \quad \Rightarrow \quad B = \frac{10-6C}{4} = -2.$$

To find the coefficients D and E, since they come from the complex root of the polynomial x^2+1 we can substitute $x = i$, from which we obtain:

$$10i + 2 = (Di + E)(i+1)(i-1)^2 \quad \Rightarrow \quad 10i + 2 = 2(D+E) + 2(D-E)i$$

Finally, we must equal the real and imaginary parts:

$$D + E = 5, \quad D - E = 1 \quad \Rightarrow \quad D = 3, \quad E = 2$$

We have obtained $A = -1$, $B = -2$, $C = 3$, $D = 3$, $E = -2$. Therefore:

$$\int \left(\frac{A}{x+1} + \frac{B}{x-1} + \frac{C}{(x-1)^2} + \frac{\mathrm{d}x + E}{x^2+1} \right) \mathrm{d}x =$$
$$= \int \left(-\frac{1}{x+1} - \frac{2}{x-1} + \frac{3}{(x-1)^2} + \frac{3x-2}{x^2+1} \right)$$
$$= -\ln|x+1| - 2\ln|x-1| - \frac{3}{x-1} + \frac{3}{2}\ln(1+x^2) - 2\arctan(x) + C.$$

Proposed Exercise 5.6. Calculate the antiderivatives of the following functions:

(a) $\displaystyle\int \frac{x-8}{x^2-x-2}\,\mathrm{d}x$

(b) $\displaystyle\int \frac{x^4-7x^3+17x^2-22x+14}{x^3-7x^2+14x-8}\,\mathrm{d}x$

(c) $\displaystyle\int \frac{4x-2}{5x^2-20x+65}\,\mathrm{d}x$

(d) $\displaystyle\int \frac{x^3+x^2+x+1}{(x-1)^5}\,\mathrm{d}x$

(e) $\displaystyle\int \frac{x^3+2x^2-4x+13}{x^4-4x^3+13x^2}\,\mathrm{d}x$

Proposed Exercise 5.7. Calculate the antiderivatives of the rational functions in Examples 5.8-5.11.

Proposed Exercise 5.8. Calculate the following antiderivatives, following the given indications:

(a) $\displaystyle\int \frac{1}{(1+x^2)^2}\,\mathrm{d}x$ change $x=\tan(u)$

(b) $\displaystyle\int \sqrt{a^2+b^2x^2}\,\mathrm{d}x$ change $bx = a\tan(t)$

(c) $\displaystyle\int \frac{\sqrt{x}+\sqrt[3]{x^2}}{\sqrt{x}}\,\mathrm{d}x$ change $x=t^6$

(d) $\displaystyle\int \arctan(x)\,\mathrm{d}x$ integration by parts

(e) $\displaystyle\int \frac{x^3+2x+1}{x^2-5x+6}\,\mathrm{d}x$

(f) $\displaystyle\int \frac{1}{(x-1)^2(x+1)}\,\mathrm{d}x$

(g) $\displaystyle\int \frac{x}{(x^2+x+1)(x+1)}\,\mathrm{d}x$

(h) $\displaystyle\int \frac{1}{3+\cos(x)}\,\mathrm{d}x$ change $x = 2\arctan(t)$

(i) $\displaystyle\int \frac{\sqrt{x^2+4}}{x}\,\mathrm{d}x$ change $x=2\tan(u)$, $x^2+4=u^2$, $u=\sqrt{x^2+4}-x$

(j) $\displaystyle\int \frac{\ln(x)}{\sqrt{x}}\,\mathrm{d}x$ change $u=\sqrt{x}$ or parts

(k) $\displaystyle\int \sin(\ln(x))\,\mathrm{d}x$ integration by parts

(l) $\displaystyle\int \tan(x)\,\mathrm{d}x$

(m) $\displaystyle\int \frac{\cos(x)}{1+\sin^2(x)}\,\mathrm{d}x$

(n) $\displaystyle\int \frac{3x-2}{x^2-4x+5}\,\mathrm{d}x$

(o) $\displaystyle\int e^{\arcsin(x)}\,\mathrm{d}x$ integration by parts

(p) $\displaystyle\int \ln(1-\sqrt{x})\,\mathrm{d}x$ change $u=\sqrt{x}$ or parts

(q) $\displaystyle\int x\cos^2(x)\,\mathrm{d}x$ integration by parts

5.4 Riemann Integral

The goal of the Riemann integral is to calculate the area under the graph of a function. It is important when reading this section to disassociate the word integral from the inverse process of differentiation. That is, we should think of the Riemann integral as a tool for calculating areas.

5.4.1 Intuitive Idea

Given a bounded and positive function $f:[a,b]\to\mathbb{R}$, we want to address the problem of determining the area bounded by the graph of f, the x-axis, and the lines $x=a$ and $x=b$. Figure 5.1 shows an example of the area to be determined.

Before introducing the basic idea of how to calculate this area, the following definitions will be useful.

Definition 5.3 (Partition). A partition p of the closed interval $[a,b]$ is an ordered and finite set of points $p=\{x_0,x_1,x_2,\dots,x_n\}$ such that:

$$a = x_0 < x_1 < x_2 < \cdots < x_{n-1} < x_n = b$$

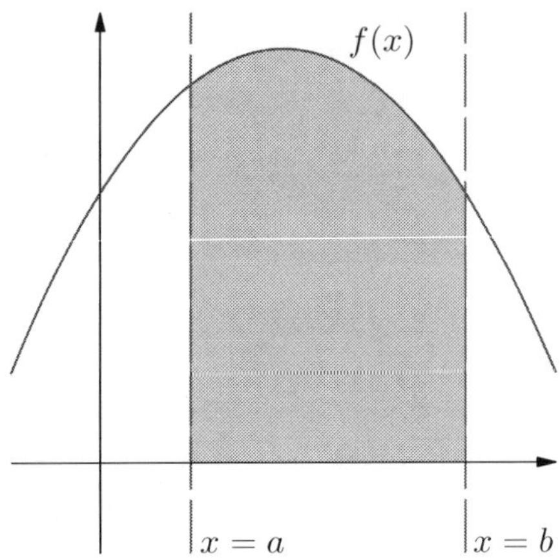

FIGURE 5.1
Area bounded by the graph of a function, the x-axis, and the lines $x = a$ and $x = b$.

Definition 5.4 (Norm of a partition). Given a partition p of the closed interval $[a, b]$, we denote by the norm of that partition, $\|p\|$, the maximum difference between any two consecutive points in the partition. That is, defining the length of the i-th interval as $\Delta x_i = x_i - x_{i-1}$ we have:

$$\|p\| = max\{\Delta x_i,\, i = 1, \ldots, n\}.$$

Definition 5.5 (Refinement of a partition). A refinement of a partition p is another partition p' that contains all the points of p and, additionally, other points, also ordered in magnitude.

Definition 5.6. We denote by $\mathcal{P}([a, b])$ the set of all possible partitions of the interval $[a, b]$.

Idea to approximate the area from below: we consider a partition $p \in \mathcal{P}([a, b])$ of our interval:

$$a = x_0 < x_1 < x_2 < \cdots < x_{n-1} < x_n = b,$$

within each subinterval $I_i = [x_{i-1}, x_i]$ we calculate the minimum value of the function:

$$m_i = \min\{f(x),\, x \in I_i\},$$

and, finally, we approximate the area under the function by the sum of the areas of the rectangles. We denote a_i as the area of the rectangle with base $\Delta x_i = x_i - x_{i-1}$ and height m_i. Then, the approximate area is:

$$s_p(f) = m_1 \Delta x_1 + m_2 \Delta x_2 + \cdots + m_n \Delta x_n = a_1 + a_2 + \cdots + a_n.$$

Thus, $s_p(f)$ is called the lower (Riemann) sum of the function f associated with the partition p and gives us a lower bound of the area we want to calculate. The subscript p indicates that this lower sum depends on the partition used. If we take a different partition, the lower sum changes.

For the case in Figure 5.2, we have an approximation $s_p = 6.5980$ when the exact area is $A = 8.933882605$.

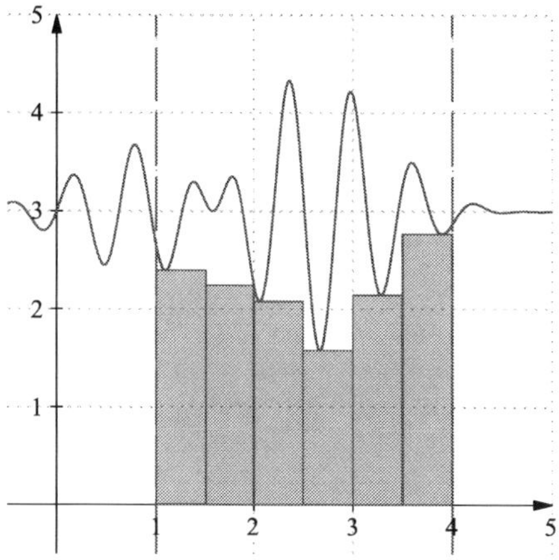

I_i	m_i	a_i
[1, 1.5]	2.3914	1.1957
[1.5, 2]	2.2402	1.1201
[2, 2.5]	2.0769	1.0385
[2.5, 3]	1.5764	0.7882
[3, 3.5]	2.1411	1.0706
[3.5, 4]	2.7698	1.3849
lower sum		6.5980

FIGURE 5.2
Calculation of the lower (Riemann) sum, $s_p(f)$, of the function f associated with p, with $f(x) = 2\sin(10x)\cos(x)e^{-(x-2)^2/2} + 3$ and $p = \{1, 1.5, 2, 2.5, 3, 3.5, 4\}$.

Idea to approximate the area from above: we will consider the previous partition, $p \in \mathcal{P}([a, b])$:

$$a = x_0 < x_1 < x_2 < \cdots < x_{n-1} < x_n = b,$$

within each subinterval $I_i = [x_{i-1}, x_i]$, we calculate the maximum value of the function:

$$M_i = \max\{f(x),\, x \in I_i\}$$

Finally, we approximate the area under the function by the sum of the areas of the rectangles. We denote A_i as the area of the rectangle with base $\Delta x_i = x_i - x_{i-1}$ and height M_i. Then, the approximate area is:

$$S_p(f) = M_1\Delta x_1 + M_2\Delta x_2 + \cdots + M_n\Delta x_n = A_1 + A_2 + \cdots + A_n.$$

Thus, $S_p(f)$ is called the upper (Riemann) sum of the function f associated with the partition p and gives us an upper bound of the area we want to calculate. Remember that the subscript p indicates that this upper sum depends on the partition used.

For the case in Figure 5.3, we have an approximation $S_p(f) = 11.4411$ when the exact area is $A = 8.933882605$.

Observation 5.6. Given a bounded function $f : [a, b] \to \mathbb{R}$ and a partition p of the interval $[a, b]$, we obtain the lower Riemann sums, $s_p(f)$, and upper Riemann sums, $S_p(f)$. If we refine the initial partition, obtaining a finer partition $\hat{p}$, then the lower sum increases $(s_{\hat{p}}(f) > s_p(f))$, and the upper sum decreases $(S_{\hat{p}}(f) < S_p(f))$. Let us see this fact in Example 5.15.

Example 5.15. *The exact area of the function in Figure 5.4 is $A = 8.933882605$. Note how refining a partition makes the rectangles better fit the shape of the function and, therefore, the lower sum increases and the upper sum decreases.*

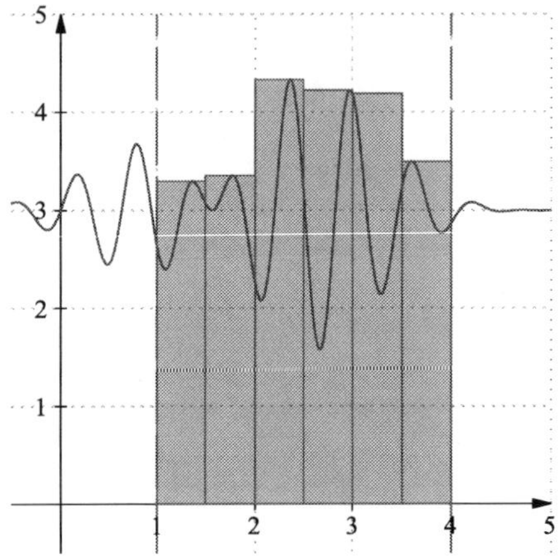

I_i	M_i	A_i
$[1, 1.5]$	3.2965	1.6482
$[1.5, 2]$	3.3523	1.6762
$[2, 2.5]$	4.3300	2.1650
$[2.5, 3]$	4.2208	2.1104
$[3, 3.5]$	4.1865	2.0933
$[3.5, 4]$	3.4961	1.7480
upper sum		11.4411

FIGURE 5.3
Calculation of the upper (Riemann) sum, $S_p(f)$, of the function f associated with p, with $f(x) = 2\sin(10x)\cos(x)e^{-(x-2)^2/2} + 3$ and $p = \{1, 1.5, 2, 2.5, 3, 3.5, 4\}$.

In this case, the lower and upper sums converge to the value of the area, as shown in Table 5.2.

5.4.2 Formal Definition

To present the definition of an integrable function in the sense of Riemann, it is necessary to recall the definition of the supremum and infimum of a subset. Recall that if B is a subset of a set A, the supremum of B is defined as the least element of A that is greater than or equal to every element of B and is denoted by $\sup(B)$. On the other hand, the infimum of B is defined as the greatest element of A that is less than or equal to every element of B and is denoted by $\inf(B)$. For example, if we take the set $B = (0, 1]$, which is a subset of the real numbers, the infimum is 0 and the supremum is 1.

In the previous section, we saw that given a function $f : [a, b] \to \mathbb{R}$ that is bounded and positive, if we refine a partition, the lower sum increases and the upper sum decreases. Thus, when for a certain function $f : [a, b] \to \mathbb{R}$, the supremum of the lower sums (taking all possible partitions of the interval $[a, b]$) coincides with the infimum of the upper sums, the function is said to be integrable in the sense of Riemann. Below is the formal definition.

Definition 5.7. If $f : [a, b] \to \mathbb{R}$ is a bounded function in its domain, f is said to be integrable in the sense of Riemann if:

$$\sup_{p\in\mathcal{P}([a,b])} \{s_p(f)\} = \inf_{p\in\mathcal{P}([a,b])} \{S_p(f)\}$$

Then, the value to which the upper and lower sums converge is called the definite integral of f on $[a, b]$ and is denoted by:

$$\int_a^b f(x)\,\mathrm{d}x.$$

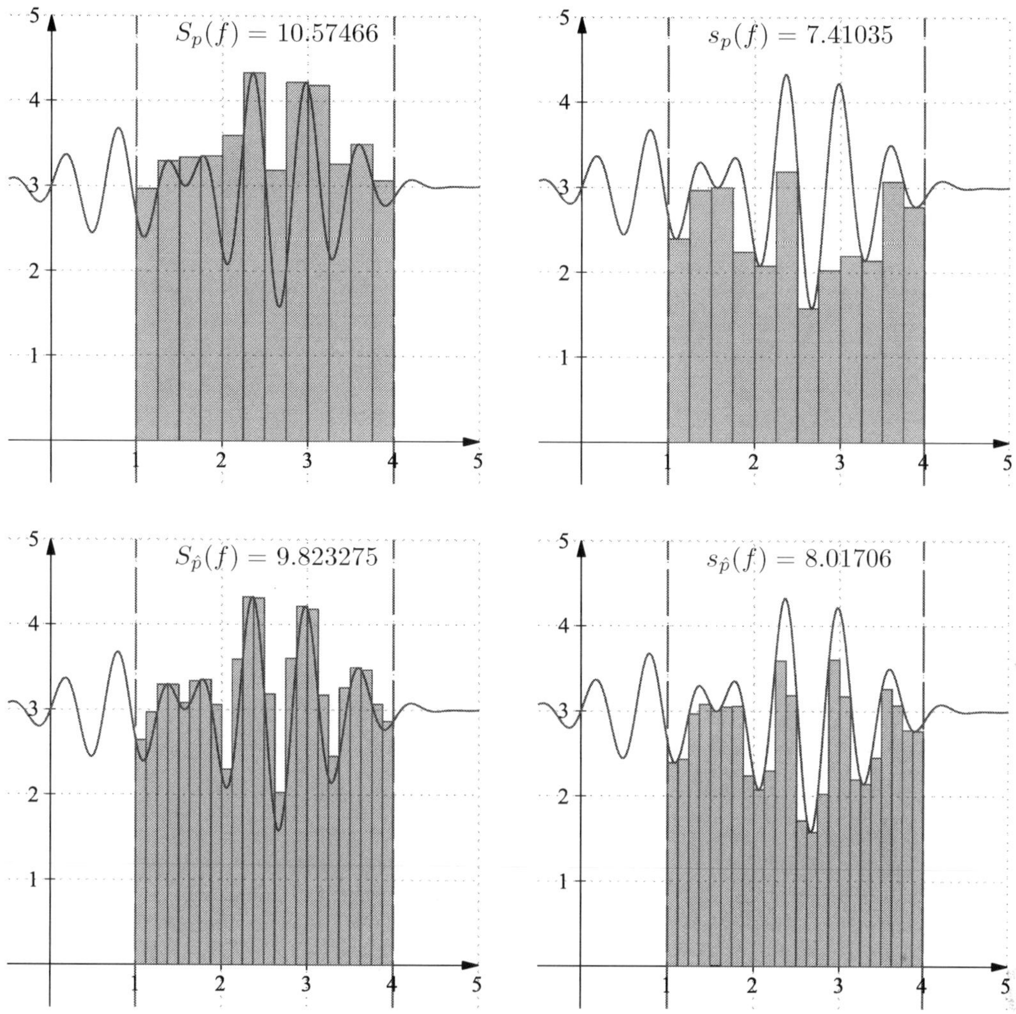

FIGURE 5.4
Calculation of the lower and upper Riemann sums of the function $f(x) = 2\sin(10x)\cos(x)e^{-(x-2)^2/2} + 3$ for a first partition p and a second finer partition $\hat{p}$.

Example 5.16. *Let's see an example of a function that is not integrable in the sense of Riemann. Consider the Dirichlet function $f : [a, b] \to \mathbb{R}$ defined by:*

$$f(x) = \begin{cases} 1, & \text{if } x \in \mathbb{Q} \\ 0, & \text{if } x \in \mathbb{R} - \mathbb{Q} \end{cases}$$

Let $p = \{x_0, x_1, x_2, ..., x_n\}$ be any partition of the interval $[a, b]$. In each subinterval I_i, the maximum value of the function is 1 and the minimum is 0, since, no matter how small the interval, there will always be rational and irrational numbers within it. Therefore, the upper sum will be $S_p(f) = 1$ and the lower sum will be $s_p(f) = 0$ for any partition p. This demonstrates that the Dirichlet function is not integrable in the sense of Riemann, as the upper and lower sums do not converge to the same value, no matter how much we refine the partition.

TABLE 5.2
Values of the lower and upper Riemann sums as finer partitions are taken.

Number of intervals	$s_p(f)$	$S_p(f)$
6	6.597950035	11.44108920
12	7.410344617	10.57466057
18	7.760077085	10.10370961
24	8.017068057	9.823274607
30	8.206761358	9.671088636
36	8.325267225	9.555041641
42	8.405479820	9.466787199
48	8.472961787	9.398836600
54	8.524642902	9.350773664
60	8.564191338	9.307486009
66	8.595828891	9.274074506
72	8.624302697	9.246079129
78	8.648065263	9.221106953
84	8.667824200	9.201460265
90	8.685040643	9.183813180
600	8.896422517	8.971374734
1200	8.915147143	8.952625076
2400	8.924513889	8.943253092
4800	8.929198023	8.938567639

$A = 8.933882605$

Example 5.17. *Let's see that the identity function, $f(x) = x$, is integrable on $[0, 1]$. Let $p_n \in \mathcal{P}([0, 1])$ be the partition into n subintervals given by:*

$$p_n = \left\{0, \frac{1}{n}, \frac{2}{n}, \cdots, \frac{n-1}{n}, \frac{n}{n} = 1\right\}.$$

Since the identity function is increasing, the minimum value of the function is always reached at the left end of the interval and the maximum value at the right end. That is, in the subinterval $I_i = [x_{i-1}, x_i] = [\frac{i-1}{n}, \frac{i}{n}]$ we have:

$$m_i = \frac{i-1}{n}, \qquad M_i = \frac{i}{n}.$$

Moreover, with the chosen partition, all subintervals have the same length:

$$\Delta x_i = \frac{1}{n}, \quad \forall i = 1, \ldots, n.$$

Thus, a lower and upper bound of the integral are given by:

$$s_{p_n}(f) = m_1 \Delta x_1 + m_2 \Delta x_2 + \cdots + m_n \Delta x_n = \left(\frac{0}{n} + \frac{1}{n} + \cdots + \frac{n-1}{n}\right)\frac{1}{n},$$

$$S_{p_n}(f) = M_1 \Delta x_1 + M_2 \Delta x_2 + \cdots + M_n \Delta x_n = \left(\frac{1}{n} + \frac{2}{n} + \cdots + \frac{n}{n}\right)\frac{1}{n}.$$

Using the formula $1 + 2 + \cdots + m = \frac{m(m+1)}{2}$ *we obtain:*

$$s_{p_n}(f) = \frac{(n-1)n}{2n^2} = \frac{1}{2}\left(1 - \frac{1}{n}\right),$$

$$S_{p_n}(f) = \frac{n(n+1)}{2n^2} = \frac{1}{2}\left(1 + \frac{1}{n}\right).$$

Since the set of partitions $\{p_n,\ n \in \mathbb{N}\}$ *is a subset of the set of all partitions* $\mathcal{P}([0,1])$, *it follows that:*

$$\frac{1}{2} = \sup_{p_n,\ n\in\mathbb{N}} \{s_{p_n}(f)\} \leq \sup_{p\in\mathcal{P}([a,b])} \{s_p(f)\},$$

as well as:

$$\frac{1}{2} = \inf_{p_n,\ n\in\mathbb{N}} \{S_{p_n}(f)\} \geq \inf_{p\in\mathcal{P}([a,b])} \{S_p(f)\}.$$

From which it is deduced that:

$$\frac{1}{2} \leq \sup_{p\in\mathcal{P}([a,b])} \{s_p(f)\} \leq \inf_{p\in\mathcal{P}([a,b])} \{S_p(f)\} \leq \frac{1}{2}$$

and, therefore, we can conclude that the identity function is integrable in the sense of Riemann, since:

$$\sup_{p\in\mathcal{P}([a,b])} \{s_p(f)\} = \inf_{p\in\mathcal{P}([a,b])} \{S_p(f)\} = \frac{1}{2}.$$

Next, we will see a necessary and sufficient condition for the existence of the Riemann integral of a bounded function.

Proposition 5.6. Let $f : [a,b] \to \mathbb{R}$ be a bounded function in its domain. Then, f is integrable in the sense of Riemann if and only if for any partition $p \in \mathcal{P}([a,b])$ the following limit exists:

$$\lim_{\|p\|\to 0} \sum_{i=1}^{n} f(t_i)\Delta x_i$$

where Δx_i is the length of the i-th interval and t_i is any point in this interval $[x_{i-1}, x_i]$. Recall that $\|p\|$ is the maximum difference between any two consecutive points of the partition. The value of this limit is called the definite integral of f on $[a,b]$ and is denoted by:

$$\int_a^b f(x)\,\mathrm{d}x.$$

Observation 5.7. If we take an equally spaced partition p, that is, where the n intervals have the same length, Δx, the previous limit can be written as:

$$\lim_{n\to\infty} \sum_{i=1}^{n} f(t_i)\Delta x$$

since Δx decreases as we increase the number of points in the partition and, therefore, the norm of the partition tends to zero.

The classical notation of the definite integral introduced by Leibniz is justified from the previous proposition. Leibniz imagined the finite sums $\sum_{i=1}^{n} f(t_i)\Delta x_i$ (called Riemann sums) becoming an infinite sum of values of the function $f(x)$ multiplied by infinitesimal widths $\mathrm{d}x$ of the subintervals. The sum symbol $\sum$ is replaced in the limit by the integral symbol $\int$, which is nothing more than an elongated S representing the word sum. The values of the function $f(t_i)$ are replaced by a continuous selection of values of the function $f(x)$. The widths of the subintervals Δx_i become the differential $\mathrm{d}x$.

Observation 5.8. If $f : [a,b] \to \mathbb{R}$ is a bounded and integrable function in the sense of Riemann, and furthermore, $f(x) \geq 0,\ \forall x \in [a,b]$, then:

$$\int_a^b f(x)\,\mathrm{d}x = A$$

where A is the area bounded by the graph of f, the x-axis and the lines with equations $x = a$ and $x = b$.

Figure 5.5 shows an example of the area bounded by the graph of a function $f : [a,b] \to \mathbb{R}$ that is bounded and integrable in the sense of Riemann, the x-axis and the lines with equations $x = a$ and $x = b$, where $f(x) \geq 0, \forall x \in [a,b]$.

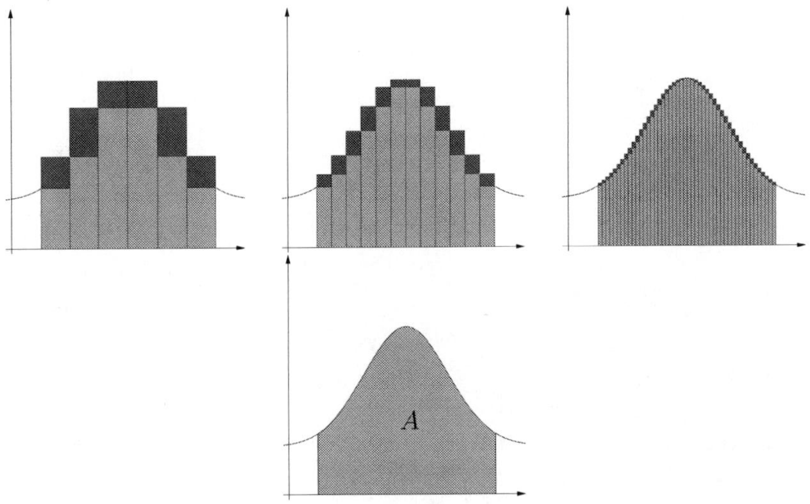

FIGURE 5.5
Area bounded by the graph of the function $f(x) = 2.5e^{-(x-2)^2} + 1$, the x-axis and the lines with equations $x = a$ and $x = b$.

Observation 5.9. If $f : [a,b] \to \mathbb{R}$ is a bounded and integrable function in the sense of Riemann, and furthermore, $f(x) \leq 0,\ \forall x \in [a,b]$, then:

$$\int_a^b f(x)\,\mathrm{d}x = -A$$

where A is the area bounded by the graph of f, the x-axis and the lines with equations $x = a$ and $x = b$.

Figure 5.6 shows an example of the area bounded by the graph of a function $f : [a,b] \to \mathbb{R}$ that is bounded and integrable in the sense of Riemann, the x-axis and the lines with equations $x = a$ and $x = b$, where $f(x) \leq 0, \forall x \in [a,b]$.

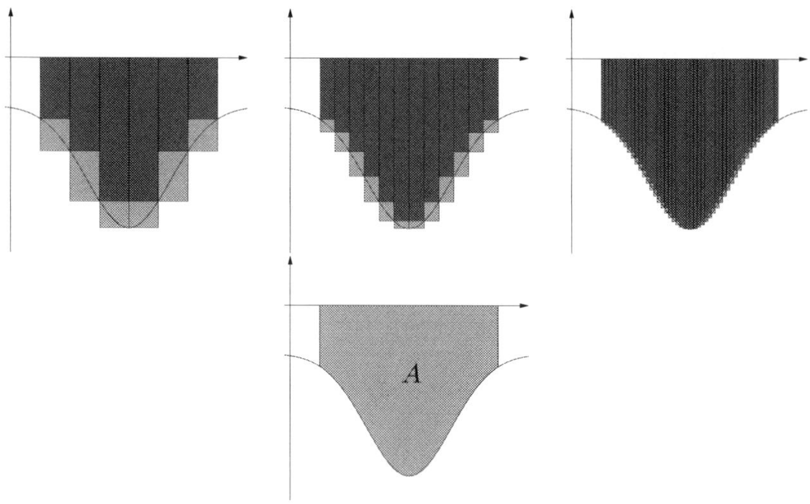

FIGURE 5.6
Area bounded by the graph of the function $f(x) = -2.5e^{-(x-2)^2} - 1$, the x-axis and the lines with equations $x = a$ and $x = b$.

Observation 5.10. If $f : [a, b] \to \mathbb{R}$ is a bounded and integrable function in the sense of Riemann, then:

$$\int_a^b f(x)\,\mathrm{d}x = A - B$$

where A is the sum of the areas bounded by the function f in the positive sections, and B is the sum of the areas bounded by the function f in the negative sections.

Figure 5.7 shows an example of the area bounded by the graph of a function $f : [a, b] \to \mathbb{R}$ that is bounded and integrable in the sense of Riemann, the x-axis and the lines with equations $x = a$ and $x = b$. In this case, the sum of the areas bounded by the function in the positive sections is $A = A_1 + A_2$. The area bounded by the function in the negative sections is B.

5.4.3 Properties of the Riemann Integral

Proposition 5.7. Given $f : [a, b] \to \mathbb{R}$ and $g : [a, b] \to \mathbb{R}$ integrable in the sense of Riemann on $[a, b]$, the following properties hold:

(a) $\displaystyle\int_a^b f(x)\,\mathrm{d}x = -\int_b^a f(x)\,\mathrm{d}x$

(b) $\displaystyle\int_a^a f(x)\,\mathrm{d}x = 0$

(c) $\displaystyle\int_a^b (f(x) + g(x))\,\mathrm{d}x = \int_a^b f(x)\,\mathrm{d}x + \int_a^b g(x)\,\mathrm{d}x$

(d) $\displaystyle\int_a^b kf(x)\,\mathrm{d}x = k\int_a^b f(x)\,\mathrm{d}x,\ \forall k \in \mathbb{R}$

(e) $\displaystyle\int_a^c f(x)\,\mathrm{d}x + \int_c^b f(x)\,\mathrm{d}x = \int_a^b f(x)\,\mathrm{d}x,\ \forall c \in [a,b]$

By defining $\int_a^b f(x)\,\mathrm{d}x$ as a limit of sums $\sum_{i=1}^n f(t_i)\Delta x_i$, we move from left to right across the interval $[a,b]$. What happens if we move from right to left, starting at $x_0 = b$ and ending at $x_n = a$? Each Δx_i in the sum would change sign, as now $x_i - x_{i-1}$ would be negative. In the final result, the sign of the integral would be modified. Since we had not previously given a meaning to backward integration, this leads us to define property (a).

On the other hand, we have only defined the integral on $[a,b]$ when $a < b$. What happens if we have an integration interval of zero width, that is, $a = b$? Since $a = b$ implies $\Delta x = 0$, as long as $f(a)$ exists, we define property (b).

While properties (a) and (b) are definitions, rules (c), (d), and (e) need to be proven. The following is a proof of rule (c). Similar proofs can be carried out to verify the other properties.

Proof. (Of property (c)) Consider $p = \{x_0, x_1, x_2, ..., x_n\}$ any partition of the interval $[a,b]$. In each subinterval:

$$I_i = [x_{i-1}, x_i] \text{ where } i = 1, \ldots, n$$

we define:

$$\begin{aligned} m_i &= \min\{f(x) + g(x),\ x \in I_i\}, \\ m_i' &= \min\{f(x),\ x \in I_i\}, \\ m_i'' &= \min\{g(x),\ x \in I_i\}. \end{aligned}$$

By the properties of the minimum regarding the sum, it holds that:

$$m_i \geq m_i' + m_i'', \quad i = 1, \ldots, n.$$

If we define the maxima M_i, M_i', M_i'' analogously, we verify:

$$M_i \leq M_i' + M_i'', \quad i = 1, \ldots, n.$$

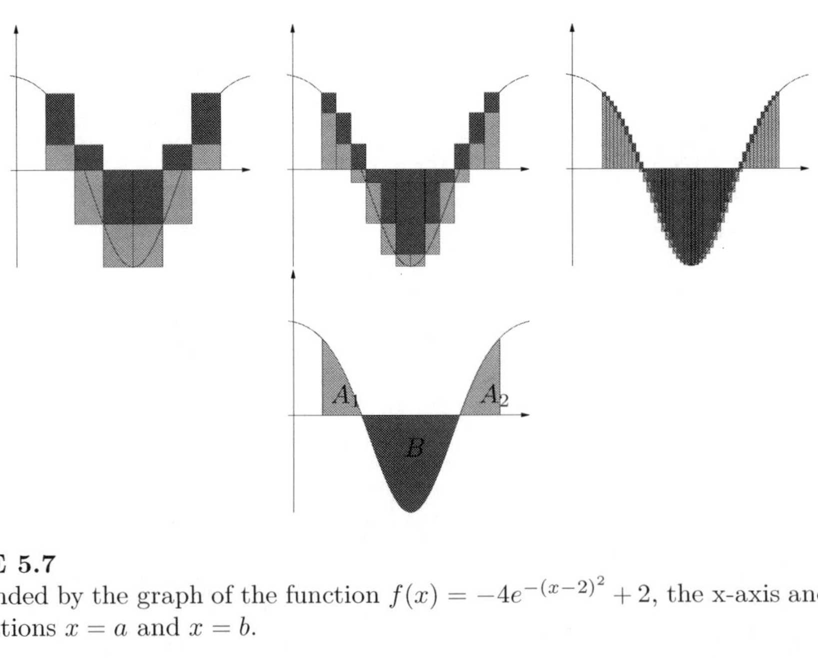

FIGURE 5.7
Area bounded by the graph of the function $f(x) = -4e^{-(x-2)^2} + 2$, the x-axis and the lines with equations $x = a$ and $x = b$.

Thus, the upper and lower sums of the functions $f+g$, f and g associated with the partition p satisfy:

$$s_p(f+g) \geq s_p(f) + s_p(g)$$
$$S_p(f+g) \leq S_p(f) + S_p(g),$$

and consequently:

$$\sup_{p\in\mathcal{P}([a,b])}\{s_p(f)\} + \sup_{p\in\mathcal{P}([a,b])}\{s_p(g)\} \leq \sup_{p\in\mathcal{P}([a,b])}\{s_p(f+g)\} \leq \inf_{p\in\mathcal{P}([a,b])}\{S_p(f+g)\}$$
$$\leq \inf_{p\in\mathcal{P}([a,b])}\{S_p(f)\} + \inf_{p\in\mathcal{P}([a,b])}\{S_p(g)\}.$$

The hypothesis of integrability of the functions f and g motivates that the first and last term are equal and, therefore, the supremum of the lower sums coincides with the infimum of the upper sums of the function $f+g$. In summary, this function is integrable and the stated property holds. □

5.4.4 Fundamental Theorem of Calculus

The formal definition of the Riemann integral (or definite integral) is difficult to use for practical integration calculations. In this section, we will see two important theorems that will be very useful for calculating integrals.

Theorem 5.1 (Fundamental Theorem of Calculus (First Part)). If f is continuous on $[a,b]$, then the function

$$F(x) = \int_a^x f(t)\,\mathrm{d}t$$

is continuous on $[a,b]$ and differentiable on (a,b) and its derivative is:

$$F'(x) = f(x).$$

Proof. Let $c \in (a,b)$. Using the formal definition of the derivative:

$$F'(c) = \lim_{h\to 0} \frac{F(c+h) - F(c)}{h}.$$

Let $h > 0$ (the case $h < 0$ is demonstrated in a similar manner and is left as an exercise for the reader), then since $F(x) = \int_a^x f(t)\,\mathrm{d}t$ we have:

$$F(c+h) - F(c) = \int_a^{c+h} f(t)\,\mathrm{d}t - \int_a^c f(t)\,\mathrm{d}t = \int_c^{c+h} f(t)\,\mathrm{d}t.$$

Next, we will obtain an upper and lower bound of the integral $\int_c^{c+h} f(t)\,\mathrm{d}t$.

Within the subinterval $[c, c+h]$ we calculate the maximum value of the function:

$$M_h = \max\{f(t),\, t \in [c, c+h]\}.$$

Thus, $M_h \cdot h$ is an upper bound of the integral $\int_c^{c+h} f(t)\,\mathrm{d}t$.

Within the subinterval $[c, c+h]$ we calculate the minimum value of the function:

$$m_h = \min\{f(t),\, t \in [c, c+h]\}.$$

Thus, $m_h \cdot h$ is a lower bound of the integral $\int_c^{c+h} f(t)\,\mathrm{d}t$. Using both bounds we can write:

$$m_h \cdot h \leq \int_c^{c+h} f(t)\,\mathrm{d}t \leq M_h \cdot h$$

and since $\int_c^{c+h} f(t)\,\mathrm{d}t = F(c+h) - F(c)$ we have:

$$m_h \leq \frac{F(c+h) - F(c)}{h} \leq M_h.$$

Finally, since f is continuous at c:

$$\lim_{h\to 0} m_h = \lim_{h\to 0} M_h = f(c),$$

and, therefore:

$$\lim_{h\to 0} \frac{F(c+h) - F(c)}{h} = f(c).$$

That is, given $c \in (a, b)$ we have seen that $F'(c) = f(c)$, as we wanted to demonstrate. □

Theorem 5.2 (Fundamental Theorem of Calculus (Second Part) or Barrow's Rule). If f is a continuous function on $[a, b]$ and $F(x)$ is an antiderivative of f, then:

$$\int_a^b f(x)\,\mathrm{d}x = F(b) - F(a).$$

The usual notation for the difference $F(b) - F(a)$ is:

$$F(x)\Big|_a^b \quad \text{or} \quad [F(x)]_a^b$$

depending on whether F has one or more terms.

Proof. Given f a continuous function on $[a, b]$ the first fundamental theorem of calculus tells us that there exists an antiderivative of this function, $G(x)$, such that:

$$G(x) = \int_a^x f(t)\,\mathrm{d}t.$$

Let $F(x)$ be another antiderivative of the function $f(x)$. Then, by Proposition 5.1, there exists a constant $C \in \mathbb{R}$ such that:

$$G(x) - F(x) = C, \quad \text{for all } x \in [a, b].$$

This can be rewritten as:

$$G(x) = F(x) + C \quad \text{for all } x \in [a, b].$$

Specifically, at $x = a$, we obtain:

$$0 = G(a) = F(a) + C,$$

that is $C = -F(a)$. Therefore:

$$G(x) = F(x) - F(a).$$

And in particular, at $x = b$, we obtain:

$$\int_a^b f(t)\,\mathrm{d}t = F(b) - F(a).$$

□

The previous proposition is very important because it relates the problem of area calculation (integral calculus) with the calculation of derivatives (differential calculus). Informally, we can say that the theorem states that differentiation and integration are inverse operations.

Example 5.18. *Calculate the integral* $\int_0^{2\pi} \cos(x)\,\mathrm{d}x$.

SOLUTION. If we consider $f(x) = \cos(x)$, we have that $F(x) = \sin(x)$ is an antiderivative of $f(x)$ and, therefore:

$$\int_0^{2\pi} \cos(x)\,\mathrm{d}x = \sin(x)\Big|_0^{2\pi} = \sin(2\pi) - \sin(0) = 0.$$

□

Example 5.19. *Calculate the definite integral:*

$$\int_0^2 \sqrt{2x+1}\,\mathrm{d}x$$

using the change of variable $u = 2x + 1$.

SOLUTION.

When calculating a definite integral using a change of variable, we can proceed in different ways:

(a) calculate an antiderivative of the function and then apply Barrow's rule (this involves applying and undoing the change of variable);

(b) apply the change of variable to the integration limits and integrate (which means it is not necessary to undo the change).

In our case, if we choose option (a) and apply the change of variable $u = 2x + 1$, which implies $du = 2\,\mathrm{d}x$, we have:

$$\int \sqrt{2x+1}\,\mathrm{d}x = \int \frac{1}{2}\sqrt{u}du = \frac{1}{3}u\sqrt{u} = \frac{1}{3}(2x+1)\sqrt{2x+1} = F(x).$$

Then:

$$\int_0^2 \sqrt{2x+1}\,\mathrm{d}x = F(2) - F(0) = \frac{5\sqrt{5}-1}{3}.$$

If we choose option (b), then the new integration limits will be:

$$u = 2x + 1|_{x=0} = 1,$$
$$u = 2x + 1|_{x=2} = 5,$$

so the definite integral is:

$$\int_{x=0}^{x=2} \sqrt{2x+1}\,\mathrm{d}x = \int_{u=1}^{u=5} \frac{1}{2}\sqrt{u}du = \frac{1}{3}u\sqrt{u}\Big|_{u=1}^{u=5} = \frac{5\sqrt{5}-1}{3}.$$

Obviously, the result in both cases is the same, but the second option can be faster by saving us from undoing a change of variable. □

Proposed Exercise 5.9. Given a curve in the plane defined by the equation $y = f(x)$, the length of this curve from $x = a$ to $x = b$ is given by:

$$L = \int_a^b \sqrt{1 + (f'(x))^2}\, dx.$$

(i) Prove the above formula.

(ii) Prove that the circumference of a circle of radius r is $2\pi r$.

Hint: calculate the length of the circumference in the first quadrant and multiply by 4.

Proposed Exercise 5.10. Consider a two-dimensional plate of constant density bounded by the curves $y_1 = g(x)$, $y_2 = f(x)$ and the lines $x = a$ and $x = b$ where $a < b$ and $f(x) > g(x)$ for all $x \in [a, b]$. Then, the center of mass of the plate is calculated as follows:

$$\bar{x} = \frac{\displaystyle\int_a^b x(f(x) - g(x))\, dx}{\displaystyle\int_a^b (f(x) - g(x))\, dx}$$

$$\bar{y} = \frac{1}{2}\frac{\displaystyle\int_a^b (f(x)^2 - g(x)^2)\, dx}{\displaystyle\int_a^b (f(x) - g(x))\, dx}$$

(i) Find the center of mass of the region bounded by the parabola $y = \sqrt{x}$, the x-axis, and the lines $x = 0$ and $x = 4$.

(ii) Find the center of mass of the region bounded by the curve $y = x^2$, the line $y = x + 2$ and the lines $x = -1$ and $x = 2$.

(iii) Find the center of mass of the semicircle of radius a located in the first and second quadrants.

5.5 Area Between Two Curves

We will now see how to calculate the area bounded between two given functions. We will start with the simplest case where both functions are positive and one is always strictly greater than the other. Then we will study the case where the functions need not be positive (but we will continue to assume that one is strictly greater than the other). Finally, we will see the most general case where we have any two integrable functions.

5.5.1 Area Bounded by Two Bounded and Integrable Functions Always Positive and One Strictly Greater Than the Other

Let f and g be two bounded and integrable functions in the Riemann sense on $[a, b]$ such that $f(x) \geq g(x) \geq 0$ $\forall x \in [a, b]$, as shown in Figure 5.8.

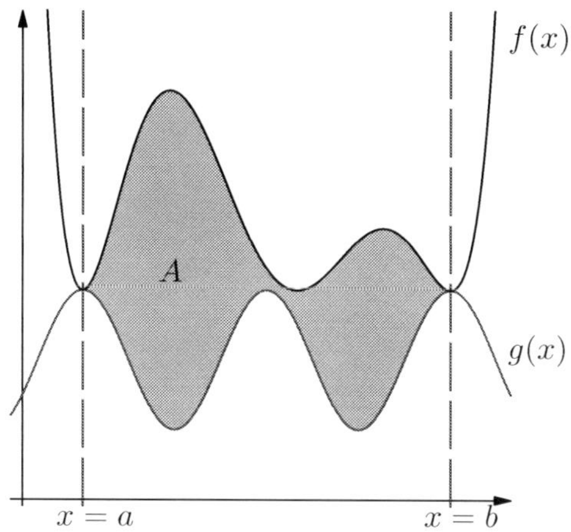

FIGURE 5.8
Example of bounded and integrable functions f and g in the Riemann sense on $[a, b]$ such that $f(x) \geq g(x) \geq 0,\ \forall x \in [a, b]$.

The area bounded between the two curves and the lines $x = a$ and $x = b$ is given by:

$$\boxed{A = \int_a^b (f(x) - g(x))\,\mathrm{d}x.}$$

Note that:

$$\int_a^b (f(x) - g(x))\,\mathrm{d}x = \underbrace{\int_a^b f(x)\,\mathrm{d}x}_{A+B} - \underbrace{\int_a^b g(x)\,\mathrm{d}x}_{B} = A,$$

as can be seen in Figure 5.9.

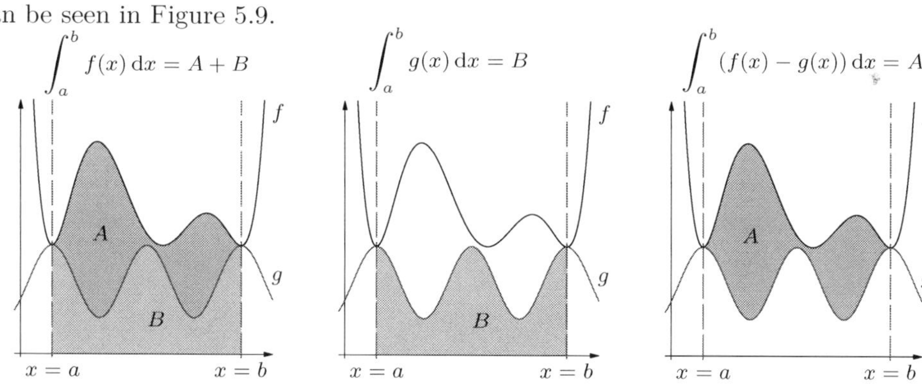

FIGURE 5.9
Area bounded by two bounded and integrable functions f and g in the Riemann sense on $[a, b]$ such that $f(x) \geq g(x) \geq 0\ \forall x \in [a, b]$.

5.5.2 Area Bounded by Two Bounded and Integrable Functions Where One is Strictly Greater Than the Other

Let f and g be two bounded and integrable functions in the Riemann sense on the interval $[a, b]$ such that $f(x) \geq g(x),\ \forall x \in [a, b]$, as shown in Figure 5.10.

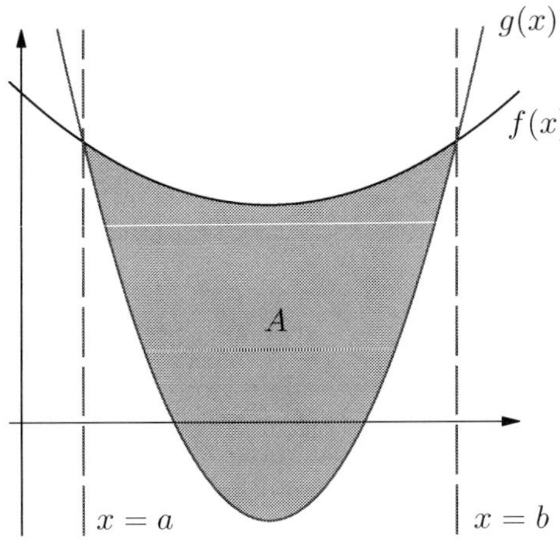

FIGURE 5.10
Example of bounded and integrable functions f and g in the Riemann sense on $[a, b]$ such that $f(x) \geq g(x),\ \forall x \in [a, b]$.

The area bounded between the two curves and the lines $x = a$ and $x = b$ is given, as in the case $f(x) \geq g(x) \geq 0$, by:

$$\boxed{A = \int_a^b (f(x) - g(x))\,\mathrm{d}x.}$$

Let's see the above equality by calculating the area represented in Figure 5.11.

$$\int_a^b f(x)\,\mathrm{d}x = B + E + D \qquad \int_a^b g(x)\,\mathrm{d}x = B - C + D \qquad \int_a^b (f(x) - g(x))\,\mathrm{d}x = E + C$$

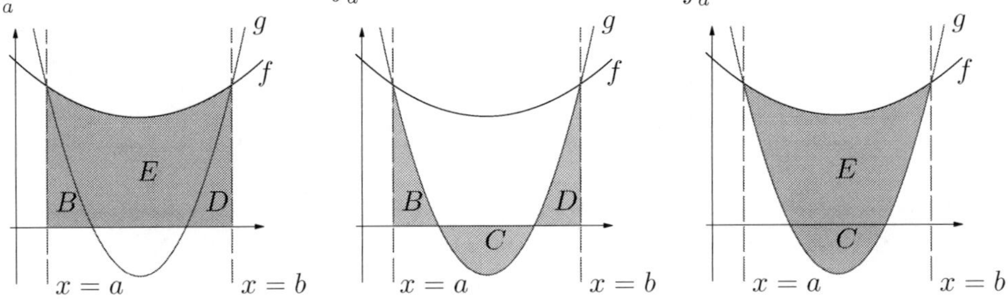

FIGURE 5.11
Area bounded by two bounded and integrable functions f and g in the Riemann sense on $[a, b]$ such that $f(x) \geq g(x),\ \forall x \in [a, b]$.

For the area represented in Figure 5.11 we have:

$$\int_a^b f(x)\,\mathrm{d}x = B + E + D$$
$$\int_a^b g(x)\,\mathrm{d}x = B - C + D.$$

Therefore:

$$\int_a^b (f(x) - g(x))\,\mathrm{d}x = B + E + D - (B - C + D) = E + C = A.$$

5.5.3 Area Bounded by Any Two Bounded and Integrable Functions

Let f and g be any two bounded and integrable functions in the Riemann sense on the interval $[a, b]$, as shown in Figure 5.12.

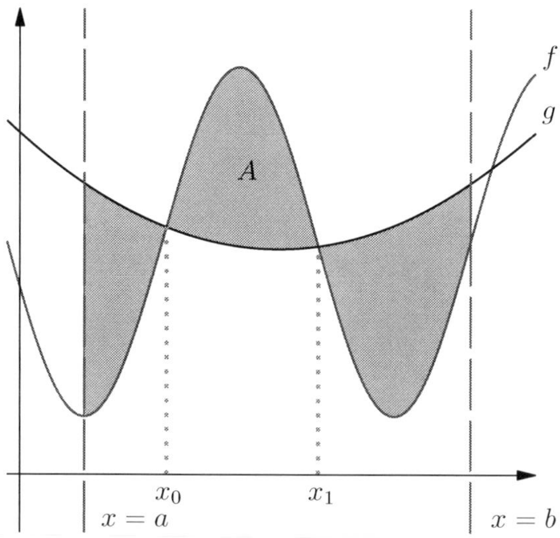

FIGURE 5.12
Example of any two bounded and integrable functions f and g in the Riemann sense on $[a, b]$.

The area bounded between the two curves and the lines $x = a$ and $x = b$ is given by:

$$\boxed{A = \int_a^b |f(x) - g(x)|\,\mathrm{d}x.}$$

Now, how do we calculate $\int_a^b |f(x) - g(x)|\,\mathrm{d}x$? Note that, to calculate the absolute value, we need to know when $f(x) - g(x) \geq 0$ and when $f(x) - g(x) \leq 0$.

Therefore, to calculate the area it is necessary to:

(1) Calculate the intersection points of the two functions in the interval $[a, b]$.

(2) Determine the subintervals where $f(x) \geq g(x)$ and the subintervals where $g(x) \geq f(x)$.

(3) The area is the sum of the integrals of $f - g$, where f is greater than g and the integrals of $g - f$, where g is greater than f.

In the case of the functions f and g in Figure 5.12:

$$\begin{aligned} A &= \int_a^b |f(x) - g(x)|\,\mathrm{d}x \\ &= \int_a^{x_0} (g(x) - f(x))\,\mathrm{d}x + \int_{x_0}^{x_1} (f(x) - g(x))\,\mathrm{d}x + \int_{x_1}^{b} (g(x) - f(x))\,\mathrm{d}x \end{aligned}$$

Example 5.20. *Given the functions $f(x) = 2\sqrt{x+1}$ and $g(x) = 3\sqrt{x}$, calculate the area of the closed region in the shape of a crescent moon that they delimit with the y-axis.*

SOLUTION.
Figure 5.13 shows the area bounded by the graphs of the functions f and g.

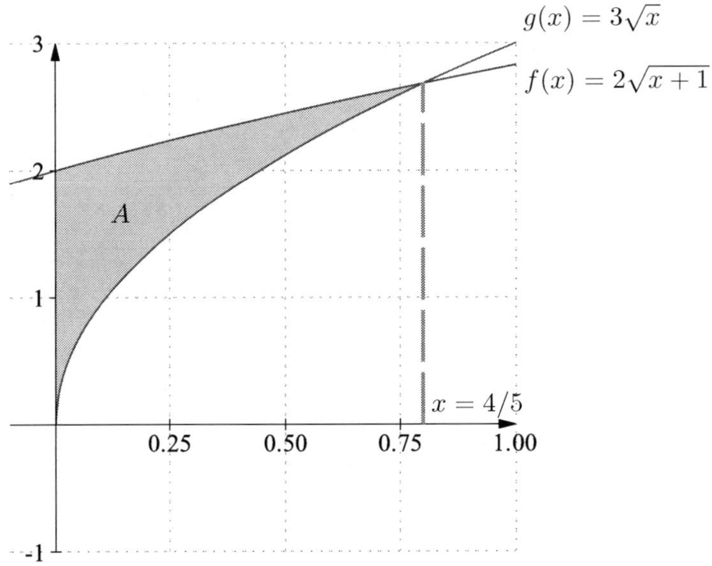

FIGURE 5.13
Area bounded by the graphs of the functions $f(x) = 2\sqrt{x+1}$ and $g(x) = 3\sqrt{x}$.

The first thing we must do is determine the intersection point, that is, the point x such that $f(x) = g(x)$:

$$\begin{aligned} 2\sqrt{x+1} = 3\sqrt{x} \quad &\Leftrightarrow \quad 4(x+1) = 9x \quad \Leftrightarrow \quad 5x - 4 = 0 \\ &\Leftrightarrow \quad x = \frac{4}{5}. \end{aligned}$$

Since in the interval $[0, \frac{4}{5}]$ we have $f(x) \geq g(x)$, the area is given by:

$$A = \int_0^{\frac{4}{5}} (f(x) - g(x))\,\mathrm{d}x = \int_0^{\frac{4}{5}} (2\sqrt{x+1} - 3\sqrt{x})\,\mathrm{d}x$$

Therefore:

$$A = \int_0^{\frac{4}{5}} \left(2\sqrt{x+1} - 3\sqrt{x}\right)\, \mathrm{d}x$$

$$= \left[\frac{4}{3}(x+1)^{\frac{3}{2}} - 2x^{\frac{3}{2}}\right]_0^{\frac{4}{5}} = \frac{4}{3}\left(\frac{9}{5}\right)^{\frac{3}{2}} - 2\left(\frac{4}{5}\right)^{\frac{3}{2}} - \frac{4}{3}$$

$$= \frac{4}{3}\frac{27}{5\sqrt{5}} - 2\frac{8}{5\sqrt{5}} - \frac{4}{3} = \frac{20}{5\sqrt{5}} - \frac{4}{3} = \frac{4}{\sqrt{5}} - \frac{4}{3} = 0.4556$$

□

Example 5.21. *Given the functions* $f(x) = 6x - x^2$, $g(x) = x^2 - 2x$ *and* $h(x) = 10 + x$, *calculate the area of the region bounded by the three curves.*

SOLUTION.
Figure 5.14 shows the area bounded by the graphs of the functions f, g and h.

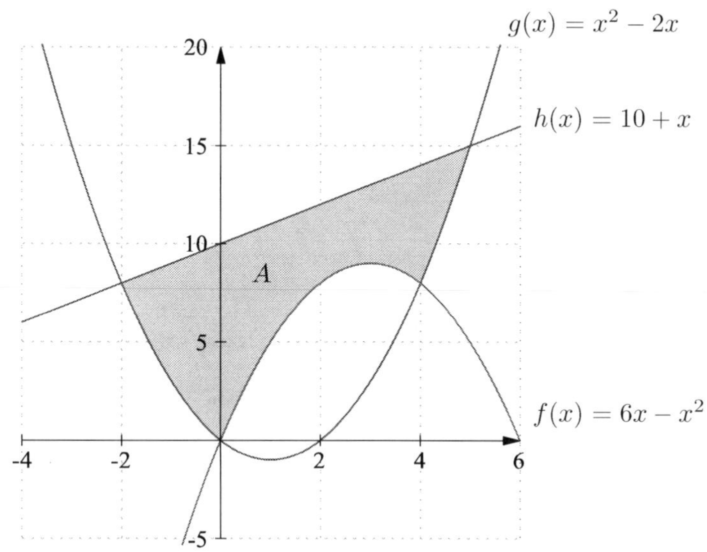

FIGURE 5.14
Area bounded by the graphs of the functions $f(x) = 6x - x^2$, $g(x) = x^2 - 2x$ and $h(x) = 10 + x$.

If we calculate the intersections of the three functions we have:

$$f(x) = g(x) \quad \Rightarrow \quad x = 0,\ 4$$

$$g(x) = h(x) \quad \Rightarrow \quad x = -2,\ 5$$

Note that to calculate the area, we have to divide the total area into three regions $x \in [-2, 0]$, $x \in [0, 4]$ and $x \in [4, 5]$, as shown in Figure 5.15.

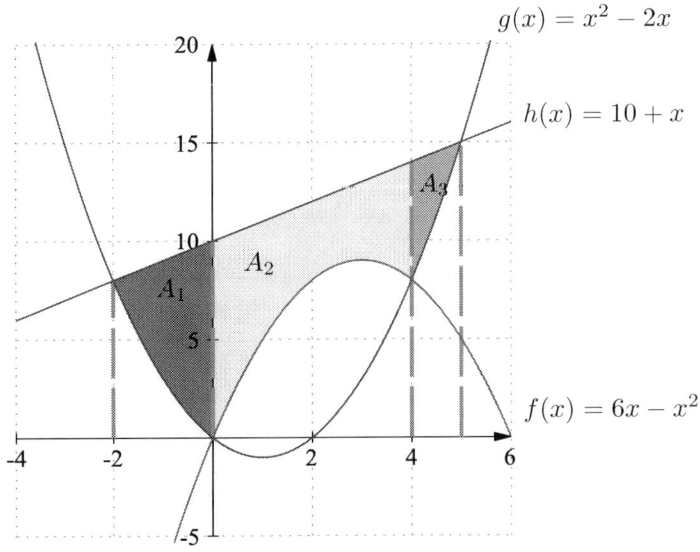

FIGURE 5.15
Division into three regions of the area bounded by the graphs of the functions $f(x) = 6x - x^2$, $g(x) = x^2 - 2x$ and $h(x) = 10 + x$.

Therefore:

$$A = \underbrace{\int_{-2}^{0} (h(x) - g(x))\,\mathrm{d}x}_{A_1} + \underbrace{\int_{0}^{4} (h(x) - f(x))\,\mathrm{d}x}_{A_2} + \underbrace{\int_{4}^{5} (h(x) - g(x))\,\mathrm{d}x}_{A_3}$$
$$= 34/3 + 64/3 + 19/6 = 215/6.$$

□

Proposed Exercise 5.11. Prove that the area of a circular sector of radius a and central angle $\alpha \in (0, \pi/2)$ is $A = \frac{1}{2}a^2\alpha$.

5.5.4 Integration with Respect to the Variable y

If the boundary curves of the region are described by functions of y, as shown in Figure 5.16, the approximation rectangles are horizontal (and not vertical). In this case, we use the formula:

$$A = \int_{c}^{d} (f(y) - g(y))\, dy$$

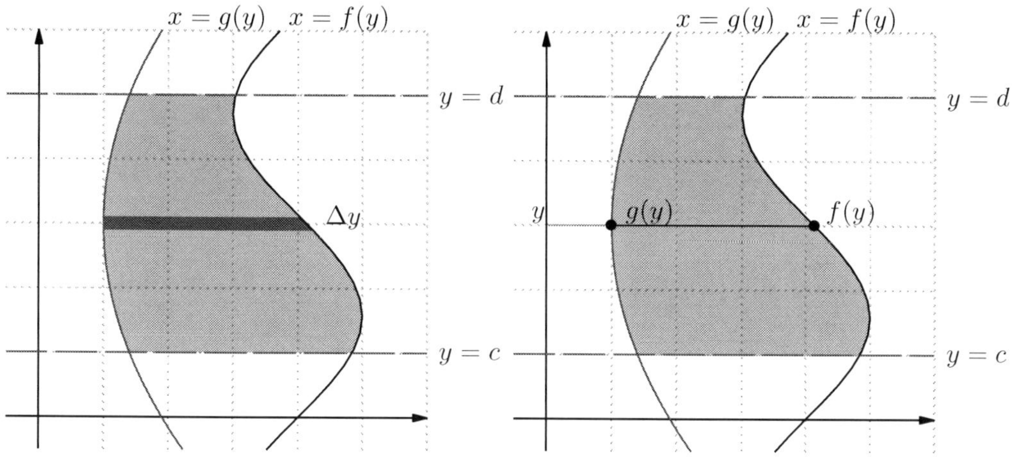

FIGURE 5.16
Integration with respect to the variable y.

Example 5.22. *Determine the area bounded by the parabola* $y^2 = 4x$ *and the line* $y = 2x-4$.

SOLUTION. Figure 5.17 shows the area bounded by the parabola $y^2 = 4x$ and the line $y = 2x - 4$.

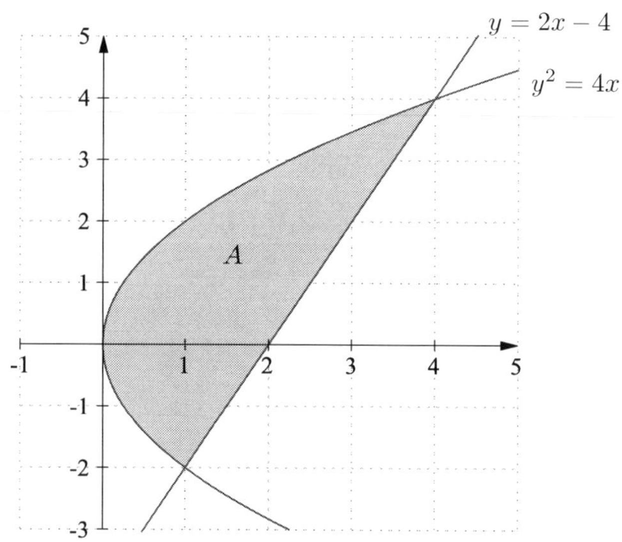

FIGURE 5.17
Area bounded by the parabola $y^2 = 4x$ and the line $y = 2x - 4$.

This problem can be approached in two ways depending on whether we integrate with respect to the variable x or the variable y, as shown in Figures 5.18 and 5.19.

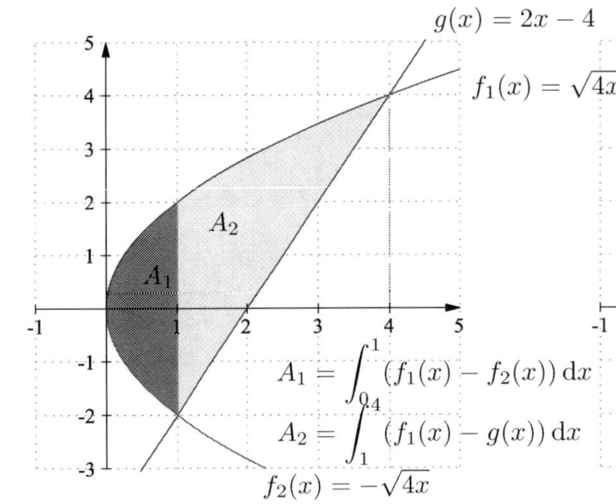

FIGURE 5.18
Calculating the area bounded by the parabola $y^2 = 4x$ and the line $y = 2x - 4$ if we integrate with respect to the variable x.

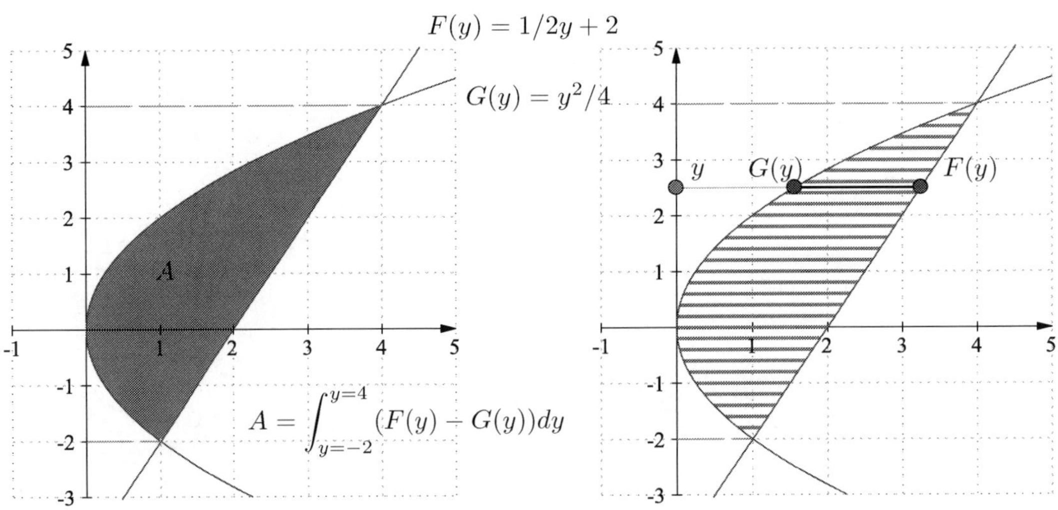

FIGURE 5.19
Calculating the area bounded by the parabola $y^2 = 4x$ and the line $y = 2x - 4$ if we integrate with respect to the variable y.

If we integrate with respect to x (Figure 5.18), two integrals are required to determine the area of this region.

$$A = \underbrace{\int_0^1 4\sqrt{x}\,\mathrm{d}x}_{A_1} + \underbrace{\int_1^4 (2\sqrt{x} - 2x + 4)\,\mathrm{d}x}_{A_2} = 8/3 + 19/3 = 9$$

In contrast, only one is needed if we integrate with respect to y (Figure 5.19).

$$A = \int_{-2}^{4} \left(\frac{1}{2}y + 2 - \frac{1}{4}y^2 \right) dy = 9$$

Obviously, the result is the same in both cases. □

5.6 Applications of the Integral

In this section, we will see two of the most important applications of the definite integral: the calculation of the average value of a continuous function and the calculation of volumes by means of cross-sections.

5.6.1 Average Value of a Function

It is easy to find the average of a set of n given numbers; we just have to sum them and divide by n. But how do we calculate the average temperature during a day?

The temperature is a function whose values rise and fall continuously and, therefore, we have infinite values of it during a day. That is, what happens if we want to find the average of an infinite number of values? We are in the presence of a type of continuous average.

Let's see how to calculate the average value of a function $f(x)$ on an interval $[a, b]$. We take a partition of the initial interval with n equal subintervals, each with length Δx. If t_i is any point of the i-th subinterval, then the average of these values of the function $f(x)$ is given by:

$$\frac{f(t_1) + f(t_2) + \cdots + f(t_n)}{n} = \frac{\sum_{i=1}^{n} f(t_i)}{n}.$$

Note that having taken equal subintervals, we have $n = \frac{b-a}{\Delta x}$. Therefore:

$$\frac{\sum_{i=1}^{n} f(t_i)}{n} = \frac{\sum_{i=1}^{n} f(t_i)\Delta x}{b-a} = \frac{1}{b-a}\sum_{i=1}^{n} f(t_i)\Delta x.$$

Note that the average is obtained by dividing a Riemann sum for f on $[a, b]$ by $(b - a)$. Therefore, if f is Riemann integrable, the average value of f on $[a, b]$ is given by:

$$\text{Avg}(f) = \frac{1}{b-a}\int_a^b f(x)\,\mathrm{d}x.$$

Example 5.23. *Suppose the current world population is 5 billion and the population within t years is given by the exponential growth law $p(t) = 5e^{0.023t}$, where t is time and $p(t)$ is expressed in billions of people. Find the average population of the earth over the next 30 years. Note that $p(0) = 5$, that is, at the present time the population is 5 billion people.*

Solution. To solve this problem, we must find the average value of the population $p(t)$ from $t = 0$ to $t = 30$.

$$\text{Avg}(p) = \frac{1}{30-0}\int_0^{30} 5e^{0.023t}\,\mathrm{d}t = \frac{1}{30-0}\left[\frac{5e^{0.023t}}{0.023}\right]_0^{30} = 7.2.$$

Thus, the average value of the world population in the next 30 years will be 7.2 billion people.

5.6.2 Calculation of Volumes

Let's see how to calculate volumes of solids through the areas of their cross-sections. A cross-section $S(\mathrm{x})$ of a solid S, as illustrated in Figure 5.20, is given by the intersection of S with a plane perpendicular to the x-axis, P_{x}, which passes through the point x in the interval $[a, b]$.

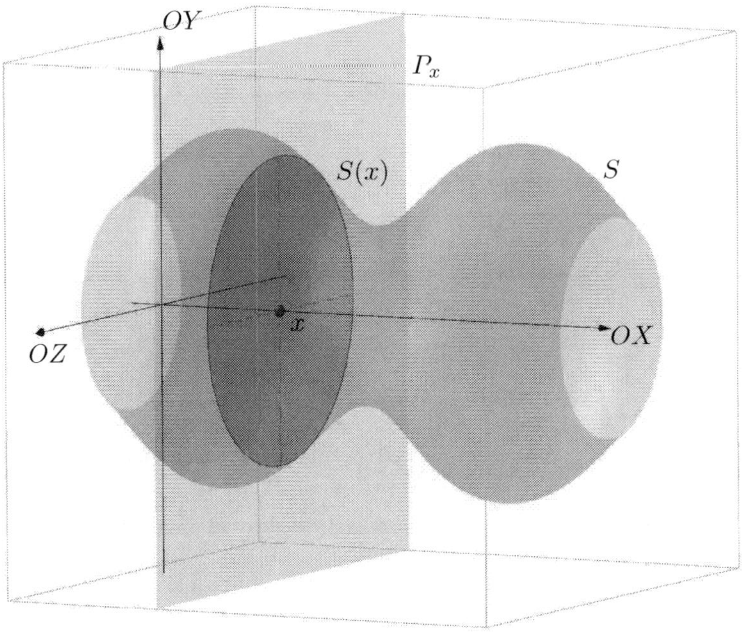

FIGURE 5.20
Cross-section $S(\mathrm{x})$ of a solid S.

The method for calculating the volume consists of *slicing* the solid as if it were a loaf of bread. We take a partition $p = \{a = x_0, x_1, x_2, ..., x_n = b\}$ of the interval $[a, b]$. The planes P_{x_i}, perpendicular to the x-axis at the points of the partition, slice the solid into thin plates (slices) as illustrated in Figure 5.21.

We approximate the volume of the plate between the planes $P_{x_{i-1}}$ and P_{x_i} by means of a cylindrical solid with base area $A(x_i)$ (area of the cross-section at x_i) and height $\Delta x_i = x_{i-1} - x_i$. The volume of this cylindrical solid is $A(x_i)\Delta x_i$, which is approximately the volume of the plate, as illustrated in Figure 5.22.

Therefore, the volume of the entire solid will be approximately:

$$V \approx \sum_{i=1}^{n} A(x_i)\Delta x_i.$$

This is a Riemann sum for the function $A(x)$ in $[a, b]$. Therefore, if $A(x)$ is a Riemann integrable function, the volume of the solid will be given by:

$$V = \lim_{\|p\|\to 0} \sum_{i=1}^{n} A(x_i)\Delta x_i = \int_a^b A(x)\,\mathrm{d}x.$$

Example 5.24. *Calculate the volume of a sphere of radius R.*

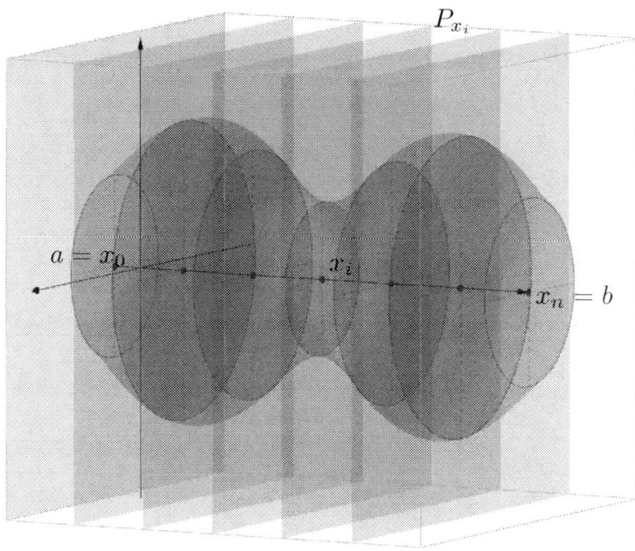

FIGURE 5.21
The planes P_{x_i}, perpendicular to the x-axis at the points of the partition, slice the solid into thin plates (slices).

SOLUTION.

First, let's imagine that the sphere is centered at the origin of coordinates. The equation of the sphere centered at the origin with radius R is given by $x^2 + y^2 + z^2 = R^2$. Therefore,

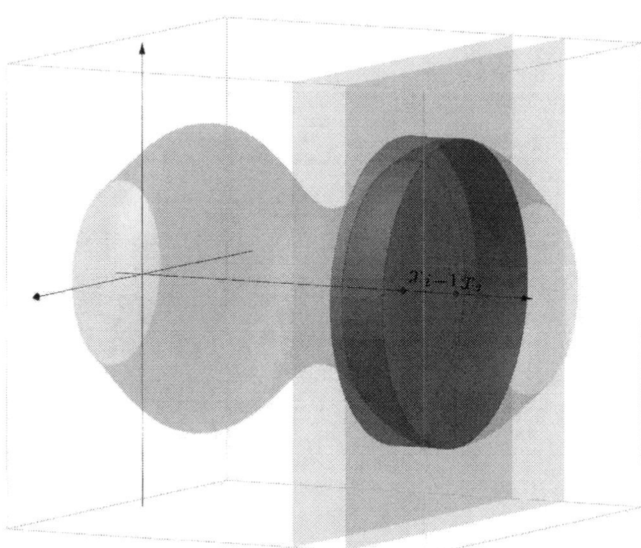

FIGURE 5.22
The volume of the plate between the planes $P_{x_{i-1}}$ and P_{x_i} is approximated by the volume of a cylindrical solid with the area of the cross-section at x_i and height $x_{i-1} - x_i$.

at a certain point x, the cross-section of the sphere is the circle $y^2 + z^2 = R^2 - x^2$. This circle has a radius of $\sqrt{R^2 - x^2}$. Thus, the area of the section will be:

$$A(x) = \pi(R^2 - x^2),$$

as illustrated in Figure 5.23.

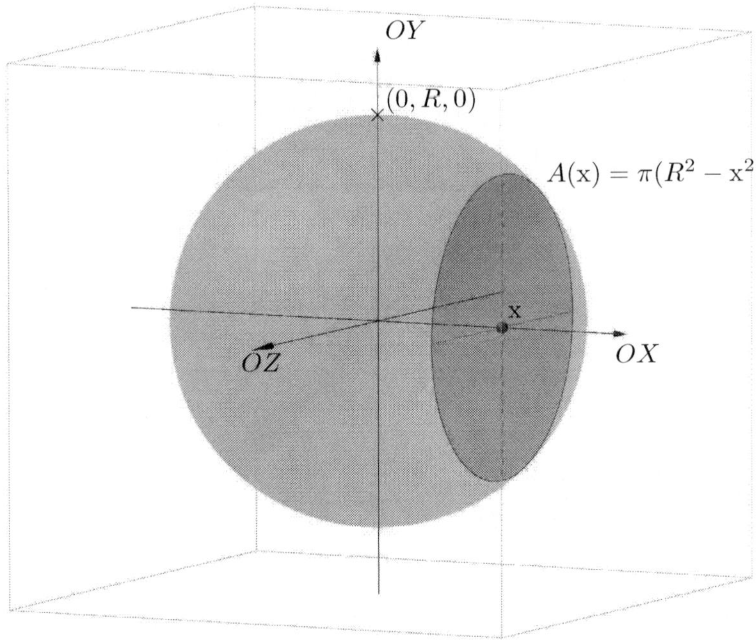

FIGURE 5.23
The area of the cross-section of a sphere centered at the origin and with radius R is equal to $\pi(R^2 - x^2)$.

Finally, the volume of the sphere will be:

$$V = \int_{-R}^{R} A(x)\,dx = \int_{-R}^{R} \pi(R^2 - x^2)\,dx =$$

$$= \pi\left[R^2 x - \frac{x^3}{3}\right]_{-R}^{R} =$$

$$= \pi(R^3 - \frac{R^3}{3}) - \pi(-R^3 + \frac{R^3}{3}) = \frac{4\pi R^3}{3}.$$

□

5.7 Improper Integrals

In the definition of a Riemann integrable function, it was required that the integration interval be bounded and that the function be bounded within that interval. When either of these conditions is no longer satisfied, the Riemann integral theory cannot be applied. This section briefly explains how to adapt Riemann's integral theory to the case where the function is unbounded and/or the integration interval is unbounded, which are called improper integrals.

5.7.1 Intuitive Idea

Does it make sense to calculate the area bounded by the function $f(x) = \frac{1}{x^2}$, the x-axis, and the line $x = 1$?

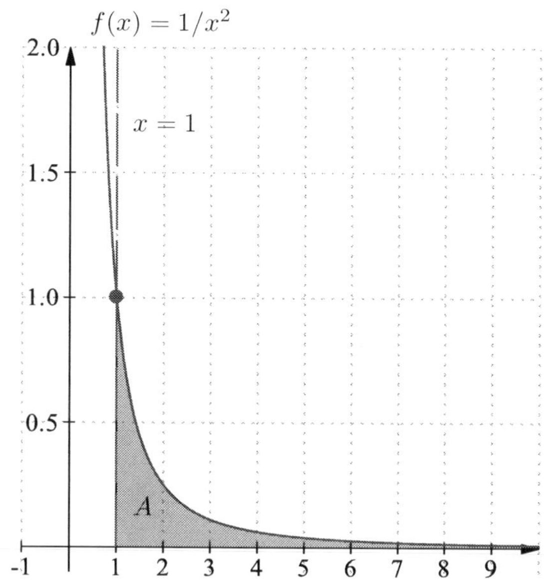

FIGURE 5.24
Area bounded by the function $f(x) = \dfrac{1}{x^2}$, the x-axis, and the line $x = 1$.

This area, shown in Figure 5.24, would be given by the integral $\int_1^{+\infty} f(x)\,\mathrm{d}x$, where one of the integration limits is infinite. In this section, we will see if it makes sense for one (or both) of the bounds of a definite integral to be positive or negative infinity.

On the other hand, does it make sense to calculate the area bounded by the same function in the interval $[-1, 1]$ and the x-axis?

In this case, the area shown in Figure 5.25, would be given by the integral $\int_{-1}^{1} f(x)\,\mathrm{d}x$, where the function $f(x)$ is not bounded within $[-1, 1]$. In this section, we will see if it makes sense to calculate the integral of an unbounded function.

5.7.2 Definitions and Classification

Definition 5.8 (Improper integral of the first kind). We say that an integral $\int_a^b f$ is *improper of the first kind* if the integration interval is unbounded ($a = -\infty, b = +\infty$, or both).

Example 5.25. *The integral* $\int_0^{+\infty} \sin(x)\,\mathrm{d}x$ *is improper of the first kind since the domain is unbounded. Figure 5.26 shows the area bounded by the function* $f(x)$ *and the x-axis.*

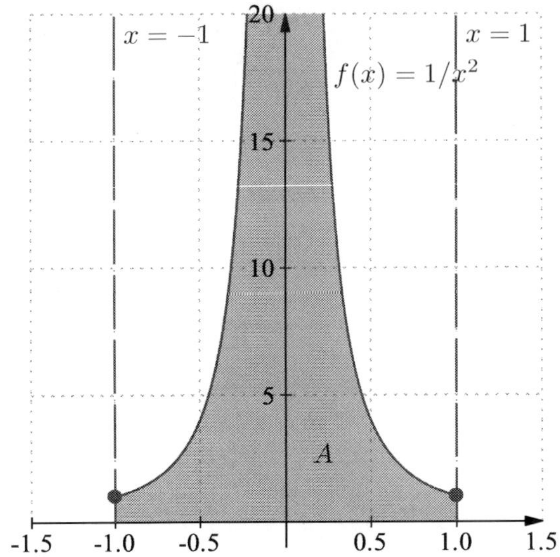

FIGURE 5.25
Area bounded by the function $f(x) = \dfrac{1}{x^2}$, in the interval $[-1, 1]$ and the x-axis.

Definition 5.9 (Improper integral of the second kind). We say that an integral $\int_a^b f$ is *improper of the second kind* if the function f is unbounded within the integration domain. Points within the interval that cause the function to be unbounded (e.g., due to vertical asymptotes) are called singular points or singularities.

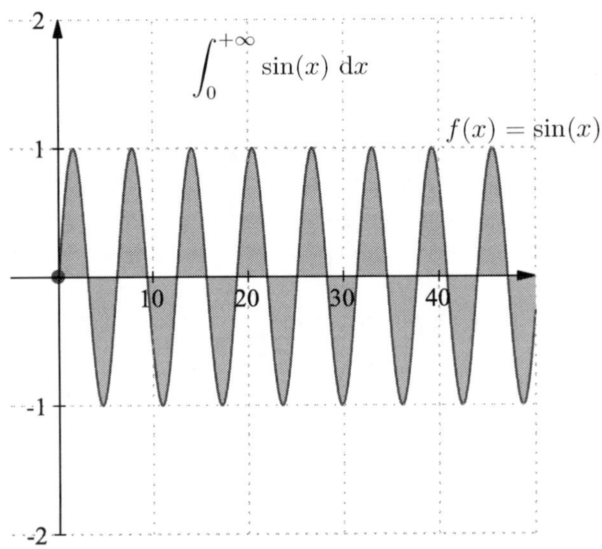

FIGURE 5.26
Area bounded by the function $f(x) = \sin(x)$ and the x-axis.

Example 5.26. *The integral $\int_0^{10} \frac{1}{x-7}\,\mathrm{d}x$ is improper of the second kind since at the point $x = 7$, which belongs to the integration interval, the function $f(x) = \frac{1}{x-7}$ is unbounded. Figure 5.27 shows the area bounded by the function $f(x)$ and the x-axis.*

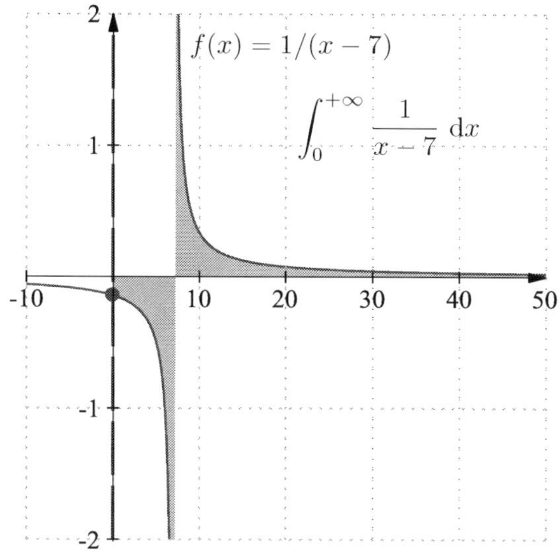

FIGURE 5.27
Area bounded by the function $f(x) = \dfrac{1}{x-7}$ and the x-axis.

Definition 5.10 (Improper integral of the third kind). We say that an integral $\int_a^b f$ is *improper of the third kind* if it is improper of both the first and second kind.

Example 5.27. *The integral $\int_{-1}^{+\infty} \frac{1}{x^2}\,\mathrm{d}x$ is improper of the third kind. Figure 5.28 shows the area bounded by the function $f(x)$ and the x-axis.*

Example 5.28. *Classify the integral:*

$$\int_0^1 \frac{\sin x}{x}\,\mathrm{d}x.$$

SOLUTION.

At first, it seems to be an improper integral of the second kind since it is not defined at $x = 0$ and it might appear that there is an asymptote at that point. However, we know that:

$$\lim_{x\to 0^+} \frac{\sin x}{x} = 1$$

and if a function has a limit at a point, there exists a neighborhood around that point where the function is bounded. Therefore, this is not an improper integral.
Figure 5.29 shows the area bounded by the function $f(x) = \frac{\sin x}{x}$ and the x-axis in the interval $[0, 1]$. □

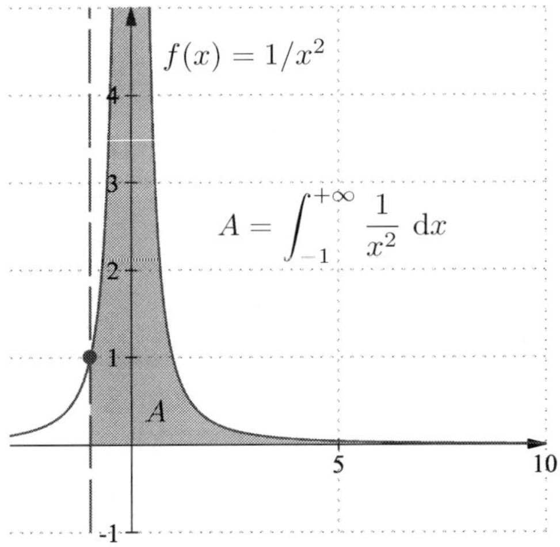

FIGURE 5.28
Area bounded by the function $f(x) = \dfrac{1}{x^2}$ and the x-axis.

Proposed Exercise 5.12. Classify the integrals:

(a) $\displaystyle\int_0^{+\infty} \frac{dx}{\ln(1+x)(x-2)^3}$

(b) $\displaystyle\int_{-\infty}^{+\infty} \frac{x^3\,dx}{e^x(x^8+x^4+1)}$

(c) $\displaystyle\int_0^1 \frac{x\cos x - \sin x}{x^3}\,dx$

(d) $\displaystyle\int_0^{+\infty} \frac{x\,dx}{x^2+x+1}$

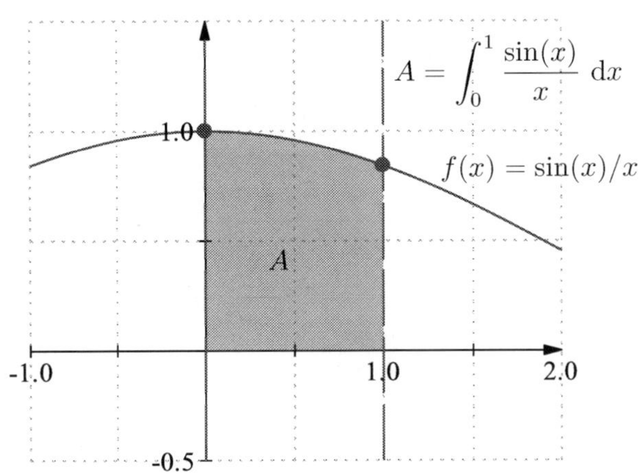

FIGURE 5.29
Area bounded by the function $f(x) = \dfrac{\sin x}{x}$ and the x-axis in the interval $[0, 1]$.

5.7.3 Improper Integrals of the First Kind

Definition 5.11 (Convergence of an improper integral of the first kind)**.** Let I be any interval of $\mathbb{R}$, $I \subset [a, +\infty)$ and $f : I \to \mathbb{R}$ a function locally integrable over I, that is, f is Riemann integrable over any interval $[a, b]$ for all $b \geq a$ contained in I. Then, we say that f has a convergent improper integral of the first kind if the real limit exists:

$$\lim_{b\to+\infty} \int_a^b f(x)\,\mathrm{d}x = \int_a^{+\infty} f(x)\,\mathrm{d}x$$

Otherwise, we say that the improper integral of the first kind is divergent.

Figure 5.30 graphically illustrates the concept of an improper integral of the first kind.

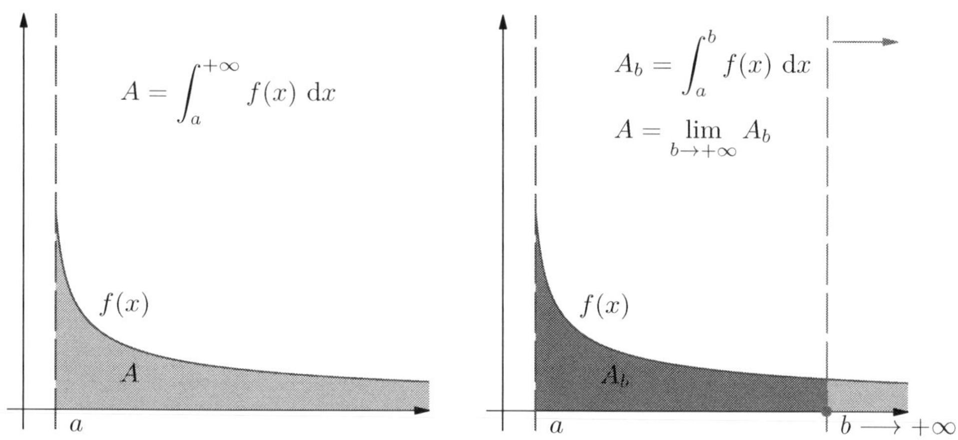

FIGURE 5.30
Graphical representation of the calculation of an improper integral of the first kind.

Remark 5.1. Similarly, we define:

$$\int_{-\infty}^{a} f(x)\,\mathrm{d}x = \lim_{b\to-\infty} \int_b^a f(x)\,\mathrm{d}x$$

Example 5.29. *Study the convergence of the integral:*

$$\int_1^{+\infty} \frac{1}{1+x}\,\mathrm{d}x.$$

SOLUTION.
This is an improper integral of the first kind.

$$\int_1^{+\infty} \frac{1}{1+x}\,\mathrm{d}x = \lim_{b\to+\infty} \int_1^b \frac{1}{1+x}\,\mathrm{d}x$$

$$= \lim_{b\to+\infty} [\ln|1+x|]_1^b = \lim_{b\to+\infty} [\ln|1+b| - \ln(2)] = +\infty.$$

The integral is divergent. □

Example 5.30. *Study the convergence of the integral:*

$$\int_{1}^{+\infty} e^{-x}\,\mathrm{d}x.$$

Solution.
This is an improper integral of the first kind.

$$\int_{1}^{+\infty} e^{-x}\,\mathrm{d}x = \lim_{b\to+\infty}\int_{1}^{b} e^{-x}$$

$$= \lim_{b\to+\infty}\left[-e^{-x}\right]_1^b = \lim_{b\to+\infty}\left[-\frac{1}{e^b}+\frac{1}{e}\right] = \frac{1}{e}.$$

The integral is convergent. □

Example 5.31. *Study the convergence of the integral:*

$$\int_{1}^{+\infty} x^{-4}\,\mathrm{d}x.$$

Solution.
This is an improper integral of the first kind.

$$\int_{1}^{+\infty} x^{-4}\,\mathrm{d}x = \lim_{b\to+\infty}\int_{1}^{b} x^{-4}\,\mathrm{d}x = \lim_{b\to+\infty}\left[-\frac{1}{3x^3}\right]_1^b$$

$$= \lim_{b\to+\infty}\left[\frac{1}{3}-\frac{1}{3b^3}\right] = \frac{1}{3}.$$

The integral is convergent. □

Example 5.32. *Study the convergence of the integral:*

$$\int_{0}^{+\infty} e^{-\lambda x}\,\mathrm{d}x$$

depending on the parameter $\lambda \in \mathbb{R}$.

Solution.
This is an improper integral of the first kind.

$$\int_{0}^{+\infty} e^{-\lambda x}\,\mathrm{d}x = \lim_{b\to+\infty}\int_{0}^{b} e^{-\lambda x}\,\mathrm{d}x = \lim_{b\to+\infty}\left[\frac{-e^{-\lambda x}}{\lambda}\right]_0^b$$

$$= \lim_{b\to+\infty}\left[\frac{-e^{-\lambda b}}{\lambda}+\frac{1}{\lambda}\right] = \frac{1}{\lambda}, \text{ if } \lambda > 0.$$

The integral is convergent if $\lambda > 0$; and divergent if $\lambda \leq 0$. □

Example 5.33. *Study the convergence of the integral:*

$$\int_{1}^{+\infty} \frac{1}{x^p}\,\mathrm{d}x$$

depending on the parameter $p \in \mathbb{R}$.

SOLUTION.
This is an improper integral of the first kind.
If $p = 1$:

$$\int_1^{+\infty} \frac{1}{x}\,\mathrm{d}x = \lim_{b\to+\infty} [\ln|x|]_1^b = +\infty$$

If $p \neq 1$:

$$\begin{aligned}\int_1^{+\infty} \frac{1}{x^p}\,\mathrm{d}x &= \lim_{b\to+\infty} \left[\frac{x^{-p+1}}{-p+1}\right]_1^b \\ &= \lim_{b\to+\infty} \frac{b^{-p+1}}{-p+1} - \frac{1}{1-p} \\ &= \frac{1}{p-1}, \text{ if } p > 1.\end{aligned}$$

Therefore, the integral is convergent if $p > 1$ and divergent if $p \leq 1$. □

Remark 5.2. In the case where the convergence of an improper integral of the first kind with both infinite bounds needs to be studied:

$$\int_{-\infty}^{+\infty} f(x)\,\mathrm{d}x,$$

it is necessary to consider any point c and separately study the convergence of the improper integrals:

$$\int_{-\infty}^{c} f(x)\,\mathrm{d}x \quad \text{and} \quad \int_{c}^{+\infty} f(x)\,\mathrm{d}x.$$

Only if both improper integrals are convergent, will the initial integral be convergent.

Proposed Exercise 5.13. Study the convergence of the integral:

$$\int_{-\infty}^{+\infty} \arctan(x)\,\mathrm{d}x.$$

Proposed Exercise 5.14. Study the convergence of the integral:

$$\int_{-\infty}^{+\infty} e^{-|x|}\sin(x)\,\mathrm{d}x.$$

5.7.4 Improper Integrals of the Second Kind

In this section, we deal with the integration of unbounded functions in bounded domains. In the following definition, we assume that f has a vertical asymptote at $x = a$.

Definition 5.12. Consider now a function $f : (a, b] \to \mathbb{R}$ that is integrable over any interval $(\mu, b]$ with $\mu \in (a, b]$, with:

$$\lim_{x\to a^+} f(x) = \infty.$$

We say that f has a convergent improper integral of the second kind if the real limit exists:

$$\lim_{\mu\to a^+}\int_{\mu}^{b} f(x)\,\mathrm{d}x = \lim_{\delta\to 0^+}\int_{a+\delta}^{b} f(x)\,\mathrm{d}x = \int_a^b f(x)\,\mathrm{d}x.$$

Otherwise, we say that the integral is divergent.

Figure 5.31 graphically illustrates the concept of an improper integral of the second kind.

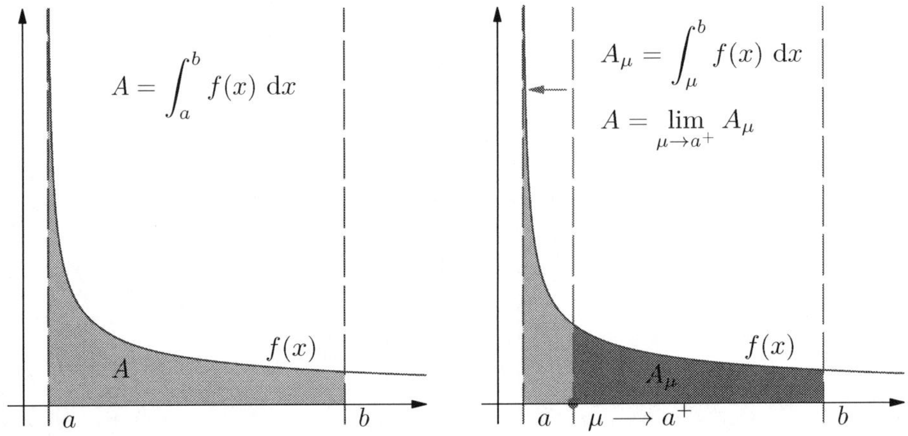

FIGURE 5.31
Graphical representation of the calculation of an improper integral of the second kind.

Remark 5.3. Let $f : [a,b]-\{c\} \to \mathbb{R}$ and $\lim_{x\to c} f(x) = \infty$. We say that $\int_a^b f$ is convergent if each of the integrals $\int_a^c f$ and $\int_c^b f$ are convergent and, in this case:

$$\int_a^b f(x)\,\mathrm{d}x = \int_a^c f(x)\,\mathrm{d}x + \int_c^b f(x)\,\mathrm{d}x.$$

In Figure 5.32, an example of an area bounded by a function $f : [a,b]-\{c\} \to \mathbb{R}$ in the interval $[a,b]$ with $\lim_{x\to c} f(x) = \infty$ and the x-axis can be observed.

Example 5.34. *Study the integral:*

$$\int_1^e \frac{1}{x(\ln(x))^{\frac{1}{3}}}\,\mathrm{d}x.$$

SOLUTION.

This is an improper integral of the second kind since for $x = 1$ the function $f(x) = \frac{1}{x(\ln(x))^{\frac{1}{3}}}$ is unbounded. To calculate it, we make the change of variable $\ln(x) = t$:

$$\int_1^e \frac{1}{x(\ln(x))^{\frac{1}{3}}}\,\mathrm{d}x = \int_0^1 t^{-\frac{1}{3}}dt = \lim_{\delta\to 0^+}\int_\delta^1 t^{-\frac{1}{3}}dt$$

$$= \lim_{\delta\to 0^+}\left[\frac{3}{2}t^{\frac{2}{3}}\right]_\delta^1 = \lim_{\delta\to 0^+}\left[\frac{3}{2}-\frac{3}{2}\delta^{\frac{2}{3}}\right] = \frac{3}{2}.$$

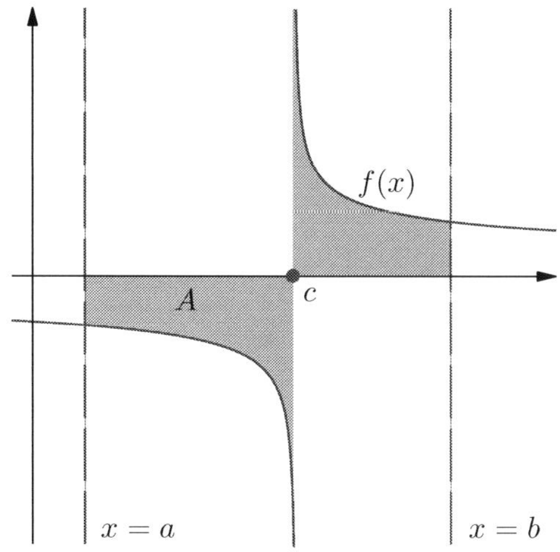

FIGURE 5.32
Area bounded by a function $f : [a, b] - \{c\} \to \mathbb{R}$ in the interval $[a, b]$ with $\lim_{x \to c} f(x) = \infty$ and the x-axis.

Therefore, this improper integral of the second kind is convergent with a value of 3/2. □

Example 5.35. *Study the convergence of the integral:*

$$\int_a^b \frac{1}{(x-a)^p} \, \mathrm{d}x, \; b > a$$

depending on the different values of $p \in \mathbb{R}$.

SOLUTION.
It is clear that for values $p \leq 0$, it is a proper integral since the function is bounded over a bounded domain.

If $p > 0$, the function $\frac{1}{(x-a)^p}$ presents a singularity at the point $x = a$. To determine the nature of the integral, we apply the definition according to the values of p.

- Case $p \neq 1$

$$\begin{aligned}\int_a^b \frac{1}{(x-a)^p} \, \mathrm{d}x &= \lim_{\delta \to 0^+} \int_{a+\delta}^b \frac{1}{(x-a)^p} \, \mathrm{d}x \\ &= \lim_{\delta \to 0^+} \left[\frac{(x-a)^{-p+1}}{-p+1} \right]_{a+\delta}^b \\ &= \frac{(b-a)^{1-p}}{1-p}, \quad \text{if } p < 1.\end{aligned}$$

  In the case $p > 1$, the last limit of the previous expression is infinite.

- Case $p = 1$

$$\begin{aligned}\int_a^b \frac{1}{x-a} \, \mathrm{d}x &= \lim_{\delta \to 0^+} \int_{a+\delta}^b \frac{1}{x-a} \, \mathrm{d}x \\ &= \lim_{\delta \to 0^+} [\ln|x-a|]_{a+\delta}^b = +\infty\end{aligned}$$

Summarizing the previous results, we can say that the integral $\int_a^b \frac{1}{(x-a)^p}\,\mathrm{d}x$ is convergent if $p < 1$, and divergent if $p \geq 1$. □

Proposed Exercise 5.15. Calculate the integral:

$$\int_{-\infty}^{+\infty} \frac{1}{x^2+2x+5}\,\mathrm{d}x$$

if it is convergent.

Proposed Exercise 5.16. Calculate the integral:

$$\int_{0}^{+\infty} x\sin(x)\,\mathrm{d}x$$

if it is convergent.

5.7.5 Improper Integrals of the Third Kind

The study of the convergence of an improper integral of the third kind is reduced, by the additivity of the integral with respect to the interval, to separately studying the convergence of one or two integrals of the first kind (depending on whether one or both ends of the integral are infinite) and one or several integrals of the second kind (depending on the singularities). The improper integral of the third kind will be convergent if and only if, all the integrals of the summands are convergent. Otherwise, the improper integral of the third kind will be divergent.

In Example 5.36, we can see how to proceed in these cases:

Example 5.36. *Determine whether the integral:*

$$\int_{-1}^{+\infty} \frac{1}{x^2}\,\mathrm{d}x$$

is convergent or divergent.

SOLUTION.

Given that at $x = 0$ the function $f(x) = \frac{1}{x^2}$ presents a vertical asymptote, we decompose the improper integral of the third kind into three improper integrals:

$$\int_{-1}^{+\infty} \frac{1}{x^2}\,\mathrm{d}x = \int_{-1}^{0} \frac{1}{x^2}\,\mathrm{d}x + \int_{0}^{1} \frac{1}{x^2}\,\mathrm{d}x + \int_{1}^{+\infty} \frac{1}{x^2}\,\mathrm{d}x.$$

The first two summands are each integrals of the second kind. The third summand is an improper integral of the first kind.

If we consider, for example, the integral of the first summand, we have:

$$\int_{-1}^{0} \frac{1}{x^2}\,\mathrm{d}x = \lim_{\delta\to 0^+} \int_{-1}^{0-\delta} \frac{1}{x^2}\,\mathrm{d}x = \lim_{\delta\to 0^+} \left[-\frac{1}{x}\right]_{-1}^{-\delta} = \lim_{\delta\to 0^+} \left(\frac{1-\delta}{\delta}\right) = +\infty$$

Since this improper integral is divergent, the initial integral of the third kind will also be divergent. □

5.7.6 Bounding Improper Integrals

Sometimes it is useful to know whether an integral is convergent without being interested in the particular value of the integral. For example, we might be interested in knowing whether the integral:

$$\int_1^{+\infty} \frac{\sin(x)+\cos(x)}{x^2+x+1}\,\mathrm{d}x$$

is finite or convergent.

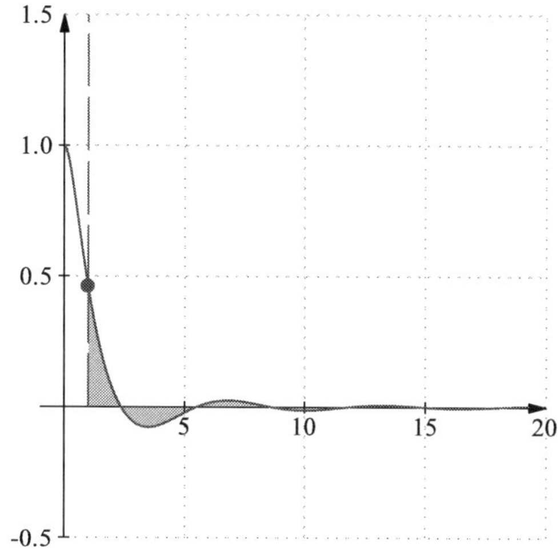

FIGURE 5.33
Area bounded by the function $f(x) = \frac{\sin(x)+\cos(x)}{x^2+x+1}$ and the x-axis.

This is equivalent to finding two real numbers L_1 and L_2 such that:

$$L_1 \leq \int_1^{+\infty} \frac{\sin(x)+\cos(x)}{x^2+x+1}\,\mathrm{d}x \leq L_2.$$

Figure 5.33 shows the area bounded by the function $f(x) = \frac{\sin(x)+\cos(x)}{x^2+x+1}$ and the x-axis.

Observation 5.11. Let $f(x)$ be integrable in the sense of Riemann in $[a, b]$, then:

$$\left|\int_a^b f(x)\,\mathrm{d}x\right| \leq \int_a^b \left|f(x)\right|\,\mathrm{d}x.$$

Furthermore, this property is also true if a and/or b are $\pm\infty$, that is, if we have an improper integral of the first kind.

From Figure 5.34, it can be seen that:

$$\left|\int_a^b f(x)\,\mathrm{d}x\right| = B+C-A-D \leq \int_a^b \left|f(x)\right|\,\mathrm{d}x = A+B+C+D$$

This observation is key to bounding integrals. Figure 5.34 graphically shows that the area bounded by the absolute value of a function and the x-axis is always equal to or greater than the area bounded by the function and the x-axis.

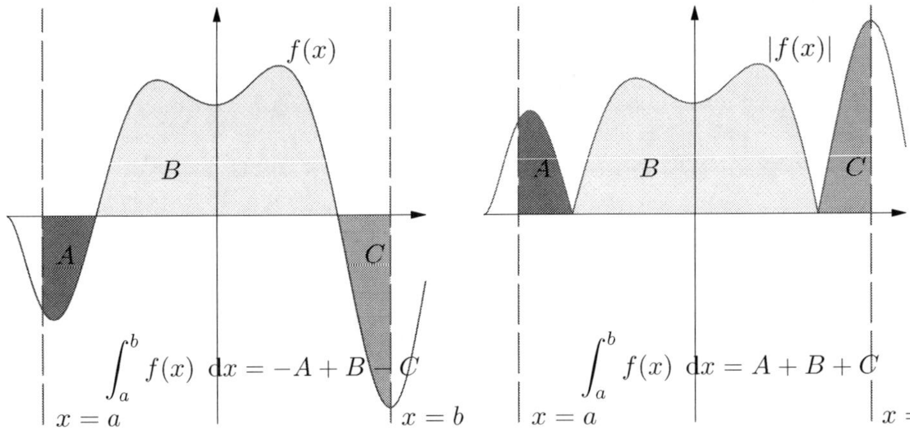

FIGURE 5.34
Area bounded by a function and the x-axis and by the absolute value of the function and the x-axis.

Example 5.37. *Determine whether the integral:*

$$\int_1^{+\infty} \frac{\sin(x) + \cos(x)}{x^2 + x + 1}\ \mathrm{d}x,$$

is convergent or divergent.

SOLUTION.
In Figure 5.35, the process of bounding the integral is schematically represented. First, we observe that:

$$\int_1^{+\infty} f(x)\ \mathrm{d}x \leq \int_1^{+\infty} |f(x)|\ \mathrm{d}x$$

and then we use a function $g(x)$ to bound the integral of the absolute value:

$$\int_1^{+\infty} |f(x)|\ \mathrm{d}x \leq \int_1^{+\infty} g(x)\ \mathrm{d}x.$$

Let $L = \displaystyle\int_1^{+\infty} \frac{\sin(x) + \cos(x)}{x^2 + x + 1}\ \mathrm{d}x$, then:

$$|L| \leq \int_1^{+\infty} \left| \frac{\sin(x) + \cos(x)}{x^2 + x + 1} \right| \mathrm{d}x \leq \int_1^{+\infty} \frac{2}{x^2 + x + 1}\ \mathrm{d}x,$$

where we have used that:

- $|\sin(x) + \cos(x)| \leq |\sin(x)| + |\cos(x)| \leq 2.$
- $|x^2 + x + 1| = x^2 + x + 1$, as we are integrating in the domain $x \geq 1$ and, therefore, x is always positive.

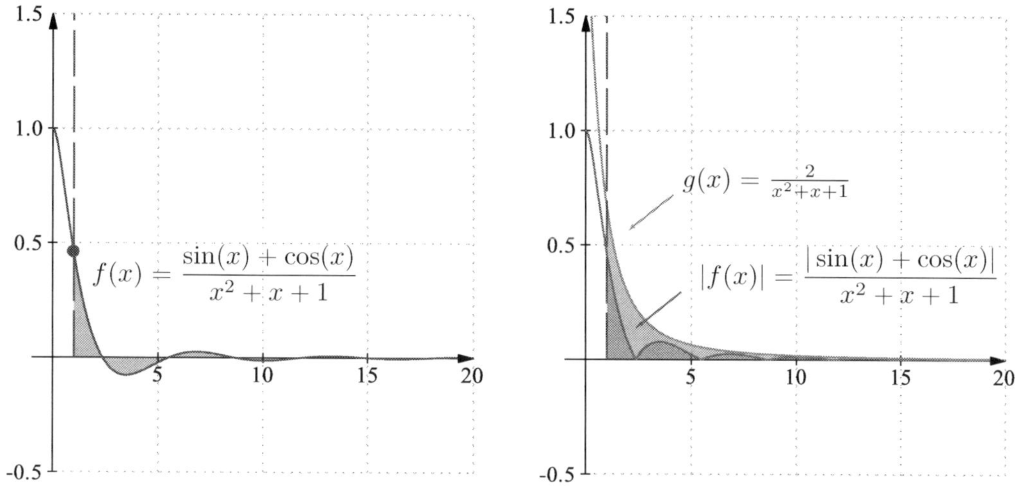

FIGURE 5.35
Graphical representation of the inequality:
$\int_1^{+\infty} f(x)\ \mathrm{d}x \le \int_1^{+\infty} |f(x)|\ \mathrm{d}x \le \int_1^{+\infty} g(x)\ \mathrm{d}x$

Finally, given that:

$$x^2 + x + 1 \ge x^2 \quad \Longrightarrow \quad \frac{1}{x^2 + x + 1} \le \frac{1}{x^2}$$

we can write:

$$\begin{aligned} |L| \ &\le \int_1^{+\infty} \frac{2}{x^2 + x + 1}\,\mathrm{d}x \le \int_1^{+\infty} \frac{2}{x^2}\,\mathrm{d}x \\ &= \lim_{b \to +\infty} \int_1^b \frac{2}{x^2}\,\mathrm{d}x = \lim_{b \to +\infty} \left[-\frac{2}{x}\right]_1^b \\ &= \lim_{b \to +\infty} \left[-\frac{2}{b} + 2\right] = 2. \end{aligned}$$

Therefore, the integral is convergent. □

Proposed Exercise 5.17. The Gamma function is defined by:

$$\Gamma(p) = \int_0^{\infty} x^p e^{-x}\,\mathrm{d}x.$$

Show that this function is defined for all $p > -1$ and that it satisfies $\Gamma(p+1) = p\Gamma(p)$.

5.8 Solved Problems (Calculation of Primitives / Definite Integration)

Problem 5.1. Calculate the following definite integrals using the method of integration by parts:

(a) $\displaystyle\int_1^e \ln(x)\,\mathrm{d}x$

(b) $\displaystyle\int_0^\pi x\cos(x)\,\mathrm{d}x$

(c) $\displaystyle\int_0^1 x^2\cos(2x)\,\mathrm{d}x$

(d) $\displaystyle\int_0^\pi e^x\sin(x)\,\mathrm{d}x$

(e) $\displaystyle\int_{-\pi/2}^{\pi/2} e^{2x}\cos(x)\,\mathrm{d}x$

(f) $\displaystyle\int_0^1 x^3e^{2x}\,\mathrm{d}x$

(g) $\displaystyle\int_{10}^{100} xe^{-x}\,\mathrm{d}x$

Solution to the problem 5.1. *(a) We will solve the definite integral using the method of integration by parts. Indeed:*

$$\int_1^e \ln(x)\,\mathrm{d}x = \int_1^e \underbrace{\ln(x)}_{u(x)}\,\underbrace{1\,\mathrm{d}x}_{v'(x)} \overset{(*)}{=} \Big[\ln(x)x\Big]_1^e - \int_1^e x\frac{1}{x}\,\mathrm{d}x$$

$$(*)\left[\begin{array}{lll} u(x)=\ln(x) & \Rightarrow & u'(x)=\frac{1}{x}\,\mathrm{d}x \\ v'(x)=1\,\mathrm{d}x & \Rightarrow & v(x)=x \end{array}\right]$$

$$= e-0-\int_1^e 1\,\mathrm{d}x = e - x\Big|_1^e = e-(e-1)=1.$$

(b) *We will solve the definite integral using the method of integration by parts. Indeed:*

$$\int_0^{\pi} x\cos(x)\,\mathrm{d}x = \int_0^{\pi} \underbrace{x}_{u(x)} \underbrace{\cos(x)\,\mathrm{d}x}_{v'(x)} \overset{(*)}{=}$$

$$(*) \begin{bmatrix} u(x) = x & \Rightarrow & u'(x) = 1\,\mathrm{d}x \\ v'(x) = \cos(x)\,\mathrm{d}x & \Rightarrow & v(x) = \sin(x) \end{bmatrix}$$

$$= \Big[x\sin(x)\Big]_0^{\pi} - \int_0^{\pi} \sin(x)1\,\mathrm{d}x$$

$$= 0 - \Big[-\cos(x)\Big]_0^{\pi} = 0 + \Big[\cos(x)\Big]_0^{\pi} = 0 + (-1-1) = -2.$$

(c) *We will solve the definite integral using the method of integration by parts twice. Indeed:*

$$\int_0^{1} x^2\cos(2x)\,\mathrm{d}x = \int_0^{1} \underbrace{x^2}_{u(x)} \underbrace{\cos(2x)\,\mathrm{d}x}_{v'(x)} \underset{(*_1)}{=}$$

$$(*_1) \begin{bmatrix} u(x) = x^2 & \Rightarrow & u'(x) = 2x\,\mathrm{d}x \\ v'(x) = \cos(2x)\,\mathrm{d}x & \Rightarrow & v(x) = \frac{1}{2}\sin(2x) \end{bmatrix}$$

$$= \Big[x^2\frac{1}{2}\sin(2x)\Big]_0^{1} - \int_0^{1} \frac{1}{2}\sin(x)\cdot 2x\,\mathrm{d}x$$

$$= \left(1\frac{1}{2}\sin(2) - 0\cdot 0\right) - \underbrace{\int_0^{1} x\sin(2x)\,\mathrm{d}x}_{\textit{por partes}}$$

The integral on the right is solved using integration by parts again:

$$\int_0^1 x^2\cos(2x)\,\mathrm{d}x = \frac{1}{2}\sin(2) - \int_0^1 \underbrace{x}_{u(x)}\underbrace{\sin(2x)\,\mathrm{d}x}_{v'(x)} \underset{(*_2)}{=}$$

$$(*_2)\left[\begin{array}{lll} u(x) = x & \Rightarrow & u'(x) = \mathrm{d}x \\ v'(x) = \sin(2x)\,\mathrm{d}x & \Rightarrow & v(x) = -\tfrac{1}{2}\cos(2x) \end{array}\right]$$

$$= \frac{1}{2}\sin(2) - \left[x\left(-\frac{1}{2}\cos(2x)\right)\right]_0^1 + \int_0^1 -\frac{1}{2}\cos(2x)\,\mathrm{d}x$$

$$= \frac{1}{2}\sin(2) - \left(1\left(-\frac{1}{2}\cos(2)\right) - 0\left(-\frac{1}{2}\cos(0)\right)\right) + \left[-\frac{1}{4}\sin(2x)\right]_0^1$$

$$= \frac{1}{2}\sin(2) + \frac{1}{2}\cos(2) + \left[-\frac{1}{4}\sin(2) - \left(-\frac{1}{4}\sin(0)\right)\right]$$

$$= \frac{1}{2}\sin(2) + \frac{1}{2}\cos(2) - \frac{1}{4}\sin(2) = \frac{1}{4}\sin(2) + \frac{1}{2}\cos(2).$$

(d) *We will solve the definite integral using the method of integration by parts and see that the integral is cyclic. Indeed:*

$$\int_0^\pi e^x\sin(x)\,\mathrm{d}x = \int_0^\pi \underbrace{e^x}_{u(x)}\underbrace{\sin(x)\,\mathrm{d}x}_{v'(x)} \underset{(*_1)}{=}$$

$$(*_1)\left[\begin{array}{lll} u(x) = e^x & \Rightarrow & u'(x) = e^x\,\mathrm{d}x \\ v'(x) = \sin(x)\,\mathrm{d}x & \Rightarrow & v(x) = -\cos(x) \end{array}\right]$$

$$= \left[-e^x\cos(x)\right]_0^\pi - \int_0^\pi -\cos(x)e^x\,\mathrm{d}x$$

$$= (e^\pi 1 - (-1)1) + \underbrace{\int_0^\pi e^x\cos(x)\,\mathrm{d}x}_{\textit{by parts}}$$

$$= (e^\pi + 1) + \int_0^\pi \underbrace{e^x}_{u(x)}\underbrace{\cos(x)\,\mathrm{d}x}_{v'(x)} \underset{(*_2)}{=}$$

$$(*_2)\left[\begin{array}{lll} u(x) = e^x & \Rightarrow & u'(x) = e^x\,\mathrm{d}x \\ v'(x) = \cos(x)\,\mathrm{d}x & \Rightarrow & v(x) = \sin(x) \end{array}\right]$$

$$= (e^\pi + 1) + \left[e^x\sin(x)\right]_0^\pi - \int_0^\pi \sin(x)e^x\,\mathrm{d}x$$

$$= (e^\pi + 1) + 0 - \underbrace{\int_0^\pi e^x\sin(x)\,\mathrm{d}x}_{\textit{original integral}}.$$

In this case, we can see that, after integrating by parts twice, we end up with an integral identical to the one we started with. We analyze the expression we have obtained:

$$\int_0^{\pi} e^x \sin(x)\,dx = (e^{\pi}+1) - \int_0^{\pi} e^x \sin(x)\,dx.$$

Solving, we obtain:

$$\int_0^{\pi} e^x \sin(x)\,dx + \int_0^{\pi} e^x \sin(x)\,dx = (e^{\pi}+1)$$

$$2\int_0^{\pi} e^x \sin(x)\,dx = (e^{\pi}+1)$$

$$\int_0^{\pi} e^x \sin(x)\,dx = \frac{1}{2}(e^{\pi}+1)$$

$$\int_0^{\pi} e^x \sin(x)\,dx = \frac{1}{2}e^{\pi}+\frac{1}{2}.$$

(e) *We will solve the definite integral using the method of integration by parts and see that the integral is cyclic. Indeed:*

$$\int_{-\pi/2}^{\pi/2} e^{2x}\cos(x)\,dx = \int_{-\pi/2}^{\pi/2} \underbrace{e^{2x}}_{u(x)}\underbrace{\cos(x)\,dx}_{v'(x)} \underset{(*_1)}{=}$$

$$(*_1)\left[\begin{array}{lll} u(x)=e^{2x} & \Rightarrow & u'(x)=2e^{2x}\,dx \\ v'(x)=\cos(x)\,dx & \Rightarrow & v(x)=\sin(x) \end{array}\right]$$

$$= \Big[e^{2x}\sin(x)\Big]_{-\pi/2}^{\pi/2} - \int_{-\pi/2}^{\pi/2} \sin(x)2e^{2x}\,dx$$

$$= (e^{\pi}1 - e^{-\pi}(-1)) - 2\underbrace{\int_{-\pi/2}^{\pi/2} e^{2x}\sin(x)\,dx}_{by\ parts}$$

$$= (e^{\pi}+e^{-\pi}) - 2\int_{-\pi/2}^{\pi/2} \underbrace{e^{2x}}_{u(x)}\underbrace{\sin(x)\,dx}_{v'(x)} \underset{(*_2)}{=}$$

$$(*_2)\left[\begin{array}{lll} u(x)=e^{2x} & \Rightarrow & u'(x)=2e^{2x}\,dx \\ v'(x)=\sin(x)\,dx\,dx & \Rightarrow & v(x)=-\cos(x) \end{array}\right]$$

$$= (e^{\pi} + e^{-\pi}) - 2\left[-e^{2x}\cos(x)\Big|_{-\pi/2}^{\pi/2} - \int_{-\pi/2}^{\pi/2} -\cos(x)2e^{2x}\,\mathrm{d}x\right]$$

$$= (e^{\pi} + e^{-\pi}) - 2\left[0 + 2\int_{-\pi/2}^{\pi/2} e^{2x}\cos(x)\,\mathrm{d}x\right]$$

$$= (e^{\pi} + e^{-\pi}) - 4\underbrace{\int_{-\pi/2}^{\pi/2} e^{2x}\cos(x)\,\mathrm{d}x}_{\textit{original integral}}.$$

In this case, we again find that, after integrating by parts twice, the integral to be solved is the same as the original. We analyze the expression we have obtained:

$$\int_{-\pi/2}^{\pi/2} e^{2x}\cos(x)\,\mathrm{d}x = (e^{\pi} + e^{-\pi}) - 4\int_{-\pi/2}^{\pi/2} e^{2x}\cos(x)\,\mathrm{d}x.$$

Solving, we obtain:

$$\int_{-\pi/2}^{\pi/2} e^{2x}\cos(x)\,\mathrm{d}x + 4\int_{-\pi/2}^{\pi/2} e^{2x}\cos(x)\,\mathrm{d}x = e^{\pi} + e^{-\pi}$$

$$5\int_{-\pi/2}^{\pi/2} e^{2x}\cos(x)\,\mathrm{d}x = e^{\pi} + e^{-\pi} = e^{\pi} + \frac{1}{e^{\pi}} = \frac{e^{2\pi}+1}{e^{\pi}}$$

$$\int_{-\pi/2}^{\pi/2} e^{2x}\cos(x)\,\mathrm{d}x = \frac{1}{5}\frac{e^{2\pi}+1}{e^{\pi}}.$$

(f) *We will solve the definite integral using the method of integration by parts three times. Indeed:*

$$\int_0^1 x^3 e^{2x}\,\mathrm{d}x = \int_0^1 \underbrace{x^3}_{u(x)}\underbrace{e^{2x}\,\mathrm{d}x}_{v'(x)} \underset{(*_1)}{=}$$

$$(*_1)\left[\begin{array}{lll} u(x) = x^3 & \Rightarrow & u'(x) = 3x^2\,\mathrm{d}x \\ v'(x) = e^{2x}\,\mathrm{d}x & \Rightarrow & v(x) = \frac{1}{2}e^{2x} \end{array}\right]$$

$$= \left[x^3\frac{1}{2}e^{2x}\right]_0^1 - \int_0^1 \frac{1}{2}e^{2x}3x^2\,\mathrm{d}x$$

$$= \frac{1}{2}e^2 - 0 - \frac{3}{2}\int_0^1 e^{2x}x^2\,\mathrm{d}x$$

$$= \frac{1}{2}e^2 - \underbrace{\frac{3}{2}\int_0^1 x^2 e^{2x}\,\mathrm{d}x}_{por\ partes} =$$

$$= \frac{1}{2}e^2 - \frac{3}{2}\int_0^1 \underbrace{x^2}_{u(x)}\underbrace{e^{2x}\,\mathrm{d}x}_{v'(x)} \underset{(*_2)}{=}$$

$$(*_2)\left[\begin{array}{lll} u(x) = x^2 & \Rightarrow & u'(x) = 2x\,\mathrm{d}x \\ v'(x) = e^{2x}\,\mathrm{d}x & \Rightarrow & v(x) = \frac{1}{2}e^{2x} \end{array}\right]$$

$$= \frac{1}{2}e^2 - \frac{3}{2}\left[x^2\frac{1}{2}e^{2x}\Big|_0^1 - \int_0^1 \frac{1}{2}e^{2x}2x\,\mathrm{d}x\right]$$

$$= \frac{1}{2}e^2 - \frac{3}{2}\left[\frac{1}{2}e^2 - \int_0^1 xe^{2x}\,\mathrm{d}x\right]$$

$$= \frac{1}{2}e^2 - \frac{3}{4}e^2 + \underbrace{\frac{3}{2}\int_0^1 xe^{2x}\,\mathrm{d}x}_{por\ partes}$$

$$= \frac{1}{2}e^2 - \frac{3}{4}e^2 + \frac{3}{2}\int_0^1 \underbrace{x}_{u(x)}\underbrace{e^{2x}\,\mathrm{d}x}_{v'(x)} \underset{(*_3)}{=}$$

$$(*_3)\left[\begin{array}{lll} u(x) = x & \Rightarrow & u'(x) = 1\,\mathrm{d}x \\ v'(x) = e^{2x}\,\mathrm{d}x & \Rightarrow & v(x) = \frac{1}{2}e^{2x} \end{array}\right]$$

$$= -\frac{1}{4}e^2 + \frac{3}{2}\left[x\frac{1}{2}e^{2x}\Big|_0^1 - \int_0^1 \frac{1}{2}e^{2x}\,\mathrm{d}x\right]$$

$$= -\frac{1}{4}e^2 + \frac{3}{2}\left[\frac{1}{2}e^2 - \frac{1}{2}\int_0^1 e^{2x}\,\mathrm{d}x\right]$$

$$= -\frac{1}{4}e^2 + \frac{3}{4}e^2 - \frac{3}{4}\cdot\frac{1}{2}e^{2x}\Big|_0^1$$

$$= \frac{1}{2}e^2 - \frac{3}{8}(e^2 - 1) = \frac{1}{2}e^2 - \frac{3}{8}e^2 + \frac{3}{8} = \frac{1}{8}e^2 + \frac{3}{8}.$$

(g) We will solve the definite integral using the method of integration by parts. Indeed:

$$\int_{10}^{100} xe^{-x}\,\mathrm{d}x = \int_{10}^{100} \underbrace{x}_{u(x)}\underbrace{e^{-x}\,\mathrm{d}x}_{v'(x)} \overset{(*)}{=}$$

$$(*)\left[\begin{array}{lll} u(x) = x & \Rightarrow & u'(x) = 1\,\mathrm{d}x \\ v'(x) = e^{-x}\,\mathrm{d}x & \Rightarrow & v(x) = -e^{-x} \end{array}\right]$$

$$\begin{aligned}
&= \Big[x\big(-e^{-x}\big)\Big]_{10}^{100} - \int_{10}^{100} -e^{-x}\,\mathrm{d}x \\
&= -100e^{-100} - \big(-10e^{-10}\big) + \int_{10}^{100} e^{-x}\,\mathrm{d}x \\
&= -100e^{-100} + 10e^{-10} - \big(-e^{-x}\big)\Big|_{10}^{100} \\
&= -100e^{-100} + 10e^{-10} - \big(e^{-100} - e^{-10}\big) \\
&= -100e^{-100} - e^{-100} + 10e^{-10} + e^{-10} \\
&= -101e^{-100} + 11e^{-10}.
\end{aligned}$$

Problem 5.2. Calculate the following indefinite integrals using the proposed variable changes.

(a) $\displaystyle\int \frac{1}{e^x + e^{-x}}\,\mathrm{d}x, \quad t = e^x$

(b) $\displaystyle\int \frac{1}{\sqrt{25 - 16x^2}}\,\mathrm{d}x, \quad x = \frac{5}{4}\sin(t)$

(c) $\displaystyle\int \frac{\sin(x)}{\cos^2 x}\,\mathrm{d}x, \quad t = \cos(x)$

(d) $\displaystyle\int \frac{\ln(x)}{x}\,\mathrm{d}x, \quad t = \ln(x)$

(e) $\displaystyle\int \frac{e^{\sqrt{x}}}{\sqrt{x}}\,\mathrm{d}x, \quad x = t^2$

(f) $\displaystyle\int \frac{\sin(x)\cos(x)}{1 + \sin(x)}\,\mathrm{d}x, \quad t = \sin(x)$

Solution to the problem 5.2. *(a) In this integral, the change $t = e^x$ is proposed, that is:*

$$(*)\left[\begin{array}{ccc} t & = & e^x \\ \mathrm{d}t & = & e^x\,\mathrm{d}x \end{array}\right]$$

We should develop the integral so that the change can be applied. In this case, we are interested in having e^x appear in the numerator so that we have $e^x\,\mathrm{d}x$ and can substitute it with $\mathrm{d}t$.

Let's start by developing the integral:

$$\int \frac{1}{e^x + e^{-x}}\,\mathrm{d}x = \int \frac{1}{e^x + \dfrac{1}{e^x}}\,\mathrm{d}x = \int \frac{1}{\dfrac{e^{2x}+1}{e^x}}\,\mathrm{d}x$$

$$= \int \frac{e^x}{e^x e^x + 1}\,\mathrm{d}x = \int \frac{1}{\underbrace{e^x}_{t}\,\underbrace{e^x}_{t} + 1}\underbrace{e^x\,\mathrm{d}x}_{\mathrm{d}t}.$$

Now we can apply the change and obtain:

$$\int \frac{1}{e^x + e^{-x}}\,\mathrm{d}x = \int \frac{1}{t^2+1}\,\mathrm{d}t$$

$$= \arctan(t) + C = \arctan(e^x) + C, \quad C \in \mathbb{R}.$$

When it is not immediately clear how to develop the integral in order to apply the change, another solution is to isolate the variable of the change in terms of the one in which the integral is expressed, find the differential by deriving, and substitute directly:

$$\int \frac{1}{e^x + e^{-x}}\,\mathrm{d}x \overset{(*)}{=} \int \frac{1}{e^{\ln(t)} + e^{-\ln(t)}}\frac{\mathrm{d}t}{t} = \int \frac{1}{t + \dfrac{1}{t}}\frac{\mathrm{d}t}{t}$$

$$(*)\left[\begin{array}{cccc} t & = & e^x & \Rightarrow x = \ln(t) \\ \mathrm{d}t & = & e^x\,\mathrm{d}x; & \mathrm{d}x = \dfrac{\mathrm{d}t}{t} \end{array}\right]$$

$$= \int \frac{t}{t^2+1}\frac{\mathrm{d}t}{t}$$

$$= \int \frac{1}{t^2+1}\,\mathrm{d}t.$$

(b) We perform the proposed variable change:

$$\int \frac{1}{\sqrt{25-16x^2}}\,\mathrm{d}x = \int \left(25-16x^2\right)^{-1/2}\mathrm{d}x \overset{(*)}{=}$$

$$(*)\left[\begin{array}{ccc} x & = & \frac{5}{4}\sin(t) \\ \mathrm{d}x & = & \frac{5}{4}\cos(t)\,\mathrm{d}t \end{array}\right]$$

$$= \int \left(25-16\left(\frac{5}{4}\sin(t)\right)^2\right)^{-1/2} \cdot \frac{5}{4}\cos(t)\,\mathrm{d}t$$

$$= \int \left(25-16\frac{25}{16}\sin^2(t)\right)^{-1/2} \cdot \frac{5}{4}\cos(t)\,\mathrm{d}t$$

$$= \int \left(25(1-\sin^2(t))\right)^{-1/2} \cdot \frac{5}{4}\cos(t)\,\mathrm{d}t$$

$$= \int \frac{1}{5}\frac{1}{\sqrt{1-\sin^2(t)}}\frac{5}{4}\cos(t)\,\mathrm{d}t$$

$$= \int \frac{1}{4\cos(t)}\cos(t)\,\mathrm{d}t = \int \frac{1}{4}\,\mathrm{d}t = \frac{1}{4}t$$

$$= \frac{1}{4}\arcsin\left(\frac{4}{5}x\right) + C, \quad C \in \mathbb{R}.$$

To undo the change, we solved for t from the proposed change:

$$x = \frac{5}{4}\sin(t) \quad \Rightarrow \quad \sin(t) = \frac{4}{5}x \quad \Rightarrow \quad t = \arcsin\left(\frac{4}{5}x\right)$$

(c) We perform the proposed variable change:

$$\int \frac{\sin(x)}{\cos^2 x}\,\mathrm{d}x = \int \underbrace{\frac{1}{\cos^2(x)}}_{t^2} \underbrace{\sin(x)\,\mathrm{d}x}_{-\,\mathrm{d}t} \overset{(*)}{=} \int -\frac{1}{t^2}\,\mathrm{d}t$$

$$(*)\left[\begin{array}{ccc} t & = & \cos(x) \\ \mathrm{d}t & = & -\sin(x)\,\mathrm{d}x \end{array}\right]$$

$$= -\int \frac{1}{t^2}\,\mathrm{d}t = -\int t^{-2}\,\mathrm{d}t = -\left(\frac{1}{-1}t^{-1}\right)$$

$$= -\left(-\frac{1}{t}\right) = \frac{1}{t} = \frac{1}{\cos(x)} + C, \quad C \in \mathbb{R}.$$

(d) *We perform the proposed variable change:*

$$\int \frac{\ln(x)}{x}\,\mathrm{d}x = \int \underbrace{\ln(x)}_{t}\,\underbrace{\frac{\mathrm{d}x}{x}}_{\mathrm{d}t} \overset{(*)}{=} \int t\,\mathrm{d}t$$

$$(*)\begin{bmatrix} t & = & \ln(x) \\ \mathrm{d}t & = & \dfrac{\mathrm{d}x}{x} \end{bmatrix}$$

$$= \frac{1}{2}t^2 = \frac{1}{2}\ln(x)^2 + C, \quad C \in \mathbb{R}.$$

(e) *We perform the proposed variable change:*

$$\int \frac{e^{\sqrt{x}}}{\sqrt{x}}\,\mathrm{d}x \overset{(*)}{=} \int \frac{e^{\sqrt{t^2}}}{\sqrt{t^2}}2t\,\mathrm{d}t = \int \frac{e^t}{t}2t\,\mathrm{d}t$$

$$(*)\begin{bmatrix} x & = & t^2 \\ \mathrm{d}x & = & 2t\,\mathrm{d}t \end{bmatrix}$$

$$= 2\int e^t\,\mathrm{d}t = 2e^t = 2e^{\sqrt{x}} + C, \quad C \in \mathbb{R}.$$

(f) *We perform the proposed variable change:*

$$\int \frac{\sin(x)\cos(x)}{1+\sin(x)}\,\mathrm{d}x = \int \frac{\sin(x)}{1+\sin(x)}\cos(x)\,\mathrm{d}x \overset{(*)}{=} \int \frac{t}{1+t}\,\mathrm{d}t$$

$$(*)\begin{bmatrix} t & = & \sin(x) \\ \mathrm{d}t & = & \cos(x)\,\mathrm{d}x \end{bmatrix}$$

We use a mathematical trick by adding and subtracting 1 to the numerator to obtain two immediate integrals (in fact, this is a quick way to obtain simple fractions in this case):

$$\begin{aligned} \int \frac{t}{1+t}\,\mathrm{d}t &= \int \frac{t+1-1}{1+t}\,\mathrm{d}t = \int \left(\frac{1+t}{1+t} - \frac{1}{1+t}\right)\mathrm{d}t \\ &= \int \left(1 - \frac{1}{1+t}\right)\mathrm{d}t = \int 1\,\mathrm{d}t - \int \frac{1}{1+t}\,\mathrm{d}t \\ &= t - \ln|1+t| = \sin(x) - \ln|1+\sin(x)| + C, \quad C \in \mathbb{R}. \end{aligned}$$

Problem 5.3. Calculate a primitive of each of the following functions, using the most suitable method in each case:

(a) $f_1(x) = \dfrac{1}{e^x + e^{-x}}$

(b) $f_2(x) = xe^{x+2}$

(c) $f_3(x) = \dfrac{1}{\tan(x)}$

(d) $f_4(x) = \dfrac{1}{3+\cos(x)}$, using the change of variable $x = 2\arctan(t)$ and taking into account that:

$$\cos(x) = \cos^2\left(\frac{x}{2}\right) - \sin^2\left(\frac{x}{2}\right)$$
$$\cos(\arctan(x)) = \frac{1}{\sqrt{1+x^2}}$$
$$\sin(\arctan(x)) = \frac{x}{\sqrt{1+x^2}}$$

Solution to the problem 5.3. *(a) This integral is almost immediate if we rewrite the function as follows:*

$$\begin{aligned}\int f_1(x)\,\mathrm{d}x &= \int \frac{1}{e^x + e^{-x}}\,\mathrm{d}x = \int \frac{1}{e^x + \frac{1}{e^x}}\,\mathrm{d}x\\ &= \int \frac{1}{\frac{e^{2x}+1}{e^x}}\,\mathrm{d}x = \int \frac{e^x}{e^{2x}+1}\,\mathrm{d}x\\ &= \int \frac{e^x}{(e^x)^2+1}\,\mathrm{d}x = \arctan(e^x) + C.\end{aligned}$$

where $C \in \mathbb{R}$. *Therefore,* $\varphi_1(x) = \arctan(e^x)$ *is a primitive.*

If we check, we have:

$$\varphi_1'(x) = \frac{1}{(e^x)^2+1}e^x = \frac{e^x}{e^{2x}+1} = \frac{1}{\frac{e^{2x}+1}{e^x}} = \frac{1}{e^x + \frac{1}{e^x}} = \frac{1}{e^x + e^{-x}}.$$

If we do not realize it is almost an immediate integral, we can make the change:

$$t = e^x \quad \Longrightarrow \quad dt = e^x\,\mathrm{d}x \quad \Longrightarrow \quad \mathrm{d}x = \frac{1}{e^x}dt = \frac{1}{t}dt$$

$$\begin{aligned}\int f_1(x)\,\mathrm{d}x &= \int \frac{1}{e^x + \frac{1}{e^x}}\,\mathrm{d}x = \int \frac{1}{t+\frac{1}{t}}\frac{1}{t}\,\mathrm{d}t\\ &= \int \frac{1}{t^2+1}dt = \arctan(t) + C = \arctan(e^x) + C\end{aligned}$$

where $C \in \mathbb{R}$.

(b) This integral is solved by integration by parts:

$$\int f_2(x)\,\mathrm{d}x = \int xe^{x+2}\,\mathrm{d}x$$

and, using the integration by parts method with the parts:

$$\left[\begin{array}{ccc} u = x & \Rightarrow & du = \mathrm{d}x \\ dv = e^{x+2}\,\mathrm{d}x & \Rightarrow & v = e^{x+2} \end{array}\right]$$

we have:

$$\int xe^{x+2}\,\mathrm{d}x = xe^{x+2} - \int e^{x+2}\,\mathrm{d}x$$

$$= xe^{x+2} - e^{x+2} + C = (x-1)e^{x+2} + C$$

where $C \in \mathbb{R}$.

Therefore, $\varphi_2(x) = (x-1)e^{x+2}$ *is a primitive.*

If we check, we have:

$$\varphi_2'(x) = 1 \cdot e^{x+2} + (x-1)e^{x+2} = e^{x+2}(1+x-1) = e^{x+2}.$$

(c) *This integral is immediate if written as:*

$$\frac{1}{\tan(x)} = \frac{\cos(x)}{\sin(x)}$$

So:

$$\int f_3(x)\,\mathrm{d}x = \int \frac{1}{\tan(x)}\,\mathrm{d}x = \int \frac{\cos(x)}{\sin(x)}\,\mathrm{d}x = \ln|\sin(x)| + C.$$

(d) *To solve this integral, we will use the help provided in the statement.*

$$\int f_4(x)\,\mathrm{d}x = \int \frac{1}{3+\cos(x)}\,\mathrm{d}x \overset{(*)}{=}$$

$$(*)\begin{bmatrix} x = 2\arctan(t) \\ \mathrm{d}x = \dfrac{2}{1+t^2}\,\mathrm{d}t \end{bmatrix}$$

$$= \int \frac{1}{3+\cos(2\arctan(t))}\frac{2}{1+t^2}\,\mathrm{d}t.$$

Using the expression of the cosine in terms of the cosine and sine of the half-angle squared, we have:

$$\cos(2\arctan(t)) = \cos^2(\arctan(t)) - \sin^2(\arctan(t))$$

and using the other relationships given in the help:

$$\cos(2\arctan(t)) = \left(\frac{1}{\sqrt{1+t^2}}\right)^2 - \left(\frac{t}{\sqrt{1+t^2}}\right)^2 = \frac{1}{1+t^2} - \frac{t^2}{1+t^2} = \frac{1-t^2}{1+t^2}.$$

Therefore:

$$3+\cos(2\arctan(t)) = 3 + \frac{1-t^2}{1+t^2} = \frac{3+3t^2+1-t^2}{1+t^2} = \frac{4+2t^2}{1+t^2}$$

and finally:

$$\int f_4(x)\,\mathrm{d}x = \int \frac{1}{3+\cos(2\arctan(t))}\frac{2}{1+t^2}\,\mathrm{d}t = \int \frac{1+t^2}{2(2+t^2)}\frac{2}{1+t^2}\,\mathrm{d}t$$

$$= \int \frac{1}{2+t^2}\,\mathrm{d}t = \frac{1}{\sqrt{2}}\int \frac{1/\sqrt{2}}{1+\left(\dfrac{t}{\sqrt{2}}\right)^2}\,\mathrm{d}t$$

$$= \frac{1}{\sqrt{2}}\arctan\left(\frac{t}{\sqrt{2}}\right) + C = \frac{1}{\sqrt{2}}\arctan\left(\frac{1}{\sqrt{2}}\tan\left(\frac{x}{2}\right)\right) + C$$

where $C \in \mathbb{R}$.

Problem 5.4. Calculate the following integrals, applying in each case the most appropriate method.

(a) $I_1 = \displaystyle\int \tan^2(x)\,\mathrm{d}x$

(b) $I_2 = \displaystyle\int \ln^2(x)\,\mathrm{d}x$

(c) $I_3 = \displaystyle\int \frac{x}{\sqrt{x^2-2}}\,\mathrm{d}x$

(d) $I_4 = \displaystyle\int \frac{1}{1+3\cos^2(x)}\,\mathrm{d}x$

(e) $I_5 = \displaystyle\int_0^1 4e^{-2x}\,\mathrm{d}x$

Help:

(a) consider the change $u = \tan(x)$ and note that:

$$\frac{u^2}{u^2+1} = 1 - \frac{1}{u^2+1};$$

(b) apply integration by parts twice;

(c) almost immediate;

(d) consider the change $t = \tan(x)$ and note that:

$$\cos(\arctan(t)) = 1/\sqrt{t^2+1}$$

Solution to the problem 5.4. *(a) The task is to calculate the integral:*

$$I_1 = \int \tan^2(x)\,\mathrm{d}x$$

This integral can be calculated very easily by considering that if we consider the function $\varphi(x) = \tan(x)$*, its derivative is:*

$$\varphi'(x) = \tan^2(x) + 1$$

Therefore, if we consider the function $\varphi(x) = \tan(x) - x$ *and derive it we have:*

$$\varphi'(x) = \tan^2(x) + 1 - 1 = \tan^2(x)$$

Therefore, a primitive of $\tan^2(x)$ *is:*

$$\tan(x) - x$$

Let's now calculate the integral using the variable change proposed in the statement.

$$I_1 = \int \tan^2(x)\,\mathrm{d}x \overset{(*)}{=}$$

$$(*)\left[\begin{array}{ll} u = \tan(x) & \\ du = (1+\tan^2(x))\,\mathrm{d}x & \Longrightarrow \quad \mathrm{d}x = \frac{1}{1+\tan^2 x}du = \frac{1}{1+u^2}du \end{array}\right]$$

$$= \int u^2 \cdot \frac{1}{1+u^2}\,du = \int \frac{u^2}{1+u^2}\,du = \int \frac{u^2+1-1}{1+u^2}\,du$$

$$= \int \left[1 - \frac{1}{1+u^2}\right] du = u - \arctan(u) + C$$

$$= \tan(x) - \arctan(\tan(x)) + C = \tan(x) - x + C, \quad C \in \mathbb{R}.$$

(b) *The task is to calculate the integral:*

$$I_2 = \int \ln^2(x)\,\mathrm{d}x$$

This integral can be solved by applying integration by parts twice.

$$I_2 = \int \ln^2(x)\,\mathrm{d}x \overset{(*_1)}{=}$$

$$(*_1)\left[\begin{array}{l} u(x) = \ln^2(x) \Longrightarrow u'(x) = 2\ln(x)\frac{1}{x} \\ v'(x) = 1 \Longrightarrow v(x) = x \end{array}\right]$$

$$= x\ln^2(x) - \int x2\ln(x)\frac{1}{x}\,\mathrm{d}x = x\ln^2(x) - 2\int \ln(x)\,\mathrm{d}x \overset{(*_2)}{=}$$

$$(*_2)\left[\begin{array}{l} u(x) = \ln(x) \Longrightarrow u'(x) = \frac{1}{x} \\ v'(x) = 1 \Longrightarrow v(x) = x \end{array}\right]$$

$$= x\ln^2(x) - 2\left[x\ln(x) - \int x\frac{1}{x}\,\mathrm{d}x\right]$$

$$= x\ln^2(x) - 2x\ln(x) + 2\int 1\,\mathrm{d}x = x\ln^2(x) - 2x\ln(x) + 2x + C, \quad C \in \mathbb{R}.$$

(c) *The task is to calculate the integral:*

$$I_3 = \int \frac{x}{\sqrt{x^2-2}}\,\mathrm{d}x$$

This integral is almost immediate. Note that the derivative of $x^2 - 2$ *is* $2x$. *Therefore, if we take:*

$$f(x) = \frac{1}{\sqrt{x}} = x^{-1/2}$$

and $g(x) = x^2 - 2$, *we have that:*

$$f(g(x)) = \frac{1}{\sqrt{g(x)}} = \frac{1}{\sqrt{x^2-2}}$$

and that:

$$I_3 = \int \frac{x}{\sqrt{x^2-2}}\,\mathrm{d}x = \frac{1}{2}\int \frac{2x}{\sqrt{x^2-2}}\,\mathrm{d}x = \frac{1}{2}\int f(g(x))g'(x)\,\mathrm{d}x = \frac{1}{2}\varphi(g(x))$$

where $\varphi(x)$ *is a primitive of* $f(x)$.
That is, since $\varphi(x) = 2\sqrt{x}$, *we have that:*

$$I_3 = \frac{1}{2}\cdot 2\sqrt{x^2-2} = \sqrt{x^2-2}.$$

Like all almost immediate integrals, it can also be solved by variable change, making the change $u = g(x) = x^2 - 2$.

$$I_3 = \int \frac{x}{\sqrt{x^2-2}}\,\mathrm{d}x \stackrel{(*)}{=}$$

$$(*)\left[\begin{array}{l} u = x^2 - 2 \\ du = 2x\,\mathrm{d}x \Longrightarrow x\,\mathrm{d}x = \frac{1}{2}du \end{array}\right]$$

$$= \int \frac{1}{2}\frac{1}{\sqrt{u}}\,du = \sqrt{u} = \sqrt{x^2-2} + C, \quad C \in \mathbb{R}.$$

(d) *The task is to calculate the integral:*

$$I_4 = \int \frac{1}{1+3\cos^2 x}\,\mathrm{d}x$$

In the help, it is recommended to use the change $t = \tan(x)$.

$$I_4 = \int \frac{1}{1+3\cos^2 x}\,\mathrm{d}x \stackrel{(*)}{=}$$

$$(*)\left[\begin{array}{l} t = \tan(x) \Longrightarrow x = \arctan(t) \\ dt = (1+\tan^2(x))\,\mathrm{d}x \Longrightarrow \mathrm{d}x = \frac{1}{1+\tan^2(x)}\,\mathrm{d}t = \frac{1}{1+t^2}\,\mathrm{d}t \end{array}\right]$$

$$= \int \frac{1}{1+3\cos^2(\arctan(t))}\cdot\frac{1}{1+t^2}\,\mathrm{d}t.$$

Also, it says that:

$$\cos(\arctan(t)) = 1/\sqrt{t^2+1} \Longrightarrow$$

$$\Longrightarrow \cos^2(\arctan(t)) = \left(1/\sqrt{t^2+1}\right)^2 = 1/(t^2+1)$$

That is:

$$\begin{aligned} I_4 &= \int \frac{1}{1+3\cos^2(\arctan(t))}\frac{1}{1+t^2}\,\mathrm{d}t \\ &= \int \frac{1}{1+\frac{3}{t^2+1}}\cdot\frac{1}{1+t^2}\,\mathrm{d}t = \int \frac{1}{1+t^2+3}dt = \int \frac{1}{t^2+4}\,\mathrm{d}t \\ &= \frac{1}{4}\int \frac{1}{(t/2)^2+1}\,\mathrm{d}t = \frac{1}{2}\int \frac{1/2}{(t/2)^2+1}\,\mathrm{d}t \\ &= \frac{1}{2}\arctan\left(\frac{t}{2}\right) = \frac{1}{2}\arctan\left(\frac{1}{2}\tan(x)\right)+C, \quad C\in\mathbb{R}. \end{aligned}$$

(e) *The task is to calculate the integral:*

$$I_5 = \int_0^1 4e^{-2x}\,\mathrm{d}x$$

This integral is immediate:

$$I_5 = \int_0^1 4e^{-2x}\,\mathrm{d}x = \left[-2e^{-2x}\right]_0^1 = -2e^{-2}+2e^0 = 2-2e^{-2}.$$

Problem 5.5. Calculate the following integrals, applying the indicated method in each case. For the integrals where no method is indicated, apply the most appropriate method.

(a) $\displaystyle\int \frac{1}{(1+x^2)^2}\,\mathrm{d}x$, applying the variable change $x=\tan(t)$

(b) $\displaystyle\int \sqrt{a^2+b^2x^2}\,\mathrm{d}x$, applying the variable changes $bx = a\tan(t)$ and $s=\frac{1}{\cos(t)}+\tan(t)$

(c) $\displaystyle\int \frac{\sqrt{x}+\sqrt[3]{x^2}}{\sqrt{x}}\,\mathrm{d}x$, applying the variable change $x=t^6$

(d) $\displaystyle\int \arctan(x)\,\mathrm{d}x$, by the method of integration by parts

(e) $\displaystyle\int \frac{x^3+2x+1}{x^2-5x+6}\,\mathrm{d}x$

(f) $\displaystyle\int \frac{1}{(x-1)^2(x+1)}\,\mathrm{d}x$

(g) $\displaystyle\int \frac{x}{(x^2+x+1)(x+1)}\,\mathrm{d}x$

(h) $\displaystyle\int \frac{1}{3+\cos(x)}\,\mathrm{d}x$, applying the variable change $x = 2\arctan(t)$

(i) $\displaystyle\int \frac{\sqrt{x^2+4}}{x}\,\mathrm{d}x$, applying one of the following variable changes: $x = 2\tan(t)$ and $s = \frac{1}{\sin(t)} + \frac{1}{\cos(t)}$, $x^2+4 = t^2$ or $t = \sqrt{x^2+4} - x$

Solution to the problem 5.5. *(a) Initially, we must replace x and $\mathrm{d}x$ with the corresponding expressions:*

$$\int \frac{1}{(1+x^2)^2}\,\mathrm{d}x \overset{(*)}{=}$$

$$(*)\left[\begin{array}{rcl} x & = & \tan(t) \\ \mathrm{d}x & = & \dfrac{1}{\cos^2(t)}\,\mathrm{d}t \end{array}\right]$$

$$= \int \frac{1}{(1+\tan^2(t))^2} \cdot \frac{1}{\cos^2(t)}\,\mathrm{d}t$$

$$= \int \frac{1}{\left(\dfrac{1}{\cos^2(t)}\right)^2} \cdot \frac{1}{\cos^2(t)}\,\mathrm{d}t$$

$$= \int \frac{1}{\frac{1}{\cos^2(t)}}\,\mathrm{d}t$$

$$= \int \cos^2(t)\,\mathrm{d}t$$

$$= \int \left(\frac{1}{2} + \frac{1}{2}\cos(2t)\right)\,\mathrm{d}t$$

$$= \frac{1}{2}t + \frac{1}{4}\sin(2t) + C, \quad C \in \mathbb{R}$$

$$= \frac{1}{2}\arctan(x) + \frac{1}{4}\sin(2\arctan(x)) + C, \quad C \in \mathbb{R}.$$

To calculate the integral, we used the following trigonometric identities:

$$\frac{1}{\cos^2(x)} = 1 + \tan^2(x)$$

$$\cos^2(x) = \frac{1}{2} + \frac{1}{2}\cos(2x).$$

To undo the change, we isolated t from the proposed change:

$$x = \tan(t) \quad \Rightarrow \quad t = \arctan(x).$$

We can develop the result by taking into account the following trigonometric identities:

$$\sin(2x) = 2\sin(x)\cos(x)$$

$$\sin(\arctan(x)) = \frac{x}{\sqrt{1+x^2}}$$

$$\cos(\arctan(x)) = \frac{1}{\sqrt{1+x^2}}.$$

So:

$$\frac{1}{2}\arctan(x)+\frac{1}{4}\sin(2\arctan(x)) + C =$$

$$= \frac{1}{2}\arctan(x) + \frac{1}{4}2\sin(\arctan(x))\cos(\arctan(x)) + C$$

$$= \frac{1}{2}\arctan(x) + \frac{1}{2}\frac{x}{\sqrt{1+x^2}}\frac{1}{\sqrt{1+x^2}} + C$$

$$= \frac{1}{2}\arctan(x) + \frac{1}{2}\frac{x}{1+x^2} + C, \quad C \in \mathbb{R}.$$

(b) *We apply the proposed change:*

$$\int \sqrt{a^2+b^2x^2}\,\mathrm{d}x = \int \sqrt{a^2+(bx)^2}\,\mathrm{d}x \overset{(*)}{=}$$

$$(*)\begin{bmatrix} bx & = & a\tan(t) \\ \mathrm{d}x & = & \frac{a}{b}\frac{1}{\cos^2(t)}\,\mathrm{d}t \end{bmatrix}$$

$$= \int \sqrt{a^2+a^2\tan^2(t)} \cdot \frac{a}{b} \cdot \frac{1}{\cos^2(t)}\,\mathrm{d}t$$

$$= \int a\sqrt{1+\tan^2(t)} \cdot \frac{a}{b} \cdot \frac{1}{\cos^2(t)}\,\mathrm{d}t$$

$$= \frac{a^2}{b}\int \sqrt{\frac{1}{\cos^2(t)}} \cdot \frac{1}{\cos^2(t)}\,\mathrm{d}t$$

$$= \frac{a^2}{b}\int \frac{1}{\cos^3(t)}\,\mathrm{d}t.$$

For the previous calculations, we used the following trigonometric identity:

$$\frac{1}{\cos^2(x)} = 1+\tan^2(x).$$

The resulting integral will be solved by the method of integration by parts:

$$\frac{a^2}{b}\int \frac{1}{\cos^3(t)}\,\mathrm{d}t = \frac{a^2}{b}\int \underbrace{\frac{1}{\cos(t)}}_{u(t)}\underbrace{\frac{1}{\cos^2(t)}}_{v'(t)}\,\mathrm{d}t \overset{(*)}{=}$$

$$(*)\left[\begin{array}{lll} u(t)=\frac{1}{\cos(t)} & \Rightarrow & u'(t)=\frac{\sin(t)}{\cos^2(t)}\,\mathrm{d}t \\ v'(t)=\frac{1}{\cos^2(t)}\,\mathrm{d}t & \Rightarrow & v(t)=\tan(t) \end{array}\right]$$

$$= \frac{a^2}{b}\left(\frac{\tan(t)}{\cos(t)} - \int \tan(t)\cdot\frac{\sin(t)}{\cos^2(t)}\,\mathrm{d}t\right)$$

$$= \frac{a^2}{b}\frac{\tan(t)}{\cos(t)} - \frac{a^2}{b}\int \tan^2(t)\cdot\frac{1}{\cos(t)}\,\mathrm{d}t$$

$$= \frac{a^2}{b}\frac{\tan(t)}{\cos(t)} - \frac{a^2}{b}\int \left(\frac{1}{\cos^2(t)}-1\right)\cdot\frac{1}{\cos(t)}\,\mathrm{d}t$$

$$= \frac{a^2}{b}\frac{\tan(t)}{\cos(t)} - \frac{a^2}{b}\int \frac{1}{\cos^3(t)}\,\mathrm{d}t + \frac{a^2}{b}\int\frac{1}{\cos(t)}\,\mathrm{d}t.$$

The integral $\int \frac{1}{\cos(t)}\,\mathrm{d}t$ *will be solved by making a variable change:*

$$\int \frac{1}{\cos(t)}\,\mathrm{d}t = \int \frac{1}{\cos(t)}\cdot\frac{\frac{1}{\cos(t)}+\tan(t)}{\frac{1}{\cos(t)}+\tan(t)}\,\mathrm{d}t$$

$$= \int \frac{\frac{1}{\cos^2(t)}+\frac{\tan(t)}{\cos(t)}}{\frac{1}{\cos(t)}+\tan(t)}\,\mathrm{d}t$$

$$= \int \frac{1}{\underbrace{\frac{1}{\cos(t)}+\tan(t)}_{s}}\cdot\underbrace{\left(\frac{1}{\cos^2(t)}+\frac{\tan(t)}{\cos(t)}\right)\mathrm{d}t}_{\mathrm{d}s} \overset{(*)}{=}$$

$$(*)\left[\begin{array}{lll} s & = & \frac{1}{\cos(t)}+\tan(t) \\ \mathrm{d}s & = & \left(\frac{1}{\cos^2(t)}+\frac{\tan(t)}{\cos(t)}\right)\mathrm{d}t \end{array}\right]$$

$$= \int \frac{1}{s}\mathrm{d}s$$

$$= \ln|s| + C, \quad C\in\mathbb{R}$$

$$= \ln\left|\frac{1}{\cos(t)}+\tan(t)\right| + C, \quad C\in\mathbb{R}.$$

Replacing this result in the previous expression:

$$\frac{a^2}{b}\int\frac{1}{\cos^3(t)}\,\mathrm{d}t = \frac{a^2}{b}\frac{\tan(t)}{\cos(t)} - \frac{a^2}{b}\int\frac{1}{\cos^3(t)}\,\mathrm{d}t + \frac{a^2}{b}\ln\left|\frac{1}{\cos(t)}+\tan(t)\right|.$$

Operating, we obtain:

$$2\frac{a^2}{b}\int \frac{1}{\cos^3(t)}\,\mathrm{dt} = \frac{a^2}{b}\frac{\tan(t)}{\cos(t)} + \frac{a^2}{b}\ln\left|\frac{1}{\cos(t)} + \tan(t)\right| + C, \quad C \in \mathbb{R}$$

$$\frac{a^2}{b}\int \frac{1}{\cos^3(t)}\,\mathrm{dt} = \frac{a^2}{2b}\frac{\tan(t)}{\cos(t)} + \frac{a^2}{2b}\ln\left|\frac{1}{\cos(t)} + \tan(t)\right| + C, \quad C \in \mathbb{R}$$

Therefore:

$$\int \sqrt{a^2+b^2x^2}\,\mathrm{d}x = \frac{a^2}{b}\int \frac{1}{\cos^3(t)}\,\mathrm{dt}$$

$$= \frac{a^2}{2b}\frac{\tan(t)}{\cos(t)} + \frac{a^2}{2b}\ln\left|\frac{1}{\cos(t)} + \tan(t)\right| + C$$

$$= \frac{a^2}{2b}\frac{\tan\left(\arctan\left(\frac{bx}{a}\right)\right)}{\cos\left(\arctan\left(\frac{bx}{a}\right)\right)} + \frac{a^2}{2b}\ln\left|\frac{1}{\cos\left(\arctan\left(\frac{bx}{a}\right)\right)} + \tan\left(\arctan\left(\frac{bx}{a}\right)\right)\right| + C$$

$$= \frac{a^2}{2b}\frac{\frac{bx}{a}}{\cos\left(\arctan\left(\frac{bx}{a}\right)\right)} + \frac{a^2}{2b}\ln\left|\frac{1}{\cos\left(\arctan\left(\frac{bx}{a}\right)\right)} + \frac{bx}{a}\right| + C$$

where $C \in \mathbb{R}$. *We develop the result using the following trigonometric identity:*

$$\cos(\arctan(x)) = \frac{1}{\sqrt{1+x^2}}.$$

So:

$$\int \sqrt{a^2+b^2x^2}\,\mathrm{d}x = \frac{a^2}{2b}\frac{\frac{bx}{a}}{\frac{1}{\sqrt{1+\left(\frac{bx}{a}\right)^2}}} + \frac{a^2}{2b}\ln\left|\frac{1}{\frac{1}{\sqrt{1+\left(\frac{bx}{a}\right)^2}}} + \frac{bx}{a}\right| + C$$

$$= \frac{1}{2}x\sqrt{a^2+b^2x^2} + \frac{a^2}{2b}\ln\left|\sqrt{1+\left(\frac{bx}{a}\right)^2} + \frac{bx}{a}\right| + C$$

where $C \in \mathbb{R}$. *To further simplify the result, we can add* $\frac{a^2}{2b}\ln|a|$ *to the expression:*

$$\frac{1}{2}x\sqrt{a^2+b^2x^2} + \frac{a^2}{2b}\ln\left|\sqrt{1+\left(\frac{bx}{a}\right)^2} + \frac{bx}{a}\right| + C =$$

$$= \frac{1}{2}x\sqrt{a^2+b^2x^2} + \frac{a^2}{2b}\ln\left|\sqrt{1+\left(\frac{bx}{a}\right)^2} + \frac{bx}{a}\right| + C + \frac{a^2}{2b}\ln|a|$$

$$= \frac{1}{2}x\sqrt{a^2+b^2x^2} + \frac{a^2}{2b}\ln\left|\sqrt{a^2+b^2x^2} + bx\right| + C, \quad C \in \mathbb{R}.$$

Therefore:

$$\int \sqrt{a^2+b^2x^2}\,\mathrm{d}x = \frac{1}{2}x\sqrt{a^2+b^2x^2} + \frac{a^2}{2b}\ln\left|\sqrt{a^2+b^2x^2}+bx\right| + C, \quad C \in \mathbb{R}.$$

(c) *We replace x and $\mathrm{d}x$ with the corresponding expressions:*

$$\int \frac{\sqrt{x}+\sqrt[3]{x^2}}{\sqrt{x}}\,\mathrm{d}x = \int \frac{x^{\frac{1}{2}}+x^{\frac{2}{3}}}{x^{\frac{1}{2}}}\,\mathrm{d}x \stackrel{(*)}{=}$$

$$(*)\left[\begin{array}{rcl} x & = & t^6 \\ \mathrm{d}x & = & 6t^5\,\mathrm{d}t \end{array}\right]$$

$$\begin{aligned} &= \int \frac{(t^6)^{\frac{1}{2}}+(t^6)^{\frac{2}{3}}}{(t^6)^{\frac{1}{2}}}\cdot 6t^5\,\mathrm{d}t \\ &= \int \frac{t^3+t^4}{t^3}\cdot 6t^5\,\mathrm{d}t \\ &= \int \left(6t^5+6t^6\right)\,\mathrm{d}t \\ &= t^6+\frac{6}{7}t^7+C, \quad C\in\mathbb{R} \\ &= (\sqrt[6]{x})^6+\frac{6}{7}(\sqrt[6]{x})^7+C, \quad C\in\mathbb{R} \\ &= x+\frac{6}{7}\sqrt[6]{x^7}+C, \quad C\in\mathbb{R}. \end{aligned}$$

To undo the change, we isolated t from the proposed change:

$$x=t^6 \quad\Rightarrow\quad t=\sqrt[6]{x}.$$

(d) *We solve the integral by the method of integration by parts:*

$$\int \arctan(x)\,\mathrm{d}x = \underbrace{\arctan(x)}_{u(x)}\cdot\underbrace{1}_{v'(x)} \stackrel{(*)}{=}$$

$$(*)\left[\begin{array}{lll} u(x)=\arctan(x) & \Rightarrow & u'(x)=\frac{1}{x^2+1}\,\mathrm{d}x \\ v'(x)=1\,\mathrm{d}x & \Rightarrow & v(x)=x \end{array}\right]$$

$$\begin{aligned} &= \arctan(x)\cdot x - \int \frac{x}{x^2+1}\,\mathrm{d}x \\ &= \arctan(x)\cdot x - \int \frac{1}{2}\frac{\overbrace{2x}^{\text{denominator derivative}}}{x^2+1}\,\mathrm{d}x \\ &= \arctan(x)\cdot x - \frac{1}{2}\ln(x^2+1)+C, \quad C\in\mathbb{R}. \end{aligned}$$

(e) *Initially, we reduce the rational function to a fraction where the degree of the numerator is less than the degree of the denominator:*

$$\frac{x^3+2x+1}{x^2-5x+6} = \frac{(x^2-5x+6)(x+5)+21x-29}{x^2-5x+6} = x+5+\frac{21x-29}{x^2-5x+6}.$$

Therefore:

$$\int\left(x+5+\frac{21x-29}{x^2-5x+6}\right)\mathrm{d}x = \frac{1}{2}x^2+5x+\int\frac{21x-29}{x^2-5x+6}\,\mathrm{d}x.$$

Next, we decompose the quotient of polynomials into simple fractions:

$$\frac{21x-29}{x^2-5x+6} = \frac{A}{x-3}+\frac{B}{x-2} = \frac{A(x-2)+B(x-3)}{(x-3)(x-2)} = \frac{(A+B)x-2A-3B}{(x-3)(x-2)}.$$

It must be satisfied:

$$\begin{aligned} A+B &= 21 \\ -2A-3B &= -29 \end{aligned}$$

Solving the system, we get $A = 34$ and $B = -13$. Therefore:

$$\begin{aligned} \int\frac{x^3+2x+1}{x^2-5x+6}\,\mathrm{d}x &= \frac{1}{2}x^2+5x+\int\frac{21x-29}{x^2-5x+6}\,\mathrm{d}x = \\ &= \frac{1}{2}x^2+5x+\int\left(\frac{34}{x-3}-\frac{13}{x-2}\right)\mathrm{d}x \\ &= \frac{1}{2}x^2+5x+34\ln|x-3|-13\ln|x-2|+C, \quad C\in\mathbb{R}. \end{aligned}$$

(f) *Initially, we decompose the numerator of the rational function into simple fractions:*

$$\begin{aligned} \frac{1}{(x-1)^2(x+1)} &= \frac{A}{x-1}+\frac{B}{(x-1)^2}+\frac{C}{x+1} \\ &= \frac{A(x-1)(x+1)+B(x+1)+C(x-1)^2}{(x-1)^2(x+1)} \\ &= \frac{(A+C)x^2+(B-2C)x-A+B+C}{(x-1)^2(x+1)}. \end{aligned}$$

It must be satisfied:

$$\begin{aligned} A+C &= 0 \\ B-2C &= 0 \\ -A+B+C &= 1 \end{aligned}$$

Solving the system, we get $A = -1/4$, $B = 1/2$ and $C = 1/4$. Therefore:

$$\begin{aligned} \int\frac{1}{(x-1)^2(x+1)}\,\mathrm{d}x &= \int\left(-\frac{1/4}{x-1}+\frac{1/2}{(x-1)^2}+\frac{1/4}{x+1}\right)\mathrm{d}x \\ &= -\frac{1}{4}\ln|x-1|-\frac{1}{2}\frac{1}{x-1}+\frac{1}{4}\ln|x+1|+C, \quad C\in\mathbb{R}. \end{aligned}$$

(g) *Initially, we decompose the numerator of the rational function into simple fractions:*

$$\frac{x}{(x^2+x+1)(x+1)} = \frac{Ax+B}{x^2+x+1} + \frac{C}{x+1}$$

$$= \frac{Ax(x+1)+B(x+1)+C(x^2+x+1)}{(x^2+x+1)(x+1)}$$

$$= \frac{(A+C)x^2+(A+B+C)x+B+C}{(x^2+x+1)(x+1)}.$$

It must be satisfied:

$$A+C=0$$

$$A+B+C=1$$

$$B+C=0$$

Solving the system, we get $A=1$, $B=1$ *and* $C=-1$. *Therefore:*

$$\int \frac{x}{(x^2+x+1)(x+1)}\,\mathrm{d}x = \int \left(\frac{x+1}{x^2+x+1} - \frac{1}{x+1}\right)\mathrm{d}x$$

$$= -\ln|x+1| + \int \frac{\frac{1}{2}(2x+1)+\frac{1}{2}}{x^2+x+1}\,\mathrm{d}x$$

$$= -\ln|x+1| + \frac{1}{2}\int \frac{2x+1}{x^2+x+1}\,\mathrm{d}x + \int \frac{\frac{1}{2}}{x^2+x+1}\,\mathrm{d}x$$

$$= -\ln|x+1| + \frac{1}{2}\ln|x^2+x+1| + \int \frac{\frac{1}{2}}{(x+\frac{1}{2})^2+\frac{3}{4}}\,\mathrm{d}x$$

$$= -\ln|x+1| + \frac{1}{2}\ln|x^2+x+1| + \frac{\frac{1}{2}}{\sqrt{\frac{3}{4}}}\arctan\left(\frac{x+\frac{1}{2}}{\sqrt{\frac{3}{4}}}\right) + C$$

$$= -\ln|x+1| + \frac{1}{2}\ln|x^2+x+1| + \frac{1}{\sqrt{3}}\arctan\left(\frac{2x+1}{\sqrt{3}}\right) + C$$

where $C \in \mathbb{R}$.

(h) Initially, we rewrite the integral based on the following trigonometric identities:

$$\cos^2(x) = \frac{1}{2} + \frac{1}{2}\cos(2x) \quad \Longrightarrow \quad \cos(2x) = 2\cos^2(x) - 1$$

$$\Longrightarrow \quad \cos(x) = 2\cos^2\left(\frac{x}{2}\right) - 1$$

$$\tan^2(x) + 1 = \frac{1}{\cos^2(x)} \quad \Longrightarrow \quad \cos^2(x) = \frac{1}{\tan^2(x) + 1}$$

Next, we replace x and $\mathrm{d}x$ with the corresponding expressions:

$$\begin{aligned}
\int \frac{1}{3+\cos(x)}\,\mathrm{d}x &= \int \frac{1}{3 + 2\cos^2\left(\frac{x}{2}\right) - 1}\,\mathrm{d}x \\
&= \int \frac{1}{2 + \frac{2}{\tan^2\left(\frac{x}{2}\right)+1}}\,\mathrm{d}x \stackrel{(*)}{=}
\end{aligned}$$

$$(*)\begin{bmatrix} x & = & 2\arctan(t) \\ \mathrm{d}x & = & \dfrac{2}{1+t^2}\,\mathrm{d}t \end{bmatrix}$$

$$\begin{aligned}
&= \int \frac{1}{2 + \frac{2}{\tan^2\left(\frac{1}{2}\cdot 2\arctan(t)\right)+1}} \cdot \frac{2}{1+t^2}\,\mathrm{d}t \\
&= \int \frac{1}{2 + \frac{2}{t^2+1}} \cdot \frac{2}{t^2+1}\,\mathrm{d}t \\
&= \int \frac{2}{2t^2+2+2}\,\mathrm{d}t \\
&= \int \frac{1}{t^2+2}\,\mathrm{d}t \\
&= \frac{1}{\sqrt{2}}\arctan\left(\frac{t}{\sqrt{2}}\right) + C, \quad C \in \mathbb{R} \\
&= \frac{1}{\sqrt{2}}\arctan\left(\frac{1}{\sqrt{2}}\tan\left(\frac{x}{2}\right)\right) + C, \quad C \in \mathbb{R}.
\end{aligned}$$

To undo the change, we isolated t from the proposed change:

$$x = 2\arctan(t) \quad \Rightarrow \quad t = \tan\left(\frac{x}{2}\right).$$

(i) We solve the integral by applying the variable changes:

$$x = 2\tan(t), \quad s = \frac{1}{\sin(t)} + \frac{1}{\tan(t)},$$

obtaining:

$$\int \frac{\sqrt{x^2+4}}{x}\,\mathrm{d}x \overset{(*)}{=}$$

$$(*)\left[\begin{array}{ccc} x & = & 2\tan(t) \\ \mathrm{d}x & = & \dfrac{2}{\cos^2(t)}\,\mathrm{d}t \end{array}\right]$$

$$= \int \frac{\sqrt{4\tan^2(t)+4}}{2\tan(t)} \cdot \frac{2}{\cos^2(t)}\,\mathrm{d}t$$

$$= \int \frac{2\sqrt{\tan^2(t)+1}}{2\tan(t)} \cdot 2(\tan^2(t)+1)\,\mathrm{d}t$$

$$= \int \frac{\frac{1}{\cos(t)}}{\tan(t)} \cdot 2(\tan^2(t)+1)\,\mathrm{d}t$$

$$= 2\int \frac{1}{\sin(t)} \cdot \frac{1}{\cos^2(t)}\,\mathrm{d}t.$$

For the previous calculations, we used the following trigonometric identity:

$$\frac{1}{\cos^2(x)} = 1 + \tan^2(x).$$

Next, we solve the resulting integral by the method of integration by parts:

$$2\int \underbrace{\frac{1}{\sin(t)}}_{u(t)} \cdot \underbrace{\frac{1}{\cos^2(t)}\,\mathrm{d}t}_{v'(t)} \overset{(*)}{=}$$

$$(*)\left[\begin{array}{lll} u(t) = \frac{1}{\sin(t)} & \Rightarrow & u'(t) = -\frac{\cos(t)}{\sin^2(t)}\,\mathrm{d}t \\ v'(t) = \frac{1}{\cos^2(t)}\,\mathrm{d}t & \Rightarrow & v(t) = \tan(t) \end{array}\right]$$

$$= 2\left(\frac{1}{\sin(t)} \cdot \tan(t) - \int \tan(t) \cdot \left(-\frac{\cos(t)}{\sin^2(t)}\right)\,\mathrm{d}t\right)$$

$$= \frac{2}{\cos(t)} + 2\int \frac{1}{\sin(t)}\,\mathrm{d}t.$$

The integral $\int \frac{1}{\sin(t)}\,\mathrm{d}t$ *we solve by making the change of variable:*

$$s = \frac{1}{\sin(t)} + \frac{1}{\tan(t)};$$

replacing:

$$\int \frac{1}{\sin(t)}\,\mathrm{d}t = \int \frac{1}{\sin(t)} \frac{\frac{1}{\sin(t)} + \frac{1}{\tan(t)}}{\frac{1}{\sin(t)} + \frac{1}{\tan(t)}}\,\mathrm{d}t =$$

$$= -\int \underbrace{\frac{1}{\frac{1}{\sin(t)} + \frac{1}{\tan(t)}}}_{s} \cdot \underbrace{\left(-\frac{1}{\sin(t)\tan(t)} - \frac{1}{\sin^2(t)}\right)\mathrm{d}t}_{\mathrm{d}s} \overset{(*)}{=}$$

$$(*)\left[\begin{array}{rcl} s & = & \frac{1}{\sin(t)} + \frac{1}{\tan(t)} \\ \mathrm{d}s & = & \left(-\frac{1}{\sin(t)\tan(t)} - \frac{1}{\sin^2(t)}\right)\mathrm{d}t \end{array}\right]$$

$$= -\int \frac{1}{s}\mathrm{d}s$$

$$= -\ln|s| + C, \quad C \in \mathbb{R}$$

$$= -\ln\left|\frac{1}{\sin(t)} + \frac{1}{\tan(t)}\right| + C, \quad C \in \mathbb{R}.$$

Therefore:

$$\int \frac{\sqrt{x^2+4}}{x}\,\mathrm{d}x = \frac{2}{\cos(t)} - 2\ln\left|\frac{1}{\sin(t)} + \frac{1}{\tan(t)}\right| + C, \quad C \in \mathbb{R}$$

$$= \frac{2}{\frac{1}{\sqrt{1+\left(\frac{x}{2}\right)^2}}} - 2\ln\left|\frac{1}{\frac{\frac{x}{2}}{\sqrt{1+\left(\frac{x}{2}\right)^2}}} + \frac{1}{\frac{x}{2}}\right| + C, \quad C \in \mathbb{R}$$

$$= \frac{2}{\frac{1}{\sqrt{1+\left(\frac{x}{2}\right)^2}}} - 2\ln\left|\frac{1}{\sin(\arctan(t))} + \frac{1}{\tan(\arctan(t))}\right| + C, \quad C \in \mathbb{R}$$

$$= \frac{2}{\frac{1}{\sqrt{1+\left(\frac{x}{2}\right)^2}}} - 2\ln\left|\frac{\sqrt{x^2+4}}{x} + \frac{2}{x}\right| + C, \quad C \in \mathbb{R}.$$

To undo the change, we isolated t from the proposed change:

$$x = 2\tan(t) \quad \Rightarrow \quad t = \arctan\left(\frac{x}{2}\right).$$

To simplify the result, we used the following trigonometric identities:

$$\sin(\arctan(x)) = \frac{x}{\sqrt{1+x^2}}$$
$$\cos(\arctan(x)) = \frac{1}{\sqrt{1+x^2}}.$$

We can also solve the integral by applying the variable change $x^2+4=t^2$*:*

$$\int \frac{\sqrt{x^2+4}}{x}\,\mathrm{d}x \stackrel{(*)}{=}$$

$$(*)\left[\begin{array}{rcl} x^2+4 &=& t^2 \qquad \Rightarrow \quad x=\sqrt{t^2-4} \\ \mathrm{d}x &=& \dfrac{t}{\sqrt{t^2-4}}\,\mathrm{d}t \end{array}\right]$$

$$= \int \frac{\sqrt{(t^2)}}{\sqrt{t^2-4}}\frac{t}{\sqrt{t^2-4}}\,\mathrm{d}t = \int \frac{t^2}{t^2-4}\,\mathrm{d}t$$
$$= \int \frac{t^2-4+4}{t^2-4}\,\mathrm{d}t = \int \left(1+\frac{4}{t^2-4}\right)\mathrm{d}t$$
$$= t + \int \frac{4}{t^2-4}\,\mathrm{d}t.$$

We decompose the denominator of the rational function into simple fractions:

$$\frac{4}{t^2-4} = \frac{4}{(t-2)(t+2)} = \frac{A}{t-2}+\frac{B}{t+2}$$
$$= \frac{A(t+2)+B(t-2)}{(t-2)(t+2)} = \frac{(A+B)t+2A-2B}{(t-2)(t+2)}.$$

It must be satisfied:

$$A+B=0$$
$$2A-2B=4$$

Solving the system, we get $A=2$ *and* $B=-2$*. Therefore:*

$$t+\int \frac{4}{t^2-4}\,\mathrm{d}t = t+\int\left(\frac{1}{t-2}-\frac{1}{t+2}\right)\mathrm{d}t$$
$$= t+\ln|t-2|-\ln|t+2|+C, \quad C\in\mathbb{R}$$
$$= \sqrt{x^2+4}+\ln\left|\sqrt{x^2+4}-2\right|-\ln\left|\sqrt{x^2+4}+2\right|+C, \quad C\in\mathbb{R}.$$

The result is identical to the one obtained with the previous change of variable, since:

$$\begin{aligned}
&\sqrt{x^2+4}+\ln\left|\sqrt{x^2+4}-2\right|-\ln\left|\sqrt{x^2+4}+2\right|+C, \quad C\in\mathbb{R}\\
&=\sqrt{x^2+4}-\ln\left|\frac{\sqrt{x^2+4}+2}{\sqrt{x^2+4}-2}\right|+C, \quad C\in\mathbb{R}\\
&=\sqrt{x^2+4}-\ln\left|\frac{\left(\sqrt{x^2+4}+2\right)^2}{\left(\sqrt{x^2+4}-2\right)\left(\sqrt{x^2+4}+2\right)}\right|+C, \quad C\in\mathbb{R}\\
&=\sqrt{x^2+4}-\ln\left|\frac{\left(\sqrt{x^2+4}+2\right)^2}{x^2}\right|+C, \quad C\in\mathbb{R}\\
&=\sqrt{x^2+4}-\ln\left|\left(\frac{\sqrt{x^2+4}+2}{x}\right)^2\right|+C, \quad C\in\mathbb{R}\\
&=\sqrt{x^2+4}-2\ln\left|\frac{\sqrt{x^2+4}}{x}+\frac{2}{x}\right|+C, \quad C\in\mathbb{R}.
\end{aligned}$$

The third variable change proposed is $t=\sqrt{x^2+4}-x$*. We solve the integral with this change:*

$$\begin{aligned}
\int\frac{\sqrt{x^2+4}}{x}\,\mathrm{d}x &= \int\left(\frac{\sqrt{x^2+4}-x}{x}+1\right)\mathrm{d}x\\
&= x+\int\frac{\sqrt{x^2+4}-x}{x}\,\frac{\sqrt{x^2+4}}{x-\sqrt{x^2+4}}\,\frac{x-\sqrt{x^2+4}}{\sqrt{x^2+4}}\,\mathrm{d}x\\
&= x+\int\frac{\overbrace{x^2+4-x\sqrt{x^2+4}}^{\frac{t^2}{2}+2}}{\underbrace{x^2-x\sqrt{x^2+4}}_{\frac{t^2}{2}-2}}\,\underbrace{\frac{x-\sqrt{x^2+4}}{\sqrt{x^2+4}}\,\mathrm{d}x}_{\mathrm{d}t} \overset{(*)}{=}
\end{aligned}$$

$$(*)\left[\begin{array}{rcl} t &=& \sqrt{x^2+4}-x \quad\Rightarrow\quad x=\dfrac{4-t^2}{2t}\\ \mathrm{d}t &=& \dfrac{x-\sqrt{x^2+4}}{\sqrt{x^2+4}}\,\mathrm{d}x \end{array}\right]$$

$$\begin{aligned}
&= \frac{4-t^2}{2t}+\int\frac{\frac{t^2}{2}+2}{\frac{t^2}{2}-2}\,\mathrm{d}t\\
&= \frac{4-t^2}{2t}+\int\frac{t^2+4}{t^2-4}\,\mathrm{d}t\\
&= \frac{4-t^2}{2t}+\int\left(1+\frac{8}{t^2-4}\right)\mathrm{d}t\\
&= \frac{4-t^2}{2t}+t+\int\frac{8}{t^2-4}\,\mathrm{d}t.
\end{aligned}$$

We decompose the denominator of the rational function into simple fractions:

$$\frac{8}{t^2-4} = \frac{8}{(t-2)(t+2)} = \frac{A}{t-2} + \frac{B}{t+2}$$

$$= \frac{A(t+2)+B(t-2)}{(t-2)(t+2)} = \frac{(A+B)t+2A-2B}{(t-2)(t+2)}.$$

It must be satisfied:

$$A+B=0$$
$$2A-2B=8$$

Solving the system, we get $A=2$ and $B=-2$. Therefore:

$$\frac{4-t^2}{2t}+t+\int \frac{8}{\sqrt{t^2-4}}\,\mathrm{d}t = \frac{4+t^2}{2t}+2\ln|t-2|-2\ln|t+2|+C$$

$$= \frac{4+t^2}{2t}+2\ln\left|\sqrt{x^2+4}-x-2\right|-2\ln\left|\sqrt{x^2+4}-x+2\right|+C$$

$$= \frac{4+2x^2+4-2x\sqrt{x^2+4}}{2\sqrt{x^2+4}-2x}-2\ln\left|\frac{\sqrt{x^2+4}-x+2}{\sqrt{x^2+4}-x-2}\right|+C$$

$$= \frac{\sqrt{x^2+4}\,(2\sqrt{x^2+4}-2x)}{2\sqrt{x^2+4}-2x}-2\ln\left|\frac{\sqrt{x^2+4}-x+2}{\sqrt{x^2+4}-(x+2)}\,\frac{\sqrt{x^2+4}+(x+2)}{\sqrt{x^2+4}+(x+2)}\right|+C$$

$$= \sqrt{x^2+4}-2\ln\left|\frac{4\sqrt{x^2+4}+8}{-4x}\right|+C = \sqrt{x^2+4}-2\ln\left|\frac{\sqrt{x^2+4}}{x}+\frac{2}{x}\right|+C$$

where $C\in\mathbb{R}$. Again, we obtain the same result.

Problem 5.6. Calculate the following integrals, applying the indicated method in each case. For the integrals where no method is specified, apply the most appropriate method.

(a) $\displaystyle\int \frac{\ln(x)}{\sqrt{x}}\,\mathrm{d}x$, applying the variable substitution $t=\sqrt{x}$ or by the method of integration by parts

(b) $\displaystyle\int \sin(\ln(x))\,\mathrm{d}x$, by the method of integration by parts

(c) $\displaystyle\int \tan(x)\,\mathrm{d}x$

(d) $\displaystyle\int \frac{\cos(x)}{1+\sin^2(x)}\,\mathrm{d}x$

(e) $\displaystyle\int \frac{3x-2}{x^2-4x+5}\,\mathrm{d}x$

(f) $\displaystyle\int e^{\arcsin(x)}\,\mathrm{d}x$, by the method of integration by parts

(g) $\displaystyle\int \ln(1-\sqrt{x})\,\mathrm{d}x$, applying the variable substitution $t=\sqrt{x}$ or by the method of integration by parts

(h) $\displaystyle\int x\cos^2(x)\,\mathrm{d}x$, applying the method of integration by parts

Solution to the problem 5.6. *(a) We solve the integral by applying the proposed substitution:*

$$\begin{aligned}\int \frac{\ln(x)}{\sqrt{x}}\,\mathrm{d}x &= \int 2\ln(x)\cdot\frac{1}{2\sqrt{x}}\,\mathrm{d}x\\ &= \int 2\ln(\underbrace{\sqrt{x}^2}_{t^2})\cdot\underbrace{\frac{1}{2\sqrt{x}}\,\mathrm{d}x}_{\mathrm{d}t} \overset{(*)}{=}\end{aligned}$$

$$(*)\left[\begin{array}{ccc} t & = & \sqrt{x}\\ \mathrm{d}t & = & \dfrac{1}{2\sqrt{x}}\,\mathrm{d}x\end{array}\right]$$

$$\begin{aligned}&= \int 2\ln(t^2)\,\mathrm{d}t\\ &= 4\int \ln(t)\,\mathrm{d}t.\end{aligned}$$

Next, we solve this integral by the method of integration by parts:

$$4\int \ln(t)\,\mathrm{d}t = 4\int \underbrace{\ln(t)}_{u(t)}\,\underbrace{1\,\mathrm{d}t}_{v'(t)} \overset{(*)}{=}$$

$$(*)\left[\begin{array}{lcl} u(t)=\ln(t) & \Rightarrow & u'(t)=\frac{1}{t}\,\mathrm{d}t\\ v'(t)=1\,\mathrm{d}t & \Rightarrow & v(t)=t\end{array}\right]$$

$$\begin{aligned}&= 4\left(\ln(t)\cdot t-\int t\cdot\frac{1}{t}\,\mathrm{d}t\right)\\ &= 4\,(\ln(t)\cdot t-t)+C,\quad C\in\mathbb{R}\\ &= 4\ln(\sqrt{x})\cdot\sqrt{x}-4\sqrt{x}+C,\quad C\in\mathbb{R}\\ &= 2\sqrt{x}\ln(x)-4\sqrt{x}+C,\quad C\in\mathbb{R}.\end{aligned}$$

The integral from the statement can also be solved directly by the method of integration by parts:

$$\int \frac{\ln(x)}{\sqrt{x}}\, dx = \int \underbrace{\ln(x)}_{u(x)} \underbrace{\frac{1}{\sqrt{x}}\, dx}_{v'(x)} \stackrel{(*)}{=}$$

$$(*) \left[\begin{array}{lll} u(x) = \ln(x) & \Rightarrow & u'(x) = \frac{1}{x}\, dx \\ v'(x) = \frac{1}{\sqrt{x}}\, dx & \Rightarrow & v(x) = 2\sqrt{x} \end{array} \right]$$

$$= \ln(x) \cdot 2\sqrt{x} - \int 2\sqrt{x} \cdot \frac{1}{x}\, dx$$

$$= 2\sqrt{x}\ln(x) - 2\int x^{-\frac{1}{2}}\, dx$$

$$= 2\sqrt{x}\ln(x) - 4\sqrt{x} + C, \quad C \in \mathbb{R}.$$

The result obtained is the same by both methods.

(b) We apply the method of integration by parts:

$$\int \sin(\ln(x))\, dx = \int \underbrace{\sin(\ln(x))}_{u(x)} \underbrace{1\, dx}_{v'(x)} \stackrel{(*)}{=}$$

$$(*) \left[\begin{array}{lll} u(x) = \sin(\ln(x)) & \Rightarrow & u'(x) = \frac{\cos(\ln(x))}{x}\, dx \\ v'(x) = 1\, dx & \Rightarrow & v(x) = x \end{array} \right]$$

$$= \sin(\ln(x)) \cdot x - \int x \cdot \frac{\cos(\ln(x))}{x}\, dx$$

$$= \sin(\ln(x)) \cdot x - \int \cos(\ln(x))\, dx.$$

Applying the method of integration by parts again:

$$\sin(\ln(x)) \cdot x - \int \underbrace{\cos(\ln(x))}_{u(x)} \underbrace{1\, dx}_{v'(x)} \stackrel{(*)}{=}$$

$$(*) \left[\begin{array}{lll} u(x) = \cos(\ln(x)) & \Rightarrow & u'(x) = \frac{-\sin(\ln(x))}{x}\, dx \\ v'(x) = 1\, dx & \Rightarrow & v(x) = x \end{array} \right]$$

$$= \sin(\ln(x)) \cdot x - \left(\cos(\ln(x)) \cdot x - \int x \cdot \frac{-\sin(\ln(x))}{x}\, dx \right)$$

$$= \sin(\ln(x)) \cdot x - \cos(\ln(x)) \cdot x - \int \sin(\ln(x))\, dx.$$

Therefore:

$$\int \sin(\ln(x))\,\mathrm{d}x = \sin(\ln(x)) \cdot x - \cos(\ln(x)) \cdot x - \int \sin(\ln(x))\,\mathrm{d}x$$

$$2\int \sin(\ln(x))\,\mathrm{d}x = \sin(\ln(x)) \cdot x - \cos(\ln(x)) \cdot x + C, \quad C \in \mathbb{R}$$

$$\int \sin(\ln(x))\,\mathrm{d}x = \frac{1}{2}\sin(\ln(x)) \cdot x - \frac{1}{2}\cos(\ln(x)) \cdot x + C, \quad C \in \mathbb{R}.$$

(c) *To solve this integral we can make the substitution* $t = \cos(x)$. *Thus:*

$$\int \tan(x)\,\mathrm{d}x = \int \frac{1}{\underbrace{\cos(x)}_{t}} \cdot \underbrace{(-\sin(x))\,\mathrm{d}x}_{\mathrm{d}t} \overset{(*)}{=}$$

$$(*)\begin{bmatrix} t & = & \cos(x) \\ \mathrm{d}t & = & -\sin(x)\,\mathrm{d}x \end{bmatrix}$$

$$= \int -\frac{1}{t}\,\mathrm{d}t$$

$$= -\ln|t| + C, \quad C \in \mathbb{R}$$

$$= -\ln|\cos(x)| + C, \quad C \in \mathbb{R}.$$

(d) *To solve this integral we can make the substitution* $t = \sin(x)$. *Thus:*

$$\int \frac{\cos(x)}{1+\sin^2(x)}\,\mathrm{d}x = \int \frac{1}{1+\underbrace{\sin^2(x)}_{t^2}} \cdot \underbrace{\cos(x)\,\mathrm{d}x}_{\mathrm{d}t} \overset{(*)}{=}$$

$$(*)\begin{bmatrix} t & = & \sin(x) \\ \mathrm{d}t & = & \cos(x)\,\mathrm{d}x \end{bmatrix}$$

$$= \int \frac{1}{1+t^2}\,\mathrm{d}t$$

$$= \arctan(t) + C, \quad C \in \mathbb{R}$$

$$= \arctan(\sin(x)) + C, \quad C \in \mathbb{R}.$$

(e) *The denominator of the rational function is an irreducible polynomial of degree 2.*

Therefore:

$$\int \frac{3x-2}{x^2-4x+5}\,\mathrm{d}x = \int \frac{\frac{3}{2}(2x-4)-2-\frac{3}{2}(-4)}{x^2-4x+5}\,\mathrm{d}x$$

$$= \int \frac{3}{2}\frac{\overbrace{2x-4}^{\textit{derivative of the denominator}}}{x^2-4x+5}\,\mathrm{d}x + \int \frac{4}{x^2-4x+5}\,\mathrm{d}x$$

$$= \frac{3}{2}\ln|x^2-4x+5| + \int \frac{4}{(x-2)^2+1}\,\mathrm{d}x$$

$$= \frac{3}{2}\ln|x^2-4x+5| + 4\arctan(x-2) + C, \quad C \in \mathbb{R}.$$

(f) *We apply the method of integration by parts:*

$$\int e^{\arcsin(x)}\,\mathrm{d}x = \int \underbrace{e^{\arcsin(x)}}_{u(x)}\underbrace{1\,\mathrm{d}x}_{v'(x)} \stackrel{(*)}{=}$$

$$(*)\left[\begin{array}{lll} u(x) = e^{\arcsin(x)} & \Rightarrow & u'(x) = \frac{e^{\arcsin(x)}}{\sqrt{1-x^2}}\,\mathrm{d}x \\ v'(x) = 1\,\mathrm{d}x & \Rightarrow & v(x) = x \end{array}\right]$$

$$= e^{\arcsin(x)}\cdot x - \int x\cdot\frac{e^{\arcsin(x)}}{\sqrt{1-x^2}}\,\mathrm{d}x.$$

We apply the method of integration by parts again:

$$e^{\arcsin(x)}\cdot x - \int x\cdot\frac{e^{\arcsin(x)}}{\sqrt{1-x^2}}\,\mathrm{d}x = e^{\arcsin(x)}\cdot x - \int \underbrace{e^{\arcsin(x)}}_{u(x)}\underbrace{\frac{x}{\sqrt{1-x^2}}\,\mathrm{d}x}_{v'(x)} \stackrel{(*)}{=}$$

$$(*)\left[\begin{array}{lll} u(x) = e^{\arcsin(x)} & \Rightarrow & u'(x) = \frac{e^{\arcsin(x)}}{\sqrt{1-x^2}}\,\mathrm{d}x \\ v'(x) = \frac{x}{\sqrt{1-x^2}}\,\mathrm{d}x & \Rightarrow & v(x) = -\sqrt{1-x^2} \end{array}\right]$$

$$= e^{\arcsin(x)}\cdot x - \left(e^{\arcsin(x)}\cdot\left(-\sqrt{1-x^2}\right) - \int -\sqrt{1-x^2}\cdot\frac{e^{\arcsin(x)}}{\sqrt{1-x^2}}\,\mathrm{d}x\right)$$

$$= e^{\arcsin(x)}\cdot x + e^{\arcsin(x)}\cdot\sqrt{1-x^2} - \int e^{\arcsin(x)}\,\mathrm{d}x.$$

Therefore:

$$\int e^{\arcsin(x)}\,\mathrm{d}x = e^{\arcsin(x)}\cdot x + e^{\arcsin(x)}\cdot\sqrt{1-x^2} - \int e^{\arcsin(x)}\,\mathrm{d}x$$

$$2\int e^{\arcsin(x)}\,\mathrm{d}x = e^{\arcsin(x)}\cdot x + e^{\arcsin(x)}\cdot\sqrt{1-x^2} + C, \quad C \in \mathbb{R}$$

$$\int e^{\arcsin(x)}\,\mathrm{d}x = \frac{1}{2}e^{\arcsin(x)}\cdot\left(x+\sqrt{1-x^2}\right) + C, \quad C \in \mathbb{R}.$$

(g) We make the substitution:

$$\int \ln\left(1-\sqrt{x}\right)\,\mathrm{d}x = \int \ln\left(1-\underbrace{\sqrt{x}}_{t}\right)\cdot 2\underbrace{\sqrt{x}}_{t}\cdot\underbrace{\frac{1}{2\sqrt{x}}\,\mathrm{d}x}_{\mathrm{d}t} \stackrel{(*)}{=}$$

$$(*)\left[\begin{array}{ccc} t & = & \sqrt{x} \\ \mathrm{d}t & = & \dfrac{1}{2\sqrt{x}}\,\mathrm{d}x \end{array}\right]$$

$$= \int \ln(1-t)\cdot 2t\,\mathrm{d}t.$$

Next, we solve this integral by the method of integration by parts:

$$\int \underbrace{\ln(1-t)}_{u(t)}\cdot\underbrace{2t\,\mathrm{d}t}_{v'(t)} \stackrel{(*)}{=}$$

$$(*)\left[\begin{array}{lll} u(t)=\ln(1-t) & \Rightarrow & u'(t)=-\frac{1}{1-t}\,\mathrm{d}t \\ v'(t)=2t\,\mathrm{d}t & \Rightarrow & v(t)=t^2 \end{array}\right]$$

$$= \ln(1-t)\cdot t^2 - \int t^2\cdot\left(-\frac{1}{1-t}\right)\,\mathrm{d}t$$

$$= \ln(1-t)\cdot t^2 + \int \frac{t^2}{1-t}\,\mathrm{d}t$$

$$= \ln(1-t)\cdot t^2 + \int \frac{(-1-t)(1-t)+1}{1-t}\,\mathrm{d}t$$

$$= \ln(1-t)\cdot t^2 + \int \left(-1-t+\frac{1}{1-t}\right)\,\mathrm{d}t$$

$$= \ln(1-t)\cdot t^2 - t - \frac{1}{2}t^2 - \ln|1-t| + C, \quad C\in\mathbb{R}$$

$$= \ln(1-\sqrt{x})\cdot\left(\sqrt{x}\right)^2 - \sqrt{x} - \frac{1}{2}\left(\sqrt{x}\right)^2 - \ln|1-\sqrt{x}| + C, \quad C\in\mathbb{R}$$

$$= x\ln(1-\sqrt{x}) - \sqrt{x} - \frac{1}{2}x - \ln|1-\sqrt{x}| + C, \quad C\in\mathbb{R}.$$

This integral can also be solved by the method of integration by parts:

$$\int \ln\left(1-\sqrt{x}\right)\,\mathrm{d}x = \int \underbrace{\ln\left(1-\sqrt{x}\right)}_{u(x)}\underbrace{1\,\mathrm{d}x}_{v'(x)} \stackrel{(*)}{=}$$

$$(*)\left[\begin{array}{lll} u(x)=\ln\left(1-\sqrt{x}\right) & \Rightarrow & u'(x)=\frac{1}{2}\frac{1}{x-\sqrt{x}}\,\mathrm{d}x \\ v'(x)=1\,\mathrm{d}x & \Rightarrow & v(x)=x \end{array}\right]$$

$$= \ln\left(1-\sqrt{x}\right)\cdot x - \int x\cdot\frac{1}{2}\frac{1}{x-\sqrt{x}}\,\mathrm{d}x$$

$$= x\ln\left(1-\sqrt{x}\right) - \frac{1}{2}\int \frac{x}{x-\sqrt{x}}\,\mathrm{d}x$$

$$= x\ln\left(1-\sqrt{x}\right) - \frac{1}{2}\int \frac{x-\sqrt{x}+\sqrt{x}-1+1}{x-\sqrt{x}}\,\mathrm{d}x$$

$$= x\ln\left(1-\sqrt{x}\right) - \frac{1}{2}\int \left(\frac{x-\sqrt{x}}{x-\sqrt{x}} + \frac{\sqrt{x}-1}{x-\sqrt{x}} + \frac{1}{x-\sqrt{x}}\right)\mathrm{d}x$$

$$= x\ln\left(1-\sqrt{x}\right) - \frac{1}{2}\int \left(1+\frac{1}{\sqrt{x}} + \frac{1}{x-\sqrt{x}}\right)\mathrm{d}x$$

$$= x\ln\left(1-\sqrt{x}\right) - \frac{1}{2}x - \sqrt{x} - \underbrace{\int \frac{1}{2}\frac{1}{x-\sqrt{x}}\,\mathrm{d}x}_{=\ln|1-\sqrt{x}|+C,\quad C\in\mathbb{R}}$$

$$= x\ln\left(1-\sqrt{x}\right) - \frac{1}{2}x - \sqrt{x} - \ln|1-\sqrt{x}| + C, \quad C\in\mathbb{R}.$$

We obtain the same result by both methods.

(h) We can simplify the integral by considering the following trigonometric identity:

$$\cos^2(x) = \frac{1}{2} + \frac{1}{2}\cos(2x).$$

Thus:

$$\int x\cos^2(x)\,\mathrm{d}x = \int \left(\frac{1}{2}x + \frac{1}{2}x\cdot\cos(2x)\right)\mathrm{d}x = \frac{1}{4}x^2 + \frac{1}{2}\int \underbrace{x}_{u(x)}\underbrace{\cos(2x)\,\mathrm{d}x}_{v'(x)} \overset{(*)}{=}$$

$$(*)\left[\begin{array}{lll} u(x) = x & \Rightarrow & u'(x) = 1\,\mathrm{d}x \\ v'(x) = \cos(2x)\,\mathrm{d}x & \Rightarrow & v(x) = \frac{1}{2}\sin(2x) \end{array}\right]$$

$$= \frac{1}{4}x^2 + \frac{1}{4}x\sin(2x) - \frac{1}{2}\int \sin(2x)\,\mathrm{d}x$$

$$= \frac{1}{4}x^2 + \frac{1}{4}x\sin(2x) + \frac{1}{4}\cos(2x) + C, \quad C\in\mathbb{R}.$$

For the previous calculations, we have used the following trigonometric identity:

$$\cos^2(x) = \frac{1}{2} + \frac{1}{2}\cos(2x).$$

Problem 5.7. Given a curve in the plane defined by the equation $y = f(x)$, the length of this curve from $x = a$ to $x = b$ is given by:

$$L = \int_a^b \sqrt{1+(f'(x))^2}\,\mathrm{d}x$$

Prove that the length of a quarter circle of radius r is $\pi r/2$.

Hint: make the substitution $x = r\sin(u)$ in the integral and use that $\arcsin(0) = 0$ and $\arcsin(1) = \pi/2$.

Solution to the problem 5.7. *The equation that defines the circle of radius r and center* $(0,0)$ *in the first quadrant is* $y = \sqrt{r^2 - x^2}$, *as the equation is:*

$$x^2 + y^2 = r^2$$

Since $f(x) = \sqrt{r^2 - x^2}$, *then:*

$$f'(x) = \frac{-2x}{2\sqrt{r^2 - x^2}}$$

Therefore, the length of a quarter of the circle is given by:

$$\begin{aligned} L &= \int_0^r \sqrt{1 + \left(\frac{-2x}{2\sqrt{r^2 - x^2}}\right)^2}\,\mathrm{d}x = \int_0^r \sqrt{1 + \frac{4x^2}{4(r^2 - x^2)}}\,\mathrm{d}x = \int_0^r \sqrt{1 + \frac{x^2}{r^2 - x^2}}\,\mathrm{d}x \\ &= \int_0^r \sqrt{\frac{r^2 - x^2 + x^2}{r^2 - x^2}}\,\mathrm{d}x = \int_0^r \sqrt{\frac{r^2}{r^2 - x^2}}\,\mathrm{d}x = r\int_0^r \frac{1}{\sqrt{r^2 - x^2}}\,\mathrm{d}x. \end{aligned}$$

Making the proposed substitution:

$$x = r\sin(u) \quad\Longrightarrow\quad u = \arcsin(x/r),$$

so we have $\mathrm{d}x = r\cos(u)du$ *and the limits of integration are now:*

$$\begin{aligned} x &= 0 \to u = \arcsin(0) = 0, \\ x &= r \to u = \arcsin(1) = \pi/2. \end{aligned}$$

$$\begin{aligned} L &= r\int_0^r \frac{1}{\sqrt{r^2 - x^2}}\,\mathrm{d}x = r\int_0^{\pi/2} \frac{r\cos(u)}{\sqrt{r^2 - r^2\sin(u)^2}}\,du = r\int_0^{\pi/2} \frac{r\cos(u)}{r\sqrt{1 - \sin(u)^2}}\,du \\ &= r\int_0^{\pi/2} \frac{\cos(u)}{\sqrt{\cos(u)^2}}\,du = r\int_0^{\pi/2} 1\,du = r\,[u]_0^{\pi/2} = r\pi/2. \end{aligned}$$

Problem 5.8. Decompose into simple fractions the rational function:

$$\frac{x^4 + 5\,x^2 + 2}{x^3 + 2\,x}$$

Solution to the problem 5.8. *Dividing the polynomial* $x^4 + 5x^2 + 2$ *by* $x^3 + 2x$, *we obtain:*

$$x^4 + 5x^2 + 2 = x \cdot (x^3 + 2x) + (3x^2 + 2).$$

Therefore:

$$\frac{x^4 + 5\,x^2 + 2}{x^3 + 2\,x} = x + \frac{3x^2 + 2}{x^3 + 2x} = x + \frac{3x^2 + 2}{x(x^2 + 2)} = x + \frac{A}{x} + \frac{Bx + C}{x^2 + 2}.$$

The coefficients A, B and C are found by solving the equation:

$$3x^2 + 2 = A(x^2 + 2) + (Bx + C)x = (A + B)x^2 + Cx + 2A,$$

that is $A + B = 3, C = 0, 2A = 2$ *and, therefore,* $C = 0, A = 1, B = 2$.
Thus, we obtain the following decomposition into simple fractions:

$$\frac{x^4 + 5\,x^2 + 2}{x^3 + 2\,x} = x + \frac{1}{x} + \frac{2x}{x^2 + 2}.$$

Problem 5.9. Decompose into simple fractions the rational function:

$$\frac{5x^2+8x+2}{x^3+2\,x^2+2\,x+4}$$

Solution to the problem 5.9. *First, we must decompose the denominator. If we notice that the polynomial in the denominator vanishes at $x=-2$, we can divide the polynomial x^3+2x^2+2x+4 by $x+2$, resulting in the polynomial x^2+2, which does not decompose further as it is always positive. So we have:*

$$x^3+2x^2+2x+4=(x+2)(x^2+2).$$

Then, we need to find the values of the parameters A, B and C that satisfy:

$$\frac{5x^2+8x+2}{x^3+2\,x^2+2\,x+4}=\frac{A}{x+2}+\frac{Bx+C}{x^2+2}.$$

The sum of the simple fractions on the right is:

$$\frac{A(x^2+2)+(Bx+C)(x+2)}{(x+2)(x^2+2)}=\frac{(A+B)x^2+(2B+C)x+(2A+2C)}{(x+2)(x^2+2)}.$$

Since:

$$5x^2+8x+2$$

must be equal to:

$$(A+B)x^2+(2B+C)x+(2A+2C),$$

we equate the coefficients:

$$A+B=5$$

$$2B+C=8$$

$$2A+2C=2$$

and solve the system, which has the solution:

$$A=1$$

$$B=4$$

$$C=0$$

Therefore:

$$\frac{5x^2+8x+2}{x^3+2\,x^2+2\,x+4}=\frac{1}{x+2}+\frac{4x}{x^2+2}.$$

Problem 5.10. Decompose into simple fractions the rational function:

$$\frac{4x^2-2x+6}{x^3+x^2-5x+3}$$

Solution to the problem 5.10. *The denominator of the fraction can be decomposed (by Ruffini) as:*

$$x^3 + x^2 - 5x + 3 = (x-1)^2(x+3).$$

Therefore, the decomposition into simple fractions will have the following structure:

$$\frac{4x^2 - 2x + 6}{x^3 + x^2 - 5x + 3} = \frac{A}{x-1} + \frac{B}{(x-1)^2} + \frac{C}{(x+3)}.$$

Adding the simple fractions and equating the denominators, we have:

$$\begin{aligned}\frac{4x^2 - 2x + 6}{x^3 + x^2 - 5x + 3} &= \frac{A}{x-1} + \frac{B}{(x-1)^2} + \frac{C}{(x+3)} \\ &= \frac{A(x-1)(x+3) + B(x+3) + C(x-1)^2}{(x-1)^2(x+3)} \\ &= \frac{(A+C)x^2 + (2A+B-2C)x - 3A + 3B + C}{(x-1)^2(x+3)}.\end{aligned}$$

Therefore, we have the following system of equations:

$$\begin{aligned} A + C &= 4 \\ 2A + B - 2C &= -2 \\ -3A + 3B + C &= 6 \end{aligned}$$

which has the solution:

$$\begin{aligned} A &= 1 \\ B &= 2 \\ C &= 3 \end{aligned}$$

Thus, the decomposition is:

$$\frac{4x^2 - 2x + 6}{x^3 + x^2 - 5x + 3} = \frac{1}{x-1} + \frac{2}{(x-1)^2} + \frac{3}{(x+3)}.$$

Problem 5.11. Decompose into simple fractions the rational function:

$$\frac{16}{(x-3)(x^2+2x+1)}.$$

Solution to the problem 5.11. *The denominator of the fraction can be decomposed as:*

$$(x-3)(x^2+2x+1) = (x-3)(x+1)^2$$

Therefore, the decomposition into simple fractions will have the following structure:

$$\frac{16}{(x-3)(x^2+2x+1)} = \frac{A}{x-3} + \frac{B}{x+1} + \frac{C}{(x+1)^2}$$

$$= \frac{A(x+1)^2 + B(x-3)(x+1) + C(x-3)}{(x-3)(x+1)^2}$$

$$= \frac{(A+B)x^2 + (2A-2B+C)x + (A-3B-3C)}{(x-3)(x+1)^2}.$$

The numerators 16 *and* $(A+B)x^2+(2A-B+C)x+(A-2B-3C)$ *must be equal, therefore, the coefficients* A, B, C *can be found by solving the linear system of equations:*

$$A + B = 0$$
$$2A - B + C = 0$$
$$A - 2B - 3C = 16$$

The solution to this system is:

$$A = 1$$
$$B = -1$$
$$C = -4$$

Finally, the decomposition into simple fractions is:

$$\frac{16}{(x-3)(x^2+2x+1)} = \frac{1}{x-3} + \frac{-1}{x+1} + \frac{-4}{(x+1)^2}.$$

5.9 Solved Problems (Calculation of Areas)

Problem 5.12. Calculate the area of the region of the plane bounded by the indicated curves:

(a) The parabolas $y^2 = 2px$ and $x^2 = 2py$;

(b) The parabola $y^2 = 2x$ and the circle $x^2 + y^2 = 8$;

(c) The sine arch $y = \sin(x)$ and the lines $x = 0$, $x = \pi$, and $y = 1/2$;

(d) The circles $x^2 + y^2 = 9\, and\, (x-3)^2 + y^2 = 9$;

(e) The curves $y = e^x$, $x = 1 - y^2$ and the x-axis;

(f) The parabolas $x = -y^2 + 2y$, $x = y^2 - 2y + 2$ and the x-axis;

(g) The curve $y^2 = x^2 - x^4$.

Solution to the problem 5.12. *(a) Find the area bounded by the two parabolas $y^2 = 2px$ and $x^2 = 2py$ between the points where they intersect.*

- *In Figure 5.36 we have represented the two curves.*

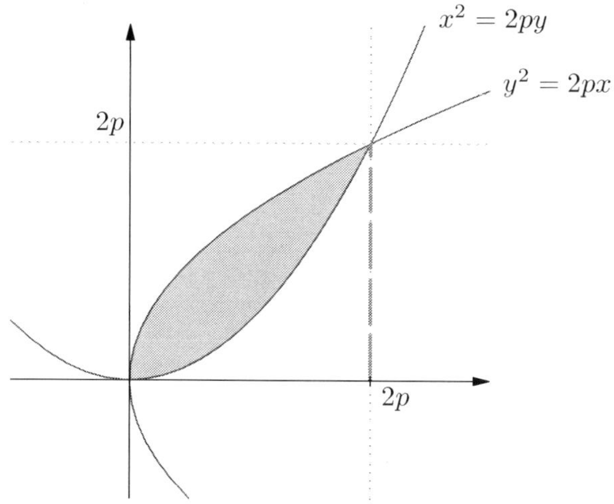

FIGURE 5.36
Graphs of the parabolas $y^2 = 2px$ and $x^2 = 2py$.

We will use:

$$y^2 = 2px \quad \Rightarrow \quad y = f(x) = \sqrt{2px}$$

$$x^2 = 2py \quad \Rightarrow \quad y = g(x) = \frac{x^2}{2p}$$

- *Points where the two functions intersect.*

  From the graph, we can see that the curves intersect at two points. To find these points, we equalize the two functions:

$$\sqrt{2px} = \frac{x^2}{2p} \quad \Leftrightarrow \quad 2px = \frac{x^4}{4p^2} \quad \Leftrightarrow \quad 8p^3 = \frac{x^4}{x} \quad \Leftrightarrow \quad 2^3p^3 = x^3$$

$$\Leftrightarrow \quad 2p = x$$

$$y = \sqrt{2p2p} = \sqrt{2^2p^2} = 2p \quad \Leftrightarrow \quad y = 2p$$

  Therefore, the two intersection points are: $(0,0)$ *and* $(2p, 2p)$.

- *We observe that in the interval* $[0, 2p]$ *the function* $f(x)$ *is greater than or equal to* $g(x)$.

- *Area bounded by the two curves.*

$$A = \int_0^{2p} (f(x) - g(x))\,\mathrm{d}x = \int_0^{2p} \sqrt{2px}\,\mathrm{d}x - \int_0^{2p} \frac{x^2}{2p}\,\mathrm{d}x$$

$$= \int_0^{2p} (2p)^{\frac{1}{2}} x^{\frac{1}{2}}\,\mathrm{d}x - \int_0^{2p} \frac{1}{2p} x^2\,\mathrm{d}x$$

$$= (2p)^{\frac{1}{2}} \int_0^{2p} x^{\frac{1}{2}}\,\mathrm{d}x - \frac{1}{2p} \int_0^{2p} x^2\,\mathrm{d}x$$

$$= (2p)^{\frac{1}{2}} \cdot \frac{2}{3} x^{\frac{3}{2}} \Big|_0^{2p} - \frac{1}{2p} \cdot \frac{1}{3} x^3 \Big|_0^{2p}$$

$$= (2p)^{\frac{4}{2}} \cdot \frac{2}{3} - \frac{1}{3}(2p)^2$$

$$= \frac{2}{3} 4p^2 - \frac{1}{3} 4p^2 = \frac{1}{3} 4p^2.$$

Thus:

$$A = \frac{1}{3} 4p^2.$$

(b) Find the area common to the parabola $y^2 = 2x$ *and the circle* $x^2 + y^2 = 8$.

- *In Figure 5.37, the graph of the parabola* $y^2 = 2x$ *and the circle* $x^2 + y^2 = 8$ *are shown.*

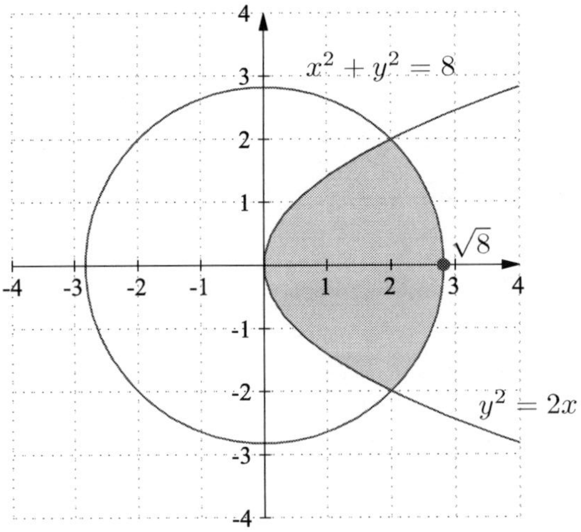

FIGURE 5.37
Graph of the parabola $y^2 = 2x$ and the circle $x^2 + y^2 = 8$

We will use:

$$y^2 = 2x \quad \Rightarrow \quad y = f(x) = \sqrt{2x}$$

$$x^2 + y^2 = 8 \quad \Rightarrow \quad y = g(x) = \sqrt{8 - x^2}$$

- *Points where the functions intersect.*

  First, we need to know the point where the circle intersects the x-axis. Since it is a circle centered at the origin, this point will coincide with the radius value:

$$x^2 + y^2 = R^2 \quad \Rightarrow \quad R^2 = 8 \quad \Rightarrow \quad R = \sqrt{8}$$

  Therefore, the point is $(\sqrt{8}, 0)$.

  Let's determine now the points where the two curves intersect by equalizing the two functions:

$$\sqrt{2x} = \sqrt{8 - x^2} \quad \Leftrightarrow \quad 2x = 8 - x^2 \quad \Leftrightarrow \quad x^2 + 2x - 8 = 0 \quad \Leftrightarrow$$

$$\Leftrightarrow \quad x = \frac{-2 \pm \sqrt{2^2 - 4 \cdot 1(-8)}}{2 \cdot 1} \quad \Leftrightarrow$$

$$\Leftrightarrow \quad x = \frac{-2 \pm 6}{2} \quad \Leftrightarrow \quad \boxed{x_1 = 2} \qquad x_2 = -4$$

$$y = \sqrt{2x} = \sqrt{2 \cdot 2} = \pm 2 \quad \Leftrightarrow \quad y = \pm 2$$

  The two points are $(2, 2)$ *and* $(2, -2)$.

- *The curves we will integrate in each interval to find the area will be:*

  - *In the interval* $[0, 2]$ *the function* $f(x)$ *is greater than or equal to* $g(x)$.
  - *In the interval* $[2, \sqrt{8}]$ *the function* $g(x)$ *is greater than or equal to* $g(x)$.

- *Area bounded by the two curves.*

  By symmetry, the area sought is twice the area bounded by the curve $f(x)$ *and the x-axis on the positive side in the interval* $[0, 2]$ *plus twice the area bounded by the curve* $g(x)$ *and the x-axis on the positive side in the interval* $[2, \sqrt{8}]$:

$$A = 2\left(\int_0^2 (f(x)\,\mathrm{d}x + \int_2^{\sqrt{8}} g(x))\,\mathrm{d}x\right)$$

$$= 2\int_0^2 \sqrt{2x}\,\mathrm{d}x + 2\int_2^{\sqrt{8}} \sqrt{8 - x^2}\,\mathrm{d}x$$

$$= 2\int_0^2 (2x)^{\frac{1}{2}}\,\mathrm{d}x + 2\int_2^{\sqrt{8}} (8 - x^2)^{\frac{1}{2}}\,\mathrm{d}x$$

$$= 2\int_0^2 (2x)^{\frac{1}{2}}\,\mathrm{d}x + 2\int_2^{\sqrt{8}} \left(8 \cdot \left(1 - \frac{x^2}{8}\right)\right)^{\frac{1}{2}}\,\mathrm{d}x$$

$$= 2\int_0^2 (2x)^{\frac{1}{2}}\,\mathrm{d}x + 2\int_2^{\sqrt{8}} \sqrt{8} \cdot \sqrt{1 - \left(\frac{x}{\sqrt{8}}\right)^2}\,\mathrm{d}x$$

$$= 2\int_0^2 (2x)^{\frac{1}{2}}\,\mathrm{d}x + 2\sqrt{8}\int_2^{\sqrt{8}} \sqrt{1 - \left(\frac{x}{\sqrt{8}}\right)^2}\,\mathrm{d}x.$$

At this point, we need to make the substitution:

$$\sin(t) = \frac{x}{\sqrt{8}}$$

to solve the second integral:

$$\cos(t)\,\mathrm{d}t = \frac{1}{\sqrt{8}}\,\mathrm{d}x \quad \Leftrightarrow \quad \sqrt{8}\cos(t)\,\mathrm{d}t = \mathrm{d}x$$

$$x_1 = 2 \rightarrow t_1 = \arcsin\left(\frac{2}{\sqrt{8}}\right),\ t_1 \in (-\pi, \pi) \rightarrow t_1 = \frac{\pi}{4}$$

$$x_2 = \sqrt{8} \rightarrow t_2 = \arcsin(1),\ t_2 \in (-\pi, \pi) \rightarrow t_2 = \frac{\pi}{2}$$

Therefore:

$$\begin{aligned}
A &= 2\int_0^2 (2x)^{\frac{1}{2}}\,\mathrm{d}x + 2\sqrt{8}\int_{\frac{\pi}{4}}^{\frac{\pi}{2}} \sqrt{1-\sin^2(t)}\sqrt{8}\cos(t)\,\mathrm{d}t \\
&= 2\int_0^2 (2x)^{\frac{1}{2}}\,\mathrm{d}x + 2\sqrt{8}\int_{\frac{\pi}{4}}^{\frac{\pi}{2}} \sqrt{\cos^2(t)}\sqrt{8}\cos(t)\,\mathrm{d}t \\
&= 2\int_0^2 (2x)^{\frac{1}{2}}\,\mathrm{d}x + 2\sqrt{8}\sqrt{8}\int_{\frac{\pi}{4}}^{\frac{\pi}{2}} \cos^2(t)\,\mathrm{d}t.
\end{aligned}$$

To solve the resulting integral, the equality must be applied:

$$\cos^2(t) = \frac{1}{2} + \frac{1}{2}\cos(2t)$$

Operating:

$$\begin{aligned}
A &= 2\int_0^2 (2x)^{\frac{1}{2}}\,\mathrm{d}x + 2\sqrt{8}\sqrt{8}\int_{\frac{\pi}{4}}^{\frac{\pi}{2}} \frac{1}{2} + \frac{1}{2}\cos(2t)\,\mathrm{d}t \\
&= 2\left[\frac{1}{3}(2x)^{\frac{3}{2}}\right]_0^2 + 16\left[\frac{\pi}{4} + \frac{1}{4}\sin(\pi) - \frac{\pi}{8} - \frac{1}{4}\sin\left(\frac{\pi}{2}\right)\right] \\
&= 2\left[\frac{1}{3}\cdot 4^{\frac{3}{2}}\right] + 16\left[\frac{\pi}{4} - \frac{\pi}{8} - \frac{1}{4}\right] \\
&= 2\cdot\frac{1}{3}\cdot 8 + 16\left(\frac{\pi}{4} - \frac{\pi}{8} - \frac{1}{4}\right) \\
&= \frac{16}{3} + 2\pi - 4 \\
&= \frac{4}{3} + 2\pi \simeq 7.6165.
\end{aligned}$$

Thus:

$$A \simeq 7.6165.$$

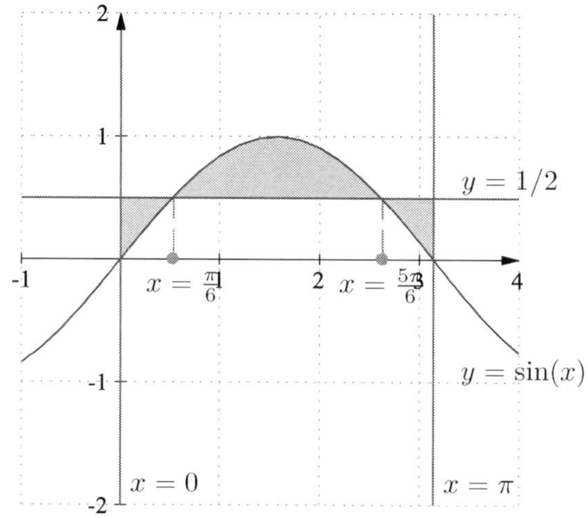

FIGURE 5.38
Graph of the function $y = \sin(x)$ and the lines $x = 0$, $x = \pi$, and $y = 1/2$.

(c) *Find the area bounded by the sine arch* $y = \sin(x)$ *and the lines* $x = 0$, $x = \pi$, *and* $y = 1/2$.

- *Figure 5.38 shows the sine arch* $y = \sin(x)$ *and the lines* $x = 0$, $x = \pi$, *and* $y = 1/2$.
- *Points of intersection.*

  First, we need to know the points where the sine arch intersects the line $y = 1/2$. *To do this, we equalize the functions* $f(x) = \sin(x)$ *and* $g(x) = \frac{1}{2}$:

$$\sin(x) = \frac{1}{2}, x \in [0, \pi] \Leftrightarrow x = \arcsin\left(\frac{1}{2}\right) = \frac{\pi}{6}$$

  The first intersection point between the function $\sin(x)$ *and the line* $y = 1/2$ *is the point:*

$$\left(\frac{\pi}{6}, \frac{1}{2}\right)$$

  Since the functions $f(x)$ *and* $g(x)$ *are symmetrical in the interval* $[0, \pi]$, *the second intersection point will be located at* $x = \pi - \frac{\pi}{6} = \frac{5\pi}{6}$, *i.e., at the point:*

$$\left(\frac{5\pi}{6}, \frac{1}{2}\right)$$

  However, it is not necessary to know the second intersection point, as by using the symmetry of the functions we can calculate the area they bound in the interval $[0, \pi]$ *as twice the area they bound in the interval* $\left[0, \frac{\pi}{2}\right]$.

- *We observe that:*
  - *In the interval* $\left[0, \frac{\pi}{6}\right]$, *the function* $g(x)$ *is greater than or equal to* $f(x)$.
  - *In the interval* $\left[\frac{\pi}{6}, \frac{\pi}{2}\right]$, *the function* $f(x)$ *is greater than or equal to* $g(x)$.

- *Area bounded by the curve and the line in the interval* $[0, \frac{\pi}{2}]$*:*

$$
\begin{aligned}
A_1 &= 2\left(\int_0^{\frac{\pi}{6}} (g(x) - f(x))\,\mathrm{d}x + \int_{\frac{\pi}{6}}^{\frac{\pi}{2}} (f(x) - g(x))\,\mathrm{d}x\right) \\
&= 2\left(\int_0^{\frac{\pi}{6}} \left(\frac{1}{2} - \sin(x)\right)\mathrm{d}x + \int_{\frac{\pi}{6}}^{\frac{\pi}{2}} \left(\sin(x) - \frac{1}{2}\right)\mathrm{d}x\right) \\
&= 2\int_0^{\frac{\pi}{6}} \left(\frac{1}{2} - \sin(x)\right)\mathrm{d}x + 2\int_{\frac{\pi}{6}}^{\frac{\pi}{2}} \left(\sin(x) - \frac{1}{2}\right)\mathrm{d}x \\
&= 2\left[\frac{1}{2}x + \cos(x)\right]_0^{\frac{\pi}{6}} + 2\left[-\cos(x) - \frac{1}{2}x\right]_{\frac{\pi}{6}}^{\frac{\pi}{2}} \\
&= 2\left[\frac{1}{2}\cdot\frac{\pi}{6} + \cos\left(\frac{\pi}{6}\right) - \cos(0)\right] + 2\left[-\cos\left(\frac{\pi}{2}\right) - \frac{1}{2}\cdot\frac{\pi}{2} + \cos\left(\frac{\pi}{6}\right) + \frac{1}{2}\cdot\frac{\pi}{6}\right] \\
&= 2\left[\frac{\pi}{12} + \frac{\sqrt{3}}{2} - 1\right] + 2\left[0 - \frac{\pi}{4} + \frac{\sqrt{3}}{2} + \frac{\pi}{12}\right] \\
&= 2\sqrt{3} - \frac{\pi}{6} - 2 \simeq 0.9405.
\end{aligned}
$$

Thus:

$$A \simeq 0.9405.$$

(d) *Find the area bounded by the circles:*

$$x^2 + y^2 = 9, \quad (x-3)^2 + y^2 = 9$$

- *Figure 5.39 shows these circles.*

  We will use:

$$
\begin{aligned}
x^2 + y^2 = 9 \quad &\Rightarrow \quad y = f(x) = \sqrt{9 - x^2} \\
(x-3)^2 + y^2 = 9 \quad &\Rightarrow \quad y = g(x) = \sqrt{9 - (x-3)^2}
\end{aligned}
$$

- *Points of intersection.*

  In this case, we have two circles of radius 3, one centered at the origin and one centered at $x = 3$*. The points of intersection are equidistant from the centers; therefore, the* x *coordinate of the intersection points must be the midpoint between the centers, that is,* $x = \frac{3}{2}$*.*

  On the other hand, we can note that the area to be calculated is symmetrical with respect to the x*-axis, and the upper half is symmetrical with respect to the line* $x = \frac{3}{2}$*. Therefore, we can obtain the total area as 4 times the area of one quarter.*

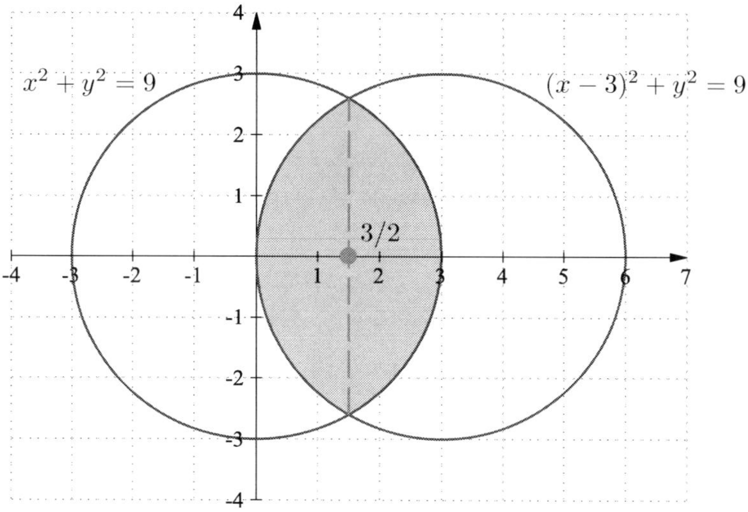

FIGURE 5.39
Representation of the circles $x^2 + y^2 = 9$ *and* $(x-3)^2 + y^2 = 9$.

- *Common area bounded by the two circles.*
  Taking the upper left quarter, it coincides with the area under the function:

$$g(x) = \sqrt{9-(x-3)^2}$$

between the points $x = 0$ *and* $x = \frac{3}{2}$. *Therefore, the total area will be:*

$$\begin{aligned} A &= 4\int_0^{\frac{3}{2}} \sqrt{9-(x-3)^2}\,\mathrm{d}x \\ &= 4\int_0^{\frac{3}{2}} \sqrt{9\left(1-\frac{(x-3)^2}{9}\right)}\,\mathrm{d}x \\ &= 4\int_0^{\frac{3}{2}} 3\sqrt{1-\left(\frac{(x-3)}{3}\right)^2}\,\mathrm{d}x. \end{aligned}$$

At this point, we need to make the substitution:

$$\sin(t) = \frac{x-3}{3}$$

to solve the integral, recalculating also the differential and the limits:

$$\cos(t)\,\mathrm{d}t = \frac{1}{3}\,\mathrm{d}x \quad \Leftrightarrow \quad 3\cos(t)\,\mathrm{d}t = \mathrm{d}x$$

$$x_1 = 0 \quad \rightarrow \quad t_1 = \arcsin(-1), t_1 \in (-\pi,\pi) \quad \rightarrow \quad t_1 = \frac{-\pi}{2}$$

$$x_2 = \frac{3}{2} \quad \rightarrow \quad t_2 = \arcsin\left(-\frac{1}{2}\right), t_2 \in (-\pi,\pi) \quad \rightarrow \quad t_2 = \frac{-\pi}{6}.$$

Therefore, the integral becomes:

$$\begin{aligned} A &= 4 \cdot 3 \int_{\frac{-\pi}{2}}^{\frac{-\pi}{6}} \sqrt{1 - (\sin(t))^2} 3 \cos(t)\, dt \\ &= 12 \int_{\frac{-\pi}{2}}^{\frac{-\pi}{6}} \sqrt{\cos^2(t)} 3 \cos(t)\, dt \\ &= 12 \cdot 3 \int_{\frac{-\pi}{2}}^{\frac{-\pi}{6}} \cos^2(t)\, dt. \end{aligned}$$

To solve the resulting integral, the equality must be applied:

$$\cos^2(t) = \frac{1}{2} + \frac{1}{2}\cos(2t).$$

Operating:

$$\begin{aligned} A &= 36 \int_{\frac{-\pi}{2}}^{\frac{-\pi}{6}} \left(\frac{1}{2} + \frac{1}{2}\cos(2t)\right) dt \\ &= 36 \left[\frac{t}{2} + \frac{1}{4}\sin(2t)\right]_{\frac{-\pi}{2}}^{\frac{-\pi}{6}} \\ &= 36 \left[\left(\frac{-\pi}{12} + \frac{1}{4}\sin\left(\frac{-\pi}{3}\right)\right) - \left(\frac{-\pi}{4} + \frac{1}{4}\sin(-\pi)\right)\right] \\ &= 6\pi - \frac{9}{2}\sqrt{3}. \end{aligned}$$

Evaluating:

$$A \simeq 11.0553.$$

Note: This area can also be calculated geometrically.

(e) *the curves $y = e^x$, $x = 1 - y^2$ and the x-axis.*

Find the area bounded by the curves $y = e^x$, $x = 1 - y^2$ and the x-axis.

- *Figure 5.40 shows the curves $y = e^x$, $x = 1 - y^2$ and the x-axis.*

  We will use:

$$y = f(x) = e^x$$

$$x = 1 - y^2 \quad \Rightarrow \quad y = g(x) = \sqrt{1 - x}$$

- *Points of intersection.*

  The area to be calculated is bounded by two curves and the x-axis. In this case, we can calculate the total area as the sum of two areas: the one below the function $f(x)$, and the one below the function $g(x)$.

  The first one is an improper integral of the first kind, since the lower limit is $-\infty$. The upper limit will be the point of intersection of the two functions. If to draw the graph we have calculated the points of intersection with the y-axis we will have seen that the two functions coincide at the point $(0, 1)$; therefore, the intersection is at $x_1 = 0$.

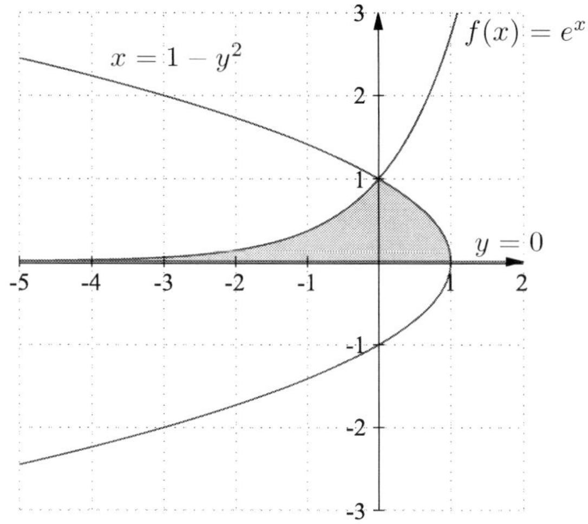

FIGURE 5.40
Representation of the curves $y = e^x$, $x = 1 - y^2$ and the x-axis.

The area below the function $g(x)$ *will have as the lower limit the intersection point* $x_1 = 0$, *and as the upper limit, the point of intersection with the x-axis, that is:*

$$g(x) = 0 \quad \Longleftrightarrow \quad \sqrt{1-x} = 0 \quad \Longleftrightarrow \quad x = 1$$

Therefore, $x_2 = 1$.

- *Area bounded by the two curves.*
  As we have mentioned, it is obtained from the sum of two areas $A = A_1 + A_2$, *where:*

$$A_1 = \int_{-\infty}^{0} e^x \, dx$$
$$A_2 = \int_{0}^{1} \sqrt{1-x} \, dx.$$

We start by solving the improper integral:

$$\begin{aligned} A_1 &= \int_{-\infty}^{0} e^x \, dx \\ &= \lim_{a \to -\infty} \int_{a}^{0} e^x \, dx \\ &= \lim_{a \to -\infty} \left[e^x\right]_a^0 \\ &= e^0 - \lim_{a \to -\infty} e^a = 1 - 0 = 1. \end{aligned}$$

We now calculate the second integral:

$$A_2 = \int_0^1 \sqrt{1-x}\,\mathrm{d}x$$
$$= \left[\frac{-2(1-x)^{\frac{3}{2}}}{3}\right]_0^1$$
$$= 0 - \left(-\frac{2}{3}\right) = \frac{2}{3}.$$

Therefore:

$$A = 1 + \frac{2}{3} = \frac{5}{3}.$$

(f) *Find the area bounded by the parabolas* $x = -y^2 + 2y$, $x = y^2 - 2y + 2$ *and the x-axis.*

- *Figure 5.41 represents the parabolas* $x = -y^2 + 2y$, $x = y^2 - 2y + 2$ *and the x-axis.*

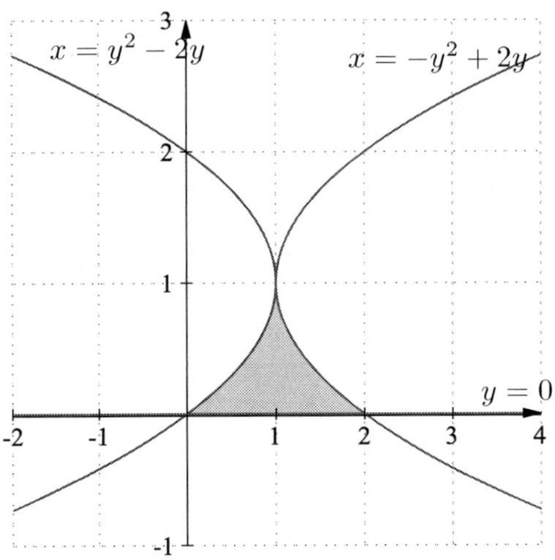

FIGURE 5.41
Representation of the parabolas $x = -y^2 + 2y$, $x = y^2 - 2y + 2$ and the x-axis.

We will use:

$$x = -y^2 + 2y \quad \Rightarrow \quad y = f(x) = \sqrt{1-x} + 1$$
$$x = y^2 - 2y + 2 \quad \Rightarrow \quad y = g(x) = \sqrt{x-1} + 1$$

- *Points of intersection.*

  To find the intersection point we equalize the two initial expressions:

$$-y^2 + 2y = y^2 - 2y + 2$$
$$2y^2 - 4y + 2 = 0 \Rightarrow y = 1$$

Substituting we find $x = 1$*; therefore, the intersection point is* $(1, 1)$.

- *Area bounded by the two parabolas.*

  In this case, it will be easier to calculate the integral if we do it with respect to the y-axis. If we establish:

$$f(y) = -y^2 + 2y$$
$$g(y) = y^2 - 2y + 2,$$

we can observe that $g(y) > f(y)$ in the entire domain. In this case, the lower limit will be $y = 0$, which corresponds to the x-axis, and the upper limit $y = 1$, corresponding to the intersection point. Therefore:

$$\begin{aligned} A &= \int_0^1 y^2 - 2y + 2 - (-y^2 + 2y)\,\mathrm{d}x \\ &= \int_0^1 2y^2 - 4y + 2\,\mathrm{d}x \\ &= \left[\frac{2}{3}y^3 - 2y^2 + 2y\right]_0^1 \\ &= \frac{2}{3} - 2 + 2 - (0) = \frac{2}{3}. \end{aligned}$$

Thus:

$$A = \frac{2}{3}.$$

(g) *Find the area defined by the curve $y^2 = x^2 - x^4$.*

- *Figure 5.42 shows the curve $y^2 = x^2 - x^4$.*

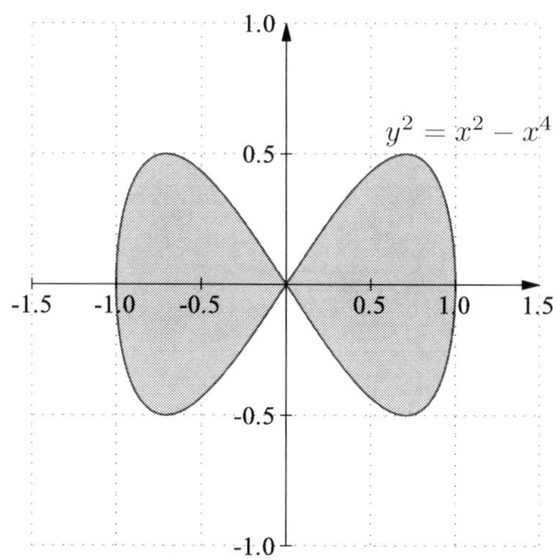

FIGURE 5.42
Representation of the curve $y^2 = x^2 - x^4$.

We will use:

$$y^2 = x^2 - x^4 \quad \Rightarrow \quad y = f(x) = \sqrt{x^2 - x^4}$$

- *Points of intersection.*

  If we calculate the intersection points of the function $f(x) = \sqrt{x^2 - x^4}$ *with the x-axis, we get* $x_1 = 0$, $x_2 = -1$ *and* $x_3 = 1$. *We observe that:*

  - *The domain of the function is* $[-1, 1]$, *since whenever* $|x| > 1$ *the radicand will be negative.*
  - *The function is symmetrical with respect to the x-axis, since it is a square root.*
  - *The function is symmetrical with respect to the y-axis, since the variable x is always raised to an even power.*

  Therefore, we can calculate the total area as 4 times the area bounded in one of the quadrants. Taking, for example, the first quadrant, the interval will be $[0, 1]$ *and we will take, therefore, the positive root.*

- *Area bounded by the curve:*

$$A = 4\int_0^1 \sqrt{x^2 - x^4}\,\mathrm{d}x = 4\int_0^1 x\sqrt{1 - x^2}\,\mathrm{d}x = 4 \cdot \frac{1}{2}\int_0^1 2x\sqrt{1 - x^2}\,\mathrm{d}x$$

  We now make the substitution $t = 1 - x^2$:

$$\mathrm{d}t = -2x\,\mathrm{d}x$$

$$x_1 = 0 \quad \Rightarrow \quad t_1 = 1$$

$$x_2 = 1 \quad \Rightarrow \quad t_2 = 0$$

  Substituting:

$$A = -2\int_1^0 \sqrt{t}\,\mathrm{d}t = -2\left[\frac{2}{3}\sqrt{t^3}\right]_1^0 = -2\left(0 - \frac{2}{3}\right)$$

  Thus:

$$A = \frac{4}{3}$$

Problem 5.13. Calculate the area formed by the graphs of the functions $y^2 = 8x$ and $y = x^2$.

Solution to the problem 5.13.

- *Figure 5.43 represents the two curves* $y^2 = 8x$ *and* $y = x^2$.

  We will use the fact that:

$$y^2 = 8x \quad \Rightarrow \quad y = f(x) = \sqrt{8x}$$

$$y = g(x) = x^2$$

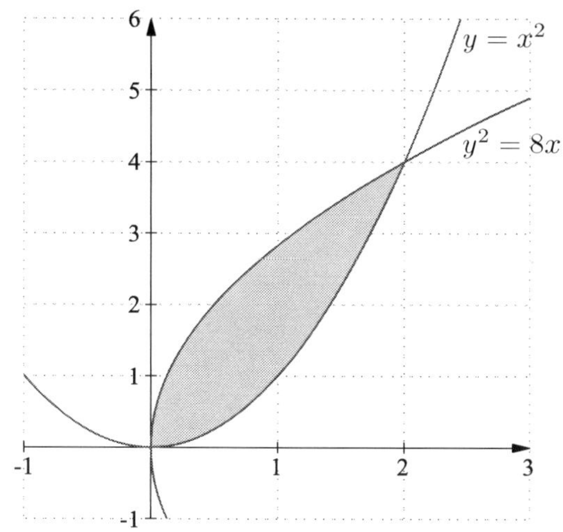

FIGURE 5.43
Representation of the curves $y^2 = 8x$ and $y = x^2$.

- *Intersection points.*

  We find the intersection points by equating the two functions:

$$f(x) = g(x) \quad \Leftrightarrow \quad \sqrt{8x} = x^2$$
$$8x = x^4 \quad \Leftrightarrow \quad x\left(x^3 - 8\right) = 0$$
$$x_1 = 0; \quad x_2 = \sqrt[3]{8} = 2$$

- *Common area bounded by the two curves.*

  Since in the interval $[0, 2]$ *the function* $f(x)$ *is greater than or equal to* $g(x)$*:*

$$A = \int_0^2 f(x) - g(x)\, \mathrm{d}x = \int_0^2 \sqrt{8x} - x^2\, \mathrm{d}x = \left[\frac{2}{3}\sqrt{8x^3} - \frac{1}{3}x^3\right]_0^2 = \frac{16}{3} - \frac{8}{3} - (0).$$

  Therefore:

$$A = \frac{8}{3}.$$

Problem 5.14. Calculate the area of the function $f(x) = e^{2x}\sin(x)$ between $x = \pi/2$ and $x = \pi$.

Solution to the problem 5.14.

- *In Figure 5.44 we can see the graph of the function* $f(x) = e^{2x}\sin(x)$.

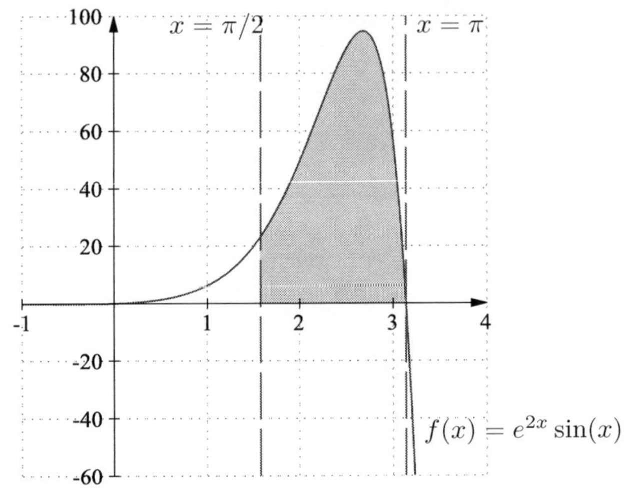

FIGURE 5.44
Graph of the function $f(x) = e^{2x}\sin(x)$.

- *Area bounded in the interval* $[\frac{\pi}{2}, \pi]$.

  To calculate the area under the curve in the given interval, we need to solve the definite integral:

$$A = \int_{\frac{\pi}{2}}^{\pi} e^{2x}\sin(x)\,\mathrm{d}x = \int_{\frac{\pi}{2}}^{\pi} \underbrace{e^{2x}}_{u(x)}\underbrace{\sin(x)\,\mathrm{d}x}_{v'(x)} \underset{(*_1)}{=}$$

$$(*_1)\left[\begin{array}{lll} u(x) = e^{2x} & \Rightarrow & u'(x) = 2e^{2x}\,\mathrm{d}x \\ v'(x) = \sin(x)\,\mathrm{d}x & \Rightarrow & v(x) = -\cos(x) \end{array}\right]$$

$$= \Big[-e^{2x}\cos(x)\Big]_{\frac{\pi}{2}}^{\pi} - \int_{\frac{\pi}{2}}^{\pi} -\cos(x)2e^{2x}\,\mathrm{d}x$$

$$= \left(e^{2\pi} - (0)\right) + \underbrace{\int_{\frac{\pi}{2}}^{\pi} 2e^{2x}\cos(x)\,\mathrm{d}x}_{by\ parts}$$

$$= \left(e^{2\pi}\right) + \int_{\frac{\pi}{2}}^{\pi} \underbrace{2e^{2x}}_{u(x)}\underbrace{\cos(x)\,\mathrm{d}x}_{v'(x)} \underset{(*_2)}{=}$$

$$(*_2)\left[\begin{array}{lll} u(x) = 2e^{2x} & \Rightarrow & u'(x) = 4e^{2x}\,\mathrm{d}x \\ v'(x) = \cos(x)\,\mathrm{d}x & \Rightarrow & v(x) = \sin(x) \end{array}\right]$$

$$= \left(e^{2\pi}\right) + \left[2e^{2x}\sin(x)\right]_{\frac{\pi}{2}}^{\pi} - 4\int_{\frac{\pi}{2}}^{\pi}\sin(x)e^{2x}\,\mathrm{d}x$$

$$= \left(e^{2\pi}\right) + \left(0 - \left(2e^{\pi}\right)\right) - 4\int_{\frac{\pi}{2}}^{\pi} e^{2x}\sin(x)\,\mathrm{d}x.$$

In this case, we can see that, after integrating by parts twice, we encounter the initial integral again. Therefore:

$$\int_{\frac{\pi}{2}}^{\pi} e^{2x}\sin(x)\,\mathrm{d}x = \left(e^{2\pi}\right) - \left(2e^{\pi}\right) - 4\int_{\frac{\pi}{2}}^{\pi} e^{2x}\sin(x)\,\mathrm{d}x.$$

Operating, we obtain:

$$5\int_{\frac{\pi}{2}}^{\pi} e^{2x}\sin(x)\,\mathrm{d}x = e^{2\pi} - 2e^{\pi}$$

$$\int_{\frac{\pi}{2}}^{\pi} e^{2x}\sin(x)\,\mathrm{d}x = \frac{1}{5}e^{2\pi} - \frac{2}{5}e^{\pi}$$

Evaluating:

$$A \simeq 97.8421.$$

Problem 5.15. Calculate the area formed in the first quadrant by the curves $x^2 + y^2 = 9$ and $y = 3 - x$.

Solution to the problem 5.15. • *Figure 5.45 represents the circle $x^2 + y^2 = 9$ and the line $y = 3 - x$.*

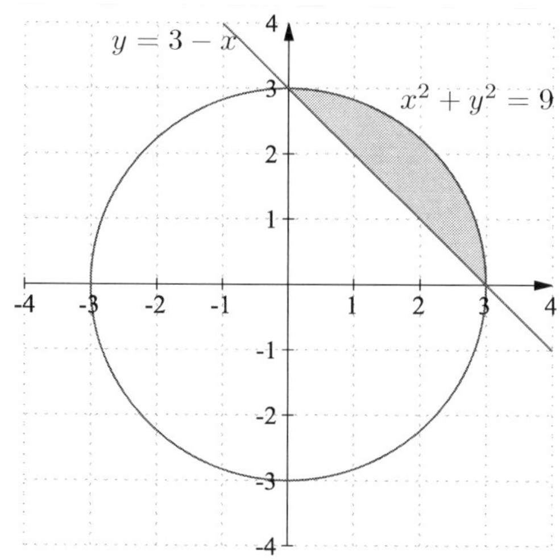

FIGURE 5.45
Graph of the circle $x^2 + y^2 = 9$ and the line $y = 3 - x$.

We will use the fact that:

$$x^2 + y^2 = 9 \quad \Rightarrow \quad y = f(x) = \sqrt{9 - x^2}$$

$$y = 3 - x \quad \Rightarrow \quad y = g(x) = 3 - x$$

- *Intersection points.*

  The line $g(x) = 3 - x$ intersects the axes at the same points as the circle with radius 3; therefore, the points are $(0,3)$ and $3,0)$.

- *Common area bounded by the circle and the line.*

  Since $g(x) \leq f(x)$ for all $x \in [0,3]$, we calculate the area:

$$\begin{aligned} A &= \int_0^3 \sqrt{9-(x)^2} - (3-x)\,\mathrm{d}x \\ &= \underbrace{\int_0^3 \sqrt{9-(x)^2}\,\mathrm{d}x}_{A_1} - \underbrace{\int_0^3 (3-x)\,\mathrm{d}x}_{A_2}. \end{aligned}$$

We initially calculate A_2:

$$\begin{aligned} A_2 &= \int_0^3 (3-x) \\ &= \left[3x - \frac{1}{2}x^2\right]_0^3 \\ &= (9 - \frac{9}{2}) - (0) = \frac{9}{2}. \end{aligned}$$

We now calculate A_1:

$$\begin{aligned} A_1 &= \int_0^3 \sqrt{9-(x)^2}\,\mathrm{d}x \\ &= \int_0^3 \sqrt{9\left(1 - \left(\frac{x}{3}\right)^2\right)}\,\mathrm{d}x. \end{aligned}$$

At this point, a change of variable $\sin(t) = \frac{x}{3}$ is needed to solve the integral:

$$\begin{aligned} \cos(t)\,\mathrm{d}t = \frac{1}{3}\,\mathrm{d}x \quad &\Leftrightarrow \quad 3\cos(t)\,\mathrm{d}t = \mathrm{d}x \\ x_1 = 0 \quad &\Longrightarrow \quad t_1 = \arcsin(0),\ t_1 \in (-\pi,\pi) \quad \Longrightarrow \quad t_1 = 0 \\ x_2 = 3 \quad &\Longrightarrow \quad t_2 = \arcsin\left(\frac{3}{3}\right),\ t_2 \in (-\pi,\pi) \quad \Longrightarrow \quad t_2 = \frac{\pi}{2}. \end{aligned}$$

Therefore, the integral is:

$$\begin{aligned} A_1 &= 3\int_0^{\frac{\pi}{2}} \sqrt{1-(\sin(t))^2} 3\cos(t)\,\mathrm{d}t \\ &= 3\int_0^{\frac{\pi}{2}} \sqrt{\cos^2(t)} 3\cos(t)\,\mathrm{d}t \\ &= 3\cdot 3\int_0^{\frac{\pi}{2}} \cos^2(t)\,\mathrm{d}t. \end{aligned}$$

To solve the resulting integral, apply the identity:

$$\cos^2(t) = \frac{1}{2} + \frac{1}{2}\cos(2t)$$

and operating:

$$\begin{aligned} A_1 &= 9\int_0^{\frac{\pi}{2}} \frac{1}{2} + \frac{1}{2}\cos(2t)\,\mathrm{d}t \\ &= 9\left[\frac{t}{2} + \frac{1}{4}\sin(2t)\right]_0^{\frac{\pi}{2}} \\ &= 9\left[\left(\frac{\pi}{4} + 0\right) - (0)\right] = \frac{9\pi}{4}. \end{aligned}$$

Therefore:

$$A = A_1 - A_2 = \frac{9\pi}{4} - \frac{9}{2}.$$

Evaluating:

$$A \simeq 2.5686.$$

Note: this area can also be calculated geometrically.

Problem 5.16. Calculate the area between $y^2 = x$ and $y^2 = 4 - x$.

Solution to the problem 5.16. • *Figure 5.46 represents the parabolas $y^2 = x$ and $y^2 = 4 - x$.*

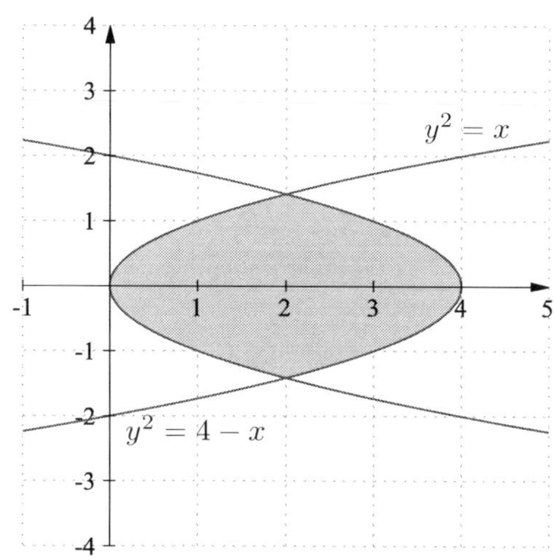

FIGURE 5.46
Representation of the parabolas $y^2 = x$ and $y^2 = 4 - x$.

We will use the fact that:

$$\begin{aligned} y^2 = x &\quad\Rightarrow\quad y = f(x) = \sqrt{x} \\ y^2 = 4 - x &\quad\Rightarrow\quad y = g(x) = \sqrt{4 - x} \end{aligned}$$

- *Intersection points.*

First, we need to find the point where the curve $g(x)$ intersects the OX axis:

$$\sqrt{4-x}=0 \quad \Rightarrow \quad 4-x=0 \quad \Rightarrow \quad x=4$$

Therefore, the point is $(4,0)$.

Now, we determine the points where the two curves intersect by equating the two functions:

$$\begin{aligned} \sqrt{x}=\sqrt{4-x} \quad &\Leftrightarrow \quad x=4-x \quad \Leftrightarrow \quad 2x=4 \quad \Leftrightarrow \\ &\Leftrightarrow \quad x=2 \\ y=\sqrt{x}=\sqrt{2} \quad &\Leftrightarrow \quad y=\pm\sqrt{2} \end{aligned}$$

The two points are $(2,\sqrt{2})$ *and* $(2,-\sqrt{2})$.

- *We observe that:*
  - *In the interval* $[0,2]$ *the function* $f(x)$ *is less than or equal to* $g(x)$.
  - *In the interval* $[2,4]$ *the function* $g(x)$ *is less than or equal to* $f(x)$.
- *Common area bounded by the two curves.*

  It is formed by twice the area between the curve $f(x)$ *and the* OX *axis in the positive part of the interval [0,2] plus twice the area between the curve* $g(x)$ *and the* OX *axis in the positive part of the interval* $[2,4]$. *Note also that the area to be calculated is symmetric with respect to the line* $y=2$.

  Therefore, we can calculate it directly as four times the area between the curve $f(x)$ *and the* OX *axis in the positive part of the interval [0,2]:*

$$\begin{aligned} A &= 4\left(\int_0^2 f(x)\,\mathrm{d}x\right) = \\ &= 4\int_0^2 \sqrt{x}\,\mathrm{d}x = \\ &= 4\int_0^2 x^{\frac{1}{2}}\,\mathrm{d}x = \\ &= 4\left[\frac{2}{3}x^{\frac{3}{2}}\right]_0^2 = \\ &= 4\cdot\frac{2}{3}\sqrt{8} = \frac{16}{3}\sqrt{2} \approx 7.5425. \end{aligned}$$

  Thus:

$$A \approx 7.5425.$$

Problem 5.17. Calculate the area bounded by the following curves (*see* Figure 5.47): $y=e^{2-x}-1$, $y=2x-4$, $y=x^2-4$ and the axes.

Solution to the problem 5.17. *Calculating the three areas separately:*

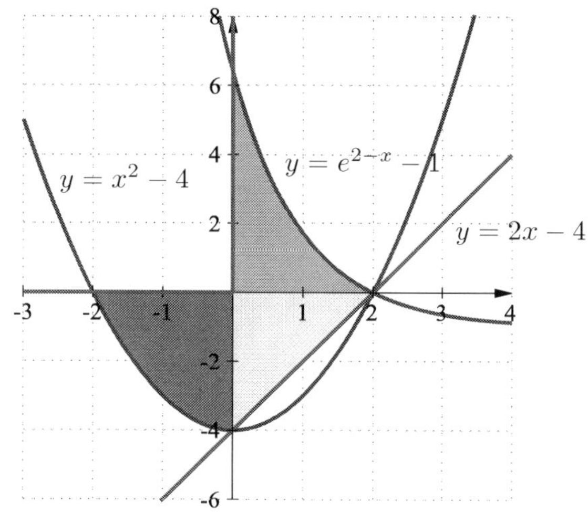

FIGURE 5.47
Representation of the curves $y = e^{2-x} - 1$, $y = 2x - 4$, $y = x^2 - 4$ and the axes.

- *The area A_1 is bounded by the curve $y = x^2 - 4$. The function $x^2 - 4$ intersects the OX axis at the points $x = \pm 2$. Furthermore, in the interval $[-2, 0]$, the function is negative.*

  Therefore, the area is given by:

$$A_1 = -\int_{-2}^{0} (x^2 - 4)\, \mathrm{d}x = -\left[\frac{x^3}{3} - 4x\right]_{-2}^{0}$$
$$= -\left(\frac{0^3}{3} - 4 \cdot 0 - \frac{(-2)^3}{3} + 4(-2)\right) = -\left(\frac{8}{3} - 8\right) = \frac{16}{3}.$$

- *The area A_2 is bounded by the line $y = 2x - 4$. The line intersects the OX axis at the point $x = 2$. Furthermore, in the interval $[0, 2]$, the function is negative.*

  Therefore, the area is given by:

$$A_2 = -\int_{0}^{2} (2x - 4)\, \mathrm{d}x = -\left[x^2 - 4x\right]_0^2 = -(4 - 8 - 0) = 4.$$

  Note that this area can also be calculated as the area of the triangle with base 2 and height 4, that is, $A_2 = 2 \cdot 4/2 = 4$.

- *The area A_3 is bounded by the curve $y = e^{2-x} - 1$. We will calculate the intersections with the OX axis.*

$$e^{2-x} - 1 = 0 \Longleftrightarrow e^{2-x} = 1 \Longleftrightarrow 2 - x = \ln(1) = 0 \Longleftrightarrow x = 2.$$

  Therefore, the function intersects the OX axis only once and is always positive.

  Thus, the area is given by:

$$A_3 = \int_0^2 \left(e^{2-x} - 1\right) \mathrm{d}x = \left[-e^{2-x} - x\right]_0^2 = -e^0 - 2 + e^2 + 0 = e^2 - 3.$$

- *Finally, the total area is:*

$$A_{TOTAL} = \frac{16}{3} + 4 + e^2 - 3 = \frac{16+3}{3} + e^2 = \frac{19}{3} + e^2.$$

Problem 5.18. We want to calculate the area bounded by the graph of the function $f(x) = e^{x^2}$, the OX axis, and the lines $x = 0$ and $x = 1$, that is, $\int_0^1 e^{x^2}\, dx$, but we do not know any primitive of the function. Therefore, we will do it approximately, substituting the function to be integrated, $f(x) = e^{x^2}$, with its Taylor polynomial of order 3 around the point $x = 0$.

(a) Calculate the Taylor polynomial of order 3 around the point $x = 0$ of the function $f(x) = e^{x^2}$.

(b) Calculate an approximation of the value of the integral $\int_0^1 e^{x^2}\, dx$ from the polynomial calculated in the previous section.

(c) Calculate the limit $\lim\limits_{x\to 0} \frac{e^{x^2}-1}{x^2}$ substituting the function $f(x) = e^{x^2}$ with the Taylor polynomial calculated in section (a).

(d) Use the Taylor polynomial calculated in section (a) to approximate the value of the number e.

Solution to the problem 5.18. *(a) We must calculate the Taylor polynomial of the function of order 3 around the point $x = 0$, therefore, we must calculate the value of the function and its first three derivatives at $x = 0$.*

$$\begin{aligned}
f(x) &= e^{x^2} &&\Longrightarrow\ f(0) = e^0 = 1\\
f'(x) &= 2xe^{x^2} &&\Longrightarrow\ f'(0) = 0\\
f''(x) &= 2e^{x^2} + 4x^2e^{x^2} = 2e^{x^2}(1+2x^2) &&\Longrightarrow\ f''(0) = 2\\
f'''(x) &= 4xe^{x^2}(1+2x^2) + 2e^{x^2}4x = 4e^{x^2}(3x+2x^3) &&\Longrightarrow\ f'''(0) = 0
\end{aligned}$$

Therefore:

$$p_3(x) = f(0) + f'(0)x + \frac{f''(0)}{2!}x^2 + \frac{f''(0)}{3!}x^3 = 1 + x^2.$$

(b) We are told to use the Taylor polynomial calculated previously to approximate the integral, that is:

$$\int_0^1 e^{x^2}\, dx \approx \int_0^1 p_3(x)\, dx = \int_0^1 (1+x^2)\, dx = x + \left.\frac{x^3}{3}\right|_0^1 = 1 + \frac{1}{3} = \frac{4}{3}.$$

(c) In this case, we are told to calculate the limit using the Taylor polynomial. Remember that since $f(x)$ is infinitely differentiable, we can use the theorem of the remainder to write:

$$f(x) = 1 + x^2 + \frac{f^4(\xi)}{4!}x^4 = 1 + x^2 + O(x^4)$$

where ξ is a value that depends on the point x where we evaluate the function, that is, $\xi(x)$.

Therefore:

$$\lim_{x\to 0}\frac{e^{x^2}-1}{x^2}=\lim_{x\to 0}\frac{1+x^2+O(x^4)-1}{x^2}=\lim_{x\to 0}\frac{x^2+O(x^4)}{x^2}$$

$$=\lim_{x\to 0}1+O(x^2)=1+0=1.$$

(d) *Note that $f(1)=e^1=e$. Therefore, we can approximate the value of the function at $x=1$ by the value of the Taylor polynomial at the point $x=1$, that is:*

$$e=f(1)\approx p_3(1)=1+1^2=2.$$

5.10 Solved Problems (Improper Integration)

Problem 5.19. Calculate the following improper integrals of the first kind:

(a) $\displaystyle\int_1^{+\infty}\frac{1}{1+x}\,\mathrm{d}x$

(b) $\displaystyle\int_1^{+\infty}e^{-x}\,\mathrm{d}x$

(c) $\displaystyle\int_1^{+\infty}x^{-4}\,\mathrm{d}x$

(d) $\displaystyle\int_0^{+\infty}e^{-\lambda x}\,\mathrm{d}x$

(e) $\displaystyle\int_1^{+\infty}\frac{1}{x^p}\,\mathrm{d}x$

Solution to the problem 5.19. (a) *We are asked to find the improper integral of the first kind* $\int_1^{+\infty}\frac{1}{1+x}\,\mathrm{d}x$.
We calculate the integral:

$$\int_1^{+\infty}\frac{1}{1+x}\,\mathrm{d}x=\lim_{b\to+\infty}\int_1^b\frac{1}{1+x}\,\mathrm{d}x$$

$$=\lim_{b\to+\infty}\left[\ln|1+x|\right]_1^b=\lim_{b\to+\infty}\left[\underbrace{\ln|1+b|}_{\to+\infty}-\ln(2)\right]=+\infty.$$

Since the real limit does not exist, the integral is divergent.

(b) *We are asked to find the improper integral of the first kind* $\int_1^{+\infty}e^{-x}\,\mathrm{d}x$.

We calculate the integral:

$$\int_1^{+\infty} e^{-x}\,\mathrm{d}x = \lim_{b\to+\infty}\int_1^b e^{-x}\,\mathrm{d}x$$

$$= \lim_{b\to+\infty}\left[-e^{-x}\right]_1^b = \lim_{b\to+\infty}\Big[\underbrace{-e^{-b}}_{\to 0}-(-e^{-1})\Big] = e^{-1}.$$

Since the real limit exists, the integral is convergent.

(c) *We are asked to find the improper integral of the first kind* $\int_1^{+\infty} x^{-4}\,\mathrm{d}x$. *We calculate the integral:*

$$\int_1^{+\infty} x^{-4}\,\mathrm{d}x = \lim_{b\to+\infty}\int_1^b x^{-4}\,\mathrm{d}x$$

$$= \lim_{b\to+\infty}\left[-\frac{1}{3}x^{-3}\right]_1^b = \lim_{b\to+\infty}\Big[\underbrace{-\frac{1}{3}b^{-3}}_{\to 0}-(-\frac{1}{3})\Big] = \frac{1}{3}.$$

Since the real limit exists, the integral is convergent.

(d) *We are asked to find the improper integral of the first kind* $\int_0^{+\infty} e^{-\lambda x}\,\mathrm{d}x$. *We calculate the integral:*

$$\int_0^{+\infty} e^{-\lambda x}\,\mathrm{d}x = \lim_{b\to+\infty}\int_0^b e^{-\lambda x}\,\mathrm{d}x = \lim_{b\to+\infty}\left[\frac{-e^{-\lambda x}}{\lambda}\right]_0^b$$

$$= \lim_{b\to+\infty}\left[\frac{-e^{-\lambda b}}{\lambda}+\frac{1}{\lambda}\right].$$

The value of this limit depends on the value of λ.
For $\lambda > 0$ *we have:*

$$\lim_{b\to+\infty}\Big[\underbrace{\frac{-e^{-\lambda b}}{\lambda}}_{\to 0}+\frac{1}{\lambda}\Big] = \frac{1}{\lambda}.$$

For $\lambda \leq 0$ *we have:*

$$\lim_{b\to+\infty}\Big[\underbrace{\frac{-e^{-\lambda b}}{\lambda}}_{\to +\infty}+\frac{1}{\lambda}\Big] = +\infty.$$

Therefore, the integral is convergent if $\lambda > 0$ *and divergent if* $\lambda \leq 0$.

(e) *We are asked to find the improper integral of the first kind* $\int_1^{+\infty} \frac{1}{x^p}\,\mathrm{d}x$. *We calculate the integral:*

$$\int_1^{+\infty} \frac{1}{x^p}\,\mathrm{d}x = \lim_{b\to+\infty}\int_1^b \frac{1}{x^p}\,\mathrm{d}x.$$

For $p = 1$ we have:

$$\begin{aligned}\lim_{b\to+\infty}\int_1^b \frac{1}{x^p}\,\mathrm{d}x &= \lim_{b\to+\infty}\int_1^b \frac{1}{x}\,\mathrm{d}x \\ &= \lim_{b\to+\infty}\left[\ln|x|\right]_1^b \\ &= \lim_{b\to+\infty}\Big[\underbrace{\ln|b|}_{\to+\infty} - \ln(1)\Big] = +\infty.\end{aligned}$$

For $p \neq 1$ we have:

$$\begin{aligned}\lim_{b\to+\infty}\int_1^b \frac{1}{x^p}\,\mathrm{d}x &= \lim_{b\to+\infty}\left[-\frac{x}{(p-1)x^p}\right]_1^b \\ &= \lim_{b\to+\infty}\left[-\frac{1}{(p-1)b^{p-1}} - \left(-\frac{1}{p-1}\right)\right].\end{aligned}$$

The value of this limit depends on the value of p.
For $p > 1$ we have:

$$\lim_{b\to+\infty}\Big[\underbrace{-\frac{1}{(p-1)b^{p-1}}}_{\to 0} + \frac{1}{p-1}\Big] = \frac{1}{p-1}.$$

For $p < 1$ we have:

$$\lim_{b\to+\infty}\Big[\underbrace{-\frac{1}{(p-1)}b^{p-1}}_{\to+\infty} + \frac{1}{p-1}\Big] = +\infty.$$

Therefore, the integral is convergent if $p > 1$ and divergent if $p \leq 1$.

Problem 5.20. Calculate the following integral:

$$I = \int_0^{+\infty} \frac{2x-1}{1+x^2}\,\mathrm{d}x$$

Hint: remember that $\lim_{x\to+\infty} \arctan(x) = \pi/2$.

Solution to the problem 5.20. *We are asked to calculate the integral:*

$$I = \int_0^{+\infty} \frac{2x-1}{1+x^2}\,\mathrm{d}x$$

We must calculate an improper integral of the first kind, therefore:

$$I = \int_0^{+\infty} \frac{2x-1}{1+x^2}\,\mathrm{d}x = \lim_{b\to+\infty}\int_0^b \frac{2x-1}{1+x^2}\,\mathrm{d}x$$

The resulting integral is an integral of a quotient of polynomials where the denominator is an irreducible polynomial, since $1+x^2 \neq 0,\ \forall x \in \mathbb{R}$.

$$\begin{aligned} I &= \int_0^{+\infty} \frac{2x-1}{1+x^2}\,\mathrm{d}x = \lim_{b\to+\infty} \int_0^b \frac{2x-1}{1+x^2}\,\mathrm{d}x = \lim_{b\to+\infty} \int_0^b \left[\frac{2x}{1+x^2} - \frac{1}{1+x^2}\right] \mathrm{d}x \\ &= \lim_{b\to+\infty} \left[\ln|1+x^2| - \arctan(x)\right]_0^b \\ &= \lim_{b\to+\infty} \left[\ln|1+b^2| - \arctan(b) - \ln|1| + \arctan(0)\right] \\ &= \lim_{b\to+\infty} \left[\ln|1+b^2| - \arctan(b)\right] = +\infty - \pi/2 = +\infty \end{aligned}$$

Problem 5.21. Consider the following integral:

$$\int_5^{+\infty} \frac{1}{x^2-16}\,\mathrm{d}x$$

Given that:

$$\frac{1}{x^2-16} = \frac{1}{8(x-4)} - \frac{1}{8(x+4)}$$

determine if it is a proper or improper integral. If it is improper, determine the type. If it is convergent, calculate its value.

Hint:

$$\ln(a) - \ln(b) = \ln(a/b)$$

Solution to the problem 5.21. *We must study the integral:*

$$\int_5^{+\infty} \frac{1}{x^2-16}\,\mathrm{d}x$$

First, let's examine the domain of the rational function. Since:

$$x^2-16 = (x-4)(x+4)$$

we have that the function $\frac{1}{x^2-16}$ *has vertical asymptotes at* $x=4$ *and* $x=-4$.

Therefore, in the interval $[5,+\infty)$, *the function is continuous and thus bounded, and the integral we need to calculate is an improper integral of the first kind (since one of the limits is infinite).*

Let's calculate the value to determine if it is convergent or divergent.

$$\int_5^{+\infty} \frac{1}{x^2-16}\,\mathrm{d}x = \lim_{b\to+\infty} \int_5^b \frac{1}{x^2-16}\,\mathrm{d}x$$

For simplicity, we will first calculate an antiderivative of the function and then substitute the value into the previous equation.

$$\begin{aligned} \int \frac{1}{x^2-16}\,\mathrm{d}x &= \int \left[\frac{A}{x-4} + \frac{B}{x+4}\right] \mathrm{d}x = \int \left[\frac{1}{8}\frac{1}{x-4} - \frac{1}{8}\frac{B}{x+4}\right] \mathrm{d}x \\ &= \frac{1}{8}\ln|x-4| - \frac{1}{8}\ln|x+4| = \frac{1}{8}\ln\left|\frac{x-4}{x+4}\right| \end{aligned}$$

since:

$$1 = A(x+4) + B(x-4) \quad \Longrightarrow \quad A = 1/8, \quad B = -1/8$$

Therefore, applying the Barrow rule we have:

$$\begin{aligned}
\int_5^{+\infty} \frac{1}{x^2-16}\,dx &= \lim_{b\to+\infty} \int_5^b \frac{1}{x^2-16}\,dx = \lim_{b\to+\infty} \left[\frac{1}{8}\ln\left|\frac{x-4}{x+4}\right|\right]_5^b \\
&= \lim_{b\to+\infty} \left[\frac{1}{8}\ln\left|\frac{b-4}{b+4}\right| - \frac{1}{8}\ln\left|\frac{5-4}{5+4}\right|\right] \\
&= \frac{1}{8}\ln|1| - \frac{1}{8}\ln\left|\frac{1}{9}\right| \\
&= -\frac{1}{8}\ln\left(\frac{1}{9}\right) = -\frac{1}{8}\left(\ln(1) - \ln(9)\right) \\
&= \frac{\ln(9)}{8} = \frac{\ln(3^2)}{8} = \frac{2\ln(3)}{8} = \frac{\ln(3)}{4}.
\end{aligned}$$

Therefore, the integral is convergent.

Problem 5.22. Consider the functions:

$$f(x) = \frac{2}{x}, \quad g(x) = \frac{2x}{x^2+1}$$

(a) Calculate the area of the region bounded by the graphs of the functions $f(x)$, $g(x)$, and the lines $x = 1$ and $x = 2$.

(b) Calculate the area of the region bounded by the graphs of the functions $f(x)$, $g(x)$ when $x \geq 1$.

Solution to the problem 5.22. *(a) We are asked to calculate the area of the region bounded by the graphs of the functions $f(x)$, $g(x)$, and the lines $x = 1$ and $x = 2$.*

- *We represent the given functions in the statement in Figure 5.48.*
- *Intersection points.*

  To find the intersection points between $f(x)$ and $g(x)$, we equal the functions:

$$f(x) = g(x) \quad \Leftrightarrow \quad \frac{2}{x} = \frac{2x}{x^2+1} \quad \Leftrightarrow \quad 2x^2 + 2 = 2x^2 \quad \Leftrightarrow \quad 2 = 0$$

  There is no point where $f(x)$ equals $g(x)$.

- *Area bounded by the functions in the interval $[1, 2]$.*

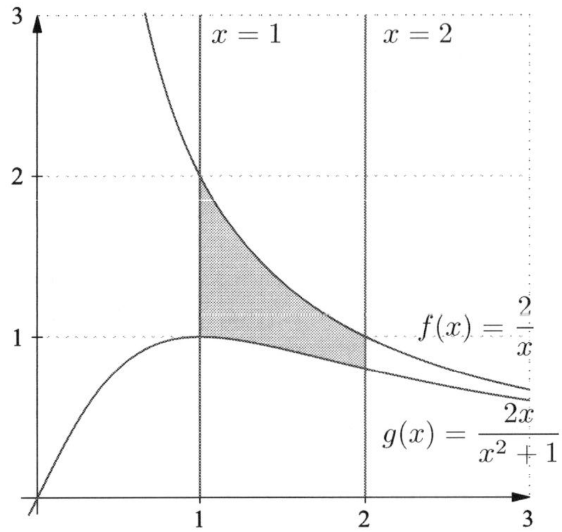

FIGURE 5.48
Graph of the functions $f(x) = \dfrac{2}{x}$ and $g(x) = \dfrac{2x}{x^2+1}$ and the lines $x = 1$ and $x = 2$.

To calculate the area bounded by the functions in the interval, we need to determine which function is greater:

$$\left.\begin{array}{l} f(1) = 2 \\ g(1) = \frac{2}{2} = 1 \end{array}\right\} \quad f(1) > g(1) \quad \Rightarrow \quad f(x) > g(x)$$

Therefore:

$$\begin{aligned}
\int_1^2 \left(\underbrace{\frac{2}{x}}_{f(x)} - \underbrace{\frac{2x}{x^2+1}}_{g(x)} \right) \mathrm{d}x &= \int_1^2 \frac{2}{x}\,\mathrm{d}x - \int_1^2 \frac{2x}{x^2+1}\,\mathrm{d}x \\
&= \Big[2\ln|x| \Big]_1^2 - \Big[\ln(x^2+1) \Big]_1^2 \\
&= 2\ln(2) - \cancel{2\ln(1)} - \ln(5) + \ln(2) \\
&= 3\ln(2) - \ln(5) \\
&= \ln(2^3) - \ln(5) = \ln\left(\tfrac{8}{5}\right) \simeq 0.47.
\end{aligned}$$

(b) *We are asked to calculate the area of the region bounded by the graphs of the functions $f(x)$ and $g(x)$ when $x \geq 1$.*

- *Figure 5.49 represents the area of the region bounded by the graphs of the functions $f(x)$ and $g(x)$ when $x \geq 1$.*

  To calculate this area, we must solve an improper integral of the first kind, as the

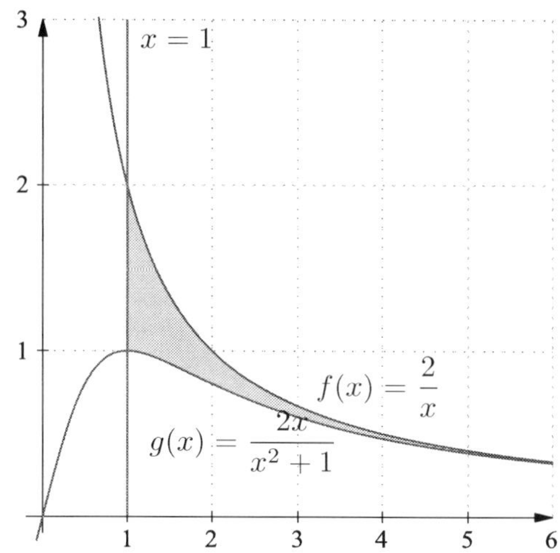

FIGURE 5.49
Area of the region bounded by the graphs of the functions $f(x)$ and $g(x)$ when $x \geq 1$.

upper limit is $+\infty$*. Thus:*

$$\int_1^{+\infty} (f(x) - g(x))\,\mathrm{d}x = \lim_{T\to\infty} \int_1^T (f(x) - g(x))\,\mathrm{d}x = \lim_{T\to\infty} \left[2\ln(x) - \ln(x^2+1)\right]$$

$$= \lim_{T\to\infty} (\underbrace{2\ln(T)}_{\ln(T^2)} - \ln(T^2+1) - \cancel{2\ln(1)} + \ln(2))$$

$$\stackrel{(*)}{=} \lim_{T\to\infty} \left(\underbrace{\ln\left(\underbrace{\frac{T^2}{T^2+1}}_{\to 1}\right)}_{\to 0} + \ln(2) \right) = \ln(2)$$

where in the penultimate step marked as (*) we used:

$$\ln(A) - \ln(B) = \ln\left(\frac{A}{B}\right)$$

Problem 5.23. Given the rational function:

$$f(x) = \frac{x^4}{x^4 - 1}$$

(a) Decompose the function $f(x)$ into partial fractions.

(b) Calculate an antiderivative of the function, i.e. $\int f(x)\,\mathrm{d}x$.

(c) Study the convergence of the improper integral $\int_1^{10} f(x)\,\mathrm{d}x$.

Solution to the problem 5.23. *(a) We are asked to decompose the function $f(x)$ into partial fractions. First, we must decompose the denominator polynomial:*

$$\frac{x^4}{x^4-1} = \frac{x^4-1+1}{x^4-1} = 1 + \frac{1}{x^4-1}$$

$$x^4 - 1 = \underbrace{(x^2+1)}_{0}\ \underbrace{(x^2-1)}_{(x+1)(x-1)} = (x^2+1)(x+1)(x-1)$$

Next, we find the coefficients A, B, C, and D that allow expressing the quotient of polynomials as a sum of quotients of lower-order polynomials:

$$\begin{aligned}\frac{1}{x^4-1} &= \frac{A}{x+1} + \frac{B}{x-1} + \frac{Cx+D}{x^2+1}\\ &= \frac{A(x-1)(x^2+1) + B(x+1)(x^2+1) + (Cx+D)(x+1)(x-1)}{(x+1)(x-1)(x^2+1)}\end{aligned}$$

By equating the numerators, we have:

$$1 = A(x-1)(x^2+1) + B(x+1)(x^2+1) + (Cx+D)(x+1)(x-1)$$

To determine the values of A, B, C, and D, we substitute into the previous equation the different values of x that make the denominator of the function to be decomposed zero:

$$x = -1, \quad 1 = -4A \quad \Leftrightarrow \quad A = \frac{-1}{4}$$

$$x = 1, \quad 1 = 4B \quad \Leftrightarrow \quad B = \frac{1}{4}$$

$$x = 0, \quad 1 = -A + B - D \quad \Leftrightarrow \quad D = \frac{-1}{2}$$

$$x = 2, \quad 1 = 5A + 15B + 6C + 3D \quad \Leftrightarrow \quad C = 0$$

Therefore:

$$\frac{x^4}{x^4-1} = 1 + \frac{1}{x^4-1} = 1 + \frac{-1/4}{x+1} + \frac{1/4}{x-1} + \frac{-1/2}{x^2+1}$$

(b) We are asked to find an antiderivative of the function $f(x)$, i.e., $\int f(x)\,dx$. For the calculation, we will use the decomposition of $f(x)$ into partial fractions carried out in part (a):

$$\begin{aligned}\int \frac{x^4}{x^4-1}\,dx &= \int \left(1 + \frac{1}{x^4-1}\right) dx\\ &= \int \left(1 - \frac{1}{4}\cdot\frac{1}{x+1} + \frac{1}{4}\cdot\frac{1}{x-1} - \frac{1}{2}\cdot\frac{1}{x^2+1}\right) dx\\ &= \underbrace{\int 1\,dx}_{I_1} - \underbrace{\frac{1}{4}\int \frac{1}{x+1}\,dx}_{I_2} + \underbrace{\frac{1}{4}\int \frac{1}{x-1}\,dx}_{I_3} - \underbrace{\frac{1}{2}\int \frac{1}{x^2+1}\,dx}_{I_4}\end{aligned}$$

$$I_1 = \int 1\,dx = x$$
$$I_2 = \int \frac{1}{x+1}\,dx = \ln|x+1|$$
$$I_3 = \int \frac{1}{x-1}\,dx = \ln|x-1|$$
$$I_4 = \int \frac{1}{x^2+1}\,dx = \arctan(x)$$

Therefore:

$$\int \frac{x^4}{x^4-1}\,dx = x - \frac{1}{4}\ln|x+1| + \frac{1}{4}\ln|x-1| - \frac{1}{2}\arctan(x) + C$$

(c) *We are asked to study the convergence of the improper integral* $\int_1^{10} f(x)\,dx$.

Since the denominator of $f(x)$ *becomes zero at* $x = 1$, *the function has a vertical asymptote at* $x = 1$. *Therefore:*

$$\int_1^{10} f(x)\,dx = \lim_{\delta\to 0^+} \int_{1+\delta}^{10} f(x)\,dx \underbrace{=}_{F'(x)=f(x)} \lim_{\delta\to 0^+} [F(10) - F(1+\delta)]$$

Using the result from part (b):

$$\lim_{\delta\to 0^+} \Bigg(10 - \frac{1}{4}\ln(11) + \frac{1}{4}\ln(9) - \frac{1}{2}\arctan(10) -$$

$$- \underbrace{1+\delta}_{\to 1} + \frac{1}{4}\underbrace{\ln(2+\delta)}_{\to \ln 2} - \underbrace{\frac{1}{4}\ln(\delta)}_{\to -\infty} + \frac{1}{2}\underbrace{\arctan(1+\delta)}_{\to \arctan(1)} \Bigg) = +\infty.$$

Therefore, the improper integral is divergent.

Problem 5.24. Prove that if f is a differentiable function such that:

$$\lim_{x\to+\infty} f(x)e^{-sx} = 0,$$

then:

$$\int_0^{+\infty} f'(x)e^{-sx}\,\mathrm{d}x = s\int_0^{+\infty} f(x)e^{-sx}\,\mathrm{d}x - f(0).$$

Hint: use the definition of improper integral and the integration by parts rule.

Solution to the problem 5.24. *We need to calculate an improper integral of the first kind; therefore:*

$$\int_0^{+\infty} f'(x)e^{-sx}\,\mathrm{d}x = \lim_{b\to+\infty} \int_0^b f'(x)e^{-sx}\,\mathrm{d}x.$$

To solve the integral from 0 to b, we will apply the integration by parts rule:

$$\int_0^b f'(x)e^{-sx}\,\mathrm{d}x \overset{(*)}{=}$$

$$(*)\left[\begin{array}{ccc} u = e^{-sx} & \Longrightarrow & u' = -se^{-sx} \\ v' = f'(x) & \Longrightarrow & v = f(x) \end{array}\right]$$

$$= \left[f(x)e^{-sx}\right]_0^b - \int_0^b \left[-se^{-sx}f(x)\right]\mathrm{d}x$$

$$= f(b)e^{-sb} - f(0)e^{-s0} + \int_0^b se^{-sx}f(x)\,\mathrm{d}x$$

$$= f(b)e^{-sb} - f(0) + s\int_0^b e^{-sx}f(x)\,\mathrm{d}x$$

Therefore:

$$\begin{aligned} \int_0^{+\infty} f'(x)e^{-sx}\,\mathrm{d}x &= \lim_{b\to+\infty}\int_0^b f'(x)e^{-sx}\,\mathrm{d}x \\ &= \lim_{b\to+\infty}\left(f(b)e^{-sb} - f(0) + s\int_0^b e^{-sx}f(x)\,\mathrm{d}x\right) \\ &= s\int_0^{+\infty} e^{-sx}f(x)\,\mathrm{d}x - f(0) \qquad \textit{since } \lim_{b\to+\infty} f(b)e^{-sb} = 0. \end{aligned}$$

Problem 5.25. Prove that the following integrals are finite and find an upper bound and a lower bound for their value:

(a) $\displaystyle\int_1^{+\infty} \frac{\cos(x)}{x^2}\,\mathrm{d}x$

(b) $\displaystyle\int_1^{+\infty} \frac{\sin(x)+\cos(x)}{x^2+x+1}\,\mathrm{d}x$

Solution to the problem 5.25. *(a) We need to prove that the integral:*

$$\int_1^{+\infty} \frac{\cos(x)}{x^2}\,\mathrm{d}x$$

is finite and find an upper bound and a lower bound for its value.
Let:

$$L = \int_1^{+\infty} \frac{\cos(x)}{x^2}\,\mathrm{d}x,$$

then $-|L| \le L \le |L|$. *Additionally:*

$$|L| = \left|\int_1^{+\infty} \frac{\cos(x)}{x^2}\,\mathrm{d}x\right| \le \int_1^{+\infty} \left|\frac{\cos(x)}{x^2}\right|\mathrm{d}x \le \int_1^{+\infty} \frac{1}{x^2}\,\mathrm{d}x,$$

since:

- $|\cos(x)| \leq 1$
- $|x^2| = x^2$

Therefore:

$$|L| \leq \int_1^{+\infty} \frac{1}{x^2}\,\mathrm{d}x = \lim_{b\to+\infty} \int_1^b \frac{1}{x^2}\,\mathrm{d}x = \lim_{b\to+\infty} \left[-\frac{1}{x}\right]_1^b$$

$$= \lim_{b\to+\infty} \Big[\underbrace{-\frac{1}{x}}_{\to 0} - (-1)\Big] = 1.$$

Thus, the integral is convergent.

(b) *We need to prove that the integral:*

$$\int_1^{+\infty} \frac{\sin(x)+\cos(x)}{x^2+x+1}\,\mathrm{d}x$$

is finite and find an upper bound and a lower bound for its value.
Let:

$$L = \int_1^{+\infty} \frac{\sin(x)+\cos(x)}{x^2+x+1}\,\mathrm{d}x,$$

then $-|L| \leq L \leq |L|$*. Additionally:*

$$|L| \leq \int_1^{+\infty} \left|\frac{\sin(x)+\cos(x)}{x^2+x+1}\right| \mathrm{d}x \leq \int_1^{+\infty} \frac{2}{x^2+x+1}\,\mathrm{d}x,$$

since:

- $|\sin(x)+\cos(x)| \leq |\sin(x)| + |\cos(x)| \leq 2$
- *since we integrate in the domain* $x \in [1, +\infty)$*, we have:*

$$x^2+x+1 \geq 0, \quad |x^2+x+1| = x^2+x+1$$

Additionally, since:

$$x^2+x+1 \geq x^2 \quad \Longrightarrow \quad \frac{1}{x^2+x+1} \leq \frac{1}{x^2}$$

we have:

$$|L| \leq \int_1^{+\infty} \frac{2}{x^2+x+1}\,\mathrm{d}x \leq \int_1^{+\infty} \frac{2}{x^2}\,\mathrm{d}x$$

$$= \lim_{b\to+\infty} \int_1^b \frac{2}{x^2}\,\mathrm{d}x = \lim_{b\to+\infty} \left[-\frac{2}{x}\right]_1^b$$

$$= \lim_{b\to+\infty} \left[-\frac{2}{b} + 2\right] = 2.$$

Thus, the integral is convergent.

Problem 5.26. Calculate the following improper integrals of the second kind:

(a) $\displaystyle\int_0^1 \frac{1}{x}\,\mathrm{d}x$

(b) $\displaystyle\int_1^e \frac{1}{x\sqrt[3]{\ln(x)}}\,\mathrm{d}x$

(c) $\displaystyle\int_0^1 x^{-1/3}\,\mathrm{d}x$

(d) $\displaystyle\int_0^1 \frac{1}{\sqrt{x}}\,\mathrm{d}x$

Solution to the problem 5.26. *(a) We need to calculate the improper integral of the second kind $\int_0^1 \frac{1}{x}\,\mathrm{d}x$. Note that the function $f(x) = \frac{1}{x}$ is not bounded at $x = 0$. Therefore:*

$$\int_0^1 \frac{1}{x}\,\mathrm{d}x = \lim_{\delta\to 0^+}\int_\delta^1 \frac{1}{x}\,\mathrm{d}x$$

$$= \lim_{\delta\to 0^+}[\ln|x|]_1^b = \lim_{\delta\to 0^+}\left[\ln(1) - \underbrace{\ln|\delta|}_{\to -\infty}\right] = +\infty.$$

Thus, the integral is divergent.

(b) We need to calculate the improper integral of the second kind $\int_1^e \frac{1}{x\sqrt[3]{\ln(x)}}\,\mathrm{d}x$. Note that the function $f(x) = \frac{1}{x\sqrt[3]{\ln(x)}}$ is not bounded at $x = 1$. Therefore:

$$\int_1^e \frac{1}{x\sqrt[3]{\ln(x)}}\,\mathrm{d}x = \lim_{\delta\to 0^+}\int_{1+\delta}^e \frac{1}{x\sqrt[3]{\ln(x)}}\,\mathrm{d}x.$$

To solve this integral, we make the substitution $t = \ln(x)$:

$$\mathrm{d}t = \frac{1}{x}\,\mathrm{d}x$$
$$x_1 = 1 \Rightarrow t_1 = 0$$
$$x_2 = e \Rightarrow t_2 = 1$$

Substituting:

$$\lim_{\delta\to 0^+}\int_{1+\delta}^e \frac{1}{x\sqrt[3]{\ln(x)}}\,\mathrm{d}x = \lim_{\delta\to 0^+}\int_\delta^1 \frac{1}{\sqrt[3]{t}}dt = \lim_{\delta\to 0^+}\left[\frac{3}{2}t^{\frac{2}{3}}\right]_\delta^1$$

$$= \lim_{\delta\to 0^+}\left[\frac{3}{2} - \underbrace{\frac{3}{2}\delta^{\frac{2}{3}}}_{\to 0}\right] = \frac{3}{2}.$$

Therefore, the integral is convergent.

(c) *We need to calculate the improper integral of the second kind* $\int_0^1 t^{-1/3}dt$. *Note that the function* $f(x) = \frac{1}{x}$ *is not bounded at* $x = 0$. *Indeed, see that this integral is the same as the one we calculated in part (b) after making the substitution. Recall that the singularity in the function from part (b):*

$$f(x) = \frac{1}{x\sqrt[3]{\ln(x)}}$$

is at $x = 1$, *and we made the substitution* $t = \ln(x)$. *Thus, the singularity in the resulting function after the substitution must be at* $t = \ln(1) = 0$, *as we observed earlier.*
Therefore, the integral $\int_0^1 t^{-1/3}dt$ *is convergent, and its value is* $\frac{3}{2}$.

(d) *We need to calculate the improper integral of the second kind* $\int_0^1 \frac{1}{\sqrt{x}}\,\mathrm{d}x$. *Note that the function* $f(x) = \frac{1}{\sqrt{x}}$ *is not bounded at* $x = 0$. *Therefore:*

$$\int_0^1 \frac{1}{\sqrt{x}}\,\mathrm{d}x = \lim_{\delta\to 0^+}\int_\delta^1 \frac{1}{\sqrt{x}}\,\mathrm{d}x$$
$$= \lim_{\delta\to 0^+}\left[2\sqrt{x}\right]_1^b = \lim_{\delta\to 0^+}\left[2\sqrt{1} - \underbrace{2\sqrt{\delta}}_{\to 0}\right] = 2$$

Thus, the integral is convergent.

Problem 5.27. Does it make sense to calculate the area bounded by the curve $f(x) = \frac{1}{x^2-4}$ from $x = -2$ to $x = 2$?

Solution to the problem 5.27. *To determine if it makes sense to calculate the area bounded by the curve* $f(x) = \frac{1}{x^2-4}$ *from* $x = -2$ *to* $x = 2$, *we need to check if the integral:*

$$\int_{-2}^{2} \frac{1}{x^2-4}\,\mathrm{d}x$$

is convergent. We can see that the function $f(x)$ *is not bounded at* $x = -2$ *or* $x = 2$; *therefore, the integral is improper of the second kind. We also note that the function is symmetric with respect to the y-axis, as* $f(x) = f(-x)$; *hence, we can calculate the area bounded by the curve in the interval [-2,2] as twice the area bounded by the curve in the interval [0,2]. Thus:*

$$\int_{-2}^{2} \frac{1}{x^2-4}\,\mathrm{d}x = 2\int_0^2 \frac{1}{x^2-4}\,\mathrm{d}x$$
$$= 2\lim_{\delta\to 0^+}\int_0^{2-\delta} \frac{1}{x^2-4}\,\mathrm{d}x$$
$$= \lim_{\delta\to 0^+} 2\left(\int_0^{2-\delta} \frac{1}{x^2-4}\,\mathrm{d}x\right).$$

To solve this integral, we decompose into partial fractions:

$$\frac{1}{x^2-4} = \frac{1}{(x+2)(x-2)}$$
$$= \frac{A}{x+2} + \frac{B}{x-2}$$
$$= \frac{Ax-2A+Bx+2B}{(x+2)(x-2)}$$

where, to ensure the sum of the polynomial fractions is equivalent to the previous fraction, $Ax-2A+Bx+2B=1$ *must hold. Thus:*

$$Ax+Bx=0 \quad \Leftrightarrow \quad A+B=0 \quad \Leftrightarrow \quad A=-B$$
$$-2A+2B=1 \quad \Leftrightarrow \quad 2B+2B=1 \quad \Leftrightarrow \quad B=\frac{1}{4} \quad \text{and} \quad A=-\frac{1}{4}$$

Then:

$$\frac{1}{x^2-4} = \frac{-\frac{1}{4}}{x+2} + \frac{\frac{1}{4}}{x-2}$$

Substituting into the integral:

$$\lim_{\delta\to 0^+} 2\left(\int_0^{2-\delta} \frac{1}{x^2-4}\,dx\right) = \lim_{\delta\to 0^+} \int_0^{2} -\delta - \frac{1}{4}\frac{1}{x+2} + \frac{1}{4}\frac{1}{x-2}\,dx$$
$$= \lim_{\delta\to 0^+} \left[-\frac{1}{4}\ln|x+2| + \frac{1}{4}\ln|x-2|\right]_0^{2-\delta}$$
$$= \lim_{\delta\to 0^+} \left[-\frac{1}{4}\ln|2-\delta+2| + \frac{1}{4}\ln|2-\delta-2| - (-\frac{1}{4}\ln|2| + \frac{1}{4}\ln|-2|)\right]$$
$$= \lim_{\delta\to 0^+} \left[-\frac{1}{4}\ln|4-\delta| + \frac{1}{4}\ln|-\delta| + \frac{1}{4}\ln|2| - \frac{1}{4}\ln|-2|)\right]$$
$$= \lim_{\delta\to 0^+} \left[-\frac{1}{4}\underbrace{\ln(4-\delta)}_{\to\ln(4)} + \frac{1}{4}\underbrace{\ln(\delta)}_{\to-\infty} + \frac{1}{4}\ln(2) - \frac{1}{4}\ln(2)\right]$$
$$= -\infty$$

Thus, the integral is divergent.

Problem 5.28. Study the convergence of

$$\int_a^\infty \frac{1}{x^\alpha}$$

depending on the values of α for $a>0$. Hint: Study separately the cases $\alpha=1$, $\alpha<1$, and $\alpha>1$.

Solution to the problem 5.28. *We need to determine if the integral* $\int_a^\infty \frac{1}{x^\alpha}$ *is convergent for different values of* α. *We will evaluate it first for* $\alpha=1$ *and then for* $\alpha<1$ *and* $\alpha>1$, *since in the last two cases the antiderivative of the function is the same.*

- *For $\alpha = 1$:*

$$\int_a^\infty \frac{1}{x^\alpha}\,\mathrm{d}x = \int_a^\infty \frac{1}{x}\,\mathrm{d}x = \lim_{b\to+\infty}\int_a^b \frac{1}{x}\,\mathrm{d}x$$
$$= \lim_{b\to+\infty}\left[\ln|x|\right]_a^b$$
$$= \lim_{b\to+\infty}\Big[\underbrace{\ln|b|}_{\to+\infty} - \underbrace{\ln|a|}_{\mathbb{R}}\Big] = +\infty.$$

- *For $\alpha < 1$ and $\alpha > 1$:*

$$\int_a^\infty \frac{1}{x^\alpha}\,\mathrm{d}x = \int_a^\infty x^{-\alpha}\,\mathrm{d}x = \lim_{b\to+\infty}\int_a^b x^{-\alpha}\,\mathrm{d}x$$
$$= \lim_{b\to+\infty}\left[-\frac{x^{-\alpha+1}}{\alpha-1}\right]_a^b$$
$$= \lim_{b\to+\infty}\left[-\frac{b^{-\alpha+1}}{\alpha-1} - \left(-\frac{a^{-\alpha+1}}{\alpha-1}\right)\right]$$
$$= \lim_{b\to+\infty}\left[\frac{b^{1-\alpha}}{1-\alpha} + \frac{a^{1-\alpha}}{\alpha-1}\right]$$

Next, we calculate the limit for each of the two cases:

  - *For $\alpha < 1$:*

$$\lim_{b\to+\infty}\Big[\underbrace{\frac{b^{1-\alpha}}{1-\alpha}}_{\to+\infty} + \frac{a^{1-\alpha}}{\alpha-1}\Big] = +\infty$$

  - *For $\alpha > 1$:*

$$\lim_{b\to+\infty}\left[\frac{b^{1-\alpha}}{1-\alpha} + \frac{a^{1-\alpha}}{\alpha-1}\right] = \lim_{b\to+\infty}\Big[\underbrace{\frac{1}{(1-\alpha)(b^{\alpha-1})}}_{\to 0} + \frac{a^{1-\alpha}}{\alpha-1}\Big]$$
$$= \frac{a^{1-\alpha}}{\alpha-1}$$

Thus, the integral is convergent only for $\alpha > 1$.

Part V

Linear Algebra

6

Linear Algebra: Matrices and Systems of Equations

«*Mathematics is the music of reason.*» James Joseph Sylvester, British mathematician (1814-1897).

In this chapter, we present the concept of a matrix, which has been used since antiquity to solve several systems of linear equations together. Although the idea of a matrix is ancient, the current term, matrix (*matrix*, in English), was not first used until 1848 by the British mathematician James Joseph Sylvester. Although matrices are closely linked to the resolution of linear systems, in this chapter we first present matrices, their types, and the fundamental concepts of rank and determinant. In the second part of the chapter, we introduce systems of linear equations and how matrix expression can help solve them easily. Finally, in this chapter, we present some notions of space geometry, such as the equation of the plane and the line, and their relationship with systems of equations.

DOI: 10.1201/9781003399735-6

6.1 Matrices

Definition 6.1 (Matrix)**.** We call a **matrix** of real numbers A of dimension $m \times n$ a set of elements of $\mathbb{R}$ arranged in m rows and n columns as follows:

$$A = \begin{pmatrix} a_{11} & a_{12} & \cdots & a_{1n} \\ a_{21} & a_{22} & \cdots & a_{2n} \\ \vdots & \vdots & \ddots & \vdots \\ a_{m1} & a_{m2} & \cdots & a_{mn} \end{pmatrix},$$

where $a_{ij} \in \mathbb{R},\ i = 1, \ldots, m,\ j = 1, \ldots, n.$

The set of all matrices with elements in $\mathbb{R}$ and of m rows and n columns is denoted by $\mathscr{M}_{m\times n}(\mathbb{R})$.

Remark 6.1. A matrix with m rows and n columns can also be represented by:

$$A = (a_{ij})\ \begin{matrix} i = 1, \ldots, m \\ j = 1, \ldots, n. \end{matrix}$$

The subscripts of each element indicate the position of the number within the matrix. The first subscript indicates the row and the second subscript, the column.

Proposed Exercise 6.1. Write the matrix $A = (a_{ij})$ of dimension 3×2, where $a_{ij} = 2i+j$. Also, write the matrices $B = (b_{ij})$ of dimension 4×3, where $b_{ij} = i/j$, $C = (c_{ij})$ of dimension 2×4 where $c_{ij} = (-1)^i + (-1)^j$ and $D = (d_{ij})$ of dimension 3×3, where $d_{ij} = j^i$.

6.1.1 Types of Matrices

Definition 6.2 (Matrix Equality)**.** We say that two matrices $A = (a_{ij})$ and $B = (b_{ij})$ are equal if:

$$a_{ij} = b_{ij}, \quad \forall\, i, j$$

Example 6.1. *The following matrices are not equal, since the elements located in the second row and second column are different:*

$$A = \begin{pmatrix} 1 & 2 \\ 3 & \boxed{4} \end{pmatrix} \neq \begin{pmatrix} 1 & 2 \\ 3 & \boxed{5} \end{pmatrix} = B$$

that is, $a_{22} \neq b_{22}$.

Remark 6.2. It is evident that for two matrices to be equal, they must have the same number of rows and the same number of columns.

Definition 6.3 (Column Vector). A **column vector** is a matrix that has only one column, that is, of dimension $n \times 1$:

$$\begin{pmatrix} x_1 \\ x_2 \\ \vdots \\ x_n \end{pmatrix}$$

Example 6.2. *The matrix:*

$$\begin{pmatrix} 1 \\ 5 \\ 3 \end{pmatrix}$$

is an example of a column vector.

Definition 6.4 (Row Vector). A **row vector** is a matrix that has only one row, that is, of dimension $1 \times n$:

$$\begin{pmatrix} x_1 & x_2 & \cdots & x_n \end{pmatrix}$$

Example 6.3. *The matrix:*

$$\begin{pmatrix} 5 & 9 \end{pmatrix}$$

is an example of a row vector.

6.1.2 Square Matrices

Definition 6.5 (Square Matrix). A **square matrix** is a matrix where the number of rows is equal to the number of columns ($m = n$). Sometimes this matrix is called a **matrix of order** n.

Definition 6.6 (Main Diagonal and Trace). The main diagonal of a square matrix consists of the elements $a_{ii}, i = 1, 2, \ldots, n$.

The sum of the elements of the main diagonal is called the **trace**:

$$\operatorname{tr}(A) = a_{11} + a_{22} + \ldots + a_{nn} = \sum_{i=1}^{n} a_{ii}.$$

Example 6.4. *Given the matrix:*

$$A = \begin{pmatrix} \boxed{1} & 2 & 0 & -1 \\ 3 & \boxed{-2} & 5 & 10 \\ 20 & 0 & \boxed{4} & 1 \\ 1 & 1 & 0 & \boxed{3} \end{pmatrix},$$

the main diagonal is formed by the numbers $1, -2, 4,$ *and* 3 *and its trace is:*

$$\operatorname{tr}(A) = 1 + (-2) + 4 + 3 = 6.$$

Definition 6.7 (Diagonal Matrix)**.** A matrix $A = (a_{ij})$ is **diagonal** if it is square and:

$$a_{ij} = 0 \text{ when } i \neq j,$$

that is, when all the elements that are not on the main diagonal are zero.

Example 6.5. *The matrix:*

$$\begin{pmatrix} 1 & 0 & 0 \\ 0 & 4 & 0 \\ 0 & 0 & -1 \end{pmatrix}$$

is diagonal.

Remark 6.3. Sometimes the diagonal matrix is written as:

$$\operatorname{diag}(a_{11}, a_{22}, \ldots, a_{nn})$$

Example 6.6. *The matrix:*

$$A = \begin{pmatrix} 1 & 0 & 0 \\ 0 & 4 & 0 \\ 0 & 0 & -1 \end{pmatrix}$$

can be represented as:

$$A = diag(1, 4, -1)$$

Definition 6.8 (Identity Matrix)**.** The **identity matrix** or **unit matrix** I is a diagonal matrix where the elements of the main diagonal are all 1 and the rest of the elements are 0; that is:

$$a_{ii} = 1,$$
$$a_{ij} = 0, \ i \neq j$$

Remark 6.4. The identity matrix of order n is usually denoted I_n.

Example 6.7. *The matrix:*

$$I_3 = \begin{pmatrix} 1 & 0 & 0 \\ 0 & 1 & 0 \\ 0 & 0 & 1 \end{pmatrix}$$

is called the unit matrix of order 3.

Definition 6.9 (Zero Matrix)**.** The zero matrix is one that has all elements equal to zero, that is:

$$a_{ij} = 0.$$

Remark 6.5. The zero matrix of order n is usually denoted $\mathbf{0}_n$.

Example 6.8. *The matrix*

$$\mathbf{0}_3 = \begin{pmatrix} 0 & 0 & 0 \\ 0 & 0 & 0 \\ 0 & 0 & 0 \end{pmatrix}$$

is called the zero matrix of order 3.

Definition 6.10 (Lower Triangular Matrix)**.** A square matrix is called **lower triangular** if all the elements that are above the main diagonal are zero, that is:

$$a_{ij} = 0, \quad j > i.$$

Example 6.9. *The matrix:*

$$L = \begin{pmatrix} 1 & \boxed{0} & \boxed{0} \\ 4 & 5 & \boxed{0} \\ 2 & 5 & 1 \end{pmatrix}$$

is lower triangular.

Definition 6.11 (Upper Triangular Matrix)**.** A square matrix is called **upper triangular** if all the elements that are below the main diagonal are zero, that is:

$$a_{ij} = 0, \quad j < i.$$

Example 6.10. *The matrix:*

$$U = \begin{pmatrix} 1 & 3 & 0 \\ \boxed{0} & 5 & 6 \\ \boxed{0} & \boxed{0} & 2 \end{pmatrix}$$

is upper triangular.

Definition 6.12 (Symmetric Matrix)**.** A square matrix $A = (a_{ij})$ is symmetric if:

$$a_{ij} = a_{ji}.$$

Example 6.11. *The matrix:*

$$S = \begin{pmatrix} 11 & 12 & 13 & 14 \\ 12 & 22 & 23 & 24 \\ 13 & 23 & 33 & 34 \\ 14 & 24 & 34 & 44 \end{pmatrix}$$

is symmetric.

Definition 6.13 (Antisymmetric Matrix)**.** A square matrix $A = (a_{ij})$ is antisymmetric if:

$$a_{ij} = -a_{ji}.$$

Example 6.12. *The matrix:*

$$A = \begin{pmatrix} 0 & 12 & 13 & 14 \\ -12 & 0 & 23 & 24 \\ -13 & -23 & 0 & 34 \\ -14 & -24 & -34 & 0 \end{pmatrix}$$

is antisymmetric.

Proposition 6.1. The elements of the main diagonal of an antisymmetric matrix are zero.

PROOF. We know that if we have an antisymmetric matrix, then:

$$a_{ij} = -a_{ji}.$$

For the elements of the main diagonal, in particular, we have that:

$$a_{ii} = -a_{ii}.$$

This implies that $2a_{ii} = 0$, or in other words, $a_{ii} = 0$. That is, the elements of the main diagonal are all zero. □

Proposition 6.2. Every square matrix Q can be expressed as the sum of a symmetric matrix S and an antisymmetric matrix A, that is:

$$Q = S + A.$$

PROOF. Let the square matrix of order n be:

$$Q = \begin{pmatrix} q_{11} & q_{12} & \cdots & q_{1n} \\ q_{21} & q_{22} & \cdots & q_{2n} \\ \vdots & \vdots & \ddots & \vdots \\ q_{n1} & q_{n2} & \cdots & q_{nn} \end{pmatrix},$$

and the matrices:

$$S = \begin{pmatrix} s_{11} & s_{12} & \cdots & s_{1n} \\ s_{12} & s_{22} & \cdots & s_{2n} \\ \vdots & \vdots & \ddots & \vdots \\ s_{1n} & s_{2n} & \cdots & s_{nn} \end{pmatrix}$$

$$A = \begin{pmatrix} 0 & a_{12} & \cdots & a_{1n} \\ -a_{21} & 0 & \cdots & a_{2n} \\ \vdots & \vdots & \ddots & \vdots \\ -a_{1n} & -a_{2n} & \cdots & 0 \end{pmatrix}$$

Since matrix A has all zeros on the main diagonal, it is evident that the diagonal of the symmetric matrix S must coincide with the diagonal of matrix Q, that is:

$$a_{ii} = 0$$

$$s_{ii} = q_{ii}.$$

On the other hand, for an element located above the diagonal and another located below it ($i < j$), we have:

$$q_{ij} = s_{ij} + a_{ij}$$
$$q_{ji} = s_{ij} - a_{ij}$$

Adding and subtracting both equations:

$$s_{ij} = \frac{q_{ij} + q_{ji}}{2}$$
$$a_{ij} = \frac{q_{ij} - q_{ji}}{2}$$

which proves the existence of these two matrices S and A. □

Definition 6.14 (Transpose)**.** The **transpose** of a matrix A is the matrix obtained by swapping rows and columns. If a matrix is of dimension $m \times n$, its transpose is of dimension $n \times m$, and is denoted by A^t. More specifically, if $A = (a_{ij})$, then $A^t = (a_{ji})$.

Proposition 6.3. Every square matrix Q can be expressed as the sum of a symmetric matrix S and an antisymmetric matrix A, that is:

$$Q = S + A.$$

In particular,

$$S = \frac{Q + Q^t}{2}$$
$$A = \frac{Q - Q^t}{2}$$

PROOF. This is immediate, considering the proof of Proposition 6.2.

Example 6.13. *Let the matrix:*

$$A = \begin{pmatrix} 1 & 0 & 2 \\ 3 & 1 & 0 \\ 0 & 4 & 5 \\ 2 & 8 & 1 \end{pmatrix}$$

whose transpose is:

$$A^t = \begin{pmatrix} 1 & 3 & 0 & 2 \\ 0 & 1 & 4 & 8 \\ 2 & 0 & 5 & 1 \end{pmatrix}$$

Proposition 6.4. Transposing a matrix twice gives the same matrix: $(A^t)^t = A$.

PROOF. If $A = (a_{ij})$, then $A^t = (a_{ji})$. In turn, if we transpose again, $(A^t)^t = (a_{ij}) = A$.
□

Example 6.14. *Let the matrix:*

$$A = \begin{pmatrix} 1 & 0 \\ 3 & 1 \\ 0 & 4 \end{pmatrix}$$

Then:

$$A^t = \begin{pmatrix} 1 & 3 & 0 \\ 0 & 1 & 4 \end{pmatrix}, \qquad (A^t)^t = \begin{pmatrix} 1 & 0 \\ 3 & 1 \\ 0 & 4 \end{pmatrix}$$

That is, $(A^t)^t = A$.

6.1.3 Echelon Matrices

Definition 6.15 (Row-Echelon Matrix)**.** A matrix A of dimension $m \times n$ is called **row-echelon**, where $r \leq m$, if the elements of the last $m - r$ rows are all zero, and if the elements below the main diagonal of the submatrix formed by the first r rows and r columns are also zero.

Example 6.15. *The matrix:*

$$\left(\begin{array}{cccccc} 1 & 3 & 0 & 3 & 1 & 3 \\ \boxed{0} & 5 & 6 & 2 & 4 & 2 \\ \boxed{0} & \boxed{0} & 2 & 0 & 2 & 0 \\ \hline 0 & 0 & 0 & 0 & 0 & 0 \\ 0 & 0 & 0 & 0 & 0 & 0 \end{array}\right)$$

is a row-echelon matrix of dimension 5×6.

6.2 Matrix Operations

In the set of matrices of real numbers of dimension $m \times n$, $\mathscr{M}_{m \times n}(\mathbb{R})$, three operations are defined:

- Matrix addition
- Multiplication of a matrix by a number
- Matrix multiplication

6.2.1 Matrix Addition

Definition 6.16 (Addition)**.** Let $A = (a_{ij})$ and $B = (b_{ij})$ be two matrices of dimension $m \times n$. The **sum** $A + B$ of the two matrices is the matrix $m \times n$:

$$A + B = (a_{ij} + b_{ij}).$$

Example 6.16 (Matrix Addition).

$$\begin{pmatrix} 1 & 2 & 0 \\ 0 & -3 & 1 \\ 2 & 4 & 0 \end{pmatrix} + \begin{pmatrix} 2 & 0 & 1 \\ 0 & 4 & 2 \\ 0 & -3 & 8 \end{pmatrix} = \begin{pmatrix} 1+2 & 2+0 & 0+1 \\ 0+0 & -3+4 & 1+2 \\ 2+0 & 4+(-3) & 0+8 \end{pmatrix} = \begin{pmatrix} 3 & 2 & 1 \\ 0 & 1 & 3 \\ 2 & 1 & 8 \end{pmatrix}$$

Property 6.1 (Matrix Addition). The sum of matrices has the following properties:

(A+) Associative property:

$$(A+B)+C = A+(B+C)$$

(C+) Commutative property:

$$A+B = B+A$$

(EN+) Existence of an additive identity:

$$\mathbf{0} + A = A$$

This additive identity is the zero matrix:

$$\begin{pmatrix} 0 & 0 \\ 0 & 0 \end{pmatrix} + \begin{pmatrix} 6 & 3 \\ 2 & 1 \end{pmatrix} = \begin{pmatrix} 6 & 3 \\ 2 & 1 \end{pmatrix}$$

(EO+) Existence of the additive inverse:

$$A + (-A) = 0$$

The additive inverse of $A = (a_{ij})$ is $-A = (-a_{ij})$:

$$\begin{pmatrix} 6 & 3 \\ 2 & 1 \end{pmatrix} + \begin{pmatrix} -6 & -3 \\ -2 & -1 \end{pmatrix} = \begin{pmatrix} 0 & 0 \\ 0 & 0 \end{pmatrix}$$

6.2.2 Multiplication by a Scalar

Definition 6.17. Let $A = (a_{ij})$ be a matrix of dimension $m \times n$, and $c \in \mathbb{R}$. Then the product of c by the matrix A is:

$$cA = (ca_{ij}).$$

Example 6.17 (Scalar Multiplication).

$$4 \begin{pmatrix} 1 & 2 & 0 \\ 0 & -3 & 1 \\ 2 & 4 & 0 \end{pmatrix} = \begin{pmatrix} 4 & 8 & 0 \\ 0 & -12 & 4 \\ 8 & 16 & 0 \end{pmatrix}$$

Property 6.2 (Properties of Scalar Multiplication). The properties of scalar multiplication are:

(a) $c(A+B) = cA + cB$, $(c+d)A = cA + dA$

(b) $(cd)A = c(dA)$

(c) $1A = A$, $0A = \mathbf{0}_{m\times n}$

where A and B are matrices of dimension $m \times n$ and $c, d \in \mathbb{R}$.

6.2.3 Multiplication of a Row Vector by a Column Vector

The product of a row vector of dimension $1 \times n$ by a column vector of dimension $n \times 1$ is a matrix of dimension 1×1, that is, a real number, which is defined as follows:

$$\begin{pmatrix} x_1 & x_2 & \cdots & x_n \end{pmatrix} \cdot \begin{pmatrix} y_1 \\ y_2 \\ \vdots \\ y_n \end{pmatrix} = x_1y_1 + x_2y_2 + \cdots + x_ny_n = \sum_{i=1}^{n} x_iy_i$$

Example 6.18 (Multiplication of a Row Vector by a Column Vector).

$$\begin{pmatrix} 1 & 2 & 3 \end{pmatrix} \begin{pmatrix} 3 \\ 1 \\ 5 \end{pmatrix} = 1 \cdot 3 + 2 \cdot 1 + 3 \cdot 5 = 20$$

This method of multiplying row vectors by column vectors is used for matrix multiplication, as we will see below.

6.2.4 Matrix Multiplication

To be able to multiply two matrices A and B, it is necessary that the number of columns of the first be equal to the number of rows of the second:

$$m\left\{\overbrace{\begin{pmatrix} & & \\ & a_{ij} & \\ & & \end{pmatrix}}^{n}\right. \cdot \quad n\left\{\overbrace{\begin{pmatrix} \\ b_{ij} \\ \\ \end{pmatrix}}^{p}\right.$$

$$(m \times n) \cdot (n \times p)$$

That is, if A is of dimension $m \times n$, B must be of dimension $n \times p$. Then the product has dimension $m \times p$ and is given by:

$$AB = (a_{ij})(b_{ij}) = (c_{ij}), \qquad c_{ij} = \sum_{k=1}^{n} a_{ik}b_{kj}.$$

That is, the element c_{ij} of the product matrix is obtained by multiplying the ith row of matrix A by the jth column of matrix B.

Example 6.19 (Matrix Multiplication).

$$\begin{pmatrix} 2 & 0 & 5 \\ 1 & 2 & 3 \end{pmatrix} \begin{pmatrix} 1 & 3 & 4 & 2 \\ 1 & 1 & 3 & 2 \\ 2 & 5 & 1 & 2 \end{pmatrix} = \begin{pmatrix} 12 & 31 & 13 & 14 \\ 9 & 20 & 13 & 12 \end{pmatrix}$$

where, for example:

$$c_{22} = \begin{pmatrix} 1 & 2 & 3 \end{pmatrix} \begin{pmatrix} 3 \\ 1 \\ 5 \end{pmatrix} = 1 \cdot 3 + 2 \cdot 1 + 3 \cdot 5 = 20.$$

Properties of Matrix Multiplication

Property 6.3 (Properties of Matrix Multiplication)**.** The properties of matrix multiplication are:

(A·) Associative property:

$$(AB)C = A(BC)$$

(EN·) Existence of an identity element with respect to multiplication:

$$IA = A$$

This identity element is the identity matrix of order equal to the number of rows of A:

$$\begin{pmatrix} 1 & 0 \\ 0 & 1 \end{pmatrix} \begin{pmatrix} 6 & 3 & 1 \\ 2 & 1 & 1 \end{pmatrix} = \begin{pmatrix} 6 & 3 & 1 \\ 2 & 1 & 1 \end{pmatrix}$$

(D) Matrix multiplication is distributive with respect to addition:

$$(A + B)C = AC + BC$$

Remark 6.6. Matrix multiplication is not commutative. That is, $AB \neq BA$. In fact, depending on the dimensions of the two matrices, once the product AB is calculated, the product BA may be impossible to calculate.

Remark 6.7 (Relationship between transposed matrices and multiplication)**.** Given two matrices A and B, we have:

$$(AB)^t = B^t A^t$$

Remark 6.8. Matrix multiplication is not cancellative. That is:

$$AB = AC \nRightarrow B = C$$

Proposed Exercise 6.2. Verify in a particular case that $(AB)^t = B^t A^t$.

Proposed Exercise 6.3. Find three matrices A, B and C such that $AB = AC$, but $B \neq C$.

6.3 Rank of a Matrix

6.3.1 Elementary Transformations

Definition 6.18 (Elementary Transformations). We call the following operations on a matrix as **elementary transformations**:

(PF) Permuting two rows of the matrix.

(PC) Permuting two columns of the matrix.

(MF) Multiplying a row by a nonzero real number.

(MC) Multiplying a column by a nonzero real number.

(SF) Adding to a row another row multiplied by a real number.

(SC) Adding to a column another column multiplied by a real number.

Remark 6.9. In this chapter, we will use the following notation for referring to the different elementary transformations:

- pf_{ij}: permuting rows i and j.
- pc_{ij}: permuting columns i and j.
- αf_i: multiplying row i by the number $\alpha \in \mathbb{R}$.
- αc_i: multiplying column i by the number $\alpha \in \mathbb{R}$.
- $f_i \leftarrow f_i + \alpha f_j$: adding to row i row j multiplied by the number $\alpha \in \mathbb{R}$.
- $c_i \leftarrow c_i + \alpha c_j$: adding to column i column j multiplied by the number $\alpha \in \mathbb{R}$.

Example 6.20 (of elementary transformations). *We start with the matrix A and, through elementary transformations, arrive at the matrix B.*

$$A = \begin{pmatrix} 1 & 5 \\ -2 & 1 \end{pmatrix} \underset{pf_{12}}{\overset{(PF)}{\longrightarrow}} \begin{pmatrix} -2 & 1 \\ 1 & 5 \end{pmatrix} \underset{pc_{12}}{\overset{(PC)}{\longrightarrow}} \begin{pmatrix} 1 & -2 \\ 5 & 1 \end{pmatrix}$$

$$\underset{2f_2}{\overset{(MF)}{\longrightarrow}} \begin{pmatrix} 1 & -2 \\ 10 & 2 \end{pmatrix} \underset{\frac{1}{2}c_2}{\overset{(MC)}{\longrightarrow}} \begin{pmatrix} 1 & -1 \\ 10 & 1 \end{pmatrix}$$

$$\underset{f_2 \leftarrow f_2 - 5f_1}{\overset{(SF)}{\longrightarrow}} \begin{pmatrix} 1 & -1 \\ 5 & 6 \end{pmatrix} \underset{c_2 \leftarrow c_2 + c_1}{\overset{(SC)}{\longrightarrow}} \begin{pmatrix} 1 & 0 \\ 5 & 11 \end{pmatrix} = B$$

6.3.2 Equivalent Matrices

Definition 6.19 (Equivalent Matrices). Given two matrices A and B, we say they are **equivalent matrices** if one can be transformed into the other through elementary transformations.

If two matrices A and B are equivalent, we denote it as $A \sim B$.

Example 6.21. *The matrices* $A = \begin{pmatrix} 1 & 2 \\ 3 & 0 \end{pmatrix}$ *and* $B = \begin{pmatrix} 2 & 1 \\ -2 & 0 \end{pmatrix}$ *are equivalent:*

$$\begin{pmatrix} 1 & 2 \\ 3 & 0 \end{pmatrix} \sim \begin{pmatrix} 2 & 1 \\ -2 & 0 \end{pmatrix}$$

since:

$$\begin{pmatrix} 1 & 2 \\ 3 & 0 \end{pmatrix} \underset{\frac{1}{3}f_2}{\sim} \begin{pmatrix} 1 & 2 \\ 1 & 0 \end{pmatrix} \underset{f_2 \leftarrow f_2 - f_1}{\sim} \begin{pmatrix} 1 & 2 \\ 0 & -2 \end{pmatrix} \underset{pc_{12}}{\sim} \begin{pmatrix} 2 & 1 \\ -2 & 0 \end{pmatrix}$$

Definition 6.20 (Rank). A matrix is said to have rank r if it is equivalent to an r-row echelon matrix. The rank of a matrix A is represented by rank(A).

Example 6.22. *Calculate the rank of the matrix:*

$$A = \begin{pmatrix} 1 & 2 & 3 \\ 3 & 0 & 1 \\ 5 & 4 & 7 \end{pmatrix}$$

SOLUTION. The rank of matrix A is 2, because:

$$\begin{pmatrix} 1 & 2 & 3 \\ 3 & 0 & 1 \\ 5 & 4 & 7 \end{pmatrix} \underset{f_3 \leftarrow f_3 - f_2 - 2f_1}{\sim} \begin{pmatrix} 1 & 2 & 3 \\ 3 & 0 & 1 \\ 0 & 0 & 0 \end{pmatrix} \underset{pc_{12}}{\sim} \begin{pmatrix} 2 & 1 & 3 \\ \boxed{0} & 3 & 1 \\ \hline 0 & 0 & 0 \end{pmatrix}$$

□

Property 6.4 (On the Rank of a Matrix). The properties of the rank of a matrix are:

- The rank of a matrix is, at most, the minimum number of rows or columns. That is, if A is a matrix of dimension $m \times n$,

$$\text{rank}(A) \leq \min\{m, n\}$$

- The rank of a matrix and its transpose are the same. That is:

$$\text{rank}(A) = \text{rank}(A^t)$$

- Equivalent matrices have the same rank. Therefore, if we perform an elementary transformation on a matrix, its rank is preserved.

6.3.3 Rank Calculation by Gaussian Method

Using elementary transformations, we can transform any matrix into an equivalent echelon form, i.e., one where the lower triangle elements are zeros. Since the rank of two equivalent matrices is the same, the rank is the number of non-zero rows left.

Example 6.23. *Calculate the rank of the matrix:*

$$\begin{pmatrix} 1 & 3 & 2 & -1 \\ 2 & 4 & 2 & 1 \\ 4 & 6 & 2 & 5 \end{pmatrix}$$

by the Gaussian method.

SOLUTION.

$$\begin{pmatrix} 1 & 3 & 2 & -1 \\ 2 & 4 & 2 & 1 \\ 4 & 6 & 2 & 5 \end{pmatrix} \underset{f_2 \leftarrow f_2 - 2f_1}{\sim} \begin{pmatrix} 1 & 3 & 2 & -1 \\ 0 & -2 & -2 & 3 \\ 4 & 6 & 2 & 5 \end{pmatrix}$$

$$\underset{f_3 \leftarrow f_3 - 4f_1}{\sim} \begin{pmatrix} 1 & 3 & 2 & -1 \\ 0 & -2 & -2 & 3 \\ 0 & -6 & -6 & 9 \end{pmatrix}$$

$$\underset{f_3 \leftarrow f_3 - 3f_2}{\sim} \left(\begin{array}{cccc} 1 & 3 & 2 & -1 \\ 0 & -2 & -2 & 3 \\ \hline 0 & 0 & 0 & 0 \end{array}\right)$$

Therefore, the rank of the matrix is 2. □

Example 6.24. *Calculate the rank of the matrix:*

$$\begin{pmatrix} 0 & 0 & 0 & 0 & 1 \\ 0 & 2 & 2 & 0 & 0 \\ 1 & 0 & 0 & 2 & 0 \\ 0 & 0 & 2 & 0 & 1 \\ 1 & 0 & 0 & 2 & 0 \\ 1 & 2 & 2 & 2 & 0 \end{pmatrix}$$

by the Gaussian method.

SOLUTION.

$$\begin{pmatrix} 0 & 0 & 0 & 0 & 1 \\ 0 & 2 & 2 & 0 & 0 \\ 1 & 0 & 0 & 2 & 0 \\ 0 & 0 & 2 & 0 & 1 \\ 1 & 0 & 0 & 2 & 0 \\ 1 & 2 & 2 & 2 & 0 \end{pmatrix} \underset{f_5 \leftarrow f_5 - f_3}{\sim} \begin{pmatrix} 0 & 0 & 0 & 0 & 1 \\ 0 & 2 & 2 & 0 & 0 \\ 1 & 0 & 0 & 2 & 0 \\ 0 & 0 & 2 & 0 & 1 \\ 0 & 0 & 0 & 0 & 0 \\ 1 & 2 & 2 & 2 & 0 \end{pmatrix}$$

$$\underset{f_6 \leftarrow f_6 - f_3}{\sim} \begin{pmatrix} 0 & 0 & 0 & 0 & 1 \\ 0 & 2 & 2 & 0 & 0 \\ 1 & 0 & 0 & 2 & 0 \\ 0 & 0 & 2 & 0 & 1 \\ 0 & 0 & 0 & 0 & 0 \\ 0 & 2 & 2 & 0 & 0 \end{pmatrix}$$

$$\underset{f_6 \leftarrow f_6 - f_3}{\sim} \begin{pmatrix} 0 & 0 & 0 & 0 & 1 \\ 0 & 2 & 2 & 0 & 0 \\ 1 & 0 & 0 & 2 & 0 \\ 0 & 0 & 2 & 0 & 1 \\ 0 & 0 & 0 & 0 & 0 \\ 0 & 0 & 0 & 0 & 0 \end{pmatrix} \underset{pf_{13}}{\sim} \begin{pmatrix} 1 & 0 & 0 & 2 & 0 \\ 0 & 2 & 2 & 0 & 0 \\ 0 & 0 & 0 & 0 & 1 \\ 0 & 0 & 2 & 0 & 1 \\ 0 & 0 & 0 & 0 & 0 \\ 0 & 0 & 0 & 0 & 0 \end{pmatrix}$$

$$\underset{pf_{34}}{\sim} \begin{pmatrix} 1 & 0 & 0 & 2 & 0 \\ 0 & 2 & 2 & 0 & 0 \\ 0 & 0 & 2 & 0 & 1 \\ 0 & 0 & 0 & 1 & 0 \\ \hline 0 & 0 & 0 & 0 & 0 \\ 0 & 0 & 0 & 0 & 0 \end{pmatrix} \underset{pc_{45}}{\sim} \begin{pmatrix} 1 & 0 & 0 & 0 & 2 \\ 0 & 2 & 2 & 0 & 0 \\ 0 & 0 & 2 & 1 & 0 \\ 0 & 0 & 0 & 1 & 0 \\ \hline 0 & 0 & 0 & 0 & 0 \\ 0 & 0 & 0 & 0 & 0 \end{pmatrix}$$

Therefore, the rank of the matrix is 4. □

Proposed Exercise 6.4. Calculate the rank of the following matrices using the Gaussian method:

(a) $\begin{pmatrix} -1 & 0 & 1 \\ 3 & 2 & 1 \\ 1 & -1 & 2 \end{pmatrix}$

(b) $\begin{pmatrix} 1 & 1 & 0 \\ 2 & -1 & 0 \\ 1 & 1 & 1 \end{pmatrix}$

(c) $\begin{pmatrix} 1 & 1 & 1 \\ 0 & 1 & 1 \\ 3 & 0 & -6 \end{pmatrix}$

(d) $\begin{pmatrix} 2 & 4 & 0 & 7 \\ 1 & 3 & 1 & 8 \\ 1 & 0 & -2 & -9 \end{pmatrix}$

(e) $\begin{pmatrix} 1 & 2 & 5 \\ 1 & -3 & -6 \\ -6 & -2 & -8 \\ 0 & 5 & 11 \\ 2 & -1 & -1 \end{pmatrix}$

(f) $\begin{pmatrix} 1 & 1 & 2 & -1 & 0 \\ 0 & -1 & 1 & 1 & 1 \\ 1 & 0 & 0 & 2 & -1 \end{pmatrix}$

(g) $\begin{pmatrix} 2 & 1 & 1 & 1 & 1 & 1 \\ 1 & 3 & 1 & 1 & 2 & 1 \\ 1 & 1 & 4 & 1 & 3 & 1 \\ 1 & 1 & 1 & 5 & 4 & 1 \end{pmatrix}$

Example 6.25. *Determine the values of a for which* $\operatorname{rank}(A) \neq 3$*, where:*

$$A = \begin{pmatrix} 1 & a & 1 \\ a+1 & 1 & -a \\ 2 & 1 & -1 \end{pmatrix}$$

SOLUTION.

$$\begin{pmatrix} 1 & a & 1 \\ a+1 & 1 & -a \\ 2 & 1 & -1 \end{pmatrix} \underset{pc_{13}}{\sim} \begin{pmatrix} 1 & a & 1 \\ -a & 1 & a+1 \\ -1 & 1 & 2 \end{pmatrix} \underset{pf_{13}}{\sim} \begin{pmatrix} -1 & 1 & 2 \\ -a & 1 & a+1 \\ 1 & a & 1 \end{pmatrix}$$

$$\underset{f_2 \leftarrow f_2 - af_1}{\sim} \begin{pmatrix} -1 & 1 & 2 \\ 0 & 1-a & 1-a \\ 1 & a & 1 \end{pmatrix}$$

$$\underset{f_3 \leftarrow f_3 + f_1}{\sim} \begin{pmatrix} -1 & 1 & 2 \\ 0 & 1-a & 1-a \\ 0 & 1+a & 3 \end{pmatrix}$$

$$\underset{pf_{23}}{\sim} \begin{pmatrix} -1 & 1 & 2 \\ 0 & 1+a & 3 \\ 0 & 1-a & 1-a \end{pmatrix} \underset{pc_{23}}{\sim} \begin{pmatrix} -1 & 2 & 1 \\ 0 & 3 & 1+a \\ 0 & 1-a & 1-a \end{pmatrix}$$

$$\underset{f_3 \leftarrow f_3 - (1-a)/3}{\sim} \begin{pmatrix} -1 & 2 & 1 \\ 0 & 3 & 1+a \\ 0 & 0 & 1-a-(1-a)(1+a)/3 \end{pmatrix}$$

This last matrix is equal to:

$$\begin{pmatrix} -1 & 2 & 1 \\ 0 & 3 & 1+a \\ 0 & 0 & 1-a-(1-a)(1+a)/3 \end{pmatrix} = \begin{pmatrix} -1 & 2 & 1 \\ 0 & 3 & 1+a \\ 0 & 0 & (3-3a-1+a^2)/3 \end{pmatrix}$$

$$= \begin{pmatrix} -1 & 2 & 1 \\ 0 & 3 & 1+a \\ 0 & 0 & (a^2-3a+2)/3 \end{pmatrix}$$

Therefore, $\operatorname{rank}(A) \geq 2$ for all values of a and $\operatorname{rank}(A) = 3 \Leftrightarrow a^2 - 3a + 2 \neq 0$, that is, if $x \neq 1$ and $x \neq 2$. □

6.4 Inverse Matrix

Definition 6.21 (Invertible Matrix). A matrix A is **invertible** if it is square and has maximum rank. In this case, its inverse is denoted by A^{-1}, and is such that:

$$AA^{-1} = A^{-1}A = I.$$

Invertible matrices are also called **regular** or **non-singular**.

Property 6.5 (On Invertible Matrices). If A and B are two regular matrices of order n, i.e., $A, B \in \mathscr{M}_{n\times n}(\mathbb{R})$, then:

(a) $\left(A^{-1}\right)^{-1} = A.$

(b) $(\lambda \cdot A)^{-1} = \frac{1}{\lambda} \cdot A^{-1}, \lambda \neq 0$

(c) $(A \cdot B)^{-1} = B^{-1} \cdot A^{-1}$

(d) $(A^t)^{-1} = (A^{-1})^t$

Proof.

(a) The matrix $\left(A^{-1}\right)^{-1}$ is, by definition, the inverse of A^{-1}. That is:

$$\left(A^{-1}\right)^{-1} \cdot A^{-1} = I.$$

Multiplying on the right by A in both cases:

$$\begin{aligned} \left(A^{-1}\right)^{-1} \cdot A^{-1}A &= IA, \\ \left(A^{-1}\right)^{-1} I &= A, \\ \left(A^{-1}\right)^{-1} &= A, \end{aligned}$$

since $A^{-1}A = I$, $IA = A$ and $\left(A^{-1}\right)^{-1} I = \left(A^{-1}\right)^{-1}$.

(b) The inverse of λA is $\frac{1}{\lambda} \cdot A^{-1}$, since:

$$\begin{aligned} \frac{1}{\lambda} \cdot A^{-1} \cdot \lambda A &= \frac{1}{\lambda}\lambda A^{-1}A \\ &= 1 \cdot I = I. \end{aligned}$$

(c) Let A and B be two regular matrices with inverses A^{-1} and B^{-1} respectively and let $C = B^{-1} \cdot A^{-1}$. The matrix C will be the inverse of the product $A \cdot B$ if:

$$C \cdot (A \cdot B) = (A \cdot B) \cdot C = I$$

Regarding the equality $C \cdot (A \cdot B) = I$:

$$\begin{aligned} C \cdot (A \cdot B) &= (B^{-1} \cdot A^{-1}) \cdot (A \cdot B) \\ &= B^{-1} \cdot (A^{-1} \cdot A) \cdot B \\ &= B^{-1} \cdot I \cdot B \\ &= B^{-1} \cdot B = I \end{aligned}$$

Regarding the equality $(A \cdot B) \cdot C = I$:

$$\begin{aligned}(A \cdot B) \cdot C &= (A \cdot B) \cdot (B^{-1} \cdot A^{-1}) \\ &= A \cdot (B \cdot B^{-1}) \cdot A^{-1} \\ &= A \cdot I \cdot A^{-1} \\ &= A \cdot A^{-1} = I\end{aligned}$$

(d) The inverse of A^t is $(A^{-1})^t$, since:

$$\begin{aligned}A^t \cdot (A^{-1})^t &= (A^{-1} \cdot A)^t \\ &= I^t = I,\end{aligned}$$

as we wanted to demonstrate. □

Proposition 6.5. If A is a square matrix of order n and invertible, then A is equivalent to the identity matrix of order n, that is, $A \sim I_n$.

Moreover, we can obtain the identity matrix from the matrix A using only row elementary transformations (PF, MF and SF).

Example 6.26. *Show that the matrix:*

$$A = \begin{pmatrix} 1 & 1 & 1 \\ 2 & 3 & 4 \\ 5 & 8 & 9 \end{pmatrix}$$

is equivalent to the identity matrix of order 3.

SOLUTION.

$$A = \begin{pmatrix} 1 & 1 & 1 \\ 2 & 3 & 4 \\ 5 & 8 & 9 \end{pmatrix} \underset{\substack{f_2 \leftarrow f_2 - 2f_1 \\ f_3 \leftarrow f_3 - 5f_1}}{\sim} \begin{pmatrix} 1 & 1 & 1 \\ 0 & 1 & 2 \\ 0 & 3 & 4 \end{pmatrix} \underset{f_3 \leftarrow f_3 - 3f_2}{\sim} \begin{pmatrix} 1 & 1 & 1 \\ 0 & 1 & 2 \\ 0 & 0 & -2 \end{pmatrix}$$

$$\underset{-\frac{1}{2}f_3}{\sim} \begin{pmatrix} 1 & 1 & 1 \\ 0 & 1 & 2 \\ 0 & 0 & 1 \end{pmatrix} \underset{f_1 \leftarrow f_1 - f_2}{\sim} \begin{pmatrix} 1 & 0 & -1 \\ 0 & 1 & 2 \\ 0 & 0 & 1 \end{pmatrix}$$

$$\underset{\substack{f_1 \leftarrow f_1 + f_3 \\ f_2 \leftarrow f_2 - 2f_3}}{\sim} \begin{pmatrix} 1 & 0 & 0 \\ 0 & 1 & 0 \\ 0 & 0 & 1 \end{pmatrix} = I$$

6.4.1 Calculation of the Inverse Matrix Using Gaussian Method

If A is an invertible matrix of order n, we can obtain the identity matrix I_n by considering only **row elementary transformations**.

Thus, A^{-1} is calculated by applying elementary row operations on the matrix $(A|I_n)$, until obtaining $(I_n|A^{-1})$. That is:

$$(A \mid I_n) \quad \underset{PF,MF,CF}{\sim} \quad \ldots \quad \sim \quad (I_n \mid A^{-1})$$

Remark 6.10. Note that given matrices A and B of dimensions $m \times n_A$ and $m \times n_B$, $(A|B)$ is the matrix of dimension $m \times (n_A + n_B)$ obtained by concatenating the columns of B to the columns of matrix A.

Example 6.27. *Calculate the inverse matrix of:*

$$\begin{pmatrix} 1 & 1 & 1 \\ 2 & 3 & 4 \\ 5 & 8 & 9 \end{pmatrix}$$

using the Gaussian method.

SOLUTION.

$$\left(\begin{array}{ccc|ccc} 1 & 1 & 1 & 1 & 0 & 0 \\ 2 & 3 & 4 & 0 & 1 & 0 \\ 5 & 8 & 9 & 0 & 0 & 1 \end{array}\right) \underset{\substack{f_2 \leftarrow f_2 - 2f_1 \\ f_3 \leftarrow f_3 - 5f_1}}{\sim} \left(\begin{array}{ccc|ccc} 1 & 1 & 1 & 1 & 0 & 0 \\ 0 & 1 & 2 & -2 & 1 & 0 \\ 0 & 3 & 4 & -5 & 0 & 1 \end{array}\right)$$

$$\underset{f_3 \leftarrow f_3 - 3f_2}{\sim} \left(\begin{array}{ccc|ccc} 1 & 1 & 1 & 1 & 0 & 0 \\ 0 & 1 & 2 & -2 & 1 & 0 \\ 0 & 0 & -2 & 1 & -3 & 1 \end{array}\right)$$

$$\underset{-\frac{1}{2}f_3}{\sim} \left(\begin{array}{ccc|ccc} 1 & 1 & 1 & 1 & 0 & 0 \\ 0 & 1 & 2 & -2 & 1 & 0 \\ 0 & 0 & 1 & -\frac{1}{2} & \frac{3}{2} & -\frac{1}{2} \end{array}\right)$$

$$\underset{f_1 \leftarrow f_1 - f_2}{\sim} \left(\begin{array}{ccc|ccc} 1 & 0 & -1 & 3 & -1 & 0 \\ 0 & 1 & 2 & -2 & 1 & 0 \\ 0 & 0 & 1 & -\frac{1}{2} & \frac{3}{2} & -\frac{1}{2} \end{array}\right)$$

$$\underset{\substack{f_1 \leftarrow f_1 + f_3 \\ f_2 \leftarrow f_2 - 2f_3}}{\sim} \left(\begin{array}{ccc|ccc} 1 & 0 & 0 & \frac{5}{2} & \frac{1}{2} & -\frac{1}{2} \\ 0 & 1 & 0 & -1 & -2 & 1 \\ 0 & 0 & 1 & -\frac{1}{2} & \frac{3}{2} & -\frac{1}{2} \end{array}\right)$$

Therefore, the inverse of the matrix:

$$A = \begin{pmatrix} 1 & 1 & 1 \\ 2 & 3 & 4 \\ 5 & 8 & 9 \end{pmatrix}$$

is:

$$A^{-1} = \begin{pmatrix} \frac{5}{2} & \frac{1}{2} & -\frac{1}{2} \\ -1 & -2 & 1 \\ -\frac{1}{2} & \frac{3}{2} & -\frac{1}{2} \end{pmatrix}$$

is the matrix:

$$A^{-1} = \begin{pmatrix} 5/2 & 1/2 & -1/2 \\ -1 & -2 & 1 \\ -1/2 & 3/2 & -1/2 \end{pmatrix}$$

□

Proposed Exercise 6.5. Calculate the inverse of the following matrices:

(a) $\begin{pmatrix} -2 & 3 \\ 1 & -1 \end{pmatrix}$

(b) $\begin{pmatrix} 1 & 3 & 1 \\ 2 & -1 & 2 \\ 3 & 2 & -3 \end{pmatrix}$

(c) $\begin{pmatrix} 1 & 2 & -3 \\ 0 & 1 & 2 \\ 0 & 0 & 1 \end{pmatrix}$

(d) $\begin{pmatrix} 2 & 4 & 3 \\ 0 & 1 & 1 \\ 2 & 2 & -1 \end{pmatrix}$

(e) $\begin{pmatrix} 2 & 1 & 1 \\ 4 & 2 & 0 \\ -3 & -1 & 1 \end{pmatrix}$

(f) $\begin{pmatrix} 0 & -1 & 0 & 1 \\ 0 & 2 & -1 & 1 \\ 1 & -2 & 1 & 1 \\ -1 & 0 & 1 & 3 \end{pmatrix}$

Proposed Exercise 6.6. Find, for the possible values of k, the inverse of the matrix:

$$\begin{pmatrix} 1 & k & k \\ 1 & 2 & 4 \\ 1 & 3 & 9 \end{pmatrix}$$

6.5 Determinant of a Matrix

6.5.1 Definition and Calculation

Definition 6.22 (Determinant). Given a square matrix $A \in \mathcal{M}_{n\times n}(\mathbb{R})$, the **determinant** of the matrix A is defined and symbolized by $\det(A)$, $|A|$ or $D(A)$, as an application that assigns to each square matrix A a real number:

$$D : \mathcal{M}_{n\times n}(\mathbb{R}) \longrightarrow \mathbb{R},$$

so that $D(A) = \det(A) = |A| \in \mathbb{R}$.

The determinant is defined inductively on the order n of the matrix A, as follows:

- If $n = 2$, then:

$$\begin{vmatrix} a_{11} & a_{12} \\ a_{21} & a_{22} \end{vmatrix} = a_{11}\cdot a_{22} - a_{12}\cdot a_{21}$$

Example 6.28. *Calculate the determinant of the matrix:*

$$\begin{pmatrix} 4 & -3 \\ -2 & 2 \end{pmatrix}$$

SOLUTION.

$$\begin{vmatrix} 4 & -3 \\ -2 & 2 \end{vmatrix} = 4\cdot 2 - (-3)\cdot(-2) = 2.$$

□

- If $n = 3$, then:

$$\begin{vmatrix} a_{11} & a_{12} & a_{13} \\ a_{21} & a_{22} & a_{23} \\ a_{31} & a_{32} & a_{33} \end{vmatrix} = a_{11}D_{11} - a_{12}D_{12} + a_{13}D_{13},$$

where:

$$D_{11} = \begin{vmatrix} a_{22} & a_{23} \\ a_{32} & a_{33} \end{vmatrix}, \; D_{12} = \begin{vmatrix} a_{21} & a_{23} \\ a_{31} & a_{33} \end{vmatrix} and D_{13} = \begin{vmatrix} a_{21} & a_{22} \\ a_{31} & a_{32} \end{vmatrix}.$$

D_{ij} is the determinant of the matrix obtained by removing row i and column j of the original matrix.

Note that we have performed what is called the *expansion by the first row*. The determinant can also be calculated, for example, by expanding by the first column. In this case, the determinant would be:

$$\begin{vmatrix} a_{11} & a_{12} & a_{13} \\ a_{21} & a_{22} & a_{23} \\ a_{31} & a_{32} & a_{33} \end{vmatrix} = a_{11}D_{11} - a_{21}D_{21} + a_{31}D_{31},$$

where:

$$D_{11} = \begin{vmatrix} a_{22} & a_{23} \\ a_{32} & a_{33} \end{vmatrix}, \ D_{21} = \begin{vmatrix} a_{12} & a_{13} \\ a_{32} & a_{33} \end{vmatrix} and D_{31} = \begin{vmatrix} a_{12} & a_{13} \\ a_{22} & a_{33} \end{vmatrix}.$$

In any case, the determinant of a 3×3 matrix can be calculated using the well-known *Sarrus rule*, which states that this determinant is:

$$\begin{vmatrix} a_{11} & a_{12} & a_{13} \\ a_{21} & a_{22} & a_{23} \\ a_{31} & a_{32} & a_{33} \end{vmatrix} = a_{11}a_{22}a_{33} + a_{12}a_{23}a_{31} + a_{13}a_{21}a_{32} -$$

$$- a_{31}a_{22}a_{13} - a_{32}a_{23}a_{11} - a_{33}a_{21}a_{12}.$$

Example 6.29. *Calculate the determinant of the matrix:*

$$\begin{pmatrix} 1 & 2 & 3 \\ 4 & 5 & 6 \\ 7 & 8 & 9 \end{pmatrix}$$

SOLUTION. Applying the Sarrus rule, we have that:

$$\begin{vmatrix} 1 & 2 & 3 \\ 4 & 5 & 6 \\ 7 & 8 & 9 \end{vmatrix} = 1 \cdot 5 \cdot 9 + 2 \cdot 6 \cdot 7 + 3 \cdot 4 \cdot 8 - 3 \cdot 5 \cdot 7 - 2 \cdot 4 \cdot 9 - 1 \cdot 6 \cdot 8 = 0.$$

□

- If $n = 4$, then:

$$\begin{vmatrix} a_{11} & a_{12} & a_{13} & a_{14} \\ a_{21} & a_{22} & a_{23} & a_{24} \\ a_{31} & a_{32} & a_{33} & a_{34} \\ a_{41} & a_{42} & a_{43} & a_{44} \end{vmatrix} = a_{11}D_{11} - a_{12}D_{12} + a_{13}D_{13} - a_{14}D_{14},$$

where:

$$D_{11} = \begin{vmatrix} a_{22} & a_{23} & a_{24} \\ a_{32} & a_{33} & a_{34} \\ a_{42} & a_{43} & a_{44} \end{vmatrix}, \quad D_{12} = \begin{vmatrix} a_{21} & a_{23} & a_{24} \\ a_{31} & a_{33} & a_{34} \\ a_{41} & a_{43} & a_{44} \end{vmatrix},$$

$$D_{13} = \begin{vmatrix} a_{21} & a_{22} & a_{24} \\ a_{31} & a_{32} & a_{34} \\ a_{41} & a_{42} & a_{44} \end{vmatrix}, \quad D_{14} = \begin{vmatrix} a_{21} & a_{22} & a_{23} \\ a_{31} & a_{32} & a_{33} \\ a_{41} & a_{42} & a_{43} \end{vmatrix}.$$

As with the case of 3×3 matrices, the calculation of the determinant of a 4×4 matrix can be done by expanding along another row or column.

Definition 6.23 (Minor). The **minor** of an element a_{ij} of a square matrix is the determinant of the matrix obtained by removing the i-th row and the j-th column of the original matrix, and is symbolized by Δ_{ij} or D_{ij}.

Definition 6.24 (Cofactor). The **cofactor** of an element a_{ij}, denoted by A_{ij}, is the minor of a_{ij} multiplied by $(-1)^{i+j}$.

Thus, we can give a general formula for calculating the determinant of a square matrix of order n, expanding, for example, by the first row:

$$\begin{vmatrix} a_{11} & a_{12} & \cdots & a_{1n} \\ a_{21} & a_{22} & \cdots & a_{2n} \\ \vdots & \vdots & \ddots & \vdots \\ a_{n1} & a_{n2} & \cdots & a_{nn} \end{vmatrix} = a_{11}A_{11} + a_{12}A_{12} + \cdots + a_{1n}A_{1n}.$$

Proposed Exercise 6.7. Calculate the determinants of the following matrices:

(a) $\begin{pmatrix} \cos(\alpha) & -\sin(\alpha) \\ \sin(\alpha) & \cos(\alpha) \end{pmatrix}$

(b) $\begin{pmatrix} 1 & 1 & 1 \\ x & y & z \\ x^2 & y^2 & z^2 \end{pmatrix}$

(c) $\begin{pmatrix} 2 & 1 & 3 & -1 \\ 3 & 1 & 2 & 1 \\ 1 & 2 & 3 & 1 \\ 1 & -1 & 1 & -1 \end{pmatrix}$

Remark 6.11. As we have said, the definition given of the determinant of order n is also called expanding the determinant by the first row:

$$\begin{vmatrix} \boxed{a_{11}} & \boxed{a_{12}} & \cdots & \boxed{a_{1n}} \\ a_{21} & a_{22} & \cdots & a_{2n} \\ \vdots & \vdots & \ddots & \vdots \\ a_{n1} & a_{n2} & \cdots & a_{nn} \end{vmatrix} = \boxed{a_{11}}A_{11} + \boxed{a_{12}}A_{12} + \cdots + \boxed{a_{1n}}A_{1n}.$$

In fact, to calculate the determinant of order n one can proceed by expanding by any row or column, i.e., summing the products of the elements of any row or any column multiplied by the respective cofactors.

Example 6.30. *Calculate the determinant of the matrix:*

$$\begin{pmatrix} 3 & 1 & -2 \\ -1 & 3 & 5 \\ 2 & 1 & 0 \end{pmatrix}$$

by expanding by the third row.

SOLUTION.

$$\begin{vmatrix} 3 & 1 & -2 \\ -1 & 3 & 5 \\ \boxed{2} & \boxed{1} & \boxed{0} \end{vmatrix} = 2\begin{vmatrix} 1 & -2 \\ 3 & 5 \end{vmatrix} - 1\begin{vmatrix} 3 & -2 \\ -1 & 5 \end{vmatrix} + 0\begin{vmatrix} 2 & 1 \\ -1 & 3 \end{vmatrix}$$
$$= 2(5+6) - (15-2) + 0 = 22 - 13 = 9.$$

Proposed Exercise 6.8. Calculate the following determinants, expanding by the most appropriate column or row:

(a) $\begin{vmatrix} 3 & -2 & 0 \\ -3 & 9 & 5 \\ 6 & 5 & 0 \end{vmatrix}$

(b) $\begin{vmatrix} 2 & 1 & -6 \\ 3 & -1 & 7 \\ 0 & 0 & 4 \end{vmatrix}$

(c) $\begin{vmatrix} 3 & -1 & 5 \\ 0 & 2 & 0 \\ 1 & 1 & -1 \end{vmatrix}$

Property 6.6. On Determinants

If two rows or two columns are interchanged, the determinant changes sign.

For example:

$$\begin{vmatrix} 1 & 2 \\ 3 & 4 \end{vmatrix} = -\begin{vmatrix} 2 & 1 \\ 4 & 3 \end{vmatrix} = -\begin{vmatrix} 3 & 4 \\ 1 & 2 \end{vmatrix}.$$

(D_2) If all elements of a row or column are multiplied by a constant, the determinant is multiplied by that constant.

For example:

$$\begin{vmatrix} 2 & 20 \\ 2 & 5 \end{vmatrix} = 2\begin{vmatrix} 1 & 10 \\ 2 & 5 \end{vmatrix} = 2\cdot 5\begin{vmatrix} 1 & 2 \\ 2 & 1 \end{vmatrix}.$$

(D_3) If a row or a column is the sum of two summands, the determinant decomposes into the sum of two determinants, each with one of the summands in the corresponding row or column and with the rest of the rows or columns being the same.

For example:

$$\begin{vmatrix} 1+3 & 3 \\ 2+7 & 4 \end{vmatrix} = \begin{vmatrix} 1 & 3 \\ 2 & 4 \end{vmatrix} + \begin{vmatrix} 3 & 3 \\ 7 & 4 \end{vmatrix}.$$

(D_4) The determinant of the identity matrix is 1. Indeed:

$$\begin{vmatrix} 1 & 0 \\ 0 & 1 \end{vmatrix} = 1.$$

(D_5) If all elements of a row or column are zero, then the determinant is 0.

For example:

$$\begin{vmatrix} 1 & 2 & 0 \\ 3 & 4 & 0 \\ 6 & 1 & 0 \end{vmatrix} = 0.$$

(D_6) If in a determinant, a row or a column is a linear combination of other rows or columns (equal to the sum of other rows or columns multiplied by a constant), the determinant is zero.

For example:

$$\begin{vmatrix} 1 & 2 & 2\cdot 1+3\cdot 2 \\ 3 & 1 & 2\cdot 3+3\cdot 1 \\ 5 & 2 & 2\cdot 5+3\cdot 2 \end{vmatrix} = 0.$$

(D_7) The determinant of a matrix is the same as that of its transpose, i.e., $|A| = |A^t|$.

For example:

$$\begin{vmatrix} 1 & 2 \\ 3 & 4 \end{vmatrix} = \begin{vmatrix} 1 & 3 \\ 2 & 4 \end{vmatrix}.$$

(D_8) The determinant of a product of matrices is equal to the product of the determinants, $|A \cdot B| = |A| \cdot |B|$.

For example:

$$\left|\begin{pmatrix} 1 & 2 \\ 3 & 4 \end{pmatrix}\begin{pmatrix} 3 & 2 \\ 7 & 1 \end{pmatrix}\right| = \begin{vmatrix} 1 & 2 \\ 3 & 4 \end{vmatrix}\begin{vmatrix} 3 & 2 \\ 7 & 1 \end{vmatrix}$$

Proposition 6.6. A square matrix A is invertible, regular or non-singular if and only if the determinant is different from zero.

Moreover, if A is invertible, $|A^{-1}| = \frac{1}{|A|}$.

PROOF. If A is invertible, then there exists a matrix A^{-1} such that $AA^{-1} = I$. Calculating determinants, we have that $|AA^{-1}| = |I|$. Then, using properties D_4 and D_8 of determinants:

$$|AA^{-1}| = |A| \cdot |A^{-1}| = |I| = 1 \Longrightarrow |A^{-1}| = \frac{1}{|A|}.$$

□

Proposed Exercise 6.9. Calculate the inverse of the matrix:

$$\begin{pmatrix} 2 & 3 \\ 1 & -1 \end{pmatrix}$$

and verify that $|A^{-1}| = 1/|A|$.

6.5.2 Calculation of the Rank of a Matrix Using Determinants

Definition 6.25 (Minor). The **minor of order** k of a matrix $A \in \mathscr{M}_{m\times n}(\mathbb{R})$ is the determinant of a square submatrix of order $k \times k$ of A, taking k rows and k columns of the matrix A.

Example 6.31. *Calculate the nine minors of order 2 of the matrix:*

$$\begin{pmatrix} 1 & 1 & -1 \\ 0 & 1 & 1 \\ 3 & 0 & -6 \end{pmatrix}$$

SOLUTION. The nine minors are:

$$D_{11} = \begin{vmatrix} 1 & 1 \\ 0 & -6 \end{vmatrix} = -6, \quad D_{12} = \begin{vmatrix} 0 & 1 \\ 3 & -6 \end{vmatrix} = -3, \quad D_{13} = \begin{vmatrix} 0 & 1 \\ 3 & 0 \end{vmatrix} = -3$$

$$D_{21} = \begin{vmatrix} 1 & -1 \\ 0 & -6 \end{vmatrix} = -6, \quad D_{22} = \begin{vmatrix} 1 & -1 \\ 3 & -6 \end{vmatrix} = -3, \quad D_{23} = \begin{vmatrix} 1 & 1 \\ 3 & 0 \end{vmatrix} = -3$$

$$D_{31} = \begin{vmatrix} 1 & -1 \\ 1 & 1 \end{vmatrix} = 2, \quad D_{32} = \begin{vmatrix} 1 & -1 \\ 0 & 1 \end{vmatrix} = 1, \quad D_{33} = \begin{vmatrix} 1 & 1 \\ 0 & 1 \end{vmatrix} = 1$$

□

Definition 6.26 (Rank). We can also define the **rank** of a matrix $A \in \mathscr{M}_{m\times n}(\mathbb{R})$ as the order of the largest non-zero minor that can be obtained from the matrix A.

Example 6.32. *Calculate the rank of the matrix:*

$$A = \begin{pmatrix} -1 & 0 & 1 \\ 3 & 2 & 1 \\ 1 & -1 & 2 \end{pmatrix}.$$

SOLUTION. The rank of matrix A is 3, since:

$$|A| = -4 - 3 - 2 - 1 = -10 \neq 0,$$

that is, there exists a minor of order 3 with a non-zero determinant. □

Example 6.33. *Calculate the rank of the matrix:*

$$B = \begin{pmatrix} 1 & 1 & -1 \\ 0 & 1 & 1 \\ 3 & 0 & -6 \end{pmatrix}.$$

SOLUTION. Since $|B| = -6 + 3 + 3 = 0$, the rank of matrix B is at most 2, that is, $\text{rank}(B) \leq 2$.

To determine if the rank is 2, we need to check if there is any non-zero minor of order two. Indeed:

$$D_{33} = \begin{vmatrix} 1 & 1 \\ 0 & 1 \end{vmatrix} = 1 \neq 0,$$

which implies $\text{rank}(B) = 2$. □

Example 6.34. *Calculate the rank of the matrix:*

$$C = \begin{pmatrix} 2 & 4 & 0 & 7 \\ 1 & 3 & 1 & 8 \\ 1 & 0 & -2 & -9 \end{pmatrix}.$$

SOLUTION. Since:

$$\begin{vmatrix} 2 & 4 \\ 1 & 3 \end{vmatrix} = 6 - 4 = 2 \neq 0,$$

then $\operatorname{rank}(C) \geq 2$. But all minors of order 3 are zero, and therefore the rank of the matrix cannot be 3 and is finally 2. □

6.6 System of Linear Equations

Definition 6.27 (Linear equation). A **linear equation** with n unknowns $x_1, x_2, \ldots, x_n$ is any expression of the type:

$$a_1x_1 + a_2x_2 + \cdots + a_nx_n = b,$$

where $a_i, b \in \mathbb{R}$. The real numbers a_i are called coefficients and b is called the **independent term**.

Definition 6.28 (System of linear equations). A system of m linear equations with n unknowns is any set of m linear equations with the same n unknowns:

$$\left\{ \begin{array}{rcl} a_{11}x_1 + a_{12}x_2 + \cdots + a_{1n}x_n & = & b_1 \\ a_{21}x_1 + a_{22}x_2 + \cdots + a_{2n}x_n & = & b_2 \\ & \vdots & \\ a_{m1}x_1 + a_{m2}x_2 + \cdots + a_{mn}x_n & = & b_m \end{array} \right.$$

Definition 6.29 (Solution). A **solution** of a system of m linear equations with n unknowns is any set of n numbers $s_1, s_2, \cdots, s_n$ such that when substituting x_1 for s_1, x_2 for s_2 and so on, the m equations of the system are verified.

Example 6.35. *Given the system of linear equations:*

$$\left\{ \begin{array}{r} 2x - y = 1 \\ x + y = 2 \end{array} \right.$$

verify that $x = 1, \ y = 1$ is a solution.

SOLUTION. The set of values $x = 1, \ y = 1$ is a solution of the system of linear equations, since:

$$2x - y = 2 - 1 = 1$$
$$x + y = 1 + 1 = 2$$

□

6.6.1 Types of Systems. Classification

The systems of linear equations are classified according to the number of solutions they have into:

- **Inconsistent system** (IS). It is a system that does not admit any solution. That is, there is no set of values that satisfies all the equations of the system at the same time.
- **Consistent system** (CS). It is a system that has at least one solution. Among the consistent systems, we distinguish two cases:
  - **Consistent and determined system** (CDS). It is a system that has exactly one unique solution.
  - **Consistent and indeterminate system** (CIS). It is a system that has an infinite number of solutions.

Example 6.36 (of inconsistent system). *The system of linear equations:*

$$\begin{cases} x + y = 1 \\ x + y = 2 \end{cases}$$

is inconsistent, since there is no set of two values $\{x, y\}$ *that can satisfy both equations at the same time.*

Example 6.37 (of consistent and determined system). *The system of linear equations:*

$$\begin{cases} x + y = 2 \\ x - y = 0 \end{cases}$$

is determined consistent, since the only solution is $\{x = 1, y = 1\}$.

Example 6.38 (of consistent and indeterminate system). *The system of linear equations:*

$$\{\ x + y = 0$$

is undetermined consistent, since the following sets are solutions: $\{x = 1, y = -1\}, \{x = -1, y = 1\}, \{x = 2, y = -2\}, \{x = -2, y = 2\}, \ldots$

6.6.2 Equivalent Systems

Definition 6.30 (Equivalent systems). Two systems are called **equivalent** if they have the same solutions.

Example 6.39. *The systems:*

$$\left.\begin{array}{r} 2x - y = 1 \\ x + y = 2 \end{array}\right\} \sim \left.\begin{array}{r} 4x - 2y = 2 \\ x + y = 2 \end{array}\right\} \sim \left.\begin{array}{r} x = 1 \\ y = 1 \end{array}\right\}$$

are equivalent, since they have the same solution $\{x = 1, y = 1\}$.

Property 6.7. Given a system of linear equations, the **elementary transformations** that allow obtaining new equivalent systems are:

(PF) Change the order of the equations.

(MF) Multiply or divide one of the equations by a real number different from 0.

(SF) Add to one equation of the system another multiplied by a real number.

Example 6.40. *Given the system of linear equations:*

$$\left.\begin{aligned} x+y+z&=1\\ 2x+3y+4z&=0\\ 5x+8y+9z&=0 \end{aligned}\right\}$$

calculate the solution of the system using elementary transformations and equivalent systems.

SOLUTION.

$$\left.\begin{aligned} x+y+z&=1\\ 2x+3y+4z&=0\\ 5x+8y+9z&=0 \end{aligned}\right\} \underset{\substack{f_2\leftarrow f_2-2f_1\\ f_3\leftarrow f_3-5f_1}}{\overset{\text{(SF)}}{\sim}} \left.\begin{aligned} x+y+z&=1\\ y+2z&=-2\\ 3y+4z&=-5 \end{aligned}\right\}$$

$$\underset{f_3\leftarrow f_3-3f_2}{\overset{\text{(SF)}}{\sim}} \left.\begin{aligned} x+y+z&=1\\ y+2z&=-2\\ -2z&=1 \end{aligned}\right\}$$

$$\underset{-\frac{1}{2}f_3}{\overset{\text{(MF)}}{\sim}} \left.\begin{aligned} x+y+z&=1\\ y+2z&=-2\\ z&=-1/2 \end{aligned}\right\}$$

$$\underset{f_1\leftarrow f_1-f_2}{\overset{\text{(SF)}}{\sim}} \left.\begin{aligned} x-z&=3\\ y+2z&=-2\\ z&=-1/2 \end{aligned}\right\}$$

$$\underset{\substack{f_1\leftarrow f_1+f_3\\ f_2\leftarrow f_2-2f_3}}{\overset{\text{(SF)}}{\sim}} \left.\begin{aligned} x&=5/2\\ y&=-1\\ z&=-1/2 \end{aligned}\right\}$$

□

6.6.3 Matrix Form of a System of Equations

Every system of equations:

$$\left\{\begin{array}{rcl} a_{11}x_1+a_{12}x_2+\cdots+a_{1n}x_n &=& b_1\\ a_{21}x_1+a_{22}x_2+\cdots+a_{2n}x_n &=& b_2\\ &\vdots& \\ a_{m1}x_1+a_{m2}x_2+\cdots+a_{mn}x_n &=& b_m \end{array}\right.$$

can be rewritten in matrix form as:

$$\begin{pmatrix} a_{11} & a_{12} & \cdots & a_{1n} \\ a_{21} & a_{22} & \cdots & a_{2n} \\ \vdots & \vdots & \ddots & \vdots \\ a_{m1} & a_{m2} & \cdots & a_{mn} \end{pmatrix} \begin{pmatrix} x_1 \\ x_2 \\ \vdots \\ x_n \end{pmatrix} = \begin{pmatrix} b_1 \\ b_2 \\ \vdots \\ b_m \end{pmatrix}$$

and is denoted by $Ax = b$.

Definition 6.31 (Matrix of a system). The **matrix of a system** is the matrix formed by all the coefficients of the unknowns. This matrix, usually symbolized as A, will have dimension $m \times n$, where m is the number of equations and n the number of unknowns.

Definition 6.32 (Augmented matrix). The **augmented matrix** of the system is obtained by adding the column of the independent terms. This matrix, usually symbolized as $(A|b)$, has dimension $m \times (n+1)$.

Example 6.41. *Given the system of linear equations:*

$$\left.\begin{aligned} 2x - y &= 1 \\ x + y &= 2 \end{aligned}\right\}$$

the matrix of the system is:

$$A = \begin{pmatrix} 2 & -1 \\ 1 & 1 \end{pmatrix}$$

while the augmented matrix is:

$$(A|b) = \left(\begin{array}{cc|c} 2 & -1 & 1 \\ 1 & 1 & 2 \end{array}\right).$$

6.6.4 Elementary Transformations in Matrix Form

The transformations PF, MF and SF of Proposition 6.7 that allow obtaining two equivalent systems, are equivalent to considering the augmented matrix $(A|b)$ and performing on it the transformations PF, MF and SF of Definition 6.18 that allowed us to obtain equivalent matrices.

Example 6.42. *Given the system:*

$$\left.\begin{aligned} x + y + z &= 1 \\ 2x + 3y + 4z &= 0 \\ 5x + 8y + 9z &= 0 \end{aligned}\right\}$$

that we have solved in Example 6.40 and whose solution is $\{x = 5/2, y = -1, z = -1/2\}$, *the augmented matrix associated with this system is:*

$$(A|b) = \left(\begin{array}{ccc|c} 1 & 1 & 1 & 1 \\ 2 & 3 & 4 & 0 \\ 5 & 8 & 9 & 0 \end{array}\right).$$

Then, by elementary row transformations, we have:

$$\left(\begin{array}{ccc|c} 1 & 1 & 1 & 1 \\ 2 & 3 & 4 & 0 \\ 5 & 8 & 9 & 0 \end{array}\right) \underset{\substack{f_2 \leftarrow f_2 - 2f_1 \\ f_3 \leftarrow f_3 - 5f_1}}{\sim} \left(\begin{array}{ccc|c} 1 & 1 & 1 & 1 \\ 0 & 1 & 2 & -2 \\ 0 & 3 & 4 & -5 \end{array}\right)$$

$$\underset{f_3 \leftarrow f_3 - 3f_2}{\sim} \left(\begin{array}{ccc|c} 1 & 1 & 1 & 1 \\ 0 & 1 & 2 & -2 \\ 0 & 0 & -2 & 1 \end{array}\right)$$

$$\underset{-\frac{1}{2}f_3}{\sim} \left(\begin{array}{ccc|c} 1 & 1 & 1 & 1 \\ 0 & 1 & 2 & -2 \\ 0 & 0 & 1 & -1/2 \end{array}\right)$$

$$\underset{f_1 \leftarrow f_1 - f_2}{\sim} \left(\begin{array}{ccc|c} 1 & 0 & -1 & 3 \\ 0 & 1 & 2 & -2 \\ 0 & 0 & 1 & -1/2 \end{array}\right)$$

$$\underset{\substack{f_1 \leftarrow f_1 + f_3 \\ f_2 \leftarrow f_2 - 2f_3}}{\sim} \left(\begin{array}{ccc|c} 1 & 0 & 0 & 5/2 \\ 0 & 1 & 0 & -1 \\ 0 & 0 & 1 & -1/2 \end{array}\right)$$

and if we return to the equations notation, we obtain the solution $\{x = 5/2, y = -1, z = -1/2\}$.

6.6.5 Rouché-Frobenius Theorem

We now present a theorem that allows us to classify the systems of linear equations based on the rank of the system matrix and the corresponding augmented matrix.

Theorem 6.1 (Rouché-Frobenius). Let $Ax = b$ be a system of m equations with n unknowns. Denoting by $r = \text{rank}(A)$ and by $r' = \text{rank}(A|b)$ then the system can be classified as:

- Inconsistent system (IS) if $r \neq r'$
- Consistent system if $r = r'$. In particular:
  - consistent and determined system (CDS) if $r = r' = n$;
  - consistent and indeterminate system (CIS) if $r = r' < n$.

Remark 6.12. In the case of an consistent and indeterminate system, the difference between the number of unknowns and the rank, $n - r$, is called the *degree of indeterminacy* or *degrees of freedom* and gives the number of parameters that will appear in the solution of the system.

Example 6.43. *Classify the system of linear equations:*

$$\left.\begin{aligned} x &= 1 \\ x &= 2 \end{aligned}\right\}$$

using the Rouché-Frobenius theorem.

SOLUTION. The matrix of the system is:

$$A = \begin{pmatrix} 1 \\ 1 \end{pmatrix}$$

and has rank 1 while the augmented matrix is:

$$(A|b) = \left(\begin{array}{c|c} 1 & 1 \\ 1 & 2 \end{array}\right)$$

To calculate the rank of the augmented matrix, we use the Gaussian method:

$$(A|b) = \left(\begin{array}{c|c} 1 & 1 \\ 1 & 2 \end{array}\right) \underset{f_2 \leftarrow f_2 - f_1}{\sim} \left(\begin{array}{c|c} 1 & 1 \\ 0 & 1 \end{array}\right),$$

which implies that $\operatorname{rank}(A|b) = 2$. Since $r = \operatorname{rank}(A) = 1$ and $r' = \operatorname{rank}((A|b)) = 2$, that is, $r \neq r'$, the system is inconsistent (IS). □

Example 6.44. *Classify the system of linear equations:*

$$\left.\begin{aligned} x + y &= 2 \\ x - y &= 0 \end{aligned}\right\}$$

using the Rouché-Frobenius theorem.

SOLUTION. The matrix of the system is:

$$A = \begin{pmatrix} 1 & 1 \\ 1 & -1 \end{pmatrix}$$

while the augmented matrix is:

$$(A|b) = \left(\begin{array}{cc|c} 1 & 1 & 2 \\ 1 & -1 & 0 \end{array}\right)$$

To calculate the rank of the matrix and the augmented matrix, we use the Gaussian method:

$$(A|b) = \left(\begin{array}{cc|c} 1 & 1 & 2 \\ 1 & -1 & 0 \end{array}\right) \underset{f_2 \leftarrow f_2 - f_1}{\sim} \left(\begin{array}{cc|c} 1 & 1 & 2 \\ 0 & -2 & -2 \end{array}\right),$$

which implies that $r = \operatorname{rank}(A) = 2$ and $r' = \operatorname{rank}(A|b) = 2$. Since:

$$r = r' = n = 2,$$

where $n = 2$ is the number of unknowns, the system is determined consistent (CDS). □

Example 6.45. *Classify the system of linear equations:*

$$\left.\begin{aligned} x + y + z &= 0 \\ y + z &= 0 \end{aligned}\right\}$$

using the Rouché-Frobenius theorem.

SOLUTION. The matrix of the system is:

$$A = \begin{pmatrix} 1 & 1 & 1 \\ 0 & 1 & 1 \end{pmatrix}$$

while the augmented matrix is:

$$(A|b) = \left(\begin{array}{ccc|c} 1 & 1 & 1 & 0 \\ 0 & 1 & 1 & 0 \end{array}\right)$$

To calculate the rank of the matrix and the augmented matrix, we use the Gaussian method:

$$(A|b) = \left(\begin{array}{ccc|c} 1 & 1 & 1 & 0 \\ 0 & 1 & 1 & 0 \end{array}\right) \underset{f_2 \leftarrow f_2 - f_1}{\sim} \left(\begin{array}{ccc|c} 1 & 1 & 1 & 0 \\ 0 & 0 & 0 & 0 \end{array}\right),$$

which implies that $r = \text{rank}(A) = 1$ and $r' = \text{rank}(A|b) = 1$. Since:

$$r = r' = 1 < n = 3,$$

where $n = 3$ is the number of unknowns, the system is undetermined consistent (CIS) with $n - r = 3 - 1 = 2$ degrees of freedom. □

Among all the systems of linear equations, there is a special type of system, the homogeneous ones. These systems are characterized because they are always consistent, as we will see below.

Definition 6.33 (Homogeneous system). A system of linear equations is called **homogeneous** if all the independent terms are zero:

$$\begin{cases} a_{11}x_1 + a_{12}x_2 + \cdots + a_{1n}x_n &= 0 \\ a_{21}x_1 + a_{22}x_2 + \cdots + a_{2n}x_n &= 0 \\ & \vdots \\ a_{m1}x_1 + a_{m2}x_2 + \cdots + a_{mn}x_n &= 0 \end{cases}$$

Remark 6.13. In any homogeneous system,

$$\text{rank}(A) = \text{rank}(A|b)$$

and, therefore, homogeneous systems are always consistent (they admit, at a minimum, the so-called *trivial solution* $x_1 = x_2 = \cdots = x_n = 0$).

6.6.6 Solving Systems Using the Gaussian Method

Gaussian rule

It is a method for solving linear systems based on obtaining equivalent systems by eliminating unknowns, until a system with an upper triangular matrix is obtained. In this case, the solution is obtained from the last unknown, by backward substitution, from which the values of the other unknowns are found.

Gauss-Jordan rule

This rule is a particular case of the previous one; instead of reaching a system whose matrix is upper triangular, a system with a diagonal matrix is reached, from which the solution is more easily obtained.

Let's see how this method works through the following example:

Example 6.46 (of solving a system by the Gauss-Jordan method)**.** *Solve the system:*

$$\left.\begin{array}{rcl} (1+a)x+y-z &=& 1+b \\ x+y+2z &=& 4 \\ 2ax-5z &=& 2b-5 \end{array}\right\}$$

depending on the parameters $a, b \in \mathbb{R}$.

SOLUTION. We want to solve a system of three linear equations with three unknowns. Consider the augmented matrix of the system:

$$\left(\begin{array}{ccc|c} 1+a & 1 & -1 & 1+b \\ 1 & 1 & 2 & 4 \\ 2a & 0 & -5 & 2b-5 \end{array}\right).$$

To solve the system, we transform the augmented matrix using elementary transformations. To simplify the calculations, we will permute the columns of the matrix (which is not an elementary transformation) so that the order of the variables instead of (x, y, z) is (y, z, x). Indeed:

$$\left(\begin{array}{ccc|c} 1 & -1 & 1+a & 1+b \\ 1 & 2 & 1 & 4 \\ 0 & -5 & 2a & 2b-5 \end{array}\right) \underset{f_2 \leftarrow f_2 - f_1}{\sim} \left(\begin{array}{ccc|c} 1 & -1 & 1+a & 1+b \\ 0 & 3 & -a & 3-b \\ 0 & -5 & 2a & 2b-5 \end{array}\right)$$

$$\underset{3f_3}{\sim} \left(\begin{array}{ccc|c} 1 & -1 & 1+a & 1+b \\ 0 & 3 & -a & 3-b \\ 0 & -15 & 6a & 6b-15 \end{array}\right)$$

$$\underset{f_3 \leftarrow f_3 + 5f_2}{\sim} \left(\begin{array}{ccc|c} 1 & -1 & 1+a & 1+b \\ 0 & 3 & -a & 3-b \\ 0 & 0 & a & b \end{array}\right)$$

The rank of A is 2 if $a = 0$ and 3 in any other case. Let's study what happens if $a = 0$ and if $a \neq 0$.

- If $a = 0$, as we said, the rank of matrix A is 2. In the case of the augmented matrix:

$$\left(\begin{array}{ccc|c} 1 & -1 & 1 & 1+b \\ 0 & 3 & 0 & 3-b \\ 0 & 0 & 0 & b \end{array}\right),$$

  its rank will also be 2 (CIS) if $b = 0$; or 3 if $b \neq 0$ (IS). In summary, if $a = 0,\ b \neq 0$, the system is inconsistent (IS). On the other hand, if $a = 0,\ b = 0$ the system is undetermined consistent (CIS) whose solution we find by converting the augmented matrix back to system form (remembering the change of order in the variables):

$$\left.\begin{array}{rcl} y-z+x &=& 1 \\ 3z &=& 3 \end{array}\right\} \rightarrow \left.\begin{array}{c} z=1 \\ y=2-x \end{array}\right\} \rightarrow \left.\begin{array}{l} x=\lambda,\ \lambda \in \mathbb{R} \\ y=2-\lambda \\ z=1 \end{array}\right\}$$

- If $a \neq 0$, the rank of A and the rank of the augmented matrix are equal to 3, which is the number of unknowns. Therefore, the system is determined consistent (CDS), whose solution we find using the Gauss-Jordan method:

$$\left(\begin{array}{ccc|c} 1 & -1 & 1+a & 1+b \\ 0 & 3 & -a & 3-b \\ 0 & 0 & a & b \end{array}\right) \underset{3f_1}{\sim} \left(\begin{array}{ccc|c} 3 & -3 & 3+3a & 3+3b \\ 0 & 3 & -a & 3-b \\ 0 & 0 & a & b \end{array}\right)$$

$$\underset{f_1 \leftarrow f_1+f_2}{\sim} \left(\begin{array}{ccc|c} 3 & 0 & 3+2a & 6+2b \\ 0 & 3 & -a & 3-b \\ 0 & 0 & a & b \end{array}\right)$$

$$\underset{f_2 \leftarrow f_2+f_3}{\sim} \left(\begin{array}{ccc|c} 3 & 0 & 3+2a & 6+2b \\ 0 & 3 & 0 & 3 \\ 0 & 0 & a & b \end{array}\right)$$

$$\underset{f_1 \leftarrow f_1 - 2f_3}{\sim} \left(\begin{array}{ccc|c} 3 & 0 & 3 & 6 \\ 0 & 3 & 0 & 3 \\ 0 & 0 & a & b \end{array}\right)$$

$$\underset{\substack{\frac{1}{3}f_1 \\ \frac{1}{2}f_2}}{\sim} \left(\begin{array}{ccc|c} 1 & 0 & 1 & 2 \\ 0 & 1 & 0 & 1 \\ 0 & 0 & a & b \end{array}\right)$$

If we convert it to equations we have:

$$\left.\begin{array}{rcl} y+x & = & 2 \\ z & = & 1 \\ ax & = & b \end{array}\right\} \longrightarrow \left.\begin{array}{l} x = b/a \\ y = 2 - b/a \\ z = 1 \end{array}\right\}$$

In summary:

- If $a = 0$ and $b \neq 0 \rightarrow$ IS;
- If $a = b = 0 \rightarrow$ SCI where $z = 1, y = 2 - x$;
- If $a \neq 0 \rightarrow$ SCD where $x = b/a, y = 2 - b/a, z = 1$.

Proposed Exercise 6.10. Consider the matrix:

$$A = \begin{pmatrix} 1 & 0 & 0 & 0 & 0 \\ 0 & 0 & 0 & a & 0 \\ 0 & 0 & 2 & 0 & 0 \\ 0 & 1 & 0 & 0 & 0 \\ 0 & 0 & 0 & 0 & a+1 \end{pmatrix}.$$

Calculate the determinant of A. Calculate the values of $a \in \mathbb{R}$ that make A singular. Calculate the inverse matrix of A when possible.

Proposed Exercise 6.11. Solve the following systems of linear equations.

(a) $\begin{cases} 3x - 2y + 2z = 4 \\ 2x + y - 3z = 1 \\ x - 3y + 4z = 2 \end{cases}$

(b) $\begin{cases} x + y - z + 2u = 0 \\ 2x - y - t = 1 \\ 3x - z + t = 1 \end{cases}$

(c) $\begin{cases} x + y + z = 3 \\ x + 2y - 4z = 1 \end{cases}$

(d) $\begin{cases} x + 2y = 1 \\ 2x - z = 1 \\ x + y - z = 0 \\ x - y + 2z = 3 \end{cases}$

(e) $\begin{cases} x + y + z = 1 \\ 2x - 2y - z = 0 \\ x + 3y + 5z = 2 \\ 5x + 3y + 6z = 4 \end{cases}$

Proposed Exercise 6.12. Classify the following systems of linear equations according to the value of the parameter $a \in \mathbb{R}$.

(a) $\begin{cases} 2x + y - az = 1 \\ 3x + 2y + z = a \\ 2x + y + z = -a \end{cases}$

(b) $\begin{cases} ax + y + 2z = 1 \\ x + ay + z = a \\ x + 2y + z = a \end{cases}$

(c) $\begin{cases} ax + 2z = 1 \\ x + y + (a+1)z = 2 \\ 2x - y + 4z = a \end{cases}$

(d) $\begin{cases} 3x - 2y + z = 1 \\ 4x + y - 2z = 2 \\ 2x - 5y - az = 3 \end{cases}$

Proposed Exercise 6.13. Classify the system of linear equations:

$$\begin{cases} ax - 2x = b + 1 \\ (2a - 4)x - y = 2b \\ (a^2 - 4)x + z - y = 0 \end{cases}$$

according to the values of the parameters $a, b \in \mathbb{R}$.

Proposed Exercise 6.14. Classify the following system of linear equations according to the values of the parameters $b, c \in \mathbb{R}$:

$$\begin{cases} s + t = 1 \\ bx + y = 1 \\ bs = c \\ bz + t = 1 \end{cases}$$

Find the solutions in cases where the system is consistent.

6.7 Geometric Approximation to Systems of Linear Equations

In this section, we will relate the geometry of space ($\mathbb{R}^3$) with systems of linear equations, establishing an equivalence between the classification of systems and the relative position of planes in space.

We will start the section by defining a plane:

Definition 6.34 (Plane/vector equation of the plane). A plane π in space $\mathbb{R}^3$ is the geometric locus of points obtained by considering a point $(x_0, y_0, z_0) \in \mathbb{R}^3$ and adding the linear combination of two vectors $\vec{u} = (u_1, u_2, u_3)$ and $\vec{v} = (v_1, v_2, v_3)$ which are non-zero and linearly independent. That is:

$$\begin{aligned} \pi : (x, y, z) &= (x_0, y_0, z_0) + \lambda\vec{u} + \mu\vec{v} \\ &= (x_0, y_0, z_0) + \lambda(u_1, u_2, u_3) + \mu(v_1, v_2, v_3), \ \ \lambda, \mu \in \mathbb{R} \end{aligned}$$

This last expression is called the *vector equation of the plane*, and the vectors $\vec{u}, \vec{v}$ are called *direction vectors* of the plane.

Definition 6.35 (Parametric equation of the plane). If we consider the vector equation of the plane and express it component-wise:

$$\left.\begin{array}{rcl} x &=& x_0 + \lambda u_1 + \mu v_1 \\ y &=& y_0 + \lambda u_2 + \mu v_2 \\ z &=& z_0 + \lambda u_3 + \mu v_3 \end{array}\right\}$$

we obtain the *parametric equation of the plane*

Now, the previous system can also be written as follows:

$$\left.\begin{array}{rcl} x - x_0 &=& \lambda u_1 + \mu v_1 \\ y - y_0 &=& \lambda u_2 + \mu v_2 \\ z - z_0 &=& \lambda u_3 + \mu v_3 \end{array}\right\} \Leftrightarrow \begin{pmatrix} x - x_0 \\ y - y_0 \\ z - z_0 \end{pmatrix} = \lambda \begin{pmatrix} u_1 \\ u_2 \\ u_3 \end{pmatrix} + \mu \begin{pmatrix} v_1 \\ v_2 \\ v_3 \end{pmatrix}$$

This last expression means that the vector:

$$\begin{pmatrix} x - x_0 \\ y - y_0 \\ z - z_0 \end{pmatrix}$$

is a linear combination of the vectors:

$$\begin{pmatrix} u_1 \\ u_2 \\ u_3 \end{pmatrix} \text{ and } \begin{pmatrix} v_1 \\ v_2 \\ v_3 \end{pmatrix}$$

which implies that the determinant of the matrix formed by the concatenation of these three vectors must be zero:

$$\begin{vmatrix} x - x_0 & u_1 & v_1 \\ y - y_0 & u_2 & v_2 \\ z - z_0 & u_3 & v_3 \end{vmatrix} = 0.$$

If we calculate this determinant and group the coefficients according to x, y, and z, we have that:

$$(u_2v_3 - u_3v_2)x + (u_3v_1 - u_1v_3)y + (u_1v_2 - u_2v_1)z +$$
$$+ (-z_0u_1v_2 + z_0u_2v_1 - x_0u_2v_3 + x_0u_3v_2 - y_0u_3v_1 + y_0u_1v_3) = 0$$

This last equation can be written more compactly as:

$$\pi : \; Ax + By + Cz + D = 0,$$

where:

$$\begin{aligned} A &= u_2v_3 - u_3v_2 \\ B &= u_3v_1 - u_1v_3 \\ C &= u_1v_2 - u_2v_1 \\ D &= -z_0u_1v_2 + z_0u_2v_1 - x_0u_2v_3 + x_0u_3v_2 - y_0u_3v_1 + y_0u_1v_3 \end{aligned}$$

This expression is called the *general, Cartesian* or *implicit* equation of a plane.

Remark 6.14. It can be seen that the coefficients A, B, C of the general equation of a plane π can be obtained as the cross product of the direction vectors $\vec{u}$ and $\vec{v}$:

$$\begin{vmatrix} \vec{i} & \vec{j} & \vec{k} \\ u_1 & u_2 & u_3 \\ v_1 & v_2 & v_3 \end{vmatrix} = A\vec{i} + B\vec{j} + C\vec{k}.$$

Proposition 6.7. The vector $\vec{n} = (A, B, C)$ is perpendicular to the plane:

$$\pi: \; Ax + By + Cz + D = 0$$

and is called the *normal vector to the plane.*

PROOF. We can consider two proofs, one based on the cross product and the other on the dot product of two vectors.

- The vector $\vec{n} = (A, B, C)$ is perpendicular because it has been calculated as the cross product of the two direction vectors of the plane, $\vec{u}$ and $\vec{v}$.
- The vector $\vec{n} = (A, B, C)$ is perpendicular to the two direction vectors of the plane, $\vec{u}$ and $\vec{v}$, since the dot products:

$$\begin{aligned} \vec{u} \cdot \vec{n} &= u_1(u_2v_3 - u_3v_2) + u_2(u_3v_1 - u_1v_3) + u_3(u_1v_2 - u_2v_1) \\ &= u_1u_2v_3 - u_1v_2u_3 + v_1u_2u_3 - u_1u_2v_3 + u_1v_2u_3 - v_1u_2u_3 = 0 \\ \vec{v} \cdot \vec{n} &= v_1(u_2v_3 - u_3v_2) + v_2(u_3v_1 - u_1v_3) + v_3(u_1v_2 - u_2v_1) \\ &= v_1u_2v_3 - v_1v_2u_3 + v_1v_2u_3 - u_1v_2v_3 + u_1v_2v_3 - v_1u_2v_3 = 0 \end{aligned}$$

  are both zero.

□

What we have just seen is that we have identified the planes in space through a linear equation with three unknowns x, y, and z.

What is, then, the geometric interpretation of a system of two linear equations with three unknowns? What will be the geometric interpretation of a system of three linear equations with three unknowns?

6.7.1 Geometric Interpretation of a System of Two Linear Equations with Three Unknowns

Given the system of equations:

$$\left.\begin{aligned} A_1x + B_1y + C_1z &= D_1 \\ A_2x + B_2y + C_2z &= D_2 \end{aligned}\right\},$$

which we can represent in matrix form as:

$$\left(\begin{array}{ccc|c} A_1 & B_1 & C_1 & D_1 \\ A_2 & B_2 & C_2 & D_2 \end{array}\right)$$

this is geometrically interpreted as the *intersection* of two planes defined by the equations that form the system. That is, the solution of the system, if it exists, is the set of points formed by the intersection of the planes:

$$\pi_1 : \; A_1x + B_1y + C_1z = D_1$$

$$\pi_2 : \; A_2x + B_2y + C_2z = D_2$$

The relative position of two planes can be:

- the two planes are the same;
- the two planes intersect in a line;
- the two planes are parallel and, therefore, do not intersect at any point.

Let's analyze each of these three cases:

- If the two planes are the same, it means that the equation defining plane π_1 is equivalent to the equation defining plane π_2, that is, we can obtain one equation from the other simply by multiplying by a scalar. Therefore, at the matrix level, the rank of the matrix is 1, the rank of the augmented matrix is also 1, and since there are 3 unknowns, the system is indeterminate with two degrees of freedom. That is, the solution is a plane, the same plane $\pi_1 = \pi_2$. This answer is logical, as the intersection of two planes that are the same is the same plane.

  For example, the planes:

  $$\pi_1 : \; x + y + z = 1$$
  $$\pi_2 : \; 2x + 2y + 2z = 2$$

  are coincident, as can be seen in Figure 6.1.

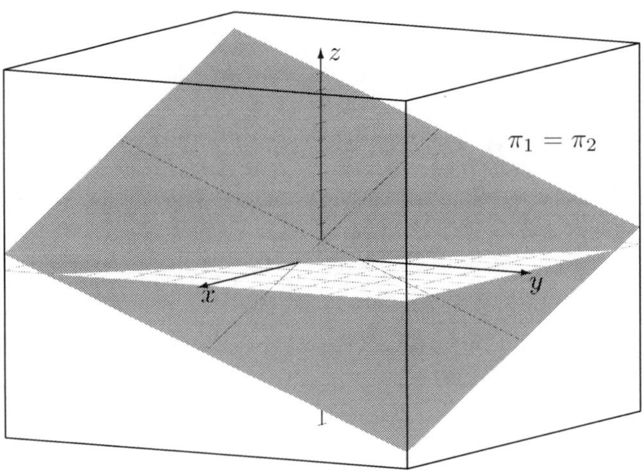

FIGURE 6.1
The planes π_1 and π_2, with equations $x + y + z = 1$ and $2x + 2y + 2z = 2$, respectively, are coincident.

- If the two planes intersect in a line, this is because the normal vectors $\vec{n}_1 = (A_1, B_1, C_1)$ and $\vec{n}_2 = (A_2, B_2, C_2)$ are not parallel, that is, they are linearly independent. Therefore, at the matrix level, the rank of the matrix is 2, the rank of the augmented matrix is also 2, and since there are 3 unknowns, the system is indeterminate with one degree of freedom. That is, the solution is a line (the solution depends on one parameter).

For example, the planes:

$$\pi_1 : \ x + y + z = 1$$

$$\pi_2 : \ -11x - 20y - 10z = 25$$

intersect in the line r with equation $y = -\frac{x}{10} - \frac{35}{10}$, as shown in Figure 6.2.

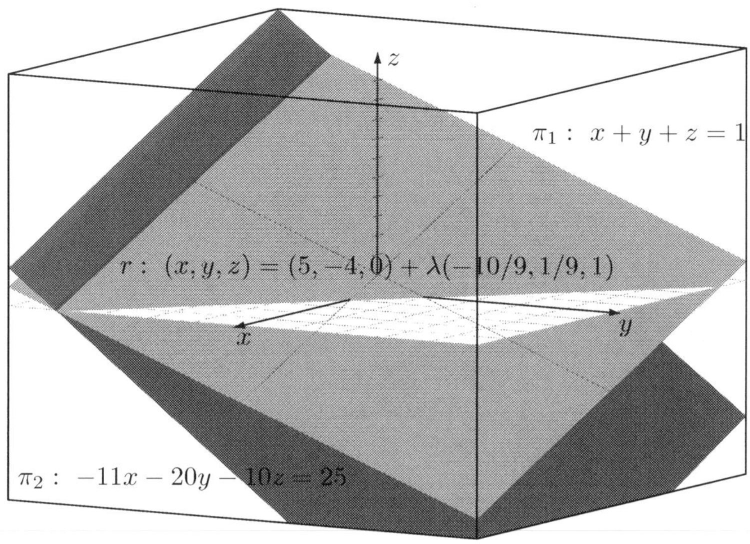

FIGURE 6.2
The planes π_1 and π_2, with equations $x + y + z = 1$ and $-11x - 20y - 10z = 25$, respectively, intersect in the line r passing through the point $(5, -4, 0)$ with direction vector $\vec{u}_r = (-10/9, 1/9, 1)$.

- If the two planes are parallel and do not intersect, this is because the normal vectors:

$$\vec{n}_1 = (A_1, B_1, C_1), \quad \vec{n}_2 = (A_2, B_2, C_2)$$

are parallel, but the planes are at different *heights*. Since the two normal vectors are parallel, the rank of the matrix is 1, but since the planes are not coincident, the rank of the augmented matrix is 2. Therefore, if the ranks are different, the system is inconsistent.

For example, the planes:

$$\pi_1 : \ x + y + z = 1$$

$$\pi_2 : \ x + y + z = 2$$

are parallel and have no common intersection point, as can be seen in Figure 6.3.

In summary, when we calculate a system of two linear equations with three unknowns, the classification of the system is related to the relative position of the planes that represent the equations:

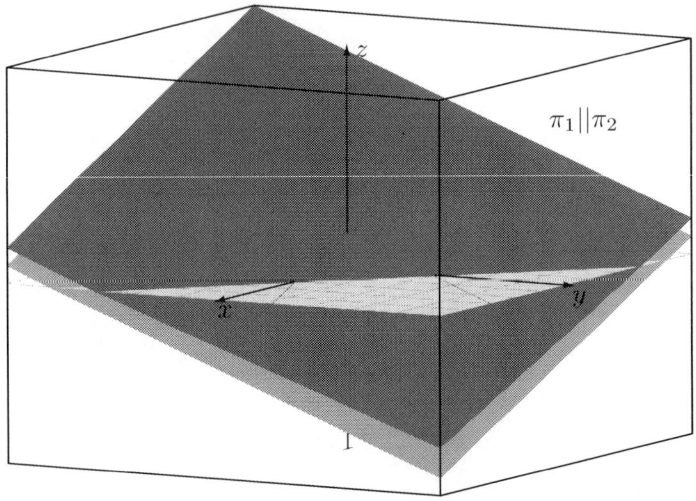

FIGURE 6.3
The planes π_1 and π_2, with equations $x + y + z = 1$ and $x + y + z = 2$, respectively, are parallel and have no common intersection point.

- The system is indeterminate (CIS) with two degrees of freedom: the two planes π_1 and π_2 are coincident (Figure 6.4, left).
- The system is indeterminate (CIS) with one degree of freedom: the two planes π_1 and π_2 intersect in a line (Figure 6.4, center).
- The system is inconsistent (IS): the two planes are parallel (Figure 6.4, right).

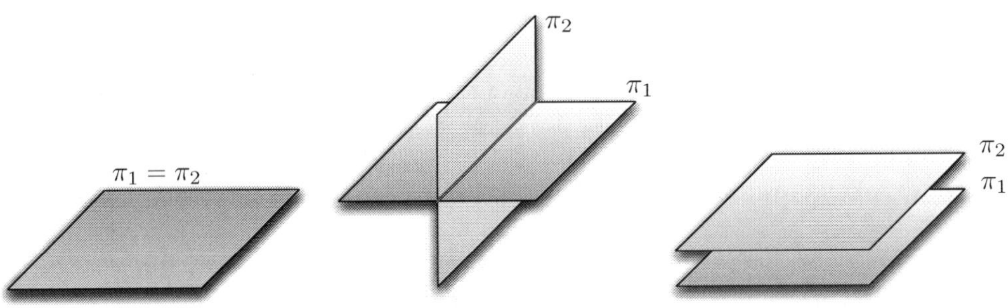

FIGURE 6.4
Relative position of two planes in space: coincident planes (left), planes intersecting in a line (center), and parallel planes (right).

6.7.2 Geometric Interpretation of a System of Three Linear Equations with Three Unknowns

Given the system of equations:

$$\left.\begin{array}{rcl} A_1x + B_1y + C_1z & = & D_1 \\ A_2x + B_2y + C_2z & = & D_2 \\ A_3x + B_3y + C_3z & = & D_3 \end{array}\right\},$$

which we can represent in matrix form as:

$$\left(\begin{array}{ccc|c} A_1 & B_1 & C_1 & D_1 \\ A_2 & B_2 & C_2 & D_2 \\ A_3 & B_3 & C_3 & D_3 \end{array}\right)$$

this is geometrically interpreted as the *intersection* of three planes defined by the equations that form the system. That is, the solution of the system, if it exists, is the set of points formed by the intersection of the planes:

$$\begin{aligned} \pi_1 &: \ A_1x + B_1y + C_1z = D_1 \\ \pi_2 &: \ A_2x + B_2y + C_2z = D_2 \\ \pi_3 &: \ A_3x + B_3y + C_3z = D_3 \end{aligned}$$

The relative position of three planes can be:

- the three planes are the same or coincident;
- the three planes are parallel and two of them are coincident;
- the three planes are parallel and distinct;
- two planes are coincident and one of them, secant;
- the three planes are secant and distinct (intersect in a line);
- two planes are parallel and the third, secant;
- the planes are secant two by two;
- the three planes intersect in a point.

Let's analyze each of these eight cases:

- If the three planes are the same, it means that the equation defining plane π_1 is equivalent to the equations defining planes π_2 and π_3, that is, we can obtain one equation from the others simply by multiplying by a scalar. Therefore, at the matrix level, the rank of the matrix is 1, the rank of the augmented matrix is also 1, and since there are 3 unknowns, the system is indeterminate with two degrees of freedom. That is, the solution is a plane, the same plane $\pi_1 = \pi_2 = \pi_3$. This answer is logical, as the intersection of three planes that are the same is the same plane.

  For example, the planes:

  $$\begin{aligned} \pi_1 &: \ x + y + z = 1 \\ \pi_2 &: \ 2x + 2y + 2z = 2 \\ \pi_3 &: \ 3x + 3y + 3z = 3 \end{aligned}$$

  are coincident.

- Suppose the three planes are parallel and two of them are coincident. Suppose the coincident planes are π_2 and π_3. This means that we can obtain the equation of π_2 by multiplying the equation of π_3 by a constant. This is equivalent to the quotient of the coefficients of the equations defining both planes being constant:

  $$\frac{A_2}{A_3} = \frac{B_2}{B_3} = \frac{C_2}{C_3} = \frac{D_2}{D_3}$$

On the other hand, if plane π_1 is parallel but not coincident, we have:

$$\frac{A_1}{A_2} = \frac{B_1}{B_2} = \frac{C_1}{C_2} \neq \frac{D_1}{D_2}$$
$$\frac{A_1}{A_3} = \frac{B_1}{B_3} = \frac{C_1}{C_3} \neq \frac{D_1}{D_3}$$

Therefore, at the matrix level, the rank of the matrix is 1, but the rank of the augmented matrix is 2, which means that the system is inconsistent.

For example, the planes:

$$\pi_1 : \ x + y + z = 2$$
$$\pi_2 : \ 2x + 2y + 2z = 2$$
$$\pi_3 : \ 3x + 3y + 3z = 3$$

are parallel with π_2 and π_3 coincident.

- Suppose the three planes are parallel and distinct. Since the three planes are parallel, the normal vectors:

$$\vec{n}_1 = (A_1, B_1, C_1)$$
$$\vec{n}_2 = (A_2, B_2, C_2)$$
$$\vec{n}_3 = (A_3, B_3, C_3)$$

are parallel. This is equivalent to the quotient of the components of these vectors being constant, but not the quotient of the terms $D_i, \ i = 1, 2, 3$. This is:

$$\frac{A_1}{A_2} = \frac{B_1}{B_2} = \frac{C_1}{C_2} \neq \frac{D_1}{D_2}$$
$$\frac{A_1}{A_3} = \frac{B_1}{B_3} = \frac{C_1}{C_3} \neq \frac{D_1}{D_3}$$
$$\frac{A_2}{A_3} = \frac{B_2}{B_3} = \frac{C_2}{C_3} \neq \frac{D_2}{D_3}$$

Therefore, at the matrix level, the rank of the matrix is 1, but the rank of the augmented matrix is 2, which means that the system is inconsistent.

For example, the planes:

$$\pi_1 : \ x + y + z = 2$$
$$\pi_2 : \ x + y + z = 2$$
$$\pi_3 : \ x + y + z = 3$$

are parallel and distinct.

- Suppose two of the planes are coincident and one secant. Suppose the coincident planes are π_2 and π_3. This means that we can obtain the equation of π_2 by multiplying the equation of π_3 by a constant. This is equivalent to the quotient of the coefficients of the equations defining both planes being constant:

$$\frac{A_2}{A_3} = \frac{B_2}{B_3} = \frac{C_2}{C_3} = \frac{D_2}{D_3}$$

On the other hand, if plane π_1 is secant, its normal vector $\vec{n}_1$ is not parallel to the vectors $\vec{n}_2 = \vec{n}_3$. Therefore, at the matrix level, the rank of the matrix is 2, and the rank of the augmented matrix is also 2 (it cannot be 3 since the last two rows are multiples of each other) and, having 3 unknowns, the system is indeterminate with one degree of freedom. That is, the solution is a line.

For example, the planes:

$$\pi_1 : \; x + y + 2z = 1$$
$$\pi_2 : \; x + y + z = 1$$
$$\pi_3 : \; 2x + 2y + 2z = 2$$

intersect in the line with equation $r : (x, y, z) = (1, 0, 0) + \lambda(-1, 1, 0)$.

- Suppose the three planes are secant and distinct (intersect in a line). The three normal vectors $\vec{n}_1, \vec{n}_2$, and $\vec{n}_3$ are not proportional two by two. However, these three vectors are located in a plane perpendicular to the intersection line of the three planes. Therefore, although the three normal vectors are not proportional two by two, it is possible to express $\vec{n}_1$, for example, as a linear combination of $\vec{n}_2$ and $\vec{n}_3$. At the matrix level, the rank of the matrix is 2. As we have said, it cannot be 3 because these three vectors are linearly dependent. The rank of the augmented matrix is also 2 and, having 3 unknowns, the system is indeterminate with one degree of freedom. That is, the solution is the line where they intersect.

  For example, the planes:

  $$\pi_1 : \; x + y = 1$$
  $$\pi_2 : \; z = 1$$
  $$\pi_3 : \; x + y + z = 2$$

  intersect in the line with equation $r : (x, y, z) = (1, 0, 1) + \lambda(-1, 1, 0)$.

- Suppose two of the planes are parallel but not coincident (π_2 and π_3) and the other secant. Then:

  $$\frac{A_2}{A_3} = \frac{B_2}{B_3} = \frac{C_2}{C_3} \neq \frac{D_2}{D_3}$$

  But the normal vectors $\vec{n}_1$ and $\vec{n}_2$ are not proportional. At the matrix level, the rank of the matrix is 2, but the rank of the augmented matrix is 3, which means that the system is inconsistent.

  For example, of the planes:

  $$\pi_1 : \; x + y + 2z = 1$$
  $$\pi_2 : \; x + y + z = 1$$
  $$\pi_3 : \; x + y + z = 2$$

  two are parallel and the third secant.

- If the three planes intersect in a point it is because the three normal vectors $\vec{n}_1, \vec{n}_2$ and $\vec{n}_3$ of the planes π_1, π_2 and π_3, respectively, are linearly independent. Therefore, the rank of the matrix is 3, the rank of the augmented matrix is also 3 and since the number of unknowns is also 3, the system is consistent and determined (CDS).

For example, the planes:

$$\pi_1: \ x+y+z=3$$
$$\pi_2: \ x+y=2$$
$$\pi_3: \ x+z=2$$

intersect at the point $(1,1,1)$.

In summary, when we calculate a system of three linear equations with three unknowns, the classification of the system is related to the relative position of the planes that represent the equations. This relationship is found in Table 6.1 and the positions of the planes can be seen in Figure 6.5.

TABLE 6.1
Relative position of three planes in space based on the rank of the matrix (r) and the rank of the augmented matrix (r').

r	r'	Position
1	1	the three planes are the same or coincident
1	2	the three planes are parallel and two of them, coincident
1	2	the three planes are parallel and distinct
2	2	two planes are coincident and one of them, secant
2	2	the three planes are secant and distinct
2	3	two planes are parallel and the third, secant
2	3	the planes are secant two by two
3	3	the three planes intersect in a point

Example 6.47. *Calculate the relative position of the planes:*

$$\pi_1: \ 2x-3y+4z-1=0$$
$$\pi_2: \ x-y-z+1=0$$
$$\pi_3: \ -x+2y-z+2=0$$

and the common intersection, if it exists.

SOLUTION. Consider the system of linear equations resulting from considering the three equations defining the planes:

$$\begin{cases} 2x-3y+4z-1=0 \\ x-y-z+1=0 \\ -x+2y-z+2=0 \end{cases}$$

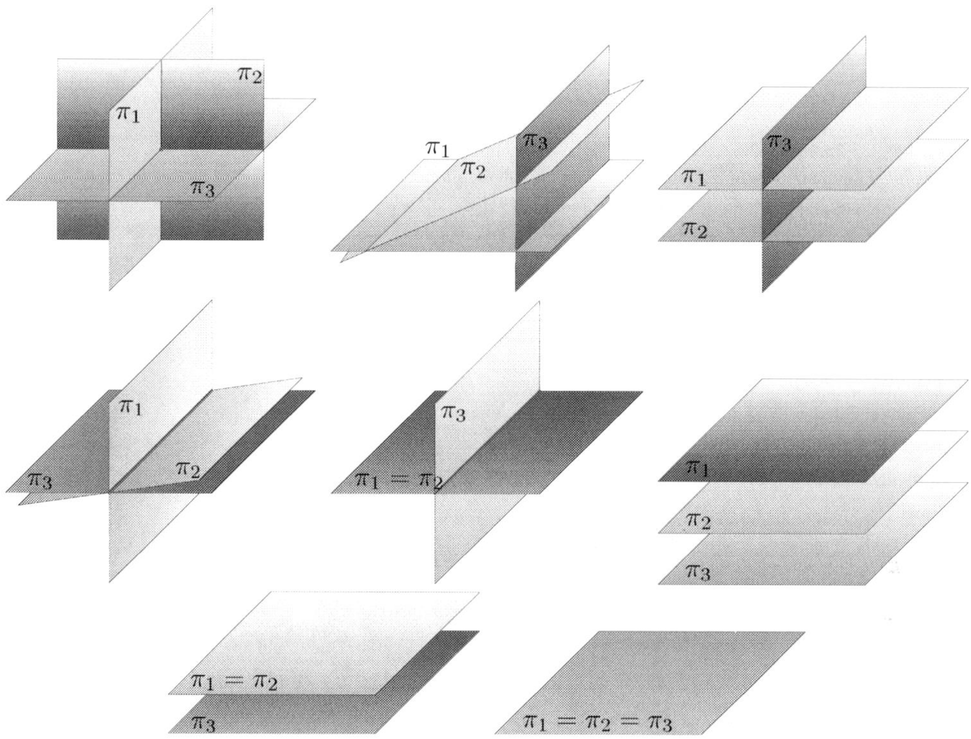

FIGURE 6.5
Relative position of three planes in space. From left to right and top to bottom: the three planes intersect in a point; the planes are secant two by two; two planes are parallel and the third, secant; the three planes are secant and distinct; two planes are coincident and one of them is secant; the three planes are parallel and distinct; the three planes are parallel and two of them, coincident; the three planes are the same or coincident.

The system matrix and the augmented matrix are:

$$A = \begin{pmatrix} 2 & -3 & 4 \\ 1 & -1 & -1 \\ -1 & 2 & -1 \end{pmatrix},$$

$$(A|\, b) = \left(\begin{array}{ccc|c} 2 & -3 & 4 & 1 \\ 1 & -1 & -1 & -1 \\ -1 & 2 & -1 & -2 \end{array}\right)$$

The determinant of A is equal to 4, which implies that the rank of the matrix is $r = 3$. The rank of the augmented matrix is also 3. Therefore, the three planes intersect in a point. The intersection point is $(-4, -3, 0)$, found by solving the system. □

Proposed Exercise 6.15. Calculate the relative position of the planes:

$$\pi_1 : \; x + y - z + 3 = 0$$

$$\pi_2 : \; -4x + y + 4z - 7 = 0$$

$$\pi_3 : \; -2x + 3y + 2z - 2 = 0$$

and the common intersection, if it exists.

Proposed Exercise 6.16. Calculate the relative position of the planes:

$$\pi_1 : \; 2x + 3y + z - 1 = 0$$

$$\pi_2 : \; x - y + z + 2 = 0$$

$$\pi_3 : \; 2x - 2y + 2z + 4 = 0$$

and the common intersection, if it exists.

Proposed Exercise 6.17. Calculate the relative position of the planes:

$$\pi_1 : \; 2x - y + 2z + 1 = 0$$

$$\pi_2 : \; -4x + 2y - 4z - 2 = 0$$

$$\pi_3 : \; 6x - 3y + 6z + 1 = 0$$

and the common intersection, if it exists.

6.7.3 Equation of a Line in Space

To complete the geometric view, we have given to systems of linear equations, we will define the different forms in which a line in space can be expressed.

Definition 6.36. Line/vector equation of the line A line r in space $\mathbb{R}^3$ is the geometric locus of points obtained by considering a point $(x_0, y_0, z_0) \in \mathbb{R}^3$ and adding the linear combination of a non-zero vector $\vec{u} = (u_1, u_2, u_3)$. That is:

$$\begin{aligned} r : (x, y, z) &= (x_0, y_0, z_0) + \lambda\vec{u} \\ &= (x_0, y_0, z_0) + \lambda(u_1, u_2, u_3) \end{aligned}$$

This last expression is called the *vector equation of the line* and the vector $\vec{u}$ is called the *direction vector* of the line.

Definition 6.37 (Parametric equation of the line)**.** If we consider the vector equation of the line and express it component-wise:

$$\left.\begin{array}{rcl} x & = & x_0 + \lambda u_1 \\ y & = & y_0 + \lambda u_2 \\ z & = & z_0 + \lambda u_3 \end{array}\right\}$$

we obtain the *parametric equation of the plane.*

Now, if the components of the direction vector $\vec{u}$ are non-zero, we can solve for the parameter λ in each of the previous equations:

$$\lambda = \frac{x - x_0}{u_1}$$

$$\lambda = \frac{y - y_0}{u_2}$$

$$\lambda = \frac{z - z_0}{u_3}$$

Definition 6.38 (Continuous equation of the line)**.** The *continuous equation* of the line results from equating the parameter λ of the parametric equation of the line:

$$r : \frac{x - x_0}{u_1} = \frac{y - y_0}{u_2} = \frac{z - z_0}{u_3}$$

From the continuous equation of the line, if we consider separately the first and second equality, we obtain the following system of equations:

$$\frac{x - x_0}{u_1} = \frac{y - y_0}{u_2}$$

$$\frac{y - y_0}{u_2} = \frac{z - z_0}{u_3}$$

which, when developed, is equivalent to:

$$u_2 x - u_1 y + u_1 y_0 - u_2 x_0 = 0$$

$$u_3 y - u_2 z + u_2 z_0 - u_3 y_0 = 0$$

Definition 6.39 (Implicit or Cartesian equation of the line)**.** The *implicit* or *Cartesian* equation of the line is the system of two linear equations:

$$r : \left\{\begin{array}{l} Ax + By + Cz + D = 0 \\ A'x + B'y + C'z + D' = 0 \end{array}\right.$$

where $A, A', B, B', C, C', D, D' \in \mathbb{R}$.

Remark 6.15. Note that, for practical purposes, a line has been identified with the intersection of two planes, provided the planes have a non-empty intersection.

Example 6.48. *Calculate the implicit equation of the line passing through the points* $P = (0,0,1)$ *and* $Q = (1,2,3)$.

SOLUTION. We seek the equation of the line passing through point $P = (0,0,1)$ and having as direction vector the vector joining points P and Q. That is:

$$\vec{u} = \vec{PQ} = Q - P = (1,2,3) - (0,0,1) = (1,2,2)$$

The vector equation is:

$$(x,y,z) = (0,0,1) + \lambda(1,2,2)$$

The parametric equation is:

$$r: \begin{cases} x = \lambda \\ y = 2\lambda \\ z = 1 + 2\lambda \end{cases}$$

The continuous equation:

$$r: \frac{x}{1} = \frac{y}{2} = \frac{z-1}{2}$$

and finally, the implicit equation:

$$r: \begin{cases} 2x - y = 0 \\ 2y - 2z + 2 = 0 \end{cases}$$

□

Example 6.49. *Calculate the direction vector of the line defined in implicit form as:*

$$r: \begin{cases} 2x - y = 0 \\ 3x - z = 0 \end{cases}$$

SOLUTION. The calculation of the direction vector is done through the resolution of the system. In fact, the problem is equivalent to finding the intersection of the two planes that define the line. In this case, being a system of two equations with three unknowns, one of the unknowns will act as the parameter. For example, let $z = \lambda$. Then, the resulting system is:

$$\begin{cases} 2x - y = 0 \\ 3x = \lambda \\ z = \lambda \end{cases}$$

whose solution is:

$$\begin{cases} x = \dfrac{2\lambda}{3} \\ y = \dfrac{\lambda}{3} \\ z = \lambda \end{cases}$$

What we have just obtained is the parametric equation of the line passing through point $P = (0,0,0)$ and whose direction vector is $\vec{u} = (2/3, 1/3, 1)$. □

6.8 Solved Problems (Matrices)

Problem 6.1. Given the matrix:

$$A = \begin{pmatrix} 2 & 1 \\ -2 & 3 \end{pmatrix}.$$

Determine the values of x, y that satisfy:

$$A^2 + xA + yI = N,$$

where I and N are the identity matrix and the null matrix of order 2.

Solution to the problem 6.1. *The matrix equation with two unknowns (x, y) in the statement contains 4 scalar equations that result from equating the 4 coefficients of the 2×2 matrices. Calculating the products of the equation we have:*

$$\begin{aligned} A^2 &= \begin{pmatrix} 2 & 1 \\ -2 & 3 \end{pmatrix} \cdot \begin{pmatrix} 2 & 1 \\ -2 & 3 \end{pmatrix} \\ &= \begin{pmatrix} 2 \cdot 2 + 1 \cdot (-2) & 2 \cdot 1 + 1 \cdot 3 \\ -2 \cdot 2 + 3 \cdot (-2) & -2 \cdot 1 + 3 \cdot 3 \end{pmatrix} = \begin{pmatrix} 2 & 5 \\ -10 & 7 \end{pmatrix} \\ xA &= x \cdot \begin{pmatrix} 2 & 1 \\ -2 & 3 \end{pmatrix} = \begin{pmatrix} 2x & x \\ -2x & 3x \end{pmatrix} \\ yI &= y \cdot \begin{pmatrix} 1 & 0 \\ 0 & 1 \end{pmatrix} = \begin{pmatrix} y & 0 \\ 0 & y \end{pmatrix} \end{aligned}$$

Therefore:

$$\begin{aligned} A^2 + xA + yI &= \begin{pmatrix} 2 & 5 \\ -10 & 7 \end{pmatrix} + \begin{pmatrix} 2x & x \\ -2x & 3x \end{pmatrix} + \begin{pmatrix} y & 0 \\ 0 & y \end{pmatrix} \\ &= \begin{pmatrix} 2 + 2x + y & 5 + x \\ -10 - 2x & 7 + 3x + y \end{pmatrix} = \begin{pmatrix} 0 & 0 \\ 0 & 0 \end{pmatrix}. \end{aligned}$$

The equations corresponding to the 12 and 21 coefficients are equivalent and allow us to calculate:

$$5 + x = 0 \quad \Leftrightarrow \quad x = -5.$$

From the equation of the 11 coefficient, we have:

$$2 + 2x + y = -8 + y = 0 \quad \Leftrightarrow \quad y = 8.$$

The values $x = -5$, $y = 8$ also verify the equation of the 22 coefficient and, therefore, they are the solution.

Problem 6.2. If $A \in \mathcal{M}_{2\times 2}(\mathbb{R})$, $A \neq 0$ and $x \in \mathbb{R}$, determine the value of x that satisfies:

$$\begin{pmatrix} 1 & 2 \\ 3 & x \end{pmatrix} \cdot A = \begin{pmatrix} 2 & 4 \\ 6 & 8 \end{pmatrix} \cdot A.$$

Solution to the problem 6.2. *Let the matrix* $A = \begin{pmatrix} a & b \\ c & d \end{pmatrix}$, *we calculate the products:*

$$\begin{pmatrix} 1 & 2 \\ 3 & x \end{pmatrix}\begin{pmatrix} a & b \\ c & d \end{pmatrix} = \begin{pmatrix} a+2c & b+2d \\ 3a+xc & 3b+xd \end{pmatrix},$$

$$\begin{pmatrix} a & b \\ c & d \end{pmatrix}\begin{pmatrix} 2 & 4 \\ 6 & 8 \end{pmatrix} = \begin{pmatrix} 2a+4c & 2b+4d \\ 6a+8c & 6b+8d \end{pmatrix}.$$

Equating the resulting matrices we obtain the following 4 equations:

$$a+2c = 2a+4c \Leftrightarrow a=-2c \tag{6.1}$$

$$b+2d = 2b+4d \Leftrightarrow b=-2d \tag{6.2}$$

$$3a+xc = 6a+8c \Leftrightarrow c(x-2)=0 \Leftrightarrow c=0 \ or\ x=2 \tag{6.3}$$

$$3b=xd = 6b+8d \Leftrightarrow d(x-2)=0 \Leftrightarrow d=0 \ or\ x=2 \tag{6.4}$$

and we have the following possibilities:

- *If* $x=2$, *then the only independent conditions are* $a=-2c$ *and* $b=-2d$ *and, therefore:*

$$A = \begin{pmatrix} -2c & -2d \\ c & d \end{pmatrix}, \quad \forall c,d \in \mathbb{R}.$$

- *If* $x \neq 2$, *then* $c=d=0$ *and from equations (1) and (2) it also follows that* $a=b=0$ *and we have that:*

$$A = \begin{pmatrix} 0 & 0 \\ 0 & 0 \end{pmatrix},$$

regardless of the value of x.

In summary, the solution is:

$$\left(x=-2 \ and\ \forall c,d \in \mathbb{R}\ A = \begin{pmatrix} -2c & -2d \\ c & d \end{pmatrix},\ x \neq 2 \ and\ A = \begin{pmatrix} 0 & 0 \\ 0 & 0 \end{pmatrix}\right).$$

Problem 6.3. Given the matrices:

$$A = \begin{pmatrix} 2 & 1 & 2 \\ 3 & 1 & 2 \end{pmatrix} \qquad B = \begin{pmatrix} 0 & 2 & -1 \\ 2 & 2 & 0 \end{pmatrix}$$

perform the following operations, if possible:

$$A+B, \quad A-B, \quad 2A-3B, \quad A\cdot B^{\mathrm{T}}, \quad A^{\mathrm{T}}\cdot B, \quad A\cdot B, \quad B\cdot A.$$

Solution to the problem 6.3. • $A+B = \begin{pmatrix} 2 & 1 & 2 \\ 3 & 1 & 2 \end{pmatrix} + \begin{pmatrix} 0 & 2 & -1 \\ 2 & 2 & 0 \end{pmatrix} = \begin{pmatrix} 2 & 3 & 1 \\ 5 & 3 & 2 \end{pmatrix}$.

- $A - B = \begin{pmatrix} 2 & 1 & 2 \\ 3 & 1 & 2 \end{pmatrix} - \begin{pmatrix} 0 & 2 & -1 \\ 2 & 2 & 0 \end{pmatrix} = \begin{pmatrix} 2 & -1 & 3 \\ 1 & -1 & 2 \end{pmatrix}$.

- *In this case:*

$$\begin{aligned} 2A - 3B &= 2\begin{pmatrix} 2 & 1 & 2 \\ 3 & 1 & 2 \end{pmatrix} - 3\begin{pmatrix} 0 & 2 & -1 \\ 2 & 2 & 0 \end{pmatrix} \\ &= \begin{pmatrix} 4 & 2 & 4 \\ 6 & 2 & 4 \end{pmatrix} - \begin{pmatrix} 0 & 6 & -3 \\ 6 & 6 & 0 \end{pmatrix} \\ &= \begin{pmatrix} 4 & -4 & 7 \\ 0 & -4 & 4 \end{pmatrix}. \end{aligned}$$

- *It is a product of matrices of dimensions 2×3 and 3×2 and, therefore, they can be multiplied:*

$$\begin{aligned} A \cdot B^{\mathrm{T}} &= \begin{pmatrix} 2 & 1 & 2 \\ 3 & 1 & 2 \end{pmatrix} \cdot \begin{pmatrix} 0 & 2 \\ 2 & 2 \\ -1 & 0 \end{pmatrix} \\ &= \begin{pmatrix} 2\cdot 0 + 1\cdot 2 + 2\cdot(-1) & 2\cdot 2 + 1\cdot 2 + 2\cdot 0 \\ 3\cdot 0 + 1\cdot 2 + 2\cdot(-1) & 3\cdot 2 + 1\cdot 2 + 2\cdot 0 \end{pmatrix} = \begin{pmatrix} 0 & 6 \\ 0 & 8 \end{pmatrix}. \end{aligned}$$

- *It is a product of matrices of dimensions 3×2 and 2×3 and, therefore, they can be multiplied:*

$$\begin{aligned} A^{\mathrm{T}} \cdot B &= \begin{pmatrix} 2 & 3 \\ 1 & 1 \\ 2 & 2 \end{pmatrix} \cdot \begin{pmatrix} 0 & 2 & -1 \\ 2 & 2 & 0 \end{pmatrix} \\ &= \begin{pmatrix} 2\cdot 0 + 3\cdot 2 & 2\cdot 2 + 3\cdot 2 & 2\cdot(-1) + 3\cdot 0 \\ 1\cdot 0 + 1\cdot 2 & 1\cdot 2 + 1\cdot 2 & 1\cdot(-1) + 1\cdot 0 \\ 2\cdot 0 + 2\cdot 2 & 2\cdot 2 + 2\cdot 2 & 2\cdot(-1) + 2\cdot 0 \end{pmatrix} = \begin{pmatrix} 6 & 10 & -2 \\ 2 & 4 & -1 \\ 4 & 8 & -2 \end{pmatrix}. \end{aligned}$$

- $A \cdot B$*: This product is not possible, as the number of columns of A is not equal to the number of rows of B.*

- $B \cdot A$*: This product is not possible, as the number of columns of B is not equal to the number of rows of A.*

Problem 6.4. Calculate the matrix $A^3 - 3A^2 - 5A + 2I$ in the cases:

(a) $A = \begin{pmatrix} 1 & 0 \\ -2 & -1 \end{pmatrix}$

(b) $A = \begin{pmatrix} 0 & 0 & 0 \\ 1 & 0 & 0 \\ -2 & -1 & 0 \end{pmatrix}$

Solution to the problem 6.4. *(a) Initially we calculate A^2 and A^3. To calculate A^3 we will first calculate A^2 and then multiply the result by A, since $A^2 \cdot A = A^3$:*

$$A^2 = \begin{pmatrix} 1 & 0 \\ -2 & -1 \end{pmatrix} \cdot \begin{pmatrix} 1 & 0 \\ -2 & -1 \end{pmatrix}$$

$$= \begin{pmatrix} 1 \cdot 1 + 0 \cdot (-2) & 1 \cdot 0 + 0 \cdot (-1) \\ -2 \cdot 1 + (-1) \cdot (-2) & -2 \cdot 0 + (-1) \cdot (-1) \end{pmatrix} = \begin{pmatrix} 1 & 0 \\ 0 & 1 \end{pmatrix}.$$

Notice that $A^2 = I_2$; therefore, it is not necessary to calculate the product:

$$A^2 \cdot A = I_2 \cdot A$$

since $I_2 \cdot A = A$. Then:

$$A^3 = \begin{pmatrix} 1 & 0 \\ -2 & -1 \end{pmatrix}.$$

Next, we simplify the initial equation based on the previous calculations:

$$A^3 - 3A^2 - 5A + 2I = A - 3I - 5A + 2I = -4A - I.$$

Therefore:

$$-4A - I = -4 \begin{pmatrix} 1 & 0 \\ -2 & -1 \end{pmatrix} - \begin{pmatrix} 1 & 0 \\ 0 & 1 \end{pmatrix}$$

$$= \begin{pmatrix} -4 & 0 \\ 8 & 4 \end{pmatrix} - \begin{pmatrix} 1 & 0 \\ 0 & 1 \end{pmatrix} = \begin{pmatrix} -5 & 0 \\ 8 & 3 \end{pmatrix}.$$

(b) Again, we start by calculating A^2 and A^3:

$$A^2 = \begin{pmatrix} 0 & 0 & 0 \\ 1 & 0 & 0 \\ -2 & -1 & 0 \end{pmatrix} \cdot \begin{pmatrix} 0 & 0 & 0 \\ 1 & 0 & 0 \\ -2 & -1 & 0 \end{pmatrix}$$

$$= \begin{pmatrix} 0 \cdot 0 + 0 \cdot 1 + 0 \cdot -2 & 0 \cdot 0 + 0 \cdot 0 + 0 \cdot 1 & 0 \cdot 0 + 0 \cdot 0 + 0 \cdot 0 \\ 1 \cdot 0 + 0 \cdot 1 + 0 \cdot -2 & 1 \cdot 0 + 0 \cdot 0 + 0 \cdot 1 & 1 \cdot 0 + 0 \cdot 0 + 0 \cdot 0 \\ -2 \cdot 0 + (-1) \cdot 1 + 0 \cdot -2 & -2 \cdot 0 + (-1) \cdot 0 + 0 \cdot 1 & -2 \cdot 0 + (-1) \cdot 0 + 0 \cdot 0 \end{pmatrix}$$

$$= \begin{pmatrix} 0 & 0 & 0 \\ 0 & 0 & 0 \\ -1 & 0 & 0 \end{pmatrix}.$$

$$A^3 = A^2 \cdot A = \begin{pmatrix} 0 & 0 & 0 \\ 0 & 0 & 0 \\ -1 & 0 & 0 \end{pmatrix} \cdot \begin{pmatrix} 0 & 0 & 0 \\ 1 & 0 & 0 \\ -2 & -1 & 0 \end{pmatrix}$$

$$= \begin{pmatrix} 0 \cdot 0 + 0 \cdot 1 + 0 \cdot -2 & 0 \cdot 0 + 0 \cdot 0 + 0 \cdot 1 & 0 \cdot 0 + 0 \cdot 0 + 0 \cdot 0 \\ 0 \cdot 0 + 0 \cdot 1 + 0 \cdot -2 & 0 \cdot 0 + 0 \cdot 0 + 0 \cdot 1 & 0 \cdot 0 + 0 \cdot 0 + 0 \cdot 0 \\ -1 \cdot 0 + 0 \cdot 1 + 0 \cdot -2 & -1 \cdot 0 + 0 \cdot 0 + 0 \cdot 1 & -1 \cdot 0 + 0 \cdot 0 + 0 \cdot 0 \end{pmatrix}$$

$$= \begin{pmatrix} 0 & 0 & 0 \\ 0 & 0 & 0 \\ 0 & 0 & 0 \end{pmatrix}.$$

Next, we simplify the initial equation based on the previous calculations:

$$A^3 - 3A^2 - 5A + 2I = -3A^2 - 5A + 2I.$$

Therefore:

$$-3A^2 - 5A + 2I = -3\begin{pmatrix} 0 & 0 & 0 \\ 0 & 0 & 0 \\ -1 & 0 & 0 \end{pmatrix} - 5\begin{pmatrix} 0 & 0 & 0 \\ 1 & 0 & 0 \\ -2 & -1 & 0 \end{pmatrix} + 2\begin{pmatrix} 1 & 0 & 0 \\ 0 & 1 & 0 \\ 0 & 0 & 1 \end{pmatrix}$$

$$= \begin{pmatrix} 0 & 0 & 0 \\ 0 & 0 & 0 \\ 3 & 0 & 0 \end{pmatrix} - \begin{pmatrix} 0 & 0 & 0 \\ 5 & 0 & 0 \\ -10 & -5 & 0 \end{pmatrix} + \begin{pmatrix} 2 & 0 & 0 \\ 0 & 2 & 0 \\ 0 & 0 & 2 \end{pmatrix}$$

$$= \begin{pmatrix} 2 & 0 & 0 \\ -5 & 2 & 0 \\ 13 & 5 & 2 \end{pmatrix}.$$

Problem 6.5. Prove that matrices of the type $\begin{pmatrix} a & b \\ b & a \end{pmatrix}$, $a, b \in \mathbb{R}$ commute under multiplication.

Solution to the problem 6.5. *We must see that* $\forall a, b, c, d \in \mathbb{R}$ *it holds that:*

$$\begin{pmatrix} a & b \\ b & a \end{pmatrix}\begin{pmatrix} c & d \\ d & c \end{pmatrix} = \begin{pmatrix} c & d \\ d & c \end{pmatrix}\begin{pmatrix} a & b \\ b & a \end{pmatrix}$$

On one hand, we have that:

$$\begin{pmatrix} a & b \\ b & a \end{pmatrix}\begin{pmatrix} c & d \\ d & c \end{pmatrix} = \begin{pmatrix} ac + bd & ad + bc \\ bc + ad & bd + ac \end{pmatrix},$$

and, on the other hand, it follows that:

$$\begin{pmatrix} c & d \\ d & c \end{pmatrix}\begin{pmatrix} a & b \\ b & a \end{pmatrix} = \begin{pmatrix} ca + db & cb + da \\ da + cb & db + ca \end{pmatrix}.$$

By the commutative property of the addition and multiplication of reals the equality is proved.

Problem 6.6. If $A, B \in \mathscr{M}_{2\times 2}(\mathbb{R})$, simplify the expression:

$$(A + B - I) \cdot (A - B + I) + (A + 2B) \cdot (B - A).$$

Solution to the problem 6.6. *To develop the expression, we use that the product of matrices is commutative with respect to the addition, that is,* $(A+B)C = AC+BC$*. Thus:*

$$
\begin{aligned}
&(A+B-I)\cdot(A-B+I)+(A+2B)\cdot(B-A)=\\
=&A(A-B+I)+B(A-B+I)-I(A-B+I)+A(B-A)+2B(B-A)\\
=&A^2-AB+\underbrace{AI}_{A}+BA-B^2+\underbrace{BI}_{B}-\underbrace{IA}_{A}+\underbrace{IB}_{B}-\underbrace{I^2}_{I}+AB-A^2+2B^2-2BA\\
=&-BA+B^2+2B-I\\
=&B(B-A+2I)-I.
\end{aligned}
$$

Therefore:

$$(A+B-I)\cdot(A-B+I)+(A+2B)\cdot(B-A)=B(B-A+2I)-I.$$

Notice that an expression of the type B^2+kB*, where* B *is a matrix and* k *a scalar, cannot be factored as* $B(B+k)$*, since a matrix cannot be added to a scalar. The correct way to factor the expression is:*

$$B^2+kB=B(B+kI)$$

since:

$$B\cdot kI=kB$$

Problem 6.7. Determine all matrices in $\mathscr{M}_{2\times 2}(\mathbb{R})$ such that their square is the identity matrix I_2.

Solution to the problem 6.7. *Let the matrix* $A=\begin{pmatrix} a & b\\ c & d\end{pmatrix}$*, then:*

$$A^2=\begin{pmatrix} a & b\\ c & d\end{pmatrix}\cdot\begin{pmatrix} a & b\\ c & d\end{pmatrix}=\begin{pmatrix} a^2+bc & ab+bd\\ ca+dc & bc+d^2\end{pmatrix}=\begin{pmatrix} 1 & 0\\ 0 & 1\end{pmatrix}\quad\Leftrightarrow$$

$$\Leftrightarrow\left\{\begin{array}{lllllll} a^2+bc &=& 1 & & & & (1)\\ ab+bd &=& 0 & \Leftrightarrow & b(a+d)=0 & \Leftrightarrow\quad b=0\ or\ a=-d & (2)\\ ca+dc &=& 0 & \Leftrightarrow & quadc(a+d)=0 & \Leftrightarrow\quad c=0\ or\ a=-d & (3)\\ bc+d^2 &=& 1 & & & & (4)\end{array}\right.$$

We then have the following possibilities:

- *Case 1:* $\boxed{b=0}$

  In this case, from equation (1) we have $a^2=1\Leftrightarrow a=\pm1$ *and from equation (4)* $d^2=1\Leftrightarrow d=\pm1$*. To verify the remaining equation (3) we then have two possibilities:*

  - *Case 1.1:* $\boxed{c=0}$

    In this case, the solutions are:

    $$\begin{pmatrix} 1 & 0\\ 0 & 1\end{pmatrix},\begin{pmatrix} -1 & 0\\ 0 & 1\end{pmatrix},\begin{pmatrix} 1 & 0\\ 0 & -1\end{pmatrix},\begin{pmatrix} -1 & 0\\ 0 & -1\end{pmatrix}$$

- *Case 1.2:* $\boxed{c \neq 0}$

  In this case, it must necessarily be $a = -d$ *and, therefore, the solutions are,* $\forall c \in \mathbb{R}, c \neq 0$:

$$\begin{pmatrix} 1 & 0 \\ c & -1 \end{pmatrix}, \begin{pmatrix} -1 & 0 \\ c & 1 \end{pmatrix}$$

- *Case 2:* $\boxed{b \neq 0}$

  To verify equation (2) it must be $a = -d$*, and then equation (3) is automatically verified. From equation (1) we have:*

$$bc = 1 - a^2 \Leftrightarrow c = (1 - a^2)/b$$

  and, therefore, the solution is:

$$\begin{pmatrix} a & b \\ (1-a^2)/b & -a \end{pmatrix}$$

In summary, the matrices in $\mathscr{M}_{2\times 2}(\mathbb{R})$ *such that their square is the identity matrix* I_2 *are* $\forall a, b, c \in \mathbb{R}, b \neq 0$:

$$\begin{pmatrix} 1 & 0 \\ 0 & 1 \end{pmatrix}, \begin{pmatrix} -1 & 0 \\ c & 1 \end{pmatrix}, \begin{pmatrix} 1 & 0 \\ c & -1 \end{pmatrix}, \begin{pmatrix} -1 & 0 \\ 0 & -1 \end{pmatrix}, \begin{pmatrix} a & b \\ (1-a^2)/b & -a \end{pmatrix}$$

Problem 6.8. Two matrices are said to commute under multiplication if it holds $AB = BA$. Find all matrices in $\mathscr{M}_{2\times 2}(\mathbb{R})$ that commute under multiplication with the matrix:

$$A = \begin{pmatrix} 1 & 2 \\ 3 & 4 \end{pmatrix}$$

Solution to the problem 6.8. *We must find* $a, b, c, d \in \mathbb{R}$ *such that:*

$$\begin{pmatrix} a & b \\ c & d \end{pmatrix}\begin{pmatrix} 1 & 2 \\ 3 & 4 \end{pmatrix} = \begin{pmatrix} 1 & 2 \\ 3 & 4 \end{pmatrix}\begin{pmatrix} a & b \\ c & d \end{pmatrix},$$

that is, such that:

$$\begin{pmatrix} a+3b & 2a+4b \\ c+3d & 2c+4d \end{pmatrix} = \begin{pmatrix} a+2c & b+2d \\ 3a+4c & 3b+4d \end{pmatrix}$$

$$\Leftrightarrow \begin{cases} a+3b = a+2c \Leftrightarrow 3b = 2c \\ 2a+4b = b+2d \Leftrightarrow 2d = 2a+3b = 2a+2c \Leftrightarrow d = a+c \\ c+3d = 3a+4c \Leftrightarrow 3d = 3a+3b \Leftrightarrow d = a+c \\ 2c+4d = 3b+4d \Leftrightarrow 3b = 2c \end{cases}$$

Therefore, the four conditions reduce to $d = a + c$ *and* $3b = 2c$*. The solutions then have two parameters, e.g.* $a, c \in \mathbb{R}$*, and can be written as:*

$$\begin{pmatrix} a & 2c/3 \\ c & a+c \end{pmatrix}.$$

Problem 6.9. Determine the rank of the following matrices:

(a) $\begin{pmatrix} 1 & 2 & 1 \\ -1 & 0 & 1 \\ 3 & 2 & 2 \end{pmatrix}$

(b) $\begin{pmatrix} 2 & 1 & 3 & -1 \\ 3 & -1 & 2 & 0 \\ 1 & 3 & 4 & -2 \\ 4 & -3 & 1 & 1 \end{pmatrix}$

(c) $\begin{pmatrix} 1 & 1 & 1 \\ 1 & -1 & 1 \\ 1 & 1 & -1 \\ 3 & 1 & 1 \end{pmatrix}$

Theorem 6.2. Problem 6.9 (cont)

$\begin{pmatrix} 1 & -1 & -2 & 1 \\ 0 & 1 & 0 & 0 \\ -2 & 0 & 4 & -1 \\ -1 & 2 & 2 & 0 \end{pmatrix}$

(d) $\begin{pmatrix} 2 & 1 & 3 & 4 \\ -1 & 0 & 1 & 2 \\ -6 & -2 & -4 & -4 \end{pmatrix}$

(f) $\begin{pmatrix} 1 & 0 & 3 & 5 \\ -1 & 0 & -3 & -5 \\ 3 & 1 & 5 & 9 \\ 2 & 2 & 1 & 1 \end{pmatrix}$

(g) $\begin{pmatrix} 2 & 3 & 5 & 1 & 3 \\ 1 & 1 & 1 & 2 & 1 \\ 2 & -1 & 3 & 1 & 4 \\ 5 & 1 & 2 & 8 & 1 \end{pmatrix}$

(h) $\begin{pmatrix} a & 1 & 1 \\ 1 & a & 1 \\ 1 & 1 & a \end{pmatrix}$

(i) $\begin{pmatrix} a & 1 & -1 \\ 1 & -a & 1 \\ 1 & 1 & a \end{pmatrix}$, $a \in \mathbb{R}$

Solution to the problem 6.9. *(a)*

$$\begin{pmatrix} 1 & 2 & 1 \\ -1 & 0 & 1 \\ 3 & 2 & 2 \end{pmatrix} \underset{f_2 \leftarrow f_2 + f_1}{\sim} \begin{pmatrix} 1 & 2 & 1 \\ 0 & 2 & 2 \\ 3 & 2 & 2 \end{pmatrix} \underset{f_3 \leftarrow f_3 - 3f_1}{\sim} \begin{pmatrix} 1 & 2 & 1 \\ 0 & 2 & 2 \\ 0 & -4 & -1 \end{pmatrix}$$

$$\underset{f_3 \leftarrow f_3 + 2f_2}{\sim} \begin{pmatrix} 1 & 2 & 1 \\ 0 & 2 & 2 \\ 0 & 0 & 3 \end{pmatrix}.$$

Therefore, the rank of the matrix is 3.

(b)

$$\begin{pmatrix} 2 & 1 & 3 & -1 \\ 3 & -1 & 2 & 0 \\ 1 & 3 & 4 & -2 \\ 4 & -3 & 1 & 1 \end{pmatrix} \underset{f_3 \leftarrow f_3 + f_2}{\sim} \begin{pmatrix} 2 & 1 & 3 & -1 \\ 3 & -1 & 2 & 0 \\ 4 & 2 & 6 & -2 \\ 4 & -3 & 1 & 1 \end{pmatrix}$$

$$\underset{f_3 \leftarrow f_3 - 2f_1}{\sim} \begin{pmatrix} 2 & 1 & 3 & -1 \\ 3 & -1 & 2 & 0 \\ 0 & 0 & 0 & 0 \\ 4 & -3 & 1 & 1 \end{pmatrix}$$

$$\underset{f_4 \leftarrow f_4 + f_1}{\sim} \begin{pmatrix} 2 & 1 & 3 & -1 \\ 3 & -1 & 2 & 0 \\ 0 & 0 & 0 & 0 \\ 6 & -2 & 4 & 0 \end{pmatrix}$$

$$\underset{f_4 \leftarrow f_4 - 2f_2}{\sim} \begin{pmatrix} 2 & 1 & 3 & -1 \\ 3 & -1 & 2 & 0 \\ 0 & 0 & 0 & 0 \\ 0 & 0 & 0 & 0 \end{pmatrix}$$

$$\underset{f_2 \leftarrow f_2 - \frac{3}{2}f_1}{\sim} \left(\begin{array}{cccc} 2 & 1 & 3 & -1 \\ 0 & -5/2 & -5/2 & -3/2 \\ \hline 0 & 0 & 0 & 0 \\ 0 & 0 & 0 & 0 \end{array}\right).$$

Therefore, the rank of the matrix is 2.

(c)

$$\begin{pmatrix} 1 & 1 & 1 \\ 1 & -1 & 1 \\ 1 & 1 & -1 \\ 3 & 1 & 1 \end{pmatrix} \underset{f_4 \leftarrow f_4 - f_1}{\sim} \begin{pmatrix} 1 & 1 & 1 \\ 1 & -1 & 1 \\ 1 & 1 & -1 \\ 2 & 0 & 0 \end{pmatrix}$$

$$\underset{f_4 \leftarrow f_4 - f_2}{\sim} \begin{pmatrix} 1 & 1 & 1 \\ 1 & -1 & 1 \\ 1 & 1 & -1 \\ 1 & 1 & -1 \end{pmatrix}$$

$$\underset{f_4 \leftarrow f_4 - f_3}{\sim} \begin{pmatrix} 1 & 1 & 1 \\ 1 & -1 & 1 \\ 1 & 1 & -1 \\ 0 & 0 & 0 \end{pmatrix}$$

$$\underset{f_3 \leftarrow f_3 - f_1}{\sim} \begin{pmatrix} 1 & 1 & 1 \\ 1 & -1 & 1 \\ 0 & 0 & -2 \\ 0 & 0 & 0 \end{pmatrix}$$

$$\underset{f_2 \leftarrow f_2 - f_1}{\sim} \begin{pmatrix} 1 & 1 & 1 \\ 0 & -2 & 0 \\ 0 & 0 & -2 \\ \hline 0 & 0 & 0 \end{pmatrix}.$$

Therefore, the rank of the matrix is 3.

(d)

$$\begin{pmatrix} 1 & -1 & -2 & 1 \\ 0 & 1 & 0 & 0 \\ -2 & 0 & 4 & -1 \\ -1 & 2 & 2 & 0 \end{pmatrix} \underset{pf_{24}}{\sim} \begin{pmatrix} 1 & -1 & -2 & 1 \\ -1 & 2 & 2 & 0 \\ -2 & 0 & 4 & -1 \\ 0 & 1 & 0 & 0 \end{pmatrix}$$

$$\underset{pc_{24}}{\sim} \begin{pmatrix} 1 & 1 & -2 & -1 \\ -1 & 0 & 2 & 2 \\ -2 & -1 & 4 & 0 \\ 0 & 0 & 0 & 1 \end{pmatrix}$$

$$\underset{pc_{12}}{\sim}\begin{pmatrix} 1 & 1 & -2 & -1 \\ 0 & -1 & 2 & 2 \\ -1 & -2 & 4 & 0 \\ 0 & 0 & 0 & 1 \end{pmatrix}$$

$$\underset{f_3 \leftarrow f_3 + f_1}{\sim}\begin{pmatrix} 1 & 1 & -2 & -1 \\ 0 & -1 & 2 & 2 \\ 0 & -1 & 2 & -1 \\ 0 & 0 & 0 & 1 \end{pmatrix}$$

$$\underset{f_3 \leftarrow f_3 - f_2}{\sim}\begin{pmatrix} 1 & 1 & -2 & -1 \\ 0 & -1 & 2 & 2 \\ 0 & 0 & 0 & -3 \\ 0 & 0 & 0 & 1 \end{pmatrix}$$

$$\underset{pc_{34}}{\sim}\begin{pmatrix} 1 & 1 & -1 & -2 \\ 0 & -1 & 2 & 2 \\ 0 & 0 & -3 & 0 \\ 0 & 0 & 1 & 0 \end{pmatrix}$$

$$\underset{f_4 \leftarrow f_4 + \frac{1}{3} f_3}{\sim}\left(\begin{array}{cccc} 1 & 1 & -1 & -2 \\ 0 & -1 & 2 & 2 \\ 0 & 0 & -3 & 0 \\ \hline 0 & 0 & 0 & 0 \end{array}\right).$$

Therefore, the rank of the matrix is 3.

(e)

$$\begin{pmatrix} 2 & 1 & 3 & 4 \\ -1 & 0 & 1 & 2 \\ -6 & -2 & -4 & -4 \end{pmatrix} \underset{pc_{12}}{\sim}\begin{pmatrix} 1 & 2 & 3 & 4 \\ 0 & -1 & 1 & 2 \\ -2 & -6 & -4 & -4 \end{pmatrix}$$

$$\underset{f_3 \leftarrow f_3 + 2f_1}{\sim}\begin{pmatrix} 1 & 2 & 3 & 4 \\ 0 & -1 & 1 & 2 \\ 0 & 0 & 2 & 4 \end{pmatrix}.$$

Therefore, the rank of the matrix is 3.

(f)

$$\begin{pmatrix} 1 & 0 & 3 & 5 \\ -1 & 0 & -3 & -5 \\ 3 & 1 & 5 & 9 \\ 2 & 2 & 1 & 1 \end{pmatrix} \underset{f_2 \leftarrow f_2 + f_1}{\sim}\begin{pmatrix} 1 & 0 & 3 & 5 \\ 0 & 0 & 0 & 0 \\ 3 & 1 & 5 & 9 \\ 2 & 2 & 1 & 1 \end{pmatrix} \underset{pf_{24}}{\sim}\begin{pmatrix} 1 & 0 & 3 & 5 \\ 2 & 2 & 1 & 1 \\ 3 & 1 & 5 & 9 \\ 0 & 0 & 0 & 0 \end{pmatrix}$$

$$\underset{pc_{12}}{\sim}\begin{pmatrix} 0 & 1 & 3 & 5 \\ 2 & 2 & 1 & 1 \\ 1 & 3 & 5 & 9 \\ 0 & 0 & 0 & 0 \end{pmatrix} \underset{pf_{12}}{\sim}\begin{pmatrix} 2 & 2 & 1 & 1 \\ 0 & 1 & 3 & 5 \\ 1 & 3 & 5 & 9 \\ 0 & 0 & 0 & 0 \end{pmatrix}$$

$$\underset{f_3 \leftarrow f_3 - \frac{1}{2}f_1}{\sim} \begin{pmatrix} 2 & 2 & 1 & 1 \\ 0 & 1 & 3 & 5 \\ 0 & 2 & 9/2 & 17/2 \\ 0 & 0 & 0 & 0 \end{pmatrix}$$

$$\underset{f_3 \leftarrow f_3 - 2f_2}{\sim} \begin{pmatrix} 2 & 2 & 1 & 1 \\ 0 & 1 & 3 & 5 \\ 0 & 0 & -3/2 & -3/2 \\ \hline 0 & 0 & 0 & 0 \end{pmatrix}.$$

Therefore, the rank of the matrix is 3.

(g)

$$\begin{pmatrix} 2 & 3 & 5 & 1 & 3 \\ 1 & 1 & 1 & 2 & 1 \\ 2 & -1 & 3 & 1 & 4 \\ 5 & 1 & 2 & 8 & 1 \end{pmatrix} \underset{pf_{12}}{\sim} \begin{pmatrix} 1 & 1 & 1 & 2 & 1 \\ 2 & 3 & 5 & 1 & 3 \\ 2 & -1 & 3 & 1 & 4 \\ 5 & 1 & 2 & 8 & 1 \end{pmatrix}$$

$$\underset{pc_{15}}{\sim} \begin{pmatrix} 1 & 1 & 1 & 2 & 1 \\ 3 & 3 & 5 & 1 & 2 \\ 4 & -1 & 3 & 1 & 2 \\ 1 & 1 & 2 & 8 & 5 \end{pmatrix}$$

$$\underset{pf_{23}}{\sim} \begin{pmatrix} 1 & 1 & 1 & 2 & 1 \\ 4 & -1 & 3 & 1 & 2 \\ 3 & 3 & 5 & 1 & 2 \\ 1 & 1 & 2 & 8 & 5 \end{pmatrix}$$

$$\underset{f_4 \leftarrow f_4 - f_1}{\sim} \begin{pmatrix} 1 & 1 & 1 & 2 & 1 \\ 4 & -1 & 3 & 1 & 2 \\ 3 & 3 & 5 & 1 & 2 \\ 0 & 0 & 1 & 6 & 4 \end{pmatrix}$$

$$\underset{f_3 \leftarrow f_3 - 3f_1}{\sim} \begin{pmatrix} 1 & 1 & 1 & 2 & 1 \\ 4 & -1 & 3 & 1 & 2 \\ 0 & 0 & 2 & -5 & -1 \\ 0 & 0 & 1 & 6 & 4 \end{pmatrix}$$

$$\underset{f_4 \leftarrow f_4 - \frac{1}{2}f_3}{\sim} \begin{pmatrix} 1 & 1 & 1 & 2 & 1 \\ 4 & -1 & 3 & 1 & 2 \\ 0 & 0 & 2 & -5 & -1 \\ 0 & 0 & 0 & 7/2 & 9/2 \end{pmatrix}.$$

Therefore, the rank of the matrix is 4.

(h)

$$\begin{pmatrix} a & 1 & 1 \\ 1 & a & 1 \\ 1 & 1 & a \end{pmatrix} \underset{pf_{13}}{\sim} \begin{pmatrix} 1 & 1 & a \\ 1 & a & 1 \\ a & 1 & 1 \end{pmatrix} \underset{f_2 \leftarrow f_2 - f_1}{\sim} \begin{pmatrix} 1 & 1 & a \\ 0 & a-1 & 1-a \\ a & 1 & 1 \end{pmatrix}$$

$$\underset{f_3 \leftarrow f_3 - a \cdot f_1}{\sim} \begin{pmatrix} 1 & 1 & a \\ 0 & a-1 & 1-a \\ 0 & 1-a & 1-a^2 \end{pmatrix}$$

$$\underset{f_3 \leftarrow f_3 + f_2}{\sim} \begin{pmatrix} 1 & 1 & a \\ 0 & a-1 & 1-a \\ 0 & 0 & -a^2-a+2 \end{pmatrix}$$

$$\underset{f_3 \leftarrow -f_3}{\sim} \begin{pmatrix} 1 & 1 & a \\ 0 & a-1 & 1-a \\ 0 & 0 & a^2+a-2 \end{pmatrix}.$$

The roots of the polynomial $a^2 + a - 2$ *are* $a = 1$ *and* $a = -2$*. Therefore:*

- *if* $a \neq 1$ *and* $a \neq -2$*, rank = 3.*
- *if* $a = 1$*, rank = 1.*
- *if* $a = -2$*, rank = 2.*

(i)

$$\begin{pmatrix} a & 1 & -1 \\ 1 & -a & 1 \\ 1 & 1 & a \end{pmatrix} \underset{pf_{13}}{\sim} \begin{pmatrix} 1 & 1 & a \\ 1 & -a & 1 \\ a & 1 & -1 \end{pmatrix}$$

$$\underset{f_2 \leftarrow f_2 - f_1}{\sim} \begin{pmatrix} 1 & 1 & a \\ 0 & -a-1 & a+1 \\ a & 1 & -1 \end{pmatrix}$$

$$\underset{f_3 \leftarrow f_3 - a \cdot f_1}{\sim} \begin{pmatrix} 1 & 1 & a \\ 0 & -a-1 & a+1 \\ 0 & 1-a & -1+a^2 \end{pmatrix}$$

$$\underset{f_2 \leftarrow f_2 - f_3}{\sim} \begin{pmatrix} 1 & 1 & a \\ 0 & -2 & -a^2+a+2 \\ 0 & 1-a & -1+a^2 \end{pmatrix}$$

$$\underset{f_3 \leftarrow f_2 \cdot \frac{1-a}{-2}}{\sim} \begin{pmatrix} 1 & 1 & a \\ 0 & -2 & -a^2+a+2 \\ 0 & 0 & (a^3-a)/2 \end{pmatrix}$$

$$\underset{2f_3}{\sim} \begin{pmatrix} 1 & 1 & a \\ 0 & -2 & -a^2+a+2 \\ 0 & 0 & a^3-a \end{pmatrix}$$

The roots of the polynomial $a^3 - a$ *are* $a = 0$, $a = -1$ *and* $a = 1$*. Therefore:*

- *if* $a \neq 0$, $a \neq -1$ *and* $a \neq 1$ *rank = 3.*
- *if* $a = 0$, $a = -1$ *or* $a = 1$ *rank = 2.*

Problem 6.10. Find, if possible, the inverse matrices of:

(a) $\begin{pmatrix} 1 & 0 & 0 \\ -1 & 1 & 0 \\ 1 & 1 & 1 \end{pmatrix}$

(b) $\begin{pmatrix} 1 & 2 & 3 \\ 1 & 0 & 1 \\ 0 & 1 & 2 \end{pmatrix}$

(c) $\begin{pmatrix} 1 & 2 & -1 & 1 \\ -1 & 1 & 2 & 1 \\ 1 & 0 & 3 & 1 \\ -1 & 2 & 1 & 4 \end{pmatrix}$

(d) $\begin{pmatrix} 1 & 1 & 0 & 0 \\ 0 & 1 & 1 & 0 \\ 0 & 0 & 1 & 1 \\ 1 & 0 & 0 & 1 \end{pmatrix}$

Solution to the problem 6.10. *(a) We calculate the inverse matrix using the Gauss method:*

$$\left(\begin{array}{ccc|ccc} 1 & 0 & 0 & 1 & 0 & 0 \\ -1 & 1 & 0 & 0 & 1 & 0 \\ 1 & 1 & 1 & 0 & 0 & 1 \end{array}\right) \underset{\substack{f_2 \leftarrow f_2 + f_1 \\ f_3 \leftarrow f_3 - f_1}}{\sim} \left(\begin{array}{ccc|ccc} 1 & 0 & 0 & 1 & 0 & 0 \\ 0 & 1 & 0 & 1 & 1 & 0 \\ 0 & 1 & 1 & -1 & 0 & 1 \end{array}\right)$$

$$\underset{f_3 \leftarrow f_3 - f_2}{\sim} \left(\begin{array}{ccc|ccc} 1 & 0 & 0 & 1 & 0 & 0 \\ 0 & 1 & 0 & 1 & 1 & 0 \\ 0 & 0 & 1 & -2 & -1 & 1 \end{array}\right).$$

Therefore, the inverse matrix is:

$$\begin{pmatrix} 1 & 0 & 0 \\ 1 & 1 & 0 \\ -2 & -1 & 1 \end{pmatrix}.$$

(b) We calculate the inverse matrix using the Gauss method:

$$\left(\begin{array}{ccc|ccc} 1 & 2 & 3 & 1 & 0 & 0 \\ -1 & 0 & 1 & 0 & 1 & 0 \\ 0 & 1 & 2 & 0 & 0 & 1 \end{array}\right) \underset{f_2 \leftarrow f_2 + f_1}{\sim} \left(\begin{array}{ccc|ccc} 1 & 2 & 3 & 1 & 0 & 0 \\ 0 & 2 & 4 & 1 & 1 & 0 \\ 0 & 1 & 2 & 0 & 0 & 1 \end{array}\right)$$

$$\underset{f_3 \leftarrow f_3 - \frac{1}{2} f_2}{\sim} \left(\begin{array}{ccc|ccc} 1 & 2 & 3 & 1 & 0 & 0 \\ 0 & 2 & 4 & 1 & 1 & 0 \\ \hline 0 & 0 & 0 & -1/2 & -1/2 & 1 \end{array}\right).$$

Since the matrix does not have full rank, it is not invertible.

(c) *We calculate the inverse matrix using the Gauss method:*

$$\left(\begin{array}{cccc|cccc} 1 & 2 & -1 & 1 & 1 & 0 & 0 & 0 \\ -1 & 1 & 2 & 1 & 0 & 1 & 0 & 0 \\ 1 & 0 & 3 & 1 & 0 & 0 & 1 & 0 \\ -1 & 2 & 1 & 4 & 0 & 0 & 0 & 1 \end{array}\right) \sim$$

$$\underset{\substack{f_2 \leftarrow f_2+f_1 \\ f_3 \leftarrow f_3-f_1 \\ f_4 \leftarrow f_4+f_1}}{\sim} \left(\begin{array}{cccc|cccc} 1 & 2 & -1 & 1 & 1 & 0 & 0 & 0 \\ 0 & 3 & 1 & 2 & 0 & 1 & 0 & 0 \\ 0 & -2 & 4 & 0 & 0 & -1 & 1 & 0 \\ 0 & 4 & 0 & 5 & 0 & 1 & 0 & 1 \end{array}\right)$$

$$\underset{\frac{1}{3}f_2}{\sim} \left(\begin{array}{cccc|cccc} 1 & 2 & -1 & 1 & 1 & 0 & 0 & 0 \\ 0 & 1 & 1/3 & 2/3 & 1/3 & 1/3 & 0 & 0 \\ 0 & -2 & 4 & 0 & 0 & -1 & 1 & 0 \\ 0 & 4 & 0 & 5 & 0 & 1 & 0 & 1 \end{array}\right)$$

$$\underset{\substack{f_1 \leftarrow f_1-2f_2 \\ f_3 \leftarrow f_3+2f_2 \\ f_4 \leftarrow f_4-4f_2}}{\sim} \left(\begin{array}{cccc|cccc} 1 & 0 & 5/3 & -1/3 & 1/3 & -2/3 & 0 & 0 \\ 0 & 1 & 1/3 & 2/3 & 1/3 & 1/3 & 0 & 0 \\ 0 & 0 & 14/3 & 4/3 & -1/3 & 2/3 & 1 & 0 \\ 0 & 0 & -4/3 & 7/3 & -1/3 & -4/3 & 0 & 1 \end{array}\right)$$

$$\underset{\frac{3}{14}f_3}{\sim} \left(\begin{array}{cccc|cccc} 1 & 0 & -5/3 & -1/3 & 1/3 & -2/3 & 0 & 0 \\ 0 & 1 & 1/3 & 2/3 & 1/3 & 1/3 & 0 & 0 \\ 0 & 0 & 1 & 4/14 & -1/14 & 2/14 & 3/14 & 0 \\ 0 & 0 & -4/3 & 7/3 & -1/3 & -4/3 & 0 & 1 \end{array}\right)$$

$$\underset{\substack{f_1 \leftarrow f_1+\frac{5}{3}f_3 \\ f_2 \leftarrow f_2-\frac{1}{3}f_3 \\ f_4 \leftarrow f_4+\frac{4}{3}f_3}}{\sim} \left(\begin{array}{cccc|cccc} 1 & 0 & 0 & 2/14 & 3/14 & -6/14 & 5/14 & 0 \\ 0 & 1 & 0 & 8/14 & 5/14 & 4/14 & -1/14 & 0 \\ 0 & 0 & 1 & 4/14 & -1/14 & 2/14 & 3/14 & 0 \\ 0 & 0 & 0 & 38/14 & -6/14 & -16/14 & 4/14 & 1 \end{array}\right).$$

Therefore, the inverse matrix is:

$$\frac{1}{38}\begin{pmatrix} 9 & -14 & 13 & -2 \\ 17 & 20 & -5 & -8 \\ -1 & 10 & 7 & -4 \\ -6 & -16 & 4 & 14 \end{pmatrix}.$$

(d) We calculate the inverse matrix using the Gauss method:

$$\left(\begin{array}{cccc|cccc} 1 & 1 & 0 & 0 & 1 & 0 & 0 & 0 \\ 0 & 1 & 1 & 0 & 0 & 1 & 0 & 0 \\ 0 & 0 & 1 & 1 & 0 & 0 & 1 & 0 \\ 1 & 0 & 0 & 1 & 0 & 0 & 0 & 1 \end{array}\right) \sim$$

$$\underset{f_4 \leftarrow f_4 - f_3}{\sim} \left(\begin{array}{cccc|cccc} 1 & 1 & 0 & 0 & 1 & 0 & 0 & 0 \\ 0 & 1 & 1 & 0 & 0 & 1 & 0 & 0 \\ 0 & 0 & 1 & 1 & 0 & 0 & 1 & 0 \\ 1 & 0 & -1 & 0 & 0 & 0 & -1 & 1 \end{array}\right)$$

$$\underset{f_4 \leftarrow f_4 + f_2}{\sim} \left(\begin{array}{cccc|cccc} 1 & 1 & 0 & 0 & 1 & 0 & 0 & 0 \\ 0 & 1 & 1 & 0 & 0 & 1 & 0 & 0 \\ 0 & 0 & 1 & 1 & 0 & 0 & 1 & 0 \\ 1 & 1 & 0 & 0 & 0 & 1 & -1 & 1 \end{array}\right)$$

$$\underset{f_4 \leftarrow f_4 - f_1}{\sim} \left(\begin{array}{cccc|cccc} 1 & 1 & 0 & 0 & 1 & 0 & 0 & 0 \\ 0 & 1 & 1 & 0 & 0 & 1 & 0 & 0 \\ 0 & 0 & 1 & 1 & 0 & 0 & 1 & 0 \\ \hline 0 & 0 & 0 & 0 & -1 & 1 & -1 & 1 \end{array}\right).$$

Since the matrix does not have full rank, it is not invertible.

Problem 6.11. Given the matrices I_4 and A:

$$I_4 = \begin{pmatrix} 1 & 0 & 0 & 0 \\ 0 & 1 & 0 & 0 \\ 0 & 0 & 1 & 0 \\ 0 & 0 & 0 & 1 \end{pmatrix}, \qquad A = \begin{pmatrix} 0 & 1 & 0 & 0 \\ 0 & 0 & 1 & 0 \\ 0 & 0 & 0 & 1 \\ 0 & 0 & 0 & 0 \end{pmatrix},$$

calculate: the inverse of $I_4 - A$, the inverse of $I_4 + A$, $(I_4 + A)(I_4 - A)^{-1}$.

Solution to the problem 6.11. *First, we calculate $I_4 - A$ and $I_4 + A$:*

$$I_4 - A = \begin{pmatrix} 1 & 0 & 0 & 0 \\ 0 & 1 & 0 & 0 \\ 0 & 0 & 1 & 0 \\ 0 & 0 & 0 & 1 \end{pmatrix} - \begin{pmatrix} 0 & 1 & 0 & 0 \\ 0 & 0 & 1 & 0 \\ 0 & 0 & 0 & 1 \\ 0 & 0 & 0 & 0 \end{pmatrix} = \begin{pmatrix} 1 & -1 & 0 & 0 \\ 0 & 1 & -1 & 0 \\ 0 & 0 & 1 & -1 \\ 0 & 0 & 0 & 1 \end{pmatrix}.$$

$$I_4 + A = \begin{pmatrix} 1 & 0 & 0 & 0 \\ 0 & 1 & 0 & 0 \\ 0 & 0 & 1 & 0 \\ 0 & 0 & 0 & 1 \end{pmatrix} + \begin{pmatrix} 0 & 1 & 0 & 0 \\ 0 & 0 & 1 & 0 \\ 0 & 0 & 0 & 1 \\ 0 & 0 & 0 & 0 \end{pmatrix} = \begin{pmatrix} 1 & 1 & 0 & 0 \\ 0 & 1 & 1 & 0 \\ 0 & 0 & 1 & 1 \\ 0 & 0 & 0 & 1 \end{pmatrix}.$$

We calculate $(I_4 - A)^{-1}$ using the Gauss method:

$$\left(\begin{array}{cccc|cccc} 1 & -1 & 0 & 0 & 1 & 0 & 0 & 0 \\ 0 & 1 & -1 & 0 & 0 & 1 & 0 & 0 \\ 0 & 0 & 1 & -1 & 0 & 0 & 1 & 0 \\ 0 & 0 & 0 & 1 & 0 & 0 & 0 & 1 \end{array}\right) \sim$$

$$\underset{f_3 \leftarrow f_3+f_4}{\sim} \left(\begin{array}{cccc|cccc} 1 & -1 & 0 & 0 & 1 & 0 & 0 & 0 \\ 0 & 1 & -1 & 0 & 0 & 1 & 0 & 0 \\ 0 & 0 & 1 & 0 & 0 & 0 & 1 & 1 \\ 0 & 0 & 0 & 1 & 0 & 0 & 0 & 1 \end{array}\right)$$

$$\underset{f_2 \leftarrow f_2+f_3}{\sim} \left(\begin{array}{cccc|cccc} 1 & -1 & 0 & 0 & 1 & 0 & 0 & 0 \\ 0 & 1 & 0 & 0 & 0 & 1 & 1 & 1 \\ 0 & 0 & 1 & 0 & 0 & 0 & 1 & 1 \\ 0 & 0 & 0 & 1 & 0 & 0 & 0 & 1 \end{array}\right)$$

$$\underset{f_1 \leftarrow f_1+f_2}{\sim} \left(\begin{array}{cccc|cccc} 1 & 0 & 0 & 0 & 1 & 1 & 1 & 1 \\ 0 & 1 & 0 & 0 & 0 & 1 & 1 & 1 \\ 0 & 0 & 1 & 0 & 0 & 0 & 1 & 1 \\ 0 & 0 & 0 & 1 & 0 & 0 & 0 & 1 \end{array}\right).$$

Therefore:

$$(I_4 - A)^{-1} = \begin{pmatrix} 1 & 1 & 1 & 1 \\ 0 & 1 & 1 & 1 \\ 0 & 0 & 1 & 1 \\ 0 & 0 & 0 & 1 \end{pmatrix}.$$

We calculate $(I_4 + A)^{-1}$ *using the Gauss method:*

$$\left(\begin{array}{cccc|cccc} 1 & 1 & 0 & 0 & 1 & 0 & 0 & 0 \\ 0 & 1 & 1 & 0 & 0 & 1 & 0 & 0 \\ 0 & 0 & 1 & 1 & 0 & 0 & 1 & 0 \\ 0 & 0 & 0 & 1 & 0 & 0 & 0 & 1 \end{array}\right) \underset{f_3 \leftarrow f_3-f_4}{\sim} \left(\begin{array}{cccc|cccc} 1 & 1 & 0 & 0 & 1 & 0 & 0 & 0 \\ 0 & 1 & 1 & 0 & 0 & 1 & 0 & 0 \\ 0 & 0 & 1 & 0 & 0 & 0 & 1 & -1 \\ 0 & 0 & 0 & 1 & 0 & 0 & 0 & 1 \end{array}\right)$$

$$\underset{f_2 \leftarrow f_2-f_3}{\sim} \left(\begin{array}{cccc|cccc} 1 & 1 & 0 & 0 & 1 & 0 & 0 & 0 \\ 0 & 1 & 0 & 0 & 0 & 1 & -1 & 1 \\ 0 & 0 & 1 & 0 & 0 & 0 & 1 & -1 \\ 0 & 0 & 0 & 1 & 0 & 0 & 0 & 1 \end{array}\right)$$

$$\underset{f_1 \leftarrow f_1-f_2}{\sim} \left(\begin{array}{cccc|cccc} 1 & 0 & 0 & 0 & 1 & -1 & 1 & -1 \\ 0 & 1 & 0 & 0 & 0 & 1 & -1 & 1 \\ 0 & 0 & 1 & 0 & 0 & 0 & 1 & -1 \\ 0 & 0 & 0 & 1 & 0 & 0 & 0 & 1 \end{array}\right).$$

Therefore:

$$(I_4 + A)^{-1} = \begin{pmatrix} 1 & -1 & 1 & -1 \\ 0 & 1 & -1 & 1 \\ 0 & 0 & 1 & -1 \\ 0 & 0 & 0 & 1 \end{pmatrix}.$$

Finally, we calculate $(I_4 + A)(I_4 - A)^{-1}$*:*

$$(I_4 + A)(I_4 - A)^{-1} = \begin{pmatrix} 1 & 1 & 0 & 0 \\ 0 & 1 & 1 & 0 \\ 0 & 0 & 1 & 1 \\ 0 & 0 & 0 & 1 \end{pmatrix} \cdot \begin{pmatrix} 1 & 1 & 1 & 1 \\ 0 & 1 & 1 & 1 \\ 0 & 0 & 1 & 1 \\ 0 & 0 & 0 & 1 \end{pmatrix}$$

$$= \begin{pmatrix} 1 & 2 & 2 & 2 \\ 0 & 1 & 2 & 2 \\ 0 & 0 & 1 & 2 \\ 0 & 0 & 0 & 1 \end{pmatrix}.$$

Problem 6.12. Demonstrate that $(A \cdot B)^{-1} = B^{-1} \cdot A^{-1}$. Verify it in the particular case:

$$A = \begin{pmatrix} 2 & -1 & 1 \\ -2 & 1 & 1 \\ 0 & 1 & 3 \end{pmatrix}, \qquad B = \begin{pmatrix} 2 & 1 & 1 \\ 6 & 3 & -5 \\ 4 & 1 & -3 \end{pmatrix}.$$

Solution to the problem 6.12. *To demonstrate that the matrix* $B^{-1} \cdot A^{-1}$ *is the inverse of* $A \cdot B$*, we use the definition of inverse, that is:*

$$(A \cdot B) \cdot (B^{-1} \cdot A^{-1}) = A \cdot (B \cdot B^{-1}) \cdot A^{-1} = A \cdot I \cdot A^{-1} = A \cdot A^{-1} = I$$

and, therefore:

$$B^{-1} \cdot A^{-1} = (A \cdot B)^{-1}.$$

Next, we verify it for the particular case given in the statement. We calculate A^{-1} *using the Gauss method:*

$$\left(\begin{array}{ccc|ccc} 2 & -1 & 1 & 1 & 0 & 0 \\ -2 & 1 & 1 & 0 & 1 & 0 \\ 0 & 1 & 3 & 0 & 0 & 1 \end{array}\right) \underset{f_2 \leftarrow f_2 + f_1}{\sim} \left(\begin{array}{ccc|ccc} 2 & -1 & 1 & 1 & 0 & 0 \\ 0 & 0 & 2 & 1 & 1 & 0 \\ 0 & 1 & 3 & 0 & 0 & 1 \end{array}\right)$$

$$\underset{pf_{23}}{\sim} \left(\begin{array}{ccc|ccc} 2 & -1 & 1 & 1 & 0 & 0 \\ 0 & 1 & 3 & 0 & 0 & 1 \\ 0 & 0 & 2 & 1 & 1 & 0 \end{array}\right)$$

$$\underset{\frac{1}{2} f_3}{\sim} \left(\begin{array}{ccc|ccc} 2 & -1 & 1 & 1 & 0 & 0 \\ 0 & 1 & 3 & 0 & 0 & 1 \\ 0 & 0 & 1 & 1/2 & 1/2 & 0 \end{array}\right)$$

$$\underset{\substack{f_1 \leftarrow f_1 - f_3 \\ f_2 \leftarrow f_2 - 3f_3}}{\sim} \left(\begin{array}{ccc|ccc} 2 & 1 & 0 & 1/2 & -1/2 & 0 \\ 0 & 1 & 0 & -3/2 & -3/2 & 1 \\ 0 & 0 & 1 & 1/2 & 1/2 & 0 \end{array}\right)$$

$$\underset{f_1 \leftarrow f_1 + f_2}{\sim} \left(\begin{array}{ccc|ccc} 2 & 0 & 0 & -1 & -2 & 1 \\ 0 & 1 & 0 & -3/2 & -3/2 & 1 \\ 0 & 0 & 1 & 1/2 & 1/2 & 0 \end{array}\right)$$

$$\underset{\frac{1}{2} f_1}{\sim} \left(\begin{array}{ccc|ccc} 1 & 0 & 0 & -1/2 & -1 & 1/2 \\ 0 & 1 & 0 & -3/2 & -3/2 & 1 \\ 0 & 0 & 1 & 1/2 & 1/2 & 0 \end{array}\right).$$

Therefore:

$$A^{-1} = \begin{pmatrix} -1/2 & -1 & 1/2 \\ -3/2 & -3/2 & 1 \\ 1/2 & 1/2 & 0 \end{pmatrix}.$$

We calculate B^{-1} using the Gauss method:

$$\left(\begin{array}{ccc|ccc} 2 & 1 & 1 & 1 & 0 & 0 \\ 6 & 3 & -5 & 0 & 1 & 0 \\ 4 & 1 & -3 & 0 & 0 & 1 \end{array}\right) \underset{\substack{f_2 \leftarrow f_2 - 3f_1 \\ f_3 \leftarrow f_3 - 2f_1}}{\sim} \left(\begin{array}{ccc|ccc} 2 & 1 & 1 & 1 & 0 & 0 \\ 0 & 0 & -8 & -3 & 1 & 0 \\ 0 & -1 & -5 & -2 & 0 & 1 \end{array}\right)$$

$$\underset{pf_{23}}{\sim} \left(\begin{array}{ccc|ccc} 2 & 1 & 1 & 1 & 0 & 0 \\ 0 & -1 & -5 & -2 & 0 & 1 \\ 0 & 0 & -8 & -3 & 1 & 0 \end{array}\right)$$

$$\underset{\frac{1}{2}f_1}{\sim} \left(\begin{array}{ccc|ccc} 1 & 1/2 & 1/2 & 1/2 & 0 & 0 \\ 0 & -1 & -5 & -2 & 0 & 1 \\ 0 & 0 & -8 & -3 & 1 & 0 \end{array}\right)$$

$$\underset{-f_2}{\sim} \left(\begin{array}{ccc|ccc} 1 & 1/2 & 1/2 & 1/2 & 0 & 0 \\ 0 & 1 & 5 & 2 & 0 & -1 \\ 0 & 0 & -8 & -3 & 1 & 0 \end{array}\right)$$

$$\underset{f_1 \leftarrow f_1 - \frac{1}{2}f_2}{\sim} \left(\begin{array}{ccc|ccc} 1 & 0 & -2 & -1/2 & 0 & 1/2 \\ 0 & 1 & 5 & 2 & 0 & -1 \\ 0 & 0 & -8 & -3 & 1 & 0 \end{array}\right)$$

$$\underset{-\frac{1}{8}f_3}{\sim} \left(\begin{array}{ccc|ccc} 1 & 0 & -2 & -1/2 & 0 & 1/2 \\ 0 & 1 & 5 & 2 & 0 & -1 \\ 0 & 0 & 1 & 3/8 & -1/8 & 0 \end{array}\right)$$

$$\underset{\substack{f_1 \leftarrow f_1 + 2f_3 \\ f_2 \leftarrow f_2 - 5f_3}}{\sim} \left(\begin{array}{ccc|ccc} 1 & 0 & 0 & 1/4 & -1/4 & 1/2 \\ 0 & 1 & 0 & 1/8 & 5/8 & -1 \\ 0 & 0 & 1 & 3/8 & -1/8 & 0 \end{array}\right).$$

Therefore:

$$B^{-1} = \begin{pmatrix} 1/4 & -1/4 & 1/2 \\ 1/8 & 5/8 & -1 \\ 3/8 & -1/8 & 0 \end{pmatrix}.$$

We calculate $B^{-1} \cdot A^{-1}$:

$$B^{-1} \cdot A^{-1} = \begin{pmatrix} 1/4 & -1/4 & 1/2 \\ 1/8 & 5/8 & -1 \\ 3/8 & -1/8 & 0 \end{pmatrix} \cdot \begin{pmatrix} -1/2 & -1 & 1/2 \\ -3/2 & -3/2 & 1 \\ 1/2 & 1/2 & 0 \end{pmatrix}$$

$$= \begin{pmatrix} \frac{1}{4}\cdot\left(-\frac{1}{2}\right)-\frac{1}{4}\cdot\left(-\frac{3}{2}\right)+\frac{1}{2}\cdot\frac{1}{2} & \frac{1}{4}\cdot(-1)-\frac{1}{4}\cdot\left(-\frac{3}{2}\right)+\frac{1}{2}\cdot\frac{1}{2} & \frac{1}{4}\cdot\frac{1}{2}-\frac{1}{4}\cdot 1+\frac{1}{2}\cdot 0 \\ \frac{1}{8}\cdot\left(-\frac{1}{2}\right)+\frac{5}{8}\cdot\left(-\frac{3}{2}\right)-1\cdot\frac{1}{2} & \frac{1}{8}\cdot(-1)+\frac{5}{8}\cdot\left(-\frac{3}{2}\right)-1\cdot\frac{1}{2} & \frac{1}{8}\cdot\frac{1}{2}+\frac{5}{8}\cdot 1-1\cdot 0 \\ \frac{3}{8}\cdot\left(-\frac{1}{2}\right)-\frac{1}{8}\cdot\left(-\frac{3}{2}\right)+0\cdot\frac{1}{2} & \frac{3}{8}\cdot(-1)-\frac{1}{8}\cdot\left(-\frac{3}{2}\right)+0\cdot\frac{1}{2} & \frac{3}{8}\cdot\frac{1}{2}-\frac{1}{8}\cdot 1+0\cdot 0 \end{pmatrix}$$

$$= \begin{pmatrix} 1/2 & 3/8 & -1/8 \\ -3/2 & -25/16 & 11/16 \\ 0 & -3/16 & 1/16 \end{pmatrix}.$$

We calculate $A \cdot B$:

$$A \cdot B = \begin{pmatrix} 2 & -1 & 1 \\ -2 & 1 & 1 \\ 0 & 1 & 3 \end{pmatrix} \cdot \begin{pmatrix} 2 & 1 & 1 \\ 6 & 3 & -5 \\ 4 & 1 & -3 \end{pmatrix}$$

$$= \begin{pmatrix} 2\cdot 2-1\cdot 6+1\cdot 4 & 2\cdot 1-1\cdot 3+1\cdot 1 & 2\cdot 1-1\cdot(-5)+1\cdot(-3) \\ -2\cdot 2+1\cdot 6+1\cdot 4 & -2\cdot 1+1\cdot 3+1\cdot 1 & -2\cdot 1+1\cdot(-5)+1\cdot(-3) \\ 0\cdot 2+1\cdot 6+3\cdot 4 & 0\cdot 1+1\cdot 3+3\cdot 1 & 0\cdot 1+1\cdot(-5)+3\cdot(-3) \end{pmatrix}$$

$$= \begin{pmatrix} 2 & 0 & 4 \\ 6 & 2 & -10 \\ 18 & 6 & -14 \end{pmatrix}.$$

We calculate $(A \cdot B)^{-1}$ using the Gauss method:

$$\left(\begin{array}{ccc|ccc} 2 & 0 & 4 & 1 & 0 & 0 \\ 6 & 2 & -10 & 0 & 1 & 0 \\ 18 & 6 & -14 & 0 & 0 & 1 \end{array}\right) \underset{\frac{1}{2}f_1}{\sim} \left(\begin{array}{ccc|ccc} 1 & 0 & 2 & 1/2 & 0 & 0 \\ 6 & 2 & -10 & 0 & 1 & 0 \\ 18 & 6 & -14 & 0 & 0 & 1 \end{array}\right)$$

$$\underset{\substack{f_2 \leftarrow f_2 - 6f_1 \\ f_3 \leftarrow f_3 - 18f_1}}{\sim} \left(\begin{array}{ccc|ccc} 1 & 0 & 2 & 1/2 & 0 & 0 \\ 0 & 2 & -22 & -3 & 1 & 0 \\ 0 & 6 & -50 & -9 & 0 & 1 \end{array}\right)$$

$$\underset{\frac{1}{2}f_2}{\sim} \left(\begin{array}{ccc|ccc} 1 & 0 & 2 & 1/2 & 0 & 0 \\ 0 & 1 & -11 & -3/2 & 1/2 & 0 \\ 0 & 6 & -50 & -9 & 0 & 1 \end{array}\right)$$

$$\underset{f_3 \leftarrow f_3 - 6f_2}{\sim} \left(\begin{array}{ccc|ccc} 1 & 0 & 2 & 1/2 & 0 & 0 \\ 0 & 1 & -11 & -3/2 & 1/2 & 0 \\ 0 & 0 & 16 & 0 & -3 & 1 \end{array}\right)$$

$$\underset{\frac{1}{16}f_3}{\sim} \left(\begin{array}{ccc|ccc} 1 & 0 & 2 & 1/2 & 0 & 0 \\ 0 & 1 & -11 & -3/2 & 1/2 & 0 \\ 0 & 0 & 1 & 0 & -3/16 & 1/16 \end{array}\right)$$

$$\underset{\substack{f_1 \leftarrow f_1 - 2f_3 \\ f_2 \leftarrow f_2 + 11f_3}}{\sim} \left(\begin{array}{ccc|ccc} 1 & 0 & 0 & 1/2 & 3/8 & -1/8 \\ 0 & 1 & 0 & -3/2 & -25/16 & 11/16 \\ 0 & 0 & 1 & 0 & -3/16 & 1/16 \end{array}\right).$$

Therefore:

$$(A \cdot B)^{-1} = \begin{pmatrix} 1/2 & 3/8 & -1/8 \\ -3/2 & -25/16 & 11/16 \\ 0 & -3/16 & 1/16 \end{pmatrix}.$$

We observe, therefore, that:

$$B^{-1} \cdot A^{-1} = (A \cdot B)^{-1}.$$

Problem 6.13. By applying the definition, calculate the determinant of the following matrices:

(a) $\begin{pmatrix} 1 & 2 \\ -5 & 3 \end{pmatrix}$

(b) $\begin{pmatrix} 1 & 2 & 3 \\ 1 & 1 & 1 \\ 0 & 2 & 4 \end{pmatrix}$

(c) $\begin{pmatrix} 2 & 1 & 1 \\ 1 & 0 & -3 \\ 3 & -2 & 1 \end{pmatrix}$

Solution to the problem 6.13. *(a) We calculate the determinant:*

$$\begin{vmatrix} 1 & 2 \\ -5 & 3 \end{vmatrix} = 1 \cdot 3 - 2 \cdot (-5) = 13.$$

(b) We calculate the determinant by applying the Sarrus rule:

$$\begin{vmatrix} 1 & 2 & 3 \\ 1 & 1 & 1 \\ 0 & 2 & 4 \end{vmatrix} = 1 \cdot 1 \cdot 4 + 2 \cdot 1 \cdot 0 + 3 \cdot 1 \cdot 2 - 1 \cdot 1 \cdot 2 - 2 \cdot 1 \cdot 4 - 3 \cdot 1 \cdot 0 = 0.$$

(c) We calculate the determinant by applying the Sarrus rule:

$$\begin{vmatrix} 2 & 1 & 1 \\ 1 & 0 & -3 \\ 3 & -2 & 1 \end{vmatrix} = 2 \cdot 0 \cdot 1 + 1 \cdot (-3) \cdot 3 + 1 \cdot 1 \cdot (-2) -$$

$$- 2 \cdot (-3) \cdot (-2) - 1 \cdot 1 \cdot 1 - 1 \cdot 0 \cdot 3 = -24.$$

Problem 6.14. Calculate the following determinants, transforming them previously into simpler ones:

(a) $\begin{vmatrix} 2 & 3 & -2 & 4 \\ 3 & -2 & 1 & 2 \\ 3 & 2 & 3 & 4 \\ -2 & 4 & 0 & 5 \end{vmatrix}$

(b)	$\begin{vmatrix} 1 & 2 & 1 & 2 & 1 \\ 0 & 0 & 1 & 1 & 1 \\ 1 & 1 & 0 & 0 & 0 \\ 0 & 0 & 1 & 1 & 2 \\ 1 & 2 & 2 & 1 & 1 \end{vmatrix}$
(c)	$\begin{vmatrix} a & 1 & 1 & 1 \\ 1 & a & 1 & 1 \\ 1 & 1 & a & 1 \\ 1 & 1 & 1 & a \end{vmatrix}$

Solution to the problem 6.14. *(a) This determinant will be calculated by expanding by the fourth row, as it has the most zeros.*

$$\begin{vmatrix} 2 & 3 & -2 & 4 \\ 3 & -2 & 1 & 2 \\ 3 & 2 & 3 & 4 \\ -2 & 4 & 0 & 5 \end{vmatrix} = a_{41}A_{41} + a_{42}A_{42} + a_{43}A_{43} + a_{44}A_{44}$$

$$= -2 \cdot (-1)^{4+1} D_{41} + 4 \cdot (-1)^{4+2} D_{42} + 0 \cdot (-1)^{4+3} D_{43} + 5 \cdot (-1)^{4+4} D_{44}$$
$$= 2D_{41} + 4D_{42} + 5D_{44},$$

where:

$$D_{41} = \begin{vmatrix} 3 & -2 & 4 \\ -2 & 1 & 2 \\ 2 & 3 & 4 \end{vmatrix}$$
$$= 3 \cdot 1 \cdot 4 + (-2) \cdot 2 \cdot 2 + 4 \cdot (-2) \cdot 3 - 3 \cdot 2 \cdot 3 - (-2) \cdot (-2) \cdot 4 - 4 \cdot 1 \cdot 2$$
$$= -62$$

$$D_{42} = \begin{vmatrix} 2 & -2 & 4 \\ 3 & 1 & 2 \\ 3 & 3 & 4 \end{vmatrix}$$
$$= 2 \cdot 1 \cdot 4 + (-2) \cdot 2 \cdot 3 + 4 \cdot 3 \cdot 3 - 2 \cdot 2 \cdot 3 - (-2) \cdot 3 \cdot 4 - 4 \cdot 1 \cdot 3$$
$$= 32$$

$$D_{44} = \begin{vmatrix} 2 & 3 & -2 \\ 3 & -2 & 1 \\ 3 & 2 & 3 \end{vmatrix}$$
$$= 2 \cdot (-2) \cdot 3 + 3 \cdot 1 \cdot 3 + (-2) \cdot 3 \cdot 2 - 2 \cdot 1 \cdot 2 - 3 \cdot 3 \cdot 3 - (-2) \cdot (-2) \cdot 3$$
$$= -58.$$

Therefore:

$$\begin{vmatrix} 2 & 3 & -2 & 4 \\ 3 & -2 & 1 & 2 \\ 3 & 2 & 3 & 4 \\ -2 & 4 & 0 & 5 \end{vmatrix} = 2 \cdot (-62) + 4 \cdot 32 + 5 \cdot (-58) = -286.$$

(b) *This determinant will be calculated by expanding by the third row, as it has the most zeros.*

$$\begin{vmatrix} 1 & 2 & 1 & 2 & 1 \\ 0 & 0 & 1 & 1 & 1 \\ 1 & 1 & 0 & 0 & 0 \\ 0 & 0 & 1 & 1 & 2 \\ 1 & 2 & 2 & 1 & 1 \end{vmatrix} = a_{31}A_{31} + a_{32}A_{32} + a_{33}A_{33} + a_{34}A_{34} + a_{35}A_{35}$$

$$= 1 \cdot (-1)^{3+1}D_{31} + 1 \cdot (-1)^{3+2}D_{32} + 0 \cdot (-1)^{3+3}D_{33} + \\ + 0 \cdot (-1)^{3+4}D_{34} + 0 \cdot (-1)^{3+5}D_{35}$$

$$= D_{31} - D_{32},$$

where:

$$D_{31} = \begin{vmatrix} 2 & 1 & 2 & 1 \\ 0 & 1 & 1 & 1 \\ 0 & 1 & 1 & 2 \\ 2 & 2 & 1 & 1 \end{vmatrix}$$

$$D_{32} = \begin{vmatrix} 1 & 1 & 2 & 1 \\ 0 & 1 & 1 & 1 \\ 0 & 1 & 1 & 2 \\ 1 & 2 & 1 & 1 \end{vmatrix}.$$

– *The minor D_{31} will be calculated by expanding by the first column:*

$$D_{31} = \begin{vmatrix} 2 & 1 & 2 & 1 \\ 0 & 1 & 1 & 1 \\ 0 & 1 & 1 & 2 \\ 2 & 2 & 1 & 1 \end{vmatrix} =$$

$$= 2 \cdot (-1)^{1+1} \begin{vmatrix} 1 & 1 & 1 \\ 1 & 1 & 2 \\ 2 & 1 & 1 \end{vmatrix} + 2 \cdot (-1)^{1+4} \begin{vmatrix} 1 & 2 & 1 \\ 1 & 1 & 1 \\ 1 & 1 & 2 \end{vmatrix},$$

where:

$$\begin{vmatrix} 1 & 1 & 1 \\ 1 & 1 & 2 \\ 2 & 1 & 1 \end{vmatrix} = 1 \cdot 1 \cdot 1 + 1 \cdot 2 \cdot 2 + 1 \cdot 1 \cdot 1 - 1 \cdot 2 \cdot 1 - 1 \cdot 1 \cdot 1 - 1 \cdot 1 \cdot 2 = 1$$

$$\begin{vmatrix} 1 & 2 & 1 \\ 1 & 1 & 1 \\ 1 & 1 & 2 \end{vmatrix} = 1 \cdot 1 \cdot 2 + 2 \cdot 1 \cdot 1 + 1 \cdot 1 \cdot 1 - 1 \cdot 1 \cdot 1 - 2 \cdot 1 \cdot 2 - 1 \cdot 1 \cdot 1 = -1.$$

Therefore:

$$D_{31} = 2 \cdot 1 - 2 \cdot (-1) = 4.$$

– *The minor D_{32} will also be calculated by expanding by the first column:*

$$D_{32} = \begin{vmatrix} 1 & 1 & 2 & 1 \\ 0 & 1 & 1 & 1 \\ 0 & 1 & 1 & 2 \\ 1 & 2 & 1 & 1 \end{vmatrix} = 1 \cdot (-1)^{1+1} \begin{vmatrix} 1 & 1 & 1 \\ 1 & 1 & 2 \\ 2 & 1 & 1 \end{vmatrix} + 1 \cdot (-1)^{1+4} \begin{vmatrix} 1 & 2 & 1 \\ 1 & 1 & 1 \\ 1 & 1 & 2 \end{vmatrix},$$

where:

$$\begin{vmatrix} 1 & 1 & 1 \\ 1 & 1 & 2 \\ 2 & 1 & 1 \end{vmatrix} = 1\cdot1\cdot1+1\cdot2\cdot2+1\cdot1\cdot1-1\cdot2\cdot1-1\cdot1\cdot1-1\cdot1\cdot2 = 1$$

$$\begin{vmatrix} 1 & 2 & 1 \\ 1 & 1 & 1 \\ 1 & 1 & 2 \end{vmatrix} = 1\cdot1\cdot2+2\cdot1\cdot1+1\cdot1\cdot1-1\cdot1\cdot1-2\cdot1\cdot2-1\cdot1\cdot1 = -1.$$

Therefore:

$$D_{32} = 1\cdot1-1\cdot(-1) = 2.$$

So:

$$\begin{vmatrix} 1 & 2 & 1 & 2 & 1 \\ 0 & 0 & 1 & 1 & 1 \\ 1 & 1 & 0 & 0 & 0 \\ 0 & 0 & 1 & 1 & 2 \\ 1 & 2 & 2 & 1 & 1 \end{vmatrix} = D_{31} - D_{32} = 4-2 = 2.$$

(c) *In this case, it doesn't matter by which row or column we expand to calculate the determinant, so we will calculate it by expanding by the first row.*

$$\begin{vmatrix} a & 1 & 1 & 1 \\ 1 & a & 1 & 1 \\ 1 & 1 & a & 1 \\ 1 & 1 & 1 & a \end{vmatrix} = a_{11}A_{11} + a_{12}A_{12} + a_{13}A_{13} + a_{14}A_{14}$$

$$= a\cdot(-1)^{1+1}D_{11} + 1\cdot(-1)^{1+2}D_{12} + 1\cdot(-1)^{1+3}D_{13} + 1\cdot(-1)^{1+4}D_{14}$$

$$= a\cdot D_{11} - D_{12} + D_{13} - D_{14},$$

where:

$$D_{11} = \begin{vmatrix} a & 1 & 1 \\ 1 & a & 1 \\ 1 & 1 & a \end{vmatrix}$$

$$= a\cdot a\cdot a+1\cdot1\cdot1+1\cdot1\cdot1-a\cdot1\cdot1-1\cdot1\cdot a-1\cdot a\cdot1 = a^3-3a+2$$

$$D_{12} = \begin{vmatrix} 1 & 1 & 1 \\ 1 & a & 1 \\ 1 & 1 & a \end{vmatrix}$$

$$= 1\cdot a\cdot a+1\cdot1\cdot1+1\cdot1\cdot1-1\cdot1\cdot1-1\cdot1\cdot a-1\cdot a\cdot1 = a^2-2a+1$$

$$D_{13} = \begin{vmatrix} 1 & a & 1 \\ 1 & 1 & 1 \\ 1 & 1 & a \end{vmatrix}$$

$$= 1\cdot1\cdot a+a\cdot1\cdot1+1\cdot1\cdot1-1\cdot1\cdot1-a\cdot1\cdot a-1\cdot1\cdot1 = -a^2+2a-1$$

$$D_{14} = \begin{vmatrix} 1 & a & 1 \\ 1 & 1 & a \\ 1 & 1 & 1 \end{vmatrix}$$
$$= 1\cdot 1\cdot 1 + a\cdot a\cdot 1 + 1\cdot 1\cdot 1 - 1\cdot a\cdot 1 - a\cdot 1\cdot 1 - 1\cdot 1\cdot 1 = a^2 - 2a + 1.$$

Therefore:

$$\begin{vmatrix} a & 1 & 1 & 1 \\ 1 & a & 1 & 1 \\ 1 & 1 & a & 1 \\ 1 & 1 & 1 & a \end{vmatrix} = a\cdot D_{11} - D_{12} + D_{13} - D_{14}$$
$$= a\cdot(a^3 - 3a + 2) - (a^2 - 2a + 1) + (-a^2 + 2a - 1) - (a^2 - 2a + 1)$$
$$= a^4 - 6a^2 + 8a - 3 = (a+3)(a-1)^3.$$

Problem 6.15. Prove that:
$$\begin{vmatrix} 1 & a & b+c \\ 1 & b & a+c \\ 1 & c & a+b \end{vmatrix} = 0.$$

Solution to the problem 6.15. *To prove that the determinant of the given matrix is zero, it is not necessary to calculate it; it is enough to consider the properties of determinants. We will use the following properties:*

(D_3) *If a row or a column is the sum of two terms, the determinant decomposes into the sum of two determinants, each of them with one of the terms in the corresponding row or column and having the rest of the rows or columns the same.*

(D_6) *If in a determinant, a row or a column is a linear combination of other rows or columns (equal to the sum of other rows or columns multiplied by a constant), the determinant is zero.*

Therefore:

$$\begin{vmatrix} 1 & a & b+c \\ 1 & b & a+c \\ 1 & c & a+b \end{vmatrix} = \begin{vmatrix} 1 & a & a+b+c-a \\ 1 & b & a+b+c-b \\ 1 & c & a+b+c-c \end{vmatrix}$$
$$= \underbrace{\begin{vmatrix} 1 & a & a+b+c \\ 1 & b & a+b+c \\ 1 & c & a+b+c \end{vmatrix}}_{=0\ (c_3=(a+b+c)\cdot c_1)} + \underbrace{\begin{vmatrix} 1 & a & -a \\ 1 & b & -b \\ 1 & c & -c \end{vmatrix}}_{=0\ (c_3=-c_2)} = 0.$$

Problem 6.16. Prove, without calculating the determinants, the equality:
$$\begin{vmatrix} 1 & a^2 & a^3 \\ 1 & b^2 & b^3 \\ 1 & c^2 & c^3 \end{vmatrix} = \begin{vmatrix} bc & a & a^2 \\ ac & b & b^2 \\ ab & c & c^2 \end{vmatrix}.$$

Solution to the problem 6.16. *To verify this equality, we will use property* (D_2), *which states that if all the elements of a row or column are multiplied by a constant, the determinant is multiplied by that constant.*
Thus:

$$\begin{vmatrix} bc & a & a^2 \\ ac & b & b^2 \\ ab & c & c^2 \end{vmatrix} = \frac{1}{abc} \begin{vmatrix} abc & a^2 & a^3 \\ abc & b^2 & b^3 \\ abc & c^2 & c^3 \end{vmatrix} = \frac{abc}{abc} \begin{vmatrix} 1 & a^2 & a^3 \\ 1 & b^2 & b^3 \\ 1 & c^2 & c^3 \end{vmatrix}.$$

Problem 6.17. Reason, without applying the Sarrus rule, that the roots of the following polynomial are 5, 7 and −12:

$$p(x) = \begin{vmatrix} x & 7 & 7 \\ 7 & x & 5 \\ 5 & 5 & x \end{vmatrix}$$

Solution to the problem 6.17. *The determinant in the statement is a third degree polynomial and, therefore, has three roots in* $\mathbb{C}$ *(that is, three values of* x *that make the determinant zero). To see that the given values are the roots of the polynomial, we only need to verify that they make the determinant zero, and, since we are given three values, these will be the three roots of the polynomial.*

- $p(5) = \begin{vmatrix} 5 & 7 & 7 \\ 7 & 5 & 5 \\ 5 & 5 & 5 \end{vmatrix} = 0$ *because* $c_2 = c_3$,
- $p(7) = \begin{vmatrix} 7 & 7 & 7 \\ 7 & 7 & 5 \\ 5 & 5 & 7 \end{vmatrix} = 0$ *because* $c_1 = c_2$,
- $p(-12) = \begin{vmatrix} -12 & 7 & 7 \\ 7 & -12 & 5 \\ 5 & 5 & -12 \end{vmatrix} = 0$ *because* $f_1 = -(f_2 + f_3)$,

where to verify that the determinants are zero we have used property (D_6), *that is, if in a determinant, a row or a column is a linear combination of other rows or columns (equal to the sum of other rows or columns multiplied by a constant), the determinant is zero.*

Problem 6.18. Reason, without calculating, whether the determinant of the following matrices will be null or not:

(a) $\begin{pmatrix} 5 & 3 \\ 1 & 3/5 \end{pmatrix}$

(b) $\begin{pmatrix} -1 & 2 & 0 \\ 3 & -1 & 5 \\ 1 & 4 & 6 \end{pmatrix}$

(c) $\begin{pmatrix} 1 & 2 & 5 & 3 \\ -3 & -1 & 4 & -1 \\ 2 & 3 & -6 & 4 \\ 0 & 4 & 3 & 6 \end{pmatrix}$

Solution to the problem 6.18. *In all three cases we will see that the determinant is null because there is at least one row or column that is a linear combination of other rows or columns (property (D_6)):*

$$(a)\ \begin{vmatrix} 5 & 3 \\ 1 & 3/5 \end{vmatrix} = 0, \qquad \text{because } f_1 = 5f_2.$$

$$(b)\ \begin{vmatrix} -1 & 2 & 0 \\ 3 & -1 & 5 \\ 1 & 4 & 6 \end{vmatrix} = 0, \qquad \text{because } c_3 = 2c_1 + c_2.$$

$$(c)\ \begin{vmatrix} 1 & 2 & 5 & 3 \\ -3 & -1 & 4 & -1 \\ 2 & 3 & -6 & 4 \\ 0 & 4 & 3 & 6 \end{vmatrix} = 0, \qquad \text{because } f_4 = f_1 + f_2 + f_3.$$

Problem 6.19. Prove, without calculating the determinants, the equality:

$$\begin{vmatrix} 1 & a^2 & a^3 \\ 1 & b^2 & b^3 \\ 1 & c^2 & c^3 \end{vmatrix} = \begin{vmatrix} bc & a & a^2 \\ ac & b & b^2 \\ ab & c & c^2 \end{vmatrix}.$$

Solution to the problem 6.19. *To verify this equality we will use property (D_2), which states that if all the elements of a row or column are multiplied by a constant, the determinant is multiplied by that constant.*
Thus:

$$\begin{vmatrix} bc & a & a^2 \\ ac & b & b^2 \\ ab & c & c^2 \end{vmatrix} = \frac{1}{abc} \begin{vmatrix} abc & a^2 & a^3 \\ abc & b^2 & b^3 \\ abc & c^2 & c^3 \end{vmatrix} = \frac{abc}{abc} \begin{vmatrix} 1 & a^2 & a^3 \\ 1 & b^2 & b^3 \\ 1 & c^2 & c^3 \end{vmatrix}.$$

Problem 6.20. Prove without expanding the determinant and applying only the properties of the determinants, that:

$$\begin{vmatrix} x & 3 & 3 & 3 \\ 3 & x & 2 & 2 \\ 2 & 2 & x & 9 \\ 1 & 1 & 1 & 1 \end{vmatrix} = (x-2)\cdot(x-3)\cdot(x-9).$$

Solution to the problem 6.20. *We see that the determinant is a third-degree polynomial with leading coefficient 1 and, therefore, can be written as:*

$$(x - r_1)(x - r_2)(x - r_3)$$

where:

$$r_1, r_2, r_3 \in \mathbb{C}$$

are the three roots of the polynomial. We only need to prove that:

$$r_{1,2,3} = 2, 3, 9$$

that is, the determinant is zero for these values.

- $\begin{vmatrix} 2 & 3 & 3 & 3 \\ 3 & 2 & 2 & 2 \\ 2 & 2 & 2 & 9 \\ 1 & 1 & 1 & 1 \end{vmatrix} = 0,$ *because* $c_2 = c_3$

- $\begin{vmatrix} 3 & 3 & 3 & 3 \\ 3 & 3 & 2 & 2 \\ 2 & 2 & 3 & 9 \\ 1 & 1 & 1 & 1 \end{vmatrix} = 0,$ *because* $c_1 = c_2$

- $\begin{vmatrix} 9 & 3 & 3 & 3 \\ 3 & 9 & 2 & 2 \\ 2 & 2 & 9 & 9 \\ 1 & 1 & 1 & 1 \end{vmatrix} = 0,$ *because* $c_3 = c_4$,

where to verify that the determinants are zero we have used property (D_6), *that is, if in a determinant, a row or a column is a linear combination of other rows or columns (equal to the sum of other rows or columns multiplied by a constant), the determinant is null.*

6.9 Solved Problems (Systems of Equations)

Problem 6.21. Discuss and solve the following systems of linear equations:

(a) $\begin{cases} x + y + z &= 6 \\ 2x - y &= 0 \\ 3y - 2z &= 0 \end{cases}$

(b) $\begin{cases} 2x + 3y + z &= 4 \\ x - 2y + z &= -2 \\ 8x + 5y + 5z &= 1 \end{cases}$

(c) $\begin{cases} x - 2y + z - t &= 1 \\ 2x + y + 3z &= 2 \\ -x + 3y - z + 4t &= -1 \end{cases}$

(d) $\begin{cases} x + 2y - z &= 1 \\ x + y + 3z &= 0 \\ x + 7z &= -1 \end{cases}$

(e) $\begin{cases} 2x + y + 3z &= 0 \\ -2x - y &= 1 \\ z &= 2 \end{cases}$

(f) $\left\{\begin{array}{rcl} x-2y+z+2t & = & 2 \\ -2x+y+3z-t & = & 0 \\ 3x-y-2z-t & = & 2 \\ -x+5y-z-3t & = & -1 \end{array}\right.$

Solution to the problem 6.21. *(a) We need to solve the system of equations:*

$$\left\{\begin{array}{rcl} x+y+z & = & 6 \\ 2x-y & = & 0 \\ 3y-2z & = & 0 \end{array}\right.$$

We write the augmented matrix of the system and determine the rank of the system matrix and the rank of the augmented matrix of the system:

$$\left(\begin{array}{ccc|c} 1 & 1 & 1 & 6 \\ 2 & -1 & 0 & 0 \\ 0 & 3 & -2 & 0 \end{array}\right) \underset{f_2 \leftarrow f_2 - 2f_1}{\sim} \left(\begin{array}{ccc|c} 1 & 1 & 1 & 6 \\ 0 & -3 & -2 & -12 \\ 0 & 3 & -2 & 0 \end{array}\right) \underset{f_3 \leftarrow f_3 + f_2}{\sim} \left(\begin{array}{ccc|c} 1 & 1 & 1 & 6 \\ 0 & -3 & -2 & -12 \\ 0 & 0 & -4 & -12 \end{array}\right).$$

In this case, $r = r' = n = 3$; therefore, it is a consistent and determined system (CDS). We solve it by applying the Gauss-Jordan rule:

$$\left(\begin{array}{ccc|c} 1 & 1 & 1 & 6 \\ 0 & -3 & -2 & -12 \\ 0 & 0 & -4 & -12 \end{array}\right) \underset{f_2 \leftarrow f_2 - \frac{1}{2}f_3}{\sim} \left(\begin{array}{ccc|c} 1 & 1 & 1 & 6 \\ 0 & -3 & 0 & -6 \\ 0 & 0 & -4 & -12 \end{array}\right)$$

$$\underset{f_1 \leftarrow f_1 + \frac{1}{3}f_2}{\sim} \left(\begin{array}{ccc|c} 1 & 0 & 1 & 4 \\ 0 & -3 & 0 & -6 \\ 0 & 0 & -4 & -12 \end{array}\right) \underset{f_1 \leftarrow f_1 + \frac{1}{4}f_3}{\sim} \left(\begin{array}{ccc|c} 1 & 0 & 0 & 1 \\ 0 & -3 & 0 & -6 \\ 0 & 0 & -4 & -12 \end{array}\right)$$

$$\underset{\substack{-\frac{1}{3}f_2 \\ -\frac{1}{4}f_3}}{\sim} \left(\begin{array}{ccc|c} 1 & 0 & 0 & 1 \\ 0 & 1 & 0 & 2 \\ 0 & 0 & 1 & 3 \end{array}\right).$$

Therefore, the solutions of the system are: $\{x = 1, y = 2, z = 3\}$.

(b) We need to solve the system of equations:

$$\left\{\begin{array}{rcl} 2x+3y+z & = & 4 \\ x-2y+z & = & -2 \\ 8x+5y+5z & = & 1 \end{array}\right.$$

We write the augmented matrix of the system and determine the rank of the system matrix and the rank of the augmented matrix of the system:

$$\left(\begin{array}{ccc|c} 2 & 3 & 1 & 4 \\ 1 & -2 & 1 & -2 \\ 8 & 5 & 5 & 1 \end{array}\right) \underset{pf_{12}}{\sim} \left(\begin{array}{ccc|c} 1 & -2 & 1 & -2 \\ 2 & 3 & 1 & 4 \\ 8 & 5 & 5 & 1 \end{array}\right)$$

$$\underset{\substack{f_2 \leftarrow f_2 - 2f_1 \\ f_3 \leftarrow f_3 - 8f_1}}{\sim} \left(\begin{array}{ccc|c} 1 & -2 & 1 & -2 \\ 0 & 7 & -1 & 8 \\ 0 & 21 & -3 & 17 \end{array}\right)$$

$$\underset{f_3 \leftarrow f_3 - 3f_2}{\sim} \left(\begin{array}{ccc|c} 1 & -2 & 1 & -2 \\ 0 & 7 & -1 & 8 \\ 0 & 0 & 0 & -7 \end{array}\right)$$

In this case, $r \neq r'$; therefore, it is an inconsistent system (IS), and does not admit any solution.

(c) *We need to solve the system of equations:*

$$\begin{cases} x - 2y + z - t &= 1 \\ 2x + y + 3z &= 2 \\ -x + 3y - z + 4t &= -1 \end{cases}$$

We write the augmented matrix of the system and determine the rank of the system matrix and the rank of the augmented matrix of the system:

$$\left(\begin{array}{cccc|c} 1 & -2 & 1 & -1 & 1 \\ 2 & 1 & 3 & 0 & 2 \\ -1 & 3 & -1 & 4 & -1 \end{array}\right) \underset{\substack{f_2 \leftarrow f_2 - 2f_1 \\ f_3 \leftarrow f_3 + f_1}}{\sim} \left(\begin{array}{cccc|c} 1 & -2 & 1 & -1 & 1 \\ 0 & 5 & 1 & 2 & 0 \\ 0 & 1 & 0 & 3 & 0 \end{array}\right)$$

$$\underset{f_3 \leftarrow f_3 - \frac{1}{5}f_2}{\sim} \left(\begin{array}{cccc|c} 1 & -2 & 1 & -1 & 1 \\ 0 & 5 & 1 & 2 & 0 \\ 0 & 0 & -1/5 & 3/5 & 0 \end{array}\right).$$

In this case, $r = r' = 3 < n = 4$; therefore, it is a consistent and undetermined system (CUS), with a degree of indeterminacy $n - r = 4 - 3 = 1$. We solve it:

$$\left(\begin{array}{cccc|c} 1 & -2 & 1 & -1 & 1 \\ 0 & 5 & 1 & 2 & 0 \\ 0 & 0 & -1/5 & 3/5 & 0 \end{array}\right) \underset{\substack{\frac{1}{5}f_2 \\ -5f_3}}{\sim} \left(\begin{array}{cccc|c} 1 & -2 & 1 & -1 & 1 \\ 0 & 1 & 1/5 & 2/5 & 0 \\ 0 & 0 & 1 & -13 & 0 \end{array}\right)$$

$$\underset{f_1 \leftarrow f_1+2f_2}{\sim} \left(\begin{array}{cccc|c} 1 & 0 & 7/5 & -1/5 & 1 \\ 0 & 1 & 1/5 & 2/5 & 0 \\ 0 & 0 & 1 & -13 & 0 \end{array}\right)$$

$$\underset{\substack{f_1 \leftarrow f_1-\frac{7}{5}f_3 \\ f_2 \leftarrow f_2-\frac{1}{5}f_3}}{\sim} \left(\begin{array}{cccc|c} 1 & 0 & 0 & 18 & 1 \\ 0 & 1 & 0 & 3 & 0 \\ 0 & 0 & 1 & -13 & 0 \end{array}\right).$$

Therefore, the solutions of the system are: $\{x = 1 - 18t,\ y = -3t,\ z = 13t,\ t = t\}$.

(d) *We need to solve the system of equations:*

$$\begin{cases} x + 2y - z &= 1 \\ x + y + 3z &= 0 \\ x + 7z &= -1 \end{cases}$$

We write the augmented matrix of the system and determine the rank of the system matrix and the rank of the augmented matrix of the system:

$$\left(\begin{array}{ccc|c} 1 & 2 & -1 & 1 \\ 1 & 1 & 3 & 0 \\ 1 & 0 & 7 & -1 \end{array}\right) \underset{\substack{f_2 \leftarrow f_2-f_1 \\ f_3 \leftarrow f_3-f_1}}{\sim} \left(\begin{array}{ccc|c} 1 & 2 & -1 & 1 \\ 0 & -1 & 4 & -1 \\ 0 & -2 & 8 & -2 \end{array}\right)$$

$$\underset{f_3 \leftarrow f_3-2f_2}{\sim} \left(\begin{array}{ccc|c} 1 & 2 & -1 & 1 \\ 0 & -1 & 4 & -1 \\ 0 & 0 & 0 & 0 \end{array}\right)$$

In this case, $r = r' = 2 < n = 3$*; therefore, it is a consistent and undetermined system (CUS), with a degree of indeterminacy* $n - r = 3 - 2 = 1$*. We solve it:*

$$\left(\begin{array}{ccc|c} 1 & 2 & -1 & 1 \\ 0 & -1 & 4 & -1 \\ 0 & 0 & 0 & 0 \end{array}\right) \underset{f_1 \leftarrow f_1+2f_2}{\sim} \left(\begin{array}{ccc|c} 1 & 0 & 7 & -1 \\ 0 & -1 & 4 & -1 \\ 0 & 0 & 0 & 0 \end{array}\right)$$

$$\underset{-f_2}{\sim} \left(\begin{array}{ccc|c} 1 & 0 & 7 & -1 \\ 0 & 1 & -4 & 1 \\ 0 & 0 & 0 & 0 \end{array}\right).$$

Therefore, the solutions of the system are: $\{x = 1 - 7z,\ y = 1 + 4z,\ z = z\}$.

(e) *We need to solve the system of equations:*

$$\begin{cases} 2x + y + 3z &= 0 \\ -2x - y &= 1 \\ z &= 2 \end{cases}$$

We write the augmented matrix of the system and determine the rank of the system matrix and the rank of the augmented matrix of the system:

$$\left(\begin{array}{ccc|c} 2 & 1 & 3 & 0 \\ -2 & -1 & 0 & 1 \\ 0 & 0 & 1 & 2 \end{array}\right) \underset{f_2 \leftarrow f_2 + f_1}{\sim} \left(\begin{array}{ccc|c} 2 & 1 & 3 & 0 \\ 0 & 0 & 3 & 1 \\ 0 & 0 & 1 & 2 \end{array}\right)$$

$$\underset{f_3 \leftarrow f_3 - \frac{1}{3} f_2}{\sim} \left(\begin{array}{ccc|c} 2 & 1 & 3 & 0 \\ 0 & 0 & 3 & 1 \\ 0 & 0 & 0 & 5/3 \end{array}\right).$$

In this case, $r = 2 \neq r' = 3$. We will see more clearly that the system matrix has rank 2 if we swap columns 2 and 3:

$$\begin{pmatrix} 2 & 1 & 3 \\ 0 & 0 & 3 \\ 0 & 0 & 0 \end{pmatrix} \underset{pc_{23}}{\sim} \begin{pmatrix} 2 & 3 & 1 \\ 0 & 3 & 0 \\ 0 & 0 & 0 \end{pmatrix}.$$

Therefore, it is an inconsistent system (IS), and does not admit any solution.

(f) *We need to solve the system of equations:*

$$\begin{cases} x - 2y + z + 2t &= 2 \\ -2x + y + 3z - t &= 0 \\ 3x - y - 2z - t &= 2 \\ -x + 5y - z - 3t &= -1 \end{cases}$$

We write the augmented matrix of the system and determine the rank of the system matrix and the rank of the augmented matrix of the system:

$$\left(\begin{array}{cccc|c} 1 & -2 & 1 & 2 & 2 \\ -2 & 1 & 3 & -1 & 0 \\ 3 & -1 & -2 & -1 & 2 \\ -1 & 5 & -1 & -3 & -1 \end{array}\right) \underset{\substack{f_2 \leftarrow f_2 + 2f_1 \\ f_3 \leftarrow f_3 - 3f_1 \\ f_4 \leftarrow f_4 + f_1}}{\sim} \left(\begin{array}{cccc|c} 1 & -2 & 1 & 2 & 2 \\ 0 & -3 & 5 & 3 & 4 \\ 0 & 5 & -5 & -7 & -4 \\ 0 & 3 & 0 & -1 & 1 \end{array}\right)$$

$$\underset{\substack{f_3 \leftarrow f_3 + \frac{5}{3} f_2 \\ f_4 \leftarrow f_4 + f_2}}{\sim} \left(\begin{array}{cccc|c} 1 & -2 & 1 & 2 & 2 \\ 0 & -3 & 5 & 3 & 4 \\ 0 & 0 & 10/3 & -2 & 8/3 \\ 0 & 0 & 0 & 5 & 1 \end{array}\right).$$

In this case, $r = r' = n = 4$; therefore, it is a consistent and determined system (CDS). We solve it by applying the Gauss-Jordan rule:

$$\left(\begin{array}{cccc|c} 1 & -2 & 1 & 2 & 2 \\ 0 & -3 & 5 & 3 & 4 \\ 0 & 0 & 10/3 & -2 & 8/3 \\ 0 & 0 & 0 & 5 & 1 \end{array}\right) \underset{\substack{-\frac{1}{3}f_2 \\ \frac{3}{10}f_3 \\ \frac{1}{5}f_4}}{\sim}$$

$$\underset{\substack{-\frac{1}{3}f_2 \\ \frac{3}{10}f_3 \\ \frac{1}{5}f_4}}{\sim} \left(\begin{array}{cccc|c} 1 & -2 & 1 & 2 & 2 \\ 0 & 1 & -5/3 & -1 & -4/3 \\ 0 & 0 & 1 & -3/5 & 4/5 \\ 0 & 0 & 0 & 1 & 1/5 \end{array}\right)$$

$$\underset{f_1 \leftarrow f_1 + 2f_2}{\sim} \left(\begin{array}{cccc|c} 1 & 0 & -7/3 & 0 & -2/3 \\ 0 & 1 & -5/3 & -1 & -4/3 \\ 0 & 0 & 1 & -3/5 & 4/5 \\ 0 & 0 & 0 & 1 & 1/5 \end{array}\right).$$

$$\underset{\substack{f_1 \leftarrow f_1 + \frac{7}{3}f_3 \\ f_2 \leftarrow f_2 + \frac{5}{3}f_3}}{\sim} \left(\begin{array}{cccc|c} 1 & 0 & 0 & -7/5 & 6/5 \\ 0 & 1 & 0 & -2 & 0 \\ 0 & 0 & 1 & -3/5 & 4/5 \\ 0 & 0 & 0 & 1 & 1/5 \end{array}\right)$$

$$\underset{\substack{f_1 \leftarrow f_1 + \frac{7}{5}f_4 \\ f_2 \leftarrow f_2 + 2f_4 \\ f_3 \leftarrow f_3 + \frac{3}{5}f_4}}{\sim} \left(\begin{array}{cccc|c} 1 & 0 & 0 & 0 & 37/25 \\ 0 & 1 & 0 & 0 & 2/5 \\ 0 & 0 & 1 & 0 & 23/25 \\ 0 & 0 & 0 & 1 & 1/5 \end{array}\right).$$

Therefore, the solutions of the system are:

$$\{x = \frac{37}{25}, \quad y = \frac{2}{5}, \quad z = \frac{23}{25}, \quad t = \frac{1}{5}\}.$$

Problem 6.22. Discuss, according to the value of $a \in \mathbb{R}$, the following systems of equations and find the solutions, if they exist.

$$\text{(a)} \quad \begin{cases} a^2x + y + z &= 3 \\ x + a^2y + z &= 4 - a \\ x + y + a^2z &= 2 + a^2 \end{cases}$$

(b) $\begin{cases} (1-a)x+y &= 0 \\ (1-a)y+z &= 0 \\ x+(1-a)z &= 0 \end{cases}$

(c) $\begin{cases} ax-2y+4z &= a+2 \\ x-y+2z &= a \\ -2x+2y+2az &= -4 \end{cases}$

(d) $\begin{cases} ax+3y+z &= 2 \\ 3x+ay+2z &= 3 \\ -x+y-z &= 1 \end{cases}$

Solution to the problem 6.22. *(a) We need to discuss the system of equations:*

$$\begin{cases} a^2x+y+z &= 3 \\ x+a^2y+z &= 4-a \\ x+y+a^2z &= 2+a^2 \end{cases}$$

according to the value of $a \in \mathbb{R}$.

We operate on the augmented matrix of the system:

$$\left(\begin{array}{ccc|c} a^2 & 1 & 1 & 3 \\ 1 & a^2 & 1 & 4-a \\ 1 & 1 & a^2 & 2+a^2 \end{array}\right) \underset{pf_{13}}{\sim} \left(\begin{array}{ccc|c} 1 & 1 & a^2 & 2+a^2 \\ 1 & a^2 & 1 & 4-a \\ a^2 & 1 & 1 & 3 \end{array}\right)$$

$$\underset{pf_{23}}{\sim} \left(\begin{array}{ccc|c} 1 & 1 & a^2 & 2+a^2 \\ a^2 & 1 & 1 & 3 \\ 1 & a^2 & 1 & 4-a \end{array}\right)$$

$$\underset{\substack{f_2 \leftarrow f_2 - a^2 f_1 \\ f_3 \leftarrow f_3 - f_1}}{\sim} \left(\begin{array}{ccc|c} 1 & 1 & a^2 & 2+a^2 \\ 0 & 1-a^2 & 1-a^4 & 3-a^4-2a^2 \\ 0 & a^2-1 & 1-a^2 & 2-a-a^2 \end{array}\right)$$

$$\underset{f_3 \leftarrow f_3 + f_2}{\sim} \left(\begin{array}{ccc|c} 1 & 1 & a^2 & 2+a^2 \\ 0 & 1-a^2 & 1-a^4 & 3-a^4-2a^2 \\ 0 & 0 & 2-a^4-a^2 & 5-a^4-3a^2-a \end{array}\right)$$

$$= \left(\begin{array}{ccc|c} 1 & 1 & a^2 & 2+a^2 \\ 0 & -(a+1)(a-1) & -(a+1)(a-1)(a^2+1) & -(a+3)(a-1) \\ 0 & 0 & -(a+1)(a-1)(a^2+2) & -(a-1)(a^3+a^2+4a+5) \end{array}\right).$$

Therefore:

- *If* $a = 1$, *the system is consistent and undetermined (CUS), as:*

$$r = r' = 1 < n = 3.$$

The solution of the system is then $\{x = 3 - y - z,\ y = y,\ z = z\}$

- *If* $a = -1$, *the system is inconsistent (IS), as* $r = 1 < r' = 3$, *and does not admit any solution.*
- *If* $a \neq \pm 1$, *then the system is consistent and determined (CDS), as* $r = r' = n = 3$. *The solution of the system is then:*

$$\{x = \frac{2a+3}{(a+1)(2+a^2)},\quad y = -\frac{a^2-2a-1}{(a+1)(2+a^2)},\quad z = \frac{a^3+a^2+4a+5}{(a+1)(2+a^2)}\}.$$

(b) *We need to discuss the system of equations:*

$$\left\{\begin{array}{rcl} (1-a)x + y & = & 0 \\ (1-a)y + z & = & 0 \\ x + (1-a)z & = & 0 \end{array}\right.$$

according to the value of $a \in \mathbb{R}$.
We operate on the augmented matrix of the system:

$$\left(\begin{array}{ccc|c} 1-a & 1 & 0 & 0 \\ 0 & 1-a & 1 & 0 \\ 1 & 0 & 1-a & 0 \end{array}\right) \underset{pf_{13}}{\sim} \left(\begin{array}{ccc|c} 1 & 0 & 1-a & 0 \\ 0 & 1-a & 1 & 0 \\ 1-a & 1 & 0 & 0 \end{array}\right)$$

$$\underset{f_3 \leftarrow f_3 - (1-a)f_1}{\sim} \left(\begin{array}{ccc|c} 1 & 0 & 1-a & 0 \\ 0 & 1-a & 1 & 0 \\ 0 & 1 & -(1-a)^2 & 0 \end{array}\right)$$

$$\underset{f_2 \leftarrow f_2 + af_3}{\sim} \left(\begin{array}{ccc|c} 1 & 0 & 1-a & 0 \\ 0 & 1 & -a^3+2a^2-a+1 & 0 \\ 0 & 1 & -(1-a)^2 & 0 \end{array}\right)$$

$$\underset{f_3 \leftarrow f_3 - f_2}{\sim} \left(\begin{array}{ccc|c} 1 & 0 & 1-a & 0 \\ 0 & 1 & -a^3+2a^2-a+1 & 0 \\ 0 & 0 & a^3-3a^2+3a-2 & 0 \end{array}\right)$$

$$= \left(\begin{array}{ccc|c} 1 & 0 & 1-a & 0 \\ 0 & 1 & -a^3+2a^2-a+1 & 0 \\ 0 & 0 & (a-2)(a^2-a+1) & 0 \end{array}\right).$$

Therefore:

- *If* $a = 2$, *then the system is consistent and undetermined (CUS), as* $r = r' = 2 < n = 3$. *In this case, the system matrix is:*

$$\left(\begin{array}{ccc|c} 1 & 0 & -1 & 0 \\ 0 & 1 & -1 & 0 \\ 0 & 0 & 0 & 0 \end{array}\right)$$

and, therefore, the solution is: $\{x = z,\ y = z\}$.

– *If* $a \neq 2$, *the system is consistent and determined (CDS), as* $r = r' = n = 3$. *The solution is then* $\{x = 0, y = 0, z = 0\}$.

(c) We need to discuss the system of equations:

$$\left\{\begin{array}{rcl} ax - 2y + 4z & = & a+2 \\ x - y + 2z & = & a \\ -2x + 2y + 2az & = & -4 \end{array}\right.$$

according to the value of $a \in \mathbb{R}$.
We operate on the augmented matrix of the system:

$$\left(\begin{array}{rrr|r} a & -2 & 4 & a+2 \\ 1 & -1 & 2 & a \\ -2 & 2 & 2a & -4 \end{array}\right) \underset{pf_{13}}{\sim} \left(\begin{array}{rrr|r} -2 & 2 & 2a & -4 \\ 1 & -1 & 2 & a \\ a & -2 & 4 & a+2 \end{array}\right)$$

$$\underset{-\frac{1}{2}f_1}{\sim} \left(\begin{array}{rrr|r} 1 & -1 & -a & 2 \\ 1 & -1 & 2 & a \\ a & -2 & 4 & a+2 \end{array}\right)$$

$$\underset{\substack{f_2 \leftarrow f_2 - f_1 \\ f_3 \leftarrow f_3 - af_1}}{\sim} \left(\begin{array}{rrr|r} 1 & -1 & -a & 2 \\ 0 & 0 & 2+a & a-2 \\ 0 & a-2 & 4+a^2 & 2-a \end{array}\right)$$

$$\underset{pf_{23}}{\sim} \left(\begin{array}{rrr|r} 1 & -1 & -a & 2 \\ 0 & a-2 & 4+a^2 & 2-a \\ 0 & 0 & 2+a & a-2 \end{array}\right).$$

Therefore:

– *If* $a = 2$, *then the system is consistent and undetermined (CUS), as* $r = r' = 2 < n = 3$. *In this case, the system matrix is:*

$$\left(\begin{array}{rrr|r} 1 & -1 & -2 & 2 \\ 0 & 0 & 8 & 0 \\ 0 & 0 & 4 & 0 \end{array}\right)$$

and, therefore, the solution is $\{x = 2 + y,\ z = 0\}$.

– *If* $a = -2$, *the system is inconsistent (IS), as* $r = 2 < r' = 3$, *and does not admit solution.*

– *If* $a \neq \pm 2$, *then the system is consistent and determined (CDS), as* $r = r' = n = 3$. *The solution of the system is then:*

$$\left\{x = -1, \quad y = -\frac{a^2 + a + 6}{a+2}, \quad z = \frac{a-2}{a+2}\right\}.$$

(d) We need to discuss the system of equations:

$$\left\{\begin{array}{rcl} ax + 3y + z & = & 2 \\ 3x + ay + 2z & = & 3 \\ -x + y - z & = & 1 \end{array}\right.$$

according to the value of $a \in \mathbb{R}$.
We operate on the augmented matrix of the system:

$$\left(\begin{array}{ccc|c} a & 3 & 1 & 2 \\ 3 & a & 2 & 3 \\ 1 & 1 & -1 & 1 \end{array}\right) \underset{pf_{13}}{\sim} \left(\begin{array}{ccc|c} 1 & 1 & -1 & 1 \\ 3 & a & 2 & 3 \\ a & 3 & 1 & 2 \end{array}\right)$$

$$\underset{\substack{f_2 \leftarrow f_2+3f_1 \\ f_3 \leftarrow f_3+af_1}}{\sim} \left(\begin{array}{ccc|c} 1 & 1 & -1 & 1 \\ 0 & a+3 & -1 & 6 \\ 0 & a+3 & 1-a & 2+a \end{array}\right)$$

$$\underset{f_3 \leftarrow f_3-f_2}{\sim} \left(\begin{array}{ccc|c} 1 & 1 & -1 & 1 \\ 0 & a+3 & -1 & 6 \\ 0 & 0 & 2-a & a-4 \end{array}\right).$$

Therefore:

- *If $a = 2$, the system is inconsistent (IS), as $r = 2 < r' = 3$, and does not admit solution.*
- *If $a = -3$ the system matrix is:*

$$\left(\begin{array}{ccc|c} 1 & 1 & -1 & 1 \\ 0 & 0 & -1 & 6 \\ 0 & 0 & 5 & -7 \end{array}\right) \underset{f_3 \leftarrow f_3+5f_2}{\sim} \left(\begin{array}{ccc|c} 1 & 1 & -1 & 1 \\ 0 & 0 & -1 & 6 \\ 0 & 0 & 0 & 23 \end{array}\right)$$

  and, therefore, the system is inconsistent (IS), as $r = 2 < r' = 3$, and does not admit solution.
- *If $a \neq 2, -3$, then the system is consistent and determined (CDS), as $r = r' = n = 3$. The solution of the system is then:*

$$\left\{x = \frac{3a-14}{(a-2)(a+3)}, \quad y = \frac{5a-8}{(a-2)(a+3)}, \quad z = \frac{4-a}{a-2}\right\}.$$

Problem 6.23. Discuss the following linear systems of equations in terms of the real parameters $a, b \in \mathbb{R}$. Solve it where possible.

(a) $$\begin{cases} ax + y + z = b \\ x + ay + z = b \\ x + y + az = b \end{cases}$$

(b) $$\begin{cases} ax + by + z = 1 \\ x + aby + z = b \\ x + by + az = 1 \end{cases}$$

(c) $\begin{cases} ax + y - z &= 1 \\ x - ay + z &= 4 \\ x + y + az &= b \end{cases}$

(d) $\begin{cases} 3ax + 4y - 2az &= 4 + b \\ -ax - y + az &= -1 \\ -ax - 2y + az &= -2 \end{cases}$

Solution to the problem 6.23. *(a) We need to discuss the system of equations:*

$$\begin{cases} ax + y + z &= b \\ x + ay + z &= b \\ x + y + az &= b \end{cases}$$

in terms of the real parameters $a, b \in \mathbb{R}$.
We operate on the augmented matrix of the system:

$$\left(\begin{array}{ccc|c} a & 1 & 1 & b \\ 1 & a & 1 & b \\ 1 & 1 & a & b \end{array}\right) \underset{pf_{13}}{\sim} \left(\begin{array}{ccc|c} 1 & 1 & a & b \\ 1 & a & 1 & b \\ a & 1 & 1 & b \end{array}\right)$$

$$\underset{\substack{f_2 \leftarrow f_2 - f_1 \\ f_3 \leftarrow f_3 - af_1}}{\sim} \left(\begin{array}{ccc|c} 1 & 1 & a & b \\ 0 & a-1 & 1-a & 0 \\ 0 & 1-a & 1-a^2 & b(1-a) \end{array}\right)$$

$$\underset{f_3 \leftarrow f_3 + f_2}{\sim} \left(\begin{array}{ccc|c} 1 & 1 & a & b \\ 0 & a-1 & 1-a & 0 \\ 0 & 0 & 2-a^2-2 & b(1-a) \end{array}\right)$$

$$= \left(\begin{array}{ccc|c} 1 & 1 & a & b \\ 0 & a-1 & 1-a & 0 \\ 0 & 0 & -(a-1)(a+2) & b(1-a) \end{array}\right).$$

Therefore:

- *If $a = 1$, then the system is consistent and undetermined (CUS), as $r = r' = 1 < n = 3$. The solution of the system is then:*

$$\{x = b - y - z, \quad y = y, \quad z = z\}.$$

– *If $a = -2$ the system matrix is:*

$$\left(\begin{array}{ccc|c} 1 & 1 & -2 & b \\ 0 & -3 & 3 & 0 \\ 0 & 0 & 0 & 3b \end{array}\right),$$

and, therefore:

* *If $b = 0$ then the system is consistent and undetermined (CUS), as $r = r' = 2 < n = 3$. The solution of the system is then:*

$$\{x = z, \quad y = z, \quad z = z\}.$$

* *If $b \neq 0$ the system is inconsistent (IS), as $r = 2 < r' = 3$, and does not admit solution.*

– *If $a \neq 1, -2$, $\forall b \in \mathbb{R}$ the system is consistent and determined (CDS), as $r = r' = n = 3$. The solution of the system is then:*

$$\left\{x = \frac{b}{a+2}, \quad y = \frac{b}{a+2}, \quad z = \frac{b}{a+2}\right\}.$$

(b) We need to discuss the system of equations:

$$\left\{\begin{array}{rcl} ax + by + z & = & 1 \\ x + aby + z & = & b \\ x + by + az & = & 1 \end{array}\right.$$

in terms of the real parameters $a, b \in \mathbb{R}$.

We operate on the augmented matrix of the system:

$$\left(\begin{array}{ccc|c} a & b & 1 & 1 \\ 1 & ab & 1 & b \\ 1 & b & a & 1 \end{array}\right) \underset{pf_{13}}{\sim} \left(\begin{array}{ccc|c} 1 & b & a & 1 \\ 1 & ab & 1 & b \\ a & b & 1 & 1 \end{array}\right)$$

$$\underset{\substack{f_2 \leftarrow f_2 - f_1 \\ f_3 \leftarrow f_3 - af_1}}{\sim} \left(\begin{array}{ccc|c} 1 & b & a & 1 \\ 0 & b(a-1) & 1-a & b-1 \\ 0 & b(1-a) & 1-a^2 & 1-a \end{array}\right)$$

$$\underset{f_3 \leftarrow f_3 + f_2}{\sim} \left(\begin{array}{ccc|c} 1 & b & a & 1 \\ 0 & b(a-1) & 1-a & b-1 \\ 0 & 0 & 2-a^2-a & b-a \end{array}\right)$$

$$= \left(\begin{array}{ccc|c} 1 & b & a & 1 \\ 0 & b(a-1) & 1-a & b-1 \\ 0 & 0 & -(a-1)(a+2) & b-a \end{array}\right).$$

Therefore:

– *If* $a = 1$ *the system matrix is:*

$$\left(\begin{array}{ccc|c} 1 & b & 1 & 1 \\ 0 & 0 & 0 & b-1 \\ 0 & 0 & 0 & b-1 \end{array}\right),$$

and, therefore:

* *If* $b = 1$ *then the system is consistent and undetermined (CUS), as* $r = r' = 1 < n = 3$. *The solution of the system is then:*

$$\{x = 1 - y - z, \quad y = y, \quad z = z\}.$$

* *If* $b \neq 1$ *the system is inconsistent (IS), as* $r = 1 < r' = 3$, *and does not admit solution.*

– *If* $a = -2$ *the system matrix is:*

$$\left(\begin{array}{ccc|c} 1 & b & -2 & 1 \\ 0 & b(a-1) & 3 & b-1 \\ 0 & 0 & 0 & b+2 \end{array}\right),$$

and, therefore:

* *If* $b = -2$ *then the system is consistent and undetermined (CUS), as* $r = r' = 2 < n = 3$. *The solution of the system is then:*

$$\{x = z, \quad y = -\frac{1}{2}(1+z), \quad z = z\}.$$

* *If* $b \neq -2$ *the system is inconsistent (IS), as* $r = 2 < r' = 3$, *and does not admit solution.*

– *If* $a \neq 1, -2$, *we consider the following two cases:*

* *If* $b = 0$ *the system is inconsistent (IS), as* $r = 2 < r' = 3$, *and does not admit solution.*
* *If* $b \neq 0$ *then the system is consistent and determined (CDS), as* $r = r' = n = 3$. *The solution of the system is then:*

$$\left\{x = \frac{a-b}{(a-1)(a+2)}, \quad y = \frac{ab+b-2}{b(a-1)(a+2)}, \quad z = \frac{a-b}{(a-1)(a+2)}\right\}.$$

(c) We need to discuss the system of equations:

$$\left\{\begin{array}{rcl} ax + y - z & = & 1 \\ x - ay + z & = & 4 \\ x + y + az & = & b \end{array}\right.$$

in terms of the real parameters $a, b \in \mathbb{R}$.

We operate on the augmented matrix of the system:

$$\left(\begin{array}{ccc|c} a & 1 & -1 & 1 \\ 1 & -a & 1 & 4 \\ 1 & 1 & a & b \end{array}\right) \underset{pf_{13}}{\sim} \left(\begin{array}{ccc|c} 1 & 1 & a & b \\ 1 & -a & 1 & 4 \\ a & 1 & -1 & 1 \end{array}\right)$$

$$\underset{\substack{f_2 \leftarrow f_2 - f_1 \\ f_3 \leftarrow f_3 - af_1}}{\sim} \left(\begin{array}{ccc|c} 1 & 1 & a & b \\ 0 & -a-1 & 1-a & 4-b \\ 0 & 1-a & -1-a^2 & 1-ab \end{array}\right)$$

$$\underset{f_2 \leftarrow f_2 - f_3}{\sim} \left(\begin{array}{ccc|c} 1 & 1 & a & b \\ 0 & -2 & 2+a^2-a & 4-+ab-b-1 \\ 0 & 1-a & -1-a^2 & 1-ab \end{array}\right)$$

$$\underset{-2f_3}{\sim} \left(\begin{array}{ccc|c} 1 & 1 & a & b \\ 0 & -2 & 2+a^2-a & 4+ab-b-1 \\ 0 & 2a-2 & 2a^2+2 & 2ab-2 \end{array}\right)$$

$$\underset{f_3 \leftarrow f_3 + (a-1)f_2}{\sim} \left(\begin{array}{ccc|c} 1 & 1 & a & b \\ 0 & -2 & 2+a^2-a & 4+ab-b-1 \\ 0 & 0 & a(3+a^2) & a^2b+3a+b-5 \end{array}\right).$$

Therefore:

- *If* $a = 0$ *the system matrix is:*

$$\left(\begin{array}{ccc|c} 1 & 1 & 0 & b \\ 0 & -2 & 2 & 4-b-1 \\ 0 & 0 & 0 & b-5 \end{array}\right),$$

and, therefore:

  * *If* $b = 5$ *then the system is consistent and undetermined (CUS), as* $r = r' = 2 < n = 3$. *The solution of the system is then* $\{4-z,\ 1+z,\ z\}$.
  * *If* $b \neq 5$ *the system is inconsistent (IS), as* $r = 2 < r' = 3$, *and does not admit solution.*

- *If* $a \neq 0$, $\forall b \in \mathbb{R}$ *then the system is consistent and determined (CDS), as* $r = r' = n = 3$. *The solution of the system is then:*

$$\left\{x = \frac{a^2+ab+4a+5-b}{a(a^2+3)}, \quad y = -\frac{4a^2-a-ab-b+5}{a(a^2+3)}, \quad z = \frac{3a-5+a^2b+b}{a(a^2+3)}\right\}$$

(d) *We need to discuss the system of equations:*

$$\begin{cases} 3ax + 4y - 2az & = & 4+b \\ -ax - y + az & = & -1 \\ -ax - 2y + az & = & -2 \end{cases}$$

in terms of the real parameters $a, b \in \mathbb{R}$.

We operate on the augmented matrix of the system:

$$\left(\begin{array}{ccc|c} 4 & 3a & -2a & 4+b \\ -1 & -a & a & -1 \\ -2 & -a & a & -2 \end{array}\right) \underset{pf_{12}}{\sim} \left(\begin{array}{ccc|c} -1 & -a & a & -1 \\ 4 & 3a & -2a & 4+b \\ -2 & -a & a & -2 \end{array}\right)$$

$$\underset{-f_1}{\sim} \left(\begin{array}{ccc|c} 1 & a & -a & 1 \\ 4 & 3a & -2a & 4+b \\ -2 & -a & a & -2 \end{array}\right)$$

$$\underset{\substack{f_2 \leftarrow f_2 - 4f_1 \\ f_3 \leftarrow f_3 + 2f_1}}{\sim} \left(\begin{array}{ccc|c} 1 & a & -a & 1 \\ 0 & -a & 2a & b \\ 0 & a & -a & 0 \end{array}\right)$$

$$\underset{f_3 \leftarrow f_3 + f_2}{\sim} \left(\begin{array}{ccc|c} 1 & a & -a & 1 \\ 0 & -a & 2a & b \\ 0 & 0 & a & b \end{array}\right)$$

$$\underset{f_1 \leftarrow f_1 + f_2}{\sim} \left(\begin{array}{ccc|c} 1 & 0 & a & 1+b \\ 0 & -a & 2a & b \\ 0 & 0 & a & b \end{array}\right)$$

$$\underset{\substack{f_1 \leftarrow f_1 - f_3 \\ f_2 \leftarrow f_2 + 2f_3}}{\sim} \left(\begin{array}{ccc|c} 1 & 0 & 0 & 1 \\ 0 & -a & 0 & -b \\ 0 & 0 & a & b \end{array}\right).$$

Therefore:

– *If* $a = 0$ *the matrix system is:*

$$\left(\begin{array}{ccc|c} 1 & 0 & 0 & 1 \\ 0 & 0 & 0 & -b \\ 0 & 0 & 0 & b \end{array}\right),$$

and, therefore:

  * *If* $b = 0$ *then the system is consistent and undetermined (CUS), as* $r = r' = 1 < n = 3$. *The solution of the system is* $\{x = 1,\ y = y,\ z = z\}$.
  * *If* $b \neq 0$ *the system is inconsistent (IS), as* $r = 1 < r' = 3$, *and does not admit solution.*

– *If* $a \neq 0$, $\forall b \in \mathbb{R}$ *then the system is consistent and determined (CDS), as* $r = r' = n = 3$. *The solution of the system is, therefore:*

$$\left\{x = 1,\quad y = \frac{b}{a},\quad z = \frac{b}{a}\right\}.$$

Kolibri 360 GmbH
Ferdinand-Jühlke-Straße 7
99095 Erfurt, Deutschland
produktsicherheit@kolibri360.de